THIRD EDITION

Teaching Mathematics in Secondary and Middle School

An Interactive Approach

James S. Cangelosi
Utah State University

Upper Saddle River, New Jersey
Columbus, Ohio

Library of Congress Cataloging-in-Publication Data

Cangelosi, James S.
Teaching mathematics in secondary and middle school: an interactive approach/James S. Cangelosi.—3rd ed.
p. cm.
Includes bibliographical references and index.
ISBN 0–13–095018–1 (pbk.)
1. Mathematics—Study and teaching (Secondary) 2. Mathematics—Study and teaching (Middle school) I. Title.

QA11.2 .C36 2003
510'.71'2—dc21

2001054440

Vice President and Publisher: Jeffery W. Johnston
Editor: Linda Ashe Montgomery
Production Editor: Mary M. Irvin
Design Coordinator: Diane C. Lorenzo
Project Coordination and Text Design: Carlisle Publishers Services
Cover Designer: SuperStock
Cover Photo: Bryan Huber
Production Manager: Pamela D. Bennett
Director of Marketing: Ann Castel Davis
Marketing Manager: Krista Groshong
Marketing Coordinator: Tyra Cooper

This book was set in New Century Schoolbook by Carlisle Communications, Ltd., and was printed and bound by Courier Kendallville, Inc., The cover was printed by Phoenix Color Corp.

Photo Credits: Donna Barry, pp. 11, 67, 68, 105, 106, 230, 279, 374, 412, 413, 436; Ted Hansen, pp. 19, 38, 52, 90, 100, 123, 140, 156, 199, 222, 257, 361, 362, 363, 364.

Pearson Education Ltd.
Pearson Education Australia Pty. Limited
Pearson Education Singapore Pte. Ltd.
Pearson Education North Asia Ltd.
Pearson Education Canada, Ltd.
Pearson Education de Mexico, S. A. de C. V.
Pearson Education—Japan
Pearson Education Malaysia Pte. Ltd.
Pearson Education, *Upper Saddle River, New Jersey*

10 9 8 7 6 5 4 3
ISBN 0–13–095018–1

To Chris

Preface

Contrary to popular belief, mathematics is an everyday human endeavor by which ordinary people construct concepts, discover relationships, invent algorithms and models, organize and communicate their thoughts in the language of mathematics, execute algorithms, and address their real-world problems. Likewise, by capitalizing on the use of common and personally relevant mathematics problems, mathematics teachers can help students learn and creatively apply mathematics to their everyday lives. Using research-based strategies, effective mathematics teachers guide students to invent and discover new mathematics perspectives and in so doing help students acquire confident attitudes and abilities in mathematics. Case studies in this text, the third edition of *Teaching Mathematics in Secondary and Middle School,* demonstrate how.

According to consistent findings in research studies cited throughout this book, students develop confident attitudes and abilities in mathematics by engaging in:

- inquiry lessons that lead them to reason inductively to construct mathematical concepts and discover mathematical relationships
- direct-instructional lessons that lead them to gain knowledge of conventions and develop algorithmic skills
- comprehension-and-communication lessons that lead them to take advantage of the special features of mathematical language
- inquiry lessons that lead them to reason deductively to devise solutions to real-life problems

However, most students do not have these experiences. Rather, they acquire a considerably different view of mathematics, perceiving it as a boring string of terms, symbols, facts, and algorithms—truly understood only by rare geniuses. Too many students are asked only to memorize mathematical content without ever discovering, inventing, or creatively applying it. The unhealthy attitudes and inabilities to extend mathematics beyond what is memorized are perpetuated by the most dominant method of teaching mathematics in our schools. Many teachers' lessons follow a tiresome sequence—lessons void of experiences whereby meaningful mathematics is discovered, invented, or applied: Students are told about a fact or the steps in an algorithm, walked through textbook examples, assigned exercises from the textbook, and given feedback on their work with the exercises.

For over a century, mathematics education specialists have encouraged teachers to deviate from common practice by applying research-based strategies. Promise for bringing typically practiced mathematics teaching in line with research-based strategies springs from the widespread dissemination and support for the National Council of Teachers of Mathematics' (NCTM) plan for school mathematics curriculum reform as articulated in *Principles and Standards for School Mathematics* (*PSSM*) (NCTM, 2000b).

Teaching Mathematics in Secondary and Middle School: An Interactive Approach is designed to lead you to develop your talent for teaching in accordance with *PSSM* so that your students eagerly construct mathematical concepts for themselves, discover mathematical relationships, develop and maintain algorithmic skills, communicate in the language of mathematics, and use mathematics to devise solutions to real-life problems.

WHERE ARE RESEARCH-BASED STRATEGIES INTEGRATED INTO THIS TEXT?

Introducing topics in a spiral fashion, this book actively involves you in researched-based learning activities throughout its 11 chapters.

Chapter 1, "Beginning a Career as a Professional Mathematics Teacher," will lead you to vicariously experience the professional activities of Casey Rudd, a beginning mathematics teacher, and expose you to some preliminary ideas for developing mathematics curricula and engaging students in meaningful lessons—ideas that you will further develop in depth as you work with chapters 2–11.

Chapter 2, "Gaining Students' Cooperation in an Environment Conducive to Doing Mathematics," will help you develop strategies that lead students to cooperate in the business of learning mathematics. From your work with other chapters you will learn how to design courses, plan lessons, conduct learning activities, and monitor students' progress toward meaningful mathematical goals. However, the success of even the best designed curricula depends on how well you establish a classroom environment in which students willingly work on-task and engage in the business of learning—an environment in which students feel free to experiment, make mistakes, raise questions, interact with you and one another, contribute ideas, and expose their thought processes without fear that they are risking embarrassment, harassment, or judgement of their self-worth.

Chapter 3, "Motivating Students to Engage in Mathematical Learning Activities," is an extension of chapter 2 that focuses on using problem-based lessons to motivate students to do mathematics, strategies for responding to students' questions, and strategies for keeping students engaged in the following types of learning activities: large-group presentations, question-discussion sessions, cooperative-learning, independent-work sessions, and homework.

Chapter 4, "Developing Mathematics Curricula," will familiarize you with *PSSM*. You will understand why *PSSM*-based curricula lead students to do meaningful mathematics whereas typical textbook-driven curricula lead students to perceive mathematics as a linear sequence of meaningless definitions, symbols, rules, and algorithms. From your work with chapter 4, you will develop an advanced organizer for designing courses that are consistent with *PSSM*-based curricula—an advanced organizer that interrelates the work you will be doing with chapters 5–11.

Chapter 5, "Leading Students to Construct Concepts and Discover Relationships," explains and illustrates how to design and conduct inquiry lessons that lead students to construct mathematical concepts and discover mathematical relationships. You will also develop strategies for conducting miniexperiments to monitor students' progress during these types of lessons and assess how well the objectives were achieved.

Chapter 6, "Leading Students to Develop Knowledge and Algorithmic Skills," is organized similarly to chapter 5, but the focus is on using direct-instructional strategies for lessons designed to lead students to acquire and remember mathematical information and develop algorithmic skills. The miniexperiments you will design for this learning level will emphasize the identification and correction of error patterns in students' execution of algorithms.

Chapter 7, "Leading Students to Communicate With Mathematics," is organized similarly to chapters 5 and 6, but the focus is on using a combination of inquiry and direct-instructional strategies for lessons designed to lead students to use mathematics for organizing and communicating ideas and to comprehend the language of mathematics. You will learn how to incorporate conversation, speaking, listening, writing, and reading in your lessons so that students do meaningful mathematics.

Chapter 8, "Leading Students to Creatively Use Mathematics," is organized similarly to chapters 5–7, but the focus is on inquiry lessons that lead students to apply mathematics to real-life situations, foster their creativity with mathematics, and develop an appreciation for and willingness to do mathematics. As with chapters 5–7, you will engage in activities that prompt you to design and field-test lessons and miniexperiments. Such activities provide critical experiences upon which you build your teaching talents. You will also collect artifacts (e.g., lesson plans) for the professional portfolio you will have organized from your work with chapter 1.

Chapter 9, "Assessing and Reporting Students' Progress With Mathematics," will lead you to (a) develop an efficient system for monitoring and evaluating students' progress and (b) apply authentic assessment strategies to make and communicate summative evaluations of your students' mathematical achievements. Furthermore, you will learn about some of the misuses of high-stakes testing and how to interpret the results of standardized and core-curriculum tests.

Chapter 10, "Technology and Resources for Teaching and Learning Mathematics," will direct you to a wide variety of (a) resources for stimulating ideas on teaching mathematics, learning mathematics, and doing mathematics and (b) technologies and mathematics curriculum materials typically available for use in middle, junior high, and high schools. You will critique curriculum materials (e.g., textbooks) and sample Internet-based resources.

Chapter 11, "Analyzing Examples of Mathematics Curricula and Instructional Practice," will lead you to further develop your talent for designing mathematics courses and conducting lessons for students by prompting you to analyze cases in which teachers implement strategies you learned from your work with chapters 1–10. With an emphasis on integrating mathematics curricula with those of other subject-content areas (e.g., biology, physical education, and social studies), you will be immersed in the thoughts and practices of teachers as they design and conduct a variety of middle, junior high, and high school mathematics courses.

HOW DOES THIS TEXT DEMONSTRATE RESEARCH-BASED TEACHING PRINCIPLES?

Teaching Mathematics in Secondary and Middle School: An Interactive Approach is an extremely unusual book. Not only does it present research-based principles for teaching mathematics, it also demonstrates each by contrasting actual classroom examples in which the principle is applied with actual examples that violate the principle. Throughout, topics (e.g., defining learning goals, designing lessons, motivating students' cooperation, and assessing learning) are integrated by the book's 189 actual classroom *cases* that follow teachers' thoughts, actions, reactions, and interactions as they engage in work of professional mathematics teachers. The book's pedagogy employs the same research-based teaching strategies it suggests you use with your students:

- You are stimulated to reason inductively to construct concepts and discover relationships as you interact with carefully selected examples and nonexamples.
- Direct instructional strategies are used to present you with information, conventions (e.g., word meanings), and processes.
- A combination of inquiry and direct-instructional strategies are employed throughout to lead you to comprehend and communicate ideas related to teaching mathematics.
- You are stimulated to reason deductively to apply concepts, relationships, and processes to your own teaching situations.

To incorporate these research-based teaching strategies, the organization of the book is much more complex than for a typical mathematics teaching methods textbook that simply presents information about teaching and examples of learning activities for teachers to use with their students. The non-linear, spiral structure of *Teaching Mathematics in Secondary and Middle School: An Interactive Approach* is not conducive to reading chapters out of order. Another consequence is that you will occasionally be prompted to visit cases and exhibits from chapters other than the one you are studying. This inconvenience benefits you by continually interconnecting chapters and building on what you are learning or on what you have already learned.

Each chapter begins with a goal defined by a set of objectives and ends with a set of *synthesis activities* and a *transitional activity.* The synthesis activities are designed to help you bring together the chapter's content, reinforce and extend what you learned, and assess what you gained from the chapter so that you can identify your areas of proficiency and the topics you need to review. Each chapter's transitional activity sets the stage for your work with the next chapter. There are also activities in which you will engage as you work with the main body of each chapter. These learning activities are designed to stimulate your ideas, lead you to clarify or expand upon what you have just read, prepare you to work with the next section of the chapter, address a problem, collaborate with colleagues, or design and use a product (e.g., a lesson plan). The numerous cases generate a considerable number of *exhibits* for you to analyze and model as you generate your own curriculum materials. Consequently, this book contains an unusually high number of exhibits (i.e., figures, tables, and illustrations): 256.

WHAT IS NEW ABOUT THIS EDITION?

This third edition of *Teaching Mathematics in Secondary and Middle School: An Interactive Approach* is considerably different from the second edition:

- The organization and writing are thoroughly reworked. The content is expanded, but the book's length is not appreciably increased because the writing is much crisper.
- This third edition addresses the following topics in much greater depth than they were addressed in the second edition: applying for faculty positions as a mathematics teacher, interacting with students (e.g., the use of naturalistic conversations), teaching mathematics from a historical perspective, communicating with mathematics, working with students as individuals, application of multicultural education strategies, working with students for whom English is not a first language, integrating mathematics curricula with those of other subject-content areas, differences among mathematics courses and age levels of students, Internet-based learning activities and resources, and high-stakes testing.
- This third edition reflects considerable updating with respect to the research literature, the publication of *PSSM,* and advances in technology.
- The third edition prompts you to organize your work in a *professional teaching portfolio.*
- Chapters have been reworked and reorganized with increased attention to mathematical modeling and mathematics as communication.
- Chapter 10 reflects new advances in technology and Web-based resources and activities.
- The text now includes a **glossary** of technical terms introduced in the book.

The instructor's manual for this third edition provides suggestions for designing mathematical teaching methods courses. It includes a course syllabus, a sequence of course activities, and exams with detailed scoring rubrics. It is available upon request from Merrill/Prentice Hall to course instructors.

ACKNOWLEDGMENTS

This text's instructional materials have been field-tested with the preservice mathematics teachers with whom I work in the field-based methods-of-teaching-mathematics course (i.e., Math 4500) at Utah State University. Math 4500 is conducted in a middle and a high school similarly to the way Professor Rice conducts her course in Case 2.33. I am particularly grateful to those preservice teachers and to the inservice mathematics teachers (e.g., Rob Hoggan, Jean Culbertson, Lori Wilson, Jessica Burch, Melanie Hall, Sarah Timmins, Darin Lentz, Monica Chase, Janiece Edginton, Fawn Groves, Russ Weeks, Tim Cybulski, Tami Britt, Michelle Hatch, Nancy Drickey, Joyce Smart, Dave Reynolds, Linda Nichols, Dan Coffin, Juliann Bales, Nan Koebbe, Mary Kirby, Scott Wright, and Heidi Hall) who work with them.

Expert reviews of the manuscript were provided by Joel Bass of Sam Houston State University, Linda Cronin Jones of the University of Florida, William Croadale of the University of Rhode Island, Ed Dickey of the University of South Carolina, John Dougherty of Lindenwood College, Jeff Frykholm of Virginia Tech University, Thomas Gibney of the University of Toledo, Jay Graening of the University of Arkansas, Boyd Holton of West Virginia University, Virginia Horak of the University of Arizona, Randy Hoover of Youngstown State University, Mark Klespis of St. Xavier (Chicago), Mary M. Lindquist of Columbus College, William L. Merrill of Central Michigan University, Nancy Minix of Western Kentucky University, E. Alexander Norman of the University of North Carolina–Charlotte, Sandra J. Olson of Winona State University, Katherine Pederson of Southern Illinois University, Tony Piccolino of Montclair State University, Ken Stillwell of Northeast Missouri State University, William K. Tomhave of Concordia College, Stephen F. West of State University of New York at Geneseo, John Wilkins of California State University–Dominguez Hills, and Earl J. Zwick of Indiana State University.

Credit for this work is shared with the professionals of Merrill/Prentice Hall (e.g., Allyson Sharp and Mary Irvin), Marilee Aschenbrenner of Carlisle Publishers Services, and Jane Parrigin, who copyedited the manuscript.

Contents

chapter

1

Beginning a Career as a Professional Mathematics Teacher

Goal and Objectives for Chapter 1

The Goal The goal of chapter 1 is to lead you to vicariously experience the professional activities of a beginning mathematics teacher and to expose you to some preliminary ideas for developing mathematics curricula and engaging students in meaningful lessons—ideas that you will further develop in depth as you work with chapters 2–11.

A Note Each chapter goal is defined by a set of specific objectives; you will note that each of these objectives is labeled by its targeted cognitive or affective learning level (i.e., "construct a concept," "discover a relationship," "simple knowledge," "algorithmic skill," "comprehension and communication," "application," "creative thinking," "appreciation," and "willingness to try"). The meanings of the learning levels will be clarified by the time you have finished working with chapter 4; chapters 5–8 will lead you to apply your understanding of these learning levels to your own teaching.

Each objective is also weighted according to its relative importance toward goal achievement. Each weight is expressed as a percentage (e.g., "25%"). You will understand the purpose of weighting objectives according to relative importance by the time you have finished working with chapter 9.

Near the end of this book is a glossary of technical words and phrases that are introduced, explained, and defined in the text. I mention this to you up front because you will be exposed to some of these words or phrases (e.g., "*Principles and Standards for School Mathematics* (*PSSM*)" (NCTM, 2000b), "construct a concept," "inquiry instructional strategy," "inductive reasoning," and "miniexperiment") before their technical meanings are adequately explained. When this happens, you can depend on your current nontechnical understanding until the technical meaning is defined in a subsequent section of the book. However, if you feel compelled to seek clarification, you can quickly refer to the glossary.

The glossary also serves another purpose: As you work with the text, you will be accumulating technical definitions of words and phrases. Sometimes you will need to recall definitions from previous chapters; it may be handier to look up a definition in the glossary than to locate it in the body of the text.

The Objectives Chapter 1's goal is defined by the following set of objectives:

A. You will distinguish between the purpose of a working professional portfolio and the purpose of a presentation professional portfolio and gain ideas for putting together your own working and presentation portfolios (construct a concept) 5%.
B. You will develop the structure for your working portfolio (willingness to try) 10%.
C. You will explain the general process by which preservice mathematics teachers become inservice mathematics teachers (comprehension and communication) 10%.
D. You will set up a computerized management system that will facilitate your work with this text and be useful to you as an inservice mathematics teacher (willingness to try) 10%.
E. You will explain why teaching mathematics in middle, junior high, or high school is an extremely complex art (construct a concept) 20%.
F. You will explain the importance of using assertive—rather than passive or hostile—communications in your role as a professional mathematics teacher (comprehension and communication) 10%.
G. You will anticipate and describe examples of the types of events and problems you are likely to confront as a professional mathematics teacher (construct a concept) 25%.

Continued

GOAL AND OBJECTIVES FOR CHAPTER 1 *(Continued)*

H. You will incorporate the following phrases into your working vocabulary: "preservice mathematics teacher," "inservice mathematics teacher," "working portfolio," "presentation portfolio," "inquiry instructional strategies," "direct instructional strategies," "withitness," "businesslike classroom," "assertive communication," "hostile communication," "passive communication," "administrative supervision of teaching," and "instructional supervision" (comprehension and communication) 10%.

PRESERVICE PREPARATION AND PROFESSIONAL PORTFOLIOS

Because you are reading this book, I assume you are either a *preservice mathematics teacher* or an *inservice mathematics teacher*. A preservice mathematics teacher is a person who is currently enrolled in a professional teacher preparation program for the purpose of becoming qualified and certified to teach mathematics in middle, junior high, and high schools. Once a person completes such a program and accepts a position as a mathematics teacher, the person is no longer classified as "preservice" but rather as an "inservice mathematics teacher."

As they engage in professional teacher preparation programs, many preservice mathematics teachers develop and maintain *working portfolios,* which are collections of selected items that reflect individual teachers' professional talents, aptitude for teaching, style of teaching, and teaching performances. A teacher selects a proper subset of the items from his working portfolio to create a *presentation portfolio* for use when applying for faculty positions (Painter, 2001). In answer to the question, "What is a professional portfolio?" Campbell, Cignetti, Melenyzer, Nettles, and Wyman (1997, pp. 3–4) stated:

> A portfolio is not merely a file of course projects and assignments, nor is it a scrapbook of teaching memorabilia. A portfolio is an organized, goal-driven documentation of your professional growth and achieved competence in the complex act called teaching. Although it is a collection of documents, a portfolio is tangible evidence of the wide range of knowledge, dispositions, and skills that you possess as a growing professional. What's more, documents in the portfolio are self-selected, reflecting your individuality and autonomy.
>
> There are actually two kinds of portfolios that you will be developing: a working portfolio and a presentation portfolio. A working portfolio is characterized by your ongoing systematic collection of selected work in courses and evidence of community activities. This collection would form a framework for self-assessment and goal setting. Later, you would develop a presentation portfolio by winnowing your collection to samples of your work that best reflect your achieved competence, individuality, and creativity as a professional educator.
>
> What is a working portfolio? A working portfolio is always much larger and more complete than a presentation portfolio. It contains unabridged versions of documents you have carefully selected to portray your professional growth. For example, it might contain entire reflective journals, complete units, unique teacher-made materials, and a collection of videos of your teaching. Working portfolios are often stored in a combination of computer disks, notebooks, and even boxes.
>
> What is a presentation portfolio? A presentation portfolio is compiled for the expressed purpose of giving others an effective and easy-to-read portrait of your professional competence. A presentation portfolio is selective and streamlined because other people usually do not have the time to review all the material in your working portfolio. In making a presentation portfolio, you will find that less is more. For example, since you would be unlikely to take to an interview all your teacher-made learning materials, you might rely on photographs. Most reviewers would not want to assess several videos of your teaching but would be interested in one well-edited and annotated video. Sample pages from a large project would replace an entire project. The two types of portfolios differ in that all documents in a presentation portfolio should be preceded by an explanation of the importance or relevance of the documents so that the reviewer understands the context of your work.

The format for professional portfolios can be electronic (e.g., on a CD-ROM or website), hard copy documents filed in a container, or a combination of both. Besides serving as a reservoir from which exhibits for your presentation portfolio can be drawn, the working portfolio can also warehouse instructional materials and resources for use when teaching.

Consider Case 1.1:

CASE 1.1

Casey Rudd is embarking on his initial year as a professional mathematics teacher. He recently graduated from college with a major in mathematics education. His university work included more than a dozen mathematics courses (e.g., multivariable calculus, linear algebra, differential equations, analysis, algebraic structures, geometry, number theory, historical foundations of mathematics, discrete structures, probability and statistics, and computer-aided mathematics for teachers) and nearly 10 professional education courses (e.g., educational psychology, special education for classroom teachers, motivation and classroom management, multicultural foundations of education, content-area reading and writing, cognitive science, assessment of student achievement, and methods of teaching mathematics in secondary and middle school). However, only a couple of his university mathematics instructors employed the kind of teaching strategies that were recommended by his methods-of-teaching-mathematics course. Although confident in his ability to do mathematics, he worries that he lacks the conceptual and application levels of understanding necessary to generate the real-life examples and nonexamples that will engage his students in the kinds of learning activities explained in chapters 5–8.

While he was taking the methods-of-teaching-mathematics course in college, Casey began developing a working portfolio. He continued to build and refine it so that by the time he completed student teaching, he had a presentation portfolio ready for use as he applied for faculty positions. Casey partitioned his working portfolio according to nine categories: (a) cognition, instructional strategies, and planning, (b) motivation, engagement, and classroom management, (c) assessment, (d) multicultural education, (e) exceptionality and accommodation, (f) technology, (g) mathematics and historical foundations of mathematics, (h) students' real-world problems, and (i) professional development. Exhibit 1.1 provides a preliminary explanation of these categories.

A semester-long student-teaching experience provided some opportunities to try out many of the ideas from the methods-of-teaching-mathematics course. However, in student teaching Casey tailored his teaching style to the curriculum already established by his cooperating teacher. The cooperating teacher was very supportive of his efforts and immensely helpful in providing learning materials and suggestions on managing behavior and organizing lessons. But opportunities to design complete units were quite limited because Casey had not been involved in planning the courses prior to the opening of the school year.

Although enthusiastic about applying for positions and embarking on his professional career, Casey is understandably nervous about succeeding with a full complement of mathematics classes for which he is solely responsible.

Throughout this textbook, you will be prompted to engage in learning activities designed to stimulate your ideas, lead you to clarify or expand upon what you have read, prepare you to work with the next section of the text, address a problem, collaborate with colleagues, or design and develop a product (e.g., a lesson plan) that you will need for subsequent work with this book. Engage in Activity 1.1:

Activity 1.1

If you did not carefully read Exhibit 1.1 when you worked with Case 1.1, then do so now.

Obtain a container to house your working portfolio or set up the structure for an electronic working portfolio. Organize the container or computer disk space so that it is ready for inserts in the nine categories listed in Exhibit 1.1.

Besides providing the structure for developing your working portfolio, your work with Activity 1.1 also serves as an advanced organizer for the work you will be doing with the rest of this textbook. This advanced organizer will help you put what you learn into the context of being a professional mathematics teacher.

SELECTING A TEACHING POSITION

Applications for Positions

The transition after graduating from a college or university teacher preparation program leading to professional teacher certification involves, of course, securing a faculty position at a middle, junior high, or high school. In Case 1.2, Casey Rudd applies for positions:

CASE 1.2

Casey realizes that a huge demand will exist for his services as a mathematics teacher when he completes student teaching and graduates from college. He finds out about faculty openings for mathematics teachers at middle, junior high, and high schools by (a) checking a website of his university's career placement office (e.g., http://www.usu.edu/~career/public.htm—a website with links to announcements for school faculty openings in school districts throughout the United States and in some foreign countries), (b) checking the bulletin boards in the offices of the Teacher Education Department and the Mathematics Department, (c) visiting booths of recruiters from school districts at career-placement fairs held at his own as well as at nearby colleges and universities, (d) searching the internet for the websites of school districts where he is interested in working, and (e) informing people (e.g., university faculty members and mathematics teachers with whom he has worked) that he is applying for positions for the upcoming school year.

Some—but not most—of the school districts with positions that interest Casey require applications to be submitted electronically and a few of those require an electronic presentation portfolio. He prepares to submit applications for openings by getting his presentation portfolio ready, obtaining several letters of recommendation from university professors and mathematics teachers with whom he has worked, and writing the resume shown by Exhibit 1.2.

Exhibit 1.1

Preliminary Explanations for the Categories of One Working Portfolio.

- **Cognition, Instructional Strategies, and Planning**

What types of exhibits fit this category?

For the *cognition* aspect of this category, select items that reflect your understanding of the cognitive or mental processes that learners employ to (a) construct mathematical concepts, (b) discover mathematical relationships, (c) acquire and retain mathematical information, (d) develop algorithmic skills, (e) communicate with mathematics, (f) apply mathematics to address real-life problems, (g) be creative with mathematics, and (h) develop attitudes about mathematics.

For the *instructional strategies* aspect of this category, select items that reflect your talent for designing lessons that target meaningful learning objectives in a way that is consistent with your understanding of cognition (i.e., how students learn). For example, you might exhibit and contrast two lesson plans: The first one would be designed to lead students to discover the Pythagorean theorem for themselves; the second one would be designed to lead students to become proficient with an algorithm for computing unknown lengths of sides of right triangles. The two contrasting lesson plans would demonstrate that you know how to use *inquiry instructional strategies* when you want students to make mathematical discoveries and *direct instructional strategies* when you want students to acquire algorithmic skills.

For the *planning* aspect of this category, select items that reflect how you go about organizing and preparing for instruction. For example, the same lesson plans you used to demonstrate how you select instructional strategies that are congruent with your understanding of cognition could illustrate the format of your planning documents (e.g., the lesson plan includes an objective and descriptions of learning activities).

Besides lessons plans, items could be course outlines, reflection papers you have written that explain the interdependence among the three elements of the category, and videotapes of you conducting a learning activity that demonstrates that the activity was thoroughly planned and based on an instructional strategy that is appropriate for the objective.

How and when do you learn how to create these exhibits?

Typically, teaching methods and educational psychology courses engage you in activities that lead you to develop your understanding of cognition, instructional strategies, and how to plan as well as lead you to create exhibits (e.g., lessons plans) for this category. Cognition, instructional strategies, and planning are the major foci of chapters 4–8 and 11 of this textbook. Your engagement in those chapters' activities will produce exhibits for this category. As you gain teaching experiences through clinical work in schools, student teaching, and as an inservice teacher, your understanding of cognition, instructional strategies, and planning will grow immensely and you will update, refine, and supplant exhibits in your working portfolio.

- **Motivation, Engagement, and Classroom Management**

What types of exhibits fit this category?

For the *motivation* aspect of this category, select items that reflect your understanding of what stimulates students to want to learn and do mathematics. For example, you might exhibit a list of problems from students' real worlds that you would incorporate in lessons so that students do mathematics in order to address problems that they consider important.

For the *engagement* aspect of this category, select items that reflect your talent for directing lessons and involving students in learning activities so that students willingly participate in them (e.g., by attending to explanations, asking questions, answering questions, interacting with one another about mathematics, and completing assignments). For example, you might exhibit (a) a videotape of students participating in a cooperative learning activity you are conducting or (b) notes on a lesson plan indicating how you intend to keep students motivated and involved.

For the *classroom management* aspect of this category, select items that reflect your talent for (a) fostering a safe, comfortable, productive learning community where students willingly cooperate with you and with one another to learn mathematics, (b) operating a smoothly running classroom, and (c) dealing with discipline problems when they occur. For example, you might exhibit (a) an essay on how you plan to organize your classroom, teach and enforce standards of conduct, and establish routine procedures or (b) notes on a lesson plan indicating how you intend to orchestrate transitions between different types of learning activities (e.g., as students move from collaborating in small subgroups to silently attending to a large-group presentation).

How and when do you learn how to create these exhibits?

Typically, professional education courses with "classroom management" in the titles are intended to help you develop your strategies for (a) teaching your students to be on task and to engage in learning activities and (b) establishing and maintaining a safe, comfortable, smoothly operating classroom community that is conducive to productive learning. Motivation, engagement, and classroom management are the major foci of chapter 2 of this textbook. Those topics are also prominent in every other chapter because the success of everything you do with students is dependent on how you motivate, engage, and manage. Your engagement in activities throughout the

Continued

Exhibit 1.1
Continued

book—especially chapters 2 and 3—will produce exhibits for this category.

- **Assessment**

  What types of exhibits fit this category?

  Select items that reflect your talents for (a) monitoring your students' progress as they engage in mathematics lessons, (b) assessing students' achievement of learning goals, and (c) interpreting and communicating the scores from core-curriculum and standardized achievement tests. For example, you might exhibit (a) a computer disk with miniexperiments that you would use to gauge students' progress with mathematics or (b) a test you designed to be relevant to students' achievement of the goal of a teaching unit.

  How and when do you learn how to create these exhibits?

  Typically, professional education courses with "assessment" or "measurement and evaluation" in the titles are intended to help you develop your strategies for assessing and monitoring student achievement and for interpreting test results. Test development, test-score interpretation, and communication of your evaluations of student achievement are the foci of chapter 9. Your work with chapters 5–9 will lead you to develop strategies for monitoring students' progress as they engage in different types of mathematics lessons. You will generate computer files of miniexperiments, a test blueprint, and a unit test.

- **Multicultural Education**

  What types of exhibits fit this category?

  Select items that reflect your talents for teaching mathematics from a multicultural perspective so that (a) the diversity among your students is valued, (b) equitable opportunities to learn mathematics exist for all students, and (c) attention to the historical foundations of mathematics is integrated into lessons so students appreciate the culturally diverse origins of mathematics. For example, you might exhibit (a) a teaching unit on {primes} that emphasizes its diverse origins and application in cryptography (an application that depended on language diversity) or (b) a video of you conducting a learning activity in which a student who speaks Spanish only and a student who speaks English only communicate with one another through the language of mathematics.

  How and when do you learn how to create these exhibits?

  Typically, professional education courses with the words "multicultural" or "diversity" in the titles are intended to help you teach from a multicultural perspective. Chapter 2's section "Working with Students as Individuals" and chapter 4's section "*PSSM*'s Equity Principle" devote particular attention to multicultural education strategies that are specific to the teaching of mathematics. Examples of teaching mathematics from a multicultural perspective are scattered throughout the book.

- **Exceptionality and Accommodation**

  What types of exhibits fit this category?

  Select items that reflect your (a) talents for designing lessons, conducting learning activities, and managing your classroom so that students with exceptionalities are included and their special needs are accommodated and (b) understanding of your legal responsibilities and rights with respect to including and accommodating students with exceptionalities. For example, you might exhibit (a) a blueprint for a unit test that includes special provisions for testing the mathematics achievement of a student with a particular learning disability or (b) an individualized education program (IEP) developed by an IEP team on which you participated.

  How and when do you learn how to create these exhibits?

  Typically, professional education courses with titles such as "Special Education for Regular Education Teachers" are intended to (a) familiarize you with your legal responsibilities and rights relative to students with exceptionalities and (b) help you fulfill those responsibilities and exercise those rights. Chapter 2's section "Accommodating the Special-Education Needs of Students" focuses on strategies for accommodating students' exceptionalities that are specific to teaching mathematics.

- **Technology**

  What types of exhibits fit this category?

  Select items that reflect how you (a) use technology to teach mathematics and (b) teach students to use technology to do mathematics. For example, you might exhibit (a) a website for your students to access and engage in web-based learning activities, (b) computer-generated mathematical simulation you developed, or (c) a lesson plan that includes a cooperative learning activity in which students use graphing calculators to discover a mathematical relationship.

  How and when do you learn how to create these exhibits?

  Typically, professional education courses in instructional technology are designed to help you develop your competencies for incorporating computers and other types of electronic technologies (e.g., video presentations) in your teaching. Specific applications of technology to the teaching and learning of mathematics is the focus of chapter 10. Examples of teachers using a wide variety of technologies to teach mathematics and students using technology to learn mathematics are included in chapters 3–11.

Continued

Exhibit 1.1
Continued

- **Mathematics and the Historical Foundations of Mathematics**

  What types of exhibits fit this category?

  Select items that reflect your (a) mathematical expertise and (b) understanding of the history of mathematics. For example, you might exhibit (a) the proof for a theorem that you originated, (b) an algorithm that you invented, or (c) an essay explaining the value of teaching mathematics from a historical perspective and how you plan to incorporate historical foundations of mathematics in courses that you teach.

  How and when do you learn how to create these exhibits?

  Obviously, the mathematics courses you have taken are supposed to lead you to build your mathematical expertise. Some professors teach upper-level mathematics courses from a historical perspective—especially number theory and analysis. A well-taught course in the history of mathematics is invaluable for mathematics teachers. Beyond your course work, you need to actively do mathematics—formulating and addressing problems, making conjectures and trying to prove them, and using mathematics to model real-life phenomena. A pleasant way to gain insights into mathematics and its history is to read some of the engaging and fascinating trade books on mathematics (e.g., *Archimedes' Revenge: The Joys and Perils of Mathematics* [Hoffman, 1988], *Pi in the Sky: Counting, Thinking, and Being* [Barrow, 1992], *Fermat's Enigma* [Singh, 1997], and *Zero: The Biography of a Dangerous Idea* [Seife, 2000]). Exhibit 7.9 contains a list of such books; all 11 chapters of this text are designed to stimulate your ideas about mathematics and its origins. If you are already an experienced mathematics teacher, then you have learned how your interactions with students—responding to their questions, formulating meaningful examples and nonexamples to lead them to make discoveries, breaking down algorithms, formulating problems, and making mathematical connections—have increased the depth of your understanding of mathematics. If you are still an inexperienced preservice teacher, prepare to be amazed at the mathematical insights you will gain from your interactions with middle, junior high, and high school students. They will stimulate you to think of mathematics in ways that you did not learn in your college mathematics courses.

- **Students' Real-World Problems**

  What types of exhibits fit this category?

  Note that this is not a category that appears in most mathematics teachers' presentation portfolios. However, because of the need to motivate students to do mathematics and to connect mathematics to their own real worlds, you are urged to include this category in your working portfolio. It will serve as a receptacle for storing descriptions of situations and problems that (a) students will be intrinsically motivated to solve and (b) can be efficiently addressed or dealt with by doing mathematics that you want your students to learn to do. For this category, select descriptions of problems and situations from articles in professional journals (e.g., *Mathematics Teacher* and *Mathematics Teaching in the Middle School*) and websites, from lesson plans and cases exhibited in this text (e.g., Case 11.2 and Exhibit 11.13), and most importantly from your interactions with students and professional colleagues.

  How and when do you learn how to create these exhibits?

  As suggested above, you build up your storehouse of problems and situations from students' real worlds as you come across them—as you attempt to address the question that mathematics teachers hear again and again: "Why do we have to learn this?"

- **Professional Development**

  What types of exhibits fit this category?

  Select items that reflect efforts you have made and will make to (a) improve your own talents for teaching mathematics, (b) collaboratively work with colleagues to improve the work of schools, and (c) contribute to the teaching profession. For example, you might exhibit (a) your National Council of Teachers of Mathematics (NCTM) membership card, (b) a reflection paper stimulated by your participation in an NCTM regional conference, (c) a proposal for making a presentation at a regional conference of the Mathematical Association of America (MAA), and (d) a professional growth plan you will follow to get feedback on your teaching, participate in inservice education workshops, and stay current mathematically and pedagogically.

  How and when do you learn how to create these exhibits?

  Opportunities for professional growth are available during student teaching and throughout one's career as a professional teacher. You need to assertively take advantage of them by (a) becoming involved in professional organizations—especially NCTM, (b) participating in school-level, district-level, and state-level committees (e.g., a committee to revise the state core-mathematics standards and mathematics core-curriculum tests), (c) participating in inservice courses and workshops for teachers, (d) submitting proposals for obtaining resources for your classroom, and (e) collaborating with your colleagues to improve one another's teaching and curricula (e.g., by integrating the teaching of mathematics with other content areas (e.g., see Cases 11.1., 11.2, and 11.14)).

Exhibit 1.2
Casey Rudd's Resume.

Resume for Casey I. Rudd

Contact Information

- Mailing Address: 70½ South 300 West, Apt. #9
  Logan, UT 84321
- Phone: 435-815-2752
- e-mail: sq563@usu.cc.edu

Educational Background*

- High school diploma from McKinley High School, Baton Rouge, LA (awarded in June, 1998)
- B.S. in Mathematics Education from Utah State University (degree to be awarded in May, 2002)

Teaching Certification

All of the requirements for Secondary Teaching Certificate (grades 6–12) with a Level IV endorsement in Mathematics have been met; official Utah teaching certificate scheduled to be received prior to July 15, 2002.

Current Position

Student teaching and all the requirements for the aforementioned B.S. have been recently completed. A faculty position as a mathematics teacher at a middle, junior high, or high school is currently being sought for the 2002–2003 school year.

Previous Experience Relevant to Teaching

- Worked for the Mathematics and Statistics Department at Utah State University as a tutor for undergraduates taking precalculus, mathematics for elementary school teachers, introductory statistics, calculus I & II, business calculus, and multivariable calculus (1999–2001).
- Worked for the Mathematics and Statistic Department at Utah State University as a test grader for calculus II, business calculus, and a course in linear algebra and differential equations.
- Served as a volunteer after-school mathematics tutor for middle and high school students in the Cache County and Logan City School Districts (2000–2001).
- Periodically, worked as a private mathematics tutor for college, middle school, and high school students (1998–2001).
- As part of the field-based methods-of-teaching-mathematics course at Utah State University, team taught two days per week in two classrooms (prealgebra and geometry) for eight weeks at Canyon Fork Middle School and team taught two days per week for two weeks in two classrooms (algebra II and precalculus) at Mountain View High School with the supervision of Cooperating Teacher Charles C. Taylor, Cooperating Teacher Fawn Grieves, and University Professor Jim Cangelosi (2001).
- Student taught (two periods of seventh grade mathematics and one period of prealgebra, and one period of geometry) with the supervision of Cooperating Teacher Felice Tau for one semester at North Point Center School in Hyrum, Utah (2002).
- Coached basketball team of 12- to 13-year olds in the Junior Tigers league for two seasons in Baton Rouge, LA (1997–1998).

Other Experiences

- Served as vice president of the Diversity Forum at Utah State University (1999–2000).
- Held a variety of part-time jobs while attending high school and summer school at Utah State University (e.g., housekeeper at a motel and server at a fast-food restaurant) (1996–2000).

Scholarly Presentations and Educational Projects

- Coauthored and copresented the paper *Teaching Mathematics from a Historical Perspective* at the Northern Utah Regional Meeting of the Utah Council of Teachers of Mathematics in Ogden, UT (November, 2001).
- Served as an actor in the video program, *Leading Students to Use their Mathematical Skills to Describe Biological Phenomena,* sponsored by the U.S. Office of Education and produced by Utah State University Multimedia and Distance Learning Services (1999).

Continued

Exhibit 1.2
Continued

Professional Organizations

- National Council of Teachers of Mathematics
- Utah Council of Teachers of Mathematics
- Mathematical Association of America

Professional References

- Ms. Felice Tau, Teacher
  North Point Center School
  821 Eugene Street
  Hyrum, UT 84319
  Phone: 435-347-6433
  e-mail: felice.tau@clay.k12.ut.us
- Dr. James S. Cangelosi, Professor
  Department of Mathematics & Statistics
  3900 Old Main Hill
  Utah State University
  Logan, UT 84322-3900
  Phone: 435-797-1415
  e-mail: jcang@math.usu.edu
- Dr. Henitia T. Vasquez, Associate Professor
  Department of Curriculum & Instruction
  2815 University Blvd.
  Utah State University
  Logan, UT 84322-2800
  Phone: 435-797-7145
  e-mail: hvasq@coe.usu.edu
- Mr. Charles C. Taylor, Teacher
  Canyon Fork Middle School
  73220 Canyon Fork Road
  Logan, UT 84321
  Phone: 435-722-2371
  e-mail: ctaylor@cf.logan.k12.ut.us

---
* - An electronic version of Casey Rudd's professional teaching portfolio is available upon request and a hard copy version is available for examination during interviews.

Because Casey is willing to relocate, he applies for nearly 100 openings and posts his resume on the website of his university's career-placement office where it can be viewed by recruiters from schools; 44 school districts invite him for either on-site or telephone interviews. He accepts nine of the invitations for on-site interviews and three to be conducted via conference phone calls.

Interviews and Decisions

Typically, several candidates for a faculty opening whose applications clear the school district office's screening process are invited for individual interviews with personnel from the school district (e.g., the school district's personnel director) and the school (e.g., the principal, the head of the Mathematics Department, and an assistant principal). Imagine yourself—having graduated from college qualified and certified to teach mathematics at middle and secondary schools—about to be interviewed for a position by the principal and Mathematics Department head. From their prior examination of your application documents, they know that you are qualified and certified to teach mathematics. What more do they need to learn about you from the interview? Assuming that they really are in the process of selecting the best available candidate for the position and the job has not already been secretly promised to the principal's son-in-law, they need to make judgments about the answers to the following questions: Will your strategies for motivating students, engaging them in lessons, and managing your classroom be effective for this particular school's population? Will you interact with students, parents, colleagues, other school personnel, and visitors to the school in a highly professional manner? Will you respond positively to instructional supervision and further develop your talents for teaching mathematics during your tenure at the school? Is there a good fit between your professional goals and this particular teaching position? Is there a good fit between your teaching style and the needs of the particular students with whom you will be working? Besides effectively teaching your mathematics classes, in what other ways will you contribute to the school (e.g., serving on faculty committees, being involved in cocurricula activities, and collaborating on special projects)? Will you teach in a manner that is consistent with the National Council of Teachers of Mathematics' *Principles and Standards for School Mathematics* (*PSSM*) (NCTM, 2000b)? Will you uphold *PSSM*'s equity principle so that you teach from a multicultural perspective and equally value every student's right to learn? Will you take care of the mundane tasks required of all faculty members (e.g., completing administrative forms)?

Besides needing to address these questions about you, they will also use the interview (a) to clarify for you the expectations of the position and (b) as an opportunity to convince you that if they offer you the position, you should take it.

That is their perspective on the interview. But what about your perspective? Of course, you need to convince them that they should offer you the position. However, you also need to use the interview to help you decide if this is the position for you. Thus, you need to use the interview as an opportunity to help you make judgments about the following questions: Is this a faculty with teachers—especially in the Mathematics Department—with whom you want to collaborate? How well will the responsibilities and expectations of the position further your professional goals? Will you receive the level and quality of administrative and supervisory support that you need to successfully teach your students? Will you have access to needed technology? Whenever you make professionally tenable decisions in accordance with stated school policies, can you depend on your supervisors and administrators to support those decisions? What structures does the school have in place (e.g., conflict-management programs, effective school-wide discipline and safety policies, and security systems) that will allow you to establish a safe, nurturing classroom community that is conducive to learning mathematics? Is the salary and benefits package associated with the position competitive with those of other positions you may be offered? What assurances do you have that if you fulfill the responsibilities of the position in a satisfactory manner, that you will have the option of retaining the position? Is the school and community environment conducive to your succeeding as a teacher at a level that meets your own professional standards? Besides the expectations listed in the formal description of the position, what are the other expectations of the person who fills the position?

Let us return to Casey Rudd:

CASE 1.3

Casey was disappointed with his first interview at a high school because the principal failed to involve any of the teachers in the process, she seemed unfamiliar with Casey's application materials (e.g., she had not bothered to open the website with Casey's electronic portfolio although he had sent her an e-mail with a link to it), she appeared more concerned with Casey's willingness to fit into the social structure of the school and community than with his strategies for teaching mathematics, and she failed to address his questions regarding administrative and supervisory support for his work. He decided that he would not accept the position if it were offered to him.

Casey was positively impressed by the second interview at Westside Middle School. He especially liked the middle education philosophy—emphasizing integrated curricula, collaboration among teachers, cooperative learning, a nurturing school community, and attention to the needs of preadolescents and adolescents—that teachers and administrators espoused and appeared to put into practice. He decided to accept this position if it were offered to him unless an even more attractive position became available first.

Before his third interview for a position at Rainbow High School, Casey familiarizes himself with the contents of the faculty and student handbooks—especially the sections pertaining to the school-wide discipline and safety programs and policies. He also studies a document labeled "South-Point School District's Personnel Evaluation Policies and Procedures" that he requested from the school-district office. Because it is July, Rainbow High School is not in regular session, but he visits the campus the day before the interview to get a feel for the place and engage in brief informal conversations with some of the teachers, staff, and students involved with summer activities. This informal preview helps him to feel more at home during the interview and allows him to raise questions and points specific to Rainbow High's unique characteristics.

The interview is held in a faculty conference room with Mathematics Teacher Vanessa Castillo, Mathematics Department Chairperson Armond Ziegler, Principal Harriet Adkins, and Associate Principal Jack Breaux, all of whom appear to be familiar with Casey's application materials. After pleasant and warm introductions, the interviewers prompt Casey to describe some of his strategies for designing lessons and for engaging students in them. Casey's portfolio serves him well as he responds to that and subsequent prompts. For example, to help him explain how he employs both inquiry and direct instructional strategies, he pulls from the portfolio his cooperating teacher's description of a learning activity Casey conducted while student teaching; see Exhibit 1.3.

After Casey uses Exhibit 1.3 to explain why the appropriate mingling of inquiry and direct instruction is necessary for adherence to *PSSM*, the following exchange occurs:

Harriet: Well I'm very happy that your teaching will be consistent with the new math standards; we need to have a standards-based curriculum so that we meet accreditation reviews. However, as the principal, I am accountable for how well our students do on the state

Exhibit 1.3

Description (by His Cooperating Teacher) of a Learning Activity Casey Rudd Conducted While Student Teaching.

Casey Rudd Conducting a Learning Activity While Student Teaching in a Geometry Class (Transcribed from a Videotape)

For this class session with his geometry class, Mr. Rudd plans to lead students to discover that the lateral surface area of a right cylinder with radius r and height h is $2\pi rh$. Because the objective requires them to reason rather than just remember, he employs inquiry instructional strategies. (Note that with *inquiry instructional strategies* the teacher engages students in activities in which they interact with information, make observations, and formulate and articulate ideas that lead them toward discovery, concept construction, or invention). If the students do discover the relationship $A = 2\pi rh$ with time remaining in the class, he plans to use direct instructional strategies to introduce them to the conventional words associated with their discovery (i.e., "right cylinder" and "surface area") and help them remember how to compute lateral surface areas of right cylinders. (Note that with *direct instructional strategies,* the teacher exposes students to the information or algorithm to be remembered and then engages them in some type of repetitive activity to commit the information or algorithm to memory.)

Mr. Rudd begins by telling the class, "With the permission of Mr. Duke (the head custodian for the school), I've asked Izar and Elaine to bring in one of the trash barrels from the school grounds." The two students, Izar and Elaine, display the barrel in front of the classroom. Mr. Rudd continues, "Mr. Duke told me he plans to repaint all these barrels, using bright colors instead of this drab gray again. Yes, Parisa, you have the floor."

Parisa: What has this got to do with us?

Mr. Rudd: Mr. Duke thought you might have some ideas on what colors he should use. Yes, Izar?

Izar: Let's draw pictures on them?

Mr. Rudd: Mr. Duke said each of the three sections of the barrels could be different colors, but nothing fancier than that. He also said he'd let us choose the colors in exchange for providing him with an estimate of the amount of paint he has to buy. If he's going to use colors other than the gray he has now, it's going to be expensive and he doesn't want to buy more than he needs.

Further discussions lead the students to recognize the need to find the lateral surface area of the barrels, although technical phrases like "lateral surface area" or "right cylinder" are not used. No one has a formula for computing it, but Mr. Rudd continues, "We may not know how to figure the area of this barrel, but there are some shapes with areas we can compute. Everyone, please write down the names of three shapes for which you have area formulas." Observing that everyone has written something, he continues, "In the order I call your names, read your lists: Nora, Mark, Eli, and Elaine."

Nora: Rectangle is the only one I put.

Mark: Rectangle, triangle, and another one I don't remember.

Eli: Square, triangle, and circle.

Elaine: Square, rectangle, and triangle.

Mark: But a square is a rectangle, so you didn't also need to list a square.

Mr. Rudd: Isn't that also true for this barrel?

Michelle: No, a barrel isn't a type of rectangle.

Mr. Rudd: Hmmm, maybe that's so. I was just thinking about the part of this barrel that needs to be painted.

Izar: Well we said only the outside needs to be painted.

Mr. Rudd: What if we just covered the outside of the barrel with colorful contact paper?

Michelle: That would never hold up.

Mr. Rudd: Yeah, too bad! Seeing our butcher paper in the back of the room gave me that idea. It seems so simple just to wrap paper around the barrel. Too bad it wouldn't hold up!

Ebony: Mr. Rudd, may I try something with our butcher paper?

Ebony wraps a section of the paper around the barrel, cuts it, unwraps it, and then displays the rectangular shape to the rest of the class. Eyes light up around the room as students begin to understand what Ebony demonstrated: The lateral surface area of the barrel is equivalent to the area of a rectangle whose dimensions can be obtained from the barrel.

Continued

mandated core-curriculum tests and *Stanford Achievement Tests.* Although I agree that our students are better off with a standards-based curriculum in which they construct concepts for themselves, discover relationships like that one for cylinders, and reason things out, I'm afraid if we spend the bulk of class time on the higher-level cognitive things, that we won't have enough time to build the lower-level memory skills that are emphasized on these high-stakes tests. What's your answer to that?

Exhibit 1.3
Continued

The lesson continues with Mr. Rudd playing upon Ebony's discovery to lead students to associate the circumference of the base of a right cylinder with a rectangle's width and the height of the cylinder with the rectangle's length. With about 15 minutes remaining in the class period, nearly all students seem to understand why the lateral surface area of a barrel is $2\pi rh$. Mr. Rudd begins employing direct instructional strategies to introduce the conventional names associated with the relationship they formulated and to use that relationship to compute lateral surface areas. Using an overhead transparency with the image shown here, he defines the words "right cylinder" and "lateral surface area" and explains, "As you can see by the formula, to find the lateral surface area of a right cylinder you multiply 2π times the radius of the base times the height. Here let me work an example for you. Suppose the cylinder's height is 20 centimeters and . . ."

After going through the steps of the algorithm, Mr. Rudd assigns the following homework exercise:

"Find three different objects located in or near your place of residence with shapes that are close to that of a right cylinder. After making the necessary measurements, compute the approximate surface area of each of those objects. For each object, bring to class a one-paragraph description of the object and what you did to compute an approximate value for its surface area."

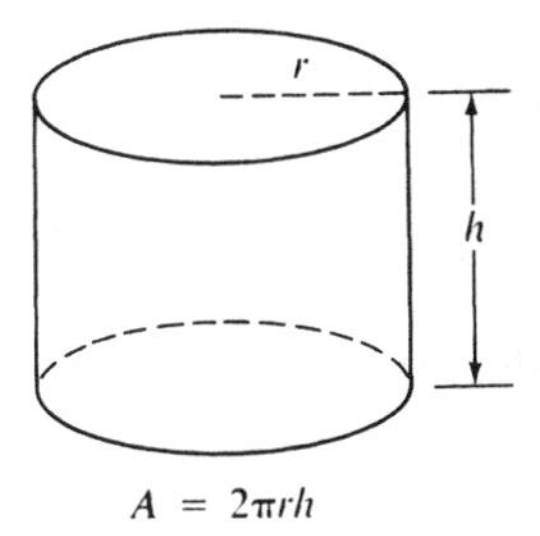

$A = 2\pi rh$

Casey: I understand that *PSSM*-based instruction emphasizes higher level learning and those tests emphasize algorithmic skills. But I'm also convinced that *PSSM*-based instruction will lead to higher rather than lower scores even on those types of tests. Allow me to illustrate with a somewhat complex example: Last year I was tutoring an eighth grader, Irene, who was taking algebra I at a middle school. She showed me this homework exercise that read, "Solve for x in the equation $8(11 - x) = 56$." So, I said, "Hmm, let's see, we need to find a number for x that will make that statement true. I wonder what it is. Okay, Irene, 8 times what number is 56?" Irene immediately answered, "7." "So," I asked, "What must $11 - x$ be?" Irene immediately answered, "7." So I asked, "What number do you subtract from 11 to get 7?" She immediately answered, "4." With a smile I asked, "So what must x be?" She immediately said, "4." "That's it," I responded excitedly. But without any sign of joy or satisfaction, she said with a frown, "That's not how we're suppose to do it." Disappointed, I thought to myself, "Okay, so her teacher wants to use a standard algorithm and display the steps along the way." I said, "Okay, let's work through the steps to rewrite this equation so that x appears all alone on the left side and a simplified constant is the right side." I then led her through these steps:

Casey writes on the conference room whiteboard as he continues.

Casey: I said to her, "Rewrite this equation so that x is by itself. We need to free x from $8(11 - x)$. What should we do to undo that multiplication by 8?" Irene answered, "Divide both sides by 8" and she does this to the equation:

$$\frac{8(11 - x)}{8} = \frac{56}{8}$$

$$11 - x = 7$$

Pleased that she seems to be following so well, but wondering why she appears so glum, I asked, "Okay, what should we do to continue to isolate x?" She says, "Subtract 11 from both sides then multiply by -1." She completes the steps:

$$11 - x = 7$$

$$11 - x - 11 = 7 - 11$$

$$-x = -4$$

$$(-1)(-x) = (-1)(-4)$$

$$x = 4$$

"So you do know how to solve these types of equations," I said with a smile. With a scowl, Irene mumbles, "But that's not the way we're supposed to do these." Becoming frustrated, I asked her to show me the section in her textbook. She opened the algebra I book to a section with the heading, "Using the Distributive Law to Solve Equations." To my horror, I followed the gruesome, illogical, inefficient algorithm laid out in the book's example:

"To solve for x in the equation

$$3(10 - x) = 15$$

First use the distributive law to get

$$3(10) - 3(x) = 15$$

Then simplify by multiplying

$$30 - 3x = 15$$

Subtract 30 from both sides

$$30 - 3x - 30 = 15 - 30$$

and further simplifying

$$-3x + 0 = -15$$

which is

$$-3x = -15$$

Now divide both sides by -3

$$\frac{-3x}{-3} = \frac{-15}{-3}$$

to get

$$x = 5$$"

Harriet: I never thought of solving that equation the way you first showed Irene. When you first stated the equation, I started multiplying without ever stopping to think why I was doing what I was doing. The book's method seems like a lot of extra trouble.

Casey: The fact that it's more trouble is not what I find most repulsive about this textbook approach. No attention is paid to students' comprehending the task at hand. Too often students are told how to find "answers" without any attention to understanding the question. For example, the textbook's algorithm for using the distributive property for solving equations begins with a step that is counter to isolating x!

Vanessa: I see what you mean. By using the distributive property you end up multiplying x by a number that makes x less isolated than before, so you have to undo that step later. It's a step you would skip if you looked at the big picture—comprehending the task of isolating x or better yet, continually asking yourself, "What does x need to be to make this statement true?"

Casey: Thanks for understanding the point. However, I haven't forgotten that Harriet prompted me to tell this story with her question concerning the effect of *PSSM*-based instruction on high-stakes test scores. I think students don't do as well as they could with these tests when they learn so-called "mathematics" using the overly focused isolated-algorithm approach that appears in textbooks. When students learn an algorithm like the one for using the distributive property to solve for equations, something like this happens: After watching his teacher solve a few equations using the textbook's algorithm, a student—call him Rasheed—begins the homework assignment, which is to solve a bunch of equations listed in a textbook exercise set appearing in a section with "Using the Distributive Property to Solve Equations" at the top of the page. Seeing "$8(14 + x) = 23$" prompts Rasheed to immediately and thoughtlessly begin grinding out the steps by multiplying 14 by 8 and so forth. At no point does Rasheed need to decide which of the hundreds of algorithms he's learned to use to solve this equation because all of the equations in this exercise set use the one that's named at the top of the page and is illustrated by the section's example. If he forgets what to do, he goes back and faithfully does what is in the example. By the way, after Rasheed "masters" this algorithm, he'll be moved on to the next section of the book to "master" another too focused isolated algorithm for another type of equation and on and on. But when Rasheed is confronted with solving an equation like $9(8 + x) = 90$ on the state core-curriculum test or a standardized test, he has trouble remembering which one of the hundreds of algorithms he supposedly mastered over the past seven months to use. Why? Because the prompts on the test are mixed up and don't have labels like "using the distributive property to solve equations."

Jack: Okay, I see why the old way of teaching causes Rasheed to mess up on those tests, but how will your style of teaching cause Rasheed to do better? He'll still have to deal with all those different kinds of problems on the test.

Casey: I won't neglect my students' algorithmic skills, but I won't need to teach them nearly as many algorithms because I'll lead my students to make connections between the different mathematical relationships that are the bases for the algorithms. Remember the description from my portfolio (i.e., Exhibit 1.3) in which students could figure the surface area of a right cylinder from their depth of understanding of the formula for finding the area of a rectangle. Students will be able to apply some key relationships that they discovered for themselves to a wide variety of mathematical problems long after they forget the rules for when to apply hundreds of algorithms that they supposedly mastered in isolation. Even more importantly, comprehension and communication are emphasized with *PSSM*-based instruction. If Rasheed approaches solving equations the way I first approached $8(11 - x) = 56$ with Irene, then he would do better on those high-stakes tests because he would attempt to comprehend the task rather than immediately trying to remember which algorithm to use. With the *PSSM*-based way, students will reason out responses to those types of test prompts without wasting time searching their memory banks for inefficient, isolated tricks.

Vanessa: I have an example that fits Casey's point. Here's a prompt I remember from one of the standardized tests we gave last year.

Vanessa writes the following on the board:

If $a + b = 10$ and $ab = 5$, then $a^2 + b^2 =$ ____

(A) 80 (B) 85 (C) 90 (D) 95 (E) 100

Vanessa: Everyone take two minutes to choose a response.

After some mild protests and nervous groans, they begin diligently figuring with pen and paper. Jack also uses a calculator.

Vanessa: Okay, time's up; what did you come up with?

Jack: I didn't have time to finish, but since there are two variables and two equations, I'm solving them simultaneously by substituting $10 - b$ for a in $ab = 5$. That gives me $(10 - b)b = 5$ which is $10b - b^2 = 5$. And that's a quadratic, which I can solve by the quadratic formula. So I rewrote it as $b^2 - 10b + 5 = 0$. But it's going to take me a while to run that through the quadratic formula.

Vanessa: And then after you solve for b you'll use one of the original equations to solve for a. Is that right?

Jack: That's right; then I'll be done.

Harriet: That may be right but by the time you get there, you won't have enough time to finish the rest of the test.

Casey: But we weren't asked to solve for a and b.

Jack: Oh, that's right! I got so caught up in solving for a and b that I forgot to use my solution to compute a value for $a^2 + b^2$.

Vanessa: Now what if instead of attacking this problem using your textbook-learned approach, you had used Casey's method with Irene to comprehend the task before diving into an algorithm? The question wasn't to solve for a and b, it was to solve for $a^2 + b^2$.

Harriet: But you need to know a and b to know $a^2 + b^2$.

Jack: Oh! No you don't; now I see what she means! Wow! If you keep the question in mind, it's a lot quicker to reason out a value for $a^2 + b^2$ without solving for a and b.

Harriet: Show me.

Jack writes the following on the board:

$$a + b = 10$$

$$a^2 + 2ab + b^2 = 100$$

$$a^2 + 2(5) + b^2 = 100 \quad \text{(because } ab = 5\text{)}$$

$$\text{So } a^2 + b^2 = 100 - 10 = 90$$

Casey: Thank you for helping me make my point about the high-stakes testing, but you also demonstrated for me how ingrained textbook-based instead of *PSSM*-based learning is in us. I was using Jack's method rather than my own to respond to that prompt.

Harriet: I understand why *PSSM*-based instruction might help students on these tests, but math courses have to cover a lot of content. How will you get through it all if you spend time getting students to discover everything? That activity with the trash barrel from your portfolio takes up a lot of class time.

Casey: Take a look at the table of contents of any of the mathematics textbooks you use here at Rainbow. At first glance, the number of topics listed appears overwhelming. But I'm going to share a secret with you that an inner circle of mathematics teachers keep as secret as the Pythagorean Brotherhood kept the existence of irrational numbers secret.

Vanessa: But because this isn't 550 B.C. in Greece, you won't be executed for heresy for giving away trade secrets. Here, I've got my prealgebra text with me; I'll pass it around so that everyone can see the table of contents.

Exhibit 1.4 displays the table of contents of the prealgebra textbook.

Harriet: That's a huge amount of content for a teacher to cover.

Casey: Here is the secret: There is not nearly as much mathematics to be learned from this text as most people think. Examine these topics carefully and you'll see that only a couple of new ideas are introduced; much of this content involves different ways of expressing the same thing over and over. Look, chapter 2 focuses on decimals, chapter 4 on fractions, and chapter 7 on ratio and proportion; chapter 10 tosses in something on percent. Typically, students perceive these four topics as if each is a new concept unrelated to the others. In reality, decimals, fractions, and percents are simply three ways of expressing exactly one concept, namely the set of rational numbers—which is in chapter 8's title.

Armond: But don't students need to learn different ways of expressing the same idea?

Casey: Absolutely, but we need to let them in on the secret by emphasizing the connections. Why, for example in this book, is the topic divisors of integers introduced in chapter 4 with no mention of multiples of integers, whereas multiples of integers is introduced in chapter 6 with no mention of divisors of integers? After all, saying "12 is a multiple of 4" is just another way of saying "4 is a divisor of 12"—just like saying "Sam is Mary's husband" is the same as saying "Mary is Sam's wife."

Armond: Well prealgebra isn't the first time students learn about divisors and multiples of integers.

Vanessa: Except that in the sixth and seventh grade textbooks divisors are called "factors."

Casey: The same content is listed over and over in mathematics textbooks from one grade to the next. How many times do we need to introduce slope? It's supposedly covered in every mathematics course from 6th grade through calculus!

Armond: But even with all that repetition students still aren't getting what they need to succeed in calculus.

Casey: To people who don't know any better, calculus is that impossible subject that's housed in a mammoth college textbook with over 1,000 pages full of strange symbols. However, if we examine those pages carefully, we see that in calculus there are only a couple of overriding concepts—like limits and continuity—and maybe a few more critical subconcepts like derivative and integral. Most of those pages are simply filled with different algorithms for applying those few concepts to different families of functions.

Armond: So you're saying that by recognizing the connections among these different topics, we could repackage the content so that we'd have less to teach. That's the "livelier and leaner" curriculum that was talked about in conjunction with the calculus-reform movement of the early 1990s.

Casey: Which would give us time to lead students to construct the key concepts and discover the key relationships for themselves.

Vanessa: Which would give students the depth of understanding they need so we could eliminate much of the overlap from course to course.

The interview continues with Casey raising his own questions about the position. Throughout, the tone of the interview is warm with occasional humorous comments, but there is never any doubt that the business at hand is very serious.

Several days after interviewing at two more schools—a junior high school and a middle school—Casey receives offers for positions at the middle school, the high school where he interviewed, and Rainbow High School. He also learns that Westside Middle School decided to hire another applicant instead of him. Disappointed about the Westside position but enthusiastic about two of the three offers that he has in hand, he decides to join the Rainbow High School faculty and cancels the remaining interviews.

A decision of which offer for a faculty position to accept should be a function of variables such as the faculty's commitment to quality teaching, the administration's style and support for the faculty's commitment to quality teaching, salary and benefits, attitude and competence of the administrators and faculty, facilities and resources, teaching assignment and load, and location of the community. To assess these variables, you need to study literature (e.g., curriculum guides and student handbooks) obtained from school district offices and schools, examine school districts' and schools' websites, consult with trusted classroom teachers and college faculty members, be extremely observant and assertively raise key questions and issues during interviews, and engage in conversations with various school personnel—especially mathematics teachers.

TEACHING LOADS AND OTHER RESPONSIBILITIES

School-year calendars vary among school districts. Most schools operate either two 18-week semesters or three 12-week trimesters per 12-month year. Year-round schools may have three 18-week sessions. Daily class-period schedules for middle, junior high, and high schools vary among districts as well as among schools within districts. For most of these

Exhibit 1.4

Prealgebra Textbook Table of Contents That Vanessa Displayed at Casey's Interview.

Chapter 1: ALGEBRAIC EXPRESSIONS WITH WHOLE NUMBERS
- 1-1 Order of Operations
- 1-2 Algebraic Expressions
- 1-3 Algebraic Expressions: Addition
- 1-4 Algebraic Expressions: Subtraction
- 1-5 Algebraic Expressions: Multiplication
- 1-6 Algebraic Expressions: Division
- 1-7 Properties
- 1-8 The Distributive Property
- 1-9 Problem Solving and Applications
- 1-10 Calculator and Computer Applications
- 1-11 Review Exercises

Chapter 2: OPERATIONS WITH DECIMALS
- 2-1 Place Value
- 2-2 Estimating and Computing Decimal Sums and Differences
- 2-3 Estimating and Computing Decimal Products
- 2-4 Estimating and Computing Decimal Quotients
- 2-5 Metric Units of Measurements
- 2-6 Scientific Notation
- 2-7 Problem Solving Using One-Step Equations
- 2-8 Significant Digits
- 2-9 Calculator and Computer Applications
- 2-10 Review Exercises

Chapter 3: SOLVING EQUATIONS USING PROPERTIES OF INTEGERS
- 3-1 Absolute Value
- 3-2 Ordering and Comparing Values
- 3-3 Using the Number Line
- 3-4 Sums of Integers
- 3-5 Differences of Integers
- 3-6 Products of Integers
- 3-7 Quotients of Integers
- 3-8 Evaluating Algebraic Expressions
- 3-9 Using Subtraction to Solve Equations
- 3-10 Using Addition to Solve Equations
- 3-11 Using Division to Solve Equations
- 3-12 Using Multiplication to Solve Equations
- 3-13 Problem Solving and Applications
- 3-14 Using Technology
- 3-15 Review Exercises

Chapter 4: FRACTIONS AND DIVISORS
- 4-1 Divisors and Monomial Expressions
- 4-2 Exponents and Powers
- 4-3 Number Theory
- 4-4 Prime Numbers
- 4-5 Cryptography
- 4-6 Prime Divisors
- 4-7 Greatest Common Divisor
- 4-8 Simplifying Fractions
- 4-9 Sums and Differences of Fractions
- 4-10 Adding Unlike Fractions
- 4-11 Products and Quotients of Fractions
- 4-12 Comparing Fractions
- 4-13 Equivalent Algebraic Fractions
- 4-14 Solving Equations Using Fractions
- 4-15 Geometric Applications
- 4-16 Problem Solving and Applications
- 4-17 Review Exercises

Chapter 5: TWO-STEP EQUATIONS AND GRAPHING
- 5-1 Solving Two-Step Equations
- 5-2 Problem Solving with Two-Step Equations
- 5-3 Ordered Pairs
- 5-4 Cartesian Coordinate System
- 5-5 Equations with Two Variables
- 5-6 Graphing Equations
- 5-7 Problem Solving with Modeling
- 5-8 Slope
- 5-9 Intercepts
- 5-10 Systems of Equations
- 5-11 Solving Systems of Equations by Graphing
- 5-12 Calculator Applications
- 5-13 Algebraic Solutions to Systems of Equations
- 5-14 Review Exercises

Chapter 6: SOLVING AND GRAPHING INEQUALITIES
- 6-1 Inequalities and the Number Line
- 6-2 Using Properties of Inequalities
- 6-3 Solving One-Step Inequalities
- 6-4 Solving Two-Step Inequalities
- 6-5 Graphing One-Variable Inequalities
- 6-6 Systems of Inequalities
- 6-7 Solving Systems of Inequalities
- 6-8 Graphing Solutions to Systems of Inequalities
- 6-9 Problem Solving and Application
- 6-10 Computer and Calculator Applications
- 6-11 Review Exercises

Chapter 7: RATIO AND PROPORTIONS: ALGEBRAIC APPLICATIONS
- 7-1 Ratios and Rates
- 7-2 Proportions
- 7-3 Solving for Proportions
- 7-4 Ratios and Estimation
- 7-5 Percent and Estimation
- 7-6 Solving Equations about Rates
- 7-7 Using Proportions to Solve Equations
- 7-8 Ratios and Geometry
- 7-9 Graphical Representations for Relating Ratios
- 7-10 Problem Solving and Application
- 7-11 Review Exercises

Continued

Exhibit 1.4
Continued

Chapter 8:	RATIONAL NUMBERS AND EXPRESSIONS
8-1	Writing Decimals and Fractions
8-2	Repeating and Terminating Decimals
8-3	Multiples
8-4	Least Common Multiples
8-5	Rational Expressions
8-6	Adding and Subtracting Rational Expressions
8-7	Multiplying Rational Expressions
8-8	Dividing Rational Expressions
8-9	Multiplicative Inverses
8-10	Solving Equations with Rational Expressions
8-11	Graphing Equations with Rational Expressions
8-12	Problem Solving and Application
8-13	Review Exercises
Chapter 9:	OPERATIONS WITH POLYNOMIALS
9-1	Monomials and Binomials
9-2	Adding and Subtracting Polynomials
9-3	Powers of Monomials
9-4	Multiplying a Polynomial by a Binomial
9-5	Multiplying Binomials
9-6	Squaring Binomials
9-7	Dividing with Polynomials
9-8	Factors of Polynomials
9-9	Calculator Applications
9-10	Review Exercises
Chapter 10:	STATISTICS, GRAPHS, AND PERCENT
10-1	Data Collection
10-2	Recording and Displaying Data
10-3	Measures of Central Tendency
10-4	Stem and Leaf Plots
10-5	Measures of Variation
10-6	Box and Whisker Plots
10-7	Data as Percentages
10-8	Histograms
10-9	Scatter Plots
10-10	Making Predictions
10-11	Deceptive Statistics
10-12	Review Exercises
Chapter 11:	DISCRETE MATH AND PROBABILITY
11-1	Counting and Tree Diagrams
11-2	Counting and Multiplication
11-3	Permutations and Combinations
11-4	Chance and Probability
11-5	Probability Tables
11-6	Probability of Independent Events
11-7	Probability of Dependent Events
11-8	Sum of Probabilities
11-9	Making Predictions
11-10	Patterns
11-11	Arithmetic Sequences
11-12	Geometric Sequences
11-13	Computer Applications
11-14	Review Exercises
Chapter 12:	ALGEBRA APPLIED TO GEOMETRY
12-1	Statements and Propositions
12-2	Logic: Conditional Statements
12-3	Venn Diagrams
12-4	Angle Relationships
12-5	Parallel Lines
12-6	Segment and Angle Constructions
12-7	Triangles
12-8	Congruent Triangles
12-9	Transformations
12-10	Similar Triangles
12-11	Quadrilaterals
12-12	Polygons
12-13	Tessellations
12-14	Problem Solving
12-15	Computer Applications
12-16	Review Exercises
Chapter 13:	DISTANCE, AREA, AND VOLUME
13-1	Perimeter and Lengths
13-2	Area of Parallelograms
13-3	Area of Triangles
13-4	Diameter, Radius, Circumference, and Pi
13-5	Area of Circles
13-6	Area and Probability
13-7	Surface Areas of Prisms
13-8	Surface Areas of Cylinders
13-9	Surface Areas of Pyramids and Cones
13-10	Volumes of Prisms
13-11	Volumes of Cylinders
13-12	Volume of Pyramids and Cones
13-13	Problem Solving and Applications
13-14	Computer Drawings
13-15	Review Exercises
Chapter 14:	ALGEBRA APPLIED TO RIGHT TRIANGLES
14-1	Square Roots
14-2	Irrational Numbers
14-3	Estimating Square Roots
14-4	The Pythagorean Relationship
14-5	Using the Pythagorean Relationship
14-6	Special Right Triangles
14-7	The Tangent Ratio
14-8	The Sine and Cosine Ratios
14-9	Applying Trigonometry
14-10	Calculator Graphs: Trigonometry
14-11	More Problem Solving
14-12	Review Exercises

Exhibit 1.5

Casey Rudd's First-Year Teaching Schedule.

First-Semester Schedule

Class Period		Assignment	Room	Course Credits per Semester
Homeroom	8:10–8:25	10th grade-B	213	—
1st	8:30–9:25	Algebra I*	213	0.5
2nd	9:30–10:25	Preparation	—	—
3rd	10:30–11:25	Geometry*	213	0.5
4th	11:30–12:25	Geometry*	108	0.5
Lunch A	12:30–12:55	Lunch supervision	Lunchroom	—
Lunch B	1:00–1:25	Free	—	—
5th	1:30–2:25	Life-Skills Mathematics**	213	0.5
6th	2:30–3:25	Precalculus**	213	0.5
Announcements	3:25–3:30	—	213	—

Second-Semester Schedule

Class Period		Assignment	Room	Course Credits per Semester
Homeroom	8:10–8:25	10th grade-B	213	—
1st	8:30–9:25	Algebra I*	213	0.5
2nd	9:30–10:25	Probability & Statistics**	213	0.5
3rd	10:30–11:25	Geometry*	213	0.5
4th	11:30–12:25	Geometry*	213	0.5
Lunch A	12:30–12:55	Free	—	
Lunch B	1:00–1:25	Lunch supervision	Lunchroom	—
5th	1:30–2:25	Preparation	—	—
6th	2:30–3:25	Probability & Statistics**	213	0.5
Announcements	3:25–3:30	—	213	—

* - Two-semester course (one group of students both semesters)

** - One-semester course (different group of students each semester)

schools, the school day is organized into either six periods of approximately 50 minutes each or five periods of about 70 minutes each. Some schools have fewer, longer class periods per day (e.g., four 90-minute periods) with each course (e.g., geometry) meeting three days one week and two days the next rather than five days per week. The daily teaching load for most secondary school teachers is either five periods of about 50 minutes each or four periods of about 70 minutes each. It is more common for mathematics teachers at smaller schools (e.g., with only 500 students) to teach a greater variety of mathematics courses per term (e.g., one section each of algebra I, geometry, life skills, pre-calculus, and algebra II) than teachers at larger schools (e.g., 3,000 students; three sections of algebra I and two sections of life-skills mathematics).

Teachers also have many other responsibilities besides teaching "their" classes. Casey Rudd's situation is not uncommon:

CASE 1.4

Exhibit 1.5 reflects Casey's year-long teaching load. Casey was somewhat concerned that for the first semester his teaching assignment includes four different courses in a five-class load. However, he thinks, "It's unrealistic to expect to keep all sections of the same course at the same pace throughout the semester; I still have to prepare differently for two different sections of the same course. Besides it's the total number of students than influences workload more than the number of courses."

Besides teaching five classes per semester, Casey is expected to do the following: (a) Serve as homeroom administrator for a group of 24 10th graders. (b) Serve as a general supervisor for students during school hours, enforcing school-wide discipline and safety policies.
(c) Serve as lunchroom monitor during one of the two lunch periods. (d) Assist in the governance of the school by responding to administrators' requests for input and management tasks and by participating in both general

faculty and Mathematics Department meetings. (e) Cooperate in the school's system for both instructional supervision and administrative supervision of his own teaching and that of other teachers. (f) Participate in professional development activities (e.g., by attending inservice workshops, taking college courses, and being involved in organizations such as the local affiliate of NCTM). (g) Represent Rainbow High School as a professional in the community.

ORGANIZING FOR THE YEAR

Weeks Before the Opening of the School Year

You gain a considerable advantage by giving yourself as much preparation time as possible between when you accept a position and the opening of the school year. Consider Case 1.5:

CASE 1.5

With the opening of school about five weeks away, Casey is issued keys for the building and his home-base classroom (Room 213), a copy of the State Education Office's *Curriculum Guide for Mathematics,* and, for each mathematics course at Rainbow, a teacher's edition of the textbook. He begins to organize for the year by surveying Room 213, for which he will be responsible, and then Room 108, the home base for a biology teacher and the classroom for fourth-period geometry during the first semester. He thinks, "Twelve computers available in my room; that's 11 for students and one exclusively for me. But there are none in Room 108 and with those fixed lab tables, 108 is just too inflexible for me to operate in there! Why can't I use my own room for fourth period? Maybe Armond can work something out and get that switched for me. For now, I should get my room arranged the way I want it. But how do I want it arranged? I guess that depends on how I organize the courses—how much students need to use the computers, and so forth—and how many students each period. Harriet said we won't get class rolls until the first week of school but that my classes should run between 20 and 30.

"So I really need to get a better idea of how I want to organize the courses before worrying much more about the room. Of course, the room will also influence the courses—having no computers available in an inflexible lab will restrict what I can do fourth period.

"Armond said that the textbooks are set for this year, but I'd have a say in the choice of some of the texts next year when there is supposed to be a turnover. Might as well start going through these texts to see what I have to work with."

Over the next few days, Casey familiarizes himself with the textbooks and state curriculum guide, taking notes to be used in planning the courses. The curriculum guide includes (a) a statement establishing NCTM's (2000b) *PSSM* as the standards to be followed and (b) goals for each state-approved mathematics course. For each of Casey's courses, the goals listed in the curriculum guide refer to mathematical topics that appear to subsume most—but not all—of the topics listed in the textbooks. He judges that the textbooks will be useful sources of exercises, examples, definitions, postulates, and theorems, but that for most topics he will not be able to depend on them for lessons targeting higher cognitive learning levels (e.g., lessons that lead students to construct concepts, discover relationships, communicate with mathematics, or apply mathematics to their own real-life situations).

Help From Colleagues

Why did you decide to become a mathematics teacher? Some teachers answer the questions with, "I love doing mathematics; a teaching certification assures me of having a job." Some others answer, "I love working with kids and mathematics is one of the areas in which there's a great need for teachers." I hope you decided to teach mathematics because you love both working with students and doing mathematics. However, for the support you need to successfully teach your students, for addressing teaching-related problems (often related to motivating students to engage in lessons and managing students' behaviors), and for adult professional stimulation with some relief from concentrated time in the company of adolescent and preadolescent students, you need to be part of a circle of trusted professional colleagues (Kramer, 2001).

Casey interacts with colleagues at Rainbow well before the school year begins:

CASE 1.6

With Armond Ziegler, Casey raises the issue of scheduling his fourth-period geometry class in Room 213 instead of 108. Understanding Casey's plight, Armond checks on the matter and finds out that 213 is the only room available during fourth period that is large enough to accommodate a health science class. Due to a new state-mandated health science requirement, a double section of more than 40 students will be using Room 213 with extra desks brought in just for fourth period; Room 108 cannot accommodate the extra desks. Casey's panicky feelings over thoughts of this fourth-period onslaught on his home-base classroom are somewhat tempered during the following conversation with Armond:

Armond: At least some of the time you can conduct class in the computer lab.

Casey: I thought teachers weren't allowed to bring whole classes into the library and technology center; that's supposed to stay open for drop-in use.

Armond: Not the technology center—the computer lab in the basement with 33 networked PCs and three printers. I didn't realize no one told you about it.

Casey: I somehow missed it when I toured the building.

Armond: I should have made a point of showing it to you. Most teachers forget it's down there, but a few—especially in business education—think of it as their exclusive domain. It's supposed to be available for individual student use from 6:30 A.M. to 8 A.M. and from 3:30 P.M. to 9 P.M. During the school day, you can schedule a whole class in there on a first-come, first-served basis. But no one class is supposed to tie it up more than two periods in any one week.

Casey: What about the technology classes; don't they need it every day?

Armond: No, there's a special technology classroom just like there's a computerized writing lab for the English Department. Mr. Tramonte in the library is in charge of scheduling classes for the basement lab.

Casey: I'd better get my courses organized so I can decide what days I'll need it because it's first come, first served.

Armond: He won't put you on the schedule before August 10, the day the faculty is officially supposed to report. But apprizing him of your intentions now will increase your chances of having it available when you need it.

Casey: Thanks.

Buoyed by thoughts of how he might take advantage of the lab, Casey begins organizing his courses. He is sitting on the floor of Room 213 with a box of mathematical manipulatives (e.g., pipes, wheels, counters, algebra tiles, measuring devices, dice, playing cards, a large wooden barrel, and string) thinking about introducing various teaching units with inquiry lessons as Vanessa Castillo and Don Delaney enter (see Exhibit 1.6):

Exhibit 1.6

Casey About to Hear Discouraging Advice From Don.

Vanessa: Casey, I'd like you to meet Don Delaney, another member of our Mathematics Department, with whom you'll enjoy working.

Casey: Hello, Don, very nice to meet you.

Don: So you're the new man on the block! Welcome to Rainbow High.

Vanessa: Don teaches algebra II and trigonometry and, like you, has a section of algebra I this year.

Don: Right. But what's all this stuff you've got here? Looks more like you teach shop than math, Casey.

Casey: These are things I've been collecting to use as manipulatives for my inquiry lessons.

Don: So you're another eager beaver fresh out of college who's going to try that discovery stuff! Look, take some advice from this veteran. I've been in this business for nearly 20 years and all that discovery stuff just wastes time and gets you in trouble.

Casey: What do you mean?

Don: You don't need to know how a car works to be able to drive it. Math works the same way. Kids can learn math correctly without knowing how it works. You spend all that time trying to get them to discover and you never get around to covering the material. Only the real smart kids are capable of understanding the whys, and they'll learn that in college.

Casey: So that's why you think the discovery stuff wastes time. But how can it get me into trouble?

Don: In two ways. One, ever try to manage a class of teenagers when they're all disorganized, running around measuring stuff? I've got perfect control of my classes; they stay in their seats with book, paper, pencil, and calculator—that's all they need.

Casey: And the second way?

Don: Look what you've got here—cards, dice, and a wine barrel. Bring that into a math class and you'll have parents coming down on you for encouraging their children to gamble and drink. You might have gotten away with that 25 years ago when teachers had some respect and authority, but now everything you do is questioned. So it's best not to try anything radical. Remember, this is a litigious society; there's a lawyer out their waiting to pick your pocket!

Casey: Ouch! I didn't think about that.

Don: Hey, we're a close-knit department here; we look out for one another. All of us love the students and love mathematics.

Casey: Well thanks for the advice.

Don: Hey, advice is cheap. Like I said, welcome aboard. You'll really like it here. Are you coming Vanessa?

Vanessa: I'll meet you in the workroom in 15 minutes, I need to talk with Casey about a few things first.

After Don leaves, Vanessa continues the conversation with Casey:

Vanessa: What did you think about Don's advice for you?

Casey: He's a bit skeptical about how I think mathematics should be taught; he doesn't exactly go for the *PSSM*-based approach.

Vanessa: That's an understatement! Don's a super guy who loves to work with kids, but he's dead wrong about how to teach mathematics. I gave up trying to argue with him years ago—he'll never change because he lacks the conceptual basis for teaching any other way than to mimic his own teachers, some of whom he idolized. If I had said anything while he was feeding you his standard line, we would have gotten into a useless debate, and I thought it was more important for you to hear what he had to say.

Casey: That I shouldn't use inquiry instructional strategies?

Vanessa: Of course you should use inquiry instructional strategies. It's the only way to lead students to higher cognitive learning. And that's not to say that there's not a place for direct instructional strategies—such as for algorithmic skills. But Don does have a message to which we should pay attention.

Casey: Which is?

Vanessa: Which is to start off conservatively. You and I know that discovery, inductive and deductive lessons, and all those other methods work—in fact, they're absolutely essential to students doing meaningful mathematics. But, as a beginning teacher, you shouldn't try to deviate too quickly from what these students are used to—at least not until you've had time to gain experience trying out different strategies and seeing what works best for you and your students. In other words, don't try to teach every lesson with every class in the "right" way. Experiment with your progressive ideas more and more as you build upon your experiences. Before you know it, you'll have built a growing arsenal of ideas and materials and be teaching confidently as you know you should. Just give yourself time; don't expect too much of yourself too soon or you'll set yourself up for failure.

Casey: That's heavy; I need to really think about that.

Vanessa: Like Don said, advice is cheap.

Casey: Then I'll ask for a bit more. I'm having a terrible time getting my courses planned and organized. I've gone through the textbooks, looked at the available materials and facilities, and all that, but I just can't get a handle on the courses from A to Z. I can put together individual lessons, but I have trouble fitting the pieces together.

Vanessa: I know just what you mean, and I have a very definitive suggestion. For each course, begin by writing a syllabus. Write it for the students to read. Being forced to describe the purposes and organization of the course for students will organize your thoughts into a coherent whole.

Planning and Organizing Courses by Writing Syllabi

Not only do syllabi lend a businesslike air to your courses and serve as guides for students, but trying to write them lends structure to the process of organizing and planning the courses. In Case 1.7, Casey follows Vanessa's advice:

CASE 1.7

At his computer, Casey outlines the syllabus for the algebra I course. He thinks, "This syllabus needs to be designed so that it sends students the message that this class is serious business and I'm serious enough about it to have it well organized and planned. The syllabi my college instructors used tended to be full of formal-looking lists of reading references, content, deadlines, and grading criteria. I need some of that here, but I can't be too specific about dates and deadlines until I get into the course and see how things go.

"I'll start with an outline of what to include: name of course, basic information such as my name and the room number, a rationale, what the course is all about, a list of materials they need, classroom standards of conduct, an idea of what they'll be doing, goals of the course, the basis for their grades. Vanessa was right: Writing this thing is going to force me to make some hard decisions that'll get me organizing the course!"

Further thought about how to format the syllabus leads Casey to an innovative idea. He decides to organize the syllabus around the questions that the document should answer for students—questions he would expect them to raise about the course. After another hour and a few false starts, he has the following list of questions, which will become the headings in the syllabus:

1. What is this course all about?
2. What is algebra?
3. Why should you learn algebra?
4. Are you ready to do algebra?

5. With whom will you be working in this course?
6. Where will you be learning algebra?
7. How will you be expected to behave in this class?
8. What materials will you need for class?
9. What will you be doing in class?
10. What will you learn from this class?
11. How will you know when you have learned algebra?
12. How will your grades for the course be determined?

Casey spends the next two days determining how to answer these 12 questions for his students and how to express the answers in the syllabus. The most taxing task is to answer the 10th question. Before he can do that, he has to determine the sequence of teaching units for the course and formulate the learning goals for each. He finds the teacher's guide to the textbook somewhat helpful in determining the units and in estimating the number of days and lessons for each. He intends to design the teaching units from a multicultural perspective and incorporate aspects from the history of mathematics. But because he will be developing the units throughout the year and still needs to work out many details, he does not try to use the syllabus to communicate the multicultural and historical features of the course. The 17 units about which he ultimately organizes the course correspond to—but do not follow—15 of the textbook's 16 chapters.

Once Casey completes the syllabus, much of his anxiety about teaching algebra I evaporates and his enthusiasm for the school year soars. Exhibit 1.7 displays his algebra I syllabus with the 17 unit titles.

In a similar fashion Casey writes syllabi for his other first-semester courses.

Engage in Activity 1.2:

Activity 1.2

Borrow a school textbook that is used for a middle school, junior high, or high school mathematics course. Examine the table of contents, familiarize yourself with the general organization of the book, and look over a sample of the sections from various chapters.

Now assume that you are responsible for teaching a course for which this book has been adopted. Develop a rough initial draft of a syllabus for that course. Keep in mind that as you work with chapters 4–11 of *Teaching Mathematics in Secondary and Middle School: An Interactive Approach* (3rd ed.), you will be making major modifications to this syllabus. But for now, write this initial draft and then (a) have colleagues who are also engaging in this activity read it and give you their feedback as you do the same for the drafts of their syllabi, (b) in light of your colleagues' feedback, modify your draft as you see fit, and (c) insert the second draft into the "Cognition, Instructional Strategies, and Planning" section of your working portfolio.

Arranging and Organizing the Classroom

No matter how fundamentally sound and clever you design courses, teaching units, and lessons that target meaningful mathematical learning objectives, it all goes for naught unless your students cooperate by being on-task during those well-designed lessons. Keeping students on-task and engaged in learning activities and constructively addressing discipline problems are at the top of most teachers' lists of concerns (Marks, 2000). These top priority concerns are the foci of chapters 2 and 3. How you arrange, equip, and organize your classroom influences (a) the efficiency of transitions between different learning activities (e.g., from a whole-class lecture session to a small-group cooperative learning session), (b) the ease with which you can move from one student to another (e.g., to deal with an off-task behavior without having to disturb the rest of the class), (c) the efficiency with which you can monitor students' activities (e.g., observe students' written responses to prompts on tasksheets), (d) how well students can see what they need to see (e.g., a visual display projected on a screen) and how well they can hear what they need to hear (e.g., your explanations), (e) students' access to technology they need to do mathematics (e.g., take a turn at a computer station), (f) the efficiency with which the special needs of students are accommodated (e.g., FM sound-field devices installed for a student with a hearing impairment), (g) the efficiency with which you can respond to and manage crisis and emergency situations (e.g., use of your telephone to signal school security personnel to come to the classroom immediately), and (h) the development of a nurturing learning community that is conducive to cooperatively doing mathematics.

Casey has these influences prominently in mind during Case 1.8:

CASE 1.8

With his syllabi in hand, Casey is anticipating the opening day of school, in two weeks. To begin organizing his classroom and working out a management plan for the year, he retrieves a copy of Exhibit 1.8's checklist (Cangelosi, 2000b, pp. 87–89) that he was introduced to while taking a classroom-management course in college. He carefully responds to each of the 15 questions under "Classroom Organization and Ongoing Routines," to the 10 questions under "One-Time-Only Tasks," and to the three questions under "Reminders for the First Week's Learning Activities."

Exhibit 1.9 is a diagram of Room 213 that Casey views. He thinks, "This just won't do. First of all, if I'm at one point in the room, I can't easily get to any one student's desk without negotiating an obstacle course and disturbing other students."

Exhibit 1.7

Casey's Syllabus for First-Period Algebra I.

Course Syllabus for First-Period Algebra I

What is this course all about?

The course is all about the following:

- Understanding how algebra is used to solve problems
- Discovering and inventing ways to use algebra to address problems and situations in our own real worlds

What is algebra?

When you learned arithmetic, you learned how to work with specific numbers. For example you learned how to divide 3.45 by 0.82. In algebra, you learn to work with *variables.* Being able to work with variables allows you to extend what you learn in one situation to many other situations. You will form your own answer to the question, "What is algebra?" during the first few weeks of the course.

Why should you learn algebra?

You should learn algebra for the following reasons:

- Using algebra will help you solve many of the problems you face right now as well as later on in your life.
- Your success in other courses you take in high school, technical school, or college depends on your understanding of algebra.
- A full credit of algebra is required for a high school diploma in this state.
- An understanding of at least some algebra is expected of literate citizens in today's society and is needed in many occupations.

Are you ready to do algebra?

Yes, as long as you can do fundamental arithmetic and want to solve problems that life tosses your way.

With whom will you be working in this course?

You will be working with Casey Rudd, who is responsible for helping you and your classmates to learn algebra. You will also be working with your classmates, each of whom will be making a unique contribution to what you get out of this course. In turn, you will contribute to what they learn by sharing your ideas, problems, discoveries, inventions, and solutions.

Where will you be learning algebra?

You will draw your understanding of algebra from your entire environment, whether at home, school, or elsewhere. Your classroom, Room 213 at Rainbow High, is the place where ideas about algebra are brought together and formalized. Room 213 is a place of business for learning algebra.

How will you be expected to behave in this class?

You and your classmates have the right to go about the business of learning algebra free from fear of being harmed, intimidated, or embarrassed. Mr. Rudd has the right go about the business of helping you and your classmates learn algebra without disruption or interference. Thus, you are required to behave in accordance with five standards of classroom conduct:

- Give yourself a complete opportunity to learn algebra.
- Do not interfere with the opportunities of your classmates to learn algebra.
- Respect the rights of all members of the class—they include you, your classmates, and Mr. Rudd.
- Follow Mr. Rudd's directions for lessons and classroom procedures.
- Adhere to the rules and policies of *Rainbow High's School-wide Discipline and Safety Code* as listed on pp. 12–15 of the *Student Handbook.*

Continued

Exhibit 1.7
Continued

What materials will you need for class?

Unless otherwise directed by Mr. Rudd, bring the following with you to every class meeting:

- The course textbook that was issued to you during the first class meeting (which you are responsible for maintaining in good condition and returning to Mr. Rudd at the end of the second semester).
- A three-section three-ring binder: (1) Section 1 is for day-by-day class notes and assignments. (2) Section 2 is for tasksheets you complete for homework and in-class assignments. (3) Section 3 is for definitions of words and symbols, postulates, and theorems.
- A scratch pad
- Pencils, pen, eraser, and a hand-held graphing calculator

You will also be issued a container for your individualized mathematics portfolio that will be secured for you in one of Mr. Rudd's cabinets. Mr. Rudd will explain how you will work with your portfolio during the first week of class.

Other supplies (e.g., rulers, compasses, computer disks, and various measuring instruments) will be available for you to use in the classroom as you need them.

What will you be doing in class?

The course is organized into 17 units of about two weeks each. During each unit you will

- Participate in class meetings that include such activities as attending to presentations by Mr. Rudd and classmates, making presentations to the class, engaging in discussions about mathematics, collaboratively doing mathematics with classmates and Mr. Rudd, raising and addressing questions, working independently, using technology (e.g., computers, manipulatives, and calculators) to do mathematics, and gauging your progress with algebra (e.g., by taking tests).
- Complete homework assignments.
- Take a unit test.

What will you learn from this class?

Each unit will either introduce you to a new algebraic topic or extend your understanding of a previous topic. During the unit you will (a) discover ideas and relationships, (b) extend your ability to communicate with the language of algebra, (c) acquire new skills or polish previously acquired skills, and (d) extend your talent for solving problems.

Here are the titles of the units:

1. Algebra and Its Language
2. Types of Numbers and Arithmetic Operations
3. Operations on Rational Numbers
4. Algebraic Inequalities
5. Powers and Roots
6. Polynomials
7. Factoring Polynomials
8. Quadratic Equations
9. Algebraic Functions
10. Extending Functions
11. Systems of Open Sentences
12. Extending Powers and Roots
13. Extending Quadratic Functions
14. Operations with Rational Polynomials
15. Extending Work with Rational Numbers
16. Special Functions with Natural Numbers
17. Extending What You've Learned

Units 1–8 are planned for the first semester; units 9–17 are planned for the second semester.

Continued

Exhibit 1.7
Continued

How will you know when you have learned algebra?

Everyone understands at least some algebra (e.g., we all manipulate variables), but no one has ever completely mastered algebra. Algebra is being discovered and invented every day. You will use what you learn in this course to develop further your ability to use algebra to solve problems and to use the language of algebra to organize and communicate your ideas. The question is not whether or not you have learned algebra; the ongoing question is the following: How well are you learning algebra?

During this course, you will be given feedback on your progress through comments Mr. Rudd makes about your work and test results. Furthermore, you will be developing an individualized portfolio that will showcase some of your many mathematical accomplishments.

How will your grades for the course be determined?

Your grades should reflect your achievement of course goals and will be based on the results from tests. As Mr. Rudd will explain to you during the first week of class, these tests are performance-based tests that build upon your homework and in-class work that involves projects, collaborative work with classmates, activities from tasksheets, and work you have done with your portfolio.

Your first semester grade will be influenced by eight unit tests, a mid-semester test scheduled between the fifth and sixth units, and a semester test as follows:

- Eight unit tests . . . . 60% (7.5% each)
- Midsemester test 15%
- Semester test 25%

Your second semester grade will be influenced by nine unit tests, a mid-semester test scheduled between the 12th and 13th units, and a semester test as follows:

- Nine unit tests . . . . 60% (6.7% each)
- Midsemester test 15%
- Semester test 25%

Will this course be enjoyable and productive?

Mr. Rudd intends it to be, but you will have to assess the answer to that question as you progress through the course.

Exhibit 1.8
Checklist From Casey's Classroom Management Course (Copied with permission from Cangelosi [2000b, pp. 87–89]).

I. *Classroom Organization and Ongoing Routines*
  - **A.** What different types of learning activities (e.g., video presentations, large-group demonstrations, small-group cooperative-learning sessions, and independent project work) do I expect to conduct this term?
  - **B.** How should the room be organized (e.g., placement of furniture, computers, resource materials, screens, and displays) to accommodate the different types of learning activities and the corresponding transition times?
  - **C.** What standards of conduct and routine procedures will be needed to maximize engagement during the different types of learning activities and to maximize on-task behaviors during transition times?
  - **D.** What standards of conduct and routine procedures will be needed to discourage disruptions to other classes or persons located in or near the school?
  - **E.** What standards of conduct and routine procedures are needed to provide a safe, secure environment in which students and other persons need not fear embarrassment, harassment, or harm?
  - **F.** How will standards of conduct and procedures be determined (e.g., strictly by me, by me with input from the students, democratically, or some combination of these)?
  - **G.** When will standards of conduct and procedures be determined (e.g., from the very beginning, as needs arise, or both)?
  - **H.** How will standards of conduct and procedures be taught to students?
  - **I.** How will standards of conduct be enforced?

Continued

Exhibit 1.8
Continued

J. What other parts of the building (e.g., detention room or other classrooms) can be utilized for separating students from the rest of the class?
K. Whom, among building personnel, can I depend on to help handle short-range discipline problems and whom for long-ranged problems?
L. How do I want to utilize the help of parents?
M. What ongoing routine tasks (e.g., reporting daily attendance) will I be expected to carry out for the school administration?
N. What events on the school calendar will need to be considered as I schedule learning activities?
O. What possible emergencies (e.g., fire, student suffering physical trauma, or violent activity) might be anticipated and, considering school policies and the school safety plan, how should I handle them?

II. *One-Time-Only Tasks*
A. How should I communicate the general school policies and school safety plan to my students?
B. What special administrative tasks will I be required to complete (e.g., identifying number of students on reduced payment lunch program and checking health records)?
C. What supplies (e.g., textbooks) will have to be distributed?
D. Are supplies available and ready for distribution in adequate quantities?
E. How will I distribute and account for supplies?
F. Are display cards with students' names ready?
G. How should I handle students who appear on the first day, but are not on the roll?
H. What procedures will be used to initially direct students into the classroom and to assigned places?
I. For whom on the student roster might special accommodations be needed for certain types of activities (e.g., students with hearing losses and students confined to wheelchairs)?
J. For whom on the student roster will I need to schedule IEP conferences and, for each, who is the relevant special education resource person?

III. *Reminders for the First Week's Learning Activities*
A. Do lesson plans for the first week call primarily for learning activities that each have:
(1) uncomplicated directions that are simple to follow, (2) challenges, but with which all students will experience success, (3) built-in positive reinforcers for engagement, and (4) all students involved at the same time?
B. Do the first week's lesson plans allow me to spend adequate time observing students, getting to know them, identifying needs, and collecting information that will help me make curricula decisions and design future learning activities?
C. Do plans allow me to be free during the first week to closely monitor student activities and be in a particularly advantageous position to discourage off-task behaviors before off-task patterns emerge, and positively reinforce on-task behaviors so that on-task patterns emerge.

IV. *Personal Reminders for Myself*
A. Am I prepared to pause and reflect for a moment on what I should say to students before I say it?
B. Am I prepared to observe exactly what students are doing and hear exactly what they are saying before making a hasty response?
C. Am I prepared to use descriptive, rather than judgmental language, as I interact with my students?
D. Am I prepared to consistently act and communicate assertively, being neither hostile nor passive?
E. Am I prepared to use a supportive response style?
F. Am I prepared to model a businesslike attitude?

Casey sits down and makes the following wish list of classroom features that will facilitate his classroom management style and the types of learning activities he plans to conduct:

- Quick and easy access between any two points in the room.
- An area for large-group presentations, discussions and questions, and individual work sessions where students are seated at desks from which they can view the whiteboard, overhead screen, and video display while listening to whomever has the floor.
- Areas in which students engage in small-group cooperative learning activities, working on tasks, discussing mathematics, or tutoring one another.
- Computer stations where students work either independently or in pairs.
- Tables where students can work alone, in pairs, or in triples with manipulatives.
- A traffic area for people entering and exiting the room that is easily monitored.
- A room where Casey can meet with individuals privately (e.g., to deal with a student's misbehavior away from the rest of the class).
- A minilibrary and quiet reading room for a few students at a time.
- Cabinets and closets for securely storing equipment and supplies out of sight.
- A secure teacher's desk at a favorable vantage point.

Exhibit 1.9
Room 213 Before Casey Modified It.

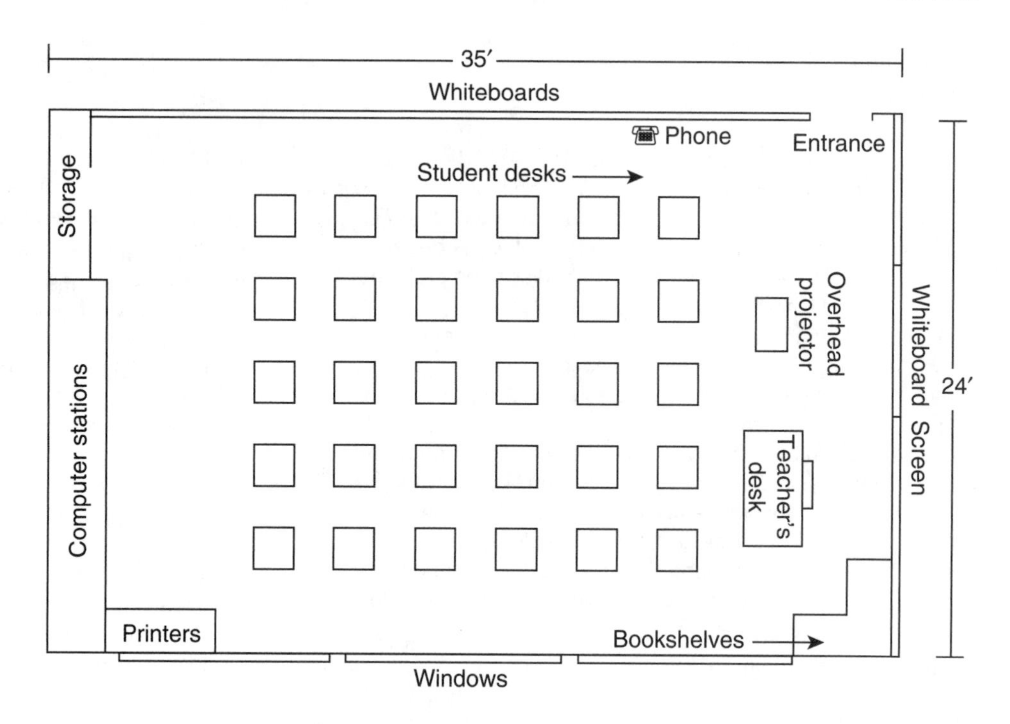

- A for-the-teacher-only computer station interlinked with the overhead projector, video monitor, students' computers, and printers.
- A two-way communication device (e.g., a telephone) with which Casey can quickly summon backup support in crisis situations.

Casey looks at his room in its present state (Exhibit 1.9) and his list and realizes that all 12 features are impossible. But he determines to get the most from what he has been handed. He takes measurements of the room and its equipment, makes some scale drawings, and designs a workable arrangement. Following a few visits to the school-district storehouse, a little trading with colleagues, and some help from a custodian, Casey begins the school year with the room arranged as indicated by Exhibit 1.10. To accommodate small-group sessions, students rearrange their desks as shown in Exhibit 1.11.

Organizing a Computerized Management System

Teachers, like Casey Rudd, begin their careers with perceptions of how they will spend their time. But teaching is full of surprises—some pleasant surprises, others not so pleasant. One of the unpleasantries teachers discover during their initial years is the overwhelming amount of paperwork they must generate, manage, and deal with (Dunican, 2000). There are three varieties of paperwork: (a) the mundane, uninteresting variety (e.g., record keeping and filling out administrative forms) that challenges your patience, (b) the nonrepetitive stimulating variety (e.g., developing unit plans, formulating mathematical problems, designing assessment devices, corresponding with parents, and writing proposals for funding) that challenges your ingenuity, and (c) the repetitive stimulating variety (e.g., scoring test papers) that requires your professional expertise but is often boring and extremely time consuming.

To prevent this paperwork from consuming all of your time, you need to set up a computerized management system prior to the opening of the school year. This is what Casey had in mind in Case 1.9:

CASE 1.9

Casey begins setting up a folder system on his computer that will serve him during the school year. He sets up space on his hard drive for the school year and then organizes that space into seven major folders:

- Courses (Under "Courses," Casey plans to use word processing to set up templates for, write, and store documents [e.g., lesson plans and tests] for each of the five courses he will be teaching.)
- Homeroom (At this point in time, Casey is unclear about the paperwork related to his homeroom period. However,

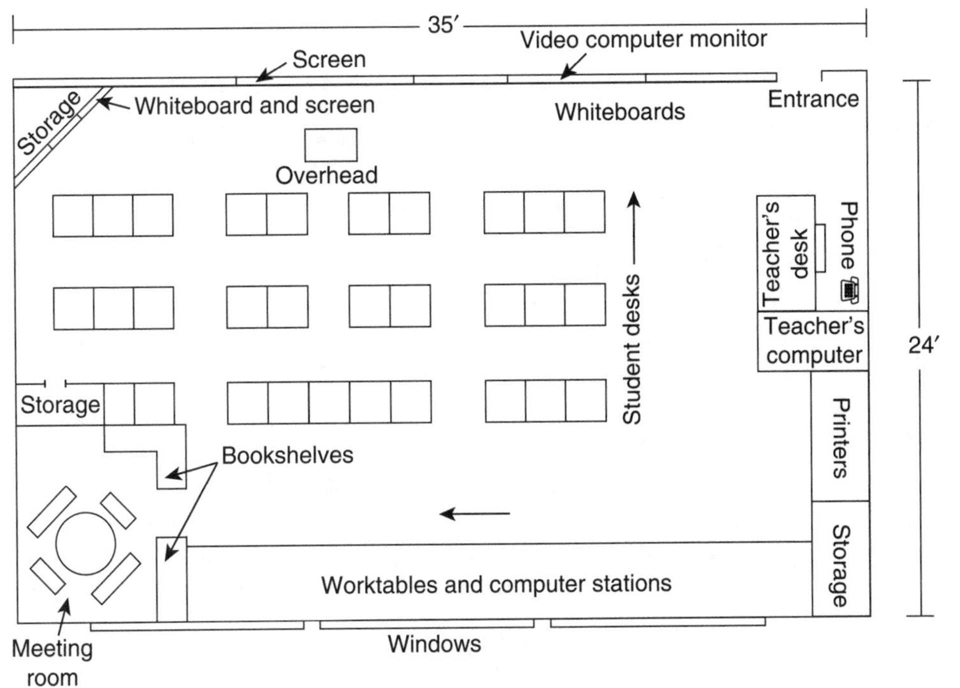

Exhibit 1.10
Room 213 (Set Up for Whole-Class Sessions) After Casey Modified It.

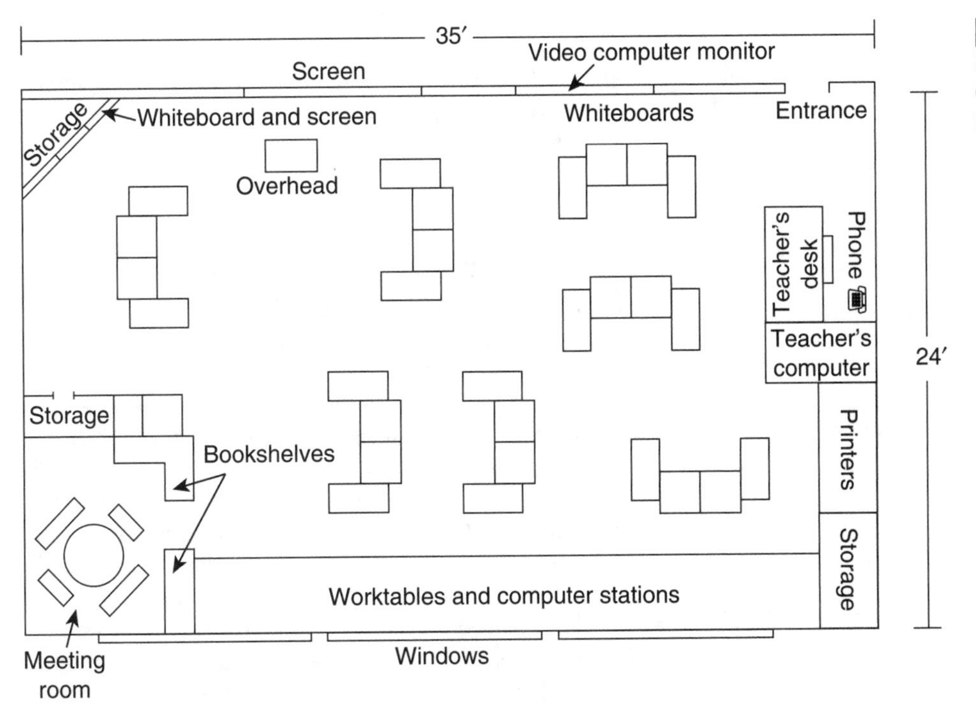

Exhibit 1.11
Room 213 (Set Up for Small-Group Sessions) After Casey Modified It.

Exhibit 1.12

Folder Structure Casey Set Up on His Computer.

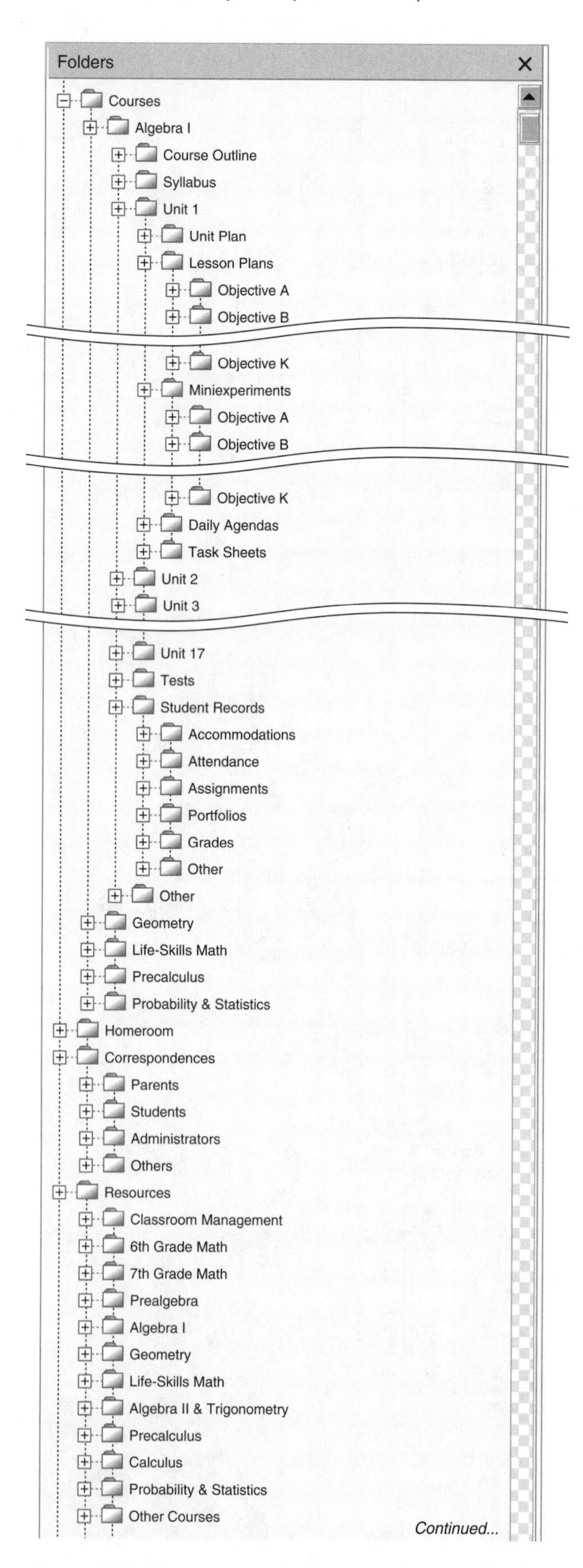

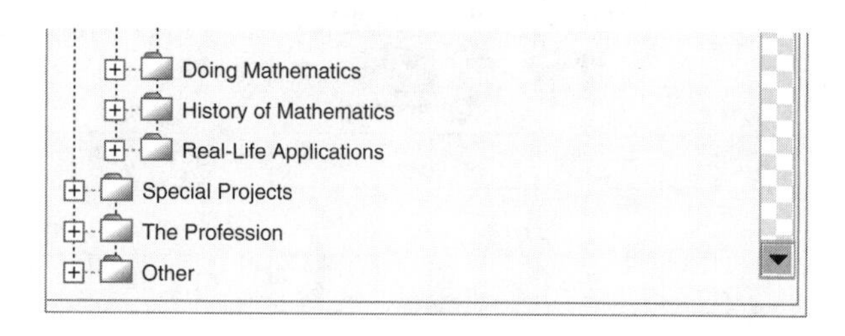

he sets up a folder in anticipation of record-keeping responsibilities.)

- Correspondences (Under "Correspondences," Casey plans to write and file memos and letters to parents, students, administrators, and others.)
- Resources (Under "Resources," Casey plans to organize and file references to teaching ideas and materials that he anticipates collecting throughout the year.)
- Special Projects (Just in case he gets involved in developing a proposal for funding, engages in an action research project [e.g., to compare the effects of two different types of learning activities], or works on a faculty-committee project, Casey sets up this folder.)
- The Profession (Casey plans to use this folder for paperwork related to his involvement in organizations such as NCTM.)
- Other (Casey sets up this folder for unanticipated work that does not fit any of the other categories.)

For each of the five courses he will be teaching during the year (i.e., algebra I, geometry, life skills, precalculus, and probability & statistics), Casey sets up a subfolder under "Courses." He then partitions each of those five subfolders with second-level subfolders as shown for "Algebra I" in Exhibit 1.12.

For algebra I, he inserts the file with the course outline and the file with the syllabus in their respective second-level subfolders. He sets up the structure for the third-level subfolders for each unit. Because he intends writing one lesson plan for each of the unit's objectives, he sets up a fourth-level subfolder for each objective's lesson. Because he has already developed the plan for Unit 1 of the algebra I course, he knows there are 11 objectives (i.e., Objectives A–K). However, for each of the other units (i.e., Units 2–17), he will not know how many lesson fourth-level subfolders to set up for the unit until he develops that unit plan during the school year.

Note: As you will better understand from your work with chapters 5–9, a "*miniexperiment* is a component of a measurement used to monitor student learning that consists of a *prompt* for students' responses and an observer's rubric for recording or quantifying those responses. Traditional test items are examples of miniexperiments." (Cangelosi, 2000a, p. G5)

For each objective under the third-level subfolder "Miniexperiments," Casey sets up a fourth-level subfolder to house the miniexperiments he will devise for that objective. He will eventually synthesize the unit test for how his students achieve Unit 1's goal by drawing miniexperiments from those 11 fourth-level subfolders.

During the two-semester school year, Casey will be conducting a total of approximately 900 55-minute class sessions for all the courses he teaches. For each daily class session, Casey intends to write an agenda that he will follow (e.g., Exhibit 11.16). Thus he sets a third-level subfolder labeled "Daily Agendas." For those class sessions and for homework assignments, Casey will be developing tasksheets to prompt students to do mathematics (e.g., Exhibit 11.22); thus, he sets up the "Tasksheets" third-level subfolder shown in Exhibit 1.12.

Identical subfolder structures are set up for all the other units.

He will use the second-level subfolder named "Tests" for each course to house files for the unit tests and other major test documents (e.g., midterm and final examinations).

For each course, the second-level subfolder "Student Records" is partitioned into six third-level subfolders: "Accommodations" will be used for documents pertaining to provisions for individualized programs for inclusion and accommodation of students with special needs (e.g., contractual agreements from students' IEPs, which you will learn more about from your work with chapter 2). "Attendance" and "Assignments" are for keeping track of individual students' records in those two areas. Casey plans to have each student develop and maintain individualized portfolios (as explained in chapter 9); "Portfolios" will help him monitor them. "Grades" will hold files with students' test scores and grades. Because of Casey's concern with anticipating the unanticipated, he throws in the "Other" third-level subfolder.

As you work with chapters 2–11, you will be prompted to produce sample documents of the "nonrepetitive stimulating" variety of paperwork (e.g., lesson). To help you manage this paperwork and access it as you need it, engage in Activity 1.3:

Activity 1.3

Follow Casey's example in Case 1.9 by setting up your own computerized management system—organized similarly to Exhibit 1.12. If you are not currently responsible for teaching mathematics courses (e.g., because you are still a preservice teacher), then under the "Courses" folder, you need to set up a subfolder structure for only one course—namely the course for which you developed an outline and a syllabus when you engaged in Activity 1.2.

Back up the folder structure for your computerized management system on a removable disk (e.g., zip disk or rewritable CD) and insert the disk in the "Technology" section of your working portfolio.

Arrangements and Acquisitions

Teaching is an extremely complex art. Besides designing curricula (e.g., units and lessons) and generating documents and instructional materials, the number and complexity of decisions teachers must continually make are unparalleled in other professions (Cangelosi, 2000a, pp. 2–10). Beginning teachers, like Casey, are particularly burdened not only by their inexperience but also by the need to produce from scratch the types of materials for which you and Casey organized a computerized management system. In his second year of teaching, Casey will be able to build on what he started during his initial year. Compounding their burdens is the fact that beginning teachers in many schools are frustrated by having to learn the ropes for managing what should be trivial matters—such as obtaining equipment and supplies for their classrooms (Duke, Cangelosi, & Knight, 1988). Each school seems to have a unique process to expedite such matters as obtaining colored board markers, replacing a burned-out lamp in an overhead projector, gaining access to a camcorder, and replacing lost books. Usually, it is a matter of identifying and befriending the staff member who knows how to get things done. Veteran teachers take such matters for granted and do not usually remember to inform beginning teachers of this informal network.

In Case 1.10, Casey continues preparing for the opening day of the school year:

CASE 1.10

Although he spends much of the week before classes begin in meetings (e.g., a six-hour orientation meeting for all new teachers in the district and meetings of the school faculty and the Mathematics Department), Casey finds time to begin learning the ropes for obtaining supplies and equipment; fortunately, Casey asked Vanessa about the process before wasting time going through formal channels.

He visits with Mr. Tramonte to work out schedules and procedures for using the computer lab. Casey reserves the lab for his fourth-period geometry class every Wednesday during the year. Mr. Tramonte also gives him a schedule of the times each week during which Casey can send as many as six students into the lab to do independent work. As Armond had indicated, the lab is also open to students and faculty for 90 minutes before school and after school until 9 P.M.

He manages to obtain a set of 20 graphing calculators for use in the classroom, along with a display model for the overhead projector. Two years ago the Mathematics Department had applied for and received a grant for obtaining calculators. Don Delaney, who hardly uses them for his own courses, was happy to accommodate Casey's request to take them off his hands. Furthermore, Casey makes arrangements with the bookstore to make graphing calculators available for students to purchase at a discount.

Most importantly, he begins interacting with teachers from outside the Mathematics Department for the purpose of establishing collaborative relationships for building integrated curricula. This effort reaps early dividends that exceed Casey's expectations:

- Mr. Bosnick, an art teacher, agrees to have students apply geometric principles they learn from Casey to art projects.
- Ms. Deere, an English teacher, and Casey plan a joint project in which students will use skills and principles developed in both life-skills and communications classes to produce videotape programs.
- Ms. Stout agrees to raise problems (e.g., involving relationships among exercise programs and physical fitness measures) in her physical education classes for Casey's algebra I and precalculus students to address.

THE BEGINNING OF AN EVENTFUL SCHOOL YEAR

Students arrive in your classroom on the first day of school with some preconceived notions about what to expect and what is expected of them. The vast majority anticipate being required to follow teachers' directions and know that antisocial behaviors such as fighting, blatant rudeness, and highly disruptive conduct such as screaming in class are not generally tolerated. But experience has taught them that teachers vary considerably relative to dedication to their work, what they tolerate, expectations, awareness of what students are doing, assertiveness, decisiveness, consistency, and their respect for students and themselves.

Thus, in students' initial encounters with you they will be filled with uncertainties. During this period of uncertainty, they will be observing your reactions, assessing your attitudes, assessing their place in the social order of the class and their relationship with you, and determining how they will behave in your classroom. Because students tend to be more attentive to your words, actions, and reactions during this feeling-out period, it is an opportune time for you to communicate some definitive messages that establish the classroom climate and set the standards of behavior for the rest of the course (Cangelosi, 2000b, pp. 84–104). Thus, it is critical for you to begin each course in a very businesslike fashion, tending to the work at hand: learning and teaching mathematics.

Being businesslike does not require you to be somber, stiff, or formal. The business of learning and teaching mathematics is best conducted in a friendly, relaxed atmosphere where hearty laughter is appreciated. Being businesslike does require that learning activities—whether enjoyable or tedious—be considered important and that matters unrelated to achievement of learning goals be dispatched efficiently.

To effect a businesslike beginning to your course, immediately involve students in a learning activity with the following features:

- *Directions for the activity are simple and unlikely to confuse anyone.* (This allows your students to get to the business of learning mathematics without experiencing bewilderment over what they are supposed to be doing. This initial experience teaches students to expect to understand your directions and enhances the chances that they will attend to them in the future. If students are confused by your initial directions, they may not be as willing to try to understand subsequent ones. Later, after they have developed a pattern of attending to your directions, you can begin introducing more complicated procedures to be followed.)
- *The activity involves them in a mathematical task that is novel for them, but one in which they are likely to succeed and experience satisfaction.* (Leaving them with the impression that they successfully learned by doing mathematics should serve to positively reinforce their engagement in learning activities you prescribe.)
- *All students are concurrently engaged in the same activity.* (Later in the course it will be advantageous to have students working on differentiated learning activities. But, initially, having all students working on the same task allows you to keep directions simple, monitor the class as a whole, and compare how different students approach a common task. Besides, until you become better acquainted with students, you hardly have a basis for deciding how to individualize.)
- *The activity is structured so that you are free to monitor student conduct and immediately stem any displays of off-task behaviors.* (Jacob Kounin [1977] demonstrated that students are more likely to cooperate with a teacher they believe is on top of things and in control of classroom activities. He coined the term "*withitness*" to refer to a teacher's awareness of what students are doing. You are in a better position to be *with-it* during activities in which you are free to move about the classroom and position yourself near students than at times when your movements are restricted [e.g., when you are stationed at a whiteboard, standing behind an overhead projector, or

sitting at your desk]. Surely you want to be especially with-it during the early phases of a course.)

Mr. Rudd has these suggestions in mind in Case 1.11:

CASE 1.11

For each class, Mr. Rudd prepares name cards for students and an opening-day questionnaire or tasksheet for them to work on as soon as they are seated. Examine the tasksheets for the four courses in Exhibits 1.13–1.16.

Mr. Rudd designed the tasksheets with the following purposes in mind:

- The students will immediately get busy doing mathematics while he takes care of administrative chores like distributing textbooks. This, he feels, will help establish a classroom learning environment that is conducive to conducting the business of teaching and learning mathematics.
- The students will be both challenged by and successful with the tasks. The algebra students, for example, already know how to add simple fractions, but most are unlikely to have ever before been prompted to describe the process they use. Also, most of the prompts involve expressing opinions, describing observations, and reporting about themselves—tasks everyone can do, but that require reasoning rather than only remembering.
- The students will begin to get the following impressions about the course they are starting under Mr. Rudd's direction:
  - The mathematics they will learn is related to their own individual interests.
  - They are expected to make judgments, and those judgments are valued.
  - They are expected to write about mathematics.
  - Their experiences in this course will be different from those they have had in previous courses.
- Students will pursue tasks they will later learn to accomplish more efficiently with mathematics planned for subsequent lessons in the course (e.g., estimating areas of irregularly shaped regions).
- Mr. Rudd will make use of information gained from responses to the final three prompts in planning lessons; he will be able to select topics for problems that will

Exhibit 1.13

Mr. Rudd's Opening-Day Tasksheet for His Algebra I Course.

1. What is your name? ____________________
2. About how many minutes does it take you to listen to a song from beginning to end? ____________________
3. About how many minutes does it take you to listen to six different songs from beginning to end? ____________________
4. Simplify each of the following; display your work:

   $\frac{3}{7} + \frac{2}{3} =$ ____________________

   $\frac{2}{5} + \frac{1}{3} =$ ____________________

   $\frac{1}{5} + \frac{3}{4} =$ ____________________

5. Use the rectangular region below to write a description of the process you used to find the three simplified sums of the fractions in number 4. Do not use the name of any specific numbers (such as 3/7, 3, or 7) in your description. Describe the general process you use for adding any two fractions.

   Your description of the process:

Continued

Exhibit 1.13
Continued

6. Fill in the blank with a number that makes the following statement true:

$$7 - \_\_\_\_\_ > 3$$

The number you put in the blank is not the only one that would have worked. But not just any number would have worked. Make up a rule for choosing a number of the blank that makes the statement true. Write your rule in the rectangular region below:

Your rule for selecting a number that will work:

7. List three things about your family, your neighborhood, or your home that you think are different from most of the other students' families, neighborhoods, or homes.

1) ______________________________

2) ______________________________

3) ______________________________

8. List three things about your family, your neighborhood, or your home that you think are the same as most of the other students' families, neighborhoods, or homes.

1) ______________________________

2) ______________________________

3) ______________________________

9. List three important questions about which you are going to have to make a decision during the next nine months.

1) ______________________________

2) ______________________________

3) ______________________________

Exhibit 1.14

Mr. Rudd's Opening-Day Tasksheet for His Geometry Course (Copied with permission from Cangelosi [2000b, pp. 230–233]).

1. What is your name? ____________________
2. Carefully look at the following two pictures of art work:

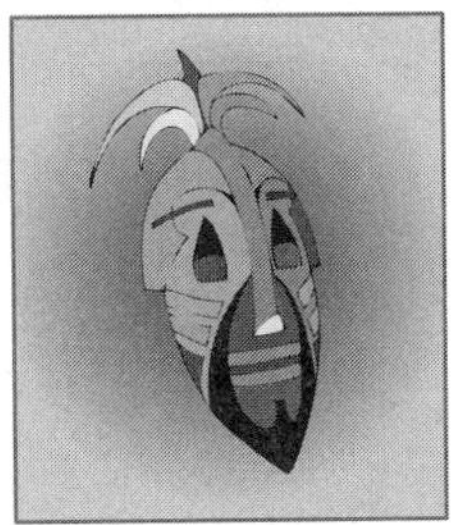

Write a description of how one of the pieces of art work looks different from the other.

____

____

____

____

____

Write a description of how one of the art work pieces looks the same as the other.

____

____

____

____

____

3. Carefully look at the clothing worn by the people in the following two pictures:

Write a description of how the clothing in one of the pictures looks different from the clothing in the other picture.

____

____

____

____

____

Continued

Exhibit 1.14
Continued

Write a description of how one of the pieces of clothing looks the same as the other.

____

____

____

____

____

4. Carefully look at the following two types of writing characters:

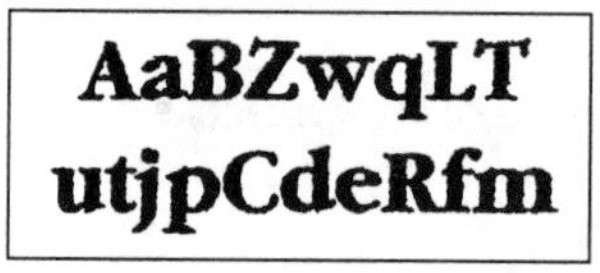

שמזוהדגבא פמלחשת

AaBZwqLT utjpCdeRfm

Write a description of how one of the types of characters looks different from the other.

____

____

____

____

____

Write a description of how one of the types of characters looks the same as the other.

____

____

____

____

____

5. Carefully look at the following two types of buildings:

Continued

Exhibit 1.14
Continued

Write a description of how one of the buildings looks different from the other.

____
____
____
____
____

Write a description of how one of the buildings looks the same as the other.

____
____
____
____
____

6. List three things about your family, your neighborhood, or your home that you think are different from most of the other students' families, neighborhoods, or homes.

   1) ____
   ____

   2) ____
   ____

   3) ____
   ____

7. List three things about your family, your neighborhood, or your home that you think are the same as most of the other students' families, neighborhoods, or homes.

   1) ____
   ____

   2) ____
   ____

   3) ____
   ____

8. List three important questions about which you are going to have to make a decision during the next nine months.

   1) ____
   ____

   2) ____
   ____

   3) ____
   ____

Exhibit 1.15

Mr. Rudd's Opening-Day Tasksheet for His Life-Skills Mathematics Course.

1. What is your name? ____________________
2. Identify a product that you bought during the month of August. ____________________

   Where did you purchase it? ____________________

   About how much did you pay for it? ____________________

   About how much do you think it cost the manufacturer or producer to make or produce it? ____________________

   How did you obtain the funds you used to make the purchase? ____________________

   Why did you decide to make the purchase? ____________________

   When you decided to buy this particular product, what other options (such as buying another brand, buying something else, or not buying anything at all) did you consider? ____________________

   Explain why you chose the option that you chose? ____________________

3. What advice do you have to help people get the most for their money when they are thinking about buying something?

4. Suppose that you are thinking about accepting a job selling some type of product. Would you rather be paid salary based on the number of hours you work or a salary based on the amount of sales you make? Explain your answer.

5. In your opinion who is the most popular entertainer in the world? ____________________

   Name another popular entertainer who one of your friends would argue is a more popular entertainer than the person you just named. ____________________

   Explain why you believe the entertainer you chose is more popular than the entertainer you think your friend would have chosen. ____________________

6. Suppose you plan to save money over the next few years to buy something very expensive. Where would you keep the money you save until you are ready to make the purchase—in a savings account, checking account, secure safe, certificate of deposit, stocks, or elsewhere? Explain how you would go about deciding where to keep the money.

Continued

Exhibit 1.15
Continued

7. List three things about your family, your neighborhood, or your home that you think are different from most of the other students' families, neighborhoods, or homes.

   1) ______________________________

   2) ______________________________

   3) ______________________________

8. List three things about your family, your neighborhood, or your home that you think are the same as most of the other students' families, neighborhoods, or homes.

   1) ______________________________

   2) ______________________________

   3) ______________________________

9. List three important questions about which you are going to have to make a decision during the next nine months.

   1) ______________________________

   2) ______________________________

   3) ______________________________

interest the students. Students' responses to these final three prompts will also provide him with information he can use to design units from a multicultural perspective.

- From both observing their behavior as they work with the tasksheet and from reading their responses, Mr. Rudd will begin to know his students as individuals.

Just before each period on the first day of class, Mr. Rudd places a name card and a tasksheet on each student's desk. The 55-minute period is spent as follows:

1. He greets students at the door as they enter and directs each to locate the desk with his name card and begin completing the tasksheet.
2. As students respond to the prompts, Mr. Rudd distributes copies of the textbook and course syllabus and checks the roll using the name tags without disturbing the students' work.
3. All students have an opportunity to work on the tasks, but no one completes them before Mr. Rudd interrupts the work and goes through the course syllabus (see Exhibit 1.7). He uses the sections of the syllabus as an advance organizer for a presentation in which he communicates expectations, introduces some classroom procedures, and discusses the nature of mathematics. Students' responses to prompts on the tasksheets are utilized in discussions stimulated by the second section of the syllabus (What is algebra, geometry, life skills, or precalculus?). For example:
   - In the algebra class he uses students' responses to set the stage for working with variables and generalizing rules of arithmetic.
   - Mr. Rudd uses students' descriptions of figures in Exhibit 1.14 to explain the focus of geometry—and to do so from a multicultural perspective.

Exhibit 1.16

Mr. Rudd's Opening-Day Tasksheet for His Precalculus Course.

1. What is your name? ______________________________
2. Suppose that you are by yourself in some unfamiliar house. You are in a hallway and come to a door. Before opening the door, you want to know what is on the other side. Fortunately, it is one of those old doors with the kind of keyhole you can look through. You peek through the keyhole and you see what appears in the picture below. *Describe* exactly what you see through the keyhole—only what you can actually see.

Your description:

____

____

____

____

Now describe what you would *infer* the room behind the door looks like, based on the limited view you had through the keyhole.

____

____

____

____

3. Now you have arrived just outside of another door in the house. Again you look through the keyhole. Again *describe* exactly what you see through the keyhole as pictured below.

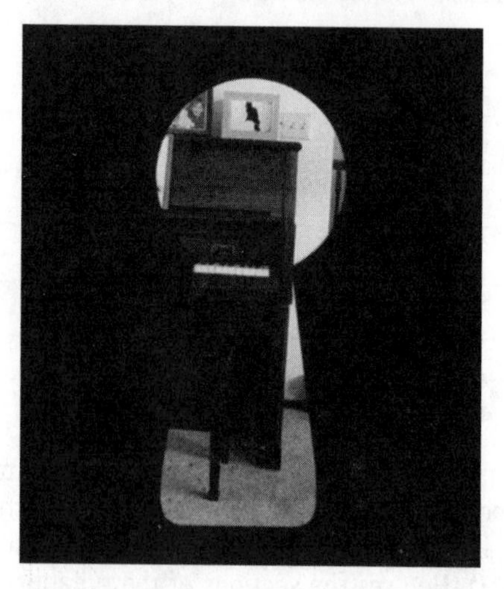

Your description:

____

____

____

____

Now describe what you would *infer* the room behind the door looks like, based on the limited view you had through the keyhole.

____

____

____

____

4. According to one hospital's records, the numbers of people treated in the emergency room for drug-induced traumas were as follows for the year 2001:

Jan: 31	Apr: 40	Jul: NA	Oct: 37
Feb: 33	May: 39	Aug: 48	Nov: 35
Mar: NA	Jun: 45	Sep: NA	Dec: 34

("NA" indicates the record was not available for that month.)

Continued

Exhibit 1.16
Continued

Do the data suggest any possible pattern relative to the relationship between frequency of drug-induced traumas and the time of the year? Explain your answer.

____

____

____

____

Estimate the total number of drug-induced traumas treated at that hospital's emergency room during 2001. Explain how you arrived at your estimate.

____

____

____

____

5. Estimate the area of the following rectangular region in terms of number of □ units:

Your estimate: _____

6. Estimate the area of the following rectangular region in terms of number of □ units:

Your estimate: _____

Continued

Exhibit 1.16
Continued

7. Estimate the area of the following rectangular region in terms of number of □ units:

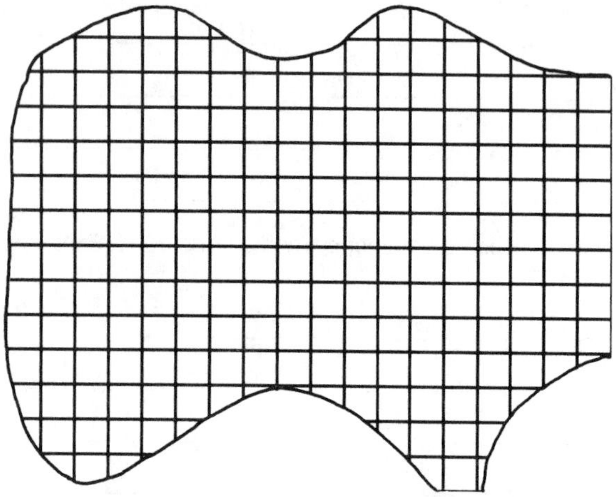

Your estimate: _____

8. List three things about your family, your neighborhood, or your home that you think are different from most of the other students' families, neighborhoods, or homes.

   1) ______________________________

   2) ______________________________

   3) ______________________________

9. List three things about your family, your neighborhood, or your home that you think are the same as most of the other students' families, neighborhoods, or homes.

   1) ______________________________

   2) ______________________________

   3) ______________________________

10. List three important questions about which you are going to have to make a decision during the next nine months.

   1) ______________________________

   2) ______________________________

   3) ______________________________

- He raises the types of personalized problems the life-skills students will be learning to solve throughout the course.
- With the precalculus class, he uses the peeking-through-a-keyhole tasks to explain what he calls the "keyhole logic" of mathematics, in which we use our intellects to expand what we know beyond what we empirically observe.

4. Mr. Rudd assigns homework that includes the completion of the tasksheet.

Mr. Rudd is generally pleased with how the first day goes. The students seem willing to cooperate, and most of them act friendly and try to follow his directions. After the first few minutes of each period, his nervousness disappears and he surprises himself with his own glibness—speaking fluently and making effective responses to students' questions and behaviors. His energy level and enthusiasm peaks in the second half of each class.

Because some students are reluctant to respond to prompts on the tasksheets without explanations and feedback from Mr. Rudd, the administrative duties take longer in every class than he had planned. Thus, no class gets through its syllabus and Mr. Rudd fears that the students are left hanging at the end of the period without understanding course expectations and the meaning of algebra, geometry, life skills, or precalculus. He is especially anxious to meet with them on the second day to finish explaining the syllabi and get back on schedule.

After the fourth period goes especially well, Mr. Rudd concludes that dealing with the disadvantages of teaching in Room 108 may be easier than he had anticipated. However, after supervising lunch and arriving back in his own room just prior to fifth period, he discovers that the fourth-period health science class has left the room in disarray. The life-skills class gets off to a rough start and he is flustered because he is unable to get the room back in order and the materials laid out by 1:30. The students do not appear surprised by the disorder. In fact, some of the students in his other classes express their surprise by how well he has things organized.

LEARNING FROM EXPERIENCES

Learning How to Apply a Complex Art

If you are a preservice teacher, then engage in Activity 1.4:

Activity 1.4

With a colleague who is also engaging in Activity 1.4, interview an inservice teacher who has at least three years experience teaching mathematics at a middle, junior high, or high school. Include the following prompts in the interview:

1. For how many years have you been working as a professional mathematics teacher?
2. How, if at all, do you teach differently now than you did during your very first year as a mathematics teacher?
3. Reflect back on that first year and talk about your experiences. What were the major surprises—both the satisfying ones and the disappointing ones?
4. How is your professional life different today than it was then?
5. We are both in the process of learning how to be professional mathematics teachers. What advice do you have for us at this stage of our careers?

Interview two other inservice teachers with at least three years experience. Without revealing anything that any of the teachers told you in confidence, compare the results of the three interviews with those from other colleagues engaging in Activity 1.4. Discuss what you learned from the interviews.

Tina and Jermain are two preservice mathematics teachers who engaged in Activity 1.4. Case 1.12 is a transcript of one of the interviews they conducted:

CASE 1.12

Tina: For how many years have you been working as a professional mathematics teacher?

Ms. Brown: This is my ninth year—four at G. W. Carver High and five here at Groves Street Middle.

Jermain: How, if at all, do you teach differently now than you did during your very first year as a mathematics teacher?

Ms. Brown: Wow! That makes me think. I don't think I've changed my philosophy of teaching. From the very start, I attempted to use the same teaching strategies that I use now—inquiry strategies to lead students to discover and invent mathematics and direct strategies to build their skills with algorithms. The huge difference is that those attempts are much more successful today than they were when I first started. They really work much better than they used to.

Jermain: Why?

Ms. Brown: Teaching is an extremely complex art. I went into teaching understanding what to do, but I needed experience in order to refine that art. What I realize now is that the teacher-preparation program I had in college provided me with the tools I needed, but it took actual experience to develop my talents for using those tools. From the start, I designed fantastic lesson plans; I knew what to do, but I wasn't always able to do it. Sometimes things went sour because I didn't communicate directions so students understood what to do or I responded to students' questions in a way that turned them off—maybe because I was too judgmental. Most of the problems stemmed from how I interacted with students and that often led to misunderstandings and then discipline problems. But my first classes taught me well. I have them to thank for being able to practice my art effectively today.

Tina: Reflect back on that first year and talk to us about your experiences. What were the major surprises—both the satisfying ones and the disappointing ones?

Ms. Brown: That was the most challenging year of my life; I worked 21 hours a day; when I was asleep I was dreaming and thinking about how to deal with classroom situations. I'm glad I don't have to go through another year like that one, but I wouldn't trade the experiences I had that year for anything! I learned so much that makes me a productive teacher today—not that I'm not still learning something new everyday. I continue to struggle with classroom management and discipline.

Tina: And the major surprises from that year—both the satisfying ones and the disappointing ones?

Ms. Brown: The most disappointing surprise was how all-consuming teaching was that first year. I hardly had time for anything else. I was also disappointed by a few of my experienced colleagues who didn't really know how to teach. They made my job harder because they reduced students' perception of mathematics to a string of mindless textbook exercises. But the satisfying surprises far outweighed the disappointments. Seeing kids turn on to real mathematics was—and still is—a real turn on for me. Those moments when you raise just the right question or pose just the right problem that leads students to discover mathematical relationships for themselves—they make up for a lot of the garbage we have to put up with. Also, I gained a lot of satisfaction from professional relationships developed with many of my fellow teachers—not the few I mentioned before. I was amazed at how much we taught one another about mathematics, classroom management, planning, communicating, and on and on—mostly when we would strategize over motivating students or dealing with discipline problems. And then, of course, there are the relationships you build with students. Don't ever let anyone convince you that teachers don't make a difference in kids' lives. Subtle difference in the seemingly little things can be huge. How you acknowledge students as you pass them in the hallway can be momentous—the difference between using a student's name in a casual greeting as opposed to no greeting or even a greeting without the name becomes important in the minds of some teenagers. When I first realized how the little things had such a major effect on how students relate to us, I felt both important and frightened—important because how I acted made a difference, frightened because of how on-my-toes I need to be every moment I'm in the company of students. That can be exhausting and that's why—as much as I enjoy interacting with students—I need time away from students and with adults who understand what it's like to be a teacher.

Jermain: How is your professional life different today than it was then?

Ms. Brown: It's easier. What I have now that I didn't have then is this storehouse of experiences and instructional materials to build on. Now I have to work only about 11 hours a day.

Tina: We're both in the process of learning how to be professional mathematics teachers. What advice do you have for us at this stage of our careers?

Ms. Brown: Take advantage of your preservice teacher preparation program. That's where you acquire those necessary tools you learn to use when you teach. And don't give up on the tools—like inquiry instructional strategies or unit planning—because you struggle to get them to work the first 100 times you try them with real students. Don't expect too much of yourself until you've had time to develop your art through experience. And, speaking of experience, seize every opportunity to work with real students—whether it be clinical experiences in conjunction with professional education courses or tutoring jobs. These preadolescent and adolescent students have so much to teach you about how to communicate, how to interact with them, and about mathematics itself. Oh! That reminds me of another major surprise—actually I was first surprised by this one during my preservice clinicals and student teaching: I'm amazed at the insights into mathematics I gained from interacting with these young people. They stimulate my brain with their questions, requests, comments, and statements such as: "Why isn't 1 prime?", "Why do we rationalize the denominators?", "So show in real life when a negative times a negative is positive.", "Do you always have to do the same thing to the numerator that you do to the denominator?", "Multiples are bigger than divisors.", "Are there always more permutations than there are combinations?", "What's the difference between a big difference and a little difference?", "Circles are one sided.", "If we figured out how to divide by zero, we wouldn't need calculus.", "We use rational numbers more than we use irrational ones.", or "Why does dividing any circumference by any diameter give you π?"

Becoming More Assertive

Santrock (2001, p. 458) stated:

> People with an *assertive style* express their feelings, ask for what they want, and say no to things they don't want. When people act assertively, they act in their own best interests. They stand for their legitimate rights, and express their views openly. Assertive individuals insist that misbehavior be corrected, and they resist being coerced or manipulated (Evertson & oth-

ers, 1997). In the view of assertiveness experts Robert Alberti and Michael Emmons (1995), assertiveness builds positive, constructive relationships.

Your communications are assertive when you send exactly the message that you want to send, being neither *passive* nor *hostile.* Your communication is passive rather than assertive if it fails to send the message you want to convey because you are intimidated or fearful of the recipient's reaction. Your communication is hostile rather than assertive if you intend it to be intimidating or insulting. (Cangelosi, 2000b, pp. 38–41, 148–150, 309–312). To gain control over your professional life, you must be assertive with students, colleagues, administrators, and parents—neither passive nor hostile. Consider Case 1.13:

CASE 1.13

Early in the school year, Mr. Rudd discovers the need to be assertive with his colleagues. For example, just before the second day's homeroom period, Mr. Rudd visits Ms. Bomgars, who teaches the fourth-period health science class in Room 213. Mr. Rudd wants to discuss the problem her class created for him yesterday, but he would rather discuss it out of earshot of her homeroom students, some of whom are milling around the room. The conversation begins:

Mr. Rudd: Hello, Ms. Bomgars, do you have a minute to talk?
Ms. Bomgars: Of course, Mr. Rudd—always glad to meet with a colleague! What can I do for you?
Mr. Rudd: It's about your fourth-period class using Room 213. Could we step away from these students to discuss a problem I had yesterday with the room?
Ms. Bomgars: Oh, don't worry about these kids; they're not interested in what we're saying. What's your problem?
Mr. Rudd: Well, it took me a long time to get the room ready yesterday for my fifth-period life-skills class.
Ms. Bomgars: I thought all you math guys needed was a piece of chalk and you're ready!
Mr. Rudd: (laughing with Ms. Bomgars) That's a common misconception, but really, I'd appreciate it if you could—

A student comes up, interrupting Mr. Rudd by asking Ms. Bomgars a question. She answers the student and they briefly converse as Mr. Rudd waits.

Ms. Bomgars: Excuse me, Mr. Rudd—it's always something. You were saying?
Mr. Rudd: I'd really appreciate it if you could have your students put things back as they were when they came into the room.
Ms. Bomgars: We took out all the extra chairs we brought in with us when we left. Were some left in?
Mr. Rudd: No, that was great. It's just that—

Another student interrupts; because the bell is about to ring, Mr. Rudd interrupts the student, to say, "Excuse me, Ms. Bomgars, I've got to get to my homeroom."

Ms. Bomgars: Thanks for coming by—I'm glad to help you out any way I can.
Mr. Rudd: Thank you, I appreciate it.

The room is in no better shape when Mr. Rudd arrives for fifth period than it was the first day; the problem persists for the remainder of the week. Realizing that he failed to tell Ms. Bomgars what he wanted out of fear of creating ill feelings, Mr. Rudd is angry at himself for not communicating assertively.

Over the weekend he phones Ms. Bomgars:

Ms. Bomgars: Hello.
Mr. Rudd: Hello, Marilyn, this is Casey Rudd. We spoke just before your homeroom Tuesday.
Ms. Bomgars: Oh, hi, Casey. How are things going?
Mr. Rudd: Some things are going very well, others aren't. *We* still have a problem to solve regarding sharing *my* classroom. I'd really like to meet with you in *my* room so *we* can work out a solution.
Ms. Bomgars: You want to meet on Monday?
Mr. Rudd: It has to be at a time when students won't be there to interrupt us. I can meet you any time today or tomorrow—or Monday at 7:15 in the morning would be okay too.
Ms. Bomgars: You seem pretty serious; I'd better meet you today. Would an hour from now be okay?
Mr. Rudd: That would be great. See you in my room in an hour. Thank you very much.
Ms. Bomgars: Good-bye, Casey.
Mr. Rudd: Good-bye, Marilyn.

Now that Ms. Bomgars understands that Mr. Rudd is serious about the two of them solving what she now perceives as her problem as well as Mr. Rudd's, she is very receptive to Mr. Rudd's explanations of just how the room should be left after fourth period. In the classroom, Mr. Rudd readily points out exactly what he expects. Mr. Rudd is pleased with the agreed upon arrangement and the way Ms. Bomgars leaves the room for the rest of the semester.

Mr. Rudd also learns to communicate assertively with parents. After school one day, Mr. Rudd passes one of his algebra students, Alphonse, walking with his father in the hallway:

Mr. Rudd: Hello, Alphonse.
Alphonse: Hi, Mr. Rudd. Mr. Rudd, this is my dad.
Mr. Rudd: Hello, Mr. Oldham, I'm Casey Rudd, very nice to meet you.
Mr. Oldham: So you're Alphonse's algebra teacher. You know, I tell my kids all the time, subjects like algebra and Latin—where you have to memorize—those are the subjects where you discipline the mind. Study algebra and you can make yourself do anything. Just

because you don't use it much, today's kids don't want to learn it. They'd rather waste time with frilly subjects—but man needs to discipline his mind with stuff like algebra. It does for the mind what football does for the body! Isn't that right, Mr. Rudd; you tell him.

Smiling broadly at Mr. Oldham but quickly thinking to himself, "If only I had stayed in my room another 20 seconds, I would have avoided this dilemma. Mr. Oldham has the best of intentions and I'm happy he's trying to encourage Alphonse to study algebra. He wants to support my efforts and I really appreciate that. But he's sending all the wrong messages about the value of algebra. I don't want Alphonse to believe what he's saying, but then I hate to contradict what a father says in front of his child!" Not wanting to appear insulting, Mr. Rudd only continues to smile and says, "I agree that algebra is important for everyone to learn. It's a pleasure meeting you, Mr. Oldham. Thank you for introducing me to your father, Alphonse."

The next day in algebra class, during a discussion about using open sentences to solve real-life problems, Alphonse says, "Yeah, but the real purpose of algebra is to train your memory. That's what my dad says, and Mr. Rudd agreed with him yesterday."

After diplomatically attempting to correct the record about his beliefs relative to the purpose of algebra, Mr. Rudd resolves to be more assertive in communicating with parents and in responding to misconceived statements about mathematics.

Fortunately, Mr. Rudd's resolve is still fresh in his mind as he meets with Ms. Minnefield about her daughter Melinda's work in geometry:

Ms. Minnefield: You know, Melinda really has a lot of respect for you; she raves about your class. That's why I thought you should be the one to help me with this problem.

Mr. Rudd: What problem?

Ms. Minnefield: Well, Melinda has got herself involved with this boy who is much too old for her. She keeps seeing him even though she knows I don't like it.

Mr. Rudd: I can see that you are concerned.

Ms. Minnefield: I thought because she likes geometry so much, I could use that to get her to break it off.

Mr. Rudd: I'm not following you.

Ms. Minnefield: Well, you know how math is harder for girls than boys. I told her if she keeps spending time with this boy, she was going to do badly in geometry. That's why girls don't do well in math because they get all wrapped up over boys. Do you see what I mean?

Mr. Rudd: Not really, ma'am, but please go on.

Ms. Minnefield: Well, I thought you could back me on this. Tell her she's going to flunk if she doesn't spend more time studying geometry and less time fooling around with that boy.

Mr. Rudd: Ms. Minnefield, I really appreciate you sharing your ideas with me, and I appreciate your concern for Melinda's welfare. You, being Melinda's parent, know far more about this situation than I. I'll confine my remarks to what I do know about. First of all, research studies clearly point out that girls do not have any more trouble learning mathematics than boys. It's a common misconception that they do, but they absolutely don't.

Ms. Minnefield: But I had always heard that.

Mr. Rudd: If you're interested, I can give you some journal articles that explain the facts about girls and women in mathematics.

Ms. Minnefield: Oh, that would be nice of you.

Mr. Rudd: Second, and more to the point here, my job is to teach Melinda and the other students mathematics as professionally and responsibly as possible. I cannot in good conscience base Melinda's grade on anything other than how well she achieves the goals of the geometry course.

Ms. Minnefield: Well, I thought if you just told her, she'd listen.

Mr. Rudd: I'm flattered that Melinda would have that much confidence in what I say. But if I start lying to her, I'll lose her confidence.

Ms. Minnefield: I don't mean for you to lie to her.

Mr. Rudd: I know, Ms. Minnefield. You want what's best for your daughter.

In his classroom, Mr. Rudd encourages more interaction among students than they had been used to in mathematics courses with previous teachers. Early in the year, some students tended to take advantage of their freedom of expression by drifting off the current topic. Mr. Rudd, not wanting to alienate students or to discourage communications about mathematics, allowed some of the discussions to waste valuable class time. For instance, during a precalculus class, the difference between "negative x" and "x is negative" is being discussed:

Opal: If x is negative, then x has to be less than 0, so negative x could be a positive.

Bernie: That's too picky. It's like in Spanish class, what's the difference between temar and temer? Mr. Waiters makes such a big deal over whether you say "ar" or "er!" Who cares?

Rita: It's so he can have something to grade you on.

Bernie: He's not fair . . .

The discussion continues in this vein for several more minutes, with Mr. Rudd worrying about the appropriateness of Mr. Waiters being the topic of conversation and about time being spent on the topic at hand.

After a few similar experiences, Mr. Rudd resolves to be more assertive in such situations, and in subsequent weeks he is more likely to respond as he does in the following instance. During a geometry lesson about applying triangle congruence theorems to real-life situations, this conversation takes place:

Eric: In basketball there's a strategy you call the three-man game, in which you form a triangle. The triangle should be equilateral.

Vontego: Who cares about basketball? Eric's always got to talk about basketball. It's stupid!

Eric: You've got to be better to play basketball than to—

Mr. Rudd: Do not debate your opinions about basketball in here today. Even those that don't like basketball can learn something about equilateral triangles from Eric's example of the three-person strategy. Please repeat your example, Eric, without defending your opinion about basketball.

It is in meeting his most challenging responsibility of keeping students on-task and responding to off-task behaviors that Mr. Rudd discovers the greatest need to be assertive. At times, he resents having to work continually to maintain students' interest and to teach students who tend to get off-task to be on-task. An expression of frustration prompts what turns out to be a productive conversation with Vanessa Castillo:

Mr. Rudd: Some days I'd just love to walk into class and discuss mathematics without having to worry about Frankie over in the corner who is going to fall asleep unless I'm either right near her or we're discussing a problem that strikes within the limited range of what she fancies! Or just once, getting through a session without having to deal with Brad's showing off or Christi's yakking with anyone who'll listen to her—wouldn't that be nice? But that's too much to ask for!

Ms. Castillo: Obviously, you're having one of those days!

Mr. Rudd: It's just that we were getting into our first formal proof in third period today, and they seemed so enthusiastic. But then they started to get a little noisy—some off-task talking. I let it go at first because I didn't want to put a damper on their enthusiasm. But then it became obvious that Christi and Livonia's conversation had nothing to do with geometry—right in the middle of my explanation!

Ms. Castillo: What did you do?

Mr. Rudd: I kept on explaining the theorem, and just moved near them and caught their eyes—that usually works for me.

Ms. Castillo: But it didn't this time?

Mr. Rudd: Oh, the two of them stopped as long as I stood there, but then other conversations broke out, and Christi and Livonia started up again as soon as I moved away. Five minutes later I'd had enough and made the mistake of threatening the class.

Ms. Castillo: What did you say?

Mr. Rudd: I told them if they didn't pipe down, they'd be sorry when the test came around. I knew that wasn't the right thing to say, but the noise just got to me and I reacted.

Ms. Castillo: Did they quiet down?

Mr. Rudd: Yes, but then Brad whispered something to Lin-Tau and she started giggling. That's when I jumped on them and called Brad a showoff. In other words, I handled it all wrong and made matters worse.

Ms. Castillo: So you weren't Mr. Perfect. You let things go too far and reacted with hostility instead of assertiveness—the way most of us react when we feel we've lost control.

Mr. Rudd: But I know better. I applied none of the stuff that's worked for me in the past—assertiveness, descriptive language, reinforcement principles—they all went out the window!

Ms. Castillo: I don't think you threw your principles out of the window; I think you waited too long to respond decisively. Most teachers wait until they are too near their threshold for tolerating noise or other annoyances before dealing with students being off-task.

Mr. Rudd: You're saying I should have stepped in and dealt with the early minor incidents before things escalated. I was passive in the beginning instead of being assertive. Then things got out of hand.

Ms. Castillo: And when things get out of hand, it's natural to be hostile.

Mr. Rudd: And that's why I'm really upset—at my own hostile behavior. I'm afraid I've lost some of the control and goodwill I've worked to build up to this point. I don't have much enthusiasm for tomorrow's class.

Ms. Castillo: What do you have planned?

Mr. Rudd: Before this, I was going to continue the session on proofs.

Ms. Castillo: I think you have two choices. You either go on with your original plan and conduct the class as if none of this happened—but have an alternative plan ready to go to as soon as they become uncooperative—you know, one where they have to work on their own while you monitor their every move. The second choice is to start the period off by expressing your feelings about what happened, even indicating that you're disappointed in your own behavior for letting things go too far and then acting with hostility as a result. But if you use this tactic, make sure to assertively demand their cooperation. If it turns out you don't get it even after clearing the air, then go immediately to the alternative learning-activity plan.

Throughout the year, Vanessa and Casey regularly share ideas on handling discipline problems, as well as other aspects of teaching. They even work out a plan by which Casey can send a student to Vanessa's classroom for custodial supervision when Casey needs to get that student out of his classroom until he has time to deal with the student's disruptive behavior. Casey reciprocates by providing the same service for Vanessa.

Benefitting From Instructional Supervision

Instructional supervision is the art of helping teachers improve their teaching performances (Cangelosi, 1991, pp. 6–7). Consider Case 1.14:

CASE 1.14

Ms. Castillo serves as an instructional supervisor for Mr. Rudd by sharing ideas on planning, managing student behavior, and other aspects of instruction. Working with Mr. Rudd not only helps his classroom effectiveness, it also benefits Ms. Castillo's teaching because conferring with Mr. Rudd causes her to analyze problem situations and reflect on instructional activities more than she would otherwise. As the year progresses and Mr. Rudd's confidence soars, Mr. Rudd more and more serves as an instructional supervisor for Ms. Castillo as well. Consistent with research findings relative to instructional supervisory practices, Casey and Vanessa's cooperative partnership provides them with the most effective type of help with their teaching (Birman, Desimone, Porter, & Garet, 2000; Cangelosi, 1991, p. 127, 158; Oliva & Pawlas, 2001, pp. 410–412).

Accustomed to sharing ideas with Ms. Castillo from his first day at Rainbow, Mr. Rudd seeks and listens comfortably to suggestions from other instructional supervisors as well (i.e., other teachers, his department head, the principal, and the district mathematics teaching supervisor). Mr. Rudd does not always agree with or follow the suggestions; however, everyone stimulates his ideas for such things as coping with individual differences among students and eliciting parents' cooperation.

Mr. Rudd's most frustrating problem is finding enough time to do what he considers necessary for optimal classroom effectiveness. The following exchange with Armond Ziegler, Mathematics Department head, helps:

Mr. Rudd: There are so many things I ought to be doing but never get around to!

Mr. Ziegler: Like what?

Mr. Rudd: I need to develop enrichment materials for some students who are ahead of the rest of the class. And there are parents I should be contacting. I haven't been entering miniexperiments into my computer folders as regularly as I should.

Mr. Ziegler: Okay, slow down! I hear you. You have to realize that it's impossible to do everything you want to or should do. Make some difficult decisions and partition your time. Prioritize from what you must do—like sleep, exercise, eat, and show up for class—to what is critical, down to what you really want to do but could put off, and then finally to what isn't all that important. Put a high priority on things that will help you save time down the road—such as keeping up with those miniexperiment folders. Neglect something else instead.

Mr. Rudd: I try to keep to a schedule, but then students come in for tutoring, and then—wham!—time I scheduled for formulating examples and nonexamples for class is used up!

Mr. Ziegler: Don't allow it to happen; take control. You wouldn't drop in on a lawyer or physician without an appointment. Don't let your clients do that to you.

Mr. Rudd: What clients?

Mr. Ziegler: Your students are your clients. Also don't let parents, colleagues, or administrators abuse your schedule either.

Mr. Rudd: What about you? Should I tell you to buzz off when you ask me to turn in my supply order while I'm in the midst of scoring tests? How about Harriet [the principal]? Can I put her off too?

Mr. Ziegler: You can tell me to buzz off anytime. And I know Harriet well enough to think she'll respect your businesslike approach and understand that you've got your time scheduled.

Mr. Rudd: But everything can't be scheduled. Unanticipated things have to be taken care of.

Mr. Ziegler: Of course they do, but get a calendar anyway and use it as an organizational tool for your own convenience—not something you blindly follow.

Mr. Rudd: Today, for example, I planned to phone a couple of parents during the B lunch period, but then we were short a lunchroom supervisor, so I had to stay that period besides my usual A lunch period.

Mr. Ziegler: Those are the times when I take a triage approach. I decide where my time can be most efficiently spent. Some crises are beyond our reasonable control and others can wait.

Mr. Rudd: So, now I have to find the time to schedule my time!

Preparing for Administrative Supervision

Whereas instructional supervision is concerned solely with improving instructional practice, *administrative supervision* is concerned with quality control (Cangelosi, 1991, pp. 163–173). Consider Case 1.15:

CASE 1.15

Rainbow High and school district administrators are responsible for determining whether or not Mr. Rudd teaches well enough to be retained as a teacher and given incentives to remain on Rainbow's faculty. The district has an administrative supervisory program in which the classroom instructional practices of each beginning teacher are evaluated three times a year. The outcomes of these evaluations hinge primarily on observational data gathered by a team composed of the school principal, the district subject supervisor, and a same-subject teacher from another school.

Ms. Castillo helps Mr. Rudd prepare for the scheduled visits from his observational team by doing the following:

- Advising him to use preobservational conferences to apprize team members of the instructional strategies they can anticipate seeing him employ.
- Suggesting that it is appropriate to utilize post-observational conferences not only for learning from the team's report but also to express his own needs regarding support services from the administration.
- Simulating a team visit, with Ms. Castillo playing the role of the team in a preobservational conference, an in-class observation (during one of her nonteaching periods), and a postobservational conference.

In anticipation of the team visits, Mr. Rudd is quite nervous. However, after the first few minutes of the first in-class observation, he relaxes and learns to enjoy the attention. It helps that he thinks of his visitors as colleagues whose goals of helping students learn mathematics are the same as his own. Although he feels the team never sees his best work, his performances receive better than satisfactory ratings and his confidence continues to rise.

SYNTHESIS ACTIVITIES FOR CHAPTER 1

The synthesis activities for each chapter are intended to (a) help you bring together the chapter's content, (b) reinforce and extend what you learned, and (c) assess what you gained from the chapter so that you can identify your areas of proficiency and the topics you need to review. Another purpose is to encourage you to articulate your thoughts about principles, strategies, and methods for teaching mathematics. Understanding is enhanced through such activities (Connolly & Vilardi, 1989; Goldin & Shteingold, 2001; Myers, 1986, pp. 167–169).

Here are the synthesis activities for chapter 1:

1. For each of the following multiple-choice prompts, select the one best response that either answers the questions or completes the statement so that it is true:

   A. Let W = {items in Casey Rudd's working portfolio}, P = {items in Casey Rudd's presentation portfolio}, $|W|$ = number of items in W, and $|P|$ = number of items in P. Which one of the following statements is true?
   a) $|P| < |W|$
   b) $|W| < |P|$
   c) $W \subseteq P$
   d) $P \cap W = \varnothing$

   B. Let R = {preservice secondary-school mathematics teachers}, I = {inservice secondary-school mathematics teachers}, and T = {secondary-school teachers}. Which one of the following statements is true?
   a) $R \subseteq I$
   b) $I \subseteq R$
   c) $R \cup I = T$
   d) $R \cap I = \varnothing$

   C. A definition for "*PSSM*" can be found in _____ of this textbook.
   a) chapter 1
   b) the glossary and chapter 4
   c) either the glossary or chapter 4

   D. Inquiry instructional strategies are intended to lead students to _____ .
   a) develop their algorithmic skills
   b) discover or invent mathematics

   E. The assertive response style is characterized by _____ .
   a) aggressiveness
   b) judgmental comments
   c) professional confidentiality
   d) sincerity

   F. According to Kounin, a teacher who is with-it _____ .
   a) deals with disruptive behavior decisively
   b) displays enthusiasm for teaching
   c) is aware of what students are doing
   d) effectively uses proximity control to manage the classroom

   G. In this chapter's section "The Beginning of an Eventful School Year," reference is made to being businesslike in the classroom. In this context, a classroom is businesslike if _____ .
   a) students adhere to formal standards of classroom conduct
   b) no attention is afforded matters unrelated to the learning of mathematics
   c) learning mathematics is afforded top priority
   d) both students and teachers follow appropriate conventions of decorum

   H. During his interview for the Rainbow High position Casey Rudd indicated that _____ .
   a) for teaching mathematics, inquiry instructional strategies are superior to direct instructional strategies
   b) for teaching mathematics, direct instructional strategies are superior to inquiry instructional strategies
   c) there is considerable redundancy of mathematical content in traditional textbook-driven mathematics curricula
   d) students need to learn algorithms for more specific situations than the more general situations presented in their mathematics textbooks

2. In a discussion with a colleague who also engaged in Synthesis Activity 1, compare your responses to the multiple-choice prompts. Also, check your choices against the following key: A–a, B–d, C–b, D–b, E–d, F–c, G–c, & H–c.
3. Locate an artifact from work you did in mathematics (e.g., the proof of a theorem that you authored, a computer program you wrote for executing an algorithm, an expository paper you wrote on some aspect of the history of mathematics, a display of your original solution to a problem, the description of a mathematical model you developed, or a disk with a computer file for a mathematical simulation you developed). Insert that artifact into the "Mathematics and the Historical Foundations of Mathematics" category in the working portfolio that you set up when you engaged in Activity 1.1.
4. Exhibits 1.13–1.16 display opening-day tasksheets that Casey Rudd developed to (a) prompt his students to succeed with challenging mathematical tasks, (b) begin setting up expectations and the tone for the courses, and (c) get an early jump on knowing his students as individuals. Suppose you are preparing to teach the course for which you developed a syllabus when you engaged in Activity 1.2. Develop an opening-day tasksheet for that course with the same purposes Mr. Rudd had in mind for his four opening-day tasksheets. Exchange your opening-day tasksheet with one from a colleague who is also engaging in this synthesis activity. Share feedback and suggestions. After modifying yours in light of the feedback, insert it into the "Motivation, Engagement, and Classroom Management" section of your working portfolio.
5. With a colleague, examine the computerized management system you set up when you engaged in Activity 1.3. In a discussion, identify which subfolders (as labeled in Exhibit 1.12) house tasks that (a) are performed only by mathematics teachers, (b) are performed by all secondary-school teachers, and (c) primarily require only clerical skills (as opposed to pedagogical talents).
6. In collaboration with a colleague, review the reported dialogue portion of the interview Casey Rudd had with Vanessa, Armond, Harriet, and Jack in Case 1.3. In light of your review, address the following questions in a discussion:
   A. What was Mr. Rudd's rationale for his belief that there really is ample room in mathematics curricula for "time-consuming" inquiry instruction?
   B. What was Mr. Rudd's rationale for his belief that *PSSM*-based curricula may actually lead to higher test scores on traditional high-stakes tests?
7. Call to mind mathematics teachers that taught you in 6th through 12th grades. Identify the one whose teaching most exemplifies Casey Rudd's approach to teaching and another whose teaching most closely exemplifies Don Delaney's approach as reflected in his comments from Case 1.6. With a colleague who is also engaging in this activity, discuss similarities and differences in the instructional methods of the four teachers the two of you identified.
8. If you are preservice teacher, then visit the classroom of an inservice mathematics teacher. Make a sketch of the room arrangement with about the same degree of detail as Exhibits 1.9–1.11. Decide how you would rearrange the room to make it more conducive to the way you visualize yourself teaching. Sketch your modified arrangement. Compare your arrangement to those of colleagues. Discuss which features from Casey Rudd's list in Case 1.8 that you were able to incorporate in your arrangements.
9. Read Case 1.16, then respond to the five prompts (A–E) that follow in light of what you read:

CASE 1.16

Shamika, one of Ms. Salzburg's students, initiates a discussion in precalculus class:

Shamika: You know those tasksheets you wanted us to turn in Friday?

Ms. Salzburg: Yes, Shamika. What about them?

Shamika: Couldn't we wait 'til Monday to give them to you?

Several other students: Oh! Yes please, please, Ms. Salzburg!

Shamika: There's a game Thursday night, and I know you want us to support the team!

Trent: You wouldn't want us to miss the game. We just want to show our school spirit, but we couldn't if you don't give us more time for that assignment.

Haikee: Be nice just this once!

Ms. Salzburg is tempted to "be nice" and enjoy the cheers she knows she'll get if she complies with their request. But she also realizes the following: (a) Delaying the assignment will cause the class to fall behind the planned lesson schedule. (b) If she does not get the tasksheets with their responses until Monday, she will not be able to read and annotate them over the weekend; consequently, she would be inconvenienced. (c) They can adjust their own schedules in order to complete the tasksheet on time without missing the game.

A. Write a *passive* response that Ms. Salzburg could make to her students in Case 1.16.
B. Write a *hostile* response that Ms. Salzburg could make to her students in Case 1.16.
C. Write an *assertive* response that Ms. Salzburg could make to her students in Case 1.16.
D. Compare what you wrote to that of colleagues. Discuss how passive, hostile, and assertive communications differ.
E. Compare what you and your colleagues wrote to the following:

A *passive* response Ms. Salzburg might make is, "Well, we really need to get these tasksheets done by Friday. I really should be going over them this weekend. I wish you wouldn't ask me to do this because I—But, okay, just this once—but only because this is an extra important game."
A *hostile* response Ms. Salzburg might make is, "You people are always trying to get out of work! Do you think your game is more important than mathematics? Understanding mathematics will take you a lot farther in life than games. Besides, if you scheduled your time better, you'd have this finished in plenty of time for the game!"
An *assertive* response Ms. Salzburg might make is, "I understand that you are worried about going to this important game and still being able to complete the tasksheet on time. You have cause for concern. Because changing the due date will mess up our schedule and because I need the weekend to analyze your responses, the tasksheets are still due on Friday."

TRANSITIONAL ACTIVITY FROM CHAPTER 1 TO CHAPTER 2

The transition activity from one chapter to the next is designed to set the stage for our work in the subsequent chapter. In a discussion with two or more of your colleagues, address the following questions:

1. Some teachers orchestrate smoothly operating classrooms where students cooperatively go about the business of learning mathematics with hardly any disruptions. Other teachers spend more time ineffectively dealing with student misbehaviors than conducting worthwhile learning activities. What do the teachers in the former group do to gain and maintain students' cooperation and to motivate students to do mathematics?
2. What are some of the strategies employed by teachers to build and maintain a classroom climate that is conducive to cooperation and engagement in meaningful mathematics?
3. What impact does the manner and style in which a teacher communicates with students and parents—how assertively the teacher speaks, the teacher's body language, and whether the teacher uses descriptive or judgmental language—have on how well students cooperate in the classroom?
4. What classroom standards of conduct and procedures for safe, efficient classroom operations should a teacher establish? How should those standards and procedures be enforced?
5. What are some effective strategies for teaching mathematics to students for whom English is not a first language and who are not proficient with English?
6. What are some effective strategies for teaching mathematics in a way that reaps the benefits of the cultural diversity in the classroom?
7. What are some effective strategies for dealing with students who are being uncooperative and off-task?
8. Chapters 4–11 of this textbook focus on the principles, strategies, and techniques for developing and delivering mathematics curricula to middle, junior high, and high school students. Why do you suppose that, prior to those chapters, a chapter with the title "Gaining Students' Cooperation in an Environment Conducive to Doing Mathematics" and a chapter with the title "Motivating Students to Engage in Mathematical Learning Activities" are inserted?

chapter

2

Gaining Students' Cooperation in an Environment Conducive to Doing Mathematics

Goal and Objectives for Chapter 2

The Goal The goal of chapter 2 is to lead you to develop strategies for gaining and maintaining your students' cooperation.

The Objectives Chapter 2's goal is defined by the following set of objectives:

A. You will explain why the success of your mathematics lessons depends on how well you teach your students to be on-task and engaged in learning activities (discover a relationship) 10%.
B. You will explain strategies for (a) communicating with students in a way that increases the likelihood that they will choose on-task instead of off-task behaviors, (b) establishing a favorable climate for learning meaningful mathematics, (c) using the cultural diversity of students to enhance opportunities for doing mathematics, (d) including and accommodating students with special needs, (e) establishing and enforcing standards of conduct and routine classroom procedures, and (f) dealing with off-task behaviors (comprehension and communication) 45%.
C. Given the responsibility for teaching a mathematics class, you will develop strategies for interrelating with students and managing the classroom to encourage cooperation and enhance learning opportunities for all students (application) 30%.
D. You will incorporate the following phrases or words into your working vocabulary: "allocated time," "transition time," "on-task," "the principle of modeling," "positive reinforcement," "destructive positive reinforcement,"-"engaged behavior," "off-task," "punishment," "disruptive behavior," "descriptive language," "judgmental language," "IRE cycle," "true dialogue," "supportive reply," "nonsupportive reply," "Individualized Education Program (IEP)." (comprehension and communication) 15%.

WELL-PLANNED LESSONS GONE AWRY

From your work with chapters 4–11, you will develop your talents for designing courses, units, and lessons that lead students to do meaningful mathematics. However, no matter how ingeniously you design a lesson, that lesson is unlikely to succeed unless you artfully apply strategies for gaining and maintaining students' cooperation as you deliver that lesson. Although the teacher in Case 2.1 developed a sound lesson plan for a meaningful objective, the lesson goes awry because she fails to interact with students effectively and practice sound classroom management principles:

CASE 2.1

As part of an algebra I unit on graphs and linear functions, Ms. Lewis planned a lesson intended to lead students to achieve the following objective:

> The student explains why the graph of $y = ax + b$, where x and y are variables and a and b are real constants, is a line (discover a relationship).

At the end of Monday's class meeting, she directs students to complete the following homework assignment:

Solve each of the following open sentences for x and y and plot the solution on a number line:

$-4 = 3x - 1$ $\quad$ $-2 = 3x - 1$

$0 = 3x - 1$ $\quad$ $3 = 3x - 1$

$4 = 3x - 1$ $\quad$ $10 = 3x - 1$

For Tuesday, she plans to have the students transfer the six number lines from their homework onto a Cartesian plane as indicated by Exhibit 2.1. Then she plans to use questions to lead them to discover that when the number lines are located where $y = -4$, $y = -2$, $y = 0$, $y = 3$, $y = 4$, and $y = 10$, respectively, the solution points are contained by a single line.

She plans to use a few more examples (e.g., $1.5 = 3x - 1$) and, for contrast, a few nonexamples (e.g., a set of equations that includes $3 = x - 7$ and $-4 = x + 5$ as well as a set of the form $y = 3x^2 - 1$) on which students are to repeat the homework task and then discuss the outcomes to discover the relationship specified by the objective.

It is now Tuesday and the bell to begin the class rings. Ms. Lewis moves to the front as some students stream into the room while others sit at their desks passing time in various ways (e.g., Joyce and Cassandra discuss plans for the evening while Randall works feverishly on his English assignment). Ms. Lewis attempts to speak above the din as she yells, "Okay, people, settle down. Hey, Patrick, did you hear me?" Patrick: "No, ma'am. What'd you say?" Ms. Lewis: "If you weren't so involved with Rolando over there, maybe you'd hear what you should be doing!" Laughter erupts around the room as Levi quips, "Wooo, Patrick is involved with Rolando!" Feeling embarrassed that once again she is struggling to gain control of the class, Ms. Lewis lets out a half-hearted chuckle at Levi's joke, but she realizes that her comment was counterproductive.

Exhibit 2.1

A Diagram on the Cartesian Plane That Ms. Lewis Expects Her Students to Plot From Their Homework Points.

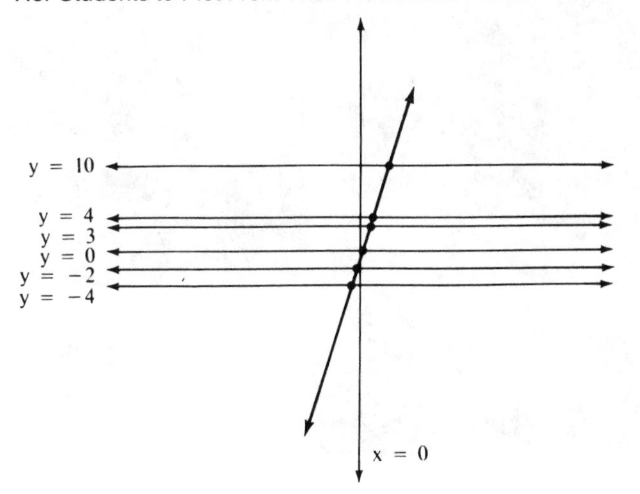

Raising her voice, Ms. Lewis calls roll:

Ms. Lewis: Genan?

Genan: Here.

Ms. Lewis: Paulette?

Paulette: Here?

Ms. Lewis: James? [pause] James, aren't you here? James, answer the roll!

James: Oh, me? Yo, I'm here.

Ms. Lewis: Answer me right away next time or you'll be marked absent.

James: Okay I will.

Ms. Lewis: Jeannie? [pause] Jeannie? Where's Jeannie?

Davilon: Oh, yeah, she's still in orchestra.

Ms. Lewis: Well what's she still doing in orchestra? Why doesn't anybody ever tell me these things?

Davilon: I just did.

Ms. Lewis: I don't mean you. The office is supposed to notify us—oh, never mind! Winston?

Seventeen minutes after the bell, the roll has been taken, the absentee list posted on the door to be picked up by the office staff, and the students are finally settled well enough for Ms. Lewis to begin implementing her lesson plan. She announces, "Take out your homework." This leads to another outbreak of quibbles:

Shauna: What homework? I didn't know you were going to pick this up!

Ms. Lewis: If you had been paying attention yesterday, you would know what you were supposed to have for today! Melissa, do you have yours?

Melissa: Right here.

Ms. Lewis: Please write it up on the board for me. I'm glad you're so dependable!

Some students pay attention while others entertain themselves in other ways as Melissa writes what appears in Exhibit 2.2 on the board.

Ms. Lewis says to Melissa, "That's not how you were supposed to do it." Turning to the class, she asks, "Did anybody do it the way I told you to?" No one volunteers. Having just been praised for being "so dependable," Melissa feels embarrassed. As Melissa sheepishly returns to her seat, Ms. Lewis says, "Let me show you what you were

Exhibit 2.2

What Melissa Wrote on the Board.

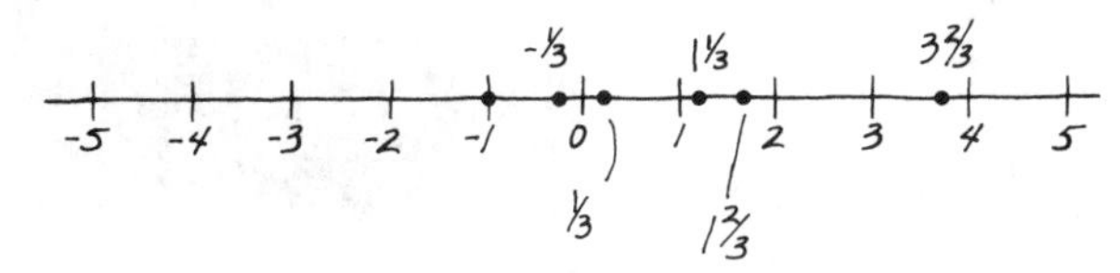

supposed to do," as she writes on the board using a separate number line to plot each of the six points (see Exhibit 2.3). As Ms. Lewis writes on the board with her back to the class, off-task conversations erupt around the room. Ms. Lewis worries that the noise might distract neighboring classes, so she yells without bothering to face the class, "Pipe down!" Moments later, she tries, "That's enough noise already!" then, "If you don't get this now, you won't know it for the test!"

With the six points plotted on six parallel number lines on the board, Ms. Lewis directs the class, "Now take out a sheet of graph paper and draw this first number line on the line where $y = -4$. Then do the second one on the horizontal line where $y = 0$, and . . ." As she continues with the directions, some students are searching for or trying to borrow graph paper. Only a few are actually listening to the directions.

Six minutes later, about two-thirds of the students have graph paper and are ready to follow Ms. Lewis' fourth repeat of the directions. She starts again, "Now this time listen! Here, I'll show you on the board graph." Midway through the explanation, Andrew enters the room and walks up to her with an admit slip for being late. Ms. Lewis: "Why are you late?" . . .

Eight minutes later, some of the students are quietly working on-task, some are waiting for Ms. Lewis to provide them individual help, some have completed the task and are thinking about other things, and some have yet to begin. Ms. Lewis tries to get to each student requesting help, but time doesn't permit.

She announces to the class, "This is what you should have done on your graph paper." She finishes what she began on the board resulting in Exhibit 2.4.

In an attempt to lead students to discover the relationship specified by the objective, she asks, "What can you say about these six points?" A number of students shout out replies.

Ms. Lewis: Pipe down! We talk one at a time in here. Okay, Eileen.
Eileen: They all line up in a row.
Ms. Lewis: That's right. Now what does that tell us?

The discussion continues, but from her position at the front of the room, Ms. Lewis is unable to detect who is actually relating the six equations to $y = 3x - 1$. Because the bell is about to ring, she aborts her plan for leading students to discover the relationship for themselves and speaks to the class as a whole as most students begin packing their bags, "Okay, this is what you're supposed to have figured out: We could express these six equations as one by writing $y = 3x - 1$ and then plotting the coordinate points (x, y) like—"

The bell rings and students rush out of the door before Ms. Lewis finishes the explanation and before she assigns homework.

Why do some teachers orchestrate smoothly operating classrooms in which students cooperatively and efficiently go about the business of learning

Exhibit 2.3
Ms. Lewis Steps in for Melissa.

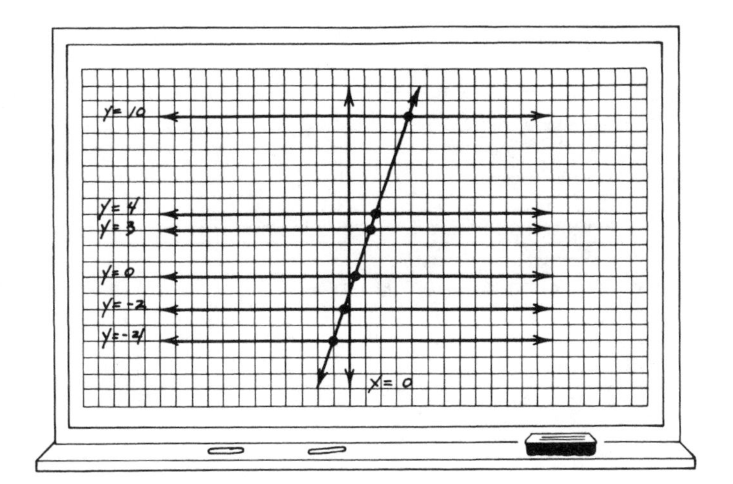

Exhibit 2.4
Ms. Lewis' Work on the Board.

mathematics with relatively few disruptions (e.g., Ms. Citerelli whom you will meet in chapter 5), whereas others (e.g., Ms. Lewis) struggle ineffectively with students' misbehaviors while trying to engage them in planned learning activities? Whether your teaching experiences are satisfying or marked by ongoing frustrating struggles in which you attempt to get your students to cooperate depends largely on how you apply fundamental research-based principles for interacting with students and managing your classroom. Ms. Lewis violated a number of those principles in Case 2.1. Overwhelmingly, teachers indicate that classroom management and discipline problems are the source of their greatest difficulties—leading to feelings of inadequacy—during their first few years of teaching (Emmer & Stough, 2001). Improper management of students is the leading cause of teacher failure (Bridges, 1986, p. 5).

Allocated and Transition Times

Examine the class meeting agenda shown in Exhibit 2.5. Note that five learning activities are planned:

1. Item 1 of the agenda indicates that the students are to work on exercises from page 124 of their texts at the very beginning of the period.
2. Items 4–6 indicate that the students are to transfer results from their previous homework onto a single Cartesian plane and then engage in a question-discussion session intended to culminate with two discoveries.
3. Item 7 indicates that students are to work independently at their places as the teacher provides one-to-one help.
4. Items 8–9 indicate that students are to engage in a question-discussion session intended to culminate in another discovery.
5. Items 10–11 indicate that students are to begin a homework assignment.

The time periods during which you plan to have your students involved in learning activities are referred to as "*allocated time.*" Thus, the day's agenda in Exhibit 2.5 provides five different allocated time periods. The time students spend between learning activities is referred to as "*transition time.*" If Exhibit 2.5's agenda is followed, transition time should occur during the following periods:

1. While students are entering the classroom, attending to the directions for the initial assignment, taking out their homework papers, and locating materials to begin the exercise from p. 124 of their textbooks.
2. As indicated by Item 3 of the agenda, when they have just stopped the first learning activity and are attending to directions for the second.
3. Just after the learning activity in Item 6 as they receive directions for and prepare to work on the tasks indicated by Item 7.
4. When the teacher calls a halt to the independent work session and directs the students into the question-discussion session indicated by Items 7–8.
5. While the teacher assigns homework.
6. Just before the class period ends when the students pack up their things and then exit the room.

Because the business of learning and doing mathematics takes place during allocated time, it is tempting to ignore the need to plan for transitions

Exhibit 2.5

Example of an Agenda for One Class Period.

Unit #6: Graphs of Linear Functions / 2nd day
Date: Tues., 10/22
Objectives:

C. The student explains why the graph of $y = ax + b$ (where x and y are real variables and a and b are real constants) is a line (discover a relationship).
D. The student states that the graph of $y = ax + b$ (where x and y are real variables and a and b are real constants) is a line (simple knowledge).

Agenda for the class session:

1. As students enter the room, direct them to begin immediately by following the assignment appearing on the overhead projector screen:
   A. Take out your homework and place it on your desk.
   B. Work Exercises 13, 14, 18, 21, and 23 from page 124 of the text.
2. As they work on the assignment, move from student to student silently taking roll and checking homework. Note whose homework you want to use as examples during explanations to the class.
3. Call a halt to the individual work. Have someone put her homework on the board as everyone else takes out a sheet of graph paper.
4. Direct the person at the board to transfer the six number lines, each with a plotted solution for x, onto the board graph where $y = -4, -2, 0, 3, 4$, and 10.
5. In a question-discussion session, lead students to discover that the six points are collinear as follows:

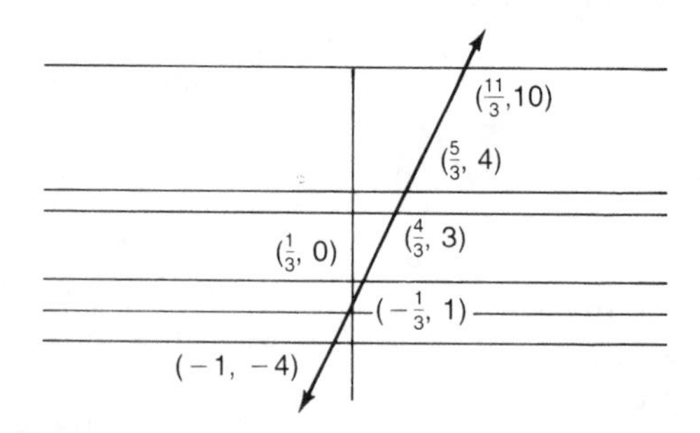

6. Use leading questions to get them to associate the six equations with $y = 3x - 1$.
7. Have the person at the board return to her place and assign everyone to repeat the procedure completed for homework and from Agenda Item #4 on each of the following sets of equations:

$-9 = 5x + 4$	$4 = -2x$	$3 = x - 7$	$-1 = 3x^2 - 1$
$-5 = 5x + 4$	$1 = -2x$	$-4 = x + 5$	$1 = 3x^2 - 1$
$0.9 = 5x + 4$	$0 = -2x$	$0 = 7x - 1$	$8 = 3x^2 - 1$
$3 = 5x + 4$	$-12 = -2x$	$13 = x - 7$	$9 = 3x^2 - 1$

   Keep the previous work on the board, so as they are working, you can provide efficient help as you circulate about the room working with individuals.
8. When at least 75% of the class has completed the first three sets and five students have completed all four sets, call a halt to the work and engage them in an inductive questioning session to lead them to discover the relationship specified by Objective C.
9. After they have articulated the relationship, have them enter it into their notes and have five people repeat it aloud.
10. Assign for homework: (a) Complete the exercises begun at the beginning of class, (b) study pp. 126–127, working through the examples, and (c) work Exercises 1, 5, 7, 9, 15, 17, 21, and 22 from pp. 127–128.
11. As time permits, have students work on the homework assignment.

between learning activities when designing lessons and daily class-meeting agendas. But how you plan for transitions has a major impact on the success of your lessons. After all, how well your students engage in a learning activity depends on how well they understood the directions for the activity you delivered to them during the preceding transition time. Furthermore, inefficient transitions waste valuable allocated time, delay the business of doing mathematics, and may affect your students' attitudes about the importance of getting down to the business of learning (Cangelosi, 2000b, pp. 23–24, 100–116).

Kounin (1977) demonstrated that student engagement in lessons depends on how smoothly teachers move from one learning activity to another and how well they maintain momentum. Consider Cases 2.2–2.4:

CASE 2.2

While other students work independently with tasksheets at their places, Mr. Rudd is at his desk with Sarah and Antonio planning a way for them to catch up on learning activities they missed while absent. When the planning is complete, Mr. Rudd suddenly announces to the class, "Okay, you may finish that tasksheet later; take out your scratchpad, straightedge, and compass and turn your textbook to page 288. [pause] Example 6-1 displays a way to construct a rhombus. Arlajean, please read the first step aloud." Some of the students who are involved with the tasksheet when Mr. Rudd suddenly announces the new activity do not listen to the directions. They continue responding to the tasksheet's prompts as Arlajean reads the first step. Others stop the tasksheet work to inquire from classmates about the page number. Only a few students engage in the construction activity when Mr. Rudd starts it.

CASE 2.3

Mr. Girardo-Jones announces to his sixth-grade mathematics class, "I see that everyone is finished with the calculator drill. Please see that your calculator is cleared and put away in the correct slot in the keeper. [pause] Okay, thank you. Now, we're going to play a game called 'What's in a composite?' I think you'll enjoy this. Everyone get out a sheet of paper and pencil. [pause] Joseph, get those other things off of your desk. [pause] Okay, as I was saying, you should have exactly one sheet of paper on your desk—that's the way to do it, Margaret. You should have only one piece of paper on your desk and a pencil—Oh! Rachel, your pencil needs sharpening. We'll take care of that in a minute. Now, this is going to be fun! At the top of your paper . . ."

CASE 2.4

With the aid of an overhead projector, Ms. Lem is demonstrating for her life-skills mathematics class an application of an algorithm for computing the cost-deferring credit-card payments. As she describes the example, she refers to "consolidation of loans," and then pauses to ask, "Have we ever before explained what 'consolidation of loans' means?"

Dandridge: No, we never did.

Lucinda: Yeah, what does that mean?

Ms. Lem: I thought we went over that before, but I guess not. Open your books to page—okay, let's see. [pause] Here it is, page 197. [pause] Do you have it yet? [pause] Good! Now read the part under heading 5-11 on "Selecting a Borrowing Plan."

Theresa: Ms. Lem, we already read this; it was a homework assignment from awhile ago.

Other students express agreement with Theresa, while only a few suggest there was no such assignment.

Ms. Lem: In that case, put your books away and let's get back to this example. Now, as I was explaining before . . ."

In Case 2.2, Mr. Rudd failed to get all his students' attention before attempting to give them directions for an upcoming learning activity. The transition between the time allocated for the tasksheet activity and the time allocated for the construction activity was not *smooth.* Because some students continued to work on the tasksheets while others were either getting ready for or had already begun the construction activity, the transition was marked by confusion. Mr. Rudd rushed through the transition without attending to details students needed to follow the directions.

The weakness of Mr. Girardo-Jones' transition in Case 2.3 is in direct contrast to the weakness created by Mr. Rudd. Mr. Girardo-Jones wasted too much time repeating details, dealing with isolated problems, and trying to convince his students how enjoyable the activity would be. Rather than putting students through a boring drawn-out transition, directly deliver the directions and get them into the activity. If they are going to enjoy the activity (some may and some may not), then they will discover that for themselves once they are engaged in it.

In Case 2.4, Ms. Lem failed to maintain momentum during the learning activities. She interrupted the activity on the application of the algorithm to begin another on the meaning of "consolidation of loans." Then she aborted the second to return to the first. Although such interruptions are sometimes necessary, be aware that they make it difficult for many students to follow what is going on. At least some of Ms. Lem's students were still thinking about who was right regarding the alleged homework assignment when she was expecting their attention to be refocused on the application of the algorithm.

Chapters 5–8 and 11 focus on designing teaching units, lesson plans, and daily agenda. Your work with those chapters will help develop your talent for planning smooth, efficient transitions.

STUDENT BEHAVIORS

On-Task Behavior

A student's behavior is *on-task* whenever the student is attempting to follow the teacher's directions during either transition or allocated time. Students' behaviors in Cases 2.5 and 2.6 are on-task:

CASE 2.5

As Joe enters the classroom for algebra I, he sees his teacher point to an assignment posted on the board. He immediately goes to his desk and begins the assignment.

CASE 2.6

During a question-discussion session about graphs of linear functions, Jolene listens to what her teacher and classmates are saying, occasionally volunteering her own questions, comments, and responses to others' questions.

Socializing with friends, eating, sleeping, partying, watching television, playing basketball, playing video games, skateboarding, and caring for family members are the kinds of behaviors that people—including your students—are ordinarily inclined to exhibit. Entering a classroom in an orderly manner, working collaboratively on a mathematics project, raising a hand before speaking, factoring polynomials, graphing inequalities, computing and comparing two standard deviations, formulating a mathematical proposition, measuring the radius and diameter of a circle and then computing the ratio of the results, responding to prompts on a mathematics test, solving quadratic equations, and finding the zeros of a function are the kinds of behaviors we expect from our students. Because such on-task behaviors tend to be contrary to how people are naturally inclined to behave, you need to teach your students to be on-task.

You teach students to be on-task by (a) assertively communicating your expectations, (b) serving as a *model* for on-task behaviors, (c) consistently positively reinforcing their being on-task, and (d) preventing punishment for their on-task behaviors.

Teaching Students to Be On-Task

Modeling

As their teacher, you cannot help but be on stage whenever you are in your students' company. They are establishing their model for appropriate behavior and enthusiasm for doing mathematics as they interact with and observe you. Students—especially those who feel a bond of mutual respect with you—are likely to imitate the behaviors and attitudes you display. The psychological *principle of modeling* states that individuals tend to imitate behaviors that they frequently observe (Martin & Pear, 1996, pp. 219–222). Because of the way Mr. Griffin behaves in Case 2.7, the principle of modeling works against students learning to be on-task; on the other hand, Mr. Rudd in Case 2.8 takes advantage of the principle of modeling to teach on-task behaviors:

CASE 2.7

Just as the bell rings signaling the beginning of probability and statistics class, Colleen, one of Mr. Griffin's students, begins explaining to him why she might go to college and major in engineering. Wanting to be encouraging, Mr. Griffin continues to discuss the matter with her, delaying the start of the planned learning activity by six minutes.

He begins to explain to the class the meaning of "conditional probability" by writing the definition on a transparency displayed by an overhead projector. What the students see is shown by Exhibit 2.6. Most students have difficulty reading what is on the screen; some do not bother trying. Carney whispers to Mai-Lin, "What's that say?" Mr. Griffin snaps, "Apparently, you two don't need to see this or you would be paying attention instead of yakking!"

Several more minutes into the explanation, Mr. Griffin notices Mr. Thibodeaux, the principal, at the doorway beckoning him over for a conversation. Mr. Griffin abruptly stops his explanation with, "Just a minute class, I have some business to take care of with Mr. Thibodeaux." In a few minutes, the class gets noisy and Mr. Griffin turns from his conversation at the doorway to yell, "Knock it off in here! I can't hear what Mr. Thibodeaux is saying; it's

Exhibit 2.6

Mr. Griffin's Display of the Definition of "Conditional Probability."

The conditional probability of A+B, given B is denoted by "P(A/B)," and is defined as follows:

$$P(A|B) = \frac{P(A \cap B)}{P(B)}$$

where $P(B) \neq 0$.

important!" Four minutes later, Mr. Griffin is ready to re-engage the students, but it takes a while for most of them to shift their thoughts back on conditional probabilities.

With his presentation back underway and nearly all students paying attention, Mr. Griffin notices Carlos staring off into space—seemingly oblivious to the explanation. Mr. Griffin interrupts his presentation to say to Carlos, "Earth to Carlos, Earth to Carlos—come in, Carlos; return to this planet for your math lesson!" Embarrassed in front of his classmates, Carlos feigns a half smile and appears to pay attention. Mr. Griffin responds, "Good! That's the way to pay attention," But Carlos is not thinking about conditional probabilities. While pretending to be attentive, his mind is focused on his embarrassment and the disrespect he perceives Mr. Griffin displayed. Furthermore, others students who were paying attention are now thinking about the incident with Carlos.

CASE 2.8

Just as the bell rings signaling the beginning of probability and statistics class, Allison, one of Mr. Rudd's students, begins explaining to him why she might go to college and major in engineering. Wanting to be encouraging but recognizing that it is time for the initial learning activity to begin, Mr. Rudd says, "I want to hear more about your interest in engineering. Check with me right after class to find a time when we can have an unrushed conversation."

He begins to explain to the class the meaning of "conditional probability" by displaying the computer-generated slide shown by Exhibit 2.7.

Several minutes into the explanation, he notices Ms. Adkins, the principal, at the doorway beckoning him over for a conversation. Rather than stop in midexplanation, Mr. Rudd acknowledges Ms. Adkins with a hand signal indicating "just a moment, until I get to a stopping point." In two minutes with Ms. Adkins still waiting by the door, Mr. Rudd reaches a stopping point, and tells the class, "Keep this thought in mind while I quickly find out what Ms. Adkins needs: For all practical purposes, the condition in a conditional probability restricts the sample space of the experiment. Hold that thought until I get right back." At the doorway, Ms. Adkins attempts to engage Mr. Rudd in a conversation about a funding proposal a faculty committee is developing. However, Mr. Rudd responds with, "I should continue with my lesson right now. I need to speak with Allison as soon as this period is over, but I could drop by your office right after fourth period to schedule a time for us to discuss the proposal." After that 33-second interruption, he makes the transition back to the learning activity with, "Thank you for waiting. Now what was that last thought I asked you to keep in mind? Okay, Maxine." Maxine: "I'm not sure what you meant, but you said that for all practical purposes . . ."

With his presentation underway again and nearly all students paying attention, Mr. Rudd notices Emerald staring off into space—seemingly oblivious to the explanation. Without missing a word in his presentation to the class, Mr. Rudd moves over to Emerald and gently touches her desktop. Emerald wakes up from her daydream and appears to attend to the explanation, which continues uninterrupted as Mr. Rudd moves about the room. Only the student seated directly behind Emerald even noticed what transpired between Mr. Rudd and Emerald.

Thirty seconds before the bell rings ending the period, Mr. Rudd meets with Allison to arrange a time to continue their conversation about her interest in engineering.

Exhibit 2.7

Mr. Rudd's Display of the Definition of "Conditional Probability."

The conditional probability of *A* and *B*, given *B*, is denoted by "P(*A*|*B*)," and is defined as follows:

$$P(A|B) = \frac{P(A \cap B)}{P(B)} \quad \text{where } P(B) \neq 0$$

Unlike Mr. Griffin, Mr. Rudd modeled behavior that reflected a commitment to spending allocated time on the business of learning mathematics. By using Exhibit 2.7 instead of Exhibit 2.6, he demonstrated that the business of learning mathematics is worth the effort to make professional, easy-to-read illustrations. Mr. Griffin made it difficult for students to be on-task by directing them to read a difficult-to-read illustration.

Unlike Mr. Griffin, Mr. Rudd managed the principal's interruption in a way that communicated to students that their engagement in learning activities is his main priority. When Mr. Rudd noticed a student drifting off-task, he cued the student back on-task without disturbing the on-task behaviors of other students. Furthermore, he did not interrupt his own explanations of conditional probability to quibble with students as Mr. Griffin did with Carney, Mai-Lin, and Carlos. Recognizing that he was being observed by students with whom he has established mutually respectful relationships, he made a concerted effort to model assertive, respectful, and polite behavior.

Positive Reinforcement

Behavioristic psychology provides a research-based foundation for teaching students to be on-task.

Particularly notable are the investigations of B. F. Skinner who examined the effects of stimuli on learning *after* a response or act. These investigations led to the following fundamental principle underlying strategies for managing behavior: Behaviors or responses that are followed by rewards (i.e., satisfying or pleasant stimuli) are more likely to be repeated than behaviors that are not. (Cangelosi, 2000b, pp. 42–45; Santrock, 2001, pp. 243–246).

A *positive reinforcer* is a stimulus presented after a response that increases the probability of that response being repeated in the future. You need to design units, plan lessons and daily agendas, and interact with students so that students' on-task behaviors are more likely to be positively reinforced than their off-task behaviors. Contrast Case 2.9 to Case 2.10:

CASE 2.9

On the first day of the school year, Mr. Boone informed his algebra I students that they are to raise their hands to be recognized before speaking aloud in class. It is now the second day and Mr. Boone asks the class, "Who can tell me what a variable is?" Amy, Ha, Tabonia, Heather, Lin-Choi, Maxwell, Bryan, Aimee, and Wanda raise their hands; but before Mr. Boone calls on anyone, J.R. speaks out, "A variable is a letter that stands for a number." Mr. Boone responds to J.R. as Tabonia, Lin-Choi, Aimee, and Wanda bring their hands down, "Where did you learn that?" Although the question was directed at J.R., Wanda responds, "That's what we learned in prealgebra." In the meantime, Amy anxiously raises her hand even higher and silently mouths, "Me-me-me!" Maxwell begins waving his hand, Bryan drops his, Tabonia reraises hers, and Jorge raises his. But Mr. Boone responds to Wanda, "In algebra we use letters to stand for variables, but the letter itself isn't actually the variable. Heights, volume, speed—those are all variables, but they aren't letters any more than your name is the same as you. For ex—" J.R. interrupts, "What's the differ—" Wanda jumps in, "Why would Mrs. Croshaw tell us that in prealgebra if it wasn't right?" Amy and Maxwell are the only ones with their hands still up as several other students begin speaking aloud. Mr. Boone raises his own voice, "Quiet! I can hear only one person at a time! We are supposed to raise our hands." Several more students besides Amy and Maxwell raise theirs. "Okay, Amy." Amy: "A variable is something that can have more than one value." "It's always an *x*," Alex says without being recognized. "No, it could be another letter," Wanda says. No longer are any students bothering to raise their hands and only a few even try to join in on the ensuing free-for-all.

Near the end of the period Mr. Boone assigns homework exercises from the textbook. That night, Amy spends 90 minutes diligently working the exercises and making the effort to not only "get the answers," but also to gain some understanding of the underlying mathematics.

The following day, Mr. Boone directs students to put their homework papers on their desks. He quickly walks by the desks checking the names of students who have "answers" for all the exercises. Those students receive five points toward their grade. Amy receives five points as do Donyell who rushed through the exercises just to get "answers" and Malinda who simply copied answers from Maxwell. Mr. Boone makes no further reference to that homework assignment.

CASE 2.10

Mr. Rudd is conducting the second class session of an algebra I course. Facing the class, he says, "I'm going to ask you a question. You are to take 30 seconds to formulate an answer silently in your mind. Then, if you want to share your answer with the rest of us, please raise your hand to be recognized. The question is, "What is a variable?" Sybil says, "It's a—" Mr. Rudd immediately faces Sybil, interrupting her with a stern look and gesture silently indicating silence. Twenty-eight seconds later, Mr. Rudd calls on Jaquilene:

Jaquilene: A variable is a letter that stands for a number.

Mr. Rudd: That's a definition that appears in some books. I'm glad you remembered it. However, before we decide on a definition that'll work for us, I want everyone to look at this open sentence.

He writes the following on the overhead: $x + 4 > 9$.

Mr. Rudd: Here's another question for everyone to answer for themselves, but I'm only going to ask Jaquilene to share her answer aloud: What is the variable in that open sentence?

Jaquilene: x.

Mr. Rudd: Raise your hand if you agree with Jaquilene's answer.

Nearly everyone—including Mr. Rudd—raises a hand.

Mr. Rudd: Here's another question for everyone, but this time you have only 15 seconds to silently come up with an answer before I call on volunteers for answers. The question is, "What is one number with which you can replace x in order to make that open sentence a true statement?"

Brad and Beverly say, "10" and "6," but Mr. Rudd ignores them and begins calling on students with hands raised and records their answers on the overhead as follows: $x = 100$, $x = 5$, $x = 9$, $x = 2\pi$, $x = 33.333$, $x = \sqrt{82}$, $x = 7$.

Mr. Rudd: Okay, we have enough for now. Yes, Beverly, you have your hand up.

Beverly: Five won't work because 5 plus 4 is 9, not greater than 9.

Mr. Rudd: (as he scratches out "$x = 5$") Thank you.

The question-discussion session continues with Mr. Rudd leading students to (a) discriminate between a letter itself and the unspecified members of a set that the letter is used to designate and (b) formulate the following definition: A variable is a quantity, quality, or characteristic that can assume more than one value.

Near the end of the period, Mr. Rudd assigns homework that is clearly relevant to the day's activities about variables. The next day, he begins the period with a short test. Students who worked diligently on the homework assignment have no trouble doing well on the test, but others find it too difficult. Furthermore, the responses students made to prompts from the homework are used during the day's learning activities.

Apparently, Amy and others in Case 2.9 wanted to share their answers to Mr. Boone's initial question; being called on would have positively reinforced their on-task behaviors of raising hands and waiting their turn to be recognized. But instead, Mr. Boone positively reinforced J. R.'s, Wanda's, and others' off-task, speaking-out-of-turn behaviors. In contrast, Mr. Rudd in Case 2.10 assertively and politely refused to allow Sybil to have the floor until she followed the on-task procedure. Mr. Rudd carefully tried to positively reinforce on-task behaviors but not off-task behaviors. After calling on Jaquilene who had her hand up, he was concerned that Jaquilene might have regretted sharing a definition that Mr. Rudd deemed unacceptable. Thus, Mr. Rudd acknowledged that Jaquilene had remembered a definition from a previous course and made sure he went right back to her with a question with an answer that follows from the definition Jaquilene initially gave. Mr. Rudd wanted to positively reinforce students' sharing answers even if those answers might be perceived as "wrong."

Unlike Mr. Boone, Mr. Rudd carefully orchestrated the homework assignment to positively reinforce diligent completion of that assignment. Also note how clearly Mr. Rudd established his expectations for the questioning session. He facilitated on-task behavior by setting up each question (e.g., "You are to take 30 seconds to formulate an answer silently in your mind.").

The idea is to teach so that your students' on-task behaviors are positively reinforced. However, in the extremely complex art of teaching, only rarely is anything simple. What one student finds rewarding, another may find aversive (e.g., some students enjoy being the focus of attention; others hate it). What is even more complicating is that your actions hardly ever impact only a single, isolated student behavior. A positive reinforcer for one behavior often has side effects on other behaviors. When those side effects are undesirable, the positive reinforcer is referred to as a "*destructive positive reinforcer*" (Cangelosi, 2000b, pp. 44–45). Consider Case 2.11:

CASE 2.11

To help her students refine their concept of sample space while practicing some memory and algorithmic skills, Ms. Ricardo introduces the game of Sumgo (Meel, 2000) during a probability unit in her seventh-grade mathematics class. To motivate them to prepare for and to put forth an effort during the game to be held the next day, she announces, "We'll have a girls' team and a boys' team. The winning team will be excused from doing tomorrow's homework assignment."

After school, Lucinda, Ilone, and Susan get together to practice for the game. The following day, their efforts are rewarded when the girls' team wins and, unlike the boys, they have no mathematics homework to do.

Apparently, the satisfaction of a victory and an evening without mathematics homework served as positive reinforcement to Lucinda's, Ilone's, and Susan's on-task behaviors. That is one of the upsides of Case 2.11 (another is the learning derived from playing Sumgo). However, there is also a downside; the positive reinforcer may have been destructive because of two undesirable side effects: Students were encouraged to (a) develop a "girls versus boys" attitude and (b) believe that mathematics homework should be avoided. An atypical—but businesslike—perspective on this is that the boys gained an advantage because by being assigned homework they had an additional opportunity to learn mathematics. Of course, you can figure out how to positively reinforce on-task behaviors while avoiding such obvious side effects, but you can not control many of the subtle, insidious side effects when you are working with 25 students in the complex environment of a classroom. Just being aware that positive reinforcers can be destructive and thinking about how to minimize undesirable side effects will help you and your students immensely. You will visit cases in which teachers address this type of problem in chapters 5–8 and 11.

Punishment

Punishment and positive reinforcement have opposite effects. *Punishment* is a stimulus presented after a response that decreases the probability of that response being repeated in the future (Cangelosi, 2000b, pp. 45–48). In Case 2.10, Mr. Rudd attempted to prevent Jaquilene's response to his initial question from being punished. In Case 2.9, on the other hand, Amy's and Maxwell's on-task behaviors were met with the discomfort of having to wait with their hands up while others participated in the discussion. Their on-task behaviors were punished. In Case 2.7, Mr. Griffin not only displayed a lack of withitness when he berated Carney and Mai-Lin, but he also punished Carney's attempt to be on-task (unless, of course, Carney didn't mind being berated by Mr. Griffin—possibly because Mr. Griffin is so disliked that "getting under his skin" is considered admirable by Carney's peers).

Of course what you need to do is to prevent on-task behaviors from being punished and cause off-task behaviors to be punished.

Punishment can have destructive side effects as can positive reinforcers. Case 2.12 is an example:

CASE 2.12

While demonstrating how to solve a textbook word problem using trigonometry, Mr. Anderson asks the class, "What sine do you think I ought to use?" Kelly (in an attempt at humor) laughs and holds up his hand with his middle finger in the air, "How about this one?" Loud laughter erupts from all around the class. Upset, Mr. Anderson retorts, "Do you need attention that badly? Don't you know a more original way to show off?" Kelly is embarrassed and wishes he had not misbehaved. But knowing that the other students' laughter could have positively reinforced what Kelly did, Mr. Anderson tries to discourage the class from laughing at inappropriate jokes again by pronouncing, "And because some of you encouraged behavior that violates school obscenity policies by laughing, I'm going to double the homework assignment for tonight. Doing the extra problems will remind you to think before you laugh!"

Kelly's embarrassment punished his inappropriate behavior, but by trying to embarrass him, Mr. Anderson modeled an immature, hostile, and disrespectful way of handling conflict. He missed an opportunity to model how to deal with the incident in a mature, assertive, and respectful way. And, of course, like Ms. Ricardo in Case 2.11, he supported the perception that mathematics homework should be avoided. As when using positive reinforcement, stop and think about possible side effects when using punishment.

Engaged Behavior

A student exhibits *engaged* behavior by being on-task during allocated time. In other words, whenever students are attempting to participate in a learning activity as planned by the teacher, the students are *engaged in the learning activity.* Jolene displayed engaged behavior in Case 2.6 as does Carol in Case 2.13:

CASE 2.13

As Mr. Rudd directed, Carol responds to the prompts on Exhibit 1.18's tasksheet.

Strategies for (a) designing learning activities that motivate students to be engaged and (b) conducting activities in a way that sustains engagement are the foci of chapter 3.

Off-Task Behavior

A student's behavior is *off-task* whenever the student fails to be on-task during either transition or allocated time. Students' behaviors in Cases 2.14 and 2.15 are off-task:

CASE 2.14

As Marlene enters the classroom for algebra I class, she ignores her teacher's directions to begin the assignment posted on the board; instead, she grabs Justin and begins arguing with him about a disagreement they had earlier.

CASE 2.15

During a question-discussion session about graphs of linear functions, Steven quietly daydreams about the truck he plans to buy.

A student's behavior is *disruptive* if it is off-task in such a way that it interferes with other students being on-task. Thus, a student who is being disruptive not only fails to cooperate during transition or allocated time but also makes it more difficult for others to behave in accordance with the teacher's plan. Marlene's behavior in Case 2.14 was disruptive as is Rudolph's in Case 2.16:

CASE 2.16

During a question-discussion session about graphs of linear functions, Rudolph interrupts others while they are speaking to make jokes that distract others from concentrating on functions and graphs.

Besides teaching students to be on-task, you need to prevent off-task behaviors from being positively reinforced and operate your classroom so that off-task behaviors are punished. Two subsequent sections of this chapter focus on teaching students to supplant off-task with on-task behaviors: "Establishing Standards for Conduct and Routine Classroom Procedures" and "Systematically Dealing with Off-Task Behaviors." For more in-depth treatises, consider sampling some of the sources of classroom and behavior management strategies listed by Exhibit 2.8.

Engage in Activity 2.1:

Activity 2.1

For one school period, observe a mathematics class in a middle, junior high, or high school. During the period, select an interval of transition time and an interval of allocated time in order to complete the following tasks:

Exhibit 2.8

Resources for Ideas on Classroom and Behavior Management Strategies.

Ballard, M., Argus, T., & Remley Jr., T. P. (1999). Bullying and school violence: A proposed prevention program. *NASSP Bulletin, 83,* 38–47.

Cangelosi, J. S. (2000). *Classroom management strategies: Gaining and maintaining students' cooperation* (4th ed.). New York: Wiley.

Canter & Associates. (1994). *Intervening safely during fights.* [Videotape]. Los Angeles: Author (www.canter.net).

Canter & Associates. (1994). *Preventing conflict and violence in the classroom.* [Videotape]. Los Angeles: Author (www.canter.net).

Canter, L. (2001). *Assertive discipline: Positive behavior management for today's classrooms.* (rev. ed.). Los Angeles: Canter & Associates.

Jones, V. F., & Jones, L. S. (1998). *Comprehensive classroom management: Creating communities of support and problem solving* (5th ed.). Boston: Allyn & Bacon.

Kerr, M. K., & Nelson, C. M. (1998). *Strategies for managing behavior problems in the classroom* (3rd ed.). Upper Saddle River, NJ: Merrill/Prentice-Hall.

Landau, B. M., & Gathercoal, P. (2000). Creating peaceful classrooms: Judicious discipline and class meetings. *Phi Delta Kappan, 81,* 450–454.

Metcalf, L. (1999). *Teaching toward solutions: Step-by-step strategies for handling academic, behavior, & family issues in the classroom.* West Nyack, NY: The Center for Applied Research in Education.

Miltenberger, R. (1997). *Behavior modification: Principles and procedures.* Pacific Grove, CA: Brooks/Cole Publishing.

Powell, R. R., McLaughlin, H. J., Savage, T. V., & Zehm, S. (2001). *Classroom management: Perspectives on the social curriculum.* Upper Saddle River, NJ: Merrill/Prentice-Hall.

Schmuck, R. A., & Schmuck, P. A. (2001). *Group processes in the classroom* (8th ed.). Boston: McGraw Hill.

Internet Sources:

Numerous websites exist that address school-wide discipline and safety issues. For example: www.adprima.com/safety.htm contains multiple links to articles and reports with school-safety information. www.looksmart.com/eus1/eus53706 has links to articles dealing with school-wide discipline policies.

Other websites address issues related to the establishment of productive, cooperative, and safe classroom communities. For example: www.angelfire.com/ks/teachme/classmanagement.html is titled "Ideas for Many Areas of Classroom Management" and includes tips for such topics as "getting organized," and "activities for the first day of school." www.ed.uiuc.edu/facstaff/m-weeks/models.html offers resources and ideas for "Positive Classroom Management." www.k-6educators.about.com/eduction/k-6educators/msub104.htm as well as www.ss.unoedu/ss/teachdevel/ClassMan/ClassManagMenu.html provide links to a variety of classroom management articles.

Websites also address issues related to preventing student disengagement and misbehavior in the classroom. For example: www.disciplinehelp.com/behindindex/main.cfm?cur_section=1 lists over 200 types of misbehavior, each with a link that includes a suggestion for dealing with the misbehavior when it occurs in the classroom. www.nwrel.org/sky/Office/Teacher and www.track0.com/canteach/index.html include a number of links with suggestions.

1. Answer the following questions about the transition time:
   A. For about how long did the transition time last?
   B. What happened immediately before the transition time began?
   C. What did the teacher do to get the students into the transition time?
   D. What happened during the transition time?
   E. What did the teacher do to get the students out of the transition time?
   F. What happened immediately after the transition time?
2. Note one student, if there were any at all, who appeared to be on-task during the transition and then complete the following tasks:
   A. Describe those aspects of the student's behavior that led you to believe that she was on-task.
   B. Describe the teacher's response to the student's apparent on-task behavior.
3. Note one student, if there were any at all, who appeared to be off-task during the transition and then complete the following tasks:
   A. Describe those aspects of the student's behavior that led you to believe that she was off-task.
   B. Describe the teacher's response to the student's apparent off-task behavior.
4. Answer the following questions about the allocated time:
   A. For about how long was the allocated time?
   B. What happened during the allocated time?
   C. What did the teacher do to get the students out of the allocated time?
   D. What happened immediately after the allocated time?
5. Note one student, if there were any at all, who appeared to be engaged during the allocated time and then complete the following tasks:
   A. Describe those aspects of the student's behavior that led you to believe she was engaged in the learning activities.
   B. Describe the teacher's response to the student's apparent engagement in the learning activities.
6. Note one student, if there were any at all, who appeared to be disengaged during the allocated time and then complete the following tasks:
   A. Describe those aspects of the student's behavior that led you to believe she was not engaged in the learning activities.
   B. Describe the teacher's response to the student's apparent lack of engagement in the learning activities.

Without revealing students' names—share, compare, and discuss your responses from this activity to those of colleagues.

ESTABLISHING A FAVORABLE CLIMATE FOR LEARNING MATHEMATICS

The success of your efforts to lead students to do meaningful mathematics in a smoothly operating classroom depends on how well you establish a learning environment in which students (a) recognize that they are in the important business of learning mathematics and (b) feel free to experiment, try and fail, make mistakes, raise questions, interact with you and their classmates, and expose their thinking processes without fear that they are risking embarrassment or harassment, or having their self-worth judged.

Students learn that learning mathematics is important business, not from your telling them so, but from the attitudes you model. In Case 1.11, Mr. Rudd took advantage of the opening of the school year to begin sending this message by being extremely organized and prepared, conducting efficient transitions, immediately involving students in thought-provoking mathematics, and modeling professional, purposeful behavior.

Establishing a learning environment in which students are comfortable experimenting and interacting with you and classmates without fear of being judged requires you to counter years of tradition that associates one's self-worth with her accomplishments. From their preschool days, most children are inundated with storybook tales, television programs, movies, and interactions with adults (including some misguided teachers who label them "good," "at-risk," "smart," or "slow") that leave them with the following unfortunate message: "The degree to which a person is loved and respected depends on that person's accomplishments." For example, Rudolph the Red-Nosed Reindeer was an object of scorn by his peers until he achieved an act of heroism one Christmas Eve, after which they began to love and respect him.

This tradition influences most students to arrive for a new school year with the idea that teachers are there to judge them. When students fear being devalued if they display a lack of understanding, knowledge, or skill, they will tend to avoid situations in which their perceived "shortcomings" could be revealed. One strategy for countering this tradition in order to establish a favorable climate for learning mathematics is to demonstrate that you do not make judgments about students themselves. Instead, you consistently focus on what they do and achieve without ever associating behaviors or achievements with their self-worth or your respect for them as people. The key is how you communicate and interact with them (Cangelosi, 2000a, pp. 115–118).

COMMUNICATING AND INTERACTING WITH STUDENTS

Descriptive Instead of Judgmental Language

Students feel less threatened, less defensive, and more willing to engage in learning activities when working with teachers who consistently use *descriptive language* than with teachers who use *judgmental language* (Manning & Bucher, 2001). Descriptive language verbally portrays a situation, behavior, achievement, or feeling. Judgmental language verbally summarizes an evaluation of a person, achievement, or behavior with a characterization or label. Judgmental language that focuses on personalities is especially detrimental to a climate of cooperation in the classroom (Manning & Bucher, 2001).

In Case 2.17, Mr. Wilcox uses judgmental language, whereas, in Case 2.18, Mr. Rudd uses descriptive language:

CASE 2.17

Amber and Abu strike up a disruptive conversation while Gail is explaining to the class how she solved a problem. Mr. Wilcox says, "Excuse me, Gail, but there are a couple of rude people in here."

After Gail's explanation, Mr. Wilcox exclaims, "Gail, you are very good at mathematics; that was ingenious!"

Later in the class period, Mr. Rudd returns Dillon's test paper with the following comment written next to one of his responses to a prompt: "You're too mechanical; you've got to be more of a thinker."

CASE 2.18

Ken and Oral strike up a dispruptive conversation while Belinda is explaining to the class how she solved a problem. Mr. Rudd says, "Excuse me, Belinda." Turning to Ken and Oral, he says, "Your talking is making it difficult for me to concentrate on what Belinda is explaining to us."

After Belinda's explanation, Mr. Rudd exclaims, "By multiplying the expression by 1 in the form of $(x - 3)/(x - 3)$, you made it obvious that the limit had to be 14. I'm glad you thought of doing that!"

Later in the class period, Mr. Rudd returns Robin's test paper with the following comment written next to one of her responses to a prompt: "All the steps in the algorithm are here without a single error. However, the final result doesn't take into account that the denominator cannot be 0."

The extra thought required to use descriptive instead of judgmental language is well worth the benefits to students' attitudes and classroom climate. Descriptive language provides richer information than judgmental language. Students gain specific feedback about their work, behavior, or situation from your descriptive comments. Judgmental comments provide only broad labels (e.g., "good") that students would be better off determining for themselves in light of specific information. Once your students learn that your comments are filled with helpful information, they are more likely to pay attention when you speak.

Descriptive language focuses on the business at hand, not on personalities. Communicating about work to be performed, rather than judgments of those performing the work enhances the businesslike atmosphere of the classroom. Comments such as "You're rude!" or "You're smart!" detract from the business of doing and learning mathematics.

Unlike judgmental language, descriptive language avoids the labeling of students and the dangerous practice of confounding mathematical achievement with self-worth. The delicate and complex relationship among students' self-concepts, need for love and acceptance, and experiences with successes and failures is a topic of extensive study (Joyce, Weil, & Calhoun, 2000, pp. 283–315). It is a common mistake to think that students will be motivated to cooperate and study diligently because their teachers praise them for appropriate behaviors and achievements and criticize them for misbehaving and failing to achieve. To the contrary, such tactics are more likely to backfire than to motivate desirable behaviors and efforts.

In Case 2.17 for example, Amber and Abu may interpret Mr. Wilcox's reference to them as "rude people" as a sign of disrespect. Thus, rather than motivating them to be polite, they may take pride in being singled out with a label that puts them at odds with a person who does not respect them. Also in Case 2.17, hearing Mr. Wilcox label Gail as "very good at mathematics," might trigger other students to think, "I didn't solve the problem, so I must not be very good at mathematics!" The praise may also have detrimental effects on Gail's attitude if she now feels pressure to live up to Mr. Wilcox's label. In time, she might protect her reputation as "very good at mathematics" by avoiding attempts at mathematical tasks at which she could fail.

True Dialogues

Compare the dialogue in Case 2.19 to that in Case 2.20:

CASE 2.19

Heidi and Shanna engage in the following conversation:

Shanna: Do you want to get something to eat before the movie?

Heidi: I'm not all that hungry, but we can go somewhere if you think we have time.

Shanna: The movie starts at 7:45, so we have over an hour to kill.

Heidi: But we still need to buy tickets. Darlene said they sold out last night before she got in.

Shanna: Oh! I didn't think of that. But maybe that was just because of opening night.

Heidi: Yeah, but I hate to take a chance—and besides I'd like to get there before the line gets too long.

Shanna: I wish I'd of grabbed a bite before I left. But if you want to get there early, I'll just get something at the snack bar.

Heidi: The snack bar is just grease and sugar. Besides, they overcharge for everything. Maybe we should stop on the way—it'd be good for me to get a little something also so I won't get hungry in the theater and be tempted to overpay for that junk.

Shanna: Why don't we run by the theater to pick up the tickets and then go to that little deli on St. Charles Street? I'll get a salad and sandwich and you can grab something light to tide you over during the movie.

Heidi: That's a plan.

CASE 2.20

As part of a unit on systems of linear functions, Ms. Cook had her algebra II students conduct experiments that produced data for comparing rates of change (e.g., Yasemin examined the speed at which two different quantities of water heated to 100°F on a burner). Ms. Cook plans to use their results to demonstrate how slopes of lines can be used to draw conclusions from data.

Yasemin's experiment resulted in the two pairs of plotted points shown by Exhibit 2.9 over which Yasemin and Ms. Cook have the following conversation:

Yasemin: Here's what I got.

Ms. Cook: Oh good! You graphed your results.

Exhibit 2.9

The Drawing Yasemin Brought to the Conversation With Ms. Cook.

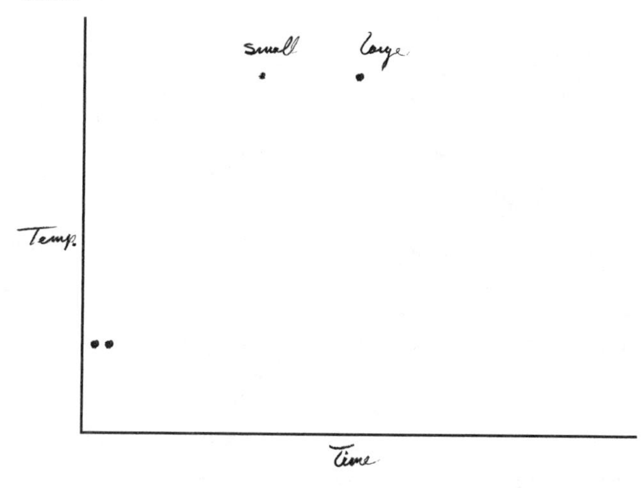

Yasemin: Were we supposed to?

Ms. Cook: Yes, this really helps us compare the two differences. Draw the two lines determined by those two pairs of points.

Yasemin produces Exhibit 2.10.

Ms. Cook: That's right; thank you. Now, how do the slopes of these two lines compare?

Yasemin: The one for the smaller pot is steeper.

Ms. Cook: Right. So, which line has a greater slope?

Yasemin: Uh, they'll both be positive, so—

Ms. Cook: That's right.

Yasemin: So the slope for the steeper line will be bigger.

Ms. Cook: Exactly. But why do you say "will be"? It *is* bigger isn't it?

Yasemin: Well I haven't figured it out yet. Were we supposed to find the slopes?

Ms. Cook: No, that's fine. My point was that lines have slopes whether we compute them or not.

Yasemin: Oh, okay.

Ms. Cook: But to get back to how slopes can help us, the size of the slopes tells us which pot heated faster.

Yasemin: Okay, but could I ask something?

Ms. Cook: Of course.

Yasemin: I didn't know if we were supposed to do this, but I also took a few in-between measurements and I don't think the temperature went straight up.

Ms. Cook: Show me.

Yasemin: I didn't bring my graph because you didn't tell us to. But can I show you what I remember?

Ms. Cook: Yes, please do.

Yasemin: The larger pot looked something like this.

Yasemin produces Exhibit 2.11 and then connects the points producing Exhibit 2.12.

Yasemin: So, what did I do wrong?

Exhibit 2.10
Yasemin Draws the Two Lines Determined by the Two Pairs of Points.

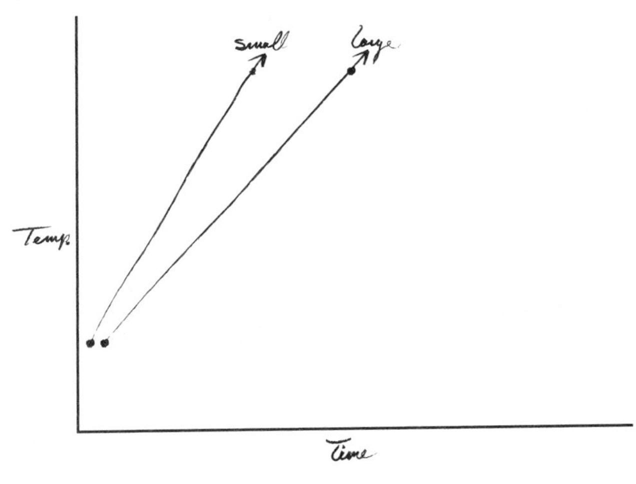

Exhibit 2.11
Yasemin Recalls "in Between" Points.

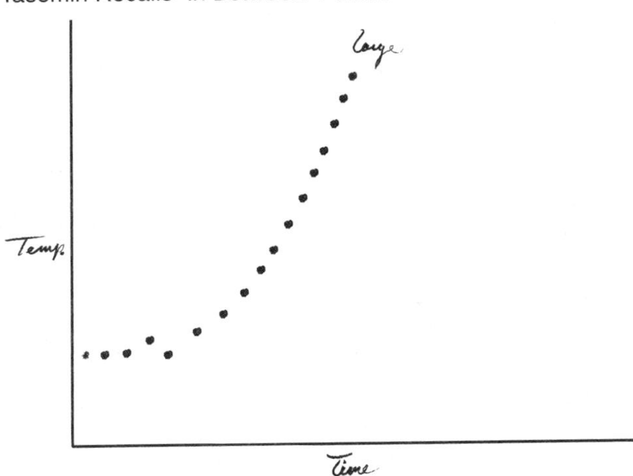

Exhibit 2.12
Yasemin Draws the Curve.

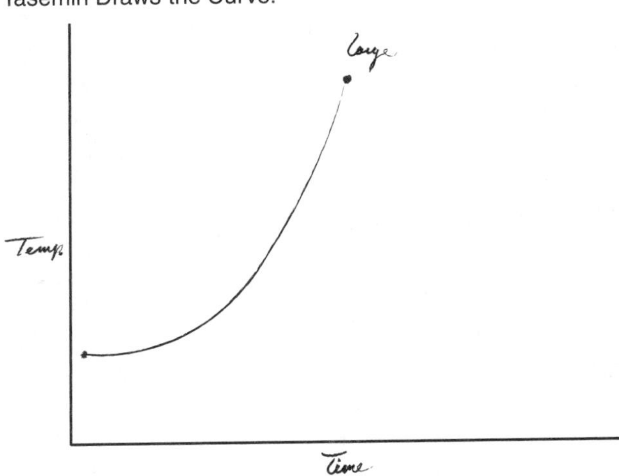

Ms. Cook: Oh, nothing. I'm glad you plotted these extra points. You see, water doesn't heat up the same number of degrees for every second. Your points not lining up in a straight line just shows that the water heated slower at first than it did later. The function isn't really linear.

Yasemin: So, how would I figure out exactly how fast it heated at the different times?

Ms. Cook: You could do that, but to get it down to the seconds, you need to use calculus.

Yasemin: I don't plan to take calculus.

Ms. Cook: You should think about taking it; it'll help you a lot.

Yasemin: I hear it's too hard.

Ms. Cook: You've done so well in algebra; calculus won't be too hard for you.

Case 2.19 is an example of the sort of natural conversation people ordinarily have in which they inform, clarify, persuade, identify problems, and address problems. Ideas evolve during these collegial interactions. As an aside, note how frequently Heidi and Shanna made quantitative or mathematical comparisons (e.g., amount of time it would take to eat with amount of time before the movie begins, waiting in a short line versus waiting in a long line, and combination of prices and nutritional value of snack bar food versus that for deli food).

Case 2.20 is an example of the kind of exchange that occurs between teacher and student in the classroom—one that does not have the same natural flow of typical outside-of-the-classroom conversations with people who are equally free to contribute ideas. Case 2.20 follows what McCormick and Pressley (1997, pp. 196–201) refer to as "initiate-response-evaluation (IRE) cycles." The teacher initiates by prompting students to respond (e.g., Ms. Cook directed students to collect data; Yasemin shows Ms. Cook her graph); the teacher evaluates the response (e.g., "Oh good!"). IRE cycles dominated the conversation. With IRE cycles, the students' role is to respond to the teacher's prompts in a way that merits a favorable evaluation. Although sometimes necessary, conversations dominated by IRE cycles discourage students from contributing their own thoughts for addressing problems; ideas do not evolve as when natural true dialogue is included (Bowers & Flinders, 1990; Cazden, 1988).

In Case 2.21, the teacher engages a student in a conversation that includes true dialogue that is not dominated by IRE cycles:

CASE 2.21

As part of a unit on systems of linear functions, Ms. Guglielmo had her algebra II students conduct experiments as did Ms. Cook's in Case 2.20. Joshua's experiment was the same as Yasemin's in Case 2.20. Looking at results similar to Exhibit 2.9, Joshua has the following conversation with Ms. Guglielmo:

Joshua: Here's what I got.
Ms. Guglielmo: So the water in both pots began with the same temperature, but it looks like you took the smaller pot off the stove first.
Joshua: No, I kept it on the stove.
Ms. Guglielmo: I thought you might have because the 100° bigger-pot point is farther to the right than the 100° smaller-pot point.
Joshua: Oh, I see why you thought that. But I just stopped testing the temperature of the big pot once the thermometer read 100°.
Ms. Guglielmo: What if we drew the two lines determined by those points to picture how much farther the temperature in the bigger pot had to travel to get to 100°?

Joshua produces exhibit 2.10.

Joshua: There, but the water didn't really travel anywhere.
Ms. Guglielmo: Yeah, I see what you mean. Maybe, it's confusing to picture a speed like this as a distance. But you've still got me thinking. Does the differences in the slopes of the two lines picture the rate of increase in temperature?
Joshua: The faster one is steeper, but you know I found out that the water didn't heat straight up like that.
Ms. Guglielmo: What do you mean?
Joshua: Well, look at this. I took some measurements in between the beginning and the end and the points lined up something like this.

Joshua produces Exhibit 2.11 and then Exhibit 2.12.

Ms. Guglielmo: I see what you mean. I wanted for you to use these results to illustrate how the slopes of lines can be used to compare rates—like the rates the water in your two pots heated. But you've got to have a line—not a curve—to have a slope.
Joshua: Right.
Ms. Guglielmo: That's too bad. Hmm, let's try something. Because two points determine a line and your curve is made up of points, pick out two points of the curve and name them something.

Joshua produces Exhibit 2.13.

Joshua: Here I called 'em *A* and *B*—Oh! I know what you're thinking; we can draw Line *AB* and it'll have a slope. So that sort of shows that the water was heating up slowly right in here because Line *AB* is flat.
Ms. Guglielmo: Wow! I see what you mean. Where do you want to pick your next two points?
Joshua: Up here at the top where the lines are gonna have a bigger slope.
Ms. Guglielmo: Should we do it for a pair of points in the middle?

Exhibit 2.13
Joshua Picks Out Two Points on the Curve.

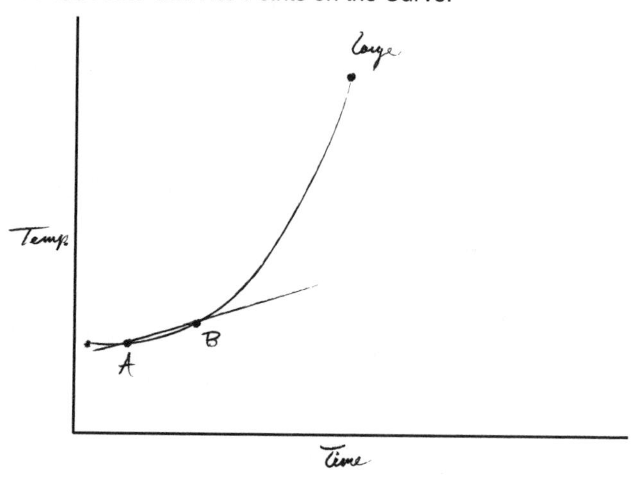

Exhibit 2.14
Joshua and Ms. Guglielmo Discuss Changing Slopes.

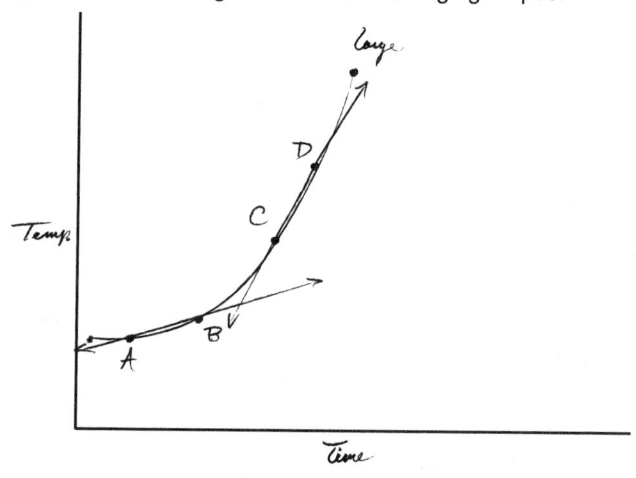

Their drawings now look like Exhibit 2.14.

Ms. Guglielmo: So are you suggesting that a curve has a a lot of different slopes?
Joshua: I'm not sure. Is that possible?
Ms. Guglielmo: Why don't you continue to mull this idea over and then let's share your idea with the class tomorrow and see if they're willing to think about expanding their idea of one slope per line to many slopes per curve.
Joshua: I don't know if I want to do that.
Ms. Guglielmo: Think about it and let me know tomorrow.

Ms. Guglielmo resisted the temptation to call Joshua "Isaac Newton" or "Gottfried Leibniz" in recognition of Joshua's beginning to develop a calculus technique; she was avoiding too many IRE cycles by not playing the role of evaluator. Ms. Guglielmo was not ready to reveal that a technique had already been invented until Joshua and the class reaped the benefits of engaging in a true dialogue leading to discovery and invention. You help establish a favorable climate for doing meaningful mathematics by engaging in real dialogues with students individually as well as in groups. Cooperative-learning activities are especially conducive to students engaging in true dialogues with one another. With true dialogues, students can share their own thoughts and try out ideas without the anxiety that each response will be met with an evaluation.

Assertiveness, Body Language, and Voice Control

In Case 1.13, Mr. Rudd discovered the advantage of communicating assertively. Studies examining traits of teachers whose students display high levels of on-task behaviors suggest that your students are more likely to cooperate with you if you consistently communicate with them in an assertive rather than in a hostile or passive manner (Canter, 2001; Santrock, 2001, pp. 457–461). As stated in chapter 1's section "Becoming More Assertive," your communications are assertive when you send exactly the message that you want to send, being neither passive nor hostile. Your communication is passive rather than assertive if it fails to send the message you want to convey because you are intimidated or fearful of the recipient's reaction. Your communication is hostile rather than assertive if you try to intimidate or insult. Review your work from Synthesis Activity 9 that appears near the end of chapter 1 (i.e., the one in which you analyzed Case 1.16).

Communicating assertively requires not only that you say what you want to say, but also that you communicate your message convincingly. Body language and voice control play a huge role in how your message is interpreted and how clearly it is understood.

Put yourself in the place of a student in Ms. Castillo's class. She is speaking to you and your classmates as pictured in Exhibit 2.15. Now visualize yourself in her class while she speaks to you and your classmates as pictured in Exhibit 2.16. Ask yourself: which of the two contrasting body languages is more likely to hold your attention?

Research suggests that students are more likely to listen to a teacher who is facing them, is making eye contact, and is nearby than one in the posture illustrated by Exhibit 2.15 (Cangelosi, 2000b, pp. 25–27, 130–131; Jones, Jones, & Jones, 2000). Ms. Castillo's body language in Exhibit 2.15 suggests that she does not take what she is saying seriously enough to face her listeners. Without saying it orally, her body language in Exhibit 2.16 clearly tells her students, "I'm speaking to you and I expect you to be listening to this important message." Your posture, body position, location in the room, use of eye contact, gestures, and facial expressions provide students with an indication of the degree to which you are in control, care for them, and expect to be taken seriously.

Get into the habit of facing and making eye contact with students to whom you are speaking. When addressing an entire class, move your eyes about the room, making eye contact with one student after another. Managing to focus your eyes on each student regularly during the course of classroom activities, and occasionally making positive expressions (e.g., a smile, wink, or thumb up) when you have caught a student's eye, helps establish an atmosphere of mutual respect. When addressing only one or two students at a time, body position can be used to clearly indicate to whom your message is intended.

However, teaching is an extremely complex art so hardly any strategy fits all circumstances. For some students from some subcultures (e.g., many American Indian subcultures), direct eye contact may be interpreted as a sign of disrespect. Some students may not make eye contact with you because

Exhibit 2.15

Ms. Castillo Speaking to You.

Exhibit 2.16

Does Ms. Castillo Expect You to Listen to Her?

they were taught at home that it is disrespectful for a child to do so with an adult (Ormrod, 2000, p. 141). Furthermore, circumstances may arise in your classroom in which directly facing a student or being close to a student is inadvisable. For example, students who are agitated may interpret Ms. Castillo's body language in Exhibit 2.16 as confrontational. In such unfortunate situations, a sideways posture at a distance may be less threatening and may help deescalate the potential for violence (Cangelosi, 2000a, pp. 395–396; Canter & Associates, 1994a, 1994b).

Which teacher displays the more effective use of body language, the one in Case 2.22 or the one in Case 2.23?

CASE 2.22

Mr. Adam's students are at their desks individually working on tasksheets as he moves about the room answering questions and providing one-to-one help. While reviewing an algorithm with Bernice, he hears Charlie and Leona talking from their desks across the room. Without turning from Bernice, he yells, "No more talking you two!" Others in the class stop their work to find out to whom he is speaking.

CASE 2.23

Ms. Petrovich's students are at their desks individually working on tasksheets as she moves about the room answering questions and providing one-to-one help. While reviewing an algorithm with Terry, she hears Moe and Bernie talking from their desks across the room. She softly tells Terry, "Excuse me, I'll be back within 35 seconds." Ms. Petrovich pivots and faces Moe and Bernie, calmly walks directly toward them, and squats down so she is at eye level. With her shoulders parallel to Moe's, she looks Moe in the eyes and softly says, "I need for you to work on these exercises without further talk." She immediately turns to Bernie, achieves eye contact and repeats the message. Standing up, she pivots and returns to Terry.

Ms. Petrovich's manner made it clear that what she had to say was meant only for Moe and Bernie. Other students did not need to stop their work to find out that her message was not directed at them. It is to your advantage to speak so that the entire class can hear you only when you expect all students to focus attention on your words. Frequent experiences stopping their work to listen to you—only to find out that you are speaking to someone else—conditions them to tune out your voice.

Your voice tone and voice control are also major variables in the mix. How well Ms. Petrovich's strategy with Moe and Bernie in Case 2.23 works is dependent on her tone of voice and the control her voice exhibits. Hostile, pleading, nervous, sing-song, screeching, or out-of-control tones influence students more than the exact words you use (Canter, 2001). Students may respond more to changes from your normal voice pitch, level, or cadence than to the actual words you say. If, for example, Ms. Petrovich sounded harsh or gruff, Moe and Bernie may only "hear" that she is angry. If she sounded distressed or used a pleading tone, they may only recognize that they "got to her." Furthermore, if you use a pleasant tone with some students, a sharp tone with others, and a sing-song tone (i.e., like some adults use with little children) with yet others, students will pick up on these differences as indicators that you like or respect some more than others. Consider recording yourself speaking to students. Privately listen to how you sound. Practice speaking in controlled, calm, adult-to-adult assertive tones.

Active Listening and Supportive Replies

The teacher in Case 2.24 tries to encourage a student to pursue a mathematical task confidently, only to reap the opposite effect:

CASE 2.24

Mr. DeCarlo is moving about his classroom as students work individually attempting to factor polynomials. As he passes Rosalie, she says, "I can't figure these out; they're too hard for me!" Mr. DeCarlo responds, "Of course you can. These are really easy for a smart lady like you! Here, I'll show you how simple they are. First you . . ."

What impact do you think Mr. DeCarlo's well-intended response had on Rosalie's thinking about doing mathematics and working with him? She said the exercises are hard; he said they are really easy for a smart lady. Besides having her feelings denied, she perceives that according to Mr. DeCarlo a smart lady would not find the exercises hard and, thus, she is not smart. Rosalie is less inclined to work with Mr. DeCarlo after he failed to understand her.

Because his response to Rosalie's expression of frustration failed to demonstrate that he understood that she was experiencing difficulty, Mr. DeCarlo's reply was *nonsupportive.* A reply to an expression of feelings—usually frustration—is *supportive* if the response clearly indicates that the feelings have been recognized and not judged to be right or wrong (Cangelosi, 2000b, pp. 132–134; Gordon, 1974, pp. 66–67; Powell, McLaughlin, Savage, & Zehm, 2001, p. 133). Case 2.25 is an example of a teacher making a supportive reply:

CASE 2.25

Mr. Marciano is moving about his classroom as students work individually attempting to factor polynomials. As he passes Seritta, he says, "I can't figure these out; they're too hard for me!" Mr. Marciano responds, "You're having difficulty identifying the common terms. That can be a real struggle."

Mr. Marciano demonstrated that he heard and understood what Seritta said. Once he lets her know he recognizes her frustration, Seritta is ready to work with him on the exercises. He listened and now she is ready to listen to him.

Responsibility for One's Own Conduct

People who consistently fail to be assertive tend to feel that others control their lives. However, except for the relatively unusual cases where one person physically accosts another, one person cannot *make* another person do something. Once students realize this and that you hold them responsible for their own conduct, they are disarmed of excuses for misbehavior. Eavesdrop on the otherwise private conversation between two teachers:

CASE 2.26

Mr. Suarez: Didn't you have Carolyn Smith in prealgebra last year?

Mr. Michelli: Yes. How's she getting along in algebra I?

Mr. Suarez: Awful! Today, I asked her why she didn't have her homework and she told me she had better blinking things to do than my blinking homework.

Mr. Michelli: Except I bet she didn't say "blinking."

Mr. Suarez: You've got that right. She used an inappropriate vulgarity in front of the class.

Mr. Michelli: What did you do?

Mr. Suarez: I was dumbfounded; I didn't know what to do. So I bought myself some time by telling her to meet me after school today.

Mr. Michelli: If she shows up, what do you plan to do?

Mr. Suarez: I had planned to take firm measures to prevent her from pulling this kind of thing again. But then, Alice—who has her for music—told me she's a victim of abuse at home and we need to give her every break. Considering some of the things Alice told me Carolyn has suffered, it's understandable for her to be so adult wary and uncooperative. How do you think I should handle the situation?

Mr. Michelli: First of all, knowing about her unfortunate circumstances helps us understand why she misbehaves. But you don't do her a favor by ever excusing misbehavior. Sure, she has it rougher than most of us, but she's still capable of conducting herself in a civil, cooperative manner in your classroom. Our job is to hold her to the same standards of classroom conduct we expect of everyone else. Because we're aware of her background, it's easier for us to respond to her misbehaviors constructively rather than angrily.

Mr. Suarez: So, I should stick with my plan for being firm with her.

Mr. Michelli: Let's hear it. You also need to come up with a strategy if she fails to show this afternoon.

Mr. Suarez: Well, first—in no uncertain terms—I plan to assure her that . . .

To lead students to understand that they are in control of their own conduct, consistently use language that is free of suggestions that one person can determine how another person chooses to behave. Purge utterances such as the following from your communications with students: "You made me lose control.", "You hurt his feelings.", "Watch out or you'll get her in trouble.", "Does he make you mad?", "If she can prove the theorem, then so can you.", and "You make me so happy!"

Replace such nonsense with remarks such as these: "It's difficult for me to control myself when you do that.", "He felt bad after you said that!", "Be careful not to encourage her to do anything she'll regret.", "Do you get mad when he does that?", "We know that the theorem can be proved; she did it.", and "I'm so happy to see how you solved that problem!"

Remind students that they are in control and responsible for their own conduct whenever they say such things as, "Sue made me do it!" or "Why blame me? I wasn't the only one!"

COMMUNICATING WITH STUDENTS' PARENTS

A Cooperative Partnership

Ideally, you, each student, and her parents form a collaborative team. Unfortunately, not all parents are able and willing to contribute to such a team. But whenever you do elicit parents' cooperation in support of your work with their children, you reap a significant advantage in managing student behavior and, thus, in helping students learn mathematics (Hargreaves, 2001). Most parents are in a position to encourage their children to cooperate with you, do homework, and attend school as well as work with you in addressing discipline problems.

The key to gaining parents' cooperation is establishing and maintaining an active, two-way channel for communications. Such a channel for each student needs to be operating before serious disruptive behavior patterns arise or summative evaluations of student achievement must be reported—especially ones involving unsatisfactory grades. You open communication channels before crises arise by keeping parents apprized of what you intend for their children to accomplish and how you plan to help them accomplish it. Parents need to be informed as to how they can contribute to this effort. Vehicles for keeping parents informed include teacher-parent-student conferences and written communiques.

Teacher-Parent-Student Conferences

Except for formal back-to-school nights held several times a year in most middle, junior high, and high schools, parents typically expect to have conferences with teachers only to be informed about their children's grades or about discipline problems. But to initiate those channels of communications, you also need conferences that focus more on *what* students are doing than on simply how well they are doing. Note in Case 2.27 how the teacher persists in steering the conversation toward what she is trying to accomplish and how she is trying to accomplish it and away from the parent's obsession with whether or not the student is being and doing "good":

CASE 2.27

Ms. Sloan teaches 148 students in five sections of mathematics. She does not have time to confer with her students' parents as frequently as she needs to. However, by routinely calling three parents every school day and limiting each phone conference to a maximum of 10 minutes, she is able to speak with a parent of each student at least once every 10 weeks. Here is her initial conversation with Redfield Breaux's mother:

Ms. Sloan: Ms. Breaux, this is Nancy Sloan, Redfield's prealgebra teacher. If you can manage the time right now, I'd like to spend five to 10 minutes talking with you about Redfield's work in mathematics. Can you do that right now or should we arrange a more convenient time to speak?

Ms. Breaux: Oh, this is fine. Is Redfield giving you some kind of trouble? Isn't he doing his work?

Ms. Sloan: He's been very cooperative with me and seems to be working very hard. The purpose of this call is to inform you about some of the things we're trying to accomplish in mathematics class.

Ms. Breaux: Is he going to pass okay? I never could do math myself so I'm not much help with his homework.

Ms. Sloan: We're just beginning a lesson on ratios and are looking at ways for determining the best price when shopping.

Ms. Breaux: That sounds more interesting than the math we had when I was in school. Will he be able to learn it?

Ms. Sloan: Yes, he should improve his skill with percentages and—more importantly—his ability to apply mathematics to his everyday life. Tomorrow, I'm going to ask the students to find newspaper ads that include things like interest rates at banks and discount sales at stores.

Ms. Breaux: It'd be better for him to read newspapers instead of spending so much of his time on the internet and playing those computer games.

Ms. Sloan: Oh! You've given me an idea. Let's capitalize on his interest in the internet to build his interest in mathematics to solve shopping problems. I'll ask him to make a record of rate-related information from advertisements that pop up on the internet. When he goes to the internet, he should take notes on prices and discounts that we can use in prealgebra class.

Ms. Breaux: I can make sure he has a pad and pencil when he's on the computer.

Ms. Sloan: That would really help; thank you.

Ms. Breaux: Anything else?

Ms. Sloan: Does Redfield have a regular time set aside for homework?

Ms. Breaux: No, but I could make him do that. How much time does he need?

Ms. Sloan: He takes six subjects. The assignments vary from subject to subject. For mathematics, he needs about 45 minutes a night. Would you please help him schedule a homework routine?

Ms. Breaux: That's a good idea. I'm working tonight, so I won't see him until late. If he's not up, I'll catch him in the morning.

Ms. Sloan: I really look forward to working with you. My time is up; I've got to phone another parent. Please feel free to call or e-mail me. I'll check back with you midway through the term, or sooner if need be.

Ms. Breaux: Thank you for calling. Good-bye.

Ms. Sloan: Good-bye, Ms. Breaux.

Regularly scheduled face-to-face teacher-parent-student conferences are common in most elementary schools. Although not as common and more difficult to schedule, more and more secondary and middle schools are also setting aside days for such conferences—especially for reporting summative evaluations of student achievement. School-sponsored teacher-parent-student conferences and, especially, individual teacher initiatives make it easier to solicit parents' help in crisis situations. Consider Case 2.28:

CASE 2.28

For two weeks, Theresa has displayed a pattern of disruptive talking in Mr. Rudd's geometry class. The second time he stops today's activities to deal with the problem, he asks her to meet him after school for yet another discussion on how they can work on breaking the pattern. Today, after school, he informs Theresa that he will arrange a conference with her parents for the purpose of devising a way to motivate her to terminate the pattern of disruptive talking.

The next day, Mr. Rudd, Theresa, and her father meet; they agree to the following plan:

> Beginning the next day, and continuing for the next two weeks, the first time Mr. Rudd detects Theresa talking disruptively in geometry class, he will issue her a warning. If she continues or disrupts the class a second time that day, he will direct her to leave class to wait for the next period in the reception room outside Ms. Slezinger's counseling office. Each day Theresa does not stay until the end of geometry class, she reports to Mr. Rudd after the day's last period to make up the missed work. On those days, she will miss her bus and her father will pick her up at 4:45 when Mr. Rudd leaves the building. At the end of the two weeks, if Theresa has stayed to the end of at least seven of the 10 class periods, she will be declared "cured" and the "treatment" will be terminated. If she has to meet after school more than three of the 10 days, then another three-way conference will be held to develop an alternative plan.

Consider the following suggestions for scheduled conferences with parents:

- Prepare an agenda for the conference that specifies the purpose of the meeting (e.g., to increase the rate at which the student completes homework assignments), a sequence of topics to be discussed, and beginning and ending times for the meeting.
- Except for special situations, invite the student to attend and participate in the conference. Healthier attitudes are more likely to emerge when the student is included as a member of the team.
- Schedule the meeting in a small conference room or other setting where distractions such as telephone calls are minimal and there is virtually no chance for outsiders to overhear the conversation.
- During the conference, concentrate remarks on description of events, behaviors, and circumstances. Focus on needs, goals, and plans for accomplishing goals. Completely avoid characterizations and judgements of personalities. Be assertive while using descriptive rather than judgmental language.
- Actively listen to the student and parents to facilitate communication among all parties and to gain ideas for working more effectively as a collaborative team.

Keep in mind that such conferences deal with privileged information that should be shared only in confidence for professional purposes with your supervisors and other teachers with whom you consult.

Written Communiques

Besides conferences with parents—which out of necessity are infrequent—some teachers send home or post on websites weekly or monthly newsletters that are designed to apprize parents of what is going on in their children's courses. Exhibit 2.17 is an example of a weekly newsletter one teacher sent to parents.

By taking the time to write such form letters, you foster the goodwill and understanding of parents. Their awareness of what you are trying to accomplish with their children will serve you well when you want to call on them for help.

Engage in Activity 2.2:

Activity 2.2

Retrieve the computerized management system that you set up when you engaged in Activity 1.3. Go to the "Parents" subfolder under the "Correspondences" folder. Create a word-processing file in which you set up a template for parents' newsletters you would conceivably write for mathematics courses you anticipate teaching. Imagine yourself teaching the course for which you wrote a syllabus

Exhibit 2.17

Example of a Weekly Newsletter One Teacher Sent to Parents.

PARENTS' NEWSLETTER FOR GEOMETRY
2nd PERIOD
From Charog Berg, Teacher
Vol. 1, No. 13, Week of November 25, 2002

Looking Back

Our last letter mentioned that we had begun a unit on quadrilaterals and polygons. I think most of the students were somewhat bored with the 1.5 days we spent reviewing and using definitions of trapezoid, parallelogram, rectangle, rhombus, square, perimeter, base, and height. I was pleasantly surprised that most already possessed a working vocabulary of these terms from their work in previous mathematics courses.

Enthusiasm picked up when we delved into some hands-on problems that led to some useful discoveries about quadrilaterals. Ultimately, the students developed some shortcut algorithms based on relations and theorems they discovered. Toward the end of the unit, we worked on applying our discoveries to real-life problem situations. This might explain why your daughter or son spent time gathering measurements from around your living space.

The results of the unit test given on November 22 proved interesting—to me anyway. The scores were somewhat higher than I had anticipated; I felt pleased about that. But what really surprised me was that according to my statistical analysis of the results, the class did far better on the parts of the test that taxed their thinking abilities than on the parts where they only had to remember something.

This Week

This week we will be working on more sophisticated problems involving parallelograms and their relations in three-dimensional space. I hope that words such as "plane" and "half plane" will creep into your son's or daughter's vocabulary as we begin examining the space about us in terms of sheets of points. One of the purposes is to get the students to analyze spatial problems systematically in a way that does not occur to most people.

Homework assignments will include: (a) Study and work selected exercises from pages 264–269 for Tuesday's class. (b) Watch the television program entitled "Spatial Fractions" from 7:00 to 8:30 on Channel 7 Tuesday night and be prepared to discuss its contents on Wednesday. (c) Begin working on the worksheet to be distributed on Wednesday and have it completed for Friday's class. (d) Study for a test on Monday, December 2.

Looking Forward

After we review the test results on Tuesday, December 3, we will tie together what we learned from these last two units with some work with mosaics and mapping three-dimensional space. This will lead us into the study of geometric similarities and proportions.

when you engaged in Activity 1.2; use the template to write a sample letter for the parents of students who are taking that course from you. Store the letter in a second file in the "Correspondences/Parents" subfolder. After getting feedback on the letter from colleagues who are also engaging in this activity and modifying the letter as you see fit, print a hard copy to insert in the "Motivation, Engagement, and Classroom Management" section of your working portfolio.

ESTABLISHING STANDARDS FOR CONDUCT AND ROUTINE CLASSROOM PROCEDURES

Necessary Standards for Conduct

Chapter 1's section "Interviews and Decisions" includes the suggestion that you raise the following question: What structures does the school have in place (e.g., conflict-management programs, effective school-wide discipline and safety policies, and security systems) that will allow you to establish a safe, nurturing classroom community that is conducive to learning mathematics? School-wide discipline and safety policies are typically published in handbooks for students, parents, and faculty. Such policies should provide for a program that encourages campus-wide civility and prevents violent, antisocial activities. Furthermore, individual teachers typically have their own sets of standards for how students are to conduct themselves in classrooms.

The standards for conduct you establish for your classroom should provide students with guidelines for their behavior while under your supervision. A *necessary* standard for classroom conduct is one that serves at least one of the following four purposes (Cangelosi, 2000b, pp. 155–162):

Exhibit 2.18

Standards for Conduct Mr. Rudd Displayed in His Classroom.

STANDARDS FOR CONDUCT

1. Respect your rights and the rights of others. (Note: All students in this class have the right to go about the business of learning mathematics free from fear of being harmed or intimidated. Mr. Rudd has the right to go about the business of teaching mathematics in the manner in which he professionally prepared without interference from others.)
2. Cooperate with your classmates and Mr. Rudd in the business of creating opportunities to learn mathematics.
3. Follow established classroom procedures as directed by Mr. Rudd.
4. Adhere to school-wide discipline and safety policies.

- Secures the safety and comfort of the learning environment
- Maximizes on-task behaviors and minimizes off-task behaviors
- Prevents the activities of the class from disturbing other classes and persons outside the class
- Maintains decorum in the classroom

A few well-understood, consistently enforced, broadly stated standards that clearly serve the four aforementioned purposes are preferable to many specific, difficult-to-remember ones (Evertson, 1989; Schmuck & Schmuck, 2001, pp. 193–221). For example, standards similar to those listed in Exhibit 2.18 may be all you need—providing that you make sure your students clearly comprehend them and they are consistently enforced.

Having such standards prominently displayed in the classroom reminds students of how you expect them to behave and helps you efficiently respond to students' disruptive behaviors as does Mr. Rudd in Case 2.29:

CASE 2.29

Mr. Rudd has Exhibit 2.18's standards displayed on the front of his classroom. While explaining an algorithm for bisecting an angle with a straightedge and compass, he notices Don lightly pricking Justin's arm with the point of the compass. Justin jerks away, turns to Don, and whispers between gritted teeth, "Cut it out!" While continuing his explanation, Mr. Rudd walks directly to Don and interrupts speaking to the class just long enough to look Don in the eye and say, "Please meet me after class today, so we can schedule a time to discuss ways to prevent you from violating Standard 1 again." Mr. Rudd immediately continues his explanation to the class.

Whereas standards like those listed in Exhibit 2.18 serve a necessary purpose, having *unnecessary* standards can distract from the business of learning mathematics. Case 2.30 is an example:

CASE 2.30

Mr. Leggio grew up with the idea that it is rude for men to wear hats indoors. Without much thought, he instituted a "no hat wearing" standard for students in his classroom. His efforts to enforce the rule have caused a number of disruptions to learning activities. On most days, Mr. Leggio stands by the doorway at the beginning of each period to check on students for such things as gum chewing and hat wearing.

Today, while Mr. Leggio is writing on the board, Mark slips on a baseball hat. Ten minutes later, Mr. Leggio notices it, stops the activity, and snaps, "I'll take that, young man!"

Mark: Why?
Mr. Leggio: You know you're not supposed to wear a hat in here.
Mark: Why?
Mr. Leggio: Because it's not polite.
Mark: Who does it hurt?
Mr. Leggio: I can't teach when you're wearing a hat!

Some students laugh as Mr. Leggio begins to feel uncomfortable. Feeling the need to assert his authority, he raises his voice, "Either you give me that hat right now, or you're going to be suspended from this class!" Mark grins as he slowly swaggers up to the front of the room and gives up his hat. Mark turns away from Mr. Leggio making a mocking face as he slowly returns to his desk. Students laugh, but Mr. Leggio is not sure why as he attempts to re-engage them in the activity.

Mr. Leggio's no-hat standard cannot be justified on the basis of either securing the safety and comfort of the learning environment, maximizing on-task behaviors and minimizing off-task behaviors, or preventing the activities of the class from disturbing other classes and persons outside the class. He might argue that the rule helps maintain acceptable standards of decorum. However, he should be careful to assure that any standard based on the need for decorum clearly does help maintain an atmosphere of politeness and cooperation rather than just imposing his personal tastes on students.

The unpleasant consequences of having unnecessary standards of conduct include the following: (a) Teachers become responsible for enforcing untenable standards. (b) When students find some standards to be unimportant, they generalize that others may be unimportant also. (c) Students who are penalized for resisting unnecessary standards may become disenchanted with school and distracted from the business of learning.

Engage in Activity 2.3:

Activity 2.3

Formulate a set of standards for classroom conduct you imagine displaying on the wall of your classroom. Now reflect how you plan to enforce those standards. Describe your "enforcement plan"—including consequences for violations—in a one- to two-page document. Obtain feedback on your work from colleagues who are also engaging in this activity. Modify your list of standards and enforcement plan as you see fit in light of the feedback.

Insert the word-processing file for this work in the "Resources/Classroom Management" subfolder that you set up when you engaged in Activity 1.3. Also insert a hard copy in the "Motivation, Engagement, and Classroom Management" section of your working portfolio.

Procedures for Smoothly Operating Classrooms

Whereas standards for conduct establish general guidelines for behavior, *classroom procedures* are the specific operational routines that students are directed to follow. How efficiently a classroom operates depends on how well procedures have been established for transitions between learning activities, movement about the room, use of equipment and supplies, large-group sessions, small-group sessions, individualized work, and administrative functions. When you examine Cases 5.8 and 5.9 in your work with chapter 5, you will note Ms. Citerelli's procedure for students speaking during question-discussion sessions. Like Mr. Rudd in Case 1.8 using Exhibit 1.8's checklist, you need to determine procedures when you organize your classroom and curricula for an upcoming school year. During the year, however, situations arise that prompt you to either modify previous procedures or develop new ones. Consider Case 2.31:

CASE 2.31

Mr. Rudd has organized his fifth-period life-skills mathematics class so that large-group sessions and collaborative-learning activities are confined to Mondays, Tuesdays, Thursday, and Fridays. Wednesdays are saved for individual, catch-up, and enrichment work. Every Wednesday, students are free to determine how they spend their time as long as they are in the classroom and independently doing mathematics in a way that does not disturb classmates.

Mr. Rudd's classroom is arranged as diagramed in Exhibit 1.10. There are 12 computers in the room to accommodate the 29 fifth-period students.

A month into the course, Mr. Rudd discovers that on Wednesdays the student demand for use of the computers exceeds the availability. Students regularly complain that they do not get to the machines because computer time is being monopolized by some of their classmates. Another complaint involves diskette abuse. The following procedure that Mr. Rudd was using for maintaining and using diskettes is apparently not working:

> Storage space is provided for each student in the file boxes kept on the computer tables. To use a computer, a student retrieves a personal diskette from the file, inserts it in an available machine, completes the work, and returns the diskette to the file.

Students complain that other students are tampering with and misfiling their diskettes. On some Wednesdays, arguments have erupted over allegations of diskette stealing and computer hogging. Mr. Rudd holds a classroom-community meeting at which students discuss ways to set up a more efficient procedure for computer use on Wednesdays and for diskette maintenance. Based on input from the students, Mr. Rudd establishes the following procedures:

> The 12 computer stations are numbered 1–12. During the last 10 minutes of class each Tuesday, Mr. Rudd will circulate among students a sign-up sheet for scheduling the next day's computer time. The sheet indicates five 10-minute blocks of time for each station. Each student may schedule up to 20 minutes of computer time the first time the sheet circulates. The sheet is recirculated, in reverse order, until either all students' computer time needs have been filled or all the 10-minute blocks are filled.

Regarding diskette use, the procedures are revised as follows:

> The file boxes will be discarded. Students are responsible for the security of their own diskettes kept in personalized portable storage cases.

Teaching Standards for Conduct and Procedures to Students

Formulating necessary standards for conduct and procedures for classroom routines will yield the results you need to conduct the business of teaching mathematics only if students comprehend them, understand exactly how to follow them, and recognize the automatic consequences for violating them. Thus, you need to apply sound pedagogical principles to deliberately teach standards and procedures just as you do to teach mathematics. The time you spend explaining and demonstrating standards and procedures will result in time saved because students will spend more time on-task and transitions will be more efficient.

WORKING WITH STUDENTS AS INDIVIDUALS

The Key: Relating to Students as Individuals

In Case 2.32 (adapted from Cangelosi, 2000b, pp. 181–182), a preservice mathematics teacher conducts a microteaching session for a group of students whom he does not know as individuals:

CASE 2.32

For an assignment in the methods-of-teaching-mathematics course he is taking at a university, Ryan Nelson develops a lesson with the objective of leading middle school students to construct the concept of probability functions. The methods course instructor, Grace Ferimundi, discusses the plan with Ryan:

Professor Ferimundi: You've built in the necessary elements for a sound lesson plan. What part of the lesson would you like to present for your field experience out in the school?

Ryan: I thought I'd do the cooperative group activity in which students play the game with the attribute blocks, using an idea from that article in *Mathematics Teaching in the Middle School* (Quinn, 2001) you showed us the other day.

Professor Ferimundi: I like your choice! Middle-education students love to work with manipulatives—especially in a game format. The key to their being engaged is having hands-on activities that keep them busy.

Through the university's field experience office, Ryan schedules his microteaching session for 30 minutes in Ms. Riggs' prealgebra class at Malaker Middle School. On the eventful day, Ryan nervously delivers the directions and distributes the attribute blocks and tasksheets for the cooperative group work as Ms. Riggs observes. The students begin playing the game as Ryan supervises, moving among groups. Nine minutes into the activity, Ms. Riggs is unexpectedly called out of the room. Just as Ryan is moving toward another group, he hears Wilson say, "This is stupid; I ain't doing this no more!" Ryan turns to see Wilson slide his desk away from the group and put his head down on his desk. Ryan goes over to Wilson and in a low but firm voice says, "Young man, I need you to return to your group and continue with the game." Burying his head even further into his arms on the desk top, Wilson mutters something that Ryan does not hear well enough to understand. Ryan: "I'm sorry, I couldn't hear what you said." Wilson picks up his head long enough to exclaim, "I don't have to listen to what your sorry white ass tells me!" Embarrassed in front of the students, Ryan is stricken with panic; he says, "You can't talk to me that way!" Wilson: "I just did!" Ryan: "Maybe so, but you'll be sorry you did!" Ryan turns to the rest of the class and says aloud, "All right, let's get back to the game in your groups. Ms. Riggs may be testing you on what you're supposed to be learning." Most students comply; Wilson puts his head back on his desk appearing to sleep. Ms. Riggs returns and Ryan relates the incident to her. Ms. Riggs apologizes for leaving and tells Ryan to finish the activity while she deals with Wilson's disruption.

The following day, Ryan discusses the incident with Professor Ferimundi and the other preservice teachers in the methods-of-teaching-mathematics class:

Ryan: Even before I said, "You can't talk to me like that," I knew that was the wrong way to respond. But not knowing what to do, I just fell into the trap of mimicking what I remember some of my teachers saying. First I was passive, then hostile. I should have been assertive, but I didn't know how. What should I have done? How should I have handled the situation?

Professor Ferimundi: Of course I wasn't there, but we might have a case of a student from a subculture in which rebelling against an authority figure representing the dominant social structure is reinforced. I don't think we should be surprised to have such a student challenge us, especially in the context of a competitive game such as the students were playing.

Ryan: So what should I have done? That kind of disrespect shouldn't be tolerated, and there's no place for that kind of language in the classroom.

Marla: Students will respect us only if we respect them.

Ryan: But I was never disrespectful to this or any other student!

Professor Ferimundi: I know that you would never intentionally be disrespectful to anyone.

Ryan: What do you mean "intentionally?"

Professor Ferimundi: Well, what is a sign of respect in one culture—such as looking at someone when you speak to her—may be interpreted as disrespectful in another culture. You know, in some Native American cultures it is disrespectful to make eye contact. I agree that the kind of language the student used is inappropriate for a classroom. However, in some cultures, language that is unacceptable to some of us is not only tolerated, but reinforced as part of posturing during competitions.

Ryan: So what was I supposed to do? Oh, never mind! Let's talk about somebody else's microteach.

Considering Ryan's question about how to handle off-task behavior, what principle factors are missing from the preceding discussion? Professor Ferimundi talked about a "dominant social culture" and cultural differences. Marla reminded everyone that she is aware of the truism, "respect breeds respect." But no one in the methods class discussed the two things that most needed to be addressed because they lacked firsthand knowledge of them: (a) the student, Wilson, and (b) the unique social environment and group dynamics of Ms. Riggs' classroom. Even with information about Wilson's ethnicity and cultural background, we know that Ryan's choice of strategies should have been based on his understanding of the unique characteristics of Wilson and the classroom situation—not on some sociological principles applicable to a culture as a whole rather than to individual members of that culture (Cangelosi, 2000b, pp. 182–183). Of course, knowledge of cultural similarities and differences is critically important to helping us explain the characteristics of individuals. However, the key to gaining and maintaining students' cooperation is recognizing and relating to them as unique individuals.

Compare Case 2.32 to Case 2.33 (adapted from Cangelosi, 2000b, pp. 183–185):

CASE 2.33

Barbara Rice is a professor at a university where she conducts a methods-of-teaching-mathematics course for preservice teachers. The course is designed to be field based so that Professor Rice holds regular class meetings with the group of preservice mathematics teachers from 7:30 A.M. to 8:20 A.M. at Westside Middle School. During the middle school's first period, which begins at 8:30 A.M., the preservice teachers work in various classrooms as follows: The 16 preservice teachers enrolled in the course are organized into eight pairs with each pair assigned to one first-period class run by one of the middle school's inservice teachers. At the very beginning of the school year, each pair spends most of its time in the classroom observing and engaging in tutoring and other types of activities that help them gain insights about the students, the social dynamics of the classroom, and the manner in which the inservice mathematics teacher conducts the class. Their interactions with students also provide opportunities for the students to get to know the preservice teachers and establish expectations, roles, and trusting relationships. As the preservice teachers learn more about teaching from the combination of work with Professor Rice and experiences in the classrooms, they begin designing and conducting lessons—first with small groups of students and then with larger groups. Whenever one member of a pair is conducting a learning activity, the other monitors students' engagement, responses, and interactions. During the first period, Professor Rice moves among the eight classrooms, depending on what the various preservice teachers are doing on that day.

Eldridge Benis and Elaine Rubenstein are the preservice teachers working in Ms. Hoggan's first-period prealgebra class. During the first week of the school year, Eldridge and Elaine observed as Ms. Hoggan applied suggestions from this chapter's section, "Establishing a Favorable Climate for Learning Mathematics." For example, on the first day, she got a jump start on getting to know the students individually by engaging them in activities similar to those Mr. Rudd used in Case 1.11. In the ensuing weeks, Eldridge and Elaine learn how Ms. Hoggan manages to accommodate students with special needs, works with students with limited English proficiency, and takes advantage of the cultural diversity in the classroom. They become familiar with Ms. Hoggan's classroom management program, participating in the establishment and enforcement of standards for classroom conduct. A poster listing the five standards for conduct (as shown by Exhibit 2.19) hangs from the wall in a large picture frame near the entrance to the classroom. Everyone in the class has learned that whenever Ms. Hoggan recognizes a student violating one of the five standards, that student is required to meet with her after school to work out a plan for preventing violations in the future.

Exhibit 2.19

Standards for Conduct Ms. Hoggan Displayed in Her Classroom.

STANDARDS FOR CONDUCT

1. Take advantage of opportunities to learn mathematics.
2. Cooperate with your classmates and Ms. Hoggan in the business of creating opportunities to learn mathematics.
3. Help create and maintain a comfortable, safe, and secure learning environment for all members of our classroom community.
4. Follow established procedures as directed by Ms. Hoggan.
5. Follow school policies.

It is now the fourth week and Eldridge is directing the students into a cooperative-group activity using attribute blocks (Quinn, 2001). Ms. Hoggan has just stepped out of the classroom, so Eldridge and Elaine are the only ones to supervise the students. In the transition period, students are arranging their desks into several circles. Margaret is following Chad as they carry their desks when Margaret trips and crashes into the back of Chad causing him to fall over his desk, banging his shin against one of the metal legs. Sprawled on the floor and in pain, Chad screams, "Ow! What are you doin'?" Margaret says, "I'm sorry," as she reaches down to help Chad. "Don't touch me, you dumb ho!" Chad screams. In the moment it takes Eldridge to move to the point of the incident, he thinks to himself, "Chad likes to play the role of the cool, in-control character in the class. He must be terribly embarrassed falling over and hurting himself like that. It's natural for him to transfer his embarrassment to anger toward Margaret. He and I have worked together before, I think he'll calm down if I calmly and firmly step in and get between the two of them. I don't think he likes to be touched, so I'll just help him turn the desk upright and move it into the circle. Once he's calm and I get the class back on-task, I can deal with the disrespectful way he spoke to Margaret." "Here, I'll help you with the desk," Eldride tells Chad. Chad suddenly leaps to his feet and gets face-to-face with Eldridge and shouts, "Help me! You're just protecting that ho! You her pimp or somethin'?!" Eldridge hears some of Chad's buddies laughing. Because Eldridge feels the class has learned to trust him, he is not embarrassed by Chad's attack; the laughter causes him to think Chad is dealing with his embarrassment by trying to impress his buddies with what he thinks are clever comebacks. Eldridge calmly turns toward the class and says, "Everyone, continue to arrange your desks and begin your group work as it's spelled out on the tasksheets." Quickly, he thinks to himself, "I see that 'help you' was the wrong phrase for me to have used with Chad. I'll put that bit of knowledge to productive use in the future. But I have to deal with the situation right now. Chad still hasn't backed off although I turned away from him. It doesn't look like he's willing to make the transition into the group activity. Elaine can supervise the activity while I deal with Chad. I can handle this myself, but it would be better if Ms. Hoggan were here in case the conflict escalates." He then beckons Jelani, a student on whom he has learned to depend, and says, "Please go down to the faculty workroom and ask one of the teachers to locate Ms. Hoggan and ask her to come in here as soon as possible." Turning to Elaine he says, "Ms. Rubenstein, would you take over for me please?" Elaine: "I'll be happy to, Mr. Benis."

Eldridge would like to speak with Chad privately away from his "audience." However, he thinks Chad is not yet ready to comply with a request to "step outside the door for a discussion." Eldridge believes that after Chad has had a few minutes to calm down he will be more likely to comply—especially if Ms. Hoggan is present. With Chad still standing by himself in the middle of the room, Eldridge takes a note pad and writes, "Chad, I need to speak with you just outside the door. Meet me there." Just as he is about to hand the note to Chad, Ms. Hoggan comes into the room and asks, "Is there a problem, Mr. Benis?" "Yes, but I believe Chad and I can work it out. Would it be all right if Chad and I step out into the hall for about five minutes?" Eldridge replies. Ms. Hoggan, knowing Eldridge well enough to trust what he will do, says, "Yes, I'll help Ms. Rubenstein work with the groups while you and Chad address the problem." Confident that Chad will follow him, Eldridge walks directly to the poster with the five standards for conduct hanging by the door, removes it, and takes it out into the hall. As Chad arrives in the hall, Eldridge leans the picture frame with the poster against the wall and they engage in the following conversation:

Chad (in a harsh tone): What?

Mr. Benis (calmly): Are you angry with me?

Chad: No!

Mr. Benis: The tone of your voice right now makes me think you're angry.

Chad: It's that dumb ho, she—

Mr. Benis: I think you should stop before you violate our third standard for conduct again. I'd like for you to read our third standard right now.

Chad silently frowns and looks down at his feet.

Mr. Benis: Do you choose to read it to me or should I read it to you?

Chad: You told me to stop, so I did.

Mr. Benis: I don't want you to stop talking, I just don't want you to use rude names in our classroom because calling Margaret a dumb ho and getting in my face and asking me if I'm a pimp are violations of Standard 3. You've already violated it today and I don't want you to do it again. I'll read it to you. Standard 3: Help—

Chad: I don't need you to read it.

Mr. Benis: Very well, please read it to me.

Chad: Help create and maintain a comfortable, safe, and secure learning environment for all members of our classroom community. So what?

Mr. Benis: What happens when we violate one of our standards of conduct?

Chad: But she dissed me.

Mr. Benis: Margaret tripped and fell into you. It hurt you.

Chad: So then she's the one that violated Standard 3. It's not safe around her.

Mr. Benis: I can understand why you got upset. When Ms. Hoggan, you, and I meet after school today, we should think of ways of remaining respectful to one another even when we're upset. Meet Ms. Hoggan and me at 2:55 right here.

Chad: Okay. Can I go back in now?

Mr. Benis: Yes. Do you want to join one of the groups or would you rather work on your own until the rest of the class finishes the group work?

Chad: I'll just go sit in my desk.

Mr. Benis: Okay, thanks for working with me.

The following morning, Eldridge and Elaine share the experience with Professor Rice and the other preservice teachers at the methods-of-teaching-mathematics course meeting. Advantages and disadvantages of various strategies for responding to such situations are discussed.

Unlike the discussion in Professor Ferimundi's class that followed the incident with Wilson in Case 2.32, Professor Rice's class in Case 2.33 was able to engage in a discussion in light of some first-hand knowledge of Chad, Margaret, and the social structure and dynamics of Ms. Hoggan's classroom. For example, the discussion took into account the roles played by three critical elements that were in place prior to the incident: (a) the relationship Eldridge had established with Chad, (2) the students' comprehension of the third standard for conduct, and (3) Ms. Hoggan's system for following up on violations of the standards.

Your knowledge of general principles regarding how people respond to situations as a function of their age, cultural backgrounds, special needs, and other circumstances is critical to your ability to develop a classroom management program that leads students to cooperate. However, the effective application of that knowledge to any unique classroom situation depends on the insights you gain and the relations you establish from your interactions with students as individuals. You must see students for who they are, not simply as a sample point of some combination of sociological categories. No two students are identical relative to the combination of needs, abilities, attitudes, experiences, background, interests, motivations, energy, genetics, chemistry, aptitudes, resources, perceptions, beliefs, and so on that they bring to your classroom. Furthermore, that combination for any one student is in a constant state of flux. Consequently, not only do you need to take into account individual differences among students as you apply classroom management strategies, you also need to be aware of how students are changing with time. As Kounin (1977) pointed out, withitness is essential for successful classroom management.

PSSM and Equity

You will become better acquainted with the *Principles and Standards for School Mathematics* (*PSSM*) (NCTM, 2000b) from your work with chapter 4. You may have already inferred from the references to it in chapter 1 that *PSSM* provides guidelines for the development and implementation of curricula that lead students in prekindergarten through grade 12 to learn meaningful mathematics. The first of *PSSM*'s six principles for high quality mathematics education is the *Equity Principle:* "Excellence in mathematics education requires equity—high expectations and strong support for *all* students." (NCTM, 2000b, p. 12)

Underlying the equity principle is the belief that all students have the right and capability to learn meaningful, nontrivial mathematics:

> All students, regardless of their personal characteristics, backgrounds, or physical challenges must have opportunities to study—and support to learn—mathematics. Equity does not mean that every student should receive identical instruction; instead, it demands that reasonable and appropriate accommodations be made as needed to promote access and attainment for all students. (NCTM, 2000b, p. 12)

Accommodating the Special-Education Needs of Students

It is essential for you to see and interact with each of your students as a unique individual regardless of whether the student is classified "regular education" or "special education." However, when students' exceptionalities interfere with their abilities to engage in activities for learning mathematics in the manner that most students are able, you will need to adjust your classroom management and instructional strategies to accommodate these students. Without such accommodations, some students will be unable to engage in your learning activities and, thus, besides violating the equity principle, you will also be faced with the discipline problems resulting from the frustration and boredom experienced by these students. Furthermore, you are legally responsible for complying with provisions of the *Individuals with Disabilities Education Act* (*IDEA*) (Turnbull & Turnbull, 1998, pp. 13–102). *IDEA* requires schools to accommodate students with special needs so that their educational opportunities are maximized in the least restrictive environment. Under this series of federal statutes, the 10 special needs categories for students aged six through 21 are as follows: mental retardation, hearing impairments, speech or language impairments, visual impairments, emotional disturbances, orthopedic impairments, autism, traumatic brain injury, other health impairments, and specific learning disabilities.

IDEA poses challenges to you regarding how you teach mathematics and operate your classroom—but not without provisions for resources for meeting these challenges. Particularly significant is the provision that each student with special education classification have an *individualized education program* (IEP) developed by a team that includes the student's parents, at least one regular education teacher (e.g., you), a special education teacher, a school administrator, and other individuals deemed appropriate. An IEP is an evolving written document agreement between the parents and the school that specifies an assessment of the student's present level of functioning, the long- and short-term goals, the services to be provided, and the plans for delivering and evaluating those services. By exerting your right to assert yourself as an active member of your students' IEP teams, you can use *IDEA* to help you accommodate individual students' needs as well as acquire needed services and equipment for your classroom. As a written contract, an IEP should not only specify instructional services you are required to provide; it should also specify the resources, services, support, and authority you require to enable you to deliver those services. Your efforts to obtain what you need from administrators, supervisors, specialists, and parents are more likely to be successful when requirements are spelled out in IEPs.

Exhibit 2.20 contains a list of references for you to use when developing strategies for accommodating students' exceptionalities as you teach mathematics and operate your classroom.

Including Students for Whom English Is Not a First Language

More than 13% of students in public school in the United States speak a non-English language in their homes; the English proficiency of over half of these students is too limited for them to fully participate in learning activities in which the English language is the primary means for communicating (McLeod, 1996). Spanish is the first language of approximately 75% of students classified as limited English proficient (LEP). Vietnamese, Hmong, Cantonese, Cambodian, or Korean are the first languages of 10% of LEP students. A variety of Native American languages are spoken by about 2.5% of LEP students. (Fleischman & Hopstock, 1993)

Typically, students who are in the process of learning English as a second language develop their conversational skills with English much faster than they develop their proficiency for profiting from academic lessons taught in English (Khisty, 1997). Thus, do not assume that because a student is able to converse with you one-to-one in English that the student can follow complex, multifaceted instruction in English requiring her to listen, read, write, speak, and employ other communication structures (e.g., interpret body language and illustrations).

Exhibit 2.20

References Relative to Accommodating Students' Exceptionalities.

Alber, S. R., Heward, W. L., & Hippler, B. J. (1999). Teaching middle school students with learning disabilities to recruit possible teacher attention. *Exceptional Children, 65,* 253–270.

Cangelosi, J. S. (2000). *Classroom management strategies: Gaining and maintaining students' cooperation.* (4th ed.). New York: Wiley (pp. 186–218).

Countinho, M. J., & Repp, A. C. (1999). *Inclusion: The integration of students with disabilities.* Belmont, CA: Wadsworth.

Feigenbaum, R. (2000). Algebra for students with learning disabilities. *Mathematics Teacher, 93,* 270–274.

Heward, W. L. (2000). *Exceptional children: An introduction to special education.* (7th ed.). Englewood Cliffs, NJ: Prentice-Hall.

Hughes, C. & Carter, E. W. (1999). *The transition handbook: Strategies for high school teachers that work.* Baltimore: Brookes.

Johnson, L. R., & Johnson, C. E. (1999). Teaching students to regulate their own behavior. *Teaching Exceptional Children, 31,* 6–10.

Lavoie, R. D. (1989). *Understanding learning disabilities: How difficult can this be? The F.A.T. city workshop* [Videotape]. Greenwich, CT: Eagle Hill Outreach (A Peter Rose Production distributed by PBS Video).

Mastropieri, M. A., & Scruggs, T. E. (2000). *The inclusive classroom: Strategies for effective instruction.* Columbus, OH: Merrill.

McLeod, B. (1996). *Exemplary school for language minority students.* Report of the Student Diversity Study by the Office of Bilingual Education. University of California, Santa Cruz, CA.

Polloway, E. A., Patton, J. R., & Serna, L. (2001). *Strategies for teaching students with special needs.* (7th ed.). Columbus, OH: Merrill.

Smith, T.E.C., Polloway, E. A., Patton, J. R., & Dowdy, C. A. (2001). *Teaching students with special needs in inclusive settings* (3rd ed.). Boston: Allyn and Bacon.

Turnbull, H. R., & Turnbull, A. P. (1998). *Free appropriate public education* (5th ed.). Denver: Love Publishing.

Yehle, A. K., & Wambold, C. (1998). An ADHD success story: Strategies for teachers and students. *Teaching Exceptional Children, 30,* 8–17.

Yesseldyke, J. E., Algozzine, B., & Thurlow, M. L. (2000). *Critical issues in special education.* (3rd ed.). Boston: Houghton-Mifflin.

Yesseldyke, J. E., & Olsen, K. (1999). Putting alternative assessments into practice. What to measure and possible sources of data. *Exceptional Children, 65,* 175–185.

Internet Sources:

Numerous websites exist that address issues related to classroom management strategies for accommodating students with exceptionalities. www.nichcy.org is a primary source. www.ced.sped.org/bk/focus/specfoc.htm is designed for special educators, but it includes links with reports and articles that should stimulate ideas for regular classroom teachers who work with students with exceptionalities. www.ida.net/users/marie/ed/cm.htm, www.chadd.org/50class.htm, and www.jaring.nmhu.edu/classman.htm all include links to articles with classroom management tips and information for regular classroom teachers who work with students with ADD/ADHD.

Of course, the success of your strategies for accommodating and including students for whom English is not a first language depends on your understanding of each student as an individual. Case 2.34 (adapted from Cangelosi, 2000b, pp. 224–228) is a continuation of Case 2.33 in which Professor Rice conducts a methods-of-teaching-mathematics course that is field based at Westside Middle School:

CASE 2.34

For Ms. Hoggan's prealgebra class, Eldridge Benis demonstrates an algorithm for using a protractor and compass to display data with circle graphs. He then directs the students to work in groups of three to practice the skill with data from experiments they previously conducted. Throughout the time Eldridge conducts the activity, his teaching partner, Elaine Rubenstein, monitors students' engagement.

The following morning in the methods-of-teaching-mathematics class meeting, Eldridge and Elaine are discussing with the other preservice teachers and Professor Rice the lesson Eldridge conducted:

Elaine: Nearly all the students really seemed to get into the lesson. I'd say 90% of the time in which Eldridge was explaining and demonstrating the algorithm, 25 of the 29 students were highly engaged. The engagement rates fluctuated more, but stayed very high, during the group work. I detected no disruptive off-task behaviors—unless you consider Cindy, Lorita, and Lena's constant, but low, talking disruptive.

Eldridge: It disrupted their own engagement, but not that of the other students because they kept to themselves and kept their voices down.

Elaine: They talked among themselves incessantly—during Eldridge's explanations as well as during the group activity—but I don't think the conversations had much to do with creating circle graphs.

Tawn: What were they talking about?

Elaine: I can't be sure because I don't understand Spanish.

Professor Rice: This would be a good time for Elaine and Eldridge to tell us about Cindy's, Lorita's, and Lena's special circumstances in Ms. Hoggan's class.

Elaine: Cindy speaks Spanish only. Lorita is bilingual with Spanish as her first language, but she's fluent with English. Lena converses well in English but she can't comprehend presentations in English like the one Eldridge gave yesterday.

Eldridge: Ms. Hoggan, their regular teacher, nearly always has the three of them working together so that Lorita can help the other two—especially Cindy—know what's going on.

Tawn: Shouldn't they learn English before we're expected to teach them?

Eldridge: Lorita functions just fine with English or Spanish, but Cindy for sure needs to be in an intensive English program.

Gabrielle: Is Cindy in Westside's bilingual education program?

Elaine: All three are so that they can experience some instruction in their native language. But this particular program—unlike the ones at some other schools—doesn't include much in the way of intensive English instruction.

Tawn: Unless we speak Spanish ourselves we can't teach students like Cindy. And even if we could, what about all the other LEP students whose first languages are Tongan, Vietnamese, Chinese, and so on. No one can expect us to teach in all those languages.

Janis: The solution is to pull Cindy out of her classes and have her concentrate on a good intensive English course. Then, after a while, she'll be ready to return to regular classes and learn math.

Elaine: So, Janis, you think we should recommend that Cindy—and possibly Lena—be referred to an intensive English course.

Janis: Right, until their English is up to speed.

Professor Rice: If Cindy were removed from the class to take intensive English, Eldridge and Elaine would no longer be faced with the classroom management dilemma of keeping her engaged in learning activities. That would solve the problem. Hmm, that makes me think about something. I've seen Cindy walking to school in the morning. There's always the possibility that she could be hit by a car crossing the street and then she wouldn't be in class tomorrow when Eldridge is scheduled to teach another lesson. That would keep us from having to face the—

Eldridge: That's a horrible thing for you to say Dr. Rice! I'm shocked that you would even imagine such a terrible event! Cindy is someone we've grown to care for over the past weeks! That's an awful thing for you to say!

Professor Rice: It was a terrible thing for me to say. But the direction this conversation was going got me to thinking that the probability of having Cindy transferred out of Ms. Hoggan's class and into an intensive English class before your next lesson is lower than the probability that a catastrophic event will keep Cindy from being in class and, thus, keep us from having to address the problem.

Eldridge: The problem being how to plan the lesson so that Cindy is included in spite of her being unable to function in English. I see your point. I apologize for attacking you; I was just so shocked at the thought of something so awful happening to Cindy.

Professor Rice: No apology needed. I would have been disappointed if you had reacted any other way. The fact of the matter is that you know Cindy as a unique individual—not simply some data point labeled "LEP"—that can be dealt with via some administrative referral process.

Tawn: So you think it's a bad idea to have her referred to an intensive English program.

Professor Rice: To the contrary, not only do I think it's a good idea to refer Cindy to an intensive English program, I think it's imperative that we do so. But that doesn't mean she has to also be excluded from mathematics class. And even if it did, I guarantee you it won't happen before Eldridge teaches tomorrow. Cindy is a living and breathing person who, if we're fortunate, is going to show up in class and Eldridge will be faced with the problem of teaching her mathematics in a way that accommodates her exceptionality. Eldridge, you've got a pedagogical and classroom management problem. How are you going to solve it?

Elaine: On one hand, I think it would be a good idea to break up the trio of Cindy, Lena, and Lorita to increase their involvement with the rest of the class. At least Lorita could be engaged without having to explain things to Cindy and Lena.

Eldridge: If only that's what she really did. But more and more they just gab away in Spanish hardly ever talking about anything to do with the lesson.

Elaine: That's why on one hand I think the group should be split up for activities. But on the other hand, Ms. Hoggan indicated that they are more comfortable in class in one another's company. Cindy, in particular, is completely left out of things without Lorita or Lena. No one else in the class knows enough Spanish to work with her.

Carrie: All of us agree that Cindy should learn to function with English. So why don't you and Eldridge come up with a plan in which you model learning in a second language yourself—namely Spanish? What if you got Cindy—with Lorita's or Lena's help—to teach some Spanish to the rest of the class?

Eldridge: That's awesome! Or what if I got Cindy with both Lorita's and Lena's help—I'm not yet ready to break up the trio—to demonstrate and explain a mathematical algorithm to the class in Spanish? Remember this is a mathematics class. We need to keep the focus on mathematics as we begin to pull Cindy into the class with an activity that uses the communication medium with which she is most comfortable. Great idea, Carrie! Thank you.

Janis: But how will the other students be able to learn an algorithm explained and demonstrated in Spanish?

Carrie: For one thing, they'll learn an appreciation for what Cindy goes through everyday and maybe they'll learn something that'll lead them to work better with Cindy and Lena.

Eldridge: But even from a learning-mathematics perspective, I think the idea has merit. One of *PSSM*'s major points is for students to communicate in the language of mathematics. Having a mathematical demonstration accompanied by explanations in a strange tongue could prompt students to raise a lot of questions leading to back and forth translations among Cindy, Lena or Lorita, and the rest of the class. This could have some real advantages relative to precision of communication with mathematics—something very important to algebraic reasoning.

Gabrielle: Won't Cindy feel strange trying this? Does she know enough mathematics to pull it off?

Elaine: She would, of course, have to be willing to give it a try. Knowing her, I think with some assertive prodding from us she'll be okay with it.

Eldridge: The demonstration and explanation would have to be very brief—not more than 10 minutes. We'd have to pick an algorithm that would be new to the class, but something we could teach Cindy to do with Lena and Lorita's help ahead of time.

Elaine: Let's see—we were going to be getting into angle relationships—but that involves a lot of new terminology—"vertical angles," "adjacent angles," and all that. That's not a good choice to try this out. What about jumping ahead to an algorithm for using a straightedge and compass to construct segments and angles. That's pretty straightforward and has a real visual aspect that can be demonstrated without a lot of technical words.

Eldridge: Cindy could conduct the demonstration and the explanations while Lena and Lorita are nearby to clarify and fill in the holes. It would turn into a mixture of Spanish and English. I love it.

Elaine: But we've got a lot of planning to do even before we run the idea by Ms. Hoggan.

Professor Rice: One question before you ask to be excused to start planning for this event: Is this going to be a one-time-only event or will you continue to put Cindy on stage for subsequent lessons?

Elaine: I think we ought to try this out with the one brief demonstration and then decide how to proceed later in light of what we learn. The whole thing could backfire.

Eldridge: I agree. What I hope will happen is that we decide to rotate the presentations among all the students with even the English-speaking students mixing in some Spanish into their explanations.

Tawn: How would that be possible?

Eldridge: Lorita and Lena could serve the same role they serve for Cindy's demonstration for other students as well—only in reverse—you know, filling in with some Spanish when the demonstration is in English just as they fill in with some English when the demonstration is presented in Spanish.

Elaine: Understand that they wouldn't be doing full translations—just plugging in a clarification here and there so that all students would be picking up both English and Spanish explanations that go along with the mathematics. Everybody can see what's going on with the demonstration and hear which Spanish or English words go along with what they see.

Tawn: I'm starting to see how this has a chance of working with two languages. But what if you've got a bunch of other kids in class whose first language varies from Korean to Russian to Italian?

Eldridge: I wouldn't know what to do. Fortunately, we don't have that problem in Ms. Hoggan's class. I guess you have to tackle each unique classroom situation head on to come up with a unique solution to each unique problem. I don't think there are any generic solutions.

Professor Rice: Elaine and Eldridge, I think you should excuse yourselves and get to work on your plan for tomorrow's activities.

Benefitting From Cultural Diversity

In Case 2.34, Eldridge and Elaine addressed a classroom management problem emanating from a difference in the language experiences of students. They realized that to gain and maintain all students' engagement they had to rethink their instructional practices. Their plan for building on Cindy's Spanish skills and Lorita's and Lena's bilingual skills turned a classroom management dilemma into a pedagogical advantage. As Eldridge envisioned the lesson stemming from Carrie's idea, the mathematical learning of all students in the class will benefit because of the difference in languages.

Besides the mathematics-education benefits of the plan, there is, of course, the advantage of the classroom community valuing Spanish—a deeply rooted part of Cindy's, Lena's, and Lorita's cultures. By conducting learning activities in ways that take advantage of language differences as well as other multicultural traits, you encourage cooperation and on-task behaviors. Students with different cultural experiences are more likely to cooperate with one another if they understand, value, and learn from one another's differences. If they do not, misunderstanding of differences leads to fear, mistrust, divisiveness, and even hostility.

Your own understanding and appreciation of cultural diversity will serve you well as you develop strategies for motivating students to be on-task and engaged. Furthermore, you are hardly in a position to elicit students' cooperation unless you are aware of differences that cause an action or communication to be perceived as a compliment within one subculture and an insult within another. The success of your teaching depends on staying abreast of multicultural issues and instructional strategies. Fortunately, a rich reservoir of instructional materials and ideas on how to implement multicultural instructional strategies exists in professional literature and media sources. A minute sample is included among the list of titles displayed by Exhibit 2.21.

In Case 2.35 (adapted from Cangelosi, 2000b, pp. 229–234), a teacher takes advantage of the cultural diversity in her geometry class:

CASE 2.35

Building on an idea from chapter 1's section "The Beginning of an Eventful School Year" and this chapter's section "Establishing a Favorable Climate for Learning Mathematics," Ms. Willis administers a tasksheet like Mr. Rudd's in Exhibit 1.14. She chose this tasksheet with the following purposes in mind:

- The students will begin to get the following impressions about the course they are starting under her direction:
  - The geometry they will be learning is related to their own individual interest, background, and experience.
  - The are expected to make judgments and those judgments are valued.
  - They are expected to write about geometry.
  - Their experiences in this course will be different from those they have had in previous mathematics courses.
- Students begin doing geometry by having to describe what they see in terms of relationships among figures imbedded in real-world images.
- Ms. Willis begins to gain some insights about their individual personalities, tastes, interests, families, neighborhoods, and backgrounds.
- Ms. Willis elicits some information that will help her begin to identify aspects of their individual cultural experiences upon which she can build and incorporate into lessons she plans to teach throughout the course.
- Their initial experience in the course exposes them to multicultural aspects of the world they will be examining from a geometric perspective.

Throughout the course Ms. Willis draws upon what she learns about students' cultural experiences to design lessons. For example, as part of a unit on geometric constructions, each student applies congruence theorems by creating an image using only a compass and a straightedge. The image is to uniquely reflect a significant aspect of the student's family, neighborhood, or home. Exhibit 2.22 displays a copy of what one student produced using only a compass and straightedge. The image reflects the fact that the student's father works in a plant that makes rocket boosters.

As part of a unit on slope, students examine the pitch on various structures in their neighborhood. One student lives with her uncle who uses a wheelchair. She reports on the angle of inclination of the ramp her uncle uses to enter and exit their apartment. This and other such units

Exhibit 2.21

References Relative to Implementing Multicultural Education Strategies.

Banks, J. A., & Banks, C. A. M. (Eds.). (1995). *The handbook of research on multicultural education.* New York: Macmillan.

Banks, J. A., & Banks, C. A. M. (Eds.). (2001). *Multicultural education: Issues and perspectives* (4th ed.). New York: Wiley.

Boaler, J. (Ed.). (2000). *Multiple persepctives on mathematics teaching and learning.* Westport, CT: Ablex Publishing.

Carnes, J. (Ed.). (1999). *Responding to hate at school: A guide for teachers, counselors, and administrators.* Montgomery, AL: Southern Poverty Law Center.

Cooney, M. P. (Ed.). (1996). *Celebrating women in mathematics and science.* Reston, VA: NCTM.

Edwards, C. A. (Ed.). (1999). *Changing the faces of mathematics: Perspectives on Asian Americans and Pacific Islanders.* Reston, VA: NCTM.

Eglash, R. (1998). Geometry in Mangbetu design. *Mathematics Teacher, 91,* 376–381.

Franklin, J. (2001). The diverse challenges of multiculturalism. *Education Update, 43,* 1, 3, & 8.

Gray, S. I. B. (2000). Mathematics in the age of Jane Austen: Essential skills of 1800. *Mathematics Teacher, 93,* 670–679.

Green, R. A., & Snyder, L. A. (2000). Primitive living structures: Tents and tipis. *Mathematics Teacher, 93,* 738–744.

Karp, K. S. (1994). Telling tales: Creating graphs using multicultural literature. *Teaching Children Mathematics, 1,* 87–91.

Karp, K. S., & Niemi, R. C. (2000). The math club for girls and other problem solvers. *Mathematics Teaching in the Middle School, 5,* 426–432.

Lamb Jr., J. F. (2000). A Chinese zodiac mathematical structure. *Mathematics Teacher, 93,* 86–91.

Lee-Chua, Q. N. (2001). Mathematics in the tribal Phillippines and other societies in the South Pacific. *Mathematics Teacher, 94,* 50–55.

Little Soldier, Lee (1997). Is there an 'Indian' in your classroom? Working successfully with urban Native American students. *Phi Delta Kappan, 78,* 650–653.

Natsoulas, A. (2000). Group symmetries connect art and history with mathematics. *Mathematics Teacher, 93,* 364–370.

Ortiz-Franco, L., Hernandez, N. G., & De La Cruz, Y. (Eds.). (1999). *Changing the faces of mathematics: Perspectives of Latinos.* Reston, VA: NCTM.

Secada, W. G. (1992). Race, social class, ethnicity, language, and achievement in mathematics. In D. A. Grouws (Ed.), *Handbook of research on mathematics teaching and learning* (pp. 623–660). New York: Macmillan.

Secada, W. G. (Ed.). (2000). *Changing the faces of mathematics: Perspectives on multiculturalism and gender equity.* Reston, VA: NCTM.

Simon, M. K. (2000). The evolving role of women in mathematics. *Mathematics Teacher, 93,* 782–786.

Strutchens, M., Johnson, M. L., & Tate, W. F. (Eds.). (2000). *Changing the faces of mathematics: Perspectives on African Americans.* Reston, VA: NCTM.

Teaching Tolerance Magazine [published by the Southern Poverty Law Center—available without cost upon request by educators from 400 Washington Ave., Montgomery, AL 36104.]

Terezinha, N. (1992). Ethnomathematics and everyday cognition. In D. A. Grouws (Ed.), *Handbook of research on mathematics teaching and learning* (pp. 557–574). New York: Macmillan.

Trentacosta, J., & Kenney, M. J. (Eds.). (1997). *Multicultural and gender equity in the mathematics classroom: The gift of diversity: 1997 Yearbook.* Reston, VA: NCTM.

Valadez, G. (2001). The gardeners' story: Teaching cross-cultural communication. *Phi Delta Kappan, 82,* 666–669.

Wiest, L. R. (2001). Selected resources for encouraging females in mathematics. *Mathematics Teacher, 94,* 14–18.

Wilson, P. S. (2001). Zero: A special case. *Mathematics Teaching in the Middle School, 6,* 300–303.

Wilson, P. S., & Chauvot, J. B. (2000). Who? How? What? A strategy for using history to teach mathematics. *Mathematics Teacher, 93,* 642–645.

Zaslavsky, C. (1998). Ethnomathematics and multicultural mathematics education. *Teaching Children Mathematics, 4,* 5.

Zaslavsky, C. (2000). The Inka quipu: Positional notation on a knotted cord. *Mathematics Teaching in the Middle School, 6,* 164–166 & 180–184.

Internet Sources:

Numerous websites exist that address issues pertaining to multicultural education and provide suggestions for teachers. The address for the Southern Poverty Law Center's web page for teachers is www.teachingtolearnce.org.

jwilson.coe.uga.edu/DEPT/Multicultural/MathEd.html is the address for Multicultural Perspectives in Mathematics Education.

www.cs.yale.edu/homes/tap/past-women.html is the address for a site focusing on women's contributions to mathematics.

www.math.buffalo.edu/mad/madhist.html is the address for a site focusing on the history of works of black mathematicians.

Exhibit 2.22

The Compass-and-Straightedge Image Produced by a Student Whose Father Works at a Rocket Production Plant (Reprinted by permission from Johnson (1997).

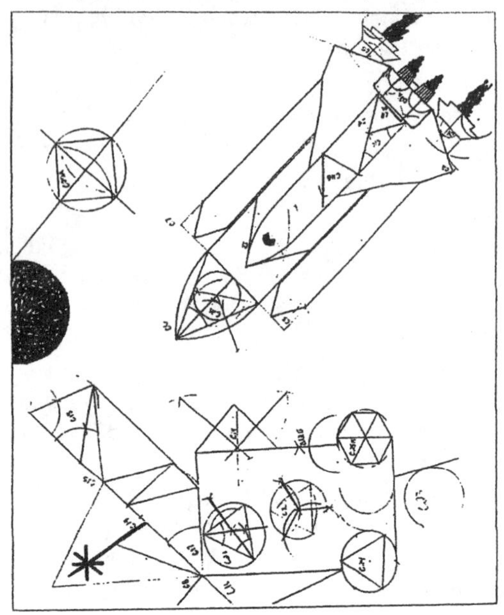

help students connect schoolwork to their family backgrounds and neighborhoods. The strategy also helps foster an understanding and appreciation for the diversity in the classroom.

SYSTEMATICALLY DEALING WITH OFF-TASK BEHAVIORS

By establishing a favorable classroom climate, communicating effectively, establishing necessary standards for conduct and routine procedures, working with students as individuals, and conducting engaging learning activities (the focus of chapter 3), you prevent most off-task behaviors from ever occurring. However, with a group of 25 or 30 adolescents, you will still have to deal with some off-task behaviors. You will be able to deal with them effectively if you (a) previously established healthy, mutually respectful relationships with students and (b) calmly apply systematic teaching strategies when confronted with the misbehaviors. It is quite natural for teachers to want to retaliate and display power over students who are infringing on the rights of those about them. But such knee-jerk reactions are virtually always counterproductive. Rather than allowing emotions to cloud her thinking, the teacher in Case 2.36 systematically and thoughtfully deals with a serious disruptive behavior pattern:

CASE 2.36

Matthew, one of Ms. Asgill's algebra I students, is working in a small-group activity with five others, playing a game called "Complete the Equation." The student conducting the game draws the next "rule" card and reads aloud, "the square root of an odd integer." The players hurriedly try out some ideas on their calculators, attempting to come up with the number that will complete the equation they have built to this point. Suddenly Oliver exclaims, "I got it—equation!" Matthew stands up and yells, "Oliver, you cheat, I was about to get mine!" With that, Matthew shoves Oliver into the game board, toppling it and scattering materials. Having observed the incident from her position across the room where she was working with another group, Ms. Asgill walks unhesitatingly between Matthew and Oliver, looks Matthew in the eye, and in a calm voice says, "Step into the hallway with me." Indicating with a gesture that he is to go first, she follows him to a point just outside the classroom. She faces him directly looking into his eyes and says in a firm, calm voice, "Please stay right here until I get back. I'm going to see if I can help Oliver; he may be hurt." She turns away before Matthew has a chance to reply. Actually, she had already noted that Oliver did not appear to be hurt, but she immediately returns to the scene of the game, where an audience has gathered around Oliver who is announcing his plans for retaliation. Ms. Asgill interrupts him with, "I'm sorry this happened, but I'm pleased that you're not hurt." Cutting off a student starting to criticize Matthew, Ms. Asgill continues, "Eric and Beatrice, please pick up this mess and set up the game again. We'll start over with Oliver conducting for three players." In a raised, but calm voice, she announces, "Everyone return to work. Thank you."

Quickly returning to Matthew standing against the hallway wall, she says, "I do not have time to deal with the way you behaved during Complete the Equation. Right now, I have a class to teach and you need to continue practicing with equations. We'll have time to discuss how to prevent these disruptions before the first bell tomorrow morning. Within three minutes after your bus arrives tomorrow, meet me at my desk. Will you remember or shall I call your house tonight to remind you?" Matthew replies, "I'll remember." Ms. Asgill responds, "Very well, it's up to you. We have only 13 more class minutes to work with equations. Go get your textbook and notebook and bring them with you to my desk." There, Ms. Asgill directs him to complete an exercise at a desk away from the other students. The exercise relates to the same skill that Complete the Equation is designed to develop.

Later in the day, when she finally has a chance to be alone, Ms. Asgill thinks, "I bought myself some time to decide what to do about Matthew's hostile outbursts. I took a chance stepping in front of him while he was still angry. Suppose he had turned on me? Then I wouldn't have him in my class anymore, and I wouldn't have to be here trying to figure out a solution.

"This is the third time he's had a disruptive outburst—but it's the first time he's gotten physically violent. Every time, it's been during some type of group activity where there's a lot of student interaction. I don't know if he's been the instigator each time, but he's been in the middle. But I'm not going to worry with who caused what, just with preventing this from happening again before somebody really gets hurt.

"Until he's learned that antisocial behavior isn't tolerated in my classroom, he'll have to be excluded from student-centered activities—nothing where he interacts with others, unless I'm right on top of things orchestrating every move.

"That takes care of the immediate goal of preventing recurrences. But if he doesn't learn to control that temper, at least in my class, another outburst will eventually occur—besides, I don't want to have to keep him separated for the rest of the year. Tomorrow, maybe I should explain my dilemma to him and ask what he would do to solve the problem if he were in my place. That tactic worked well before with Janice. But no, Matthew isn't ready for that; he's far too defensive. He'd start trying to tell me how it's so unfair, that he's always picked on. Here's what I'll try: (a) Tomorrow, I will not even attempt to explain my reasons for what I'm doing. If I do, he will try to argue with those reasons and I don't need that. I will simply tell him what we're going to do and not try to defend the plan. (b) Whenever he would normally be in a group activity that I'm not personally directing, I will assign work for him to do by himself at a desk away from the others. As far as possible, his assignment will target the same objective as the group activity. (c) I'll watch for indicators that he is progressing toward willingness to cooperate in activities with other students. (d) As I see encouraging indications, I will gradually work him back in with the other students. But I will begin very slowly and only with brief, noncompetitive activities.

"Now to prepare for this. I'd better come up with a contingency plan in case he doesn't show for our meeting tomorrow morning."

Note how Ms. Asgill addressed the problem of eliminating the undesirable behavior pattern as she would a problem of how to help a student achieve a learning objective. She focused her time, energy, and thought on strategies for leading Matthew to choose to be on-task rather than disruptive. She did not try to moralize to Matthew about the evils of fighting or attack him personally for fighting. She put a halt to the altercation and then went to work on preventing a recurrence.

Teachers who do not systematically focus on the behavior to be altered tend to compound difficulties by dwelling on irrelevant issues. For example, in Case 2.1, Ms. Lewis called attention to Patrick being "involved with Rolando" when she should have been focusing attention on efficiently dispensing with the attendance report. Later she asked Andrew, "Why are you late?" instead of quickly directing him to his seat and continuing with the business of teaching mathematics. In Case 2.7, Mr. Griffin dwelt on Carlos' daydreaming to make a joke at Carlos' expense.

Do not interpret your students' off-task behaviors as a personal attack on you. It is annoying to have your plans disrupted, your efforts ignored, and your authority questioned by adolescents. But they do it out of naivete, boredom, frustration, or for other reasons that do not threaten your personal worth. Keeping this in mind helps you maintain your wits well enough to take decisive, effective action that terminates misbehavior and reduces the probability of it recurring.

Of course how to develop effective systematic strategies to address behavior management problems is a complex topic itself. You may find Exhibit 2.8's list of references for addressing such problems helpful.

SYNTHESIS ACTIVITIES FOR CHAPTER 2

1. For each of the following multiple-choice prompts, select the one response that either completes the statement so that it is true or accurately answers the question:
   A. Time for students to achieve learning objectives increases as _____ .
      a) more time is spent on developing students' algorithmic skills instead of higher cognitive abilities
      b) more time is spent on developing students' higher cognitive abilities instead of algorithmic skills.
      c) allocated time decreases
      d) transition time decreases
   B. Duane listens intently to his classmates and occasionally volunteers his own opinions during large-group discussion sessions about applications of combinations and permutations to real-life situations. Duane's behavior appears to be _____ .
      a) disruptive but engaged
      b) off-task and disruptive
      c) on-task and engaged
      d) on-task but not engaged
   C. Pat quietly rearranges her desk as directed by her teacher as the class moves from a

large-group to a small-group session. Pat's behavior appears to be _____ .
a) disruptive but engaged
b) off-task but not disruptive
c) on-task and engaged
d) on-task but not engaged

D. Dawn speaks out, interrupting Erin during a class discussion session on the origins of certain mathematical terms. Dawn's behavior appears to be _____ .
a) disruptive but engaged
b) off-task but not disruptive
c) off-task and disruptive
d) on-task but not engaged

E. When a student is rewarded for an off-task behavior, the reward serves as _____ .
a) positive reinforcement for an off-task behavior
b) punishment for an off-task behavior
c) punishment for an on-task behavior
d) destructive positive reinforcement

F. Larry interrupts his teacher in class. The teacher says, "Larry, what makes you think we care what you have to say? Sit and listen without interrupting!" Larry feels so embarrassed he silently vows not to interrupt the teacher again. He begins doubting that his peers really want to hear his opinions. In this example, the teacher's remarks serve as _____ .
a) destructive positive reinforcement
b) destructive punishment
c) positive reinforcement of an off-task behavior
d) punishment of an on-task behavior

G. An assertive communication style is characterized by _____ .
a) aggressiveness
b) judgmental comments
c) professional confidentiality
d) sincerity

H. A classroom with a businesslike atmosphere is characterized by _____ .
a) democratic decision making
b) authoritarian decision making
c) a highly formalized structure
d) purposeful activity

I. Which one of the following contributes to a businesslike classroom conducive to learning and doing mathematics?
a) use of descriptive language
b) use of judgmental language
c) maximizing transition time
d) consistent use of corporal punishment for unbusinesslike behaviors

J. A supportive reply to a student tends to communicate _____ .
a) recognition of feelings
b) passiveness
c) assertiveness
d) value judgments

K. Students tend to be most receptive to communications about your expectations for them _____ .
a) following tests
b) near the very beginning of a course
c) during the last few days of a course
d) when you are delivering a lecture

L. By "withitness," Kounin refers to how _____ .
a) well teachers maintain students on-task and engaged in learning activities
b) aware teachers are of what is going on in their classrooms
c) well teachers display enthusiasm for teaching and learning
d) assertively teachers conduct themselves with students

2. In a discussion with a colleague who also engaged in Synthesis Activity 1, compare your responses to the multiple-choice prompts. Also, check your choices against the following key: A–d, B–c, C–d, D–c, E–a, F–b, G–d, H–d, I–a, J–a, K–b, L–b.
3. Now that you have studied this chapter's suggestions for managing students—positively reinforcing on-task behaviors, conducting efficient transitions, focusing on the business of learning mathematics, using descriptive instead of judgmental language, and effectively responding to off-task behaviors, reread Case 2.1 and identify examples in which Ms. Lewis failed to heed these suggestions. Discuss with colleagues what Ms. Lewis could have done differently to improve her classroom climate and encourage her students to be on-task.
4. Imagine having just directed your students to devise proofs independently for a theorem you have just stated. Although you were quite clear that they were to work silently by themselves, you notice that Haywood and Howard are talking. You walk over to them and realize that they are discussing how to prove the theorem. State an example of a *descriptive* comment you could make to them. State an example of a *judgmental* comment you could make to them. With colleagues discuss the relative advantages and disadvantages of making the first instead of the second comment.
5. After you have directed students to devise a proof for a theorem, Delcima exclaims to you, "I could

never make up my own proof!" State an example of a *supportive* comment you could make to Delcima. State an example of a *nonsupportive* comment you could make. With colleagues discuss the relative advantages and disadvantages of making the first instead of the second comment.

6. By his comments in Cases 2.37, a teacher violates a principle you studied in this chapter. What is the principle?

CASE 2.37

Mr. Zebart confronts Jackie, Fred, and Lamont with evidence that they cheated on a test. "What have you got to say for yourselves?" he asks. Lamont: "I didn't steal the test; Jackie already had it before I even knew about it!" Mr. Zebart: "So, Jackie, you not only cheated, but you got these other two to cheat also!"

Compare your responses on this activity to that of a colleague as well as to the following:

The principle is that students should be held responsible for their own behaviors. In Case 2.37, Mr. Zebart failed to communicate that by indicating that one student could cause another to cheat.

7. With several colleagues, critique and discuss Professor Rice's provocative comment designed to lead the preservice mathematics teachers in Case 2.34 to focus on Cindy as an individual rather than an abstract LEP statistic, and to confront the immediate problem Eldridge faced.
8. Reread Cases 2.20 and 2.21 in the section "True Dialogues." In collaboration with a colleague, write scripts for two different dialogues in which you engage your students in a conversation about some mathematical problem—one that is different from the one addressed in Cases 2.20 and 2.21. IRE cycles should dominate the first conversation. The second should be a natural conversation with true dialogue. Explain in a paragraph the pedagogical advantages of engaging students in the second conversation instead of the first. Insert the word-processing file for this work in the "Resources/Classroom Management" subfolder that you set up when you engaged in Activity 1.3. Also insert a hard copy in the "Motivation, Engagement, and Classroom Management" section of your working portfolio.

TRANSITIONAL ACTIVITY FROM CHAPTER 2 TO CHAPTER 3

In a discussion with two or more of your colleagues, address the following questions:

1. Why are some students bored by the mathematics lessons of one teacher but enthusiastically energized by the mathematics lessons of another teacher?
2. What strategies do teachers employ to help students understand and follow directions for learning activities?
3. What strategies do teachers employ to keep students engaged in the following types of learning activities?
   - large-group presentations
   - cooperative learning
   - question-discussion
   - independent work
4. What are the preferable methods for motivating students to do homework assignments?
5. What are appropriate strategies for responding to questions raised by students during mathematics lessons?

chapter

3

Motivating Students to Engage in Mathematical Learning Activities

GOAL AND OBJECTIVES FOR CHAPTER 3

The Goal The goal of chapter 3 is to lead you to develop strategies for conducting learning activities so that students willingly and enthusiastically engage in them.

The Objectives Chapter 3's goal is defined by the following set of objectives:

A. You will distinguish between intrinsic and extrinsic motivation to do mathematics (construct a concept) 10%.
B. You will explain the problem-based strategy for designing mathematics lessons so that students are intrinsically motivated to engage in them (comprehension and communication) 20%.
C. You will develop strategies for responding to students' reasoning-level questions and strategies for responding to their memory-level questions (application) 20%.
D. You will develop strategies for initiating and maintaining student engagement in the following types of learning activities: large-group presentations, question-discussion sessions, cooperative-learning sessions, independent-work sessions, and homework (application) 40%.
E. You will incorporate the following phrases or words into your working vocabulary: "intrinsic motivation," "extrinsic motivation," "problem-based approach teaching unit," "reasoning-level question," "memory-level question," and "cooperative learning" (comprehension and communication) 10%.

MOTIVATING STUDENTS TO DO MATHEMATICS

Intrinsic and Extrinsic Motivation

Eavesdrop on three teenagers discussing their algebra class:

CASE 3.1

Adam: Today, I learned to love tornadoes! Perfect timing for a surprise tornado drill—got us out of Mitchell's algebra class!

Jalynie: Borrrring! Yea—like I'm really gonna need that crap!

Xiao: They keep trying to sell us how we're going to need this in college. If math is so useful, why do they spend so much time trying to sell it to us? Last year, Iancenti kept telling us how proofs were going to teach us how to think logically, but all we ever did was memorize proofs out of the geometry book.

Jalynie: Look at these so-called "application problems" Mitchell gave us for homework.

Jalynie points to the following word problem in his algebra textbook:

> Jose's sister is three fifths his age. Six years ago, Jose was three times older than his sister. How old is Jose now? How old is his sister?

Jalynie: When in real life would I ever know how many times one person was older than another six years ago without knowing his age today?

Adam: The only math I'll ever need is how to balance my checkbook—and I guess with computers, I really don't even need that!

Xiao: Anything this boring should at least be useful!

Does the attitude reflected in Case 3.1 ring familiar? Often students are not motivated to engage in activities that are supposed to teach them mathematics because they find the activities both uninteresting and disconnected from what they consider important (Battista, 1999). Mathematical learning activities cannot always be entertaining; learning mathematics is often hard work. Nevertheless, there are strategies for designing mathematics lessons that stimulate enthusiastic student engagement. Such designs depend on the intrinsic value of mathematics to students' everyday lives.

Students are *intrinsically motivated* to engage in learning activities if they recognize that by experiencing the activity they will satisfy a need. Intrinsically motivated students value engagement as directly beneficial. The learning activity itself is perceived to be valuable. (Cangelosi, 2000b, p. 238; Santrock, 2001, pp. 397–401). The students in Case 3.2 are intrinsically motivated to engage in mathematical learning activities:

CASE 3.2

Miranda connects her interest in developing her artistic talent to what Mr. Rudd is teaching in geometry class. In particular, she engages in any of Mr. Rudd's activities that she can apply to perspective drawings.

Salmonie enthusiastically participates in Mr. Rudd's lesson on solving systems of linear relationships because he believes he can apply this area of mathematics to his interest in developing strategies for winning video games.

Aimee's biology teacher presented the class with a challenging population-growth problem involving flour beetles. Aimee wonders if she can apply what she is learning about quadratic functions in Mr. Rudd's class to solve the problem from biology class. Thus, she intently engages in Mr. Rudd's activities.

Allison realizes that her chance of becoming an engineer depends on how well she learns mathematics. Thus, she diligently attempts Mr. Rudd's homework assignments.

Unlike those in Case 3.2, students are *extrinsically motivated* to engage in learning activities not because they recognize value in experiencing the activity, but because they desire to receive the rewards that have been artificially associated with engagement or want to avoid consequences artificially imposed on those who are off-task (Cangelosi, 2000b, pp. 238–239; Santrock, 2001, pp. 397–401). The students in Case 3.3 are extrinsically motivated to engage in mathematical learning activities:

CASE 3.3

Ms. Nightingale assigns her prealgebra class to solve a set of textbook word problems related to volumes of prisms. James works on the assignment so he will earn the 10 points Ms. Nightingale awards for it. James wants to please his dad by making high grades.

Jaymie listens intently to Ms. Nightingale's explanation on when to use various formulas for computing volumes because she fears that if she does not listen, she will be embarrassed by not knowing answers when asked questions in class.

Desire for high grades, fear of low grades, competition for academic honors, the need to graduate, desire for praise and awards from parents, and fear of criticism and punishment from parents are just some of the incentives teachers and school administrators use to extrinsically motivate students to engage in learning activities. Typically, such incentives are far more motivating for students who have a history of academic success than for students who have the greatest need for motivation. Students with histories of academic failures are hardly likely to participate in academic competitions.

Connecting Mathematics to Students' Interests: The Problem-Based Approach

Extrinsically motivating students to engage in learning activities (e.g., by "participation points" toward grades) is superior to not motivating students at all but inferior to students being intrinsically motivated (Hidi & Harackiewicz, 2000). Students can be intrinsically motivated to do mathematics once they discover the connection between the mathematics and their own needs and interests (McLeod, 1992; NCTM, 2000b, pp. 64–66, Noble, Nemirovsky, Wright, & Tierney, 2001).

Compare Case 3.4 to Case 3.5:

CASE 3.4

Mr. DeMarco teaches his prealgebra class about ratios by (a) giving a brief lecture on the importance of learning about ratios, (b) using a textbook word problem to demonstrate how ratios can be useful (e.g., "The fish commission is stocking fish in City Park Lake. Suppose 800 bass and 2,800 perch are put into the lake. Express the ratio of bass to total fish stocked as a fraction in simplest form."), (c) conducting an independent-work session in which students solve textbook word problems on ratios as he provides individual help, (d) assigning similar textbook word problems for homework, and (e) answering questions on the homework assignment the next day in class.

CASE 3.5

Throughout her prealgebra course, Ms. Castillo monitors her 24 students' interests and achievement of objectives. Often she combines intraclass grouping techniques with problem-solving activities to accommodate the variety of achievement levels and interests among the students. She is now planning a lesson for the following objective:

> When confronted with a real-life problem, the student determines whether or not computing a ratio will facilitate a solution of the problem (application).

She thinks: "I'll partition the class into four collaborative groups that focus on different interest areas but all of which require them to decide when to use and when not to use ratios. Shamika, Todd, Clovis, and Brevin are all interested in basketball, but are not naturally inclined to do mathematics. I'll have them work on a basketball-related problem along with Sharee and Ping who are motivated to do mathematics but have little knowledge about basketball. Those who are disinclined to do mathematics will be enticed by the basketball, whereas Sharee and Ping will extend their mathematical talents by having to apply them to an unfamiliar topic. All six students will contribute to the effort with some serving as basketball consultants and others as math consultants."

Using similar reasoning, Ms. Castillo identifies students for the other three groups—one involving nutrition, another involving music CDs, and the final group involving crime statistics. She develops the four tasksheets shown by Exhibit 3.1.

During the lesson, each group has 18 minutes to answer its tasksheet's questions and prepare a six-minute oral report to be presented to the rest of the class. After the reports, Ms. Castillo conducts a question-discussion session in which students articulate the differences between problems to which ratios apply and problems to which ratios do not apply.

With the typical approach to teaching mathematics that Mr. DeMarco's used in Case 3.4, topics are presented in isolation—disconnected from their real-world applications that interest most students (Battista, 1999; Cangelosi, 2001; O'Brien, 1999). By consistently employing the problem-based learning strategy she used in Case 3.5, Ms. Castillo will lead her students to connect mathematics to what already interests them (Cangelosi, 2000b, pp. 240–245). Recognizing the need to intrinsically motivate students who are not motivated by traditional extrinsic incentives, the authors of *PSSM* (NCTM, 2000), included the "Connections Standard," among their five "Process Standards":

> When students can connect mathematical ideas, their understanding is deeper and more lasting. They can see mathematical connections in the rich interplay among mathematical topics, in contexts that relate mathematics to other subjects, and in their own interests and experience. Through instruction that emphasizes the interrelatedness of mathematical ideas, students not only learn mathematics, they also learn about the utility of mathematics. (p. 64)

You will revisit *PSSM*'s Connection Standard in greater detail when you work with chapter 4. Furthermore, cases in which teachers plan and conduct problem-based units and lessons that meet the Connection Standard so that students are intrinsically motivated to do mathematics are included in chapters 5–8 and 11.

DIRECTING STUDENTS INTO LEARNING ACTIVITIES

Directness, Explicitness, and Specificity

Indirect and inexplicit communications are appropriate for inquiry-learning activities to stimulate your students to reason, discover, or create. Case 3.6 is an example.

CASE 3.6

As part of a unit on the behavior of trigonometric functions, Mr. Koebbe is trying to lead students to discover certain relationships among various components of a

Exhibit 3.2
Maria Examines the Behavior of Trigonometric Functions.

Exhibit 3.1
A Different Tasksheet for Each of Ms. Castillo's Four Collaborative Groups.

Group A

DIRECTIONS: Answer the questions based on the box score from Tuesday's Flyers-Tigers basketball game. Prepare a six-minute oral report to the class that explains what problem-solving strategies you used to answer the eight questions at the bottom.

Flyers 79, Tigers 75

Flyers	min.	fgm-fga	ftm-fta	reb.	ass.	pf	tp
Champagne	23	4–7	2–2	8	1	4	10
Noto	28	9–19	6–8	4	2	2	26
Kora	21	3–4	4–9	11	3	4	10
Guillory, K.	15	2–7	2–3	2	0	0	8
Demouy	32	4–6	0–0	3	10	3	8
Guillory, T.	11	0–3	0–0	0	2	1	0
Miller	20	7–13	1–1	1	1	0	15
Knight	5	0–0	0–0	0	0	2	0
Losavio	5	1–3	0–0	0	0	0	2
TOTALS	160	30–62	15–23	29	19	16	79

Flyers	min.	fgm-fga	ftm-fta	reb.	ass.	pf	tp
Cassano	32	11–19	0–0	4	3	4	22
Silva	30	2–9	5–8	7	0	4	9
Burke	14	0–6	1–2	2	1	2	1
Parino	32	12–27	5–5	9	7	3	29
Price	22	4–5	0–0	4	7	5	8
Weimer	19	3–6	0–4	6	2	1	6
Bowman	11	0–4	0–1	1	1	0	0
TOTALS	160	32–76	11–20	33	21	19	75

1. By how much did the Flyers win?
2. Which team made more three-point goals?
3. Who made more field goals, Noto or Parino?
4. Who was the more accurate field-goal shooter, Noto or Parino?
5. Who scored more points for the amount of time he played in the game, Noto or Parino?
6. Whereas the Tigers made more field goals than the Flyers, how did the Flyers manage to score more points than the Tigers?
7. If Kora had played the whole game, rebounding at the same rate as he did for the time he was in the game, what would his rebound total have been?
8. For which ones of the preceding seven questions did you use ratios to help determine your answers?

Continued

function and its graph (e.g., how b in $y = a\sin(b\phi + c) + d$ influences the period of the graph). At one point in the lesson, Maria shows him her calculator with the display shown in Exhibit 3.2 and says, "I see what happens to the period if you multiply the angle, but what if you divide it?" Mr. Koebbe responds, "Hmmm, I wonder. Let's think about it. Would that exaggerate the influence of the angle? Or maybe not [pause] mmm [pause] before you try some examples on your calculator, let's see if we can predict what will happen.

Mr. Koebbe knew the answer to Maria's question, but instead of answering it directly, he probed with another question. His indefinite, evasive communication was appropriate in Case 3.6 because Mr. Koebbe's objective was to stimulate Maria to reason, not simply to know the answer. On the other hand, when you give students *directions* for an upcoming learning activity, your communications need to be *explicit, specific,* and *precise* so that transition time is minimized and allocated time is maximized because students understand how to engage in the learning

Exhibit 3.1
Continued

Group B

DIRECTIONS: Answer the questions based on the labels from the two soup cans. Prepare a six-minute oral report to the class that explains what problem-solving strategies you used to answer the eight questions at the bottom.

CHICKEN WITH LETTERS

Size: 10.6 oz (298 g) *Price:* 88¢

Nutritional Information

Serving size (condensed) .... 4 oz
Serving size (prepared) .... 8 oz
Per serving:
Calories 60
Protein (grams) 3
Simple sugars (grams) 1
Complex carbohydrates (grams) 6
Fat (grams) 2
Cholesterol (mg) 10
Sodium (mg) 870

Percentage of DV (Daily Value) based on a 2,000-calorie diet:

Protein	4	Riboflavin	2
Vitamin A	8	Niacin	4
Vitamin C	*	Calcium	*
Thiamine	2	Iron	4

*Contains less than 2% of the DV of this nutrient.

NOODLES AND CHICKEN

Size: 10.6 oz (298 g) *Price:* 88¢

Nutritional Information

Serving size (condensed) .... 4 oz
Serving size (prepared) .... 8 oz
Per serving:
Calories 80
Protein (grams) 3
Simple sugars (grams) 1
Complex carbohydrates (grams) 8
Fat (grams) 3
Cholesterol (mg) 15
Sodium (mg) 960

Percentage of DV (Daily Value) based on a 2,000-calorie diet:

Protein	4	Riboflavin	2
Vitamin A	25	Niacin	6
Vitamin C	*	Calcium	*
Thiamine	4	Iron	4

*Contains less than 2% of the DV of this nutrient.

1. With which soup do you get more grams for the money?
2. Which soup has more vitamins?
3. Which soup has more cholesterol?
4. Which soup has more vitamins per calorie?
5. Which soup is the best buy considering only the amount of iron it provides?
6. Which appears to be more expensive, calories or complex carbohydrates?
7. How many grams of sodium are contained in the entire can of Noodles and Chicken?
8. For which ones of the preceding seven questions did you use ratios to help determine your answers?

Continued

Exhibit 3.1
Continued

Group C

DIRECTIONS: Answer the questions based on the lists of songs from two CD albums. Prepare a six-minute oral report to the class that explains what problem-solving strategies you used to answer the eight questions at the bottom.

RABID DOG IN CONCERT

Price: $12.98

Every Rose Ain't a Flower (2:55), *Feelin' Too Much Pain* (3:06), *Three-Woman Dog* (4:20), *Too Fine to Be Mine* (2:30), *Too Young* (3:11), *Lucy the Lucky* (2:05), *It Lasts Forever* (8:09), *After the Pigs Come Home* (3:01), *Smilin' 'Stead of Cryin'* (4:44), *My Kind of Party* (3:15), *Allison* (2:02).

SENSATIONAL SCREAMERS

Price: $11.79

Help Is on the Way (2:10), *Ib's Sticks* (9:04), *Mamma Said It's Okay* (3:29), *Comfort* (4:04), *One More Time* (3:35), *One More Time Again* (5:17), *Reasons to Live* (3:11), *Not Much More to Say* (5:12), *No-Mo-Dough* (2:45), *Mindy, Mindy* (1:55), *Not Too Much Longer* (3:45), *In Your Way* (3:00).

1. Which of the two CDs has more songs?
2. Which of the two CDs has more songs for the money?
3. Which of the two CDs has more minutes of music?
4. Which of the two CDs has more minutes of music for the money?
5. Which of the two CDs do you like better?
6. What is the average length of a song on "Rabid Dog in Concert"?
7. What is the difference in time between the longest and shortest song on the two CDs?
8. For which ones of the preceding seven questions did you use ratios to help determine your answers?

Group D

DIRECTIONS: Answer the questions based on the data given about crime in our city during the years 2000 and 2001. Prepare a six-minute oral report to the class that explains what problem-solving strategies you used to answer the eight questions at the bottom.

Year	2000	2001
Population	320,000	323,000
Murders	34	37
Rapes	99	112
Robberies	704	691
Aggravated assaults	1,001	1,120
Burglaries	3,092	3,212
Thefts by larceny	7,344	7,360
Motor vehicle thefts	1,611	1,786

1. Did crime increase or decrease from 2000 to 2001?
2. Which type of crime is the most common?
3. Relative to the size of the population, did the rate of murders go up or down from 2000 to 2001?
4. Relative to the size of the population, did the rate of burglaries go up or down from 2000 to 2001?
5. The rate relative to population of which type of crime increased the least?
6. The rate relative to population of which type of crime increased the most?
7. By how much did the population grow over the course of one year?
8. For which ones of the preceding seven questions did you use ratios to help determine your answers?

activity. Compare Case 3.7 to Case 3.8. In which one are the teacher's directions clear and to the point? In which will student engagement in the upcoming learning activity be impaired because the directions do not communicate exactly what is to be done?

CASE 3.7

As part of a lesson on pattern recognition, Ms. Bey addresses her prealgebra class as follows: "I have three books here. This one is called *Mathematical Cavalcade* (Bolt, 1992), this one is *Puzzles, Paradoxes and Brain Teasers* (Gibilisco, 1990), and this one is *Mathemagics* (Benjamin & Shermer, 1993). These books are filled with enjoyable ideas, games, and experiments to do with mathematics. There are some real surprises in store in these pages! Anyway, I want each of you to read about one of these things in here and teach the rest of us about it. Any questions? Yes, Shelton."

Shelton: Can I do the one on magic? Magic is awesome!

Ms. Bey: Yes, you may choose something out of *Mathemagics*.

Eicho: Oh, we get to pick which one we want! Suppose I want the same one as—

Ms. Bey: Please, Eicho, don't interrupt. Everyone will get a chance to choose.

Eicho: But suppose—

Ms. Bey: I'll pass these books around and you can write down the project of your choice.

Ms. Bey continues to give directions for the activity, but students are more attentive to the books being passed around. Some are reading from the books while others beckon them or pass them on. Only a few follow Ms. Bey's directions on what they are to do and the schedule for completing the activity.

CASE 3.8

In preparation for an activity on pattern recognition, Ms. Culbertson selects 28 experiments or demonstrations (one for each student in her prealgebra class) from the same three trade books mentioned in Case 3.7. She duplicates the relevant pages for each (see Exhibit 3.3).

She then develops the tasksheet displayed in Exhibit 3.4 and duplicates one for each student. She also makes a transparency slide of the tasksheet as well as the two sample entries shown in Exhibit 3.3. Based on her prior assessments of students' interests and needs, she assigns one selection to each of the 28 students. For each student she makes up a packet consisting of a copy of the selection from the books and the tasksheet.

The next day, she puts a packet in each student's desk just before the beginning of the prealgebra period. She addresses the class: "I placed a packet inside your desk—No, Rascheed, not yet. Leave it there until I tell you it's time. Thank you. Your individual packet contains a copy of a page or two out of one of these three books." She displays the three books and continues, "The pages in your packet are different from everyone else's. They explain either an experiment, a demonstration, or a game that you will teach yourself and then teach to the rest of us in the class. Here's an example." She turns on the overhead projector and quickly displays the two samples from Exhibit 3.3, keeping them on just long enough to give the students an idea of what they will be reading, but not long enough for them to read more than several words.

Ms. Culbertson then goes through the directions step by step, displaying each line from Exhibit 3.4 as she explains the step. Only after she has completed the instructions does she direct them to examine their packets.

Eight Points about Directions

Ms. Culberton's explicit directions more efficiently communicated exactly what students were to do during learning activities than did Ms. Bey's. Ms. Culbertson took advantage of the following:

- Students in classes where teachers model businesslike attitudes are more likely to efficiently follow directions than do those whose teachers seem lackadaisical and less organized. Ms. Culbertson appeared to know exactly what tasks students were expected to complete and had well-organized plans for accomplishing those tasks.
- Because giving directions is a frequent, routine occurrence in a classroom, teachers can minimize transition time, streamline communication procedures, model a businesslike attitude, and reduce the amount of teacher-talk in classrooms by establishing signals or cues that nearly instantaneously communicate certain recurring expectations to students. Ms. Culbertson's use of the overhead projector helped focus students' attention.
- Students who have learned that their teacher tends to say things only once tend to listen the first time the teacher speaks. Sometimes teachers make the mistake of saying, "I'm going to say this only once," but then end up repeating themselves because their initial directions were vague.
- Students are more likely to listen carefully to the directions of teachers who restrict their remarks to exactly what students need to know to successfully engage in the upcoming learning activity. Ms. Culbertson did not mix uninformative, inane words with directions. You do not need to make comments like, "You're really going to like this!"
- When teachers are giving directions, they are not conducting an inquiry lesson. Efficiently communicated directions in which transition time is

Exhibit 3.3
Two Examples Ms. Culbertson Chose From *Mathematical Cavalcade.*

76 Calculator golf

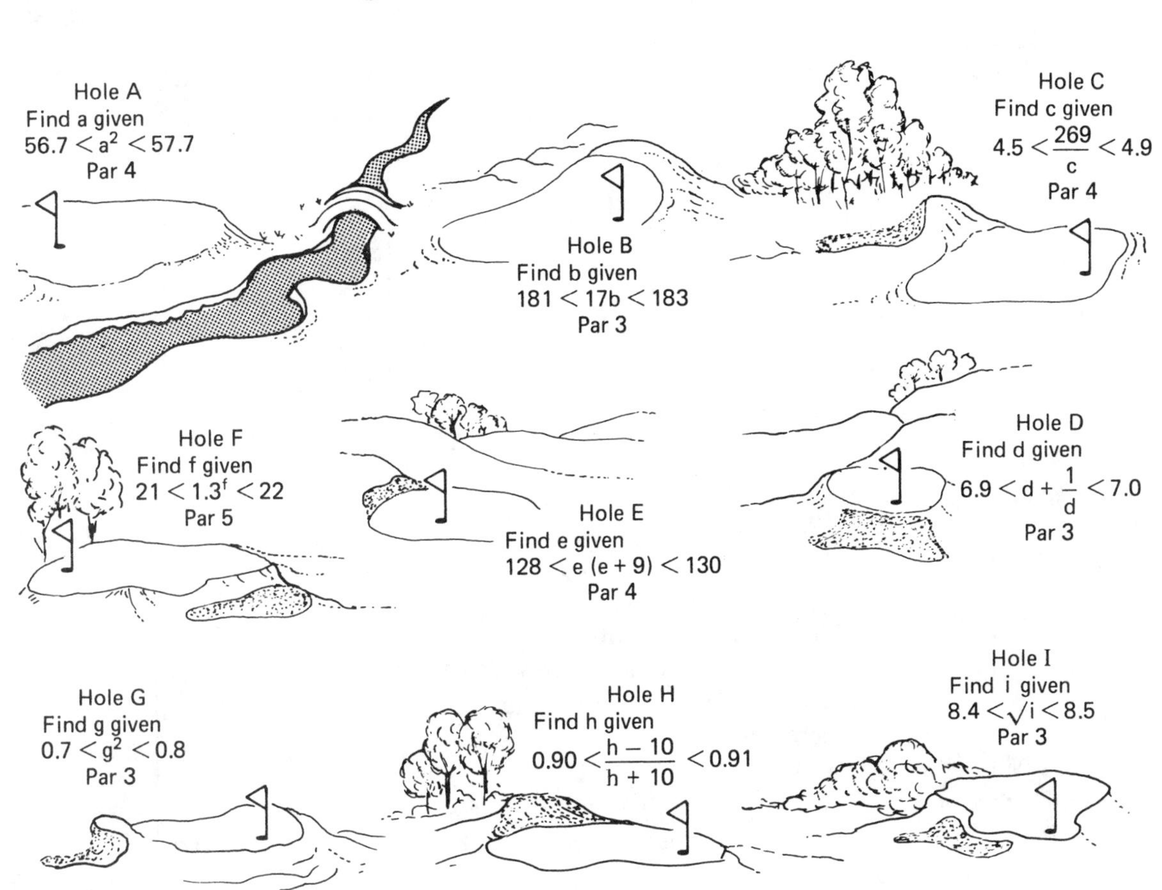

This is a game based on your ability to estimate. You will need a calculator and a means of recording your estimates. To play a hole, make an estimate of a number for the letter and use your calculator to work out the value of the calculation indicated. If this number lies between the limits indicated you will have 'holed in one'. This is unlikely, so record your estimate and the result of the calculation. From this you should be in a position to make a better estimate and get nearer to the hole.

Make further estimates and test their accuracy with your calculator until your estimate lands you in the hole between the limits. The number of estimates you require for a hole is your score.

Above is a nine-hole course with suggested pars for each hole. The par is the estimated standard score for the hole that a good player should make. Can you match par or better? Make up a similar course yourself and challenge your friends.

Continued

Source: Reprinted with permission from B. Bolt, *Mathematical Cavalcade,* pp. 40, 61

Exhibit 3.3
Continued

117 This tetraflexagon has six faces

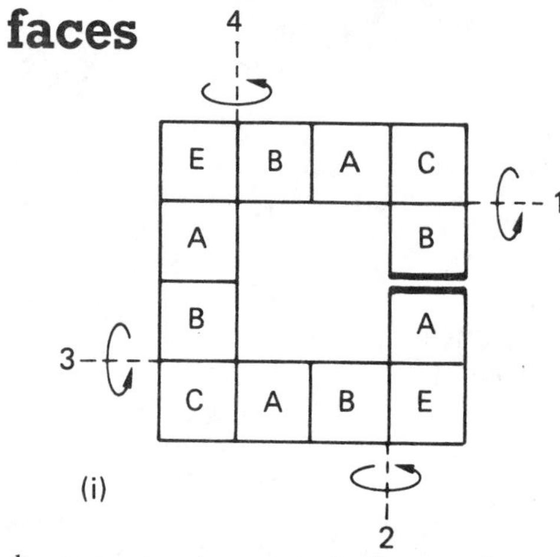

(i)

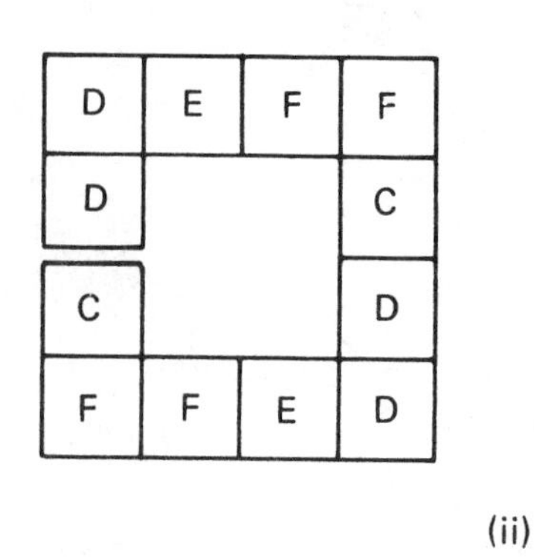

(ii)

Four-sided paper structures made from strips of paper which can be folded to expose different surfaces are known as tetraflexagons. The one discussed here can be folded to expose any one of six faces denoted here by A, B, C, D, E, F. Made with care it gives an intriguing object to manipulate. Instead of the same letter repeated four times on each exposed surface a message of four letter words can be sent, such as:

LILY WITH MUCH LOVE FROM JOHN

To make this flexagon carefully draw the band of twelve squares shown in figure (i) on plain paper. (A 3-cm edge to each square is appropriate.)

Letter the band as shown in (i), cut it out, turn it over and letter the underside as shown in (ii). Where the band has been cut between A and B the cut edges have been marked with a double line, for they will later be stuck together.

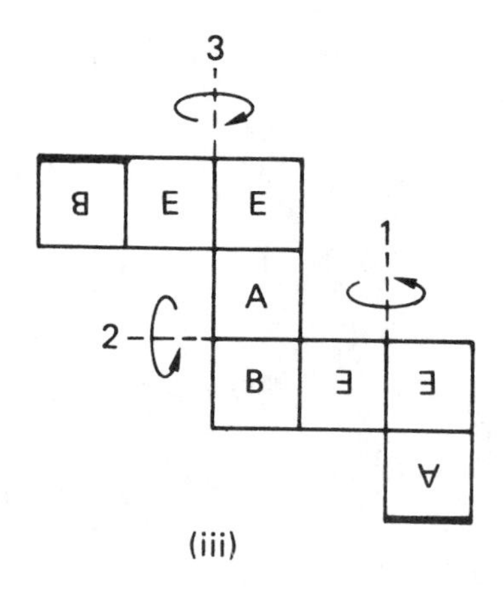

(iii)

Carefully fold the band along each of the edges of the squares so that it flexes easily. (If card is used then score all the fold lines.) Return the band to the configuration shown in (i) and fold it along the dotted lines in the order indicated. In each case fold under. The result should look like (iii).

Now fold as indicated in (iii) where the first and third folds are up over, and the second fold is under. At the third fold tuck C behind A so the result looks like (iv) and the reverse side like (v).

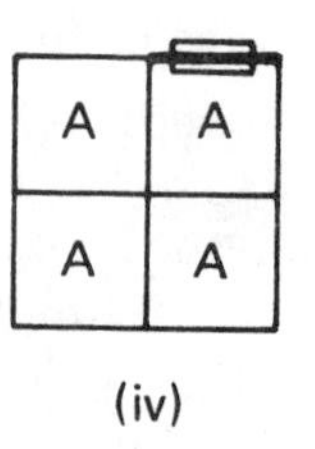

(iv)

Use a piece of sticky tape to join the top edge of the A to the edge of the B behind as shown. These should be the edges marked in (i). By folding along the horizontal mediator you will now be able to expose any of A, B, C, D and by folding along the vertical mediator find E and F.

Have fun!

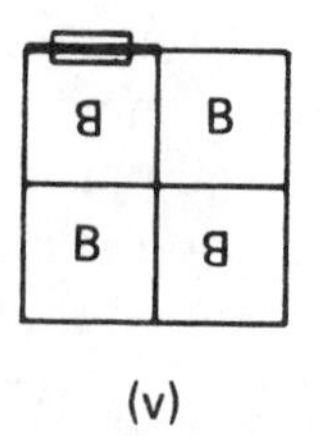

(v)

Exhibit 3.4
The Tasksheet Ms. Culbertson Uses to Communicate Directions.

You are responsible for teaching the class the experiment, demonstration, or game explained in your packet. Here is what you should do next to meet this responsibility:

1. Carefully read the page or pages in the packet. Teach yourself to do the experiment, perform the demonstration, or play the game. This will require you to do it yourself more than once. This task is to be completed by the following date: __________ .
2. Your partner's name is ______________________________ .

   During the last five minutes of today's class, meet your partner to decide on a time when the two of you can get together outside of class. You will practice teaching your experiment, demonstration, or game to your partner and your partner will practice teaching you. The task of teaching one another is to be completed by the following date: __________

   Here is a place to write down the date, time, and place you and your partner will meet:

   ____
3. Meet on the day and the time you scheduled with your partner. Practice teaching one another as indicated above. Give one another suggestions on how each of you can teach your experiments, demonstrations, or games to the whole class.
4. Plan to teach our class about your experiment, demonstration, or game. The lesson should take between 10 and 15 minutes. Use some visuals or illustrations in your lesson. Also get the class involved in doing some mathematics themselves. Be ready to explain your plan to Ms. Culbertson on the following date: __________ . You and Ms. Culbertson will schedule the date you will teach your lesson to the class. Here is a place to write it down:

   ____
5. After discussing your plan with Ms. Culbertson, teach your experiment, demonstration, or game to the class on the scheduled date.
6. After your lesson, the class will discuss with you what they learned.

minimized do not normally allow time for students to debate the pros and cons of what is to be done. Unlike Ms. Bey in Case 3.7, Ms. Culbertson did not open an opportunity to argue about the selection.

- The more senses (seeing, hearing) through which directions are communicated, the more likely students are to understand them. Besides telling her students what to do, Ms. Culbertson displayed the written tasksheet.

VARIETY OF LEARNING ACTIVITIES

Exhibit 1.3 describes part of a lesson Casey Rudd conducted on surface area. He engaged students in various types of learning activities: large-group presentation, question-discussion session, independent work, and homework. The more types of learning activities in a lesson, the more transitions you need to manage. However, today's students need the variety to remain engaged (Marks, 2000). The remainder of this chapter presents ideas for keeping students engaged during various types of sessions.

ENGAGING STUDENTS IN LARGE-GROUP PRESENTATIONS

Student Engagement During Large-Group Presentations

To be engaged in a large-group presentation, students must listen attentively to try to follow the teacher's thought pattern. Thus, they are expected to be cognitively active but physically inactive. Note taking may also be necessary. Such behavior is not readily achieved for students of any age and is virtually impossible to sustain for younger students. Presentations that continue uninterrupted for more than 10 minutes are ill-advised learning activities. In Case 3.9, the teacher's presentation style is not likely to maintain students' attention. In contrast, the teacher in Case 3.10 utilizes presentation techniques designed to obtain and maintain student engagement.

CASE 3.9

Mr. Johnson's 26 seventh-grade mathematics students are sitting at their desks. Nine have paper and pen poised to take notes; others are involved with their own thoughts as he begins, "Today, class, we're going to study a measure of central tendency called the 'arithmetic mean.' Some of you may have already heard of it." Turning away from the class to write on the whiteboard, he continues to speak, "The arithmetic mean of *N* numbers equals the sum of the numbers divided by *N*." Turning his side to the class, he says, "For example, to compute the mean of these numbers—15, 15, 20, 0, 13, 12, 25, 40, 10, and 20—we would first add the numbers to find the sum. Right?" He looks at the class but does not notice whether students appear to respond to his question, turns back, and adds the numbers on the board. Turning to the class, he says, "So the sum is 170. Now, because we have 10 numbers, *N* in the formula is 10 and we divide 170 by *N*, or 10. And what does that give us? It gives us 17.0. So, the arithmetic mean of these numbers is 17.0. Is that clear?"

Mr. Johnson stares at the class momentarily, notices Armond nodding and softly saying, "Yes." With a smile, Mr. Johnson quickly says, "Good! Okay, everybody, the arithmetic mean is an important and useful statistic. Suppose, for example, I wanted to compare the following group of numbers (from his notes he copies the following numbers on the board: 18, 35, 30, 7, 20) to these over here." He points to the previous data sequence.

Mr. Johnson: What could we do, Ramon?

Ramon: Compute the arithmetic mean you told us about.

Mr. Johnson: That's right! We could compute the arithmetic mean. $18 + 35 + 30 + 7 + 20 = 110$, and $110 \div N$ - which in this case is 5—okay?—is 22.0. Okay? Now that means this second data set has a higher average than the first, even though the first sequence has more numbers. Any questions? [pause] Good! [pause] Oh, okay, Angela?

Angela: Why do you write "17.0" and "22.0" instead of "17" and "22"? Aren't they the same?

Mr. Johnson: Good question! Hmmm, can anybody help Angela out? [pause] Well, you see in statistics the number of decimal places indicates something about the accuracy of the computations, and for that matter, the data-gathering device. So that one decimal point indicates that the statistics are more accurate than if we had written just "17" and "22" and not as as accurate if we had written, say "17.00000" or "22.00000." Got it? That was a good question. Do you understand now?

Angela: I guess so.

Mr. Johnson: Good! Now, if there are no more questions, there's some time left to get a head start on your homework.

CASE 3.10

Ms. Erickson's 27 seventh-grade mathematics students are quietly sitting at their desks, each ready with paper and pen. She has previously taught them how to take notes during large-group presentations so that they record information during the session on paper and then, after the session, organize the notes and transfer them into their required notebooks.

After distributing a copy of Exhibit 3.5's tasksheet to each student, she faces the class from a position near the overhead projector and says, "I'm standing here looking at you people and I just can't get one question out of my mind." Very deliberately, she walks in front of the fourth row of students and quickly, but demonstratively, looks at their feet (see Exhibit 3.6). Then she moves in front of the first row and repeats the odd behavior with those students. "I just don't know!" she says shaking her head as she returns to her position at the overhead.

She turns on the overhead, displaying the first line of Exhibit 3.5, and says, "In the first blank on your copy please write: Do the people sitting in the fourth row have bigger feet than those in the first row?" She moves closer to the students to monitor how they follow directions. Back at the overhead as they complete the task, she says, "Now, I've got to figure a way to gather data that will help me answer that question." Grabbing her head with a hand and closing her eyes, she appears to be in deep thought for a few seconds and then suddenly exclaims, "I've got it! We'll use shoe sizes as a measure. That'll be easier than using a ruler on smelly feet!" Some students laugh, and one begins to speak while two others raise their hands. But Ms. Erickson quickly says, "Not now please, we need to collect some data." She flips an overlay off the second line of the transparency, exposing "Data for Row 4."

Ms. Erickson: "Those of you in the fourth and first rows, quickly jot down your shoe size on your paper. If you do not know it, either guess or read it off your shoe if you can do it quickly. Starting with Jasmine in the back and moving up to Lester in front, those of you in the fourth row call out your shoe sizes one at a time so we can write them down in this blank at our places." As the students call out the sizes, she fills in the blank on the transparency as follows: 6, 10.5, 8, 5.5, 6, 9. Exposing the next line, "Data for Row 1," on the transparency she asks, "What do you suppose we're going to do now, Pauline?" Pauline: "Do the same for Row 1." Ms. Erickson says, "Okay, you heard Pauline; Row 1, give it to us from the back so we can fill in this blank." The numbers 8.5, 8, 7, 5.5, 6.5, 6.5, 9, and 8 are recorded and displayed on the overhead.

Exhibit 3.5
Tasksheet Ms. Erickson Uses During a Large-Group Presentation.

An Experiment

Question to be answered: ____________________

____

Data for Row 4: ____________________

Data for Row 1: ____________________

Treatment of Data for Row 4:

Treatment of Data for Row 1:

Treatment to compare the two data sequences:

Results: ____________________

____

Conclusions: ____________________

____

Exhibit 3.6
Why Is Ms. Erickson Looking at Students' Shoes?

Ms. Erickson: "Now, I've got to figure out what to do with these numbers to help me answer the question." Several students raise their hands, but she responds, "Thank you for offering to help, but I want to see what I come up with." Pointing to the appropriate numerals on the transparency, she seems to think aloud saying, "It's easy enough to compare one number to another. Jasmine's 6 from Row 4 is less than Rolando's 8.5 from Row 1. But I don't want to compare just one individual's number to another's. I want to compare this whole bunch of numbers (circling the set of numbers from Row 4 with an overhead pen) to this bunch (circling the numbers from Row 1). I guess we could add all the Row 4 numbers together and all the Row 1 numbers together and compare the two sums—the group with the greater sum would have the bigger feet."

A couple of students try to interrupt with, "But that won't work because—" but Ms. Erickson motions them to stop speaking and asks, "What's the sum from Row 4, Lau-Chou?"

Lau-Chou: 45.

Ms. Erickson: Thank you. And what is the sum for Row 1, Stace?

Stace: 59.

Ms. Erickson: Thank you. So Row 1 has bigger feet because 59 is greater than 45. She writes, "59 > 45" on the transparency.

Ms. Erickson: I'll pause to hear what some of you with your hands up have to say. Evangeline?

Evangeline: That's not right; it doesn't work.

Ms. Erickson: You mean 59 isn't greater than 45, Evangeline?

Evangeline: 59 is greater than 45, but there are more feet in Row 1.

Ms. Erickson: All the people in Row 1 have two feet each just like the ones in Row 4. I carefully counted. Now that we've taken care of that concern, how about other comments or questions? Brooke.

Brooke: You know what Evangeline meant! There are more people in Row 1. So what you did isn't right.

Ms. Erickson: Alright, let me see if I now understand Evangeline's point. Evangeline said we don't want our indicator of how big the feet are to be affected by how many feet—just the size of the feet. So, I've got to figure a way to compare the sizes of these two groups of numbers when one has more numbers than the other. I'm open for suggestions. [pause] Kip?

Kip: You could drop the two extra numbers from Row 1; then they'd both have six.

Ms. Erickson: That seems like a reasonable approach. I like that. But first let's hear another idea—maybe one where we use all the data. Myra?

Myra: Why not do an average?

Ms. Erickson: What do you mean?

Myra: You know, divide Row 4's total by 6 and Row 1's by 8.

Ms. Erickson: How will that dividing help? Seems like just an unnecessary step. Tom.

Tom: It evens up the two groups.

Ms. Erickson: Oh, I see what you people have been trying to tell me! Dividing Row 4's sum of 45 by 6 counts each number 1/6. And dividing Row 1's sum of 59 by 8 counts each number by 1/8. And that's fair, because

Topic: Proof by induction
Date: 3/7

Presentation Outline

I. Review of familiar methods of proving theorem
  A. Direct
  B. By contradiction
II. Types of theorems to which proof by induction applies
III. Logic of a proof by induction
  A. Sequential cases
  B. Is it true for one case?
  C. If it is true for one case, will it be true for the next case?
IV. Some everyday examples of the induction principle
  A. Playing music
  B. On soccer field
  C. In the kitchen
  D. Eating food
  E. Computer programming
V. An example with an arithmetic series, $\sum_{i=1}^{n} = n(n+1)/2$
VI. Formalizing the process
  A. Show the statement is true for $i = a$.
  B. Show that if the statement is true for some value of i, then it must also be true for $i + 1$.
  C. Draw a conclusion.
VII. Proof of the following theorem:

$$\sum_{i=1}^{n} i^2 = (n/6)(n+1)(2n+1)$$

VIII. Summary

Exhibit 3.7
Example of an Outline Distributed to Students for a Large-Group Presentation.

six one sixths is a whole just as eight one eighths is a whole. How am I doing, Jasmine?
Jasmine: A lot better than you were.

Removing another overlay on the transparency, Ms. Erickson displays the next two lines from Exhibit 3.5 and continues.

Ms. Erickson: Let's write: "The sum of Row 4's numbers is 45." 45 ÷ 6 is what, Lester?
Lester: 7.5.
Ms. Erickson: Thanks. And on the next line we write "59 ÷ 8." Which is what, Sandy?
Sandy: 7.375.
Ms. Erickson: Since 7.5 is greater than 7.373, I guess we should say that the feet in Row 4 are larger than the feet in Row 1. That is, of course, if you're willing to trust this particular statistic—which is known as the "MEB." Any questions? Yes, Evangeline.
Evangeline: Why the MEB?
Ms. Erickson: Because I just named it that after its three inventors, Myra, Evangeline, and Brooke. They're the ones who came with the idea of dividing the sum.

Ms. Erickson shifts to direct instruction to help students remember the formula, practice using it, and remember its more conventional name, "arithmetic mean."

Ten Points About Large-Group Presentations

Consider the following thoughts when designing large-group presentations:

- Students are more likely to be engaged during a presentation if the teacher has provided clear directions for behavior. Students need to have learned how to attend to a presentation. Questions about how to take notes, if at all, should be answered before the presentation begins.
- Some sort of advanced organizer to direct students' thinking helps them actively listen to the speaker. Ms. Erickson used Exhibit 3.5's tasksheet to focus students' attention and structure the activity. Consider taking that idea a step further by having an outline of the presentation (see Exhibit 3.7) or a session agenda (see Exhibit 3.8) in the hands of students or displayed on an overhead transparency. You can then use it to direct attention and provide a context for topics and subtopics. Having such advanced organizers in students' hands facilitates their note taking and helps you monitor their engagement (e.g., by sampling what they write on the form in Exhibit 3.9). By using transitional remarks such as "Let's move on to Item 4" in

Exhibit 3.8
Example of a Class Meeting Agenda Distributed to Students.

Meeting Agenda for 10/18
Algebra II, 4th Period

1. Hello (0.50–0.75 minutes)
2. Formative quiz and roll (15–16 minutes)
3. Review quiz items and discuss some subset of the following questions, as needed according to the quiz review (10–20 minutes):
   a. What is a proportion?
   b. What types of problems does setting up proportions help solve?
   c. What is the so-called proportion rule?
   d. What are some efficient strategies for estimating solutions to problems involving proportions?
4. Lecture presentation on further applying our understanding of proportions to problems involving direct and indirect variations (10–15 minutes)
5. Homework assignment (2–4 minutes)
6. Head start on homework (7–21 minutes)
7. Prepare for dismissal (1–1.75 minutes)
8. Be kind to yourself (1310 minutes)

conjunction with an outline or agenda, you help students maintain their bearings during sessions.

- Signals, especially nonverbal ones, can efficiently focus students' attention during a presentation. Ms. Erickson's curious behavior in Case 3.10—deliberately staring at feet—encouraged students to take notice and wonder, "What is she going to do next?" Her deliberate movements in the first part of the presentation established cues she took advantage of in the remainder of the session. For example:

  > After distributing the form, she walked directly to a point near the overhead projector and faced the class. From that position she spoke to the students. Silently, she walked directly to a point in front of the fourth row, then to a point in front of the first row, and then back to the position near the overhead projector where she once again spoke to the class. When she wanted students to look at an illustration, she switched on the overhead projector. When she wanted them to stop looking at it, she turned the overhead projector off or controlled what they could see with transparency overlays.

  These movements cued students to associate her location and movements with what they should be doing throughout the session (e.g., listen attentively when she is by the overhead projector).
- Presentations are useful learning activities when teachers want to have a group of students all following a common thought pattern. Lectures (e.g., Ms. Erickson's) that are designed to do more than just feed information to students run the risk of becoming question-discussion sessions. Thus, some means for staying on track should be considered when planning the presentation. One method is to have signals worked out with students so that they clearly discriminate between times when you are only lecturing and other times when discussions or questions are welcomed. While standing by the overhead projector, Ms. Erickson presented the task to be addressed, collected information, and focused thoughts on examining information. During this first part, she had students speak, but they did not enter into a discussion session. They simply provided information used in the presentation.
- Voice volume, inflection, pitch, rhythm, and pace should be strategically modulated according to the message you want to send and to the level of students. Even when the message itself is important and exciting, monotone speech is a recipe for boredom. Punctuate key sentences with voice variations. Follow key statements and questions with strategic pauses. Pauses indicate points to ponder. Pace your speech so that sessions move briskly but still allow students enough time to absorb your messages and take notes. The type of lesson you are teaching should, of course, influence pace. A lecture for an inquiry-learning activity would ordinarily proceed at a slower pace than one using direct instruction. Quina (1989, p. 143) suggested that between 110 and 130 words per minute is optimal.
- Students are more likely to follow presentations that utilize professional-quality media and technology. Computerized multimedia presentations are now cost effective for everyday classroom use.
- At least five advantages can be gained by videotaping presentations ahead of time: (a) Videotaped presentations avoid some of the interruptions in thought that occur when students make comments or ask questions. (b) The teacher can more attentively monitor students' behavior and

Exhibit 3.9
Example of a Note-Taking Form Distributed to Students for Use During a Large-Group Presentation.

Topic: Developing the quadratic formula
Date: 1/23

Main Ideas	Margin Notes
The need for a general method:	
Completing-the-square method:	
The need for an easier method:	
Completing-of-the-square example:	
Generalizing completing-the-square method:	
Reforming expressions to obtain the formula:	
Examples of equations solved with the formula: 1.	
2.	
3.	
Summary:	
What to do next:	

effectively respond to indications of disengagement. (c) Flaws in the presentation can be corrected and improvements made before the presentation is played for the class. (d) Recorded presentations can be edited and reused. (e) With videotape and other record-and-play devices, teachers can easily start, interrupt, replay, and pause presentations.

- Entertaining is not teaching. However, interjecting a bit of humor or other attention-getting devices helps keep students more alert than does a straight monologue. But be careful that attention-getting devices do not distract attention away from the objective of the lesson.
- Mind wandering and daydreaming are major causes of student disengagement during presentations. By making eye contact with individual students and incorporating students' names into your presentations, you cue students to pay attention. Punctuating explanations with focused questions encourages students to stay alert and helps you to monitor their comprehension. Asking students vague questions like, "Do you understand?" is unlikely to produce responses that are indicative of what students understand or misunderstand. But focused questions (e.g., "Why did we decide to divide the sum by 12?") provide you with formative feedback to help you regulate your presentation.
- You are in a far better position to detect what students are doing and thinking when you are circulating among them than when you are confined to a single area of the classroom (e.g., at the board or behind an overhead projector). Thus, consider organizing large-group presentations so that you are free to move about the classroom as you conduct the session. During Ms. Snyder's presentation in Case 3.11, she is hardly in a position to monitor her students' engagement and comprehension. By contrast, Mr. Heaps makes a presentation in Case 3.12 in a way that he readily monitors his students' engagement and learning:

CASE 3.11

Ms. Snyder has designed an integrated mathematics and science lesson for the purpose of leading her sixth-grade students to achieve the following objective:

> Students discover that the speed of a moving object can be measured by dividing the distance the object traveled by the time it took the object to travel the distance (discover a relationship).

The lesson plan calls for Ms. Snyder to use an overhead transparency to display a path with equally spaced grid lines and to move a tiny image of a car at a slow, steady speed along the path as depicted by Exhibit 3.10. She will have a student use a stopwatch to time how long it takes the image to cross a number of grid lines. Then she will repeat the demonstration several times, but with each trial she either moves the image at a different steady speed or varies the number of grid lines the image crosses. All along, she plans to raise questions and provide explanations that will lead students to discover the relationship specified by the objective. By recording the number of lines crossed (i.e., the distance) and the time it took for each trial on the board, she hopes to get students to make comparisons demonstrating that the faster the image moves, the greater the ratio of distance to time.

Ms. Snyder is now standing in front of the class conducting the planned demonstration on the overhead projector. See Exhibit 3.11. "Now count how many lines the car crosses this time," she directs the class as Yoshi operates the stopwatch. [pause] "There, eight lines that time. What was the time, Yoshi?" she asks. Yoshi: "A little over 12 seconds." Ms. Snyder: "So that time the car took about 12 seconds to cross eight lines. How does that compare to the last trip when it took about 17 seconds to cross eight lines?" Manny quickly answers loudly, "This time the car went faster." "I agree with Manny. The car went faster this time because it took less time to cross the same number of lines," Ms. Snyder says as she turns to add the results of this latest trial to the data already on the whiteboard. Based on Manny's response, she thinks that he answered "faster" because he recognized that the car crossed just as many lines in less time. However, while she is writing on the board and starting the next trial, she does not hear the following exchange between Manny and

Exhibit 3.10
Ms. Synder Uses an Overhead Transparency to Demonstrate the Speed of a Moving Object.

Frank who is sitting next to him off to one side of the room:

Frank: How did you know the answer?
Manny: Couldn't you see her move her hand faster that time. I didn't count no lines. I don't know what she's talking about.

Occupied at her place in the front of the room, Ms. Snyder also did not notice Rosalina in the back of the room starting to raise her hand when Manny blurted out his answer. If Rosalina had the opportunity to respond, she would have said, "The time before, the car took 17 seconds to go eight lines. That's about 2 seconds for one line. But this time, the car took only 12 seconds to go eight lines; that's like 1 1/2 seconds for one line."

Consequently, Ms. Snyder makes an erroneous formative judgment about Manny's progress and misses an opportunity for feedback relevant to Rosalina's understanding. More importantly, the whole class would have profited from a discussion stimulated by Rosalina's thoughts had she shared them.

After completing all the trials using the overhead, she summarizes:

Ms. Synder: So you see we find out the distance by counting the number of lines crossed and the time with our watch. Then the rate of speed is the distance, or number of lines crossed, divided by the time. When the car crossed 15 lines in 5 seconds, its speed was 3 lines per second. What would its speed be if it took 10 seconds to go 15 lines?
Mindy: 1.5 lines per second.
Ms. Snyder: Does everyone understand how Mindy found the speed?"

Some students nod positively; others stare straight ahead or look down.

Ms. Snyder: Any questions? Yes, Parker.
Parker: I see how Mindy got 1.5, but I don't know why.
Ms. Snyder: Because the rate of speed is the distance, or number of lines crossed, divided by the time. Do you understand, now?
Parker: Uhh,—okay, yeah.

Throughout the activity, Ms. Snyder was so busy from her place by the overhead projector and the whiteboard that she missed a myriad of indicators of who was on-task, who was off-task, who had insightful comments to make, who was following the discussion, and who got lost in the process.

CASE 3.12

Mr. Heaps devises a lesson plan for his sixth-grade students that targets the same objective as Ms. Snyder's in Case 3.11. His approach is similar to hers. But instead of standing in the front of the classroom by the overhead

Exhibit 3.11
Ms. Snyder Misses Some Important Opportunities for Unplanned Measurements.

projector and whiteboard, he has a student, Nakisha, operate the overhead projector with the gridded pathway and image of a car. Another student, Jorge, records data from the trials on the whiteboard while Jason operates the stopwatch as Yoshi did in Case 3.11.

Because he is not busy at the overhead in front of the room, Mr. Heaps is free to move among the students as shown by Exhibit 3.12. He gives step-by-step directions to Nakisha on how to perform each trial. Everyone in the classroom also hears the directions, thus, making it easier for them to understand the procedure. Furthermore, he has the students at their places respond to the prompts on the tasksheet shown by Exhibit 3.13.

At one point in the activity, Mr. Heaps directs Nakisha: "For this third trial, start the car at the second line on the path and move it at a steady pace for 10 seconds. Wait until Jason gives you the signal to start. The rest of us will count how many lines the car crosses." As the trial is performed, Mr. Heaps observes who is following the procedure; then he directs the class to fill in the data for the third trial on the tasksheets. He quickly surveys a sample of the students' responses, noting that most wrote down "18" and "10 seconds," as did Jorge on the whiteboard, but that Brett wrote nothing for Trial #3 and Mitsuko wrote "10" for the number of lines crossed and "18" for the time. Quickly and discreetly, he puts his finger on the blanks on Brett's tasksheet and then points to the data on the whiteboard. Quietly, he sends Mitsuko a hand signal indicating that she should transpose the two numbers. He notes that they both comply with his directions—directions that hardly anyone else in the class noticed.

As students work on the tasksheet after the fourth trial, Mr. Heaps walks around the room sampling their responses and watching them use their calculators in response to the last four-part prompt of the tasksheet. Exhibit 3.14 contains three of the completed tasksheets he read.

At this stage of the lesson, he expects only a few students to have discovered that speed can be measured by dividing the number of lines crossed by the number of seconds. He sees that Woofa has used her calculator to compute the actual number of lines per second as indicated by the numbers at the bottom of her tasksheet. But he also interprets her explanations of why the speed was fastest in the third trial and her choice of Trials #1 and #2 for the slowest as indicative of a misconception. Apparently, she only considered number of lines crossed without considering how much time it took to cross those lines. But yet, she computed the correct rates.

From Julie's explanations of her choices for fastest and slowest trials, Mr. Heaps judges that she has developed very sophisticated insights about the targeted relationship. However, he is confused by the numbers written at the bottom of her sheet. Instead of the expected higher numbers for higher speeds, she has lower numbers for higher speeds.

Exhibit 3.12
Mr. Heaps Conducts the Large-Group Presentation So That He Is Free to Monitor Students' Engagement.

Trial #1:

Number of lines crossed __________

Time in motion __________

Trial #2:

Number of lines crossed __________

Time in motion __________

Trial #3

Number of lines crossed __________

Time in motion __________

Trial #4

Number of lines crossed __________

Time in motion __________

During which of the four trials did the car travel fastest? __________

Why do you think your answer is correct? __________

During which of the four trials did the car travel slowest? __________

Why do you think your answer is correct? __________

Write a number that indicates the speed of the car during:

Trial #1 __________ Trial #2 __________

Trial #3 __________ Trial #4 __________

Exhibit 3.13
Tasksheet Mr. Heaps' Students Use as They Follow the Demonstration Jorge Conducts in Front of the Room.

Saul's responses suggest to Mr. Heaps that Saul reasoned that the greater the number of seconds, the faster the speed. Because he judges Saul's progress with the lesson's objective to be behind the progress of the vast majority of the class, he decides not to discuss Saul's responses at the moment, but rather to work with Saul individually at a more convenient time.

He begins a class discussion by having Julie read her explanations for her choices of Trial #3 as the fastest and Trial #1 as the slowest. A comparison of Julie's responses to those of some other students leads the class to agree that speed is a function of distance and time. Then Mr. Heaps asks Julie to explain what he considers mysterious numbers at the bottom of her sheet. She says, "My numbers tell us the average number of seconds the car took to cross one line." Now Mr. Heaps realizes that her computations are perfectly consistent with her explanations. Each of the numbers is a ratio of time to distance rather than the ratio of distance to time as he had anticipated.

A very productive discussion ensues with most students recognizing that either Julie's method or the more conventional method provide equally accurate measures of speed. Rather than raise general vague questions, such as, "Does everybody understand?", Mr. Heaps directs focused queries at specific students to gauge

Exhibit 3.14
Woofa's, Julie's, and Saul's Tasksheets.

Woofa's

Trial #1:

Number of lines crossed 8

Time in motion 15.4 sec

Trial #2:

Number of lines crossed 8

Time in motion 7 1/2 sec

Trial #3

Number of lines crossed 18

Time in motion 10.0 sec

Trial #4

Number of lines crossed 15

Time in motion 9.0 sec

During which of the four trials did the car travel fastest? #3

Why do you think your answer is correct? Because it crossed the most lines.

During which of the four trials did the car travel slowest? #1 and #2

Why do you think your answer is correct? #1 and 2 because

Write a number that indicates the speed of the car during:

Trial #1 .52 Trial #2 1.1

Trial #3 1.8 Trial #4 1.7

Continued

their progress. For example, he asks, "Whereas Woofa's numbers are different from Julie's, how could they both be equally good indicators of how fast the cars were going?"

RESPONDING TO STUDENTS' QUESTIONS

Memory-Level and Reasoning-Level Questions

Questions that can be answered by remembering previously learned responses are *memory-level questions.* Questions that require respondents to reason and make judgments to answer are *reasoning-level questions.*

Engage in Activity 3.1:

Activity 3.1

Analyze the cognitive processes a student would use to answer the following questions; classify each as either *reasoning level* or *memory level:*

A. Who wrote the book *Fermat's Enigma?*
B. What is a polyhedron?

Exhibit 3.14
Continued

Julie's

Trial #1:

Number of lines crossed 8

Time in motion 15.4 seconds

Trial #2:

Number of lines crossed 8

Time in motion 7.5 seconds

Trial #3

Number of lines crossed 18

Time in motion 10 seconds

Trial #4

Number of lines crossed 15

Time in motion 9 seconds

During which of the four trials did the car travel fastest? 3

Why do you think your answer is correct? Because it took only about half a second to cross a line. The others took longer for a line. #4 took just a little longer #2 took about 1 second and #1 took about 2 seconds

During which of the four trials did the car travel slowest? #1

Why do you think your answer is correct? Because of what I said above

Write a number that indicates the speed of the car during:

Trial #1 1.92 Trial #2 .93

Trial #3 .55 Trial #4 .6

Continued

C. Would it be quicker for you to find the real roots of the equation, $15x^2 + 2x + 8 = 0$, by factoring, completing the square, or the quadratic formula?
D. In chapter 5's Exhibit 5.11, how are the examples alike? How do they differ from the nonexamples?
E. What type of graph is best for comparing the first-year salaries of college graduates with mathematics education majors who accept various positions outside of teaching to those who accept teaching positions? (The National Commission on Mathematics and Science Teaching for the 21st Century, 2000, p. 36)
F. Have you ever proven the division algorithm (i.e., $(\forall\ (a, b) \ni (a \in \{\text{integers}\}$ and $b \in \{\text{natural numbers}\})\ \exists!$ (an ordered pair of integers $(q, r) \ni a = bq + r$ with $0 \leq r < b$)?
G. If you have not already done so, how long would it take you to develop a proof for the division algorithm?
H. Does $x^2 + 4x = 18$ have rational roots?
I. Why are so few women mentioned in historical accounts of mathematics prior to the 20th century?
J. Who was Maria Gaetana Agnesi?

Compare your classifications with those of a colleague. Resolve any differences in a discussion.

Exhibit 3.14
Continued

Saul's

Trial #1:

Number of lines crossed 8

Time in motion 15.4 seconds

Trial #2:

Number of lines crossed 8

Time in motion 7.5 seconds

Trial #3

Number of lines crossed 18

Time in motion 10 seconds

Trial #4

Number of lines crossed 15

Time in motion 9 seconds

During which of the four trials did the car travel fastest? 1

Why do you think your answer is correct? 15.4 is the most times

During which of the four trials did the car travel slowest? 9

Why do you think your answer is correct? 9 seconds is the shortest

Write a number that indicates the speed of the car during:

Trial #1 ? Trial #2 ____

Trial #3 ? Trial #4 ____

In Case 3.13, two preservice mathematics teachers discuss their classifications of Activity 3.1's questions:

CASE 3.13

Tina: I classified questions A, B, F, H, and J as memory level; the others as reasoning level.

Jermain: Yours agree with mine except for H. Why don't you think determining if a quadratic equation has rational roots involves reasoning? You don't just remember whether it does or not; you have to figure it out.

Tina: True, but I walked through the process I used to answer the question as, "No, it has two irrational roots." And what I did was to simply remember that one method for determining the nature of the roots of a quadratic equation is to compute the discriminant. I did that by remembering the expression $b^2 - 4ac$, and going through the algorithm for calculating its value for $a = 1$, $b = 4$, and $c = -18$. That gave me 88, which isn't a perfect square. And I remembered the rule: If the discriminant is positive, but not a perfect square, the equation has two irrational roots. I just used an algorithmic skill—not any reasoning.

Jermain: I guess it all depends on where you're coming from. I can see that for experienced people—like us—who have addressed the question for numerous quadratic equations repeatedly, there's virtually no conscious reasoning. So maybe I classified question H as reasoning level because I was thinking of a less experienced person who had to think through why the discriminant rule works. But even for us—who do this fairly mindlessly—didn't we use at least some deductive reasoning to determine the discriminant applied to this case—you know, the syllogism thing?

Tina: You've got a point, but it just seems so automatic. We can agree that for most people—surely the inexperienced algebra student—Question H is reasoning level.

Jermain: Some of these other questions might also be reclassified if you viewed them from different backgrounds.

Tina: Yeah, like Question I: If all one did was to remember an opinion expressed by a mathematical historian, there wouldn't be much reasoning.

Your strategies for addressing students' reasoning-level questions should differ from those for answering memory-level questions. If you are in the midst of an inquiry lesson for a higher cognitive reasoning-level objective (i.e., construct a concept, discover a relationship, comprehension and communication, application, or creative-thinking objective as explained in chapters 4, 5, 7, and 8) when students raise a reasoning-level question, you may want to lead students to develop their own answers to the questions rather than answering the question yourself. On the other hand, you may be inclined to be more direct when students raise memory-level questions—especially during lessons for memory-level objectives (i.e., simple knowledge or algorithmic skill as explained in chapters 4 and 6).

Student-Initiated Reasoning-Level Questions

Imagine yourself as the teacher in Case 3.14:

CASE 3.14

You are conducting a class session as part of an application-level lesson on compound probability. One student, Jennifer, asks you a question:

Jennifer: In biology class, we learned about this genetic thing—a defect where if the parents had it, there's a 25% chance that a kid they have will also have it. What if they have two kids, what are the chances that one or both will have the defect? Is it 50% or is it still 25%?

Jennifer's question in Case 3.14 is answerable via a cognitive process—namely deductive reasoning—that involves more than just recalling a response. Being in the middle of an application-level lesson dealing with the mathematical content of Jennifer's question (i.e., compound probability), you want to respond in a way that will advance the cause of the lesson. You have at least four options: (a) *answer the question directly,* (b) *use a think-aloud strategy,* (c) *probe back to the student,* and (d) *probe and redirect the question to other students.*

Answering the question directly is the easiest and quickest option. Give the answer and, because it is a reasoning-level question, explain the rationale for your answer. Case 3.14.1 is an example:

CASE 3.14.1

You: I think the probability would be somewhere between 25% and 50%. Think of the probability of neither of the two children inheriting the defect. The chance of the first not having it is 75%; the chance of the second is also 75%. It's a compound probability of two independent events, so we should multiply .75 by .75. And that would be about .56. So, if there is about 56% chance of neither having the defect, there must be about 44% chance that one or both will have it.

Rather than explaining the answer directly, you may choose to spend more time demonstrating what goes though your mind when formulating a solution by employing a *think-aloud strategy.* For Jennifer's question, you would be modeling the application-level thinking you want the students to learn. Case 3.14.2 is an example:

CASE 3.14.2

You: Wow! That's a challenging question, Jennifer. Let's see, how should I go about solving that one? There will be two children, an older one and then a second. The probability that the first will have the defect is 25%. Isn't that what you said? [pause] Thanks. So, what's the probability that the second child will inherit the defect? Let see, [pause] that's not affected by whether or not the older sibling has the defect so, the probability of the second child inheriting the defect is also 25%. With two children the likelihood that at least one gets the defect must be greater than the likelihood of one getting the defect if there is only one child. So, the probability is going to be greater than 25%. So for two children, doubling the probability of one child having the defect gives us a probability of 50%. Is that right? With that reasoning the probability for three children would be 75%, 100% for four children, and 125% for five children. Ouch! That can't be right. No matter how many children there will be, I can't believe that we have certainty or the impossibility of more than 100%

probability. I've got to rethink this question case by case. What's the probability of the first inheriting the defect and the second not? The chances are .25 for the first to inherit and .75 for the second not to inherit it. The two events are independent, so the multiplication rule applies. $.25 \times .75 \approx .19$. That takes care of one case. Now, what about the first not having it, but the second one does—that' another .19. What's left? Both of them having it. That's $.25 \times .25 \approx .06$. Those are all the cases in which at least one child inherits the defect. Because the three cases are mutually exclusive, we add the three probabilities and get about 44%. Does that seem right? I think so [pause] Oh! I just thought of a quicker way to come up with that answer! We could have just computed the probability of neither child inheriting the defect. That would be $.75 \times .75 \approx .56$. So subtracting 56% from 100% gives us the probability of at least one child getting the defect. And that comes out to the same 44% that we computed when we did it the longer way.

Even more time consuming, but possibly more beneficial for the student, is to *probe back to the student*. With this strategy you respond to the student's question with a sequence of your own questions that prompts the student to engage in the type of reasoning demonstrated by Case 3.14.2. In Case 3.14.3, you probe back to Jennifer:

CASE 3.14.3

You: Why do you think it might be 50%?

Jennifer: There's a 25% chance the first one will have it, right?

You: Right.

Jennifer: And 25% for the second one. So, that's another 25% chance—giving you a total of 50%.

You: Then what would the probability of at least one child having the defect if the family were going to have three children?

Jennifer: 75%.

You: What if they were going to have four children?

Jennifer: 100%.

You: And if they were going to have five children?

Jennifer: Ahh [pause] $.25 \times 5$ is [pause] that's impossible; 125% is impossible!

You: So, do we need to rethink our approach?

Jennifer: We have to.

You: What about a case-by-case strategy?

Jennifer: Like what?

You: Here; come up to the overhead projector to illustrate your thinking for the whole class.

You continue with your questions to guide Jennifer's thought processes.

To involve more students in the activity, a fourth option is to *probe and redirect the question to other students*. With this strategy, you respond with a sequence of leading questions but direct them to other students as well as the one who raised the initial question. You use this strategy for Jennifer's question in Case 3.14.4:

CASE 3.14.4

You: Which seems like the probability—25% or 50%, Wade?

Wade: It's got to be more than 25% because that's the probability if they had only one child.

You: So does that mean the probability is 50%, Agnes?

Agnes: That's all that's left.

You: Okay, class, assuming we agree with Agnes, compute the probability of at least one child having the defect if the family is going to have three children instead of only two children. And then compute the probability if they're going to have four children. Then compute the probability if they're going to have five children. Hold on to your results until I see that everyone has had an opportunity to finish.

You walk around the room as the students compute. Seeing that nearly all are ready to continue the questioning session, you get the class' attention:

You: Share your results with us, Jennifer.

Jennifer: 75%, 100%, and 125%—but that's impossible because you can't have 1.25 for a probability.

You: So should we try another strategy? Okay, Agnes.

Agnes: Why don't we . . .

Student-Initiated Memory-Level Questions

Imagine yourself as the teacher in Case 3.15:

CASE 3.15

You are conducting a class session as part of an application-level lesson on compound probability. One student, Janet, asks you a question:

Janet: What does "mutually exclusive" mean?

Janet's question in Case 3.15 is answerable by recalling from memory a response—in this case, a definition. Your options, of which there are at least five, are simpler to implement than those for reasoning-level questions: (a) *answer the question directly,* (b) *use a how-might-we-find-that-out* strategy, (c) *refer the student to a source to be used right away,* (d) *refer the student to a source to be used later,* and (e) *redirect the question to another student.*

In Case 3.15.1, you answer Janet's memory-level question from Case 3.15 *directly:*

CASE 3.15.1

You: Two events are mutually exclusive if they are so related that one cannot happen if the other one does. For example, you cannot get an A and a D on Thursday's test. An A cannot also be a D and a D cannot also be an A. The two events are mutually exclusive.

If, on the other hand, you are interested in teaching students how to use reference sources for themselves, consider the *how-might-we-find-that-out* strategy that you illustrate in Case 3.15.2:

CASE 3.15.2

You: We really need to know what "mutually exclusive" means. Where should we go to look it up?
Janet: It's probably in the book, but I don't know where.
You: Try the index; that should give us some page numbers.

Besides wanting to teach students how to look up information, you may also feel the class needs an answer to the memory-level question right away. If so, *referring the student to a source to be used right away* is an option—an option you employ in Case 3.15.3:

CASE 3.15.3

You: Quickly, get the mathematical dictionary off the shelf. Look up the definition and read it to the class.

If the class does not need an answer to the student's question immediately, you might choose to *refer the student to a source to be used at a later time*—as you do in Case 3.15.4:

CASE 3.15.4

You: As soon as we finish this discussion, please look up the definition either in your textbook or the mathematical dictionary on the shelf. We'll take a minute at the beginning of tomorrow's class for you to share the definition with the rest of us.

Finally, you might simply want to *redirect the question to another student* as you do in Case 3.15.5:

CASE 3.15.5

You: Please raise your hand if you remember the defintion of "mutually exclusive." Yes, Salvador.
Salvador: It's when two things can't happen at the same time.

ASKING STUDENTS QUESTIONS

Student Engagement During Questioning Sessions

For students to be engaged in a questioning session, they must attentively listen to each question asked by the teacher, attempt to formulate responses, and either express their responses in a manner prescribed by the teacher or listen to others express their responses.

Recitation Sessions for Memory-Level Questions

Recitation is one type of questioning session in which teachers raise memory-level questions to help students review information (e.g., definitions) or practice remembering steps in algorithms. Case 3.16 is an example:

CASE 3.16

Mr. Winn: What will the graph of $3x - 7 = 10y + 1$ look like, Deanna?
Deanna: A line.
Mr. Winn: How do you know that?
Deanna: Because it is a first-degree function.
Mr. Winn: Thank you. Now, what does the graph of $y = x^2 - 16x + 5$ look like, Gail?
Gail: Because it's a . . .

Recitation sessions can be awfully boring. Yet television quiz shows and board games such as Trivial Pursuit are immensely popular although they are forms of recitations addressing memory-level questions. Thus, consider occasionally tailoring some of your recitation sessions along the lines of some popular quiz games. Case 3.17 is an example:

CASE 3.17

Ms. Saucony periodically has her algebra II students play a game with the following rules:

1. One student is selected as the game conductor, another as the game scorekeeper.
2. The rest of the class is divided into two teams, X and Y.
3. Six members of each team are selected to serve as that team's panel.
4. Panels X and Y sit facing the rest of the class at separate tables in front of the room.

5. The game conductor randomly draws the name of a non-panelist member of one of the teams. That student then selects one of the following categories for the first question of the opposing team:
   - Geometric terms, symbols, and expressions
   - Algebraic terms, symbols, and expressions
   - Statistical terms, symbols, and expressions
   - Trigonometric terms, symbols, and expressions
   - Geometric relationships
   - Algebraic relationships
   - Statistical relationships
   - Trigonometric relationships
   - Mathematical history
   - Potpourri
6. The game conductor randomly draws a question card from the selected category and asks the question to the panel (Panel A if the student selecting the category is from B and vice versa).
7. The panel members have 15 seconds to confer and answer the question. If they answer correctly, their team is awarded two points. If they fail, then the panel members call on a nonpanelist from their team for the answer. If the team member correctly answers the question, the team gets one point. If that member does not answer correctly, then the conductor asks the question to the other panel, and the team goes through the same process. If no correct answer is forthcoming, then the conductor announces the correct answer and no points are awarded.
8. A second nonpanelist is randomly chosen to select a category and steps 6 and 7 are repeated with the following exceptions: (a) A new category has to be selected; no previously selected category can be chosen until all categories have been used once. (b) Each time a panel calls on one of its nonpanelist team members to answer a question, she must select a student who has not been called on earlier in the game.
9. The game continues along these lines—repeating the cycles established in steps 6–8—until a prespecified number of questions have been asked.

Question-Discussion Sessions for Reasoning-Level Questions

Generally more interesting than recitation sessions are question-discussion sessions in which the teacher raises reasoning-level questions as part of a lesson using inquiry instructional strategies. Such questions are designed to stimulate students to think, discover, and reason. Case 3.18 is an example:

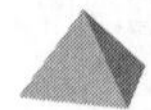

CASE 3.18

Mr. Grimes is conducting an inquiry-learning activity designed to help his 28 algebra I students achieve the following objective:

The student distinguishes between examples of variables and examples of constants (construct a concept).

With the overhead projector he displays the list shown by Exhibit 3.15 and initiates the following exchange:

Exhibit 3.15
Mr. Grimes' Overhead Transparency for His Inquiry Activity on Variables and Constants.

■ GARY ABOUD
● RECTANGLE
● RATIONAL NUMBER
■ 13
■ THE OVERHEAD PROJECTOR SCREEN YOU SEE RIGHT NOW
● PEOPLE IN THIS ROOM
● AGE
■ 15 YEARS OLD

Mr. Grimes: Notice that some of the elements in this set are marked with square regions and others with circular regions. What do the ones marked with squares have in common? And what do the ones marked with circles have in common?

Akeem, Jardine, and Sharon eagerly raise their hands. Immediately, Mr. Grimes calls on Jardine.

Jardine: The ones with the blocks are all in this room. Gary is here, the screen is there, and most of us are 15 years old.

Akeem and Sharon are waving their hands trying to convince Mr. Grimes to call on them. Akeem blurts out, "No, no!" as three other students raise their hands.

Mr. Grimes: Easy, Akeem. What's the matter?
Akeem: That can't be right because I don't see any 13s in here. And besides, one of the ones with a circle is "people in this room." I think the ones with circles have something to do with numbers.
Mr. Grimes: Okay, Sharon, before you fall out of your desk.
Sharon: The ones with the squares or blocks are exact and the—
Mr. Grimes: Excuse me, Sharon, but the procedure is that you must counter the previous hypothesis before you give your own.
Sharon: Oh, I'm sorry. What was it again?
Akeem: I said the ones with squares have something to do with numbers, but I see that's not right because there's a circle by 13. I thought it was a square at first.
Mr. Grimes: Okay, Akeem just countered his own hypothesis, so go ahead with yours, Sharon.
Sharon: Like I was saying, the ones with the squares are exact things—like Gary, not just people in the room—

or 13, not just any rational number. See what I'm saying?

Mr. Grimes: Do you understand what Sharon said, Vesna?

Vesna: Oh! Ahh, sure.

Mr. Grimes: Fine! Yes, Jardine, what is it?

Jardine: I don't get it.

Sharon: Look at the ones with the circles. There can be all different kinds of rectangles. A rational number can be 13, 15, or 100.75. People in this room can be any of us. And there's more than one age.

Jardine: Yea, but Gary is a person in this class and 13 is a rational number. So, why don't they have blocks instead of balls?

Akeem: It's like Sharon said. They're just one of a kind. Gary is only one person in this . . .

The session continues toward closure, with Mr. Grimes defining a constant and then a variable with input from Akeem, Jardine, and Sharon.

What do you think of Mr. Grimes' use of questioning or Socratic methods for stimulating students to reason in Case 3.18? The session was probably very valuable for Akeem, Jardine, and Sharon. But what about the other 25 students; what did they gain? Mr. Grimes seemed to know how to effectively utilize questioning strategies, but only a relatively small portion of the class seemed involved. For reasoning-level questioning sessions to be effective for all students, students must do more than passively listen to their classmates' responses. They do not necessarily have to be recognized and state their answers aloud for the group, but they do need to formulate and articulate their own answers for themselves.

Because Mr. Grimes allowed Jardine to answer aloud immediately after that initial question was raised, most students did not have time to formulate their own answers. They quit thinking about their own answers to listen to Jardine's. Only the students who were outspoken and quick to respond engaged in the learning activity as they should. Mr. Grimes did not pause long enough after each question before allowing a student to answer aloud. The overall, average time that teachers wait for students to respond to in-class questions is less than two seconds (Arnold, Atwood, & Rogers, 1974; Tobin, Tippins, & Gallard, 1994). After experiencing a few sessions in which they are asked questions they do not have the opportunity to answer, most students will not even attempt to formulate their own responses. Some will politely listen to the responses of the few, others entertain themselves with off-task thoughts, and others—if allowed—entertain themselves with disruptive behaviors (Cangelosi, 2000b, p. 269).

Mr. Grimes should not discard Socratic methods, but he needs to reorganize his question sessions and apply strategies for leading all students to address questions raised. One alternative is to preface questions with directions for all students to answer each question in their minds without answering aloud or volunteering answers until you ask them to do so. In Case 3.19 Mr. Smart applies that strategy to Mr. Grimes' situation.

CASE 3.19

Mr. Smart is conducting an inquiry-learning activity designed to help his 28 algebra I students achieve the same objective Mr. Grimes targeted in Case 3.18.

With the overhead projector he displays the list shown by Exhibit 3.15 and initiates the following exchange:

Mr. Smart: I am going to ask you some questions, but I don't want anyone to answer aloud until I call on him. Just answer the question in your mind. Here is the first one for you to take three minutes to reason out an answer: Notice that some of the elements in this set are marked with square regions and others with circular regions. What do the ones marked with squares have in common? And what do the ones marked with circles have in common?

Two students immediately raise their hands and say aloud, "Oh, Mr. Smart!" He is tempted to call on them to positively reinforce their enthusiasm, but resists—instead cuing them to follow directions with a gesture and a stern look. He waits, watching students' faces. Convinced after three minutes that nearly all have thought through the question and most have an answer ready, he asks, "Joyce, do you have an answer?" Joyce nods.

Mr. Smart: I'm glad that you do! How about you, Curtis?

Curtis: Yes.

Mr. Smart: Fine! Are you ready, Melissa?

Melissa: Not yet.

Mr. Smart: Just think aloud for us. Let's hear your thoughts about the two kinds of elements in the list.

Melissa: I don't see how they are different. Some of the ones with the round things are all about mathematics, but then so are some of those with the little squares.

Mr. Smart: You've made an important observation. Now I'd like some volunteers to share their answers with us. Okay, Ruth.

Ruth: I was thinking like Melissa, and then I . . .

Another possibility is to direct students to write answers on tasksheets as you circulate about the room quietly reading responses (e.g., as Mr. Heaps did in Case 3.12). Case 3.20 is another example of a teacher applying this option to Mr. Grimes's situation.

Exhibit 3.16
The Tasksheet Ms. Cramer Distributed During the Inquiry Activity on Variables and Constants.

■ WILSON McCARDLE
● RECTANGLE
● RATIONAL NUMBER
■ 13
■ THE OVERHEAD PROJECTOR SCREEN YOU SEE RIGHT NOW
● PEOPLE IN THIS ROOM
● AGE
■ 15 YEARS OLD

How are the ■'s alike but different from the ●'s?
How are the ●'s alike but different from the ■'s?

____
____
____
____
____
____
____
____

COMPARISON

____
____
____

CASE 3.20

Ms. Cramer is conducting an inquiry-learning activity designed to help her 28 algebra I students achieve the same objective Ms. Grimes targeted in Case 3.18.

She distributes the tasksheet shown by Exhibit 3.16 and says, "At the bottom of this form, please write one paragraph describing why you think the elements marked by square regions go together and the ones with the circular regions go together. In what ways are all the ones with square regions alike but different from the ones with circular regions? You have nine minutes for this task. I'll be around to read answers, but just keep working. Please don't talk to me during this time. Now, begin." She demonstratively sets the timer on the chronograph on her wrist. As students think and write, Ms. Cramer moves about the room, reading over shoulders. Some students are slow to start until she passes their desks and gently taps their tasksheets. As she samples what they write, she notes to herself whose answers she wants to use and which answers she wants to avoid in the upcoming large-group discussion. At the nine-minute mark, she moves to the front of the room, displays Exhibit 3.17 on the overhead projector, and initiates the following discussion:

Ms. Cramer: Please read yours, Simon.
Simon: I didn't know exactly what—
Ms. Cramer: Just read what you have on your paper.
Simon: The ones with squares are alike because first of all they have the squares by them. And they are all things you can see or that are numbers. The ones with circles all have circles. They have one geometry thing and two that are kinds of numbers and another that's people. They both seem pretty mixed up, alike in some ways and different in others.

Ms. Cramer writes "Simon" under "Person," "can see or is a number" under the ■ heading, and "types of things" under the ● heading on the transparency being displayed.

Ms. Cramer: Read yours, Megan.
Megan: Those with balls are kinds of things, not exactly things. The ball ones are things you study about. The block ones are more exact. The blocks have a weird thing, but I think he's cool anyway.

The class laughs as Ms. Cramer writes "Megan" under "Person," "actual and exact," under the ■, and "kinds of things" under the ● heading.

Person	■	●

Exhibit 3.17
Ms. Cramer's Overhead Transparency for Her Inquiry Activity on Variables and Constants.

Ms. Cramer: On two lines of your tasksheet under the heading "Comparison," quickly write one or two sentences describing how Simon's and Megan's answers are similar. You've got one minute.

She quickly moves about the room glancing at responses as the session continues toward closure.

Also consider having students formulate and discuss answers in collaborative groups and then report their findings and conclusions to the class. Engaging students in these types of cooperative-learning sessions is a topic for a subsequent section of this chapter.

Six Points About Question-Discussion Sessions

Consider the following thoughts when designing question-discussion sessions:

- Unlike recitation sessions, student engagement during reason-level questioning sessions requires students to take time to ponder and think about questions posed by teachers before expressing answers. Consequently, your students are hardly able to engage in reasoning-level questioning sessions unless you provide for periods of silent thinking between the time when questions are asked and when answers are expected.
- Prompting students to write out responses to questions has a least four advantages: (a) Articulating a response in writing requires students to organize their thoughts in a way that often enhances their understanding of the mathematics being studied (Connolly & Vilardi, 1989; Goldin & Shteingold, 2001; Myers, 1986, pp. 167–169). (b) Providing time for students to write serves as a silent period for all students during which they can be thinking about how to respond to questions. (c) Written responses make it possible for you to preview students' responses and decide which ones should be read to the class. (d) Having written responses available to read to the class avoids some of the stammering and grasping for words that is typical of students answering aloud in front of their peers.
- If you direct a question to a particular student before articulating the question itself, other students may be disinclined from carefully listening to that question. In most cases asking, "What has the Pythagorean relationship got to do with the distance formula? [pause] Traci." is preferable to "Traci, what has the Pythagorean relationship got to do with the distance formula?" With the latter phrasing, students other than Traci may not bother to listen to or think about how to answer the question because they know it is not directed at them.
- Moving quickly from one student to another allows more students to express answers aloud. However, with reasoning-level questions, responses should often be complex, thus needing to be discussed in some detail; answers are not simply right or wrong. To involve more students and yet have some answers fully discussed, use the responses of some students to formulate subsequent questions for other students. Ms. Cramer in Case 3.20, for example, directed the class to compare Simon's answer to Megan's.
- Students are more likely to engage in question-discussion sessions in which (a) questions relate to one another and focus on a central theme or problem and (b) questions are specific rather than vague. Vague questions such as "Do you understand?" hardly focus thought as well as, "How would reducing the denominator affect the value of y?" If you want to assess students' understanding, confront them with a specific mathematical task and observe their responses.
- Students learn the importance of engaging in question-discussion sessions when the sessions culminate in problem resolutions that are applied in subsequent learning activities. When you work with chapter 5, note how Ms. Citerelli in Cases 5.8–5.10 has students use what they gain from question-discussion sessions in other types of activities.

ENGAGING STUDENTS IN COOPERATIVE-LEARNING SESSIONS

Students Learning From One Another

Cooperative-learning activities in which students learn from one another have proved to be quite successful (Johnson & Johnson, 1999; Joyce, Weil, & Calhoun, 2000, pp. 119–122). Students can engage in cooperative-learning activities in large-group settings, but small task groups are particularly well suited for students teaching one another.

Intraclass grouping arrangements provide greater opportunities than whole-class activities for students to interact with one another, for tasks to be tailored to special interests or needs, and for a variety of tasks to be addressed concurrently. A variety of intraclass task-group patterns are commonly used to facilitate cooperative learning. They include (a) *peer instruction groups,* (b) *practice groups,* and (c) *interest or achievement-level groups.*

Peer Instruction Groups

In a peer instruction group, one student or a team of students teaches others—either presenting a brief lesson, tutoring, or providing help with specified tasks. Traditionally, this type of activity involves a student with a relatively advanced level of achievement of an objective helping others achieve that objective. Case 3.21 is an example:

CASE 3.21

Mr. Jackson notes the following from the results of a unit test he recently administered to the 21 students in his probability and statistics class:

- Regarding test promts 3, 7, 8, and 16, all of which are relevant to students' comprehension of the central limit theorem, Anita responded correctly to all four, whereas none of the following students got more than two correct: Bernie, Deborah, Amalya, Francine, and Jay.
- Regarding test prompts 5, 6, and 11, all of which are relevant to how well students construct the concept of a normal distribution, Benju responded correctly to all three, but none of the following students got any of the three correct: Bernie, Amalya, Don, Steve, and Malcom.

Mr. Jackson thinks that (a) Anita's and Benju's insights into the content would be enhanced by experiences teaching their peers, (b) the other eight students will not succeed in the next unit until they better achieve certain objectives, and (c) Bernie and Amalya need to construct the concept of a normal distribution before they can comprehend the central limit theorem. Thus, he decides to conduct a session in which the class is subdivided into three groups:

- Group I: Anita will explain the central limit theorem and how she responded to test prompts 3, 7, 8, and 16 to Deborah, Francine, and Jay.
- Group II: Benju will explain normal distributions and how he responded to test prompts 5, 6, and 11 to Bernie, Amalya, Don, Steve, and Malcom.
- Group III: The other 11 students will engage in an activity with a tasksheet in an independent-work session.

But peer instruction does not always involve mentor students who display more advanced achievement levels than their peers. Consider Case 3.22:

CASE 3.22

Ms. Harris integrates historical topics into most of the units for her algebra class of 25 students. As part of the unit on numbers and numeration, she subdivides the class into five groups of five students, with groups assigned to historical topics as shown by Exhibit 3.18.

She then conducts an hour-long task-group session, with each of the five groups studying and discussing its topic using references from Ms. Harris' resource library (e.g., Ifrah, 2000; McLeish, 1991; NCTM, 1989b; Seife, 2000). For homework, each student prepares a 15-minute lesson on her group's topic to be presented to four students from the other groups.

Exhibit 3.18
Ms. Harris' Peer Instruction Groups.

Group A

Topic: Origins of the Hindu-Arabic numeration system
Students: Osprey, Byron, Bryce, Chris, and Nadine

Group B

Topic: Origins of our beliefs about prime numbers
Students: Marion, Joe, Charlene, Jennifer, and Dominica

Group C

Topic: Discoveries of π
Students: Patti, Scott, Jan, Chen-Pai, and Garth

Group D

Topic: The history of perfect, deficient, and abundant numbers
Students: Crystal, Henry, Cinny, Jason A., and Jason T.

Group E

Topic: Origins of a dangerous idea: zero
Students: Julie, Eian, John, Rich, and Willie

Over the next two days, Ms. Harris conducts additional task-group sessions in which the five students from each group present their 15-minute lessons to groups of four students from other groups. For example, Osprey presents her lesson on the origins of Hindu-Arabic numbers to Marion, Patti, Crystal, and Julie. Marion presents his on the origins of our beliefs about prime numbers to Osprey, Patti, Crystal, and Julie. Patti presents hers on the discovery of π to Osprey, Marion, Crystal, and Julie. Crystal presents hers on the history of perfect, deficient, and abundant numbers to Osprey, Marion, Patti, and Julie. And Julie presents hers on the origins of zero to Osprey, Marion, Patti, and Crystal.

The other four groups for rounds of the lesson are: Byron, Joe, Scott, Henry, and Eian; Bryce, Charlene, Jan, Cinny, and John; Chris, Jennifer, Chen-Pai, Jason A., and Rich; Nadine, Dominica, Garth, Jason T., and Willie.

Practice Groups

Large-group recitation sessions like those in Cases 3.16 and 3.17 do not always provide highly efficient ways for students to review, drill, and receive feedback for memory-level objectives. With small-group arrangements, several can recite concurrently. Students, for example, could work in groups of three each. One student reads questions about vocabulary, symbols, and relationships from a pack of cards. The other two answer and the questioner gives feedback. The role of questioner rotates. Another possibility is for students to play mathematics memory games as was done in Case 3.17. But to increase individual involvement and the number of trials per student, students play in groups of five rather than as a whole class.

Interest or Achievement Groups

Intraclass groups may be organized around interests as Ms. Castillo did in Case 3.5 or achievement levels as Ms. Rosenberg does in Case 3.23:

CASE 3.23

To help her deal with the wide range of student achievement levels in her seventh-grade general mathematics class, Ms. Rosenberg conducts virtually the whole course with the 32 students partitioned into three groups:

- The green group—consisting of eight students who average about 3.5 weeks per teaching unit.
- The blue group—consisting of 15 students who average 2.5 weeks per teaching unit.
- The gray group—consisting of nine students who average about 1 week per teaching unit.

For all practical purposes, she conducts three courses in one. She manages the configuration by staggering large-group, small-group, and individual-work sessions among the groups.

Guidance and Structure for Maintaining Engagement

Research studies that examine how students spend their time in classrooms indicate that they are likely to have poor engagement levels in small-group learning activities unless the teacher is actively involved in the session (Evertson, 1989; Fisher, Berliner, Filby, Marliave, Cahen, & Dishaw, 1980). But a teacher cannot be in the middle of several groups at once and often subgroups fail to address their tasks due to a lack of guidance. Consider Case 3.24:

CASE 3.24

Ms. Clay has her prealgebra students organized into four subgroups. She directs them, "I want each group to discuss when we use ratios to solve real-life problems. Okay, go ahead and get started."

After six minutes discussing what they are supposed to be doing, the students in one group no longer bother with ratios and socialize with one another. Ms. Clay hardly notices that they are off-task because she is busy explaining to another group what she meant for them to do. A third group becomes quite noisy; Ms. clay raises her voice from her position with the second group and announces, "Better keep it down in here. You won't learn how to apply ratios unless you get your discussion going." In the fourth group, Magdalina dominates the first five minutes telling the others about ratios. She stimulates Ann's interest and the two of them engage in a conversation in which Magdalina reviews what Ms. Clay explained in a previous session about computing ratios. The other three members of the group are talking but not about the topic.

After managing to get the second group on track, Ms. Clay moves to the noisy third group, saying, "You people aren't following directions; you are supposed to be discussing problems that use ratios." She then tells the group what she had hoped they would discover for themselves.

After spending nine minutes with the second group, Ms. Clay calls a halt to the activity and announces, "Okay, class let's rearrange our desks back. Now that you understand when to apply ratios, let's move on to . . ."

Ms. Clay failed to obtain satisfactory student engagement because her directions did not spell out what tasks each subgroup was to accomplish and just how to go about completing them. Without an advanced organizer (e.g., Ms. Castillo's in Exhibit 3.1) to focus students' attention, it was difficult for her to provide guidance efficiently to one group while monitoring the others. Furthermore, her students are less likely to engage diligently in her next intraclass group session because they failed to achieve closure on this one. Unlike Ms. Castillo in Case 3.5, she did not take what the subgroups did and use it in a subsequent activity.

In Case 3.24, Ms. Clay also demonstrated another common misapplication of cooperative-learning strategies by stepping in and doing the second group's work for it. Sometimes a teacher gets so frustrated when a subgroup fails to maintain its focus that she breaks into an explanation instead of asking leading questions to prompt students back on track.

Nine Points About Cooperative-Learning Sessions

Consider the following when designing cooperative-learning sessions:

- Expect the same sort of off-task behaviors Ms. Clay's students exhibited in Case 3.24 unless you clearly and explicitly spell out tasks for each group as well as individual responsibilities for each group member (as indicated by Ms. Gaudchaux's lesson plan in Exhibit 5.19, which you will see when you work with chapter 5).
- All group members should be jointly accountable for completing the shared task with each member responsible for fulfilling an individual role.
- Efficient procedures for transitions into and out of small-group activities avoid the time-wasting chaos following a teacher's directions such as, "Let's move our desks so that we have four groups of six each." When you read chapter 5's Case 5.13, you will note that Ms. Smith uses an especially efficient and attention-getting method for making a transition into an intraclass group activity. Case 3.25 illustrates another idea:

CASE 3.25

Prominently displayed on the front wall of Mr. Rudd's classroom is a brightly colored 3' by 3.5' poster depicted in Exhibit 3.19. He moves the arrow to the symbol on the poster to indicate if students should be working alone ("1"), in pairs ("2"), . . ., in groups of seven ("7"), or as a single group ("Whole class").

Exhibit 3.19
The Poster Mr. Rudd Displays to Indicate Grouping Arrangements.

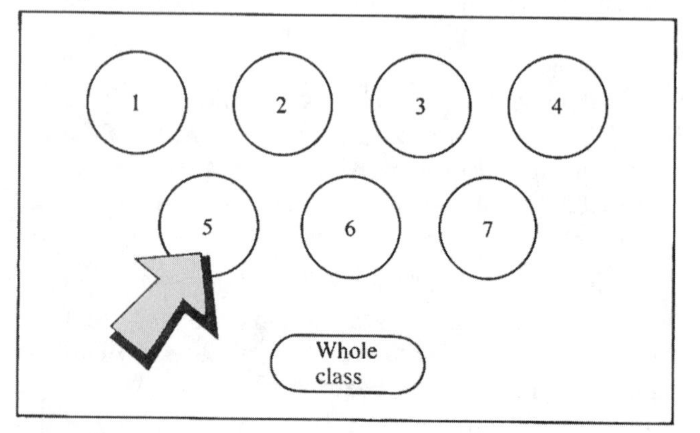

- Tasksheets (e.g., Exhibit 3.1) direct students' focus and provide them with an overall picture of what they are expected to accomplish in their groups.
- You need to monitor groups' activities and provide guidance as needed without usurping students' responsibilities for designated tasks. Move from one group to another, cuing students to be on-task without actually becoming a member of any one group. Ms. Castillo applies this suggestion in Case 3.26—a continuation of Case 3.5:

CASE 3.26

Ms. Castillo stops to sit with Group C's members as they struggle with question 4 from Exhibit 3.1. One student addresses Ms. Castillo, "I like Rabid Dog better. Doesn't that make a difference?" Ms. Castillo turns to the group, "What does *more* minutes of music mean?" Noticing the discussion is on track, she moves over to Group B, continuing to keep an eye on all the groups.

- Formative feedback should be used to regulate activities. Engaged behaviors during intragroup sessions are observable; students should be involved in discussions and working on specified tasks. Thus, formative feedback for regulating activities is readily obtained.
- Positively reinforcing closure points are needed for lengthy sessions. Ms. Castillo's sequence of questions on Exhibit 3.1's tasksheets provides closure points leading up to the report for the whole class.
- Subgroup work should be followed up and utilized during subsequent learning activities. You will note from chapter 5 that Ms. Gaudchaux's lesson plan in Exhibit 5.19 provides for large-group activities to be dependent on the outcomes of the work of the subgroups.
- Model descriptive language, true dialogues, assertiveness, and active listening techniques (as explained in chapter 2's section "Communicating and Interacting With Students") for your students. These are communication skills that facili-

tate collaboration. In Case 3.26—a continuation of Case 3.25—Ms. Castillo models active listening techniques for students:

CASE 3.26

Ms. Castillo stops to sit with Group B as they struggle with question 6 from Exhibit 3.1. As Emily is commenting, John attempts to engage Ms. Castillo with his own private question, "Ms. Castillo, this doesn't make--" But Ms. Castillo uses a frown and a hand motion to cue John to be quiet and then says to the group, "Excuse me, Emily, would you repeat that last part about adding the two prices? I missed what you said about that." Emily repeats and finishes her comment. Ms. Castillo says, "Thank you. That should shed some light on John's concern. John, raise your concern with the group." John (this time to the group): "To me, the question ought to be . . ."

Later in the large-group session, Ms. Castillo uses the comments that different students made in their subgroups as she raises issues and questions and delivers explanations.

ENGAGING STUDENTS IN INDEPENDENT-WORK SESSIONS

Student Engagement During Independent-Work Sessions

The importance of students interacting with one another in cooperative-group activities is emphasized in *PSSM* (NCTM, 2000b, pp. 60–63). But a lesson's group activities need to be interspersed with some quiet thinking time during which students engage in independent work. You will note, for example, that Ms. Gaudchaux's lesson plan in Exhibit 5.19 indicates that she will integrate small cooperative-group, question-discussion, independent-work, and homework sessions in her lesson.

Engagement in independent-work sessions requires a student to complete an assigned task without disturbing others. Typically, students work individually while the teacher is available for help. Independent-work sessions are appropriately used in a variety of ways:

- As an integral part of an inquiry lesson, students work independently on tasks that were specified previously during a large-group session. The work from the independent-work sessions is subsequently used in follow-up activities. You will note from Cases 5.8–5.10 that Ms. Citerelli begins a lesson to lead students to construct the concept of arithmetic sequences with an independent-work session in which students complete Exhibit 5.11's tasksheet. Their work with the tasksheet is used in a large-group question-discussion session in which students explain and reflect on that independent work.
- As an integral part of a direct instruction lesson, students independently practice an algorithm that has just been explained to them in a large-group presentation. The teacher coaches students individually as they practice. Students receive feedback on their independent work during a subsequent large-group question-discussion session, a cooperative-group session, or by the teacher collecting and then annotating their work with the algorithm. When you read Case 6.13 as you work with chapter 6, you will note that after Ms. Allen explains an algorithm for differentiating algebraic functions in a large-group presentation, she directs students to complete a related exercise independently; she circulates about the room monitoring work and coaching as needed.
- Students use part of a class period to begin a homework assignment in an independent-work session. This very common practice allows them to begin the assignment while the teacher is still available to provide individual guidance.

Independent-work sessions are essential to most teaching units. However, they need to be integrated with other types of learning activities, monitored, and guided. Rosenshine (1987, p. 261) stated:

> Studies have shown that when students are working alone during seatwork they are less engaged than when they are being given instruction by the teacher. Therefore, the question of how to manage students during seatwork, in order to maintain engagement, becomes of primary interest.
>
> One consistent finding has been the importance of a teacher (or another adult) monitoring the students during seatwork. Fisher et al. (1980) found that the amount of substantive teacher interaction with students during seatwork was positively related to achievement and that when students have contacts with the teacher during seatwork their engagement rate increases by about 10 percent. Thus it seems important that teachers not only monitor seatwork, but that they also provide academic feedback and explanations to students during their independent practice. However, the research suggests that these contacts should be relatively short, averaging 30 seconds or less. Longer contacts would appear to pose two difficulties: the need for a long contact suggests that the initial explanation was not complete and the more time a teacher spends with one student, the less time there is to monitor and help other students.
>
> Another finding of Fisher et al. was that teachers who had more questions and answers during group work had more engagement during seatwork. That is,

another way to increase engagement during seatwork was to have more teacher-led practice during group work so that students could be more successful during seatwork.

A third finding (Fisher et al. 1980) was that when teachers had to give a good deal of explanation during seatwork, then student error rates were higher. Having to give a good deal of explanation during seatwork suggests that the initial explanation was not sufficient or that there was not sufficient practice and correction before seatwork.

Another effective procedure for increasing engagement during seatwork was to break the instruction into smaller segments and have two or three segments of instruction and seatwork during a single period.

Six Points About Independent-Work Sessions

Consider the following ideas when designing and conducting independent-work sessions:

- By providing explicit directions that clearly specify the tasks in the transition into an independent-work session, you avoid many of the nagging questions about what to do that typically plague independent work. Your time during independent-work sessions should be spent conferring with and prompting individual students' progress through the task—not reiterating directions in response to questions such as, "What are we supposed to be doing?"
- Having artifacts (e.g., notes, a worked-out example, or a list of the steps in an algorithm) from the previous group-learning activity still visible to students on the whiteboard or in their notebooks during an independent-work session provides a reference that facilitates your providing efficient individual help. Consider Case 3.27:

CASE 3.27

Ms. Towers has just conducted a large-group presentation that included an explanation of an algorithm for factoring algebraic polynomials, a demonstration of the algorithm with an example, and the class working through an example with Ms. Towers' guidance.

Leaving the two completed examples and the step-by-step outline of the algorithm on the whiteboard, Ms. Towers directs the students into an independent-work session in which they are to practice the algorithm on six exercises from the textbook.

In a few minutes, Phyllis and Juaquin raise their hands. Silently acknowledging Juaquin with a wink and hand gesture, Ms. Towers moves to Phyllis' desk; Phyllis says, "I don't know where to start." Ms. Towers, seeing that Phyllis has nothing written on her paper, says "Read the first two steps from the board and then look at what we did first for the example on the right. I'll be back within 30 seconds." She moves directly to Juaquin's desk and Juaquin says, "Is this right?" Detecting that he has begun the first exercise by factoring out only 2 when he could have factored out 6, she responds, "It's not wrong, but you could make it easier on yourself if you repeat Step 2 before continuing—see if there's still another factor that's common." She walks back to Phyllis as Juaquin blurts out, "Oh, yeah! The 3—" Realizing that he's being disruptive, Juaquin grabs his mouth, muffling the rest of his sentence.

Back at Phyllis' desk, Ms. Towers sees that Phyllis is now started and softly tells her, "Just keep following the steps on the board and checking how that example is worked—one step at a time. I'll check with you every few minutes." By this time, three others have their hands up. By referring to the outline and examples and raising pointed questions, Ms. Towers has all three students back on track within a minute. She continues moving about the room, responding to students' requests for help and volunteering guidance as she sees fit from observing their work. At no time does she spend more than 20 seconds at a time with any one student.

- You can provide efficient individual guidance and help to students, but only if you organize for it prior to the session and communicate assertively during the session.
- Cooperative-learning activities can be incorporated into independent-work sessions to increase opportunities for students to receive help. Ms. Strickland in Case 3.28 manages to do this in an unusual way:

CASE 3.28

At the beginning of the school year Ms. Strickland constructs a flag-raising device. Each device is supplied with yellow, red, and green flags, which can be raised one at a time. She then produces enough such devices to install them on the corners of the students' desks in her classroom.

She then establishes a procedure for independent-work sessions whereby students display (a) a yellow flag as long as they are progressing with the work and do not feel a need for help, (b) a red flag to indicate a request for help, and (c) a green flag to indicate that they have finished the work and are willing to help others. See Exhibit 3.20.

As Ms. Strickland monitors an independent-work session, she responds to a red flag either by conferring with the student herself or by signaling a student who has raised a green flag to provide the consultation.

Exhibit 3.20
Flag Raising During One of Ms. Strickland's Independent-Work Sessions.

Ms. Strickland believes her system has four distinct advantages over the conventional hand-raising procedure: (a) Cooperative learning among students is encouraged. (b) When waiting for help, students can continue doing some work without having to be burdened with holding up a hand. (c) Students who finish the task before others have something to do that will not only help their peers, but also be a learning experience for them. (d) The systematic air of the procedure enhances the businesslike environment of the classroom.

- To avoid having students who finish early idly waiting for others to complete a task, you can sequence independent-work sessions so they are followed by individual activities with flexible beginning and ending times. Students work at varying paces. Unless the task can be completed by all in less than 10 minutes, you need to manage independent-work sessions to accommodate students' finishing at varying times. One solution is to schedule a subsequent activity that early finishers can start (e.g., begin a homework assignment) but that can be interrupted conveniently when you are ready to halt the independent-work session. Case 3.29 is an example:

CASE 3.29

Ms. Wharton has established a routine in which each teaching unit includes a long-range assignment that requires students to use computers to complete. The assignment is given at the beginning of the unit and is not due until the day of the summative unit test. Ordinarily it involves either completing some programmed exercises related to the unit's objectives or writing programs for executing algorithms learned during the unit. Ms. Wharton's classroom is equipped with nine computers, but she has no fewer than 21 students in any one of her classes.

For independent-work sessions, she established a routine by which students who finish the task before the end of the session work on their computer assignment. Those who may have already completed the computer assignment have the option of either beginning their homework assignment or using the computers in other ways (e.g., to play games).

- As with other types of learning activities, students need feedback to correct errors, reinforce correct responses, and positively reinforce engagement. Formative feedback is facilitated during independent-work sessions because each student's efforts are reflected by a product (e.g., written responses on a tasksheet). You will note when working with chapter 6 how Ms. Allen utilizes formative feedback during the independent-work session she conducts near the end of Case 6.13.

MOTIVATING STUDENTS TO COMPLETE HOMEWORK ASSIGNMENTS

Appropriate Use of Homework

Homework provides students with opportunities to work alone at their own pace. The crowded social

Exhibit 3.21
Mr. Greene's Overhead Transparency for Recording Data Students Collected at Home.

Person	Object	Circumference (C)	Diameter (D)	C/D

setting of a classroom is not particularly conducive to concentrated, undisturbed thinking that is essential for doing some types of reasoning-level mathematics. Furthermore, school schedules do not always allow adequate allocated time to practice exercises essential to the development of some algorithmic skills. Posamentier and Stepelman (1999, p. 18) suggested that for many students, ". . . classroom instruction may serve as a forum for exposure to new material, whereas the time spent on homework provides the genuine learning experience." Learning activities via homework assignments should complement what students do in the classroom as preparation for, extension of, or follow-up to in-class activities.

Use of Homework as Preparation for In-Class Activities

One way to lead students to connect mathematics with their own real worlds is to engage them in learning activities using numbers and other mathematical variables from their outside-of-school environments. Rather than always working with numerals printed in textbooks or data that you bring to class, sometimes assign homework in which students collect data or other information for use in classroom activities. Mr. Greene does this in Case 3.30:

CASE 3.30

Mr. Greene directs his students near the end of a class period, "For your homework assignment, locate three circles determined by objects in or near your home. Anything that determines a circle will be fine, such as, the base of a light, a bicycle wheel, a dinner plate, or the top of your little brother's head. After locating each of your three circles, measure its diameter and circumference. Write the measurements down and bring in the three pairs of numbers to class tomorrow."

The next day, Mr. Greene displays Exhibit 3.21 on the overhead projector and directs students, one at a time, to call out circumference/diameter pairs. Each student contributes one pair as Mr. Greene completes the form in Exhibit 3.22. Mr. Greene then conducts a reasoning-level question-discussion session leading students to discover that the ratio of the circumference of any circle to its diameter is a constant that is slightly greater than 3.

You may sometimes use a homework assignment to expose students to a problem or task that stimulates them to direct their thoughts toward a topic you plan to introduce during the next class session. Cases 3.31 and 3.32 are examples:

CASE 3.31

Ms. McKnight conducts an inquiry lesson on permutations, followed by direct instruction on computing permutations by the usual formula:

$$_nP_r = \frac{n!}{(n-r)!}$$

For homework, she assigns nine word problems. Solutions to the first eight problems require computations of permutations, but the ninth requires a combination to be computed. Ms. McKnight does not plan to teach about combinations until the next day after the homework is due.

The following day, the homework is reviewed in class. Some students automatically used the permutation formula for the ninth problem. Others realized that the ninth problem was different from the others but did not devise a solution. One student correctly solved the ninth problem via a tedious examination of all possible cases. Students complain to Ms. McKnight that the ninth problem was a

Person	Object	Circumference (C)	Diameter (D)	C/D
Barbara	ring	6.8 cm	2.2 cm	3.09
Andrea	barrel bottom	32"	10"	3.20
Jerry	jar	19"	6"	3.17
Oral	base of light fixture	78.74 cm	25 cm	3.15
Glenn	wheel	81.5"	26"	3.13
Karel	top of head	7.75"	2.94"	2.94

Exhibit 3.22
Mr. Greene Records One Pair of Measurements from Each Student.

"trick problem." Acknowledging her "error," Ms. McKnight uses their discussion of that problem as a springboard into the day's lesson, which is about combinations—an inquiry lesson in which they develop the following formula:

$$_nC_r = \frac{n!}{(n-r)!r!}$$

CASE 3.32

Mr. Rudd's students know how to plot points on an algebraic function in a Cartesian plane. However, except for linear functions they are unaware of any methods for sketching a graph without plotting numerous points.

The day before he plans to introduce shortcuts for finding key features of graphs of quadratic functions (e.g., the vertex and line of symmetry), Mr. Rudd assigns three quadratic functions to be graphed for homework.

The next day students have the graphs for class, with varying degrees of accuracy. Some complain about how boring it was to plot so many points. Mr. Rudd jumps on the students' expressed need for some shortcuts and embarks on his planned inquiry lesson to discover some useful relationships.

Some skills or knowledge needed for participation in an upcoming in-class activity may be efficiently acquired during a homework assignment. Cases 3.33 and 3.34 are examples:

CASE 3.33

Mr. Triche assigns some background reading on the Pythagoreans for homework (e.g., Calinger, 1999, pp. 68–79). The next day, he draws on the knowledge of mathematics history the students acquired from the reading for a lesson on polyhedra.

CASE 3.34

Ms. Carion distributes a list of mathematical words (e.g., "polynomial") and shorthand symbols (e.g., " $\sum_{i=a}^{n} p(i)$ ") she plans to use in a large-group presentation the next day.

For homework, the students are directed to look up and familiarize themselves with the definitions of those words and symbols.

Use of Homework as Extensions of In-Class Activities

One of the disadvantages of independent-work sessions is that students complete assigned tasks at different times. Often, there is inadequate class time allocated for all students to complete the tasks. Furthermore, practice exercises to polish algorithmic skills may be too time consuming to schedule during class. Homework assignments relieve at least some of the pressure of trying to squeeze necessary work into class periods.

Use of Homework as a Follow-Up to In-Class Activities

For some objectives, students need solitary time to analyze mathematical content at their own pace. Consider Case 3.35:

CASE 3.35

Ms. Hundley designs a lesson to help students apply a problem-solving strategy that you will read about in Case 8.3 of chapter 8. Ms. Hundley believes that students will learn to apply the strategy only by trying it out on their

Exhibit 3.23
Ms. Hundley's Solution Blueprint Form.

Solution Blueprint

Question(s) posed by the problem:

____

Principal variable(s):

____

Other variables affecting results:

____

Possible approaches to consider:

____

Delimited principal variable(s):

Solution plan:

Overall design

____

Measurements to make

____

Relations to establish

____

Algorithms

____

Results:

____

Conclusions:

____

How you would solve this type of problem in the future?

____

own in numerous situations. Many students, she notes, lack the confidence to trust their own thinking; they would rather be told a solution than to devise one. Consequently, she wants them to experience their initial successes with this problem-solving strategy by themselves rather than in a social setting.

Thus, she structures the lesson so that she spends nearly an entire class period explaining how to do the homework assignment. She begins by saying, "Your homework assignment for tonight is to solve three problems on the tasksheet I will distribute to you at the end of today's class. When you solve these problems, please follow the procedure outlined on the solution blueprint form that I'm about to give you." She distributes copies of Exhibit 3.23.

She then spends the next 45 minutes explaining how to use the form and the nine-stage strategy for solving a problem. The day's classroom activities present the strategies to the students that they will learn to apply when they do the homework.

A more common type of homework assignment that follows up in-class activities is homework used to provide formative feedback on what was learned in class. In class, students engage in an activity to help them progress toward an objective. The results from the homework are used to reinforce what was learned and to identify areas in need of remediation.

Four Suggestions About Homework Assignments

Unlike most other types of learning activities, students allocate their own time for engaging in homework assignments. Sometimes students doing homework have parents nearby encouraging them to be on-task. However, parental supervision of students' homework is extremely variable and depends on circumstances in homes, ages of students, and a myriad of other factors. Engagement in a homework assignment usually requires students to (a) understand the directions for the assignment, (b) schedule time away from school for the assignment, (c) resist outside of school distractions while completing the assigned task, and (d) deliver a report of the completed work in class by a specified deadline.

To motivate your students to engage in homework you assign, consider the following suggestions:

- Every day for the first few weeks of a course, assign clearly defined, specific tasks for homework. Spend class time during that early part of the course teaching students to schedule time for homework and efficient ways of completing it. Follow up on every assignment. Until you make a concerted effort to teach them, your students are unlikely to (a) know how to schedule time for homework—especially considering that they have assignments from other courses besides yours; (b) discriminate whether content relative to an assignment should be memorized, figured out by themselves, or found out from an outside source; (c) know how to study; and (d) know how you expect results of their homework to be reported. Your extra efforts along these lines during the first few weeks of a course will pay dividends in time saved and completed assignments once students learn your routine and expectations. Students will develop a behavior pattern of engaging in homework if initial efforts are positively reinforced.
- Keep in mind that students have other assignments besides yours. Fewer, but well-chosen, exercises tend to be more productive than a lengthy assignment that is more time consuming for you and the students. Long-range assignments that are expected to take days for the students to complete should be broken out into a sequence of shorter assignments, with due dates that serve as progress points toward completion. Consistently heeding this suggestion in the early stages of a course encourages students to get into a routine for doing homework. Otherwise, they tend to be overwhelmed by lengthy assignments and to put them off until just before they are due.
- To elicit parents' cooperation in encouraging and supervising homework, utilize ideas from chapter 2's section "Communicating With Students' Parents."
- Positively reinforce engagement in homework and punish failure to do homework by designing units so that success in classroom activities—especially tests—depends on homework efforts. Students can be motivated to engage in homework faithfully without your having to resort to awarding points toward grades for turning in assignments. Consider the system Ms. Goldberg changed to in Case 3.36.

CASE 3.36

Ms. Goldberg uses a procedure in which each geometry student's grade is determined by the number of points accumulated during a semester. Her students have two ways to earn points: (a) Half of the total possible points is based on their test scores. (b) The other half is awarded for homework that, when turned in on time, is scored according to the number of correct responses.

Ms. Goldberg discovers that a number of students receive high scores on their homework but low scores on their test papers. Under her system, such students are able to pass the course. After analyzing the situation, she realizes that these students are either copying their homework from others or having others do it for them. Thus, she decides to change her grading procedures. She will annotate students' homework to provide them with feedback, but she will not grade their homework so that it influences their semester reports. Ms. Goldberg begins to make a concerted effort to assign homework and design tests so that completing homework will clearly be an effective way to prepare for tests.

To begin conditioning her students to the new system, she assigns homework one day and then on the next day administers a test that covers that same objectives as the homework.

Beginning class periods with short tests that include prompts similar to those from the previous homework assignments teaches students the importance of doing homework far better than preaching to them about the importance of doing homework or threatening to lower grades if they do not do it. In Case 3.37, Mr. Heidingfelder demonstrates a businesslike attitude to motivate students to do homework:

CASE 3.37

Mr. Heidingfelder carefully examines the 30 algebraic inequalities from the exercises in the class textbook. From the 30, he selects the 11 exercises he thinks will provide students with the most useful practice for the different problem-solving situations they will encounter in a subsequent lesson.

The day after the assignment he collects the work and performs a quick error-pattern analysis (Ashlock, 2001). The papers are returned to the students with a clear indication of exactly which steps in the algorithm they did correctly and which they did incorrectly. Because he had carefully selected the exercises, he was able to do the analysis more efficiently than if they had been selected with less thought.

While the rest of the students are correcting their work and beginning an independent-work session, he calls aside the five students, Angela, Donna, Pruitt, Pam, and Carl, who did not complete the homework and initiates the following conversation:

Mr. Heidingfelder: I'm sorry you didn't give me an opportunity to provide you with feedback on how to solve the inequalities we had for homework."

Pam: I would have done it, but—

Mr. Heidingfelder (interrupting): I need to figure out when you can get this done so I can get my analysis back to you before you leave school today. You need that from me before you will be able to go on to our next lesson.

Pruitt: I forgot—

Mr. Heidingfelder (interrupting)**:** Please let me think how to help you. I've got it! Here's what we'll do. I'll meet you in here as soon as final announcements are completed this afternoon. As soon as you've finished the 11 exercises, I'll give you my analysis.

Engage in Activity 3.2:

Activity 3.2

When you engaged in Activity 1.2, you developed a syllabus listing units for the course. Retrieve a copy of the syllabus from the "Cognition, Instructional Strategies, and Planning" section of your working portfolio. Write an objective that conceivably could be one of the objectives that contributes to the goal for that unit. Now, imagine yourself teaching a lesson for that objective—a lesson that includes a homework assignment. Develop the homework assignment. Now design a brief test that is relevant to your students' progress with the objective that you conceivably would administer at the beginning of the class period at which the homework assignment is due. Design the test so that diligent completion of the homework assignment is positively reinforced.

Discuss your work with a colleague who is also engaging in Activity 3.2; provide one another with feedback. Revise your work in light of the feedback. Insert a copy of what you produced from this activity into the "Motivation, Engagement, and Classroom Management" section of your working portfolio.

SYNTHESIS ACTIVITIES FOR CHAPTER 3

1. For each of the following examples, discuss with a colleague whether the student is intrinsically or extrinsically motivated to engage in a mathematical learning activity:

    A. Harold carefully listens to his teachers' explanation of how to bisect an angle with a compass and straightedge because he fears feeling embarrassed if he is called on to demonstrate the algorithm for the class and he does not know how to do it.

    B. Because Nicole's mother told her she may borrow the car next weekend if she receives at least an A− on her next algebra II test, Nicole diligently works on her homework assignment.

    C. To keep up his grade point average so that he can remain eligible to play basketball, Horace diligently follows his geometry teacher's directions in class.

    D. To develop her ability to apply principles of logic in everyday decision making, Emily makes an extra effort to attend to her geometry teacher's lessons on truth tables and proof strategies.

    E. Because he wants to better understand how to solve problems he confronts in physics class, Nazia pays careful attention during calculus class.

    F. While playing Ms. Saucony's math-review game explained in Case 3.17, Kasee carefully listens to the game conductor's questions because she wants her team to win.

    G. After engaging in Ms. Castillo's learning activity on the application of ratios that is explained in Case 3.5, Brevin diligently works on Ms. Castillo's homework assignment because he now connects his ability to do mathematics with his ability to analyze basketball-related situations.

    H. Believing that she needs to learn how to better manage and invest money, Abby raises questions in her life-skills mathematics class.

    Did you and your colleague infer that students are intrinsically motivated in Cases D, E, G, and H and those in Cases A, B, C, and F are extrinsically motivated?

2. Discuss with a colleague the advantages and disadvantages—relative to keeping students engaged—of using a problem-based approach to teaching mathematics as opposed to a more traditional follow-the-textbook approach.

    Did you include the following points in your discussion?

    In general, students tend to be intrinsically motivated to engage in problem-based learning activities because they discover they can accomplish some of the things that are important to them by learning mathematics. Because problem-based learning activities tend to be more student centered and indirect than the activities in more traditional mathematics classrooms, students may initially consider them unusual and unstructured. Consequently, they may tend to get off-task until they learn how to take responsibility for their own learning. Initially, some students may resist the need to experience the perplexity that always accompanies real problem-solving activities. Consequently, at the beginning of a school year, introduce students to problem-based activities gradually.

3. Suppose Kwanso, one of your prealgebra students, asks you, "Why aren't there any negative prime numbers?" Formulate an answer to Kwanso's question using a how-might-we-find-that-out strategy for responding to memory-level questions. Discuss your answer with a colleague who is also engaging in this activity.
4. Suppose Sharyce, one of your prealgebra students, asks you, "Is zero odd, even, or neither?" Formulate an answer to Sharyce's question using a think-aloud strategy for responding to reasoning-level questions. Discuss your answer with a colleague who is also engaging in this activity.
5. Compare Mr. Smart's method of engaging students in a question-discussion session in Case 3.19 to Ms. Cramer's in Case 3.20. Which one of the two methods do you expect to use more often? Explain why. Discuss your responses with a colleague.
6. Observe a mathematics class session in a middle, junior high, or high school. For each allocated time period, classify the type of learning activity as (a) a large-group presentation; (b) a question-discussion session, (c) a cooperative-learning session, (d) an independent-work session, or (e) other type of session. Select one of the sessions and design an alternative type of learning activity. Explain the relative advantages and disadvantages of the two approaches for the given circumstances. Discuss your work on this activity with a colleague. Insert the resulting document in either the "Cognition, Instructional Strategies, and Planning" or "Motivation, Engagement, and Classroom Management" section of your working portfolio.
7. Relative to keeping students engaged, discuss with a colleague the advantages and disadvantages of a teacher using videotaped presentations instead of making the presentation entirely "live."
8. Relative to keeping students engaged, discuss with a colleague the advantages and disadvantages of a teacher using preprepared tasksheets for small-group cooperative-learning sessions.
9. Swap stories with a colleague about mathematics homework assignments that include examples of efficient, productive assignments, as well as examples of useless or counterproductive assignments.
10. Revisit Case 3.17 to refamiliarize yourself with Ms. Saucony's math-review game. For each of the game's 10 categories (e.g., geometric terms, symbols, and expressions), create three question cards with answers. Insert the cards—with a brief explanation as to how you would use them—into the "Motivation, Engagement, and Classroom Management" section of your working portfolio.

TRANSITIONAL ACTIVITY FROM CHAPTER 3 TO CHAPTER 4

In a discussion with two or more of your colleagues, address the following questions:

1. Why are people so mystified by mathematics? Why do so many students consider mathematics harder to learn than other school subjects?
2. Exactly what is a mathematics curriculum?
3. Who controls the design of mathematics curricula?
4. What is the book *Principles and Standards for School Mathematics* or *PSSM* (NCTM, 2000b)? Who wrote it? How does it influence mathematics curricula? Why should you be familiar with it?
5. In what ways should the mathematics lessons you design for your students differ from the lessons you experienced as a student? In what ways should they be the same?

chapter

4

Developing Mathematics Curricula

GOAL AND OBJECTIVES FOR CHAPTER 4

The Goal The goal of chapter 4 is to lead you to better understand how you control mathematics curricula and how you need to design and implement your mathematics curriculum so that your students learn meaningful mathematics.

The Objectives Chapter 4's goal is defined by the following set of objectives:

A. You will explain how teachers control curricula as they design teaching units and conduct lessons (discover a relationship) 10%.
B. You will distinguish between a traditional textbook-driven mathematics curriculum and a mathematics curriculum that leads to meaningful learning as recommended by the *Principles and Standards for School Mathematics* (NCTM 2000b) (construct a concept) 20%.
C. You will explain why traditional textbook-driven mathematics curricula tend to leave students to perceive mathematics as a linear sequence of meaningless definitions, symbols, rules, and algorithms (comprehension and communication) 20%.
D. You will distinguish between mathematical concept constructions, discoveries and mathematical inventions (construct a concept) 10%.
E. You will explain a nine-stage process by which people apply mathematics to solve real-life problems (comprehension and communication) 10%.
F. You will explain how mathematics courses are organized by teaching units (comprehension and communication) 15%.
G. You will incorporate the following phrases or words into your working vocabulary: "school curriculum," "mathematics curriculum," "course curriculum," "teaching unit," "unit goal," "learning objective," "*Principles and Standards for School Mathematics* (*PSSM*)" (NCTM, 2000b), "mathematical content of an objective," "learning level of an objective," "cognitive domain," "affective domain," and "meaningful learning of mathematics," (comprehension and communication) 15%.

A CURRICULUM

Many teachers and school administrators perceive a curriculum as the subject content (e.g., mathematical topics) listed in adopted textbooks or mandated state or local curriculum guides. Definitions range from that very narrow view to the broad, all-encompassing view that a curriculum is whatever persons experience in a setting (Oliva, 2001, p. 1–21).

Herein, the meaning of "curriculum" falls between those two extremes (Cangelosi, 2000a, p. 44):

- A *school curriculum* is a system of planned experiences (e.g., coursework, school-sponsored social functions, and contacts with school-sponsored services [e.g., the library]) designed to educate students.
- A *course curriculum* is a sequence of *teaching units* designed to provide students with experiences

that help them achieve specified learning goals. A *teaching unit* consists of (a) a learning goal defined by a set of specific objectives, (b) a planned sequence of lessons—each consisting of learning activities designed to lead students to achieve the lesson's objective, (c) mechanisms for monitoring student progress and using formative feedback to guide lessons, and (d) a summative evaluation of student achievement of the learning goal.

- A *mathematics curriculum* is a sequence of mathematics courses as well as other school-sponsored functions (e.g., a mathematics club) for the purpose of encouraging students to do mathematics.
- A *school district curriculum* is the set of all school curricula within the school district.
- A *state-level curriculum* is the set of all school district curricula within a state.

State-level, district-level, school-level, and mathematics curricula guidelines are articulated in documents housed in the files of virtually every school. The consistency between official curricula guidelines and actual school curricula varies considerably (Goodlad & Su, 1992; Oliva, 2001, pp. 438–589, Trafton, Reys, & Wasman, 2001). Obviously, a school's curriculum can be no more in line with official guidelines than the composite of the course curricula developed by its teachers.

Because mathematics is widely misunderstood to be a linear sequence of skills to be mastered one at a time in a fixed order, some people think teaching mathematics is a matter of following a prescribed curriculum guide or mathematics textbook. In reality, there are three reasons you must creatively develop curricula to succeed as a mathematics teacher. First of all, although state-level and district-level guidelines typically list competencies for mathematics courses (e.g., "Understand real numbers, rational and irrational, ways of representing numbers, relationships among numbers, and number systems" [Utah State Office of Education, 2001]), you are responsible for designing teaching units that lead students to develop those competencies. Textbooks present information and exercises on mathematical topics, but typical textbook presentations are pedagogically unsound from a constructivist perspective (American Association for the Advancement of Science, 2000; Battista, 1999; Ward, 2001) and inconsistent with *PSSM*'s Curriculum Principle (NCTM, 2000b, pp. 15–16). Thus, textbooks should be used only as references and sources of exercises—not religiously followed page by page. Furthermore, you need to tailor units to the unique characteristics of your students.

Second, although understanding of one mathematical topic (e.g., solving first-degree equations) is requisite to the understanding of another (e.g., solving quadratic equations), there is no fixed linear sequence that is optimal for all groups of students. Effective teaching requires teachers to arrange topics in response to feedback on their students' progress, diagnoses of the students' needs, and the students' interests (National Council of Teachers of Mathematics, 1991, pp. 110–119; Romberg, 1992).

Third, how you design and conduct lessons usually influences what your students learn about mathematics more than which mathematical topics are addressed in the lessons. Compare Cases 4.1, 4.2, 4.3, and 4.4:

CASE 4.1

Mr. Jackson's algebra II students have learned to solve quadratic equations by factoring, providing the left side of the equation in standard form can be factored easily (e.g., $x^2 - 6x - 16 = 0$). To teach them to solve any quadratic equation (e.g., $3x^2 + 5x + 1 = 0$), he introduces the quadratic formula by displaying it on an overhead transparency and saying, "Here is a formula for finding the complex roots of any quadratic equation. For example, suppose we want to solve for $3x^2 + 2x = 7$. Watch how much easier it is to use the formula than to try to factor the polynomial. First we rewrite the equation in standard form, and then . . ."

Mr. Jackson continues by working through several examples and assigning some equations for students to solve with the quadratic formula.

CASE 4.2

Ms. Youklic's algebra II students have learned to solve quadratic equations by factoring, providing the left side of the equation in standard form can be factored easily (e.g., $x^2 - 12x + 32 = 0$). She wants to teach them to solve any quadratic equation (e.g., $x^2 + 6x + 4 = 0$) using the quadratic formula. But instead of simply stating the formula, she introduces a real-world problem whose solution requires finding roots to quadratic equations. Because of some of her students' interest in sports, she takes an idea from a *Mathematics Teacher* (Eisner, 1986) and uses sports-related situations to establish a need for solving quadratic equations. Most of the equations are not easily factored, so she leads them through the completing-the-square algorithm for solving them.

Using inquiry instructional strategies to prompt inductive reasoning (explained in chapter 5), she leads students to generalize the quadratic formula from their experiences completing the square. Ms. Youklic then uses direct instructional strategies to help students develop their algorithmic skills with the quadratic formula.

CASE 4.3

Ms. Estrada tells her students, as she lists the rules for multiplying signed numbers on the whiteboard, "The product of two positive numbers is positive. The product of a positive number and a negative number is negative. The product of two negative numbers is positive. The product of any number and zero is zero. Do you understand?"

She directs the students to complete a tasksheet at their desks as she circulates among them looking at their responses, correcting errors, and answering questions. She notices the following on Bonita's paper:

$17 \times 10 = 170$ $\quad$ $.3347 \times 0 = 0$

$-4.1 \times 3 = -12.3$ $\quad$ $\left(\frac{1}{4}\right)\left(-\frac{2}{3}\right) = -\frac{2}{12}$

$-20 \times 9 = -180$ $\quad$ $(-5)(-10) = -50$

$(0)(-19) = 0$ $\quad$ $(-11)\left(-\frac{1}{17}\right) = -\frac{11}{17}$

$4\pi(8) = 32\pi$

They engage in the following conversation:

Ms. Estrada: Bonita, what is −5 times −10?
Bonita: Minus −50.
Ms. Estrada: But a negative times a negative is a positive.
Bonita: Why?
Ms. Estrada: Because that's the rule. See, I have it listed on the board and it's right here on page 23 of the text.

CASE 4.4

With an overhead projector, Mr. Cocora displays Exhibit 4.1 to his class and says, "A friend of mine works for the city's Traffic Control Department. She asked if we could help her solve a problem. I told her we were willing to try. Here's the situation. The department needs to estimate when and where city traffic is likely to be congested. My friend is going to be collecting data at the observation point marked here on the screen. The point is located at Highway 30 at the west edge of the city. She tells me that most of the traffic entering from west of the city passes this observation point.

"Here's the deal. Using a radar gun, the observer measures the direction and rate in miles per hour of a vehicle traveling on Highway 30. She needs some rules for using this one observation to estimate where the vehicle will be or was at any point in time. Do you think we can help?"

A discussion ensues in which the problem is clarified and Mr. Cocora explains that at traffic control (a) travel into the city is coded as a positive (+) number of miles per hour, (b) travel out of the city is coded as a negative (−) number of miles per hour, (c) locations on the city side of the observation point (i.e., to the east) are coded as a positive (+) number of miles, (d) locations outside of the city to the west of the observation point are coded as a negative (−) number of miles, (e) time in the future is coded as a positive (+) number of hours, and (f) time in the past is coded as a negative (−) number of hours.

Moving a toy car over the highway on the transparency, Mr. Cocora confronts the students with each of the following questions:

1. Where will a red Chevrolet that is headed into the city at 60 miles per hour be located 6 minutes from now?
2. Where will a dump truck that is headed out of the city at 60 miles per hour be located 6 minutes from now?
3. Where was the red Chevrolet 6 minutes ago?
4. Where was the dump truck 6 minutes ago?
5. Where is a green Toyota that is passing the observation point right now?
6. Where will a yellow van that is broken down and not moving in front of the observation point be in 5 minutes?

Mr. Cocora directs the class into collaborative subgroups that are to answer the six questions and then to generalize rules for the Traffic Control Department to use. The subgroups answer the questions by applying the relationship rate × time = distance as follows:

1. $(+60)(+0.1) = +6$ (6 miles east of the observation point)
2. $(-60)(+0.1) = -6$ (6 miles west of the observation point)
3. $(+60)(-0.1) = -6$ (6 miles west of the observation point)
4. $(-60)(-0.1) = +6$ (6 miles east of the observation point)
5. $(?)(0) = 0$ (in front of the observation point)
6. $(0)(+5) = 0$ (in front of the observation point)

After further discussion, the students settle on rules for the Traffic Control Department that are tantamount to the usual rules for multiplying signed numbers.

Exhibit 4.1
Mr. Cocora's Overhead Transparency for His Inquiry Lesson on Rules for Multiplying Signed Numbers.

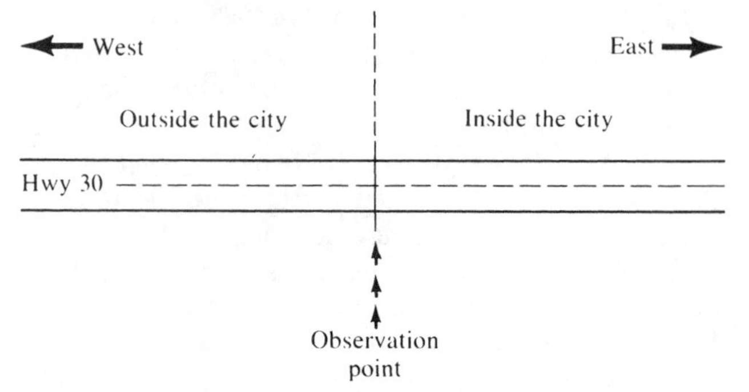

The next day Mr. Cocora restates the rules devised in the activity into more conventional textbook form. He then uses direct instruction to help them remember the rules.

Both Mr. Jackson and Ms. Youklic taught lessons on the quadratic formula, so their students learned about a different mathematical topic than Ms. Estrada's and Mr. Cocora's students. However, Mr. Jackson's and Ms. Estrada's lessons were similar in that both their classes learned to think of mathematics as a set of rules to be remembered, but not necessarily understood. Both Ms. Youklic's and Mr. Cocora's students discovered and invented mathematics. Thus, although according to a curriculum guideline or textbook table of contents, Mr. Jackson and Ms. Youklic were "covering" the same topic, the impact on students was vastly different.

Exhibit 4.2

One of the *TIMMS*-R Summaries Reported by the National Commission on Mathematics and Science Teaching for the 21st Century (2000, p. 10).

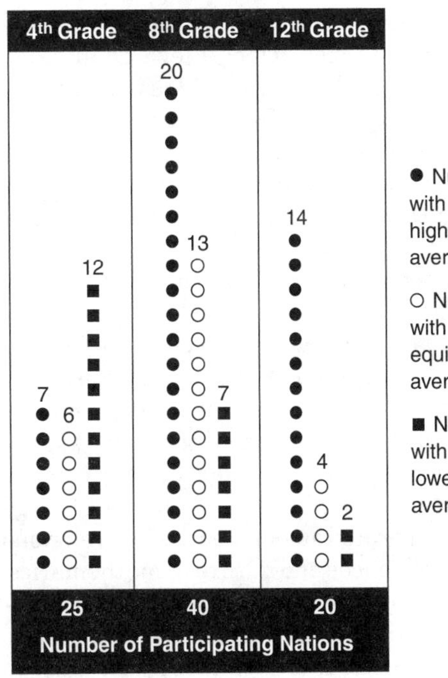

THE GAP BETWEEN RESEARCH-BASED MATHEMATICS CURRICULA AND TYPICAL PRACTICE

The Mystification of Mathematics

NAEP and TIMSS Results and Perceptions of Mathematics

Highly publicized comparisons of U.S. students' mathematics test scores to those of students in other countries have raised major concerns about the quality of mathematics education in U.S. schools (National Commission on Mathematics and Science Teaching for the 21st Century, 2000). In particular, periodic reports of the National Assessment of Educational Progress (NAEP) and the Third International Mathematics and Science Study (TIMSS) are prompting calls for reforms of mathematics curricula (Heddens & Speer, 2001, pp. 3–4). Exhibit 4.2 is one summary outcome of the TIMSS-R reported in *Before It's Too Late: A Report to the Nation from the National Commission on Mathematics and Science Teaching for the 21st Century* (2000, p. 10). For detailed NAEP and TIMSS results and updates go to the website of the National Center for Educational Statistics (http://nces.ed.gov/) and follow links to NAEP and TIMSS.

Besides the perception that U.S. students are not being taught mathematics as well as their counterparts in some other countries (e.g., Japan and Germany) (Mueller, 2001; NCTM, 2001b), you are well aware of how mystifying many people find mathematics. For example, did you ever attend a social gathering where it became known that you are or plan to be a mathematics teacher? If so, you probably heard comments such as, "Math?! So, you're some kind of a genius!", "Gosh, I avoid every math course possible!", "Math! How can you learn all that stuff? It's impossible for me!", "I can't imagine teaching math! How can you learn to work all those problems? I can't even balance my checkbook!", "Really, mathematics? My roommate is smart too. He does math all the time; he's an accounting major."

Why is mathematics commonly thought of as a mystifying subject that most people perceive as too difficult for them to learn? There are at least four contributing factors: (a) a failure to link mathematics to its historical origins, (b) a failure to comprehend the language of mathematics, (c) the fragmentation of mathematical topics into seemingly disconnected subtopics—especially by textbook-driven curricula, and (d) the failure of students to construct concepts and discover relationships for themselves.

Misunderstandings of the Historical Foundations of Mathematics

Mathematical vocabulary, symbols, concepts, relationships, and algorithms are typically presented in schools without references to their origins. Without an understanding of who, why, when, and where a concept (e.g., rational number) was defined, a relationship (e.g., area of a circular region = πr^2) was discovered, an algorithm (e.g., the sieve of Eratosthenes) was invented, a symbol (e.g., "$n!$") became conventional, or a theorem (e.g., $\sqrt{2} \notin \{rational\}$) was proven, mathematics is perceived as some sort of magic. Most students' exposure to the origins of mathematics is limited to either a passing mention of ancient Greeks (e.g., Pythagoras) in a geometry text, captioned pictures (e.g., of Sophie Germain) inserted—but not integrated—in a textbook, or a wall poster with captioned pictures of a few "geniuses of mathematics" (e.g., Carl Friedrich Gauss). Some students confuse the ancient Greek mathematicians with the Greek gods (see Exhibit 4.3) they read about in language arts courses (Cangelosi, 2001a). The fact that today's mathematics is a living, dynamic area for discoveries and inventions by modern and otherwise ordinary people—not only men and not only geniuses—is lost.

Of course, mathematics is a powerful tool for addressing problems, explaining phenomena, discovering relationships, and—because of the precision of its language—communicating and organizing ideas. However, mathematics was developed and continues to be developed by imperfect, fallible human beings who sometimes operated under political or religious pressures that influenced the nature of mathematics. Furthermore, because of Kurt Gödel's (1943) theorems of undecidablity, we now know that a complete and consistent mathematical system is impossible. To truly understand the power as well as the weaknesses of mathematics, one needs to understand its history and logical underpinnings. Consider Case 4.5:

Exhibit 4.3
Many People Confuse the Origins of Mathematics With Ancient Greek Mythology.

CASE 4.5

Mr. Edginton follows the lesson plan shown by Exhibit 4.4 to lead his prealgebra class to construct the concept of prime number.

He then introduces the following definition:

> An integer is *prime* if and only if it is greater than 1 and has no positive integer factors other than 1 and itself.

One student, Charog, asks, "Why can't 1 be prime? You can make only one rectangle out of 1 square just like you can make only one rectangle out of 2, 3, 5, 7, 11, and so forth. One seems even more prime than 2, 3, and the others because it has even fewer factors. And what about 0 and negative numbers like −1, −2, −3, −5, −7, −11, and so on? Why can't they be prime too?"

Mr. Edginton assumes that Charog's questions have a logical answer, although he's not really sure what it is. Believing in the importance of students' understanding why things are the way they are in mathematics, he doesn't want to do what Ms. Estrada did in Case 4.3 and simply say, "That's the rule." Thus, Mr. Edginton treats Charog's questions as reasoning-level questions and attempts to lead students to use reasoning to answer them. Thus, he initiates the following conversation with a think-aloud response:

Mr. Edginton: Wow, Charog, those are stimulating questions. I'm glad you asked. I'm not really sure why, but let's try to figure out why we don't consider integers less than 2 prime. Hmm, maybe because it's so obvious that 1 can't be factored, we don't need to call 1 prime. And 0 shouldn't be prime because—when you think about it—every number is a factor of 0. Seven is because $7 \times 0 = 0$ and 19 is because 19×0 is 0. Yeah, 0 should definitely not be prime.

Charog: But what about −11? −11 has the exact same factors as 11. So if 11 is prime, −11 should be prime.

Mr. Edginton: I see what you're saying. Does anyone have any ideas as to why a prime can't be negative? Okay, Heather.

Heather: The definition says primes have only two positive factors. Negative numbers have negative factors too.

Charog: Positive numbers have negative factors too. The Factors of 11 are 1, 11, −11, and −1.

Mr. Edginton: I hear what Charog is saying. I guess we could think of −11 as prime, but that's not the way it's usually defined.

Jahini: Maybe it's like square roots. Nine has two square roots, but radical nine stands for only +3, not −3.

Because of Mr. Edginton's belief in the invincibility of mathematics, he searched his brain for a logical, reasoned answer to Charog's question. What he needed to understand is that there is a historical an-

Exhibit 4.4
Mr. Edginton's Lesson Plan for Leading Students to Construct the Concept of Prime Number.

Objective: The student constructs the concept of prime number (construct a concept).

Stage 1 of the lesson:

Organize the class into the following triples of students. Give each triple four sets of plastic square regions—a red set, a yellow set, a green set, and a blue set. The numbers of square regions in the sets and the individual student's roles in each triple (i.e., Group I, II, . . ., VIII) are as follows:

Triple	Chairperson	Manager of the Squares	Data Recorder	Number of Red Squares	Number of Yellow Squares	Number of Green Squares	Number of Blue Squares
I	Shambay	Tricia	Ivory	2	10	18	26
II	Bernard	Magnella	Tanisha C.	3	11	19	27
III	Alicia	Tyler	Allison	4	12	20	28
IV	Yolanda	Byron	Tanisha L.	5	13	21	29
V	Rosalie	Brittany	Soo-Chen	6	14	22	30
VI	Ping	Wansoo	Roberta	7	15	23	31
VII	Seritta	Kellie	R. J.	8	16	24	32
VIII	Peggy	Saddam	Roberto	9	17	25	33

Direct each group to do the following:

A. Make as many red rectangular regions, yellow rectangular regions, green rectangular regions, and blue rectangular regions as possible using the following rules: (a) each rectangular region should be made from all of the squares of a single color, and (b) spatial orientation is irrelevant (e.g., count a 3 by 4 rectangular region as the same as a 4 by 3).
B. Use tally marks to record the number of different regions you are able to make with the four different numbers of square regions on the following chart:

Number of Squares	Tallies for Number of Rectangular Regions

Continued

swer to Charog's question rather than one that can be reasoned logically. Had he understood this, Case 4.5 might have gone as the one in Case 4.5.1:

CASE 4.5.1

Mr. Edginton: Thanks for raising those questions, Charog. We got the idea of prime number from the ancient Greeks from about 500 B.C. There was a really weird cult of followers of Pythagoras—known as the Pythagoreans. They believed that all of creation was mathematical and could be explained using the positive integers. It would have been contrary to their religion to conceive of nothingness—so they had no concept of 0. They visualized numbers as lengths or distances, so they had no negative numbers. That explains why their obsession with prime numbers would not include numbers less than 1. But why didn't they consider 1 prime? From what I've read about the Pythagoreans, I get the idea that 1—which they referred to as the Greek letter "α"—was set apart from other numbers because it was their unit. So 3—which they called "γ"—was 3α—all of their numbers except α were multiple units. I don't think anyone is sure why they didn't include 1, but that's the best explanation I've heard.

Heather: So, we could call 1 prime if we wanted?

Mr. Edginton: I don't want to do that, because the rest of today's mathematical world sticks with the Pythagorean idea of primes just like we accept the meaning of other words from their historical origins.

Charog: Just like we say "love" instead of zero in tennis. It's just a tradition.

Mr. Edginton: Thank you for that example. I hadn't thought of that before. It's just a tradition based on the history of primes to begin with 2.

Exhibit 4.4
Continued

Stage 2 of the lesson:

Reorganize the class into a large-group session. Have each triple clarify what it did for the whole class and then report its results. Have a student record the results from all the triples on the following transparency displayed for the whole class:

Number of Square Regions	Number of Rectangular Regions (tally)	Number of Square Regions	Number of Rectangular Regions (tally)
2		18	
3		19	
4		20	
5		21	
6		22	
7		23	
8		24	
9		25	
10		26	
11		27	
12		28	
13		29	
14		30	
15		31	
16		32	
17		33	

Stage 3 of the lesson:

In a large-group question-discussion session, lead students to discriminate between sets of rectangular regions that produced more than one square region and those that produced only one. Prompt students to articulate what is special about {2, 3, 5, 7, . . ., 31} as compared to {4, 6, 8, 9, 10, 12, 14, 15, . . ., 33}. If students appear ready, prompt them to develop a definition of numbers that would belong to {2, 3, 5, 7, . . ., 31, . . .}.

Stage 4 of the lesson:

For homework, have students test out their definition of what we will later refer to as "{primes}." Give them several selected other integers (e.g., 51) to test for whether or not they belong to {2, 3, 5, 7, . . ., 31, . . .}. The following day, have students discuss their results and refine their definition if warranted. Then make transition into a direct lesson to formalize definitions for "prime" and "composite.

Jarron: So why were the Pythager people so weird? What was wrong with them?

Mr. Edginton: That's a question that might be fun for us to answer. Jarron, there's a book on the shelf behind my desk. Grab it for me. It's the little black one called *Fermat's Enigma* (Singh, 1997). I'd like you to read a couple of paragraphs from—let's see—here, on page 6—start here.

Jarron: "In the 6th century B.C., Pythagoras of Samos was one of the most influential and yet mysterious figures . . ."

After the class session, Mr. Edginton decides that before introducing {primes} to a class again, he will have students do some background reading or watch a videotape (e.g., a segment from *Early History of Mathematics* [California Institute of Technology, 2000]) that lays the historical groundwork for the concept.

The historical perspective students gained in Case 4.5.1 removed some of the magic from their perceptions of mathematics while adding a bit of intrigue and human interest.

Exhibit 4.5
A Sample of Conventional Mathematics Shorthand Symbols.

$(A \cap B)'$ $\prod_{i=a}^{n} f(i)$ $\int_{-\infty}^{\infty} f(x)\, d_x$

$\forall x \in Z, \exists ! y \in Z \ni x + y = 0$

$\overrightarrow{AB}$ $\pm\sqrt{x}$ $\bigcup_{i=a}^{n} x_i$

$f: A \xrightarrow[\text{onto}]{1:1} B$ $\sin^{-1}\theta$

$x \equiv y \pmod{m}$ $S_1 \ V \ S_2$

$\Delta ABC \cong \Delta DEF$ $(-\infty, 74]$

Exhibit 4.6
An Algorithm for Integrating a Function Involves Only Adding, Subtracting, Multiplying, and Dividing Whole Numbers.

$$\int_1^2 \left(2x^3 - \frac{3}{x^2}\right) dx$$
$$= \int_1^2 (2x^3 - 3x^{-2})dx$$
$$= 2\int_1^2 x^3 dx - 3\int_1^2 x^{-2} dx$$
$$= 2\left(\frac{x^4}{4}\right)\Bigg|_1^2 - 3\left(\frac{x^{-1}}{-1}\right)\Bigg|_1^2$$
$$= 2\left(\frac{16}{4} - \frac{1}{4}\right) - 3\left(\frac{2^{-1}}{-1} - \frac{1}{-1}\right)$$
$$= 2\left(4 - \frac{1}{4}\right) - 3\left(-\frac{1}{2} + 1\right)$$
$$= 6$$

Miscomprehension of the Language of Mathematics

A glance through virtually any mathematics textbook reveals shorthand notations and symbols that appear strange to the uninitiated. Exhibit 4.5 contains a relatively minute number of examples. The meanings of such expressions are a mystery to those who have not learned the rules for translating them.

Consider the following shorthand expression:

$$\int_1^2 \left(2x^3 - \frac{3}{x^2}\right) dx$$

As you learned from your study of calculus and as illustrated by Exhibit 4.6, the expression simply denotes the number 6. Note that the algorithm for simplifying the expression can be boiled down to nothing more than permutations of the four fundamental operations of addition, subtraction, multiplication, and division with whole numbers. Most people do not think of adding, subtracting, multiplying, and dividing whole numbers as difficult, but integrating a function is mystifying to them. Why? Because they have not learned or do not remember how to translate the shorthand expression.

From their preschool days, students are taught to read texts presented in rectangular arrays by following words and symbols from the very top of each array progressing toward the bottom row by row, moving their eyes from left to right. It is a struggle for most primary-grade students to develop this cognitive and psychomotor skill (Cunningham, Moore, Cunningham, & Moore, 1995, pp. 46–48). After being conditioned to read text in this manner, students are confronted with mathematical expressions that are often meant to be read from bottom up, right to left, and diagonally as illustrated by Exhibit 4.7.

Negotiating shorthand symbols and expressions is only part of the struggle people have with the language of mathematics—or at least the way that language is presented in typical textbooks. Often, students can do the mathematics required to solve a problem or respond to a task presented in a textbook, by a teacher, or on a test, but they fail to solve it or respond to the prompt correctly because they do not comprehend the question they are being asked (Cangelosi, 2001). In Case 1.3 in which Casey Rudd interviewed for his position at Rainbow High School, he and Vanessa Castillo argued that with typical textbook-driven curricula, students do not learn to comprehend the mathematical questions before diving into algorithms for supposedly answering questions they failed to comprehend in the first place. They illustrated this with Casey's example "to solve for x in the equation $3(10 - x) = 15$" and Vanessa's example of the multiple-choice prompt that begins, "If $a + b = 10$ and $ab = 5$, then $a^2 + b^2 =$ what?" Casey also used Exhibit 1.4's table of contents from a prealgebra textbook to illustrate that people are tricked into thinking that one mathematical idea expressed multiple ways represents multiple mathematical ideas. This misperception stems not only from miscomprehension of the language, but also from fragmentation of mathematical topics in mathematics curricula.

Exhibit 4.7
Reading Mathematical Expressions Does Not Proceed in the Usual Left-to-Right, Top-to-Bottom Order.

The expression $\frac{7}{x} - (5 - 3x)^2$ is read in this order:

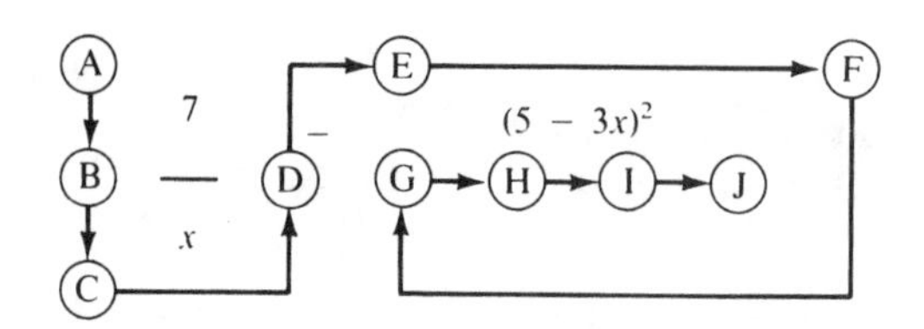

Begin at Ⓐ, reading "7," go down to Ⓑ to read "divide by"; then continue down to Ⓒ to read "x" and up to Ⓓ for "subtract." Go over to Ⓔ for "quantity of," up and over to Ⓕ for "square the quantity," down to Ⓖ for "5," right to Ⓗ to "subtract," right to Ⓘ to "three times," and right to Ⓙ for "x."

Fragmentation of Mathematical Topics

As Principal Harriet Adkins expressed in Case 1.3 when she viewed Exhibit 1.4, the number of topics listed in the table of contents of most mathematics textbooks appears overwhelming. But Casey pointed out redundancies in the list with related topics or even the same topic expressed in two different ways (e.g., divisors and multiples) being fragmented and disconnected in the minds of students because of typical textbook presentations (Cangelosi, 1989b; 2001).

Multiple expressions for the same concept or relationship and multiple algorithms for reaching the same result are ubiquitous in mathematics. Virtually everything in calculus revolves around one concept: limit. Yet a typical sequence of college calculus courses requires students to work with between 800 and 1,200 pages of text. The major thrust of the so-called "calculus-reform movement" was to move toward a "leaner and livelier" curriculum by reconnecting traditionally fragmented topics with the emphasis on depth instead of breadth (Cole, 1993; Culotta, 1992; Rowley, 1996).

Failure to Construct Concepts and Discover Relationships for Oneself

Learning mathematics is *meaningful* to students if students apply that mathematics to situations they consider important (Coxford, 1995; NCTM, 1989a; 2000b, pp. 15–16). From a constructivist perspective, how well students are able to apply mathematics to new, meaningful situations depends on their having constructed certain key concepts and discovered key relationships in their own minds (Cangelosi, 2000a, pp. 211–247; Heddens & Speer, 2001, pp. 11–16; Kamii & Warrington, 1999; Parsons, Hinson, & Sardo-Brown, 2001, pp. 418–445; Steen, 1999). However, in the most commonly conducted type of mathematics lessons (e.g., Mr. Jackson's in Case 4.1 and Ms. Estrada's in Case 4.3), concepts (e.g., rational number) and relationships (e.g., the product of two negative reals is positive) are presented without providing students with the experiences that lead them to construct the concepts or discover the relationships for themselves (Cangelosi, 2001a; Jesunathadas, 1990; Post & Cramer, 1989).

Demystifying Mathematics

Human Concept Constructions, Discoveries, and Inventions

Mathematics seems far less mystifying once one understands that it originates and continues to develop from the work of mortal, mistake-prone human beings attempting to solve problems and explain the world in which they live. Learning mathematics from a historical perspective may begin with the distinction between mathematical *concept constructions, discoveries,* and *inventions.*

A *concept* is an abstract category people mentally construct from a set of two or more elements that share some common attributes. Mathematical concepts have been and continue to be identified, studied, and defined. For example:

- Prehistoric people began constructing quantitative concepts by discriminating between collections with only a few objects and collections with many objects.
- Because of their belief that everything is governed by natural numbers and that they could elevate themselves to a more divine state by better understanding these numbers, the Pythagoreans (late sixth century B.C.) identified the concept of perfect number (i.e., a number [e.g., as 6, 28, 496, 8128, and 33550336] that equals the sum of its proper divisors) (Calinger, 1999, p. 72; Hoffman, 1988, pp. 8–25).
- In Case 4.5, Mr. Edginton's students constructed the concept of prime number.
- Prior to the beginnings of recorded history, people constructed the concept of right triangle.

- Girolamo Cardano (1501–1576) and Albert Girard (1593–1632) constructed the concept of imaginary numbers while examining the roots of equations such as $x^4 - 4x + 3 = 0$ (Baumgart, 1989; Burton, 1999, pp. 295–302).

A *relationship* is a particular association between (a) concepts, (b) a concept and a specific, or (c) specifics. Mathematical relationships have been and continue to be discovered. For example:

- Euclid (323–285 B.C.) discovered that $2^{n-1}(2^n - 1)$ is perfect whenever $2^n - 1$ is prime (Hoffman, 1988, pp. 9–14).
- Prompted by a teacher to explore the proper divisors of powers of 2 (i.e., $\{2^t : t \in$ {natural numbers} as it appears Euclid did [Singh, 1997, pp. 11–13]), Amanda Cangelosi discovered that each power of 2 is 1 greater than the sum of its proper divisors (i.e., [$\sum_{i=1}^{t} 2^{i-1} = 2^t - 1$).
- More than 1,000 years before the birth of Pythagoras, Babylonians discovered that $a^2 + b^2 = c^2$ where c is the length of a right triangle and a and b are the lengths of the triangle's legs (as evident from the deciphering of clay tablet known as "Plimpton 322" by Neugebauer and Sachs in 1945) (Burton, 1999, pp. 68–75).
- The French jurist Pierre de Fermat (1601–1665) conjectures that $\nexists\, a, b, c, \in$ {integers} $\ni a^n + b^n = c^n$ for any integer $n > 2$ (i.e., what we now call "Fermat's Last Theorem").
- Peter Roth in 1608 or Albert Girard in 1692 discovered the relationship we now refer to as "The Fundamental Theorem of Algebra" (Burton, 1999, pp. 498–500; Western, 1989): Every n^{th}-degree polynomial equation with $n \geq 1$ has at least one complex root and no more than n complex roots.
- From 1939 to 1945, a team of British and American cryptologists, number theorists, logicians (e.g., Alan Turing), language specialists, and others at England's Government Code and Cypher School at Bletchley Park, Buckinghamshire discovered mathematical relationships that allowed them to break the German enigma code, giving the Allies a decided advantage over the Nazis during World War II (Singh, 1999, pp. 143–189).
- In Case 4.4, Mr. Cocora's students discovered that the product of two negative real numbers is positive.

An *invention* is a particular device contrived by one or more persons. Mathematical techniques, proofs, algorithms, language structures, conventions, and technology have been and continue to be developed. For example:

- To keep track of possessions (e.g., goats in a herd), an enumeration algorithm was developed and incorporated into the Papuan language of New Guinean people that used a sequence of pointing to one's own body parts (Ifrah, 2000, pp. 3–22).
- Euclid (323–285 B.C.) devised a proof demonstrating that $2^{n-1}(2^n - 1)$ is perfect whenever $2^n - 1$ is prime (Dence & Dence, 1999, pp. 258–259).
- Eratosthenes of Cyrene (276–194 B.C.) devised an algorithm for identifying primes that is known today as the "sieve of Eratosthenes" (Burton, 1997, pp. 46–48; Tattersall, 1999, pp. 79–81).
- One of the earliest of some 400 known proofs of the Pythagorean theorem was devised by Chou-pei Suan-ching of China in about 250 B.C. (see Exhibit 4.8). In 1881, U.S. President James Garfield contributed his own proof (Exhibit 4.9) that was also known in Arabia and India in the seventh century (Veljan, 2000).
- Carl Friedrich Gauss (1777–1855) proved and named the Fundamental Theorem of Algebra (Fraleigh, 1999, p. 407; Western, 1989).
- Various forms for today's shorthand symbol for percent ("%") have been conventional in business computations since the end of the 15th century. But its origins are associated with the Roman emperor Augustus' tax levies on all goods sold at

Exhibit 4.8
Figure Illustrating Chou-pei Suan-ching's Proof of the Pythagorean Theorem (ca. 250 B.C.).

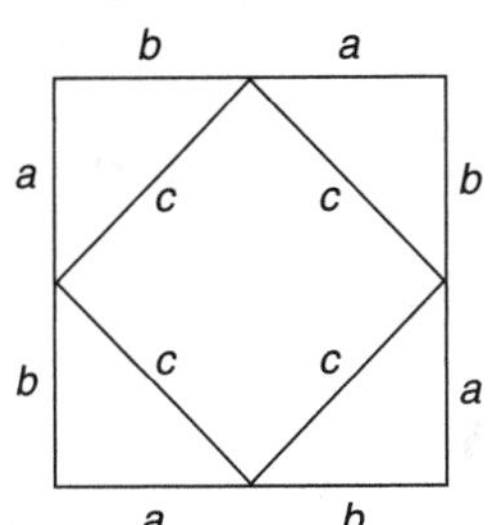

$$(a + b)^2 = c^2 + 4 \cdot \frac{ab}{2} \Rightarrow a^2 - b^2 = c^2$$

Exhibit 4.9
Figure Illustrating James Garfield's Proof of the Pythagorean Theorem (1881).

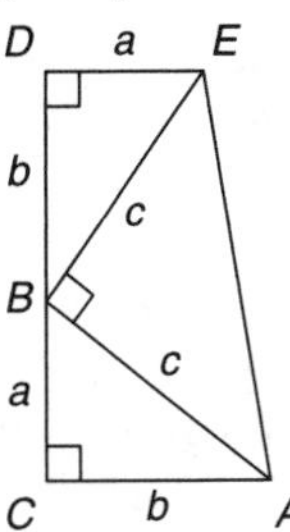

$$\frac{a + b}{2} \cdot (a + b) = 2 \cdot \frac{ab}{2} + \frac{1}{2}c^2 \Rightarrow a^2 + b^2 = c^2$$

auctions at the rate of 1/100 near the year A.D. 1 (Amundson, 1989).

- In 1811, Carl Friedrich Gauss represented imaginary numbers geometrically by representing them as ordered pairs of real numbers in the complex plane. René Descartes (1596–1605) originated the terms "real" and "imaginary." Leonhard Euler (1707–1783) introduced the letter "i" for $\sqrt{-1}$ (Baumgart, 1989; Burton, 1999, p. 486).
- The symbol "π" to represent the ratio of the circumference of a circle to its diameter was first introduced by English writer William Jones in 1706 (von Baravelle, 1989).
- Max Newman, a mathematician working at Bletchely Park on breaking the German enigma code during World War II, invented a way to mechanize the cryptanalysis of the Nazis' most sophisticated codes. An engineer, Tommy Flowers, built the so-called "Colossus" machine Newman designed—the first programmable computer (Singh, 1999, pp. 243–245).
- Building upon the work of hundreds of others (e.g., Sophie Germain, Évariste Galois, Yutaka Taniyama, Goro Shimura, Gerhard Frey, Ken Ribet, Nick Katz, and Richard Taylor), Andrew Wiles, a Princeton University mathematics professor, proved Fermat's Last Theorem in 1995—357 years after Fermat proposed it (Singh, 1997).

Once your students have a realistic idea of how mathematics is developed, they are more likely to do real mathematics—contributing to the history of mathematics by constructing concepts, discovering relationships, and inventing mathematics for themselves. Consider Case 4.6:

CASE 4.6

Several days after Casey Rudd conducted the lesson on lateral surface area of right cylinders as indicated by Exhibit 1.3 of Case 1.3, several students describe the shape of several different size fish bowls that one of them, Norton, brought to class (see Exhibit 4.10). Thus, they constructed the concept of this particular three-dimensional figure. They invented a name for it: "Norton."

Noting the attributes of a Norton (e.g., two surfaces with the same size circular regions), Mr. Rudd's students discover the following relationship: The surface area of a Norton $\approx 2\pi(2s^2 + 2sw - t^2)$ where s = the radius of the circular side panels, w = the width of one of the curved sides, and t = the radius of the opening at the top.

The relationship is the basis for an algorithm they invent to compute surface areas of Nortons.

Because Mr. Rudd had previously taught these students mathematics from a historical perspective, they were inclined to attempt to construct the concept, discover the relationship, and invent the name and algorithm alluded to in Case 4.6. That is what doing real mathematics is all about.

Exhibit 4.10
A Norton.

Exhibit 4.11 is a relatively minute list of available references and resources for learning the history of mathematics and teaching mathematics from a historical perspective.

Engage in Activity 4.1:

Activity 4.1

From your working professional portfolio, select a unit from the mathematics course outline for which you developed a syllabus when you engaged in Activity 1.2. Select a mathematical topic that you would conceivably teach from a historical perspective. Teach yourself something about the history of that topic by studying one or more sources such as those from Exhibit 4.11.

Write several paragraphs explaining how you would consider incorporating what you learned about the history of that mathematical topic into a lesson for students. Discuss your work with a colleague who is also engaging in Activity 4.1. Refine your paper, if you think you should, in light of your discussion. Insert the paper into one of the following sections of your working portfolio: (a) Cognition, Instructional Strategies, and Planning, (b) Multicultural Education, or (c) Mathematics and Historical Foundations of Mathematics.

Principles and Standards for School Mathematics

In the early 1900s, experimental studies of teaching and learning (James, 1890; Thorndike & Woodworth,

Exhibit 4.11
A Relatively Minute Sample of Resources for Learning about the History of Mathematics and for Teaching Mathematics From a Historical Perspective.

A&E Network. (1998). *Sir Issac Newton* [Videotape]. New York: Author.

Barrow, J. D. (1992). *Pi in the sky: Counting, thinking, and being.* Boston: Back Bay Books.

Bell, E. T. (1965). *Men of mathematics.* New York: Simon and Schuster.

Boyer, C. G. (1991). *A history of mathematics* (2nd ed.). New York: Wiley.

Burton, D. M. (1999). *The history of mathematics: An introduction* (4th ed.). Boston: WCB/McGraw-Hill.

Cajori, F. (1985). *A history of mathematics* (4th ed.). New York: Chelsea Publishing.

California Institute of Technology. (2000). *Early history of mathematics* [Videotape]. Pasadena, CA: Author.

Calinger, R. (1999). *A contextual history of mathematics.* Upper Saddle River, NJ: Prentice-Hall.

Dunham, W. (1994). *The mathematical universe: An alphabetical journey through the great proofs, problems, and personalities.* New York: Wiley.

Dunham, W. (1999). *Euler: The master of us all.* Washington: Mathematical Association of America.

Eves, H. (1983). *Great moments in mathematics after 1650.* Washington: Mathematical Association of America.

Eves, H. (1983). *Great moments in mathematics before 1650.* Washington: Mathematical Association of America.

Gullberg, J. (1997). *Mathematics: From the birth of numbers.* New York: W. W. Norton.

Hoffman, P. (1988). *The man who loved only numbers: The story of Paul Erdös and the search for practical truth.* New York: Hyperion.

Hoffman, P. (1998). *Archimedes' revenge: The joys and perils of mathematics.* New York: Fawcett Columbine.

Ifrah, G. (2000). *The universal history of numbers: From prehistory to the invention of computers.* New York: John Wiley & Sons.

Katz, V. J. (1998). *A history of mathematics: An introduction* (2nd ed.). New York: Addison-Wesley/Longman.

Katz, V. J. (Ed.) (2000). *Using history to teach mathematics: An international perspective.* Washington: The Mathematical Association of America.

Lewinter, M., & Widulski, W. (2002). *The saga of mathematics: A brief history.* Upper Saddle River, NJ: Prentice-Hall.

Lynch, J. (1997). *The proof* [Videotape]. Boston: WBGH Boston Video.

Mathematical Association of America. (2000). *Early history of mathematics* [Videotape]. Washington: Author.

Mathematical Sciences Research Institute. (1993). *Fermat's last theorem* [Videotape]. Berkeley: Author.

McLeish, J. (1991). *Number: The history of numbers and how they shape our lives.* New York: Fawcett Columbine.

National Council of Teachers of Mathematics. (1989b). *Historical topics for the classroom.* Reston, VA: Author.

National Council of Teachers of Mathematics. (2000). *Mathematics Teacher, 93* [The November, 2000 issue focuses on mathematics history for the classrooms; it includes the following articles: "Mathematic in Search of History" by D. T. Barry; "A Visit from Pythagoras—Using Costumes in the Classroom" by L. H. Shirley; "The Mathematics of Levi ben Gershon" by S. Simonson; "Word Histories: Melding Mathematics and Meanings" by R. N. Rubenstein & R. K. Schwartz; "Mathematics in the Age of Jane Austen: Essential Skills of 1800" by S. I. B. Gray; "Kepler and Wiles: Models of Perseverance" by P. G. Shotsberger; "Correspondences from Mathematicians" by J. Horn, A. Zamierowski, & R. Barger; "The Evolutionary Character of Mathematics" by R. M. Davitt; "Mathematicians Are Human Too" by J. E. Lightner; "From the Top of the Mountain" by D.W. Smith; "The Role of History in Mathematics Class" by G. L. Marshall & B. S. Rich; "Benoit Mandelbrot: The Euclid of Fractal Geometry" by D. R. Camp.]. Reston, VA: NCTM.

Reid, C. (1996). *Julia: A life in mathematics.* Washington: Mathematical Association of America.

Salem, L., Testard, F., & Salem, C. (1992). *The most beautiful mathematical formulas.* New York: Wiley.

Seife, C. (2000). *Zero: The Biography of a Dangerous Idea.* New York: Viking.

Singh, S. (1997). *Fermat's enigma.* New York: Walker and Company.

Singh, S. (1999). *The code book: The science of secrecy from ancient Egypt to quantum cryptography.* New York: Anchor Books.

Stein, S. (1999). *Archimedes: What did he do besides cry Eureka?* Washington: Mathematical Association of America.

Veljan, D. (2000). The 2500-year-old Pythagorean theorem. *Mathematics Magazine, 73,* 259–272.

WBGH Boston Video (1999). *Decoding Nazi secrets* [Videotape]. Boston: Author.

Continued

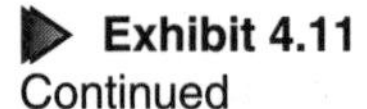

Exhibit 4.11
Continued

Internet Sources:

As you will discover by entering "mathematics history" with a search engine, there are thousands of web pages related to the historical foundations of mathematics and famous mathematical problems. The following are examples:

http://archives.math.utk.edu/topics/history.html

http://aleph0.clarku.edu/~djoyce/mathhist/mathhist.html

http://forum.swarthmore.edu/~isaac/mathhist.html

http://www.cs.yale.edu/homes/tap/past-women.html

http://www.seanet.com/~ksbrown/ihistory.htm

http://www.math.buffalo.edu/mad/madhist.html

1901) undermined the faculty psychology and formal discipline principles upon which the prevailing mathematics curricula of the day were based (Strom, 1969, pp. 147–208). Numerous curriculum-reform efforts attempted to bring the teaching of mathematics more in line with research-based principles indicating that students need experience in discovering and inventing mathematics and utilizing mathematics to solve real-life problems [The 1908 Committee of Fifteen on the Geometry Syllabus (Kinney & Purdy, 1952, pp. 22–23)]; The Joint Commission to Study the Place of Mathematics in Secondary Education (NCTM, 1940); The School Mathematics Study Group (Begle, 1958); *An Agenda for Action: Recommendations for School Mathematics of the 1980s* (NCTM, 1980); *Educating Americans for the 21st Century* (National Science Board Commission on Precollege Education in Mathematics, Science, and Technology, 1983); *The Mathematical Science Curriculum K–12: What Is Fundamental and What Is Not* (Conference Board of the Mathematical Sciences, 1983a); *New Goals for Mathematical Sciences Education* (Conference Board of the Mathematical Sciences, 1983b)]. Advances in cognitive science emphasized the need for constructivist-based instructional strategies for meaningful learning (Cangelosi, 2000a, pp. 211–247). Unfortunately, the impact of these curriculum-reform projects on actual mathematics curricula was disappointing. The gap between typical mathematics instruction and how research-based principles indicate mathematics should be taught continued into the 1990s (Battista, 1999; Brophy, 1986; Romberg, 1992; Smith, 2001, pp. 1–36). Jesunathadas (1990) described a tiresome pattern for the most commonly practiced method of teaching mathematics:

> The teacher introduces a topic by stating a rule or definition and then demonstrating it with textbook examples on the chalkboard or on overhead transparencies. Students work on exercises at their seats, with the teacher providing individual help to those experiencing difficulties. Similar exercises are completed for homework and checked as "right" or "wrong" at the beginning of the next class period. Homework exercises that were particularly troublesome are worked out for the class, either by the teacher or student volunteers.
>
> Emphasis is almost entirely on algorithmic skills. Neither understanding of why rules and algorithms work nor applications to real-life problems are stressed. The teacher's expectations for all but a few students are very low. The pace of lessons is slow, with plodding, repetitive exercises.

Research-based principles do not suggest that these expository, drill, and review lessons should be eliminated, but rather that they should not continue to be the dominant form of instruction. Acquiring algorithmic skills should not be the primary goal of school mathematics curricula.

The publication of the *Curriculum and Evaluation Standards for School Mathematics* (NCTM, 1989a) followed by a decade filled with projects (e.g., numerous articles in NCTM's journals *Mathematics Teaching in the Middle School* and *Mathematics Teacher*, NCTM's *Addenda Series, Assessment Standards for Teaching Mathematics* [NCTM, 1995], *MathFinder Sourcebook: A Collection of Resources for Mathematics Reform* [Kreindler & Zahm, 1992], and federally funded workshops and video productions for teachers [Cangelosi, 1990]) providing resources and inservice for teachers to implement those standards served to narrow the gap between research-based practice and typical practice (Thompson & Senk, 2001). However, the gap is still significant and the tiresome sentences of its preface (NCTM, 2000b, p. ix):

> *Principles and Standards for School Mathematics* is intended to be a resource and guide for all who make decisions that affect the mathematics education of students in prekindergarten through grade 12. The

recommendations in it are grounded in the belief that all students should learn important mathematical concepts and processes with understanding. *Principles and Standards* makes an argument for the importance of such understanding and describes ways students can attain it.

A vision for school mathematics is described (NCTM, 2000b, p. 3):

> Imagine a classroom, a school, or a school district where all students have access to high-quality, engaging mathematics instruction. There are ambitious expectations for all, with accommodations for those who need it. Knowledgeable teachers have adequate resources to support their work and are continually growing as professionals. The curriculum is mathematically rich, offering students opportunities to learn important mathematical concepts and procedures with understanding. Technology is an essential component of the environment. Students confidently engage in complex mathematical tasks chosen carefully by teachers. They draw knowledge from a wide variety of mathematical topics, sometimes approaching the same problem from different mathematical perspectives or representing the mathematics in different ways until they find methods that enable them to make progress. Teachers help students make, refine, and explore conjectures on the basis of evidence and use a variety of reasoning and proof techniques to confirm or disprove those conjectures. Students are flexible and resourceful problem solvers. Alone or in groups and with access to technology, they work productively and reflectively, with the skilled guidance of their teachers. Orally and in writing, students communicate their ideas and results effectively. They value mathematics and engage actively in learning it.

Resources for designing and implementing *PSSM*-based curricula will be available (NCTM, 2000b, p. xii):

> The [National Research] Council has established a task force to develop a series of materials, both print and electronic, with the work title of Navigations to assist and support teachers as they work in realizing the *Principles and Standards* in their classrooms, much as the Addenda series did following the release of the *Curriculum and Evaluation Standards for School Mathematics.*

PSSM is available as a 401-page book as well as on a CD-ROM. The full text can also be accessed electronically through NCTM's website (*www.nctm.org*). It includes six principles for school mathematics (i.e., principles for equity, curriculum, teaching, learning, assessment, and technology), five content standards and expectations (i.e., for number and operations, algebra, geometry, measurement, and data analysis and probability), and five process standards (i.e., for problem solving, reasoning and proof, communication, connections, and representation).

PSSM'S EQUITY PRINCIPLE

From chapter 2, reread "*PSSM* and Equity."

No longer is it acceptable in the mathematics education community to concentrate our efforts primarily on the so-called "mathematically talented" (a misnomer) with their stereotypical "clean-cut" images. Other students are just as mathematically talented (whatever that means) but not everyone always responds well to typical follow-the-textbook instruction nor have their mathematical achievements been accurately assessed (Kohn, 2001; Thompson, 2001).

Furthermore, students who are not inclined toward the traditional mathematically oriented fields (e.g., engineering, physics, or computer science) also need to do mathematics for real-life decision making (NCTM, 2000b, p. 20).

Because of increased emphasis on multicultural education and focus on the diversity of our school populations, it is popular for "teacher's wraparound" editions of school mathematics textbooks to include inserts labeled "Multicultural Education" and "Limited English Proficiency." Such inserts should aid you to uphold *PSSM*'s Equity Principle. Unfortunately, with some textbooks, this is not the case. The explanation of the "Multicultural Education" inserts in one very popular prealgebra textbook states (in its entirety), "MULTICULTURAL EDUCATION highlights how other people and cultures have influenced mathematics or presently use it." Keeping the words "other people" fresh in your mind, read the first "Multicultural Education" insert from that textbook: "Tom Fuller, an African-American enslaved in Virginia from 1724 until his death in 1790, possessed very amazing mental skills. Among these skills was the ability to mentally multiply any two nine-digit numbers and announce the product almost instantly."

Two things bother me about that insert: First, the inference is that African-American students are "other people." Second, is it being suggested that it is extraordinary for an African-American to possess such skills? Other inserts point out the baseball accomplishments of Jackie Robinson and the tennis accomplishments of Althea Gibson with no mention of any work in mathematics.

The explanation for the teacher regarding the "Limited English Proficiency" insert is, "LIMITED ENGLISH PROFICIENCY features methods to reach and teach students for whom English is not their primary language." Here is a typical insert—not exactly in-depth help for the teacher: "Provide students with additional practice using the Mini-Lab materials. Give them several simple equations and allow them an unrestricted amount of time to solve the equations.

Be sure students understand the terms *addition, subtraction, multiplication,* and *division,* and they understand the effects of inverse operations."

The point here is not to be critical of these textbooks; the point is that you cannot depend on such inane, simplistic bits of lip service to teach consistently with *PSSM*'s Equity Principle. Multicultural education and accommodating students need to be incorporated in your mathematics curriculum in a serious and thoughtful way. Use your knowledge of your students as individuals and devise meaningful and, sometimes, innovative strategies as did the teachers in Cases 2.34 and 2.35. You will study additional examples of teachers incorporating sound multicultural education principles in the remainder of this book. Make use of Exhibit 2.21's references.

PSSM'S CURRICULUM PRINCIPLE

According to *PSSM*'s Curriculum Principle, "A curriculum is more than a collection of activities: it must be coherent, focused on important mathematics, and well articulated across the grades." (NCTM, 2000b, p. 14)

The following explication of the Curriculum Principle (NCTM, 2000b, pp. 14–16) guided the curriculum-development strategies put forth for your understanding and consideration throughout this book:

> A school mathematics curriculum is a strong determinant of what students have an opportunity to learn and what they do learn. In a coherent curriculum, mathematical ideas are linked to and build on one another so that students' understanding and knowledge deepens and their ability to apply mathematics expands. An effective mathematics curriculum focuses on important mathematics—mathematics that will prepare students for continued study and for solving problems in a variety of school, home, and work settings. A well-articulated curriculum challenges students to learn increasingly more sophisticated mathematical ideas as they continue their studies.
>
> Mathematics comprises different topical strands, such as algebra and geometry, but the strands are highly interconnected. The interconnections should be displayed prominently in the curriculum and in instructional materials and lessons. A coherent curriculum effectively organizes and integrates important mathematical ideas so that students can see how the ideas build on, or connect with, other ideas, thus enabling them to develop new understandings and skills.
>
> Curricular coherence is also important at the classroom level. Researchers have analyzed lessons in the videotape study of 8th-grade mathematics classrooms that was part of the Third International Mathematics and Science Study (Stigler & Hiebert, 1999). One important characteristic of the lessons had to do with the internal coherence of the mathematics. The researchers found that typical Japanese lessons were designed around one central idea, which was carefully developed and extended; in contrast, typical American lessons included several ideas or topics that were not closely related and not well developed.
>
> In planning individual lessons, teachers should strive to organize the mathematics so that fundamental ideas form an integrated whole. Big ideas encountered in a variety of contexts should be established carefully, with important elements such as terminology, definitions, notation, concepts, and skills emerging in the process. Sequencing lessons coherently across units and school years is challenging. And teachers also need to be able to adjust and take advantage of opportunities to move lessons in unanticipated directions.
>
> School mathematics curricula should focus on mathematics content and processes that are worth the time and attention of students. Mathematics topics can be considered important for different reasons, such as their utility in developing other mathematical ideas, in linking different areas of mathematics, or in deepening students' appreciation of mathematics as a discipline and as a human creation. Ideas may also merit curricular focus because they are useful in representing and solving problems within or outside mathematics.
>
> Foundational ideas like place value, equivalence, proportionality, function, and rate of change should have a prominent place in the mathematics curriculum because they enable students to understand other mathematical ideas and connect ideas across different areas of mathematics. Mathematical thinking and reasoning skills, including making conjectures and developing sound deductive arguments, are important because they serve as a basis for developing new insights and promoting further study. Many concepts and processes, such as symmetry and generalization, can help students gain insights into the nature and beauty of mathematics. In addition, the curriculum should offer experiences that allow students to see that mathematics has powerful uses in modeling and predicting real-world phenomena. The curriculum also should emphasize the mathematics processes and skills that support the quantitative literacy of students. Members of an intelligent citizenry should be able to judge claims, find fallacies, evaluate risks, and weigh evidence (Price, 1997).
>
> Although any curriculum document is fixed at a point in time, the curriculum itself need not be fixed. Different configurations of important mathematical ideas are possible and to some extent inevitable. The relative importance of particular mathematics topics is likely to change over time in response to changing perceptions of their utility and to new demands and possibilities. For example, mathematics topics such as recursion, iteration, and the comparison of algorithms are receiving more attention in school mathematics because of their increasing relevance and utility in a technological world.
>
> Learning mathematics involves accumulating ideas and building successively deeper and more refined understanding. A school mathematics curriculum should provide a road map that helps teachers guide

students to increasing levels of sophistication and depths of knowledge. Such guidance requires a well-articulated curriculum so that teachers at each level understand the mathematics that has been studied by students at the previous level and what is to be the focus at successive levels. For example, in grades K–2 students typically explore similarities and differences among two-dimensional shapes. In grades 3–5 they can identify characteristics of various quadrilaterals. In grades 6–8 they may examine and make generalizations about properties of particular quadrilaterals. In grades 9–12 they may develop logical arguments to justify conjectures about particular polygons. As they reach higher levels, students should engage more deeply with mathematical ideas and their understanding and ability to use the knowledge is expected to grow.

Without a clear articulation of the curriculum across all grades, duplication of effort and unnecessary review are inevitable. A well-articulated curriculum gives teachers guidance regarding important ideas or major themes, which receive special attention at different points in time. It also gives guidance about the depth of study warranted at particular times and when closure is expected for particular skills or concepts.

PSSM'S TEACHING PRINCIPLE

According to *PSSM*'s Teaching Principle, "Effective mathematics teaching requires understanding what students know and need to learn and then challenging and supporting them to learn it well." (NCTM, 2000b, p. 16)

Coherent, well-articulated mathematics curricula targeting appropriate learning objectives with carefully planned lessons designed in accordance with research-based instructional strategies are essential to students doing meaningful mathematics. However, the success of even the best designed curriculum depends on the pedagogical and mathematical talent, professional attitude, and interpersonal skills of the teacher (i.e., you) in the classroom who implements that curriculum with students (Cangelosi, 2001b; National Commission on Teaching and America's Future, 1996; Tate & Johnson, 1999). In chapters 1–3, you were introduced to the professionalism of Casey Rudd and classroom management strategies for establishing an environment conducive to learning meaningful mathematics, interacting with students, motivating engagement in learning activities, and teaching students to be on-task before getting into the details of designing curricula and implementing specific instructional strategies of chapters 4–11. Why? Because—as illustrated by Case 2.1 in the section "Well-Planned Lessons Gone Awry"—the success of curricula depends on how individual teachers work with students.

PSSM's teaching principle focuses attention on the complete dependence of students' success on how you practice the extraordinarily complex art of teaching, the depth of your preparation in both mathematics and pedagogy, and how well your efforts are supported by school administrators and instructional leaders (NCTM, 2000b, pp. 16–18). The criticality of the continual decisions you make and your efforts to grow professionally are emphasized (NCTM, 2000b, pp. 18–19):

> Teachers make many choices each day about how the learning environment will be structured and what mathematics will be emphasized. These decisions determine, to a large extent, what students learn. Effective teaching conveys a belief that each student can and is expected to understand mathematics and that each will be supported in his or her efforts to accomplish this goal.
>
> Teachers establish and nurture an environment conducive to learning mathematics through the decisions they make, the conversations they orchestrate, and the physical setting they create. Teachers' actions are what encourage students to think, question, solve problems, and discuss their ideas, strategies, and solutions. The teacher is responsible for creating an intellectual environment where serious mathematical thinking is the norm. More than just a physical setting with desks, bulletin boards, and posters, the classroom environment communicates subtle messages about what is valued in learning and doing mathematics. Are students' discussion and collaboration encouraged? Are students expected to justify their thinking? If students are to learn to make conjectures, experiment with various approaches to solving problems, construct mathematical arguments and respond to others' arguments, then creating an environment that fosters these kinds of activities is essential.
>
> In effective teaching, worthwhile mathematical tasks are used to introduce important mathematical ideas and to engage and challenge students intellectually. Well-chosen tasks can pique students' curiosity and draw them into mathematics. The tasks may be connected to the real-world experiences of students, or they may arise in contexts that are purely mathematical. Regardless of the context, worthwhile tasks should be intriguing, with a level of challenge that invites speculation and hard work. Such tasks often can be approached in more than one way, such as using an arithmetic counting approach, drawing a geometric diagram and enumerating possibilities, or using algebraic equations, which makes the tasks accessible to students with varied prior knowledge and experience.
>
> Worthwhile tasks alone are not sufficient for effective teaching. Teachers must also decide what aspects of a task to highlight, how to organize and orchestrate the work of the students, what questions to ask to challenge those with varied levels of expertise, and how to support students without taking over the process of thinking for them and thus eliminating the challenge.
>
> Effective teaching involves observing students, listening carefully to their ideas and explanations,

having mathematical goals, and using the information to make instructional decisions. Teachers who employ such practices motivate students to engage in mathematical thinking and reasoning and provide learning opportunities that challenge students at all levels of understanding. Effective teaching requires continuing efforts to learn and improve. These efforts include learning about mathematics and pedagogy, benefitting from interactions with students and colleagues, and engaging in ongoing professional development and self-reflection.

Opportunities to reflect on and refine instructional practice—during class and outside class, alone and with others—are crucial in the vision of school mathematics outlined in Principles and Standards. To improve their mathematics instruction, teachers must be able to analyze what they and their students are doing and consider how those actions are affecting students' learning. Using a variety of strategies, teachers should monitor students' capacity and inclination to analyze situations, frame and solve problems, and make sense of mathematical concepts and procedures. They can use this information to assess their students' progress and to appraise how well the mathematical tasks, student discourse, and classroom environment are interacting to foster students' learning. They then use these appraisals to adapt their instruction.

Reflection and analysis are often individual activities, but they can be greatly enhanced by teaming with an experienced and respected colleague, a new teacher, or a community of teachers. Collaborating with colleagues regularly to observe, analyze, and discuss teaching and students' thinking or to do "lesson study" is a powerful, yet neglected, form of professional development in American schools (Stigler & Hiebert, 1999). The work and time of teachers must be structured to allow and support professional development that will benefit them and their students.

The computer management system and working professional portfolio that you organized and will continue to build are mechanisms for structuring your professional-growth efforts.

PSSM'S LEARNING PRINCIPLE

According to *PSSM*'s Learning Principle, "Students must learn mathematics with understanding, actively building new knowledge from experience and prior knowledge." (NCTM, 2000b, p. 20)

The gap between teaching strategies based on how we now understand students construct concepts, discover relationships, and develop algorithmic skills leading to meaningful learning and too often used follow-the-textbook strategies is recognized by the *PSSM* Writers Group (NCTM, 2000b, p. 20):

> The vision of school mathematics in *Principles and Standards* is based on students' learning mathematics with understanding. Unfortunately, learning mathematics *without* understanding has long been a common outcome of school mathematics instruction. In fact, learning without understanding has been a persistent problem since at least the 1930s, and it has been the subject of much discussion and research by psychologists and educators over the years (e.g., Brownell (1947); Skemp (1976); Hiebert and Carpenter (1992)).

The rationale for applying the teaching strategies you will develop as a result of your work with chapters 5–8 are included in the explication of the Learning Principle:

> In recent decades, psychological and educational research on the learning of complex subjects such as mathematics has solidly established the important role of conceptual understanding in the knowledge and activity of persons who are proficient. Being proficient in a complex domain such as mathematics entails the ability to use knowledge flexibly, applying what is learned in one setting appropriately in another. One of the most robust findings of research is that conceptual understanding is an important component of proficiency, along with factual knowledge and procedural facility (Bransford, Brown, & Cocking, 1999).
>
> The alliance of factual knowledge, procedural proficiency, and conceptual understanding makes all three components usable in powerful ways. Students who memorize facts or procedures without understanding often are not sure when or how to use what they know, and such learning is often quite fragile (Bransford, Brown, & Cocking, 1999). Learning with understanding also makes subsequent learning easier. Mathematics makes more sense and is easier to remember and to apply when students connect new knowledge to existing knowledge in meaningful ways (Schoenfeld, 1988). Well-connected, conceptually grounded ideas are more readily accessed for use in new situations (Skemp, 1976).
>
> The requirements for the workplace and for civic participation in the contemporary world include flexibility in reasoning about and using quantitative information. Conceptual understanding is an essential component of the knowledge needed to deal with novel problems and settings. Moreover, as judgments change about the facts or procedures that are essential in an increasingly technological world, conceptual understanding becomes even more important. For example, most of the arithmetic and algebraic procedures long viewed as the heart of the school mathematics curriculum can now be performed with handheld calculators. Thus, more attention can be given to understanding the number concepts and the modeling procedures used in solving problems. Change is a ubiquitous feature of contemporary life, so learning with understanding is essential to enable students to use what they learn to solve the new kinds of problems they will inevitably face in the future.
>
> A major goal of school mathematics programs is to create autonomous learners, and learning with understanding supports this goal. Students learn more and learn better when they can take control of their learn-

ing by defining their goals and monitoring their progress. When challenged with appropriately chosen tasks, students become confident in their ability to tackle difficult problems, eager to figure things out on their own, flexible in exploring mathematical ideas and trying alternative solution paths, and willing to persevere. Effective learners recognize the importance of reflecting on their thinking and learning from their mistakes. Students should view the difficulty of complex mathematical investigations as a worthwhile challenge rather than as an excuse to give up. Even when a mathematical task is difficult, it can be engaging and rewarding. When students work hard to solve a difficult problem or to understand a complex idea, they experience a very special feeling of accomplishment, which in turn leads to a willingness to continue and extend their engagement with mathematics.

PSSM'S ASSESSMENT PRINCIPLE

According to *PSSM*'s Assessment Principle, "Assessment should support the learning of important mathematics and furnish useful information to both teachers and students." (NCTM, 2000b, p. 22)

The principle calls attention to the need for assessment to provide the basis for formative judgments that guide how you teach and how students learn as well as for summative evaluations used to report mathematical achievements (Cangelosi, 2000a, pp. 1–175). But for assessment to enhance learning for students, you must practice it with the following in mind (NCTM, 2000b, pp. 22–24):

- Continual monitoring of students' progress as you teach is essential for successful lessons (Black & Wiliam, 1998).
- Your students determine what mathematical tasks and what types of cognitive activities are important by how you assess their learning (Stiggins, 1988).
- To reflect students' progress with stated goals, assessment activities (e.g., conducting miniexperiments as explained in chapter 5) should focus on the same objectives as lessons and often mimic the lessons' learning activities.
- By involving students in assessment activities (e.g., by becoming familiar with and using your scoring rubrics to measure their own learning), students learn to learn and are encouraged to take responsibility for their own learning.
- A wide variety of measurement devices (e.g., traditional tests, ongoing informal observations, structured performance observations, portfolios, and product examinations) are needed.
- Making accommodations for the special needs of students (e.g., providing prompts orally to a student with a particular learning disability) is necessary to accurately assess their progress with mathematics; furthermore, you are legally required to do so (Cangelosi, 2000a, pp. 278–280).

Because of the inextricable interrelation between monitoring students' progress and conducting lessons, strategies for implementing *PSSM*'s Assessment Principle are integrated with strategies for designing lessons in chapters 5–8. Chapter 9 focuses entirely on measuring and evaluating your students' mathematical achievements.

PSSM'S TECHNOLOGY PRINCIPLE

According to *PSSM*'s Technology Principle, "Technology is essential in teaching and learning mathematics; it influences the mathematics that is taught and enhances students' learning." (NCTM, 2000b, p. 24)

Chapters 5–8 and 11 include cases in which technology is used to teach, learn, and do mathematics. Strategies and resources for incorporating technology in your teaching are the foci of chapter 10. The following explication of the Technology Principle (NCTM, 2000b, pp. 24–27) is the rationale for the technology-related aspects of those chapters:

> Electronic technologies—calculators and computers—are essential tools for teaching, learning, and doing mathematics. They furnish visual images of mathematical ideas, they facilitate organizing and analyzing data, and they compute efficiently and accurately. They can support investigation by students in every area of mathematics, including geometry, statistics, algebra, measurement, and number. When technological tools are available, students can focus on decision making, reflection, reasoning, and problem solving.
>
> Students can learn more mathematics more deeply with the appropriate use of technology (Boers-van Oosterum 1990; Dunham and Dick 1994; Groves 1994; Rojano 1996; Sheets 1993). Technology should not be used as a replacement for basic understandings and intuitions; rather, it can and should be used to foster those understandings and intuitions. In mathematics-instruction programs, technology should be used widely and responsibly, with the goal of enriching students' learning of mathematics.
>
> The existence, versatility, and power of technology make it possible and necessary to reexamine what mathematics students should learn as well as how they can best learn it. In the mathematics classrooms envisioned in Principles and Standards, every student has access to technology to facilitate his or her mathematics learning under the guidance of a skillful teacher.
>
> Technology enhances mathematics learning. Technology can help students learn mathematics. For example, with calculators and computers students can examine more examples or representational forms than are feasible by hand, so they can make and explore conjectures easily. The graphic power of technological tools affords access to visual models that are powerful but

that many students are unable or unwilling to generate independently. The computational capacity of technological tools extends the range of problems accessible to students and also enables them to execute routine procedures quickly and accurately, thus allowing more time for conceptualizing and modeling.

Students' engagement with, and ownership of, abstract mathematical ideas can be fostered through technology. Technology enriches the range and quality of investigations by providing a means of viewing mathematical ideas from multiple perspectives. Students' learning is assisted by feedback, which technology can supply: drag a node in a Dynamic Geometry® environment, and the shape on the screen changes; change the defining rules for a spreadsheet, and watch as dependent values are modified. Technology also provides a focus as students discuss with one another and with their teacher the objects on the screen and the effects of the various dynamic transformations that technology allows.

Technology offers teachers options for adapting instruction to special student needs. Students who are easily distracted may focus more intently on computer tasks, and those who have organizational difficulties may benefit from the constraints imposed by a computer environment. Students who have trouble with basic procedures can develop and demonstrate other mathematical understandings, which in turn can eventually help them learn the procedures. The possibilities for engaging students with physical challenges in mathematics are dramatically increased with special technologies.

The effective use of technology in the mathematics classroom depends on the teacher. Technology is not a panacea. As with any teaching tool, it can be used well or poorly. Teachers should use technology to enhance their students' learning opportunities by selecting or creating mathematical tasks that take advantage of what technology can do efficiently and well—graphing, visualizing, and computing. For example, teachers can use simulations to give students experience with problem situations that are difficult to create without technology, or they can use data and resources from the Internet and the World Wide Web to design student tasks. Spreadsheets, dynamic geometry software, and computer microworlds are also useful tools for posing worthwhile problems.

Technology does not replace the mathematics teacher. When students are using technological tools, they often spend time working in ways that appear somewhat independent of the teacher, but this impression is misleading. The teacher plays several important roles in a technology-rich classroom, making decisions that affect students' learning in important ways. Initially, the teacher must decide if, when, and how technology will be used. As students use calculators or computers in the classroom, the teacher has an opportunity to observe the students and to focus on their thinking. As students work with technology, they may show ways of thinking about mathematics that are otherwise often difficult to observe. Thus, technology aids in assessment, allowing teachers to examine the processes used by students in their mathematical investigations as well as the results, thus enriching the information available for teachers to use in making instructional decisions.

Technology not only influences how mathematics is taught and learned but also affects what is taught and when a topic appears in the curriculum. With technology at hand, young children can explore and solve problems involving large numbers, or they can investigate characteristics of shapes using dynamic geometry software. Elementary school students can organize and analyze large sets of data. Middle-grades students can study linear relationships and the ideas of slope and uniform change with computer representations and by performing physical experiments with calculator-based-laboratory systems. High school students can use simulations to study sample distributions, and they can work with computer algebra systems that efficiently perform most of the symbolic manipulation that was the focus of traditional high school mathematics programs. The study of algebra need not be limited to simple situations in which symbolic manipulation is relatively straightforward. Using technological tools, students can reason about more general issues, such as parameter changes, and they can model and solve complex problems that were heretofore inaccessible to them. Technology also blurs some of the artificial separations among topics in algebra, geometry, and data analysis by allowing students to use ideas from one area of mathematics to better understand another area of mathematics.

Technology can help teachers connect the development of skills and procedures to the more general development of mathematical understanding. As some skills that were once considered essential are rendered less necessary by technological tools, students can be asked to work at higher levels of generalization or abstraction. Work with virtual manipulatives (computer simulations of physical manipulatives) or with Logo can allow young children to extend physical experience and to develop an initial understanding of sophisticated ideas like the use of algorithms. Dynamic geometry software can allow experimentation with families of geometric objects, with an explicit focus on geometric transformations. Similarly, graphing utilities facilitate the exploration of characteristics of classes of functions. Because of technology, many topics in discrete mathematics take on new importance in the contemporary mathematics classroom; the boundaries of the mathematical landscape are being transformed.

PSSM'S CONTENT STANDARDS AND EXPECTATIONS

For prekindergarten through high school *PSSM* includes two types of standards, content and process (NCTM, 2000b, p. 29):

> What mathematical content and processes should students know and be able to use as they progress

> through school? *Principles for School Mathematics* presents NCTM's proposal for what should be valued in school mathematics education. Ambitious standards are required to achieve a society that has the capability to think and reason mathematically and a useful base of mathematical knowledge and skills.
>
> The ten Standards . . . describe a connected body of mathematical understanding and competencies—a comprehensive foundation recommended for all students, rather than a menu from which to make curricular choices. Standards are descriptions of what mathematics instruction should enable students to know and do. They specify the understanding, knowledge, and skills that students should acquire from prekindergarten through grade 12. The Content Standards—Number and Operations, Algebra, Geometry, Measurement, and Data Analysis and Probability—explicitly describe the content that students should learn. The Process Standards—Problem Solving, Reasoning and Proof, Communication, Connections, and Representation—highlight ways of acquiring and using knowledge.

PSSM's Number and Operations Standard focuses on, ". . . deep and fundamental understanding of and proficiency with counting, numbers, and arithmetic, as well as understanding of number systems and their structures. The concepts and algorithms of elementary arithmetic are part of number and operations, as are the properties and characteristics of the classes of numbers that form the beginnings of number theory." (NCTM, 2000b, p. 32)

Exhibit 4.12 lists the expectations for middle and secondary school instructional programs regarding *number and operations* (NCTM, 2000b, p. 393).

PSSM's Algebra Standard, ". . . emphasizes relationships among quantities, including functions, ways of representing mathematical relationships, and the analysis of change." (NCTM, 2000b, p. 37). Exhibit 4.13 lists the expectations for middle and secondary school instructional programs regarding *algebra* (NCTM, 2000b, p. 395).

Regarding *PSSM*'s Geometry Standard, "Through the study of geometry, students will learn about geometric shapes and structures and how to analyze their characteristics and relationships. Spatial visualization—building and manipulating mental representations of two- and three-dimensional objects and perceiving an object from different perspectives—is an important aspect of geometric thinking. Geometry is a natural place for the development of students' reasoning and justification skills, culminating in work with proof in the secondary grades. Geometric modeling and spatial reasoning offer ways to interpret and describe physical environments and can be important tools in problem solving." (NCTM, 2000b, p. 41)

Exhibit 4.14 lists the expectations for middle and secondary school instructional programs regarding *geometry* (NCTM, 2000b, pp. 397).

You make a *measurement* by observing with one or more of your empirical senses (i.e., sight, hearing, smell, taste, and touch) and then recording results from that observation as a quantity. Numbers are extracted from the environment through measurement algorithms resulting in the assignment of numerical values to variables such as time, location, length, area, volume, set cardinality, monetary value, heat, velocity, weight, density, strength, and sound. The examples in Exhibit 4.15 illustrate relationships among observable variables, measurement algorithms, measurement instruments, and measurement units.

Students construct the critical concept of measurement error by collecting data for themselves. Numerals appearing in textbooks seem exact and precise—fostering the misconception that mathematics is an "exact science" with "answers" that are either precisely "right" or "wrong." To connect mathematics to their real lives, students need to do mathematics with numbers (i.e., data) they collect from their own environment—not only manipulate numerals printed in textbooks. *PSSM*'s Measurement Standard (NCTM, 2000b, pp. 44–47) addresses this need. Exhibit 4.16 lists the expectations for middle and secondary school instructional programs regarding *measurement* (NCTM, 2000b, pp. 399).

PSSM's Data Analysis and Probability Standard, ". . . recommends that students formulate questions that can be answered using data and addresses what is involved in gathering and using data wisely. Students should learn how to collect data, organize their own or others' data, and display the data in graphs and charts that will be useful in answering their questions. This Standard also includes learning some methods for analyzing data and some ways of making inferences and conclusions from data. The basic concepts and applications of probability are also addressed, with an emphasis on the way that probability and statistics are related." (NCTM, 2000b, p. 48)

Exhibit 4.17 lists the expectations for middle and secondary school instructional programs regarding *data analysis and probability* (NCTM, 2000b, p. 401).

PSSM'S PROCESS STANDARDS

Cognitive Processes

Whereas *PSSM*'s Content Standards focus on the mathematical topics with which students need to work, *PSSM*'s Process Standards focus on the various ways students' minds should work with those mathematical topics. For example, discovering a relationship (e.g., $x^3 \leq x \; \forall \, x \in [0, 1]$) is a different cognitive process than remembering that relationship. As you will better understand from subsequent work with this chapter and with chapters 5–8, you will need to be very

Exhibit 4.12

PSSM's Number and Operations Standards and Expectations for Middle and Secondary School Grades (Reprinted with permission from Principals and Standards for School Mathematics, p. 393, copyright 2000 by The National Council of Teachers of Mathematics).

Number and Operations Standards

Standard 1: Instructional programs from prekindergarten through grade 12 should enable all students to understand numbers, ways of representing numbers, relationships among numbers, and number systems.

Grades 6–8 Expectations: In grades 6–8 all students should—

- work flexibly with fractions, decimals, and percents to solve problems;
- compare and order fractions, decimals, and percents efficiently and find their approximate locations on a number line;
- develop meaning for percents greater than 100 and less than 1;
- understand and use ratios and proportions to represent quantitative relationships;
- develop an understanding of large numbers and recognize and appropriately use exponential, scientific, and calculator notation;
- use factors, multiples, prime factorization, and relatively prime numbers to solve problems;
- develop meaning for integers and represent and compare quantities with them.

Grades 9–12 Expectations: In grades 9–12 all students should—

- develop a deeper understanding of very large and very small numbers and of various representations of them;
- compare and contrast the properties of numbers and number systems, including the rational and real numbers, and understand complex numbers as solutions to quadratic equations that do not have real solutions;
- understand vectors and matrices as systems that have some of the properties of the real-number system;
- use number-theory arguments to justify relationships involving whole numbers.

Standard 2: Instructional programs from prekindergarten through grade 12 should enable all students to understand meanings of operations and how they relate to one another.

Grades 6–8 Expectations: In grades 6–8 all students should—

- understand the meaning and effects of arithmetic operations with fractions, decimals, and integers;
- use the associative and commutative properties of addition and multiplication and the distributive property of multiplication over addition to simplify computations with integers, fractions, and decimals;
- understand and use the inverse relationships of addition and subtraction, multiplication and division, and squaring and finding square roots to simplify computations and solve problems.

Grades 9–12 Expectations: In grades 9–12 all students should—

- judge the effects of such operations as multiplication, division, and computing powers and roots on the magnitudes of quantities;
- develop an understanding of properties of, and representations for, the addition and multiplication of vectors and matrices;
- develop an understanding of permutations and combinations as counting techniques.

Standard 3: Instructional programs from prekindergarten through grade 12 should enable all students to compute fluently and make reasonable estimates.

Grades 6–8 Expectations: In grades 6–8 all students should—

- select appropriate methods and tools for computing with fractions and decimals from among mental computation, estimation, calculators or computers, and paper and pencil, depending on the situation, and apply the selected methods;
- develop and analyze algorithms for computing with fractions, decimals, and integers and develop fluency in their use;
- develop and use strategies to estimate the results of rational-number computations and judge the reasonableness of the results;
- develop, analyze, and explain methods for solving problems involving proportions, such as scaling and finding equivalent ratios.

Grades 9–12 Expectations: In grades 9–12 all students should—

- develop fluency in operations with real numbers, vectors, and matrices, using mental computation or paper-and-pencil calculations for simple cases and technology for more complicated cases.
- judge the reasonableness of numerical computations and their results.

Exhibit 4.13
PSSM's Algebra Standards and Expectations for Middle and Secondary School Grades (NCTM, 2000b, p. 395).

Algebra Standards

Standard 1: Instructional programs from prekindergarten through grade 12 should enable all students to understand patterns, relations, and functions.

Grades 6–8 Expectations: In grades 6–8 all students should—

- represent, analyze, and generalize a variety of patterns with tables, graphs, words, and, when possible, symbolic rules;
- relate and compare different forms of representation for a relationship;
- identify functions as linear or nonlinear and contrast their properties from tables, graphs, or equations.

Grades 9–12 Expectations: In grades 9–12 all students should—

- generalize patterns using explicitly defined and recursively defined functions;
- understand relations and functions and select, convert flexibly among, and use various representations for them;
- analyze functions of one variable by investigating rates of change, intercepts, zeros, asymptotes, and local and global behavior;
- understand and perform transformations such as arithmetically combining, composing, and inverting commonly used functions, using technology to perform such operations on more complicated symbolic expressions;
- understand and compare the properties of classes of functions, including exponential, polynomial, rational, logarithmic, and periodic functions;
- interpret representations of functions of two variables.

Standard 2: Instructional programs from prekindergarten through grade 12 should enable all students to represent and analyze mathematical situations and structures using algebraic symbols.

Grades 6–8 Expectations: In grades 6–8 all students should—

- develop an initial conceptual understanding of different uses of variables;
- explore relationships between symbolic expressions and graphs of lines, paying particular attention to the meaning of intercept and slope;
- use symbolic algebra to represent situations and to solve problems, especially those that involve linear relationships;
- recognize and generate equivalent forms for simple algebraic expressions and solve linear equations.

Grades 9–12 Expectations: In grades 9–12 all students should—

- understand the meaning of equivalent forms of expressions, equations, inequalities, and relations;
- write equivalent forms of equations, inequalities, and systems of equations and solve them with fluency—mentally or with paper and pencil in simple cases and using technology in all cases;
- use symbolic algebra to represent and explain mathematical relationships;
- use a variety of symbolic representations, including recursive and parametric equations, for functions and relations;
- judge the meaning, utility, and reasonableness of the results of symbol manipulations, including those carried out by technology.

Standard 3: Instructional programs from prekindergarten through grade 12 should enable all students to use mathematical models to represent and understand quantitative relationships.

Grades 6–8 Expectations: In grades 6–8 all students should—

- model and solve contextualized problems using various representations, such as graphs, tables, and equations.

Grades 9–12 Expectations: In grades 9–12 all students should—

- identify essential quantitative relationships in a situation and determine the class or classes of functions that might model the relationships;
- use symbolic expressions, including iterative and recursive forms, to represent relationships arising from various contexts;
- draw reasonable conclusions about a situation being modeled.

Continued

Exhibit 4.13
Continued

Standard 4: Instructional programs from prekindergarten through grade 12 should enable all students to analyze change in various contexts.

Grades 6–8 Expectations: In grades 6–8 all students should—

- use graphs to analyze the nature of changes in quantities in linear relationships.

Grades 9–12 Expectations: In grades 9–12 all students should—

- approximate and interpret rates of change from graphical and numerical data.

Exhibit 4.14
PSSM's Geometry Standards and Expectations for Middle and Secondary School Grades (NCTM, 2000b, p. 397).

Geometry Standards

Standard 1: Instructional programs from prekindergarten through grade 12 should enable all students to analyze characteristics and properties of two- and three-dimensional geometric shapes and develop mathematical arguments about geometric relationships.

Grades 6–8 Expectations: In grades 6–8 all students should—

- precisely describe, classify, and understand relationships among types of two- and three-dimensional objects using their defining properties;
- understand relationships among the angles, side lengths, perimeters, areas, and volumes of similar objects;
- create and critique inductive and deductive arguments concerning geometric ideas and relationships, such as congruence, similarity, and the Pythagorean relationship.

Grades 9–12 Expectations: In grades 9–12 all students should—

- analyze properties and determine attributes of two- and three-dimensional objects;
- explore relationships (including congruence and similarity) among classes of two- and three-dimensional geometric objects, make and test conjectures about them, and solve problems involving them;
- establish the validity of geometric conjectures using deduction, prove theorems, and critique arguments made by others;
- use trigonometric relationships to determine lengths and angle measures.

Standard 2: Instructional programs from prekindergarten through grade 12 should enable all students to specify locations and describe spatial relationships using coordinate geometry and other representational systems.

Grades 6–8 Expectations: In grades 6–8 all students should—

- use coordinate geometry to represent and examine the properties of geometric shapes;
- use coordinate geometry to examine special geometric shapes, such as regular polygons or those with pairs of parallel or perpendicular sides.

Grades 9–12 Expectations: In grades 9–12 all students should—

- use Cartesian coordinates and other coordinate systems, such as navigational, polar, or spherical systems, to analyze geometric situations;
- investigate conjectures and solve problems involving two- and three-dimensional objects represented with Cartesian coordinates.

Standard 3: Instructional programs from prekindergarten through grade 12 should enable all students to apply transformations and use symmetry to analyze mathematical situations.

Grades 6–8 Expectations: In grades 6–8 all students should—

- describe sizes, positions, and orientations of shapes under informal transformations such as flips, turns, slides, and scaling;
- examine the congruence, similarity, and line or rotational symmetry of objects using transformations.

Continued

Exhibit 4.14
Continued

Grades 9–12 Expectations: In grades 9–12 all students should—

- understand and represent translations, reflections, rotations, and dilations of objects in the plane by using sketches, coordinates, vectors, function notation, and matrices;
- use various representations to help understand the effects of simple transformations and their compositions.

Standard 4: Instructional programs from prekindergarten through grade 12 should enable all students to use visualization, spatial reasoning, and geometric modeling to solve problems.

Grades 6–8 Expectations: In grades 6–8 all students should—

- draw geometric objects with specified properties, such as side lengths or angle measures;
- use two-dimensional representations of three-dimensional objects to visualize and solve problems such as those involving surface area and volume;
- use visual tools such as networks to represent and solve problems;
- use geometric models to represent and explain numerical and algebraic relationships;
- recognize and apply geometric ideas and relationships in areas outside the mathematics classroom, such as art, science, and everyday life.

Grades 9–12 Expectations: In grades 9–12 all students should—

- draw and construct representations of two- and three-dimensional geometric objects using a variety of tools;
- visualize three-dimensional objects and spaces from different perspectives and analyze their cross sections;
- use vertex-edge graphs to model and solve problems;
- use geometric models to gain insights into, and answer questions in, other areas of mathematics;
- use geometric ideas to solve problems in, and gain insights into, other disciplines and other areas of interest such as art and architecture.

Exhibit 4.15
Examples of Variables and Related Measurement Algorithms, Instruments, and Units.

Examples

VARIABLE	MEASUREMENT ALGORITHM OR INSTRUMENT	MEASUREMENT UNIT
Length	rulers, caliper, odometer, sextant,	centimeters, meter, inch, mile, kilometer,
Mass	scale	gram, pound, ton
Set Cardinality	counter, counting	1
Body Fat	skinfold caliper, hydrostatic weighing	percentage
Temperature	thermometer, touch,	Fahrenheit, Kelvin, Celsius
Air Pressure	barometer,	torr
Price	reading price tag	dollar, cents, lyre, peso, centavo
Angle Size	protractor, compass, sextant	radian, degree,
Location	sight, sextant	coordinate point
Speed	speedometer	km/hr, mi/hr
Direction	motion sensor	vector, + or −
Acceleration	radar detector, motion sensor	vector, coordinate point
School Achievement Score	teacher-made test, observation, standardized test, core test	raw score, percentile, stanine, scaled score, *z*-score, NCE, T-score
Time or Age	calendar, clock, stopwatch,	hour, minute, year, month, day, week
Frequency of Occurrence	counter	1
Volume	graduated cylinder, beakers, pipette	cubic centimeter, gallon, liter, quart
Sound	oscilloscope	decibel
Light	eyes, spectrophotometer	Hertz, nanometer, candela
Acidity	pH meter, taste buds	pH

Exhibit 4.16

PSSM's Measurement Standards and Expectations for Middle and Secondary School Grades (NCTM, 2000b, p. 399).

Measurement Standards

Standard 1: Instructional programs from prekindergarten through grade 12 should enable all students to understand measurable attributes of objects and the units, systems, and processes of measurement.

Grades 6–8 Expectations: In grades 6–8 all students should—

- understand both metric and customary systems of measurement;
- understand relationships among units and convert from one unit to another within the same system;
- understand, select, and use units of appropriate size and type to measure angles, perimeter, area, surface area, and volume.

Grades 9–12 Expectations: In grades 9–12 all students should—

- make decisions about units and scales that are appropriate for problem situations involving measurement.

Standard 2: Instructional programs from prekindergarten through grade 12 should enable all students to apply appropriate techniques, tools, and formulas to determine measurements.

Grades 6–8 Expectations: In grades 6–8 all students should—

- use common benchmarks to select appropriate methods for estimating measurements;
- select and apply techniques and tools to accurately find length, area, volume, and angle measures to appropriate levels of precision;
- develop and use formulas to determine the circumference of circles and the area of triangles, parallelograms, trapezoids, and circles and develop strategies to find the area of more complex shapes;
- develop strategies to determine the surface area and volume of selected prisms, pyramids, and cylinders;
- solve problems involving scale factors, using ratio and proportion;
- solve simple problems involving rates and derived measurements for such attributes as velocity and density.

Grades 9–12 Expectations: In grades 9–12 all students should—

- analyze precision, accuracy, and approximate error in measurement situations;
- understand and use formulas for the area, surface area, and volume of geometric figures, including cones, spheres, and cylinders;
- apply informal concepts of successive approximation, upper and lower bounds, and limit in measurement situations;
- use unit analysis to check measurement computations.

Exhibit 4.17

PSSM's Data Analysis and Probability Standards and Expectations for Middle and Secondary School Grades (NCTM, 2000b, p. 401).

Data Analysis and Probability Standards

Standard 1: Instructional programs from prekindergarten through grade 12 should enable all students to formulate questions that can be addressed with data and collect, organize, and display relevant data to answer them.

Grades 6–8 Expectations: In grades 6–8 all students should—

- formulate questions, design studies, and collect data about a characteristic shared by two populations or different characteristics within one population;
- select, create, and use appropriate graphical representations of data, including histograms, box plots, and scatterplots.

Grades 9–12 Expectations: In grades 9–12 all students should—

- understand the differences among various kinds of studies and which types of inferences can legitimately be drawn from each;
- know the characteristics of well-designed studies, including the role of randomization in surveys and experiments;
- understand the meaning of measurement data and categorical data, of univariate and bivariate data, and of the term variable;
- understand histograms, parallel box plots, and scatterplots and use them to display data;
- compute basic statistics and understand the distinction between a statistic and a parameter.

Continued

Exhibit 4.17
Continued

Standard 2: Instructional programs from prekindergarten through grade 12 should enable all students to select and use appropriate statistical methods to analyze data.

Grades 6–8 Expectations: In grades 6–8 all students should—

- find, use, and interpret measures of center and spread, including mean and interquartile range;
- discuss and understand the correspondence between data sets and their graphical representations, especially histograms, stem-and-leaf plots, box plots, and scatterplots.

Grades 9–12 Expectations: In grades 9–12 all students should—

- for univariate measurement data, be able to display the distribution, describe its shape, and select and calculate summary statistics;
- for bivariate measurement data, be able to display a scatterplot, describe its shape, and determine regression coefficients, regression equations, and correlation coefficients using technological tools;
- display and discuss bivariate data where at least one variable is categorical;
- recognize how linear transformations of univariate data affect shape, center, and spread;
- identify trends in bivariate data and find functions that model the data or transform the data so that they can be modeled.

Standard 3: Instructional programs from prekindergarten through grade 12 should enable all students to develop and evaluate inferences and predictions that are based on data.

Grades 6–8 Expectations: In grades 6–8 all students should—

- use observations about differences between two or more samples to make conjectures about the populations from which the samples were taken;
- make conjectures about possible relationships between two characteristics of a sample on the basis of scatterplots of the data and approximate lines of fit;
- use conjectures to formulate new questions and plan new studies to answer them.

Grades 9–12 Expectations: In grades 9–12 all students should—

- use simulations to explore the variability of sample statistics from a known population and to construct sampling distributions;
- understand how sample statistics reflect the values of population parameters and use sampling distributions as the basis for informal inference;
- evaluate published reports that are based on data by examining the design of the study, the appropriateness of the data analysis, and the validity of conclusions;
- understand how basic statistical techniques are used to monitor process characteristics in the workplace.

Standard 4: Instructional programs from prekindergarten through grade 12 should enable all students to understand and apply basic concepts of probability.

Grades 6–8 Expectations: In grades 6–8 all students should—

- understand and use appropriate terminology to describe complementary and mutually exclusive events;
- use proportionality and a basic understanding of probability to make and test conjectures about the results of experiments and simulations;
- compute probabilities for simple compound events, using such methods as organized lists, tree diagrams, and area models.

Grades 9–12 Expectations: In grades 9–12 all students should—

- understand the concepts of sample space and probability distribution and construct sample spaces and distributions in simple cases;
- use simulations to construct empirical probability distributions;
- compute and interpret the expected value of random variables in simple cases;
- understand the concepts of conditional probability and independent events;
- understand how to compute the probability of a compound event.

concerned with both the *mathematical content* as well as the *learning levels* of objectives when you formulate them and design lessons to lead your students to achieve them. For now, keep in mind that *PSSM*'s Content Standards concern the mathematical content of your objectives whereas *PSSM*'s Process Standards concern the learning levels of your objectives.

PSSM includes five Process Standards: Problem Solving, Reasoning and Proof, Communication, Connections, and Representation.

Problem Solving

The Role of Problem Solving

Students need to identify, articulate, and address problems in order to apply mathematics as well as to learn mathematics (NCTM, 2000b, p. 52):

> Problem solving means engaging in a task for which the solution method is not known in advance. In order to find a solution, students must draw on their knowledge, and through this process, they will often develop new mathematical understandings. Solving problems is not only a goal of learning mathematics but also a major means of doing so. Students should have frequent opportunities to formulate, grapple with, and solve complex problems that require a significant amount of effort and should then be encouraged to reflect on their thinking.
>
> By learning problem solving in mathematics, students should acquire ways of thinking, habits of persistence and curiosity, and confidence in unfamiliar situations that will serve them well outside the mathematics classroom. In everyday life and in the workplace, being a good problem solver can lead to great advantages.
>
> Problem solving is an integral part of all mathematics learning, and so it should not be an isolated part of the mathematics program. Problem solving in mathematics should involve all the five content areas described in these Standards. The contexts of the problems can vary from familiar experiences involving students' lives or the school day to applications involving the sciences or the world of work. Good problems will integrate multiple topics and will involve significant mathematics.

Real-Life Problem Solving

Recall from chapter 3's section "Motivating Students to Do Mathematics" that the problem-based approach used by Ms. Castillo helps students to connect mathematics to what is important to them—thus, making mathematics meaningful. Ultimately, you want your students to apply mathematics to their own *real-world problems.* But before developing strategies for designing and implementing a curriculum that leads students to develop their talents for real-world problem solving, examine the process by which the student in Case 4.7 develops a solution to a problem from her real world (see Exhibit 4.18):

CASE 4.7

Fifteen-year-old Brenda and her two brothers have just been told by their dad that he is tired of them leaving lights on in the house while he has to pay "outrageous" electric bills. Consequently, he will charge them 25¢ each time he catches one of them leaving lights on unnecessarily. Brenda retires to her room and thinks to herself: "Twenty-five cents just for leaving a light on isn't fair! It doesn't cost that much to burn a light bulb, or does it? I ought to be able to figure what it costs and show Dad that 25¢ just isn't fair. Let's see, for a problem in school, Mr. Martinez (her mathematics teacher) has us write down the question we want to answer. Okay, here it is: How much does it cost to leave on a light bulb? Then from that

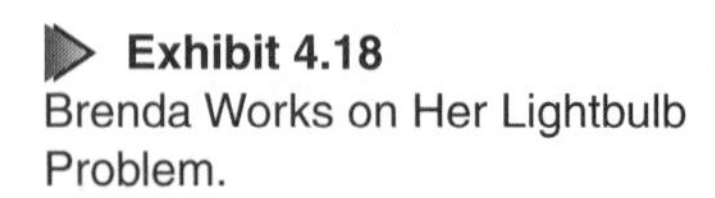

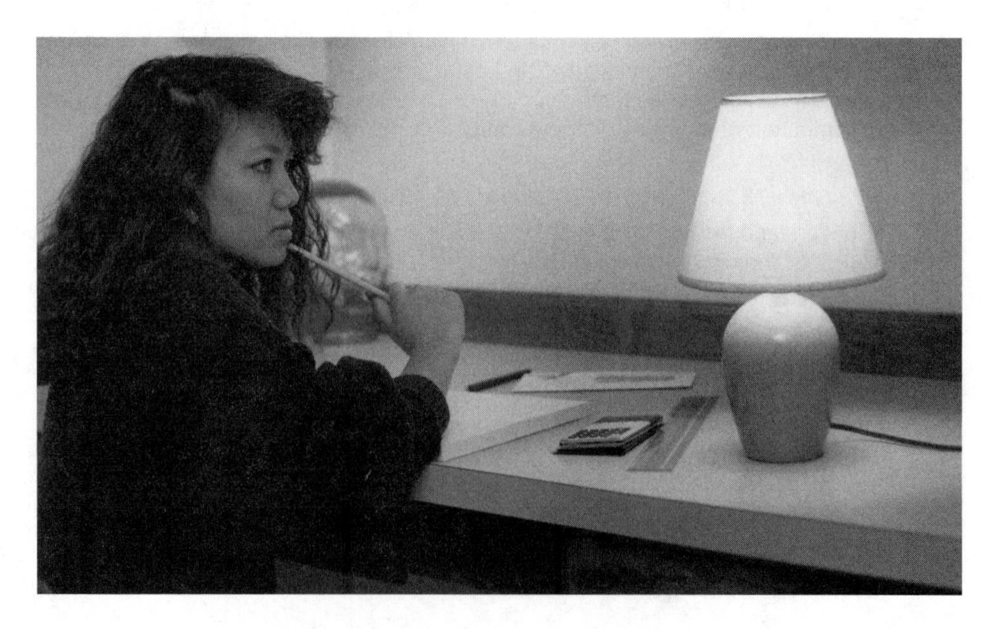

Exhibit 4.18
Brenda Works on Her Lightbulb Problem.

CITY OF LOGANA
Electric & Utilities Company

Billing date: 5/24/02 Account no.: 32-260-01-2

Electric Meter Billing Date: 5/20/02	Reading: 75698	Kilowatt-hours: 733
Charge for electricity	\$53.51	
Charge for sewer:	\$11.00	
Waste charge:	\$4.75	
Utility tax:	\$3.24	

TOTAL AMOUNT DUE: \$72.52
DATE DUE: 6/14/02

Exhibit 4.19
Brenda's Data Source.

I should be able to identify the main variable to solve. The variable is the number of dollars the electric company charges us for burning one bulb.

"So how should I solve for that variable? I could do an experiment. On the first of next month, I'll put a brand new bulb in my lamp and keep it on for a whole month. The next month, I'll keep the lamp off. Then I can find the difference between the electric bill for the month with the lamp on and the one for when it was off. Oh! What size bulb should I use? That's a variable that'll influence the results. Let's see, it's got a 60-watt one now. I'll stick with that and do the experiment for 60-watt bulbs, so size of the bulb will be constant rather than another variable to deal with.

"Oh! Oh! This experiment will take more than two months and I'll have to convince Dad not to charge me for leaving the lamp on all month. He should agree because it's necessary for the experiment and he's charging us only for unnecessary lights. But two months is too long to make my point. There must be some way to shorten this.

"I've got it! I could use last month's bill for the 'off' month and then I'd only have to wait for the 'on' month. That'd get me the results in half the time! Naw, that's not going to work because I used the lamp for some of the time last month. This is going to have to take two months. But even with the two months, there are still too many variables to worry about, like differences in how much we run the heat and electric fans and stuff. I've got to find a better way to do this. I don't know, but maybe if I looked at some old electric bills I might get some ideas."

Brenda collects some past electric bills like the one in Exhibit 4.19; she examines them at her desk and thinks: "This gives me an idea! Maybe I don't have to do an experiment after all. Last month, they charged us \$53.51. And it says we used 733 kilowatt-hours. From that I ought to be able to figure how much they charge us by the kilowatt-hour. Let's see, the rate must be less than 25¢. I'll divide 733 by 53.51. Here's my calculator; okay, 733 divided by 53.51 is [pause] What? \$13.70! That can't be! Electricity can't be that expensive. Oh, I know! I should've divided the dollar amount by the kilowatt-hours instead of into them. So, it's 53.51 divided by 733 equals 0.073. So the rate is 7.3¢." Brenda's computation of the electric rate is shown in Exhibit 4.20.

Brenda continues to think: "Maybe from this I can solve for my variable, the cost of burning one 60-watt bulb, without having to do an experiment. But what's a kilowatt-hour? It's in my science book, better look it up. [pause] It's, uhh, 'power consumption of 1,000 watts for 1 hour.' This bulb is 60 watts, so it burns [pause] 60 divided by 1,000 watts per hour. So, if the rate is 7.3¢ per kilowatt-hour, it costs 0.06 times 0.073 dollars to burn the bulb for an hour. That's about \$0.0044 every hour.

"So to leave one light bulb on all day would be 24 × 0.0044 or 0.1056, which isn't much more than a tenth of a cent. Gosh! Is that all it cost? It doesn't seem right for Dad to charge us 25¢. Oh, no! I've been doing this in dollars, so that's 0.1056 of a dollar, not a cent. So, it's really about 10½¢. And that's for only one 60-watt bulb." Brenda's computation is shown in Exhibit 4.21. She continues, "Let's see, my overhead light has three 60-watt bulbs. Leaving lights on can get pretty expensive. But 25¢ is still too high for him to charge us. I'll show him these figures and see if we can negotiate this down."

Brenda hurries out her room but quickly returns to shut off the lights before speaking to her father.

Exhibit 4.20
Brenda's Computation of the Electric Rate.

Kilowatt-hours = 733
Cost to customer = \$53.51
~~733~~ ÷ ~~53.51~~ = ~~13.70~~
53.51 ÷ 733 = 0.073
Rate = \$0.073 per Kilowatt-hour

Exhibit 4.21
Brenda's Computation of the Cost of Burning One 60-Watt Bulb for a Day.

A Kilowatt-hour =
the power consumption of 1000 watts for one hour.
The light bulb is 60 watts.
60 ÷ 1000 = 0.06
Rate = $0.073 per Kilowatt-hour.
0.06 x 0.073 = 0.0044.
It cost $0.0044 each hour to burn a 60 watt light bulb.
24 x 0.0044 = 0.1056.
~~It cost about 1.05 cents to burn a 60 watt bulb for a day.~~
It cost about 10.5 cents to burn a 60 watt bulb for a day.

Most people do not confidently address problems, nor do they systematically formulate solutions as Brenda did in Case 4.7 (Cangelosi, 2001a; Schoenfeld, 1985). Brenda's confident, systematic pursuit of a solution grew out of her experiences with a mathematics teacher who consciously taught her a nine-stage problem-solving process:

1. *The person is confronted with a puzzling question or questions about how to do something or explain a phenomenon.* In Brenda's case, the overall question was: Is it fair for her dad to charge her and her brothers 25¢ each time they leave a light bulb on unnecessarily?
2. *The person clarifies the question or questions posed by the problem, often in terms of more specific questions about quantities.* Brenda refined the overall question about the fairness of 25¢ per incident to the more mathematical question: How much does it cost to leave on a light bulb?
3. *The principal variable or variables to be solved are identified.* Brenda inferred the variable—number of dollars the electric company charges for burning one bulb—from the question: How much does it cost to leave on a light bulb?
4. *The situation is visualized so that relevant relationships involving the principal variable or variables are identified and possible solution designs are considered.* Brenda thought about how a bulb's wattage rating affects the cost of burning the bulb and how burning a bulb impacts the total monthly electric bill. This led her to consider experimenting with one 60-watt bulb for two months. After judging her plan impractical, she identified a relationship involving the rate the company charges for electricity. From there, she decided she could determine the rate from a previous monthly bill and figure the cost of burning the 60-watt bulb.
5. *The solution plan is finalized, including (a) selection of measurements (that is, how data are to be collected), (b) identification of relationships to establish, and (c) selection of algorithms to execute.* For this stage, Brenda decided to use a previous bill as the data source for the dollars charged and the kilowatt-hours consumed. Those figures were used to calculate the rate and then the cost of burning the 60-watt bulb.
6. *Data are gathered or measurements taken.* Brenda read the relevant information from the bill.
7. *Algorithms are executed with the data.* Brenda completed the computations leading to the figures of $0.073 per kilowatt-hour and 10.56¢ per day for burning a 60-watt bulb.
8. *Results of the executions of algorithms are interpreted to shed light on the original question or questions.* Brenda compared the 10.5¢ per day figure to the 25¢ figure per incident her father was charging.
9. *The person makes a value judgment regarding the original question or questions.* Brenda decided she had a reasonable chance of using her findings to negotiate successfully with her father.

Real-life problem solving requires the use of *deductive reasoning,* which you will better understand as well as learn to design lessons that stimulate students to reason deductively in your work with chapter 8.

Requisite Attitudes, Skills, and Abilities

Students' success in applying the nine-stage process to solving problems from their own real worlds depends on their acquisition of five attitudes, abilities, and skills:

- *Confidence and willingness to pursue solutions to problems* (The dogged pursuit of problem solutions requires [a] confidence in the potential for success, [b] freedom from fear of the consequences of failure, and [c] an appreciation for truth. "Appreciation for truth" may appear corny, but Brenda probably would not have systematically persisted in addressing the light-bulb problem had she been more interested in fooling her father into changing his mind than in presenting him with accurate information. Such persistence and appreciation are learned behaviors. Your work with chapters 2–3 should help you establish a classroom environment that promotes such attitudes; chapter 8's section "Influencing Students' Attitudes about Mathematics" should also prove helpful.)

- *Conceptual-level understanding of the mathematical concepts, relationships, and algorithms from which problem solutions are built* (Unless students understand how examples and nonexamples of concepts differ and why relationships and algorithms work, they will not recognize how to apply them in novel situations with real-world clutter, that is, when information irrelevant to the problem or its solution is present. In Case 4.7, for example, Brenda had never before tried to solve for an electric rate, but because she had previously constructed the concept of *rate* in her mind and discovered why relationships [e.g., rate × time = distance and rate × principal = interest] work, she was able to associate that understanding with the question about leaving on lights. [Davis, Maher, & Noddings, 1990; Kami & Warrington, 1999; Tang & Ginsburg, 1999]. *PSSM*'s Reasoning and Proof Standard focuses on this requisite to doing meaningful mathematics, as does chapter 5.)
- *Skills in recalling or retrieving formulas and executing algorithms* (Accuracy of solutions depends not only on selecting appropriate mathematics but also on correctly executing algorithms by hand, calculator, or computer. Brenda did not remember the meaning of kilowatt-hour, but she knew where to look it up. Also, she knew how to compute with her calculator and how to use estimation for monitoring answers. When you work with chapter 6, you will develop strategies for leading students to develop algorithmic and memory-level skills.)
- *Comprehension of necessary language and structural conventions for organizing, retaining, and relating mathematics to problem solving* (Language and structural conventions provide mechanisms for [a] organizing and storing concepts constructed, relationships discovered, and algorithms invented and [b] communicating about them. In Case 4.7, for example, Brenda retrieved mathematics she had conceptualized and stored under such labels as "rate" and "multiplication of rational numbers" for use in solving her problem. Her knowledge of conventions helped her organize what she understands, use tools such as a calculator, and report her findings to her father. *PSSM*'s Communication Standard and Representation Standard focus on this requisite to doing meaningful mathematics, as does chapter 7.)
- *Ability to discriminate between appropriate and inappropriate mathematical concepts, relationships, and algorithms depending on problem situations* (How to decide when to use the mathematics one understands is a learned ability that is the focus of *PSSM*'s Problem Solving Standard. In Case 4.7, for example, Brenda recognized certain features of her problem situation that helped her decide to use one type of mathematics rather than another. Your work with chapter 8 will help you to lead your students to develop this application-level ability.)

Textbook Word Problems

Word problems from mathematics textbooks provide convenient exercises for students to experience some—but not all—aspects of real-life problem solving. Engage in Activity 4.2:

Activity 4.2

Access a mathematics textbook that is used in middle or secondary schools. Select a word problem to analyze (e.g., appearing under the heading "Law of Sines" is the following word problem [Foster, Winters, Gordon, Rath, & Gell, 1992, p. 767]: "A ship is sighted at sea from two observation points on the coastline that are 30 miles apart. The angle between the coastline and the line between the ship and the first observation point measures 34°. The angle between the coastline and the line between the ship and the second observation point measures 43°34′. How far is the ship from the second observation point?")

Solve the word problem yourself, but as you do, make note of the decisions you make. Describe the cognitive activity the task prompted in your mind.

Now, compare the stages you went through to solve the textbook word problems to the nine stages of real-world problem solving enumerated shortly after Case 4.7.

Discuss your work with this activity with a colleague who is also engaging in it.

If you expressed your comparison in writing, edit the paper and insert it in the "Cognition, Instructional Strategies, and Planning" section of your working portfolio.

Some of the competencies needed to solve textbook word problems are similar to those for real-world problems; others are dissimilar:

1. With a real-life problem, students are confronted with puzzling questions they want to answer. Textbook word problems (e.g., "What is the height of a tree if it casts a 75-foot shadow when the angle of elevation of the sun is 27°?") present puzzling questions, but rarely are they questions students feel a need to answer.
2. To solve a real-life problem, students must clarify the questions posed by problems, often in terms of more specific questions about quantities. With textbook word problems, the specific

questions involving quantities (e.g, "What is the maximum possible area of the patio?") are typically articulated for students. Thus, instead of taxing students' abilities to formulate questions, what is taxed is the students' reading comprehension skill.

3. With both real-world and textbook word problems, principal variables to be solved must be inferred from the questions about quantities.
4. To solve real-life problems, students must visualize situations so that relevant relationships involving the principal variables are identified and possible solution designs are considered. Typically, this step is unnecessary for solving textbook word problems because all relevant data are given and open sentences can be formulated by simply following the pattern established by the examples and other word problems in the text section where the problem appears.
5. The solution plan for real-life problems needs to be finalized, but this is unnecessary for textbook word problems, as alluded to in point 4.
6. Numbers used in real-life problem solving result from measurements. Numbers used in textbook word problems are usually given.
7. Both real-life and textbook word problems require students to execute algorithms.
8. Results from algorithms are interpreted to shed light on the original questions posed by real-life problems. Simply obtaining the results—not interpreting them—is all that is typically required for textbook word problems (e.g., finding the height of the tree that casts the 75-foot shadow is sufficient; it is not necessary to decide if it is practical to move the tree).
9. Unlike real-life problems, solving textbook word problems does not require students to make value judgments.

Exhibit 4.22 summarizes some of the principal similarities and differences between solving real-life problems and textbook word problems.

Strategies for modifying textbook word problems so that they are more representative of real-world problems are explained and illustrated in chapter 8. Case 4.8 is an example of a teacher taking a step in the right direction:

CASE 4.8

Mr. Pepper thinks to himself as he plans on solving first-degree open sentences: "Before demonstrating methods for solving these algebraic sentences, I should first have them analyze some application-level problems. Let's see, what kind of word problems does the textbook offer? Looks as if they involve the mathematics I want to teach, but they're not very motivating for my ninth graders. I'll rewrite them so the mathematics stays the same but the situations are more in line with their interests. Okay, the first one reads:

Exhibit 4.22
Solving Real-Life Problems Compared to Solving Textbook Word Problems.

Characteristic	Real Problems	Textbook Word Problems
Problem is personalized with student having felt a need to solve it.	Yes	No
Questions posed by the problem are clarified and articulated for the student.	No	Yes
Reading comprehension skills are likely to be taxed.	No	Yes
Principal variables must be inferred from questions posed by the problem.	Yes	Yes
Type of problem is categorized for the student according to the type of mathematics needed.	No	Yes
Student is likely to have to select the type of mathematics to be used.	Yes	No
Measurement procedures must be selected and data collected.	Yes	No
Irrelevant data and information (i.e. clutter) are present.	Yes	No
Student needs to execute formulas, processes, or algorithms.	Yes	Yes
Solutions have pat answers.	No	Yes
Student must interpret results and make value judgments.	Yes	No

> An exotic tribe has a rule that the number of guards protecting the tribe must be at least one tenth the number in the tribe, less 50. According to this rule, how many tribespeople can be protected by 40 guards?

"Great! An exotic tribe—just what my students can identify with! To what should I make this relate? Something where one number depends on another. Phil has experience racing dirt bikes. Maybe I could do something with the relation between tire size and power instead of guards and tribespeople. Oh, I've got it! Most of these kids love to go to music concerts—rap, rock, whatever they like. I'll change the guards in the word problem to security people for a concert and the tribespeople to concertgoers. I've got to work this out. Okay, here's the problem:

> The number of security guards working at a rock concert depends on the number of people expected to attend. One rule of thumb stipulates that the number of security guards must be at least one tenth the number of concertgoers, less 50. According to the rule, how many people can attend a concert with 40 security guards?

"But I don't know if that rule of thumb is all that realistic. It doesn't make any more sense to me than the one for the exotic tribe. Oh, another brilliant idea! Naomi loves to read those rock magazines. I'll bet she could be our resource person for coming up with the rule. There's bound to be something about that in her magazines. This is perfect! Naomi isn't real fond of mathematics, but this time she can be the one to provide the formula we use in the word problem. Okay, I'll make this next one relate to something different. Let's see . . ."

Reasoning and Proof

Students' need to construct concepts, discover relationships, and invent mathematics is the focus of *PSSM*'s Reasoning and Proof Standard (NCTM, 2000b, p. 56):

> Mathematical reasoning and proof offer powerful ways of developing and expressing insights about a wide range of phenomena. People who reason and think analytically tend to note patterns, structure, or regularities in both real-world situations and symbolic objects; they ask if those patterns are accidental or if they occur for a reason; and they conjecture and prove. Ultimately, a mathematical proof is a formal way of expressing particular kinds of reasoning and justification.
>
> Being able to reason is essential to understanding mathematics. By developing ideas, exploring phenomena, justifying results, and using mathematical conjectures in all content areas and—with different expectations of sophistication—at all grade levels, students should see and expect that mathematics makes sense. Building on the considerable reasoning skills that children bring to school, teachers can help students learn what mathematical reasoning entails. By the end of secondary school, students should be able to understand and produce mathematical proofs—arguments consisting of logically rigorous deductions of conclusions from hypotheses—and should appreciate the value of such arguments.
>
> Reasoning and proof cannot simply be taught in a single unit on logic, for example, or by "doing proofs" in geometry. Proof is a very difficult area for undergraduate mathematics students. Perhaps students at the postsecondary level find proof so difficult because their only experience in writing proofs has been in a high school geometry course, so they have a limited perspective (Moore, 1994). Reasoning and proof should be a consistent part of students' mathematical experience in prekindergarten through grade 12. Reasoning mathematically is a habit of mind, and like all habits, it must be developed through consistent use in many contexts.

Concept construction and relationship discovery resulting in conjectures require the use of *inductive reasoning,* which you will better understand as well as learn to design lessons that stimulate students to reason inductively from your work with chapter 5. Of course, deductive reasoning is used to formulate proofs—a topic for chapter 8.

Communication

You noted the need to teach students to comprehend the language of mathematics when you worked with this chapter's section "Miscomprehension of the Language of Mathematics." This need as well as the need for students to engage in discourse with and about mathematics are the foci of *PSSM*'s (NCTM, 2000b, pp. 60–61) Communication Standard:

> Communication is an essential part of mathematics and mathematics education. It is a way of sharing ideas and clarifying understanding. Through communication, ideas become objects of reflection, refinement, discussion, and amendment. The communication process also helps build meaning and permanence for ideas and makes them public. When students are challenged to think and reason about mathematics and to communicate the results of their thinking to others orally or in writing, they learn to be clear and convincing. Listening to others' explanations gives students opportunities to develop their own understandings. Conversations in which mathematical ideas are explored from multiple perspectives help the participants sharpen their thinking and make connections. Students who are involved in discussions in which they justify solutions—especially in the face of disagreement—will gain better mathematical understanding as they work to convince their peers about differing points of view (Hatano & Inagaki, 1991). Such activity also helps students develop a language for expressing mathematical ideas and an appreciation of the need for precision in that language. Students who have opportunities, encouragement, and

support for speaking, writing, reading, and listening in mathematics classes reap dual benefits: they communicate to learn mathematics, and they learn to communicate mathematically.

Because mathematics is so often conveyed in symbols, oral and written communication about mathematical ideas is not always recognized as an important part of mathematics education. Students do not necessarily talk about mathematics naturally; teachers need to help them learn how to do so (Cobb, Wood, & Yackel, 1994). As students progress through the grades, the mathematics about which they communicate should become more complex and abstract. Students' repertoire of tools and ways of communicating, as well as the mathematical reasoning that supports their communication, should become increasingly sophisticated. Support for students is vital. Students whose primary language is not English may need some additional support in order to benefit from communication-rich mathematics classes, but they can participate fully if classroom activities are appropriately structured (Silver, Smith, & Nelson, 1995).

Students need to work with mathematical tasks that are worthwhile topics of discussion. Procedural tasks for which students are expected to have well-developed algorithmic approaches are usually not good candidates for such discourse. Interesting problems that "go somewhere" mathematically can often be catalysts for rich conversations. Technology is another good basis for communication. As students generate and examine numbers or objects on the calculator or computer screen, they have a common (and often easily modifiable) referent for their discussion of mathematical ideas.

Students gain insights into their thinking when they present their methods for solving problems, when they justify their reasoning to a classmate or teacher, or when they formulate a question about something that is puzzling to them. Communication can support students' learning of new mathematical concepts as they act out a situation, draw, use objects, give verbal accounts and explanations, use diagrams, write, and use mathematical symbols. Misconceptions can be identified and addressed. A side benefit is that it reminds students that they share responsibility with the teacher for the learning that occurs in the lesson (Silver, Kilpatrick, & Schlesinger, 1990).

Communication with and about mathematics is encouraged in classrooms where descriptive rather than judgmental language prevails and students engage in true dialogues that are not dominated by IRE cycles (recall your work with two of chapter 2's sections: "Descriptive Instead of Judgmental Language" and "True Dialogues").

Connections

The popularity of precision teaching, direct instruction, and mastery learning (Joyce, Weil, & Calhoun, 2000, pp. 317–357) has encouraged teachers and curriculum designers to present topics in small, fragmented segments with easier skills preceding those more difficult to learn. Some textbooks, for example, explain multiplication of fractions before addition of fractions. Presumably, multiplication is treated first because the algorithm is simpler than that for addition in which one must bother with finding common denominators. When the two are presented as unrelated algorithms, remembering one tends to interfere with the learning of the other (Parsons, Hinson, & Sardo-Brown, 2001, pp. 252–258). Consequently, the following error pattern (Ashlock, 2001) is likely to emerge in representative samples of students' work:

$$\text{Seventh grader: } \frac{17}{10} + \frac{4}{7} = \frac{21}{17}$$

$$\text{Ninth grader: } \frac{x+1}{x-1} + \frac{3}{x+7} = \frac{x+4}{2x+6}$$

If addition of fractions is treated first and multiplication is then presented simply as a special case of addition, students are more likely to relate the two so that the learning of one enhances the learning of the other. Methods for connecting and integrating topics so that students have fewer concepts and relationships with which to contend are explained and illustrated throughout this book. But to give you some semblance of the idea, consider Case 4.9:

CASE 4.9

Mr. Sanchez has just used direct instruction to lead his students to develop their skills adding two algebraic fractions by the algorithm used in the following example:

$$\frac{x-9}{4} + \frac{2x+3}{x-5}$$

$$= \frac{(x-9)(x-5)}{4(x-5)} + \frac{(2x+3)(4)}{4(x-5)}$$

$$= \frac{(x-9)(x-5) + (2x+3)(4)}{4(x-5)}$$

$$= \frac{(x^2-14x+45) + (8x+12)}{4x-20}$$

$$= \frac{x^2-6x+57}{4x-20}$$

Now, to introduce multiplication of algebraic fractions, he tells his class, "Consider multiplying these two numbers." He writes on the board:

$$\frac{2x-1}{x+3} \cdot \frac{4}{x-2}$$

He continues, "Because we already know how to add algebraic fractions, let's turn this into an addition task.

Multiplication is repeated addition, so we can write the following:"

$$\frac{2x-1}{x+3}\cdot\frac{4}{x-2}$$

$$=\underbrace{\frac{2x-1}{x+3}+\frac{2x-1}{x+3}+\frac{2x-1}{x+3}+\ldots+\frac{2x-1}{x+3}}_{\frac{4}{x-2}\text{ times}}$$

Mr. Sanchez: But "(4/(x – 2))times" doesn't make a lot of sense to everyone. So, let's rework what we've done to this point with a constant in place of *x*. Pick an odd whole number for us, Gretchen.

Gretchen: 7.

Mr. Sanchez extends what is on the board:

$$\underbrace{\frac{2(7)-1}{7+3}+\frac{2(7)-1}{7+3}+\frac{2(7)-1}{7+3}+\ldots+\frac{2(7)-1}{7+3}}_{\frac{4}{7-2}\text{ times}}$$

$$\underbrace{\frac{13}{10}+\frac{13}{10}+\frac{13}{10}+\ldots+\frac{13}{10}}_{\frac{4}{5}\text{ times}}$$

Mr. Sanchez: I understand adding 13/10 to itself 4 times but what in the world does adding 13/10 to itself 4/5 times mean? Cam-Loi.

Cam-Loi: You add 1/5 of 13/10 four times.

Mr. Sanchez: Oh! So you mean:

He writes:

$$\left(\frac{13}{10}\div 5\right)+\left(\frac{13}{10}\div 5\right)+\left(\frac{13}{10}\div 5\right)+\left(\frac{13}{10}\div 5\right)$$

$$=\frac{13}{50}+\frac{13}{50}+\frac{13}{50}+\frac{13}{50}$$

$$=\frac{13+13+13+13}{50}=\frac{(13)\,(4)}{50}=\frac{52}{50}$$

Mr. Sanchez: Now let's use the same process Cam-Loi suggested to work out an answer in variable form.

Together, they complete the following:

$$\frac{2x-1}{x+3}\cdot\frac{4}{x-2}=$$

$$\underbrace{\frac{2x-1}{x+3}+\frac{2x-1}{x+3}+\frac{2x-1}{x+3}+\ldots+\frac{2x-1}{x+3}}_{\frac{4}{x-2}\text{ times}}$$

Which is,

$$\left(\frac{2x-1}{x+3}\div(x-2)\right)(4)$$

Or simply,

$$=\frac{(2x-1)(4)}{(x+3)(x-2)}=\frac{8x-4}{x^2+x-6}$$

Mr. Sanchez: By the way, what is $(8x-4)/(x^2+x-6)$ if *x* is 7?

The lesson continues with the usual algorithm for multiplying algebraic fractions spelled out and practice exercises assigned.

As indicated by the quote from *PSSM* (NCTM, 2000b, p. 64) from chapter 3's section "Connecting Mathematics to Students' Interests: The Problem-Based Approach," the Connections Standard focuses on the need to reconnect the fragments of traditional follow-the-textbook mathematics instruction.

Representation

PSSM's Representation Standard is especially related to the Communication Standard (NCTM, 2000b, p. 67):

> The ways in which mathematical ideas are represented is fundamental to how people can understand and use those ideas. Consider how much more difficult multiplication is using Roman numerals (for those who have not worked extensively with them) than using Arabic base-ten notation. Many of the representations we now take for granted—such as numbers expressed in base-ten or binary form, fractions, algebraic expressions and equations, graphs, and spreadsheet displays—are the result of a process of cultural refinement that took place over many years. When students gain access to mathematical representations and the ideas they represent, they have a set of tools that significantly expand their capacity to think mathematically.
>
> The term "representation" refers both to process and to product—in other words, to the act of capturing a mathematical concept or relationship in some form and to the form itself. . . . The graph of $f(x) = x^3$ is a representation. Moreover, the term applies to processes and products that are observable externally as well as to those that occur "internally," in the minds of people doing mathematics. All these meanings of representation are important to consider in school mathematics.
>
> Some forms of representation—such as diagrams, graphical displays, and symbolic expressions—have long been part of school mathematics. Unfortunately, these representations and others have often been taught and learned as if they were ends in themselves.

> Representations should be treated as essential elements in supporting students' understanding of mathematical concepts and relationships; in communicating mathematical approaches, arguments, and understandings to one's self and to others; in recognizing connections among related mathematical concepts; and in applying mathematics to realistic problem situations through modeling. New forms of representation associated with electronic technology create a need for even greater instructional attention to representation.

DESIGNING MATHEMATICS COURSES

Chapter 11 serves as a resource of sample course outlines, teaching units, lesson plans, assessment instruments, daily agendas, and cases of teachers engaging students in learning activities and managing classrooms for different mathematics courses designed to be consistent with *PSSM*'s recommendations. Before continuing with this section, spend at least 10 minutes thumbing through chapter 11 pausing long enough with some of the exhibits so that you have a picture of a variety of planning documents in your head as you finish your work with chapter 4. Note from chapter 11's sample of course outlines (i.e., Exhibits 11.2, 11.5, 11.33, 11.37, and 11.40) that each course is a sequence of teaching units.

Engage in Activity 4.3:

Activity 4.3

When you engaged in Activities 1.2 and 1.3, you outlined at least one mathematics course and organized subfolders in your computerized folder structure to accommodate that course or those courses. Carefully develop the outline for another mathematics course and organize the needed subfolders for it. Discuss your outline with a colleague who is also engaging in this activity. If you think you should, modify your work in light of the discussion. Insert a hard copy of the course outline into the "Cognition, Instructional Strategies, and Planning" section of your working portfolio.

TEACHING UNITS

Components of a Teaching Unit

Each teaching unit listed in a course outline consists of (a) a learning goal, (b) a set of objectives that define the learning goal, (c) a planned sequence of lessons—each consisting of learning activities designed to help students achieve the objective as well as mechanisms for monitoring student progress during the lesson, and (d) a summative evaluation of student achievement of the learning goal.

Note chapter 11's sample of unit plans (i.e., Exhibits 11.10, 11.34, and 11.41).

The Learning Goal

The *learning goal* is the overall purpose of the teaching unit. It indicates what students are expected to gain if the teaching unit is successful. For example, the learning goal of an algebra II unit 6, titled "Systems of Linear Equations and Inequalities," might conceivably be as follows:

> Students can (i) formulate and efficiently use systems of linear equations and inequalities to solve real-life problems and (ii) explain interrelationships within those systems that facilitate problem solving.

The Objectives

The Need for Specificity

The learning goal provides direction for designing the teaching unit by identifying the overall student outcomes. However, teaching is a complicated art. Leading students from where they are to where they can "formulate and efficiently use systems . . . that facilitate problem solving" involves a complex set of different learning stages requiring varying teaching strategies. For students to achieve a learning goal such as the one for the unit on systems of linear equations and inequalities, they must acquire a number of specific skills, abilities, and attitudes. Thus, the learning goal is defined by a set of *specific objectives,* each indicating the particular skill, ability, or attitude that is a necessary but insufficient component of learning-goal achievement. The union of the objectives equals the learning goal. For example, the aforementioned learning goal might be defined by the objectives listed in Exhibit 4.23. Keep in mind that the terminology and mathematical notations used in the statement of the objectives are for the teachers' benefit and do not necessarily reflect the terminology or notations to which the students will be exposed.

Specifying the Mathematical Content

You write each objective to specify a mathematical content so that you clearly identify the mathematical topics students are to learn. In Exhibit 4.23, for example, Objectives C, D, and E specify the algorithms students will execute to solve systems of linear equations (i.e., substitution, graphing, and addition method).

Objectives need to be stated so that you know what types of sets, operations, algorithms, terms, and so forth are to be dealt with in the unit's lessons. How those lessons should be designed depends upon—among other things—the type of mathematical content specified in the objectives. As you will better understand from your work with chapters 5–8, you design lessons differently for teaching about one type of content (e.g., a concept) than for another

Exhibit 4.23
Sample Set of Objectives Defining the Goal for Unit on Systems of Linear Equations and Inequalities.

Unit Goal:

Students can (i) formulate and efficiently use systems of linear equations and inequalities to solve real-life problems and (ii) explain interrelationships within those systems that facilitate problem solving.

Unit Objectives:

A. Given a problem whose solution is facilitated by solving for an equation of the form $f_1(x) + f_2(x) + f_3(x) + \ldots + f_n(x) = 0$, the student explains why the solution is also facilitated by a system of equations with the following form:

$$
\begin{aligned}
a_{1,1}x_1 + a_{1,2}x_2 + a_{1,3}x_3 + \ldots + a_{1,n}x_n &= 0 \\
a_{2,1}x_1 + a_{2,2}x_2 + a_{2,3}x_3 + \ldots + a_{2,n}x_n &= 0 \\
a_{3,1}x_1 + a_{3,2}x_2 + a_{3,3}x_3 + \ldots + a_{3,n}x_n &= 0 \\
&\vdots \\
a_{n,1}x_1 + a_{n,2}x_2 + a_{n,3}x_3 + \ldots + a_{n,n}x_n &= 0
\end{aligned}
$$

(discover a relationship) 15%.

B. The student describes situations, both real life and mathematical, that are reflected by each of the types of linear equations: (i) simultaneous, (ii) inconsistent, and (3) equivalent (construct a concept) 15%.

C. Given a system of *n* *n*-variable linear equations or inequalities, the student solves for the *n* variables via the substitution algorithm (algorithmic skill) 5%.

D. Given a system of *n* *n*-variable linear equations or inequalities, the student solves for the *n* variables via the graphing algorithm (algorithmic skill) 5%.

E. Given a system of *n* *n*-variable linear equations or inequalities, the student solves for the *n* variables via the addition algorithm (algorithmic skill) 5%.

F. Given a system of *n* *n*-variable linear equations or inequalities, the student solves for the *n* variables by using matrices (algorithmic skill) 5%.

G. The student explains the difference between problems with solutions that are efficiently facilitated by linear programming and those that are not (application) 15%.

H. The student solves linear programming problems (algorithmic skill) 10%.

I. Given a problem, the student (i) determines whether or not the solution is facilitated by formulating and solving for a system of linear equations or inequalities, and, if so, (ii) formulates the system (application) 25%.

type (e.g., an algorithm). Thus, before designing a lesson for leading students to achieve a particular objective, you need to consider whether the content is (a) a *concept* (e.g., rational number), (b) a *discoverable relationship* (e.g., rate × time = distance), (c) a *convention* (e.g., "$|x|$" is read "the absolute value of x"), or (d) an *algorithm* (e.g., using Cramer's rule to solve for a system of linear equations).

Specifying the Learning Level

Compare the following five objectives for similarities and differences:

A. The student willingly attempts to develop a general formula for making it easier than either factoring or completing the square to solve quadratic equations (willingness to try).

B. The student explains why the quadratic formula yields the roots of any one-variable quadratic equation with real coefficients (discover a relationship).

C. The student states the quadratic formula (simple knowledge).

D. Given a one-variable quadratic equation with real coefficients, the student finds the roots of the equation using the quadratic formula (algorithmic skill).

E. Given a real-life problem, the student determines how, if at all, a solution to that problem is facilitated by setting up and solving for a quadratic equation (application).

All five of these objectives specify the same mathematical content—namely the quadratic formula. However, no two of the five objectives are the same. The objectives differ in the way students are expected to think about and deal with the quadratic formula. Objective A focuses on students' willingness to develop the formula. Objective B focuses on understanding why the formula works. Objective C is for students to remember the formula. Objective D is for them to compute the formula. Objective E is for the students to apply it in real-life situations. These objectives differ in the *learning levels* they specify:

> An objective's *learning level* is the manner in which students will mentally interact with the objective's mathematical content once the objective is achieved.

Just as an objective's mathematical content influences how you go about teaching to that objective (e.g., you teach about the quadratic formula differently than you teach about conditional probabilities), so should how you teach depend on the objective's learning level.

Familiarity with one of the published schemes for classifying objectives according to their targeted learning levels will help you clarify your own objectives. The scheme suggested to you in this text is especially adapted from a variety of sources (Bloom, 1984; Cangelosi, 1980; 1982, pp. 90–95; 2000a, pp. 211–247; Guliford, 1959; Krathwohl, Bloom, & Masia, 1964) for teaching mathematics in harmony with the *PSSM* (NCTM, 2000b). It takes into account the need for inquiry instruction from the constructivist perspective (Davis, Maher, & Noddings, 1990; Guzzetti, Snyder, Glass, & Gamas, 1993), as well as for direct instruction for skill building (Joyce, Weil, & Calhoun, 2000, pp. 191–214, 323–345).

Two learning domains are included: *affective* and *cognitive.* If the intent of the objective is for students to develop a particular attitude or feeling (e.g., a desire to prove a theorem or willingness to work toward the solution of a problem), the learning level of the objective falls within the *affective domain.*

If the intent of the objective is for students to be able to do something mentally (e.g., remember a formula or deduce a method for solving a problem), the learning level of the objective falls within the *cognitive domain.*

The scheme for classifying the learning levels under the two domains is explained and its uses illustrated in chapters 5–9. However, to give you a preview Exhibit 4.24 provides an outline of the scheme.

The Planned Sequence of Lessons

The paramount component of a teaching unit is the sequence of lessons you design and conduct for the purpose of achieving the stated objectives. Each lesson is designed to lead students to achieve an objective. The examples of unit plans in chapter 11 as well Exhibit 4.23 list objectives in the approximate order in which lessons for those objectives should be taught—at least with respect to what research in cognitive science suggests for learning to be meaningful (NCTM, 2000b, pp. 20–21). I say "approximate" because the lesson for one objective often overlaps lessons for other objectives—often sharing common learning activities. "The Overall Plan for Lessons" section in chapter 11's unit plans also indicates how objectives, and thus lessons, should be sequenced. Following are some general principles to keep in mind when you sequence objectives for a unit; how to apply these principles is explained and illustrated in chapters 5–8:

- Ordinarily, conventional names for concepts or relationships should not be introduced before students have engaged in inquiry lessons leading them to construct the concepts or discover the relationships for themselves. Memorizing words (e.g., "sample space" or "Pythagorean theorem") to attach to a concept (e.g., sample space) or relationship (e.g., the Pythagorean theorem) before the concept or relationship is conceptualized is meaningless for students.
- Comprehension-and-communication objectives relative to certain messages (e.g., the proof of the side-angle-side theorem as presented in a textbook) or use of certain technical expressions (e.g, "$\sum_{i=a}^{b} f(i)$") should be taught before conducting learning activities that depend on those messages or technical expressions. Students experience considerable difficulty engaging in lessons that include either messages they have yet to comprehend or technical expressions they have not learned to use.
- Ordinarily, the statement of a relationship (e.g., "$a^2 + b^2 = c^2$" or "the product of two negative integers is positive") should be committed to memory *after* students discover why that relationship exists.
- Students are ready to engage in inquiry lessons for an application-level objective only after they have (a) constructed concepts and discovered relationships underlying the mathematical content of the objective, (b) acquired relevant comprehension-and-communication skills, and (c) acquired relevant algorithmic skills.
- If creative-thinking or affective objectives are targeted by a unit, then lessons for them are ordinarily scattered throughout the unit and integrated with lessons for other objectives. Both creative and affective behaviors are usually acquired by experiences that extend over the entire course of a unit rather than tending to appear near the beginning (as with construct-a-concept or

Exhibit 4.24
Scheme for Categorizing Learning Levels Specified by Objectives.

I. **Cognitive Domain**

A. **Construct a Concept**

Students achieve an objective at the construct-a-concept learning level by using inductive reasoning to distinguish examples of a particular concept from nonexamples of that concept.

For example:
- The student (i) discriminates between relations that are functions and relations that are not and (ii) explains why the functions are functions and the other relations are not (construct a concept).
- Given a geometric figure (either concrete or abstract), the student discriminates between its surface area and other quantitative characteristics (e.g., height, volume, and angle size) (construct a concept).
- The student distinguishes between examples and nonexamples of each of the following: sample space, compound event, independent events, dependent events, mutually exclusive events, empirical sample space, theoretical sample space, complement of an event, and conditional probability (construct a concept).

B. **Discover a Relation**

Students achieve an objective at the discover-a-relationship learning level by using inductive reasoning to discover that a particular relationship exists or why the relationship exists.

For example:
- The student explains why the area of a rectangle equals the product of its length and width (discover a relationship).
- The student explains why a subset of integers with a lower bound has a least element whereas a subset of reals with a lower bound does not necessarily have a least element (discover a relationship).
- The student explains the rationale underlying the following formula for calculating the accumulated amount, *A,* in a compound interest savings plan:

$A = \left(P + \frac{r}{f}\right)^{kn}$ where *P* is the principal, *r* is the annual rate, *k* is the number of times per year the interest is compounded, and *n* is the number of years

(discover a relationship).

C. **Simple Knowledge**

Students achieve an objective at the simple-knowledge learning level by remembering a specified response (but not multiple-step process) to a specified stimulus.

For example:
- The student states the definitions of the six trigonometric functions (simple knowledge).
- The student associates the notation "$a|b$" where *a* and *b* are integers with the statements "*a* is a divisor of *b*" and "*b* is a multiple of *a*" (simple knowledge).
- The student states that the ratio of the circumference of any circle to its diameter is π (simple knowledge).

D. **Comprehension and Communication**

Students achieve an objective at the comprehension-and-communication level by (i) extracting and interpreting meaning from an expression, (ii) using the language of mathematics, and (iii) communicating with and about mathematics.

For example:
- The student explains—in his own words—the ϵ-δ definition of limit of a sequence (comprehension and communication).
- The student explains how to translate the summation notation (i.e., $\sum_{i=a}^{n} f(i)$) (comprehension and communication).
- The student explains the logic of the argument of the proof of the angle-side-angle congruence theorem as presented in the textbook (comprehension and communication).

E. **Algorithmic Skill**

Students achieve an objective at the algorithmic-skill level by remembering and executing a sequence of steps in a specific procedure.

For example:
- Given the dimensions of a triangle, the student computes the area of its interior (algorithmic skill).
- The student uses the chain rule to compute the derivative of $f(g(x))$ where *f* and *g* are algebraic functions such that *g* has a derivative at *x* and *f* has a derivative at $g(x)$ (algorithmic skill).
- The student bisects any Euclidean angle with a straightedge and compass (algorithmic skill).

Continued

Exhibit 4.24
Continued

F. **Application**

Students achieve an objective at the application level by using deductive reasoning to decide how to utilize, if at all, a particular mathematical content to solve problems.

For example:

- Given a real-life problem, the student decides how, if at all, graphing a nonlinear relationship and determining certain critical values (e.g., zeros or intercepts) would help solve the problem (application).
- Given a real-life problem, the student decides how, if at all, a solution to the problem is facilitated by using the following relationship:

  $A = \left(P + \frac{r}{f}\right)^{kn}$ where P is the principal, r is the annual rate, k is the number of times per year the interest is compounded, and n is the number of years (application).
- Given the task of proving a proposition, the student develops a plan for proving it, deciding what definitions, axioms, or prior theorems to use (application).

G. **Creative Thinking**

Students achieve an objective at the creative-thinking learning level by using divergent reasoning to view mathematical content from unusual and novel ways.

For example:

- The student develops a novel paradigm for illustrating the following relationship: ($\forall a, b, c \in \mathbb{R}$) ($x$ is a real variable and $f(x) = ax^2 + bx + c) \Rightarrow f'(x) = 2ax + b$) (creative thinking).
- The student generates novel and unusual conjectures about straightedge-and-compass constructions of angles (creative thinking).
- The student generates novel and unusual paradigms for multiplying signed numbers (creative thinking).

II. **Affective Domain**

A. **Appreciation**

Students achieve an objective at the appreciation learning level by believing the mathematical content specified in the objective has value.

For example:

- The student believes that an understanding of systems of linear equations can help solve problems about which he cares (appreciation).
- The student prefers to formulate open sentences when solving word problems rather than having someone else set them up for him (appreciation).
- The student recognizes the advantages of being able to analyze the behavior of a function without having to plot very many points on its graph (appreciation).

B. **Willingness to Try**

Students achieve an objective at the willingness-to-try learning level by choosing to attempt a mathematical task specified by the objective.

For example:

- The student attempts to formulate algebraic open sentences to solve word problems before turning to someone else to set them up (willingness to try).
- When executing an algorithm for using trigonometric functions to solve a problem, the student lists sequential results from the process neatly and in an orderly manner so that the work can readily be checked for errors (willingness to try).
- The student chooses to use graphing calculators or computer software to explore the behavior of functions (willingness to try).

discover-a-relationship learning), the middle (as with comprehension-and-communication, simple-knowledge, and algorithmic-skill learning), or the end (as with application learning).

When you plan a teaching unit, you design its lessons. However, because teaching is such a complex, mathematically chaotic art, you need to routinely monitor students' progress throughout the unit. The formative feedback from your assessments guides the pace of lessons and influences the design of learning activities.

A Summative Evaluation of Student Achievement of the Learning Goal

As a teacher, you are expected to make periodic reports to communicate to students, their parents, and your supervisors how well your students are achieving learning goals. Consequently, most teaching units terminate with measurements of students' achievement of the goal. Your judgments of students' success are referred to as "*summative evaluations.*" Strategies for measuring student achievement for purposes of making summative evaluations is a focus of chapter 9.

DAILY PLANNING

A unit's overall plan for lessons provides only general ideas about how the lessons will be taught. You develop a detailed plan for a lesson a day or two before you actually plan to begin it. A single lesson may take only a few minutes (e.g., for a simple-knowledge objective) or it may take a week (e.g., for a discover-a-relationship objective). A lesson for one objective may overlap with lessons for other objectives. The time it takes you to deliver lessons varies considerably. Then, of course, you must develop a daily plan or agenda for each class period. You will examine examples of all such planning documents as you work with chapters 5–8 and 11.

SYNTHESIS ACTIVITIES FOR CHAPTER 4

1. Look over a mathematics textbook currently being used in a middle, junior high, or high school. Select two topics from different chapters of the text. Examine how each topic is presented. Categorize aspects of each topic as to whether they originated as a concept construction, discovery of a relationship, or invention. For example, here is how I labeled some aspects of two topics from the prealgebra text whose table of contents appears in Exhibit 1.4:

- *Topic: Formula for perimeter*
  - The concept perimeter exists in nature and was constructed by people from their experiences. The name for that concept, "perimeter," is an invented convention.
  - The following method of depicting the dimensions of the perimeter of a rectangle is an invention:

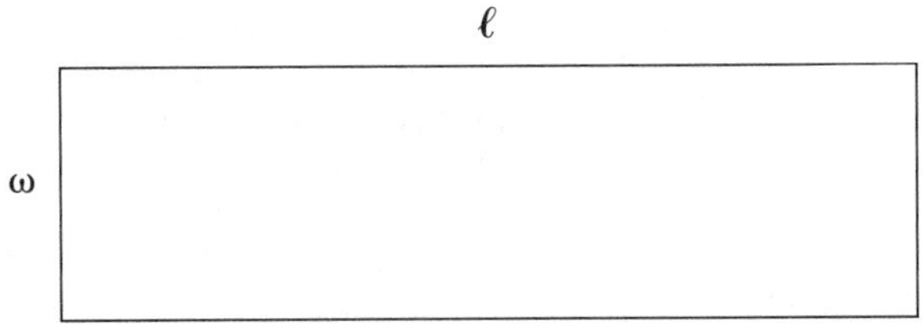

  - The concepts rectangle, length, and width are constructed. Their names, "rectangle," "length," and "width" are inventions.
  - The relationship $P = 2\ell + 2\omega$ is a discovery, but the expression of that relationship is an invention.
- *Topic: Graphing equations*
  - The Cartesian plane is an invention for illustrating relationships.
  - The following relationship is a discovery: The graph of any linear equation $ax + by = c$ (where a, b, and c are real constants and x and y are real variables) is a line.
  - The algorithm for graphing a linear equation is an invention.

Now, in a brief paragraph, describe one way the two topics you examined are related. For example, here is what I wrote for the two topics I chose:

The perimeter of a rectangle is dependent on two variables: length and width. One type of problem that the formula can be used to solve involves determining the possible dimensions of a rectangle with a given perimeter (e.g., in the case of a fixed amount of fencing available for a garden). For such situations, the perimeter formula is of the form $ax + by = c$ where c is the given perimeter and $a = b = 2$. Thus the possibilities for the length and width can be illustrated via the graph of a linear equation.

2. Following are two passages, one by Stewart (1992b, pp. 9–10), the other by Fowler (1994, p. 12). Read one of the passages yourself while a colleague reads the other. Then summarize the two selections for one another. Discuss strategies for narrowing the gap between "schoolmath" and "real mathematics."

Stewart's passage:
One of the biggest problems of mathematics is to explain to everyone else what it is all about. The technical trappings of the subject, its symbolism and formality, its baffling terminology, its apparent delight in

lengthy calculations: these tend to obscure its real nature. A musician would be horrified if his art were to be summed up as "a lot of tadpoles drawn on a row of lines"; but that's all that the untrained eye can see in a page of sheet music. The grandeur, the agony, the flights of lyricism and discords of despair: to discern them among the tadpoles is no mean task. They are present, but only in coded form. In the same way, the symbolism of mathematics is merely its coded form, not its substance. It too has its grandeur, agony, and flights of lyricism. However, there is a difference. Even a casual listener can enjoy a piece of music. It is only the performers who are required to understand the antics of the tadpoles. Music has an immediate appeal to almost everybody. But the nearest thing I can think of to a mathematical performance is the Renaissance tournament, where leading mathematicians did public battle on each other's problems. The idea might profitably be revived; but its appeal is more that of wrestling than of music.

Music can be appreciated from several points of view: the listener, the performer, and the composer. In mathematics there is nothing analogous to the listener; and even if there were, it would be the composer, rather than the performer, that would interest him. It is the creation of new mathematics, rather than its mundane practice, that is interesting. Mathematics is not about symbols and calculations. These are just tools of the trade-quavers and crotches and five-finger exercises. Mathematics is about *ideas.* In particular, it is about the way that different ideas relate to each other. If certain information is known, what else must necessarily follow? The aim of mathematics is to understand such questions by stripping away the inessentials and penetrating to the core of the problem. It is about understanding why an answer is possible at all, and why it takes the form that it does. Good mathematics has an air of economy and an element of surprise. But, above all, it has *significance.*

Fowler's passage:
What is mathematics? The answer you will get to this question depends on whom you ask. What most people in the United States have experienced is not mathematics—it is not even an elementary or student version of mathematics. It is an entirely different subject—one that could be called "schoolmath." Schoolmath has its own terminology. The rational numbers between 0 and 1 are called The Fractions. The multiplicative inverses of these numbers are called The Improper Fractions. Students are taught that the Improper Fractions exist in a sort of unstable equilibrium and should always be reduced to Mixed Numbers. [An article by Hassler Whitney (1987) got me thinking of this.]

Schoolmath has its own protocol ranging from the Story Problem to the Two-Column Proof, the latter taught in a course called Geometry. And schoolmath has its own set of beliefs about the working world—for example, beliefs about how people in all professions use schoolmath to solve problems. A typical belief is that carpenters spend a great deal of time sawing long boards into many short boards, and must therefore be skilled at dividing one mixed number by another. The truth about carpenters is that their work requires highly developed spatial visualization skills—skills they don't get from schoolmath at any grade level, and certainly not Geometry. The cognitive psychologist Robert Schank (1987) compiled a collection of similar beliefs in an article titled "Let's Eliminate Math from Schools."

Although many people survive schoolmath and may even become mathematicians, most people end up—as Keith Devlin (1993) has pointed out in an article in the *Notices of AMS*—thinking of what they studied as "mathematics."

Is schoolmath a necessary step to learning mathematics? No. In fact, schoolmath prevents most people from doing mathematics in any real way, just as contrived "schoolmusic" pieces discourage most people from being musicians. During the period 1962–72 in the U.S.A., mathematicians and some mathematics educators attempted to replace schoolmath with a version of mathematics that was generally called "new math." This did not work for several reasons; in fact, it inspired a "back-to-basics" movement that returned to schoolmath with a vengeance.

In 1972, Seymour Papert introduced Logo, and argued in a paper called "Teaching Children To Be Mathematicians Versus Teaching About Mathematics" that children could be mathematicians in an authentic sense, just as children could be artists or musicians. Papert was not talking about prodigies. His attempt was not to produce little Bourbakists, as many of the new math people had seemed inclined to do, nor was it to set up phony "discovery" situations where children somehow always came to the conclusion the teacher had previously prepared. Papert used the example of a child trying to construct a Logo procedure for some form that the child wanted the turtle to draw, such as a "squiral," and argued that children engaged in such activities were doing mathematics. In some cases, Logo is now taught as Papert intended. In other cases, teachers are suggesting that Logo be replaced by a "better drawing program," which suggests that the original purposes for Logo were not universally understood.

The mathematics reform movement, whose constitutional document is the NCTM *Curriculum and Evaluation Standards* (NCTM, 1989a), offers a rational plan for evolving from schoolmath, first to better schoolmath, and eventually to a version of real mathematics. This plan is, in fact, the official agenda for mathematics education.

3. Recall a real-life problem you once solved. Analyze the process by which you solved it. Describe exactly what you did for each of the nine stages listed after Case 4.7 in this chapter's section "Real-Life Problem Solving." Exchange your descriptions with a colleague and discuss the similarities and differences in the ways you, your col-

league, and Brenda in Case 4.7 addressed the respective problems.

4. Obtain a copy of *PSSM* (NCTM, 2000b)—either the book, CD, or from NCTM's website (www.nctm.org). Thumb through Chapters 6 ("Standards for Grades 6–8") and 7 ("Standards for Grades 9–12") so that you become aware of how the contents can serve as a resource as you develop curricula by designing mathematics courses, units, lessons, and assessment procedures.
5. In a one-to-one interview raise the following prompts with an adolescent student:
   A. If it were not a required subject, would you choose to study mathematics? Why or why not?
   B. Is mathematics more or less interesting than other school subjects you take? Why?
   C. Is mathematics more or less difficult to learn than other school subjects? Why?
   D. From where do you think the mathematics we study in school comes? Explain why you believe what you just told me.
   E. What do the words "mathematical research" bring to your mind?
   F. Tell me what you know about any one of the following people: Pythagoras, Euclid, Hypatia, René Descartes, Maria Agnesi, Issac Newton, or Andrew Wiles.
   G. What is the easiest thing about learning mathematics?
   H. What is the hardest thing about learning mathematics?
   I. What is the most interesting thing about doing mathematics?
   J. What is the most boring thing about doing mathematics?
   K. Do you read your mathematics textbook differently from the way you read textbooks for your other subjects?
   L. How often and for what reasons do you use mathematics outside of school-related work?
6. By engaging in Activities 1.2, 1.3, and 4.3, you outlined at least two mathematics courses—enumerating a sequence of teaching units for each course. Select one of those teaching units, formulate its goal, and define that goal with a sequence of objectives. Although you do not comprehend Exhibit 4.24's scheme for classifying learning levels as well now as you will after you have worked with chapters 5–8, specify each objective's learning level by labeling it as either "construct a concept," "discover a relationship," "simple knowledge," "algorithmic skill," "comprehension and communication," "application," "creative thinking," "appreciation," or "willingness to try." As you formulate and sequence these objectives, keep in mind the principles for leading students to learn meaningful mathematics that are listed in this chapter's section "The Planned Sequence of Lessons." Note that you are developing an initial draft of a goal and objectives that you will be prompted to modify and refine while working with subsequent chapters. Discuss your draft and those of colleagues—giving one another feedback. Store this unit goal and objectives in the appropriate subfolder of your computerized folder structure. Also, insert a hard copy as a "work in progress" in the "Cognition, Instructional Strategies, and Planning" section of your working portfolio.

TRANSITIONAL ACTIVITY FROM CHAPTER 4 TO CHAPTER 5

In a discussion with two or more of your colleagues, address the following questions:

1. What strategies should teachers employ to lead their students to construct mathematical concepts for themselves? How should lessons for construct-a-concept objectives be designed? What are some strategies for monitoring students' progress during lessons for construct-a-concept objectives?
2. What strategies should teachers employ to lead their students to discover mathematical relationships for themselves? How should lessons for discover-a-relationship objectives be designed? What are some strategies for monitoring students' progress during lessons for discover-a-relationship objectives?
3. What are some strategies for teachers to use to help students who display misconceptions about mathematics?

chapter

5

Leading Students to Construct Concepts and Discover Relationships

Goal and Objectives for Chapter 5

The Goal The goal of chapter 5 is to lead you to develop strategies for designing lessons that lead students to construct mathematical concepts and to discover mathematical relationships.

The Objectives Chapter 5's goal is defined by the following set of objectives:

A. You will distinguish between examples of (a) concepts, (b) discoverable relationships, and (c) other types of mathematical content specified by learning objectives (construct a concept) 5%.

B. You will describe the inductive reasoning process by which students construct concepts for themselves (comprehension and communication skills) 10%.

C. For a given group of middle or secondary school students, you will formulate construct-a-concept objectives that are consistent with *PSSM* (NCTM, 2000b) (application) 5%.

D. You will design lessons for construct-a-concept objectives (application) 20%.

E. You will design miniexperiments that are relevant to your students' achievement of construct-a-concept objectives (application) 10%.

F. You will describe the inductive reasoning process by which students discover relationships for themselves (comprehension and communication skills) 10%.

G. For a given group of middle or secondary school students, you will formulate discover-a-relationship objectives that are consistent with *PSSM* (NCTM, 2000b) (application) 5%.

H. You will design lessons for discover-a-relationship objectives (application) 20%.

I. You will design miniexperiments that are relevant to your students' achievement of discover-a-relationship objectives (application) 10%.

J. You will incorporate the following phrases or words into your working vocabulary: "specific," "concept," "concept attribute," "example noise," "subconcept," "inductive reasoning," "miniexperiment," "prompt," "observer's rubric," and "discover a relationship" (comprehension and communication) 5%.

MATHEMATICAL CONCEPTS

Conceptualizing

To lead you to refine your notion of a *concept,* engage in Activity 5.1:

Activity 5.1

Collect any 15 objects from your immediate environment and place them in a box. Write a list of the names of the 15 objects. Each of the 15 entries should be unmistakably unique. For example, to list the book you are now reading, do not simply write "a book," because there are not only many other book titles but also other copies of the title *Teaching Mathematics in Secondary and Middle School: An Interactive Approach.* Write, for example, "the copy of the book from which I read the directions for this activity."

Now make a second list with exactly three entries; include each of the 15 objects named on the first list, but do not necessarily list them separately.

Although you may be confused by the directions for this activity, make your two lists following your best guess as to what these directions are intended to mean.

Now, compare your two lists to those of colleagues, and in light of those comparisons, help one another comprehend the directions and revise your lists so the intended directions have been followed.

In Case 5.1, Tina (whom you first met in Case 1.12) engages in Activity 5.1:

CASE 5.1

After collecting 15 objects from her room, Tina thinks to herself: "Okay, first of all, there are these three pencils I dumped out of my pencil box. But I'm supposed to list each separately so that it is unmistakably unique. It sure would be easier to write, 'three pencils,' but . . ."

Six minutes later, after realizing that describing unique entities is not as simple a task as she initially perceived, Tina completes the following list:

1. The yellow pencil I'm using to write this list that was in the cardboard box on my floor a minute ago
2. The never-before-sharpened pencil that was in my pencil box on my desk just before I dumped the pencils into the cardboard box that's on my floor right now
3. The green pencil that was in my pencil box 10 minutes ago
4. My copy of the book *Another Fine Math You've Got Me Into*
5. The copy of the book *Pi in the Sky: Counting, Thinking, and Being* that I borrowed from Bill
6. My copy of the *Random House College Dictionary*
7. The paper clip in the cardboard box on my floor
8. My pencil box that's now in the cardboard box on the floor of my room right now
9. The rubber band that's farthest to the left side in the cardboard box on my floor
10. The rubber band that's farthest to the right side in the cardboard box on my floor
11. The rubber band that's second from the left in the cardboard box on my floor
12. The rubber band that's second from the right in the cardboard box on the floor of my room
13. The pair of scissors in the cardboard box on my floor
14. The copy of *Investigating Statistics with the TI-92* that I checked out of the Math Department library
15. The brown, shiny-looking, oddly shaped, odorless thing that's about two centimeters long that I picked up off the floor of my room and put in the cardboard box on my floor

She thinks, "Now my task is to collapse some of these specifics so I end up with three categories. Let's see, I've got four pencils, and . . ." Quickly she collapses the list of 15 into the following:

1. Three pencils
2. A pencil box
3. Four books
4. Four rubber bands
5. A paper clip
6. A pair of scissors
7. The brown, shiny-looking, oddly shaped, odorless thing that's about two centimeters long that I picked up off the floor of my room and put in the cardboard box on my floor

"That brown, shiny thing must have broken off of something; I have no idea what it is or what to call it. I don't see a way to group it with any of the other stuff—*stuff!* That's a category that includes all of these things. But I'll follow the directions and get these down to exactly three," she thinks before finalizing the following lists:

1. Office supplies
2. Books
3. The brown, shiny-looking, oddly shaped, odorless thing that's about two centimeters long that I picked up off the floor of my room and put in the cardboard box on my floor

Our real world is composed of *specifics* we detect with our empirical senses. Specifics are far too numerous for us to think about each one as a unique entity. Thus, we categorize and sub-categorize specifics according to certain commonalities or attributes. The categories provide a mental filing system for storing, retrieving, and thinking about information. The process by which a person groups specifics to construct a mental category is referred to as "*conceptualizing.*"

Exhibit 5.1
A Concept Relates to a Subconcept as a Set Relates to Its Subsets.

A polygon is a concept; it is a set with more than one element.	polygon
Each of the following concepts is a subconcept of a polygon: triangle, quadrilateral, pentagon, hexagon, heptagon.	triangle
Special types of triangles (e.g., isosceles), quadrilaterals (e.g., rectangles), and pentagons (e.g., regular) are subconcepts of triangle, quadrilateral, and pentagon, respectively (i.e., subsets of subsets).	isosceles triangle
A specific example of a concept is not a concept but a constant; it is the specified element of a set.	The unique isosceles triangle determined by the following three points:

The category itself is a *concept.* Here are two definitions to keep in mind:

- A *specific* is a unique entity, something that is not abstract.
- A *concept* is a category people mentally construct by creating a class of specifics possessing a common set of characteristics. In other words, a concept is an abstraction.

Of the three items in Tina's final list in Case 5.1, two refer to more than one specific thing. Thus, those two items, office supplies and books, are concepts. Exhibit 5.1 illustrates how concepts can relate to one another, with broader concepts including narrower subconcepts.

The task of listing 15 specific objects for Activity 5.1 was facilitated by having conventional names to call concepts that we recognize (e.g., "pencil," "book," "box," and "rubber band"). You have never seen Tina's pencil box, but you have a vision of what it might be like because you have seen other pencil boxes. On the other hand, the 15th item on her initial list did not fit any concept for which we know a conventional name—other than "thing", which does not distinguish it from anything else. Thus, Tina used more words to describe the 15th entry than for the ones that fit preconceived categories.

In the second phase of Activity 5.1 you clustered specific examples of a concept (e.g., office supply) and reported them as a single category. That required a higher level of thought but was not as tedious as listing each one specifically; it also provided an easier list to read. Of the three sets in Tina's final list, two contain more than one element. Thus, those two sets, the office supplies in the box and the books in the box, are *variables*. The set that contains only the brown, shiny-looking . . . thing is a one-element set and thus is not a variable but a *specific*.

The terms "variable" and "concept" denote the same idea—namely, a set with more than one element or value. The terms "constant" and "specific" both refer to a unique entity. Consider the following learning objective for a probability and statistics unit:

> The student distinguishes between examples and nonexamples of a sample space for an experiment (construct a concept).

Note that the mathematical content specified by the objective is sample space. The following, for example, is an experiment:

> On a basketball court, Joseph and Emily each shoot two foul shots. The number of successful shots for each is recorded.

And the following is one possible sample space for that experiment:

> {possible ordered pairs (j, e) where j = the number of successful foul shots Joseph makes and e = the number of successful foul shots that Emily makes} or, in other words, {(0, 0), (1, 0), (2, 0), (0, 1), (1, 1), (2, 1), (0, 2), (1, 2), (2, 2)}

Because experiments differ, sample spaces vary. Furthermore, more than one sample space can be set up for a single experiment. Thus, the idea of sample space is a concept.

To continue refining your notion of a concept, engage in Activity 5.2:

Activity 5.2

Examine the following list and (a) determine which entries are concepts and which are specifics (i.e., constants), (b) for each specific identify those concepts on the list for which it is an example, and (c) identify pairs of concepts in which one is a subconcept of the other:

1. number
2. geographic feature
3. 9.013
4. rational number
5. irrational number
6. mountain
7. mile
8. former baseball player
9. Mount Olympus
10. line segment
11. the line segment determined by the following two points: (a) the most extreme top corner of the textbook page you are now reading and (b) the most extreme point at the right top corner of the same textbook page
12. limit of a function
13. former baseball player who was inducted into the Baseball Hall of Fame
14. $7(x + 4)(3x)$ where $x \in$ {reals}
15. degree
16. standard measurement unit
17. rectangle
18. Jackie Robinson
19. real number
20. square (geometric)
21. derivative of a function
22. polynomial
23. meter

Compare how you categorized the 23 items to the following:

1. number: a concept
2. geographic feature: a concept
3. 9.013: a specific that is an example of the concepts listed as Items 1, 4, 12, and 19
4. rational number: a concept that is a subconcept of Items 1, 12, and 19
5. irrational number: a concept that is a subconcept of Items 1, 12, and 19
6. mountain: a concept that is a subconcept of Item 2
7. mile: a concept that is a subconcept of Item 16
8. former baseball player: a concept
9. Mount Olympus: a specific that is an example of concepts listed as Items 2 and 6
10. line segment: a concept
11. the line segment determined by the following two points: (a) the most extreme top corner of the textbook page you are now reading and (b) the most extreme point at the right top corner of the same textbook page: a specific that is an example of Item 10
12. limit of a function: a concept
13. former baseball player who was inducted into the Baseball Hall of Fame: a concept that is a subconcept of Item 8
14. $7(x + 4)(3x)$ where $x \in$ {reals}: a concept that is a subconcept of Items 1 and 22
15. degree: a concept that is a subconcept of Item 16
16. standard measurement unit: a concept
17. rectangle: a concept
18. Jackie Robinson: a specific that is an example of Items 8 and 13
19. real number: a concept that is a subconcept of Items 1 and 12
20. square (geometric): a concept that is a subconcept of Item 17
21. derivative of a function: a concept that is subconcept of Items 1 and 12

22. polynomial: a concept that is a subconcept of Item 1
23. meter: a concept that is a subconcept of Item 16

Concept Attributes

Whether or not a specific is an example of a particular concept depends on whether that specific possesses the defining *attributes* of the concept. For example, are you an example of the concept preservice mathematics teacher? As you recall from your work with chapter 1's section "Preservice Preparation and Professional Portfolios," a preservice mathematics teacher is a human being who is currently enrolled in a professional teacher-preparation program for the purpose of becoming qualified and certified to teach mathematics in middle or secondary schools. Thus, the attributes of a preservice mathematics teacher are (a) human being, (b) enrolled in a professional program to prepare mathematics teachers, and (c) not an inservice mathematics teacher. If you possess all three of these characteristics, you are an example of the concept preservice mathematics teacher. By definition:

> A *concept attribute* is a characteristic common to all examples of a particular concept; a concept attribute is a necessary requirement for a specific to be subsumed within a concept.

The attributes define the concept. An even number, for example, is an integer that is a multiple of 2. Thus, the attributes of the concept even number are (a) integer and (b) multiple of 2. Any specific that meets these two requirements (e.g., 10, $-8^{\frac{1}{3}}$, 0, 33 + 47, −3.3 − 2.7, and 96) is an even number. Any specific that does not (e.g., 11, $8^{\frac{1}{2}}$, $\frac{3}{4}$, −0.001, you, and the book you are now reading) is not an even number.

Example Noise

Although you may be a preservice mathematics teacher, you possess a myriad of other characteristics that are not attributes of a preservice mathematics teacher. Not all preservice mathematics teachers have the same color eyes as you, nor are they your age, nor do they have the same likes and dislikes. Similarly, 10 ∈ {even numbers}, but not all even numbers are multiples of 5, less than 34, positive, and the principal square root of 100, as is 10.

To enhance your concept of concept attributes and example noise, engage in Activity 5.3:

Activity 5.3

With a colleague, examine the following two examples:

A. Until it is won, the amount of money each hour in a radio giveaway jackpot that begins with \$100 and increases by \$20 each hour

Exhibit 5.2
The Brick Arrangement for Activity 5.3's Example B.

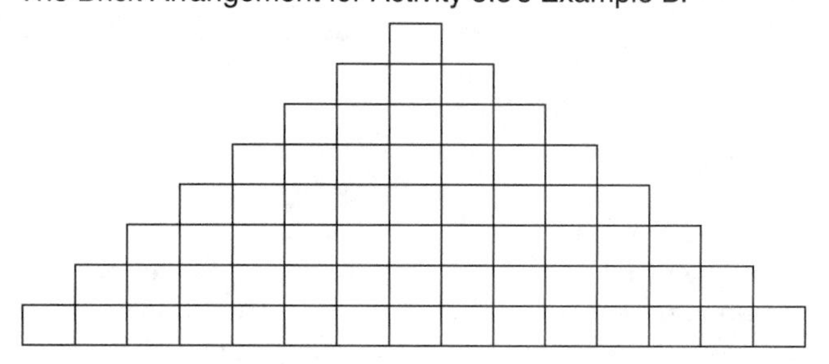

B. The number of bricks in the 1st, 2nd, 3rd, . . ., 8th row with the bricks arranged as shown by Exhibit 5.2

Accept the following definition:

$((a_1, a_2, a_3, \ldots, a_n)$ is a *finite arithmetic sequence*$) \Leftrightarrow (a_1, a_2, a_3, \ldots a_n \in \{\text{reals}\} \ni \exists\, d \in \{\text{reals}\} \ni a_{i+1} - a_i = d \;\forall\, i < n)$.

In other words, a finite arithmetic sequence is an ordered subset of {reals} such that the differences between any two consecutive numbers of the sequence is the same. Thus, Examples A and B are examples of arithmetic sequences. Now complete the following tasks:

1. List what you observed about Examples A and B that led you to believe that both are finite arithmetic sequences.
2. List characteristics of Example A that make it different from Example B.
3. List characteristics of Example B that make it different from Example A.

In Case 5.2, Tina and Jermain discuss their analyses from Activity 5.3:

CASE 5.2

Tina: Practically all we did in our discrete math course was work with sequences. An arithmetic sequence always looked like *a*, *a* + *d*, *a* + 2*d*, *a* + 3*d*, and so on until *a* + *kd* if it's finite.

Jermain: Yeah, like 4, 7, 10, 13, 16, 19. But this radio jackpot thing and this stack of bricks don't look like any sequences of numbers we ever studied. The bricks are in a sequence alright, but the elements are bricks, not numbers. The definition of an arithmetic sequence requires the elements to be real numbers.

Tina: That's what I thought when I first looked at them. Arithmetic sequences are usually expressed in textbooks as strings of numerals separated by commas. But when I reread these examples, I focused on the "amount of money" in the first and "number of bricks" in the second.

Jermain: Oh! So, it's not the dollars or bricks; it's amount or number as elements of the sequence. Okay, the elements really are real numbers and obviously each progresses by a common difference.

Tina: Let's express that in the more familiar textbook way. That'll make them easier to analyze.

Jermain: Example A is 100, 120, 140, 160, and so on until the jackpot is won. So let h be the number of hours before it's won and you get 100, 120, 140, 160, . . ., $100 + 20h$.

Tina: And the brick-related one is 1, 3, 5, 7, . . ., 15.

Jermain: Or you could start counting from the bottom and write 15, 13, 11, . . ., 1. Either way, we have a finite arithmetic sequence.

Tina: Okay we've convinced ourselves that both are examples of finite arithmetic sequences. That brings us to the three "list" tasks at the bottom.

Jermain: We've really already done the first one.

Tina: I agree; we matched characteristics of the examples to the definition of arithmetic sequence. We've just got to write them down now. The first attribute we talked about is that the elements are real numbers.

Jermain: Then we expressed them in the standard textbook form for finite arithmetic sequences, $a, a + d, a + 2d, a + 3d, \ldots, a + kd$. Being able to do that shows the other attribute of the concept.

Tina: So let's go to the second "list" task, the characteristics of Example A that make it different from Example B.

Jermain: One difference is that the numbers for the sequence in Example A are generated by dollars in a radio jackpot rather than by the bricks in the rows of a stack.

Tina: For Example A, the a of $a, a + d, a + 2d, a + 3d, \ldots, a + kd$ is 100 as opposed to 1 or 15, depending on which direction we start for Example B.

Jermain: And the d for Example A is 20 instead of 2 or -2 for Example B. The k in Example A is unknown, but for Example B it's the number of rows of bricks which is—ahh [pause] 8.

Tina: Actually, because the number of rows of bricks is 8, $k = 7$. k is the number of rows less 1.

Jermain: What? [pause] Oh! I see, thanks.

Tina: Now for the third "list" task. [pause] Oh! We already did it when we did the second.

When you, like Tina and Jermain, listed characteristics of the two examples verifying that both are finite arithmetic sequences, you were, of course, listing attributes with respect to the concept. When you listed other characteristics of those two examples—characteristics distinguishing one example of a finite arithmetic sequence from others—you were identifying *example noise.* By definition:

> *Example noise* is any characteristic of an example of a concept that is not an attribute of that concept.

All finite arithmetic sequences have a first element—that is an attribute of the concept. But in a finite arithmetic sequence, the first element does not have to be 100 (as in Example A). Thus, $a_1 = 100$ is a bit of noise for Example A. The fact that you are reading this book right now is a bit of example noise with regard to your being an example of a mathematics teacher because not all mathematics teachers are reading this book right now.

The noise in the examples you use in lessons for leading students to construct mathematical concepts plays a key role in how well they conceptualize. The noise in the examples you choose will influence whether or not your students construct misconcepts. Subsequent sections of this chapter suggest how you should design lessons so that example noise helps rather than hinders learning.

Revisit Exhibit 1.3 of Case 1.3. Casey Rudd used the trash barrel as an example of a right cylinder. What are some of the characteristics of the trash barrel that are bits of example noise? Unless Mr. Rudd's students are also exposed to other examples of right cylinders—examples with noise that are not characteristic of a trash barrel—he runs the risk of students misconstructing the concept of right cylinder by attributing noise (e.g., having a painted surface, serving as a container, and larger than a thimble) as characteristics of all right cylinders. Thus, he includes different examples (e.g., a section of a water hose and a thimble) with different kinds of noise in subsequent learning activities of this lesson.

Concepts in Mathematics Curriculum

Constructing concepts in our minds enables us to extend what we understand beyond the specific situations we have experienced in the past. Concepts are the building blocks of mathematical knowledge, but concepts are not the only type of mathematical content included in curricula. There are also discoverable relationships, conventions, and algorithms. Exhibit 5.3 lists a relatively minute subset of concepts typically included in middle and secondary school mathematics curricula.

INDUCTIVE REASONING

To construct a concept, students use *inductive reasoning* distinguishing examples from nonexamples of the concept. By definition:

> *Inductive reasoning* is generalizing from encounters with specifics. It is the cognitive process by which people discover commonalities among specific examples, thus leading them to formulate abstract categories (i.e., concepts) or discover abstract relationships.

Exhibit 5.3
A Relatively Minute Subset of the Concepts Typically Included in Middle and Secondary School Mathematics Curricula.

number	variable	point
natural number	set	line
whole number	intersection	line segment
prime number	union	plane
integer	finite	space
rational number	infinite	ray
irrational number	sequence	angle
real number	relation	polygon
complex number	function	distance
root	1:1 function	area
factor	permutation	volume
polynomial	limit	weight
interval	continuity	measurement
tangent line	derivative	perimeter
rate	integral	circle
vector	countable	fractal

Students use inductive reasoning in Cases 5.3, 5.4, and 5.5:

CASE 5.3

Rubin encounters variables in his psychology, chemistry, and mathematics courses. He notices differences and similarities among those variables. Without a conscious effort to do so, he creates a dichotomy between two types:

1. variables such as (a) aptitudes people have for learning, (b) temperature fluctuations, and (c) real-number intervals (e.g., $[-6, \infty)$)
2. variables such as (a) different types of emotional disorders, (b) atomic numbers of chemical elements, and (c) subsets of integers (e.g., $\{-6, -5, -4, \ldots\}$)

Rubin finds it more difficult to deal with the first type than the second because, as he says, "Those kinds are too packed in to list things that are next to one another." Apparently, Rubin is constructing the concept of continuous data and the concept of discrete data. He does not, however, know the concepts by those names.

CASE 5.4

After completing a homework assignment for which she used a protractor to measure angles of six triangles her teacher put on a tasksheet, Robin looks at the resulting six triples: (100°, 45°, 35°), (82°, 60°, 39°), (30°, 60°, 90°), (142°, 15°, 25°), (60°, 60°, 60°), and (30°, 31°, 120°).

She thinks, "Anytime there is a big angle, the other two are small." She then attempts to draw a triangle with two "big" angles and finds it impossible. Experimenting with triangles that are nearly equiangular, she finds all their angles are near 60°. Curious about the phenomenon, she measures the angles of 16 different triangles determined by concrete objects (e.g., two edges of a mirror). She then thinks, "Adding up the degrees of the three angles of any triangle will be about 180."

CASE 5.5

While exploring different functions with the aid of a computer program, Christi notices that $f(i)$ for $i = 0, 1, 2, \ldots, 25$ is always a prime number when $f(i) = i^2 - i + 41$. Christi concludes she has discovered a function from {integers} into {primes}.

In Case 5.3, Rubin organized variables into two categories, thus constructing two concepts. In Case 5.4, Robin formed a conjecture from her experiences with specific triangles, thus discovering a relationship. As illustrated by Case 5.5, inductive reasoning sometimes results in conjectures that can be disproven with counterexamples (e.g., $i = 41$). But disproving the conclusion does not discredit the reasoning.

CONSTRUCT-A-CONCEPT OBJECTIVES

The Construct-a-Concept Learning Level

An objective for students to use inductive reasoning to distinguish between examples and nonexamples of a mathematical concept is at the *construct-a-concept learning level.*

Review the examples of construct-a-concept objectives in Exhibit 4.24.

Stating Construct-a-Concept Objectives

How a teacher states objectives is a matter of individual style and preference. The mathematical content of a construct-a-concept objective is necessarily a concept or concepts. As long as your statement clearly specifies the mathematical concept you intend and you label it "construct a concept," the statement sufficiently conveys the intent of the lesson.

To continue refining your own concept of construct-a-concept objectives, examine Exhibit 5.4's examples of objectives taken from four different teachers' unit plans.

LESSONS FOR CONSTRUCT-A-CONCEPT OBJECTIVES

Challenging but Critical to Teach

Designing inquiry lessons for construct-a-concept objectives will tax your understanding of your students,

- The student sorts examples of a wide variety of geometric figures into the following categories (but not necessarily in association with these conventional names): (a) one dimensional, (b) convex polygon, (c) nonconvex polygon, (d) two-dimensional nonpolygon, (e) prism, (f) nonprism polyhedron, and (g) three-dimensional nonpolyhedron (construct a concept).
- The student distinguishes between relations (both real life as well as mathematical) that are functions and relations that are not functions (construct a concept).
- The student distinguishes between examples of geometric sequences and examples of sequences that are not geometric (construct a concept).
- The student discovers how countably infinite sets are like one another but different from other infinite sets (construct a concept).

Exhibit 5.4
Examples of Construct-a-Concept Objectives Taken From Four Different Teaching Units.

pedagogical principles, and mathematics. Coming up with choice examples, nonexamples, problems, and leading questions that will stimulate students to use inductive reasoning to form concepts is challenging to say the least. However, students' conceptualizations provide the basis for subsequent meaningful learning of mathematics. For example, students who have themselves constructed the concept of limit of a function will more easily comprehend the ϵ-δ definition of limit of a function than students who do not understand the interplay among ϵ and δ and other variables used in the wording of the definition.

The failure of many students to develop healthy attitudes about mathematics, algorithmic skills, comprehension and communications skills with mathematics, and application-level abilities to do mathematics to solve problems is well publicized (Brenner, Herman, Ho, & Zimmer, 1999; Heddens & Speer, 2001, pp. 3–4). Many of these failures can be traced to conceptual gaps in their learning (Ball, 1988; Goldin & Shteingold, 2001; Hiebert & Carpenter, 1992; Shuell, 1990). Such gaps are hardly surprising in light of the fact that many teachers never even consider conducting lessons for construct-a-concept objectives (Cangelosi, 2001a; Jesunathadas, 1990; NCTM, 1991; Smith, 2001, pp. 1–5).

Designing Lessons for Construct-a-Concept Objectives

Four Stages

Inquiry learning activities stimulating students to reason inductively to construct a concept are embedded in a lesson with four stages: (1st) *sorting and categorizing,* (2nd) *reflecting and explaining,* (3rd) *generalizing and articulating,* and (4th) *verifying and refining.*

Stage 1: Sorting and Categorizing

In Stage 1, you present students with a task requiring them to sort and categorize specifics. While orchestrating the activity, managing the environment, and providing guidance, you allow students to complete the task themselves.

Stage 2: Reflecting and Explaining

In Stage 2, students explain their rationales for categorizing the specifics as they did. You raise leading questions, stimulate thought, and clarify students' expressions.

Stage 3: Generalizing and Articulating

In Stage 3, students describe the concept in terms of attributes (i.e., what sets examples of the concept apart from nonexamples). They may also develop a definition for the concept; however, it is not necessary for the conventional name of the concept to be used.

Stage 4: Verifying and Refining

In Stage 4, the description or definition is tested with additional specifics—that the students already know to be examples and thus should fit—and with additional nonexamples—that students know should not fit. Further verification is pursued depending on your judgment of the situation. The description or definition of the concept is modified in light of the outcome of the tests. Prior stages are revisited as you judge necessary.

Selecting Examples and Nonexamples

One aspect of designing construct-a-concept lessons that you may find particularly challenging is producing appropriate examples and nonexamples for students to categorize. As you create or identify examples and nonexamples of a concept, your attention to concept attributes and example noise is critical. Case 5.6 illustrates the point with a kindergarten teacher attempting to lead students to develop their concepts of set cardinality:

CASE 5.6

Mr. Edwards is using questioning strategies with individual kindergarten students to stimulate them to reason

Exhibit 5.5
Which Two Sets Are Alike?

A B C

inductively to construct the concept of set cardinality. He displays Exhibit 5.5 to Stacie and asks her, "Which two groups are alike?" Stacie points to sets *A* and *B* and replies:

Stacie: These.
Mr. Edwards: Why?
Stacie: 'Cause they're not round like those.
Mr. Edwards: Thank you. Now, can you see a way this group of round things is like one of the other groups?
Stacie: No, it's different.

Mr. Edwards shows Stacie Exhibit 5.6 and asks, "Which two of these groups are alike? Stacie points to A and C and says, "'Cause they have as many." "Thank you," Mr. Edwards responds, "Let's go back and look at these again." He displays Exhibit 5.5.

Note how Mr. Edwards manipulated examples to control for example noise. Had Stacie readily categorized by cardinality upon seeing Exhibit 5.5, Mr. Edwards would have moved to a noisier situation (e.g., Exhibit 5.7) in which the noise varies more. But because Stacie experienced difficulty with Exhibit 5.5, he reduced the noise by moving to Exhibit 5.6. Eventually, Stacie needs to recognize similarities and differences with respect to cardinality even in high-noise situations because the real world is quite noisy.

In general, how well students construct a concept depends on how well they distinguish concept attributes from example noise. Discriminating between examples and nonexamples in high-noise situations is indicative of more sophisticated concept construction than in low-noise situations.

Exhibit 5.6
Examples Used by Mr. Edwards With Less Variability of Example Noise Than Exhibit 5.5.

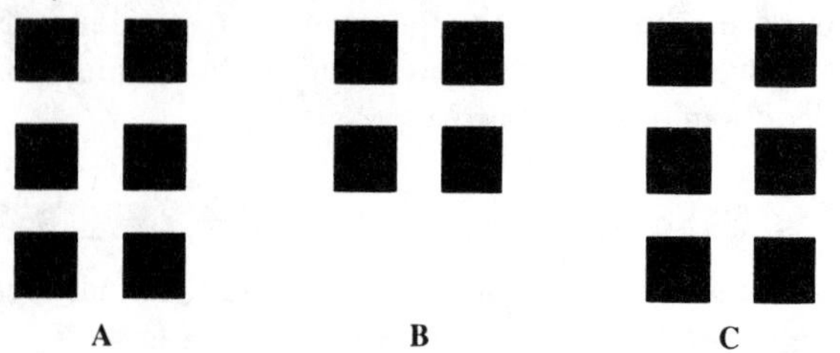

Exhibit 5.7
Examples Used by Mr. Edwards With Greater Variability of Example Noise Than Exhibit 5.5.

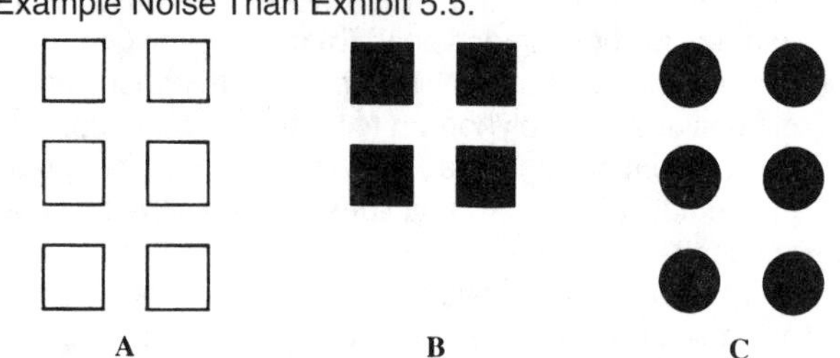

Incorporating the Four Stages Into Construct-a-Concept Lessons

In Case 5.7, note how the teacher plans for Stage 1 of a construct-a-concept lesson by making use of her understanding of concept attributes and example noise:

CASE 5.7

Working from a unit plan that includes the goal and objectives listed in Exhibit 5.8, Ms. Citerelli thinks as she designs the initial lesson, "Objective A is the key to the whole unit. All the others depend on it. The learning level is construct a concept and the mathematical content is arithmetic sequence. That means I've got to stimulate them to reason inductively beginning with a sorting and categorizing task. First, I've got to come up with the arithmetic and nonarithmetic sequences for them to ponder. If I start with a list of arithmetic sequences, I can match each with a nonarithmetic sequence that has similar example noise. I don't want them to abstract the noise as part of the concept."

With the help of the course textbook and further thought, she develops the two lists shown by Exhibit 5.9.

She thinks, "Not bad. The noise in the examples varies so they're unlikely to develop too narrow a concept. The characteristics of the nonexamples match the noise in the examples, so they should be able to sort out the attributes from the noise. Oh! I should also toss in a few real-life sequences so they'll maintain the connection between mathematics and the real world. That's going to complicate things in a necessary way. I'll think of a few." Six minutes later, she adds Exhibit's 5.10's list to Exhibit 5.9's list.

She thinks, "I'll try starting the lesson off with the whole class in one large group. I'll present them with a tasksheet with a single list of mixed-up examples and nonexamples and have them try to pull out one group according to some common attributes. But I'm afraid that would take a really long time and might get us into some other concepts I don't want to take time for. That's likely with this list; it's pretty complex. Of course, they already know what a sequence is; that won't be a problem. But having them all see uniform differences between

Exhibit 5.8
Goal and Objectives for Ms. Citerelli's Unit on Arithmetic and Geometric Sequences.

Unit 8: Arithmetic and Geometric Sequences (Intermediate Algebra)

Goal: Students explore arithmetic and geometric sequences so that they discover some fundamental relationships, invent some useful algorithm, and apply them to real-life situations.

Objectives:

A. The student distinguishes between examples and nonexamples of arithmetic sequences, explains the defining attributes, and formulates a definition (construct a concept) 10%.

B. The student explains why the following relationship holds:

$$a_n = a_1 + (n - 1)(a_2 - a_1)$$

where a_n is the nth member of the arithmetic sequence $(a_1, a_2, a_3, \ldots, a_n)$ (discover a relationship) 8%.

C. The student states the definition of "arithmetic sequence" and the formula for finding the nth member of an arithmetic sequence (simple knowledge) 5%.

D. The student interprets and uses fundamental shorthand notations conventionally used with sequences and series:

$$\text{“}(a_1, a_2, a_3, \ldots, a_n)\text{,” “}(a_n)_{i=1}^{n}\text{,” and “}\sum_{i=a}^{n} f(i)\text{”}$$

(comprehension and communication) 6%.

E. The student explains why the following relationship holds:

$$\sum_{i=1}^{n} a_i = \frac{n}{2}(a_1 + a_n)$$

where $(a_1, a_2, a_3, \ldots, a_n)$ is an arithmetic sequence (discover a relationship) 9%.

F. The student executes algorithms based on the relationships listed in Objectives B and E to find an unknown member and the sum of first n members of an arithmetic sequence (algorithmic skill) 5%.

G. Given a real-life problem, the student determines how, if at all, a solution to that problem is facilitated by using relationships and algorithms involving arithmetic sequences (application) 10%.

H. The student distinguishes between examples and nonexamples of geometric sequences, explains the defining attributes, and formulates a definition (construct a concept) 10%.

I. The student explains why the following relationship holds:

$$a_n = a_1\left(\frac{a_2}{a_1}\right)^{n-1}$$

where a_n is the nth member of the geometric sequence $(a_1, a_2, a_3, \ldots, a_n)$ (discover a relationship) 8%.

J. The student states the definition of "geometric sequence" and the formula for finding the nth member of a geometric sequence (simple knowledge) 5%.

K. The student explains why the following two relationships hold:

$$\sum_{i=1}^{n} a_i = \frac{a_1\left(1-\left(\frac{a_2}{a_1}\right)^{n}\right)}{1-\frac{a_2}{a_1}}$$

where $(a_1, a_2, a_3, \ldots)$ is a geometric sequence

and

$$\sum_{i=1}^{\infty} a_i = \frac{a_1}{1-\frac{a_2}{a_1}}$$

where $(a_1, a_2, a_3, \ldots)$ is a geometric sequence with $\left|\frac{a_2}{a_1}\right| < 1$ (discover a relationship) 9%.

L. The student executes algorithms based on the three relationships listed in Objectives I and K to find unknown members and sums of geometric sequences (algorithmic skill) 5%.

M. Given a real-life problem, the student determines how, if at all, a solution to that problem is facilitated by using relationships and algorithms involving geometric sequences (application) 10%.

Exhibit 5.9
The Initial List of Examples and Nonexamples Ms. Citerelli Developed.

Examples:
- 3, 3.1, 3.2, 3.3, 3.4, 3.5, 3.6
- −13, −2, 9, ..., 130, 141, 152
- 15, 10, 5, 0, −5, −10, ...
- 22.6, 22.6, 22.6, 22.6, 22.6, 22.6

Nonexamples:
- 7, 0.7, 0.07, 0.007, 0.0007, 0.00007
- ..., 46, 6.78, 2.60, 1.61, ..., 1.02, 1.00, 1.00, 1.00
- 1, 2, 4, 8, 16, ..., 1048576
- 1, 1, 2, 3, 5, 8, 13, 21, ...

consecutive elements is problematic unless I use a simple list—limiting the examples to sequences like 5, 10, 15, 20, . . . and the nonexamples to ones like 5, 6, 8, 2, 0, 10. But I hate to simplify it and have them develop an overly limited concept of arithmetic sequences.

"I know! Instead of making the list so that it's easier to distinguish examples from nonexamples, I'll give them a head start on the task by having them already grouped. Then it'll be their job to decide and explain how the examples are like one another but different from the nonexamples. That way I can keep the same variety of example noise without the danger of the activity being drawn out and heading out on a tangent. This'll give me more control over the situation."

Ms. Citerelli rearranges the list and adds the work space at the bottom as shown in Exhibit 5.11. She makes a copy for each student, one for her to use in class, and one for her file for use with subsequent classes.

She plans the rest of the learning activities for the final three stages of the lesson. The lesson plan is shown by Exhibit 5.12.

Exhibit 5.10
The Real-Life Examples Ms. Citerelli Added to Exhibit 5.9's List.

Examples:
- Until it is won, the amount of money each hour in a radio giveaway jackpot that begins with \$100 and increases by \$20 each hour
- The amount of money in Betty's jar each Saturday if she starts with \$60 and puts in exactly \$7.50 each Friday night (She never removes money or puts any in at any other time.)
- The number of bricks in the 1st, 2nd, 3rd,. . ., 8th row with the bricks arranged as shown by Exhibit 5.2

Nonexamples:
- The monthly savings account balance of a person who deposits \$750 in the first month and leaves it there, allowing it to collect interest at the rate of 4% compounded monthly
- The ages of all the people in our class, listed in alphabetical order

In Case 5.8, Ms. Citerelli conducts Stage 1 (sorting and categorizing) and Stage 2 (reflecting and explaining) of the lesson she planned in Case 5.7:

CASE 5.8

On the first day of the unit, Ms. Citerelli distributes copies of Exhibit 5.11's tasksheet. She directs the students to take 13 minutes to examine the two lists silently and then write out their first conjectures as to how the examples are like one another but different from the nonexamples.

After each student has written something in the first blank, she engages the class in a question-discussion session. From engaging in such sessions before, the students have learned that whenever Ms. Citerelli calls on someone and says, "Keep it going," the following procedure is in effect until Ms. Citerelli interrupts with, "Excuse me": One student at a time has the floor. Anyone wanting to speak raises her hand to request the floor from the student who is speaking.

The question-discussion session begins:

Ms. Citerelli: Read your first conjecture, Bill.

Bill: The examples are alike because they're all in the same column. The nonexamples are in a different column.

Ms. Citerelli: That surely can't be contradicted. Now, let's hear an idea that might explain what I had in mind when I grouped the examples on the left and the nonexamples on the right. Okay, Mavis, you start; keep it going.

Mavis: The first thing I noticed is that both columns contain sequences. But those on the left have more of a pattern to them. Jeannie.

Jeannie: The ones on the right have patterns also. So just having a pattern can't be it. Okay, Mark.

Mark: Not all of them. Look at the second one.

Jeannie: Sure it does—

Ms. Citerelli: Excuse me. Let's allow Mark and Jeannie to hash this out in a two-way conversation while the rest of us listen.

Jeannie: The numbers in the second one are getting smaller; that's a pattern.

Mark: Not the last two numbers.

Jeannie: They might if we saw more decimal places. I think it has something to do with taking square roots. I played around with roots on my calculator and there's something related to roots of 46 with those numbers.

Ms. Citerelli: Excuse me. Mark, do you agree that at least some of the nonexample sequences have predictable patterns?

Exhibit 5.11

Tasksheet Ms. Citerelli Uses on the First Day of Her Construct-a-Concept Lesson on Arithmetic Sequences.

Examples	Nonexamples
• 3, 3.1, 3.2, 3.3, 3.4, 3.5, 3.6 • −13, −2, 9, . . ., 130, 141, 152 • Until it is won, the amount of money each hour in a radio giveaway jackpot that begins with $100 and increases by $20 each hour • 15, 10, 5, 0, −5, −10,. . . • The amount of money in Betty's jar each Saturday if she starts with $60 and puts in exactly $7.50 each Friday night (She never removes money or puts any in at any other time.) • 22.6, 22.6, 22.6, 22.6, 22.6, 22.6 • The number of bricks in the 1st, 2nd, 3rd,. . ., 8th row with the bricks arranged as follows:	• 7, 0.7, 0.07, 0.007, 0.0007, 0.00007 • . . ., 46, 6.78, 2.60, 1.61,. . ., 1.02, 1.00, 1.00, 1.00 • The monthly savings account balance of a person who deposits $750 in the first month and leaves it there, allowing it to collect interest at the rate of 4% compounded monthly • The ages of all the people in our class, listed in alphabetical order • 1, 2, 4, 8, 16,. . ., 1048576 • 1, 1, 2, 3, 5, 8, 13, 21,. . .

How are the examples alike? How do they differ from the nonexamples?

Write your first conjecture here: ____________________

____

____

____

Write your second conjecture here: ____________________

____

____

____

Write your third conjecture here: ____________________

____

____

____

Write your fourth conjecture here: ____________________

____

____

____

Exhibit 5.12

Ms. Citerelli's Lesson Plan for Objective A of Exhibit 5.8.

Objective: The student distinguishes between examples and nonexamples of arithmetic sequences, explains the defining attributes, and formulates a definition (construct a concept)

The Four-Stage Lesson Plan:

1. Solving and categorizing: With the class organized in a single large group, I'll direct them to take 13 minutes to analyze the examples and nonexamples on the first day's tasksheet (Exhibit 5.11) and individually develop a first conjecture.
2. Reflecting and explaining: After everyone has written something for a first conjecture, I'll conduct a large-group question-discussion session in which they will share and discuss a sample of their conjectures. I'll raise questions that lead them to verbalize their rationales.
3. Generalizing and articulating: Continuing with the large-group question-discussion session, I'll lead them to compare their conjectures with each other, emphasizing similarities and differences in intent as well as in wording. We will work toward developing a commonly worded generalization. To do this, I'll need to develop a test for concept attributes.
4. Verifying and refining: At the end of the question-discussion session, I'll assign the homework tasksheet (Exhibit 5.13). On the second day, I'll direct several students to display their responses to Item I of the homework on the board. I'll then engage in a question-discussion session in which I'll scrutinize—for validity—those definitions, as well as others that surface during the session. The intent is to lead everyone to refine their own definitions until we converge on one that's consistent with the conventional meaning of "arithmetic sequence" and that they both comprehend and own.

Following our routine for making transitions from large-group to cooperative-group sessions, I'll direct them into six cooperative task groups (Reggie's, Hu-Li's Kathleen's, Choime's, Tabby's, and James') with the individual roles that were established on Monday. Within each group, they will review one another's responses to Items II, III, and IV of the homework tasksheet.

The transition will then be made into the lesson for Objective B.

Exhibit 5.13

Ms. Citerelli's Homework Assignment After the First Day of Her Construct-a-Concept Lesson on Arithmetic Sequences.

I. Using "*a*" to represent the first member, "*d*" for the common difference between consecutive members, and "*n*" for the number of members, define what we called in class today an "addition sequence," and which your textbook calls an "arithmetic sequence."

II. Carefully read section 13.2 on pages 474–475 of your textbook. Be sure to familiarize yourself with the meanings of the following members: "arithmetic sequence," "progression," "common difference," and "arithmetic mean of two numbers." Add those members and their definitions to the glossary in your notebook.

III. Determine whether each of the following sequences is or is not arithmetic. If it is, then solve for *a, d,* and *n,* according to your definition from Item I of this assignment. If the sequence is not arithmetic, then prove that by illustrating that the difference between one pair of consecutive members does not equal the difference between some other pair of consecutive members:

A. $\sqrt{3}, -\sqrt{3}, -3\sqrt{3}, -5\sqrt{3}, -7\sqrt{3}$
B. $-1, 0, -1, 0, -1, 0, -1, 0, -1, 0, -1, 0$
C. $7.1 + \pi, 7.1 + 2\pi, 7.1 + 3\pi, \ldots, 7.1 + 1000\pi$
D. All whole numbers arranged consecutively in ascending order
E. All integers arranged consecutively in ascending order
F. The sequence of digits in the numeral π (i.e., 3, 1, 4, 1, 5, 9, 2, 6, 5, ...)
G. The amount of money you spent each month beginning with January of last year
H. $f(1), f(2), f(3), f(4), \ldots$ where $f(x) = x^2 - 5$
I. $g(1), g(2), g(3), g(4), \ldots$ where $g(x) = x - 5$
J. The accumulated total, by inning, of outs the visiting team makes in one regular nine-inning baseball game
K. The accumulated total, by inning, of outs both teams make in a regular nine-inning baseball game

IV. Work the following exercises from page 476 of your textbook: 1, 7, 8, 9, 19, & 22.

Mark: Sure.

Ms. Citerelli: Those that agree, raise your hands. [pause] It looks like that's one thing we agree on. Bill, would you please come up to the board and help keep track of the points on which we agree?

As the discussion continues, Bill takes notes on the board.

In Case 5.9, Ms. Citerelli leads the class through Stage 3 of the lesson, generalizing and articulating:

CASE 5.9

Ms. Citerelli: Mavis, do you want to modify your conjecture?

Mavis: No, but I'll withdraw it because we agree that some nonexamples have patterns also.

Ms. Citerelli: Thank you. Read yours, Inez, and then keep it going.

Inez: You get the next number by adding something to the one before. Okay, Chico.

Chico: That's not right because look at the last nonexample. You add something to get to the next one. 1+1=2, 1+2=3, 2+3=5, 3+5=8, and so on. So, if Inez is right, then that one should be on the left side. Okay, Luis.

Luis: Besides, if Inez is right that you just added something, then the numbers should all be going up, but two of 'em go down and one stays the same.

Ms. Citerelli: Excuse me. Let's hear from Inez because it's her conjecture we're discussing.

Inez: Fist of all, Luis, if you add negatives, they go down. Also, I'm changing my conjecture so that it's this: You get the next number by adding the *same* amount to each one.

Ms. Citerelli: I see Bill has written down Inez's revised conjecture. Bill, would you please write the first example under that.

Bill writes, "3, 3.1, 3.2, 3.3, 3.4, 3.5, 3.6."

Ms. Citerelli: Thanks. Now write the first nonexample over here.

Bill writes, "7, 0.7, 0.07, 0.007, 0.0007, 0.00007."

Ms. Citerelli: Now, if Inez's conjecture holds, what should be true about the two sequences Bill just wrote? [pause] Okay, Bill.

Bill: You should be able to add the same thing to each of these to get the next one and you shouldn't be able to do it over here.

Ms. Citerelli: I need a volunteer to use Bill's test. Okay, Habebe. Now, if Habebe shows that Inez's conjecture fails Bill's test, what will we know? Mavis.

Mavis: That the conjecture is wrong.

Ms. Citerelli: And what if it passes? James.

James: I don't know. What's the question?

Ms. Citerelli: What is the question, Bill?

Bill: I wrote it down right here.

James: Then the conjecture is right.

Ms. Citerelli: Do you agree, Anita? Keep it going.

Anita: Well the conjecture is right; I agree with it. Chico.

Chico: But that's not what Ms. Citerelli asked. She asked if passing the test proves it's true and the answer to that question is no. Luis.

Luis: It's got to work for all of them and I see one it doesn't work for—

Ms. Citerelli: Excuse me. Let's have Habebe perform Bill's test.

Habebe writes down the following:

$3 + 0.1 = 3.1$

$3.1 + 0.1 = 3.2$

$3.2 + 0.1 = 3.3$

$3.3 + 0.1 = 3.4$

$3.4 + 0.1 = 3.5$

$3.5 + 0.1 = 3.6$

Habebe: But it doesn't work for this other sequence because to get from 7 to 0.7 you have to add some negative number that's different than the one you add to get from 0.7 to 0.07.

Ms. Citerelli: While Bill and Habebe do the same at the board, I'd like the rest of you to find the difference for each of these at your place. Bill, you may use the calculator on my desk. Habebe, I'll get yours for you.

A few minutes after the following list is on the board, everyone agrees that Inez's conjecture passes the first test:

$7 + (-6.3) = 0.7$

$0.7 + (-0.63) = 0.07$

$0.07 + (-0.063) = 0.007$

$0.007 + (-0.0063) = 0.0007$

$0.0007 + (-0.00063) = 0.00007$

Ms. Citerelli then directs the students to quietly perform Inez's test on the remainder of the examples and nonexamples. During the activity, she circulates among them, looking at their work and occasionally whispering a probing question to individuals.

Twelve minutes later, the class returns to a question-discussion session and agrees to Inez's conjecture. Then Ms. Citerelli directs them to individually formulate two more example sequences before they engage in the following conversation:

Ms. Citerelli: Mavis, Luis, and J.J., please write one of your examples on the board.

Mavis writes: "5, 10, 15, 20, . . ."
Luis writes: "0, 0, 0, 0, 0, 0, 0"
J.J. writes: "$\frac{1}{2}$, 0, $-\frac{1}{2}$, -1, $-1\frac{1}{2}$, -2"

Pointing to the three examples on the board, Ms. Citerelli continues:

Ms. Citerelli: Think about what just went through your mind as you determined the third number of your sequence. [pause] What told you what it should be, when you went from here to here? [pause] J.J.

J.J.: Just make up a number; it doesn't make any difference what you add until you get to the second one.

Ms. Citerelli: How did you pick the first number, Alysia?

Alysia: I don't know.

Ms. Citerelli: Read your sequence.

Alysia: 9, 10, 11, 12, 13.

Ms. Citerelli: Why 9?

Alysia: I just like 9; my birthday is on the ninth.

Ms. Citerelli: And why 1?

Alysia: What do you mean?

Ms. Citerelli: Mavis chose a difference of 5, Luis chose 0, and J.J. picked $-\frac{1}{2}$. Why did you create a difference of 1?

Alysia: I don't know; it just seemed simple.

Ms. Citerelli: So, Alysia is telling us that you get to pick whatever you want for the first number and whatever you want for the difference, but then after that everything else is determined for you. Is that right, Mark?

Mark: Yes.

Ms. Citerelli: Is that right, Jeannie?

Jeannie: Yes.

Ms. Citerelli: Is that right, Tabby?

Tabby: No.

Ms. Citerelli: Why not?

Tabby: Because you wouldn't keep asking the question over and over if you agreed with the answers.

Ms. Citerelli: What have you got to say about all this, Inez? Keep it going.

Inez: Once you've got the first number and the difference between them, the pattern is set. Jacelyn.

Jacelyn: But how do you know when to stop the sequence?

Hue-Li: So that's a third thing you have to decide.

Ms. Citerelli: Excuse me. Write down the three variables that determine any example sequence. Quickly, right now, on the paper in front of you! [pause] Okay, what have you got Tabby? Keep it going.

Tabby: The first number, the second number, and how many numbers. Inez.

Inez: Instead of the second number, it should be the difference between the numbers. Mavis.

Mavis: But if you know the first and second, you know the difference. So, it's the same thing.

Inez: Oh, okay.

Ms. Citerelli: Excuse me. Summarize what Tabby, Inez, and Mavis concluded about sequences that fit with our examples, Reggie.

Reggie: All you need is the first number, the difference—which is the same between numbers, and how many numbers.

Ms. Citerelli: And what was Mavis' point, Reggie?

Reggie: Oh yeah! You can figure the difference from the first two numbers.

Ms. Citerelli: Yes, Jeannie.

Jeannie: Shouldn't we give these sequences a name—like we've done with other math ideas we've come up with before?

Ms. Citerelli: Pick someone to make up a name for us, Linda.

Linda: I think Bill should; he did all the writing on the board.

Ms. Citerelli: Bill?

Bill: I'm thinking. [pause] How about "addition sequence," because you add the same thing to go from one to the next.

Ms. Citerelli: I like that. So, for now, let's call them "addition sequences." Yes, Tabby.

Tabby: You said "for now." That usually means there's already a name. What's the real name?

Ms. Citerelli: The real name is whatever we decide. But you'll find there's a conventional name when you do your homework assignment from the text [pause] not now, don't get your books out yet. Let's get back to defining an addition sequence. As Tabby said, we've got three determining variables: the first number, the difference between consecutive numbers, and the number of numbers.

Ms. Citerelli turns on the overhead projector; the students recognize that as a cue to take notes. Ms. Citerelli uses the projector to highlight what is said:

Ms. Citerelli: If we let a be the first number of an addition sequence, what will the second be? [pause] Okay, J.J.

J.J.: a plus whatever the difference is.

Ms. Citerelli: Name the difference for us, Helen.

Helen: 10.

Ms. Citerelli: Thank you. Now, give the difference a more general name so we can choose it to be 10, -7π, or any other number we want. Helen.

Helen: How about "d"?

Ms. Citerelli: So, the second number is what, Choime?

Choime: $a + d$.

Ms. Citerelli: And the third, Habebe?

Habebe: $a + d + d$.

Ms. Citerelli: Simplify Habebe's expression for us, James.

James: Uhh——What?

Ms. Citerelli: Read this line from the overhead—right here, James.

James: Uhh, $a + d + d$.

Ms. Citerelli: Now, simplify that expression.
James: Do you want $a + 2d$?
Ms. Citerelli: Thank you. And the fourth number, Inez?
Inez: $a + 3d$.
Ms. Citerelli: And the fifth number, Reggie?
Reggie: $a + 4d$.
Ms. Citerelli: And the nth number, Jacelyn?
Jacelyn: $a + 5d$.
Ms. Citerelli: And the $(n + 1)^{th}$ number, Linda?
Linda: Wait a minute! Jacelyn said, "$a + 5d$," but you didn't ask her for the sixth number. You asked her for the nth number. So, it's not necessarily $a + 5d$, because n could be anything.
Ms. Citerelli: Oh! I should have been more specific. I failed to define n. Thank you. Now, what did we want n to be?

Fourteen students raise their hands to be recognized, but Ms. Citerelli says, "Just a moment, let me reconstruct my thoughts before I lose them. Lets see [pause] " The following is now displayed on the overhead projector screen:

1st a

2nd $a + d$

3rd $a + 2d$

4th $a + 3d$

5th $a + 4d$

⋮ ⋮

n^{th} $a + ??$

Ms. Citerelli: Oh, I remember now. I was getting tired of listing each of these one at a time, so I thought maybe we could just come up with a general n^{th} member. So what is n now? Tabby.
Tabby: n is the one you want.
Ms. Citerelli: Please say that for me in other words, Hu-Li.
Hue-Li: n is the position of any number in the sequence. So the nth number is $a + nd$.

As Ms. Citerelli writes "$a + nd$" in place of "$a + ??$," 13 hands are raised as students are eager to correct the error.

Ms. Citerelli: Okay, what's wrong now? Jacelyn.
Jacelyn: It's not nd; it should be $(n - 1) \times d$.

Ms. Citerelli replaces "$a + nd$" on the transparency with "$a + (n - 1)d$."

Ms. Citerelli: Is this what you mean?
Jacelyn: Right.
Ms. Citerelli: But why?
Jacelyn: Look at the others. What you multiply d by is always 1 less than the position of the number.
Ms. Citerelli: Raise your hand if you agree with Jacelyn? Everybody! Okay, we're about out of time. Here's the assignment that's due at the start of class tomorrow. Please use the rest of today's period to get started.

She distributes Exhibit 5.13's tasksheet.

In Case 5.10, Ms. Citerelli leads the class through Stage 4 of the lesson, Verifying and Refining:

CASE 5.10

On the second day of the lesson, Ms. Citerelli has Leon, Jeannie, James, Luis, and Candice write their definitions from Item I of the homework tasksheet on the whiteboard:

Leon's: "An arithmetic sequence has the same difference between any two numbers. The first number a, the difference d, the last one n."

Jeannie's: "A sequence is an addition sequence if and only if the first number is a, the next is $a + d$, the next is $a + 2d$, the next is $a + 3d$, and so forth until the nth number, which is $a + (n - 1)d$. (d can be positive, negative, or 0.)"

James': "An arithmetic sequence or arithmetic progression is a sequence in which each term after the first is obtained by adding a constant, d, to the preceding term. In the sequence above the common difference is 5."

Luis': "An addition sequence is one that has the same difference, d, between any members like a, all the way to n."

Candice's: "Addition, or arithmetic, sequences or progressions have an nth term of $a + (n - 1)d$."

Mr. Citerelli: Please spend the next 12 minutes comparing your definition to the five definitions on the board. Decide which ones, if any, are equivalent to yours. Which define an addition or arithmetic sequence as we described it yesterday? How about your own definition? Okay, we've got 11.5 minutes. [pause]
Ms. Citerelli: Okay, time is up. Let's look at Leon's first. Read it aloud for us, Eldon.
Eldon: An arithmetic sequence has the same difference between any two numbers. The first number a, the difference d, the last one n.
Mr. Citerelli: Share your analysis of Leon's definition, Mavis. Keep it going.
Mavis: Well, it's not the same as mine, but it's the same idea. I think it's right. Okay, Jeannie.
Jeannie: But he's got n as a member of the sequence. n isn't in the sequence. It's ——
Mark: Sure it is; it's the ——
Jeannie: Hey, you don't have the floor; I didn't call on you! As I was saying, n is the place of any one member—not a member itself. Okay, Mark.
Mark: It's the same thing. The place is part of the sequence. Jeannie.

Jeannie: I'm going to ask you a question, Mark, but only if you give me the floor right back. Okay? Which seat in this row are you in right now? Mark.

Mark: You can see I'm in the fourth. Jeannie.

Jeannie: Okay. Fourth is your place, but it's not you! Mark.

Mark: Okay, I get it. Uhh, Habebe.

Habebe: There's something wrong with Leon's definition. It says *d* is the difference between *any* two numbers. Leon.

Leon: That's the whole idea of an arithmetic sequence; there has to be the same difference between the numbers. Habebe.

Habebe: Look at what I write on the board:

5, 10, 15, 20, 25, 30

Is this an arithmetic sequence? [pause] You agree, but the difference between 25 and 10 is 15 and the difference between 25 and 20 is 5. See, it's not the same between every two numbers. Leon.

Leon: You know what I mean. By "the difference between two numbers," I mean numbers next to each other.

Ms. Citerelli: Excuse me. What adjective might Leon use to clear up this matter? Mavis, keep it going.

Mavis: "Next to"? Linda.

Linda: "Neighboring." J.J.

J.J.: That's what I was going to say. Okay, Habebe.

Habebe: I like "consecutive." Leon.

Leon: So, if I said, arithmetic sequence has the same difference between *consecutive* numbers, would it be all right?

Ms. Citerelli: Excuse me. Everyone, locate their work for the sequence for B of III of the homework tasksheet (Exhibit 5.13). [pause] Take 45 seconds to examine that sequence and then think about Leon's question: Do we have a satisfactory definition if we say, "An arithmetic sequence is a sequence of real numbers such that the difference is the same between any two consecutive members"? [pause] Okay, Inez, keep it going.

Inez: I said B wasn't arithmetic because to get from the first to the second, *d*, must be 1, but from the second to the third, *d* must be -1. It makes a difference which way you're going. I think it's easier to write the definition kind of like Jeannie's. Leon.

Leon: I'm going to change mine. Let's talk about somebody else's definition.

Ms. Citerelli: Excuse me. Read Jeannie's definition, Bill, and keep it going.

Bill: A sequence is an addition sequence if and only if the first number is a, the next is $a + d$, the next is $a + 2d$, the next is $a + 3d$, and so forth until the nth number, which is $a + (n - 1)d$. d can be positive, negative, or 0. Tabby.

Tabby: Do you have to say that last part? Isn't that understood as long as you know *d* is a number? Jeannie.

Jeannie: You're right; just scratch that. Anybody else?

Ms. Citerelli: Nobody wants the floor to argue with Jeannie's definition? Okay, let's look at James'. Choime.

Choime: He didn't use the *a, d,* and *n* like we were supposed to. Mark.

Mark: What's that part about the common difference being 5? James, put your hand up so I can call on you. James.

James: I don't know; that's what the book said.

Ms. Citerelli: Excuse me, if you're going to copy straight out of the book, it's a good idea to credit the authors with a reference. Jeannie.

Jeannie: I checked the book and I see where James got the difference of 5 from. It's about an example that's not in the definition.

Ms. Citerelli: I suggest we accept Jeannie's revised definition for an arithmetic sequence and move on. Yes, Mavis.

Mavis: Why "revised"?

Ms. Citerelli: Because Jeannie deleted the part about *d* being positive, negative, or zero. Does anyone have an objection or suggestion? [pause] Okay, remember, we can always call a definition into question later on. Right now, let's look at Item III on the homework. Yes, Inez.

Inez: I was confused by E.

Ms. Citerelli: Please read it aloud.

Inez: All integers arranged consecutively in ascending order.

Ms. Citerelli: List all the integers in ascending order on the board for us, Candice.

Candice writes: . . ., -2, -1, 0, 1, 2, . . .

Ms. Citerelli: Raise your hand if you claim that sequence is arithmetic. [pause] Thank you. Now, raise your hand if you claim it's not arithmetic. [pause] Interesting! Why do you claim it's arithmetic, Chico?

Chico: Because there's the same difference between any two consecutive numbers.

Ms. Citerelli: Why did you say it wasn't, Sydna?

Sydna: Because there's no *a*. If it's an addition sequence, it's got to have an *a*.

Ms. Citerelli: Inez, you didn't raise your hand either way. Why not?

Inez: Like Sydna said, our definition says you've got to have a starting number. The book says "first term." But I also agree with Chico. You can't disprove it's arithmetic by the way you told us. Any two consecutive numbers have the same difference. So what are we supposed to do?

Ms. Citerelli: Helen.

Helen: Let's just throw that one out; it's a bad question.

Ms. Citerelli: It may be a bad homework task, but I don't want to throw out that sequence. We have a decision to make. We must decide if we want to keep our current definition and restrict arithmetic sequences to those with a first member, or revise our definition to include sequences that are *open* on the left side. Tabby.

Tabby: Which one is right?

Ms. Citerelli: Which one is consistent with our textbook's definition, Jacelyn?

Jacelyn: The book's says there's a first term—but we've been calling it "member" or "number."

Ms. Citerelli: Okay, in contrast to our textbook's definition, here's one from the *CRC Standard Mathematical Tables* (Beyer, 1987, p. 8): "An arithmetic progression is a sequence of numbers such that each number differs from the previous number by a constant amount called the common difference." Now, it goes on to mention the first member, but it's not part of the definition. What do you think, Alysia?

Alysia: Let's stick with our textbook and Jeannie's definition and have a starting number.

Ms. Citerelli: Raise your hand if you vote in favor of Alysia's proposal. [pause] So be it. But remember that in another place and time, and textbook, it may be defined to include sequences without first members.

Ms. Citerelli organizes the class into collaborative groups and continues with the plan outlined in Exhibit 5.12.

To deepen your insights into how lessons for construct-a-concept objectives should be designed, engage in Activity 5.4:

Activity 5.4

Write a construct-a-concept objective that specifies one of the concepts from the following list as its subject content:

geometric sequence	prime number
irrational number	sample space
Diophantine equation	variable
continuous variable	countably infinite set
integral	derivative
area	slope
conic section	prism
function	volume

Develop a plan for a lesson that is designed to lead two or three students to achieve the objective. Discuss your plan with and get feedback from colleagues. Refine your plan as you see fit.

Implement your lesson plan with two or three students. Seek feedback from the students regarding what they learned and how you might make the lesson more engaging for them.

Share your experiences with colleagues who are also engaging in this activity.

Store the lesson plan—annotated with comments about the experience—in the "Cognition, Instructional Strategies, and Planning" section of your working portfolio.

MINIEXPERIMENTS RELEVANT TO STUDENTS' ACHIEVEMENT OF CONSTRUCT-A-CONCEPT OBJECTIVES

In Cases 5.8, 5.9, and 5.10, Ms. Citerelli listened to students' comments, questions, and answers; she observed how students responded to prompts from Exhibit 5.11's and Exhibit 5.13's tasksheets. Such activities provided opportunities to monitor students' progress with Objective A. To make formative judgments for regulating lessons and to make summative evaluations near the end of the unit, Ms. Citerelli *measures* student achievement relative to each objective. By definition:

> A *measurement* is a process by which data or information are gathered via empirical observations and those data or information are recorded or remembered.

The results of your measurements of student achievement influence your formative judgments and summative evaluations. Measurements are composed of *miniexperiments* (Cangelosi, 2000a, p. G5):

> A *Miniexperiment* is a component of a measurement used to monitor student learning that consists of a prompt for students' responses and an observer's rubric for recording or quantifying those responses. Traditional test items are examples of such miniexperiments. Herein, "miniexperiment" is used instead of the more conventional "test item," "measurement item," or "test question" for the following reasons: (a) The author's experiences working with teachers and teacher-educators suggest that the conventional terms connote only traditional item formats (e.g., multiple choice and essay). (b) "Miniexperiment" reminds teachers that in order to gather evidence of student learning they need to create situations in which students behave in observable ways that are either consistent or inconsistent with achievement of a learning objective (i.e., the teachers are performing an experiment). (c) "Miniexperiment" reminds teachers that they are free to design an information-gathering device that may look much different than anything they've ever before seen.

A *prompt of a miniexperiment* is the aspect of the miniexperiment that stimulates students to behave or respond in a manner that is indicative of what the miniexperiment is designed to measure. A miniexperiment's *observer's rubric* is the set of rules, key, or procedures a teacher or other observer follows to record an analysis of a student's response to the miniexperiment's prompt.

Exhibit 5.14 is an example of a miniexperiment designed to be relevant to a construct-a-concept objective that specifies *function* as its mathematical content.

Designing miniexperiments for monitoring your students' achievement of construct-a-concept objectives typically requires more creative thought on

Exhibit 5.14
An Example of a Miniexperiment Relevant to Students' Construction of the Concept *Function.*

Prompt:

Let *R* be the relation *is-the-biological-child-of.* Is *R* a function from {living people} into {people who ever lived}? After answering "yes" or "no," write a paragraph proving that your answer is correct.

Observer's Rubric:

Maximum score is 4 points distributed according to the following criteria:

A. +1 for answering "no."
B. +1 for proving *R* is not a function by explaining that a living person is the biological child of two people who ever lived.
C. +1 for demonstrating a comprehension of at least one attribute of a function (e.g., by scoring +1 for Criterion B or pointing out that every living person is a biological child of a person who once lived).
D. +1 if nothing extraneous or erroneous is included in the response.

your part than miniexperiments for other types of objectives. To stimulate your thoughts about how objectives' learning levels influence miniexperiment designs and how to design miniexperiments for construct-a-concept objectives, engage in Activity 5.5:

Activity 5.5

In collaboration with a colleague, design two miniexperiments that are relevant to students' achievement of Objective C from Exhibit 5.8. Then design two more miniexperiments, but this time, they should be relevant to students' achievement of Objective A from Exhibit 5.8.

Store the four miniexperiments—labeled "works in progress"—in the "Assessment" section of your working portfolio.

In Case 5.11, preservice mathematics teachers, Jermain and Tina, engage in Activity 5.5:

CASE 5.11

Jermain: Why are we starting with Objective C instead of A?

Tina: My guess is because it should be easier to come up with miniexperiments for C than for A because C is simple knowledge.

Jermain: And A is a construct-a-concept objective. We haven't gotten into simple-knowledge objectives yet, but they seem to be just memory-level learning.

Tina: So the miniexperiments for a simple-knowledge objective should require students only to remember mathematical content.

Jermain: Like for Objective C, to remember the definition of "arithmetic sequence."

Tina: Let's look up the definition of "arithmetic sequence."

Jermain: The students' final, or at least their most recent, version of the definition should be stated somewhere in the verifying and refining stage of the lesson—that would be Case 5.10.

Tina: The class decided on a modified version of Jeannie's definition. It's equivalent to something like this: A sequence is arithmetic if and only if there exist real numbers a and d such that the first member is a, the second is $a + d$, the third is $a + 2d$, and so on, with the nth member being $a + (n - 1)d$.

Jermain: So one of our miniexperiments for Objective C could present them with the task of showing that they remember that definition or one that's equivalent to it. Why not just ask the student, "What's the definition of arithmetic sequence?" And then just see if they include all the attributes—nothing more, nothing less in the answer.

Tina: The rubric could have one point for each attribute and maybe subtract a point for anything in the response that's not an attribute—like "the numbers increase from the first to the second to the third and so on by the same amount."

Jermain: That response would merit a point for indicating the numbers changed by the same amount, but lose a point for saying it had to be an increasing sequence.

Tina: I can see that even simple-knowledge miniexperiments might be a little more difficult to design than one might first imagine—that is, if you want valid ones.

Jermain: Okay, for a second miniexperiment on Objective C, we could ask them to list the attributes of any arithmetic sequence.

Tina: Just to toss in some variety, why don't we make this one a multiple choice.

Exhibit 5.15 displays the two miniexperiments Tina and Jermain designed to be relevant to students' achievement of Objective C.

Tina: Okay, now for the really tough one, Objective A. I don't have a clue as to how to start!

Jermain: Objective A says, "The student distinguishes between examples and nonexamples of arithmetic

Exhibit 5.15
Two Miniexperiments Tina and Jermain Designed for Objective C of Exhibit 5.8.

One Miniexperiment

Prompt:

Use the following page area to write the definition of *arithmetic sequence:*

Observer's Rubric:

Maximum score is 6 points distributed according to the following criteria:

A. +1 if the response is written as a definition.
B. +1 for each of the following attributes included in the definition:
  i. Sequence of numbers
  ii. Has a first element
  iii. The difference between any two consecutive members is the same.
C. +1 if nothing extraneous or erroneous is included in the definition.

A Second Miniexperiment

Prompt:

(Multiple Choice) One and only one of the following is true for all *arithmetic sequences.* Which one is it? Show your answer by circling the letter in front of your choice:

A. The numbers in the sequence increase from 1st to 2nd, from 2nd to 3rd, and so forth.
B. The numbers in the sequence decrease from 1st to 2nd, from 2nd to 3rd, and so forth.
C. The sequence is finite.
D. The sequence is infinite.
E. The difference between any two consecutive numbers in the sequence is the same.

Observer's Rubric:

+1 for circling "E" and only "E."

sequences, explains the defining attributes, and formulates a definition—construct a concept." That's what they did in Citerelli's class. Don't we know that at least the students who participated in the development of the definition achieved the objective?

Tina: Making observations during the lesson would seem to be the most efficient way of measuring a construct-a-concept objective.

Jermain: Like reading what they write on Ms. Citerelli's tasksheets and just listening to them in class.

Tina: Exactly. But what about trying to measure achievement of this objective for summative evaluation purposes at the end of the unit? Is there any way to do that?

Jermain: Let's just take a piece of Objective A. The first part of the objective says "distinguishes between examples and nonexamples of arithmetic sequences." What if one of our miniexperiments presents students with a list of sequences—some arithmetic, some not? Then we direct the students to circle the arithmetic ones.

Tina: That seems like the obvious thing to do until you realize that by the time the students are through this unit they should be able to execute an algorithm to test if a sequence is arithmetic or not. And maybe they could do the test without ever having conceptualized arithmetic sequences.

Jermain: So if we give them sequences like −3, −3.1, −3.2, −3.3, and one like −3, −3.1, −3.12, −3.123, −3.1234, and ask them to label each as being an example or not, you're saying they might respond correctly by just remembering to see if the differences between each pair of consecutive members is the same.

Tina: Exactly.

Jermain: But suppose we ask them to explain why that test works. Wouldn't that be getting at conceptualization?

Tina: Maybe so. Or would that just be showing that they comprehend the definition—not that they've actually constructed the concept in their heads? Sure is complicated!

Jermain: If we ask them to explain why my example is arithmetic and my nonexample isn't, that begins to measure conceptualization.

Tina: Depending on whether they were formulating their own explanation or just recalling one they had heard in class.

Jermain: Speaking of complicated, I think we're more likely to get at conceptualization instead of just comprehension or remembering an explanation from class if we choose examples and nonexamples that are complicated—ones with example noise that are really different from what they are exposed to in class.

Tina: Like what?

Jermain: Like—remember Ms. Citerelli's thinking when she came up with those two real-world examples for the tasksheet in Exhibit 5.11?

Tina: The examples about the radio giveaway and the stack of bricks.

Jermain: Right! If we come up with some sequences described in ways that sort of veil their algebraic characteristics—you know, really noisy—wouldn't that get closer to testing how well they constructed the concept of arithmetic sequence in their heads?

Tina: You may have hit on something! You also gave me another thought. What about getting the students to come up with their own examples and nonexamples—complicated, noisy ones. Oh, I've got it! What if we put them in teams, and one team tries to stump the other team by developing examples and nonexamples where it is difficult to determine whether or not they're arithmetic?

Jermain: That's not exactly the kind of miniexperiment you see on the typical math test, but let's try it.

Tina and Jermain developed three miniexperiments for Objective A; they are displayed in Exhibit 5.16.

Note that Tina and Jermain included both parts to each of their miniexperiments: prompt and observer's rubric. The prompt, of course, is the miniexperiment component that appears on the measurement or test to which the students respond. The observer's rubric (or simply "rubric" for short) is the teacher's scoring key.

To further develop your talent for designing miniexperiments for construct-a-concept objectives, engage in Activity 5.6:

Activity 5.6

Design a miniexperiment to measure the two or three students' achievement of the construct-a-concept objective for which you designed and implemented a lesson when you engaged in Activity 5.4.

Try out your miniexperiment with the students. Get feedback from them on how it prompted them to think and on the clarity of the directions.

Discuss the experience with a colleague who is also engaging in this activity. Refine the miniexperiment in light of what you learn.

Store the resulting miniexperiment—along with your notes regarding the results of the experiment—as either (a) an attachment to the annotated lesson plan you stored in the "Cognition, Instructional Strategies, and Planning" section of your working portfolio or (b) in the portfolio's "Assessment" section.

DISCOVERABLE MATHEMATICAL RELATIONSHIPS

Discoverable Relationships in Mathematics Curricula

A *relationship* is an association between either (a) concepts (e.g., {irrationals} $\subseteq$ {reals}), (b) a concept and a specific (e.g., $x^2 > -4 \; \forall \; x \in$ {reals}), (c) a specific and a concept (e.g., 5,981 is prime), or (d) specifics (e.g., $\sqrt{13} \geq 1.1$). Unlike concepts that are expressed by a word or phrase (e.g., "rational number"), a relationship is expressed as a complete statement (e.g., "$a^2 + b^2 = c^2$") or the name of a complete statement (e.g., "the Pythagorean theorem").

A relationship is *discoverable* if one can use reasoning or experimentation to find out that the relationship exists. Exhibit 5.17 lists a relatively minute sample of discoverable mathematical relationships.

Not all relationships specified as mathematical content by learning objectives are discoverable. There are also relationships that are a matter of *convention,* having been established through tradition or agreement. For example, the use of the numeral "$\sqrt{3}$" to mean "the positive number that when squared equals 3" is a matter of convention. Logic does not dictate that the symbol for "principal square root of three" should look like that. By reading Miller (1989), you will learn something about the history of how the radical symbol evolved into common usage. However, the fact that the relationship "$\sqrt{3}$" means "the positive number that when squared equals 3" is not one you would be expected to discover for yourself by using reasoning.

How to use direct instruction to inform students about conventional relationships is addressed in chapter 6. This chapter focuses on inquiry instruction, where you lead students to construct concepts and discover discoverable relationships.

Discovering Relationships for Oneself

As with constructing concepts, students need to reason inductively to discover relationships. Review Cases 5.4 and 5.5 for examples of students discovering a relationship via inductive reasoning. Note that in both cases, students formed hypotheses or formulated a proposition from their experiments with specifics.

Exhibit 5.16
Three Miniexperiments Tina and Jermain Designed for Objective A of Exhibit 5.8.

One Miniexperiment

Prompt:

Is *S* an arithmetic sequence, given the following information about *S*? (Circle Yes or No):

Bonnie eats exactly two bananas a day for a week beginning on a Sunday. *S* is the following sequence: The bananas she ate that week by the end of the day on Sunday, the bananas she ate that week by the end of the day on Monday, the bananas she ate that week by the end of the day on Tuesday, . . ., the bananas she ate that week by the end of the day on Saturday.

YES NO

If you circled "YES," write a paragraph explaining exactly why *S* is arithmetic. If you circled "NO," write a paragraph explaining exactly why *S* is not arithmetic.

Observer's Rubric:

Maximum score is 2 points distributed as follows:

A. + 1 for circling "NO" and explaining that the members of *S* are bananas rather than real numbers.
B. +1 for including a one-paragraph explanation that does not include anything extraneous or erroneous.

A Second Miniexperiment

Prompt:

Is *T* an arithmetic sequence, given the following information about *T*? (Circle Yes or No):

Bonnie eats exactly two bananas a day for a week beginning on a Sunday. *T* is the following sequence: The number of grams of banana she consumed that week by the end of the day on Sunday, the number of grams of banana she consumed that week by the end of the day on Monday, the number of grams of banana she consumed that week by the end of the day on Tuesday, . . ., the number of grams of banana she consumed that week by the end of the day on Saturday.

YES NO

If you circled "YES," write a paragraph explaining exactly why *T* is arithmetic. If you circled "NO," write a paragraph explaining exactly why *T* is not arithmetic.

Observer's Rubric:

Maximum score is 2 points distributed as follows:

A. + 1 for circling "NO" and explaining that it is unlikely that all 14 bananas had the same weight.
B. + 1 for including a one-paragraph explanation that does not include anything extraneous or erroneous.

A Third Miniexperiment

Prompt:

Students will be paired off into two-person teams to play the "Is It an Arithmetic Sequence?" game. The directions for the game are as follows:

A. One team (call it "Team A") will make up four examples of sequences. Team A should design the sequences so that it is not easy to determine if they are or are not arithmetic.
B. A second team (call it "Team B") then examines Team A's sequences and decides which are arithmetic and which are not. Team B must explain its rationale for the decision.
C. A third team (call it "Team C") serves as the referees for the game. Team C manages the game, hears the explanations, and awards points as follows:

+ 1 to Team A for each incorret explanation given by Team B.
+ 1 to Team B for each correct explanation.
− 1 to Team A for each sequence that Team B demonstrates Team A has mislabeled.

Observer's Rubric:

See Team C's role in the rules for the game.

Exhibit 5.17
A Relatively Minute Sample of Discoverable Relationships Typically Included in Mathematics Curricula.

Rate × time = distance.

Area of a rectangle = length × width.

$\forall \Delta ABC$, $m\angle A + m\angle B + m\angle C = 180$.

The ratio of the circumference of any circle to its diameter is π.

Give f:{reals} → {reals}, the graph of $f(x) = x^2$ is a parabola.

The graph of a linear function on {reals} is a line.

$\binom{n}{q} = \dfrac{n!}{q!(n-q)!}$.

$\sum_{i=1}^{n} (2i-1) = n^2$.

If ΔABC and ΔDEF are such that $AB = DE$, $m\angle B = m\angle E$, and $BC = EF$, then $\Delta ABC \cong \Delta DEF$.

The law of sines

The binomial expansion theorem

Cramer's rule

DISCOVER-A-RELATIONSHIP OBJECTIVES

The Discover-a-Relationship Learning Level

Review the definition of the discover-a-relationship learning level and the examples of discover-a-relationship objectives in Exhibit 4.24.

The mathematical content of a discover-a-relationship objective is necessarily a discoverable relationship. Some discover-a-relationship objectives target only students' discovering that the relationship does in fact exist. For example:

> From experimenting with various circles, the student concludes that the ratio of the circumference of any circle to its diameter is about 3.1 (discover a relationship).

Others target students' discovering why relationships exist. For example:

> The student develops her own explanations as to why the ratio of the circumference of any circle to its diameter is about 3.1 (discover a relationship).

The Importance of Discover-a-Relationship Learning Level

Consider the following four objectives:

A. The student explains why the Pythagorean relationship exists (discover a relationship).
B. The student states the Pythagorean theorem (simple knowledge).
C. Given the length of two sides of a right triangle, the student solves for the length of the third side using the relationship $a^2 + b^2 = c^2$ where c is the length of the hypotenuse and a and b are the lengths of the other two sides (algorithmic skill).
D. Confronted with a real-life problem, the student determines how, if at all, the Pythagorean relationship can be used to solve the problem (application).

By first discovering the relationship themselves (achieving Objective A), students are more likely to remember it (Objective B) and make sense out of the algorithm specified by Objective C. Furthermore, they can hardly achieve Objective D, which is to deduce when and how the relationship is applicable to solve real-life problems they have never before confronted, without having first achieved Objective A (Davis, Maher, & Noddings, 1990; Hiebert & Carpenter, 1992).

LESSONS FOR DISCOVER-A-RELATIONSHIP OBJECTIVES

Designing Discover-a-Relationship Lessons

Four Stages

Inquiry learning activities stimulating students to reason inductively to discover a relationship are embedded in a lesson with four stages: (1st) *experimenting,* (2nd) *reflecting and explaining,* (3rd) *hypothesizing and articulating,* and (4th) *verifying and refining.*

Stage 1: Experimenting

In Stage 1, students experiment with various specific situations (e.g., measure the three angles of a variety of triangles or compare a figure that they do not know how to compute its area [e.g., a right cylinder] to one that they do [e.g., a rectangle]). You orchestrate the activity, manage the environment, and provide guidance.

Stage 2: Reflecting and Explaining

In Stage 2, students analyze outcomes of their experiments. Prompted by your questions, they discuss and explain their analyses.

Stage 3: Hypothesizing and Articulating

During Stage 3, students articulate propositions about possible relationships. With your guidance, they test, analyze further, and massage their stated hypotheses, conjectures, or propositions.

Stage 4: Verifying and Refining

In Stage 4, the students attempt to verify or disprove their statements about the relationship. The level of verification may range from "seems intuitively clear," to a failure to produce a counterexample, to a formal deductive proof. If "holes" are found in the stated proposition, then the statement is modified until students agree to an acceptable proposition about the relationship.

Incorporating the Four Stages Into Discover-a-Relationship Lessons

In Case 5.12, a teacher plans a discover-a-relationship lesson:

CASE 5.12

Ms. Gaudchaux's algebra I students have already learned to plot points by hand for the graphs of simple algebraic functions. They also have experience using graphing calculators—not only as calculating tools but also for experimenting with relationships. As part of a unit on linear functions, Ms. Gaudchaux develops the lesson plan shown in Exhibit 5.18 for the following objective:

> The student explains how the values of a and b affect the graph of the linear function $f(x) = ax + b$ (discover a relationship).

In Case 5.13, Ms. Smith plans for her students to discover a relationship via a lesson in which they experiment with problem-solving strategies during Stage 1:

CASE 5.13

In the nine years that Ms. Smith has been teaching mathematics, she has noticed that most students who learn about rate relationships in one context (e.g., as simple interest on savings $= prt$) do not recognize rate relationships in other contexts (e.g., as distance $= rt$). Thus, when she planned the prealgebra course she is currently teaching, she decided to attempt a general unit on rate relationships rather than treating interest rates on savings, rates of motion, interest rates on loans, discount rates, and other types of rate problems separately.

Now, she is in the process of designing the lesson for Objective A of that unit:

> The student explains why the accumulative effect of the application of rate per unit over an observed frequency of that unit is given by the product of the rate and the frequency (discover a relationship).

She thinks, "It was difficult for me to put the relationship that's the mathematical content of this objective into words. But I know what I mean. It's the general relationship that is the foundation of formulas for distance, interest rate, pricing rate, heart rate, body fat, and anything else that's so much per unit. The learning level is discover a relationship, so I need inductive learning activities beginning with the class experimenting with sample problems without benefit of knowing the relationship ahead of time.

"I think I'll begin with students in small collaborative teams, each working on a different problem in different real-world contexts. Maybe I can assign the problems according to their own personal interest. After seven weeks with this class, I know most of them pretty well. Okay, I'll begin with a list of different types of problems to which rate relationships are applicable. Let's see [pause] there's distance when traveling, and bank interest for savings and for loans. Then there are markups and discounts at stores. Oh, yeah! Sales commission is another. But except for distance ones, these all have something to do with money. I need more variety, or else they'll always think of rates as either money or distance. I also need to make these germane to their own lives. Interest on savings is more relevant to them today than bank loans. What kinds of problems does the text have? They're scattered throughout the book, but not many of them will tease my students' interests. [pause] I can modify some of these to make them more like real life.

"I should do this systematically and start with a list of areas from which to draw problems. Distance is one for sure and one on interest on savings. How about one for sports fans in the class—commissions for sports agents would be good. Oh, that reminds me! Rachael was discussing greeting cards from one of those mail-order firms. How does that work? They get so many prizes for selling so many dollars' worth of cards. That's like a commission. Maybe that could produce a problem for Rachael and a couple of others to work on."

With further thought, Ms. Smith develops Exhibit 5.20's list.

Looking over Exhibit 5.20's list, she thinks, "How many problems do I actually need? Each team should have at least three people and no more than five. With 31 in

Exhibit 5.18

Ms. Gaudchaux's Lesson Plan for a Discover-a-Relationship Objective.

Objective: The student explains how the values of a and b affect the graph of the linear function $f(x) = ax + b$ where x is a real variable and a and b are real constants (discover a relationship).

Note 1: During Stage 1, I will organize the class into five collaborative teams as follows:

Team	Student	Role*
I	Lawanda	m/o
	Juan L.	sup
	Chen	com
	Pauline	rec/t
	James	rep

Team	Student	Role*
II	Angel	m/o
	Haeja	sup
	Willard	com
	Eiko	rec/t
	Cynthia	rep

Team	Student	Role*
III	Orlando	m/o
	Al	sup
	Brenda	com
	Fred	rect/t
	Bryce	rep

Team	Student	Role*
IV	Ellen	m/o
	Edgardo	sup
	Juan T.	com
	Paige	rec/t
	Kim	rep

Team	Student	Role*
V	Wanda	m/o
	Suzanne	sup
	Salinda	com
	Brookelle	rec/t & rep

*-Roles are defined as follows:

m/o Designates the *manager and organizer* who chairs the meeting for the team and is responsible for maintaining group focus and reminding team members to stay on-task.

sup Designates the *calculator and materials supervisor* who takes custodial care of, distributes, and returns the team's calculators and other materials.

com Designates the *communicator* who communicates and clarifies the directions for the team.

rec/t Designates the *recorder/timer* who keeps track of time, making sure the team keeps to its schedule. She also maintains a record of the data collected by the group.

rep Designates the *reporter* who summarizes the team's findings and conclusions in writing and presents them to the rest of the class.

Note 2: Each team's recorder/timers will use one of the five following tasksheets as indicated during Stage 1 of the lesson:

TEAM I TASKSHEET

i	$f_1(x)$	Picture of graph on calculator screen	Description of how this graph compares to the one just above
1	x		**(Leave this cell blank.)**
2	$x + 1$		
3	$x + 2$		
4	$x + 5$		
5	$x + 10$		
6	$x + 13$		

Continued

Exhibit 5.18
Continued

TEAM II TASKSHEET

i	$g_1(x)$	Picture of graph on calculator screen	Description of how this graph compares to the one just above
1	x		**(Leave this cell blank.)**
2	$2x$		
3	$3x$		
4	$6x$		
5	$10x$		
6	$100x$		

TEAM III TASKSHEET

i	$h_1(x)$	Picture of graph on calculator screen	Description of how this graph compares to the one just above
1	x		**(Leave this cell blank.)**
2	$x - 1$		
3	$x - 2$		
4	$x - 5$		
5	$x - 10$		
6	$x - 13$		

TEAM IV TASKSHEET

i	$t_1(x)$	Picture of graph on calculator screen	Description of how this graph compares to the one just above
1	x		**(Leave this cell blank.)**
2	$-x$		
3	$-2x$		
4	$-6x$		
5	$-10x$		
6	$-.5x$		

Continued

Exhibit 5.18
Continued

TEAM V TASKSHEET

i	$u_1(x)$	Picture of graph on calculator screen	Description of how this graph compares to the one just above
1	x		**(Leave this cell blank.)**
2	$.1x$		
3	$.4x$		
4	$.01x$		
5	$.9x$		
6	$-.5x$		

The Four-Stage Lesson Plan:

1. Experiment: I'll direct the class to organize into the five collaborative teams with the calculator and materials supervisors making sure everyone in their respective teams has a graphing calculator ready to run and each recorder/timer has a copy of her team's tasksheet. In the meantime, I'll meet with the five communicators to explain the following directions to them:
   A. Use the calculators to graph the function listed on the 1st row of your tasksheet. Make sure everyone gets the same graph; quickly sketch it in the indicated cell of the tasksheet.
   B. Do the same for the function listed in the 2nd row. Also, in the indicated cell, describe how the 2nd graph compares to the first.
   C. Repeat what you did for the 2nd function for the 3rd, 4th, 5th, and 6th functions.
   D. After all six graphs are simultaneously displayed on each calculator, discuss how the graphs are affected by what operations are used with x. The discussion should take about 15 minutes, with another 8 minutes allowed to pull the report together and summarize the group's findings.
2. Reflecting and explaining: After making the transition into a single group, I'll direct the reporters to spend 5 minutes explaining each group's findings. As the reporter speaks, I'll use the graphing overhead-projector display to exhibit graphs (see Exhibit 5.19) from the reporting group. Next, I'll engage the class in a question-discussion session in which we'll tie the five reports together. This should lead them to begin to make conjectures about how a and b influence the graph of $y = ax + b$.
3. Hypothesizing and articulating: For homework, I'll direct the students to individually propose a hypothesis describing how the location and angle of the line (what they'll later refer to as "slope") is influenced by a and b. The following day, we'll return to the same five collaborative teams and have them share their conjectures and come to a team consensus on an all-encompassing general conjecture. I'll encourage them to experiment further with various functions and the graphing calculators as they articulate their statements.
4. Verifying and refining: I'll engage them in an independent-work session with Exercises 3, 4, 12, 17, 22, 25, and 33 from pp. 231–232 of the textbook, completing the task for homework. The following day, I'll engage them in a question-discussion session in which they'll reflect on their homework. This should lead them to either accept (for now) their conjecture as true or refine it so that we're convinced the new conjecture is true.

the class, that means eight teams—so eight problems. Well, I'm expected to cover motion and interest problems because they'll be on standardized tests. So, . . ."

After formulating a travel problem and two interest problems, she thinks: "Five more to go. Some of the students might remember distance and bank interest formulas from fifth and sixth grades. It'd be too bad if they just recall the formulas rather than discover them for themselves. But none of them will have pat formulas for any of the other areas. So I'll make sure to assign students who are likely to recall the formulas to the other areas. Oh! Now that I think about it, it wouldn't be so bad if one or two of the groups recalled a formula; then we could compare it to how the other problems are solved. That way we could illustrate that all the solutions are based on the same relationship. [pause] Oh, another idea! When I direct them into the experimenting stage, I'll require each group to develop a visual diagram of what's happening in the

Team I's Graphs of $f_1(x) = x$, $f_2(x) = x + 1$, $f_3(x) = x + 2$, $f_4(x) = x + 5$, $f_5(x) = x + 10$, $f_6(x) = x + 13$

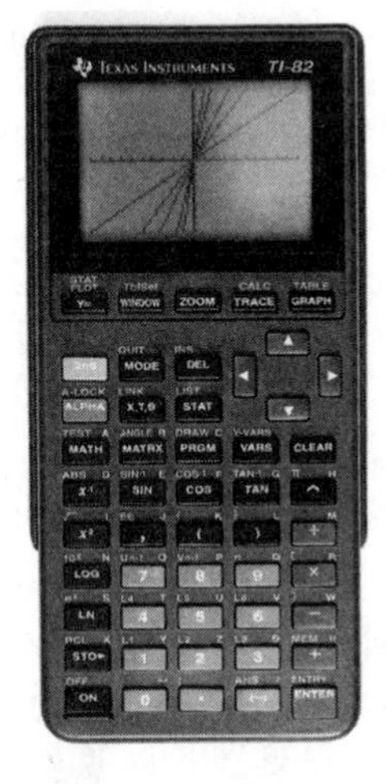

Team II's Graphs of $g_1(x) = x$, $g_2(x) = 2x$, $g_3(x) = 3x$, $g_4(x) = 6x$, $g_5(x) = 10x$, $g_6(x) = 100x$

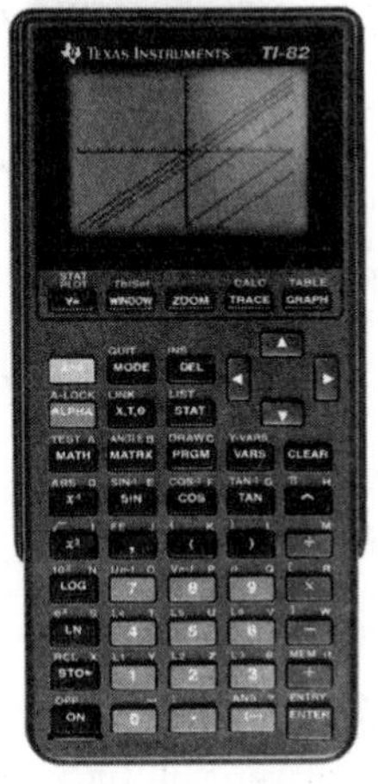

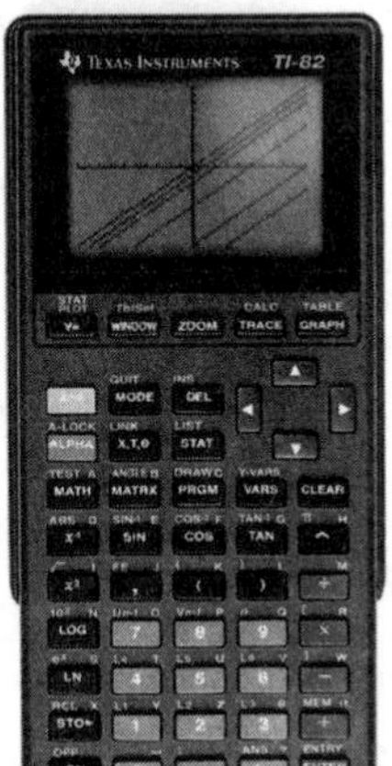

Team III's Graphs of $h_1(x) = x$, $h_2(x) = x - 1$, $h_3(x) = x - 2$, $h_4(x) = x - 5$, $h_5(x) = x - 10$, $h_6(x) = x - 13$

Team IV's Graphs of $t_1(x) = x$, $t_2(x) = -x$, $t_3(x) = -2x$, $t_4(x) = -6x$, $t_5(x) = -10x$, $t_6(x) = -.5x$

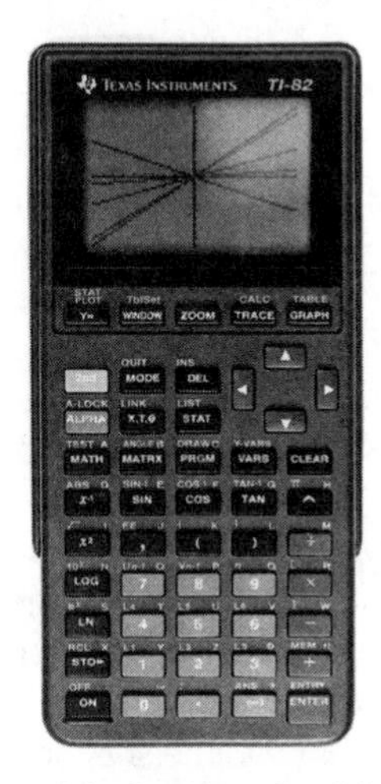

Team V's Graphs of $u_1(x) = x$, $u_2(x) = .1x$, $u_3(x) = .4x$, $u_4(x) = .01x$, $u_5(x) = .9x$, $u_6(x) = - .5x$

Exhibit 5.19
What Ms. Gaudchaux Expects the Calculators in the Cooperative Groups to Display Upon Completion of Stage 1 of the Lesson Outlined by Exhibit 5.18.

Exhibit 5.20
Ms. Smith's List of Types of Problems Involving Rates.

Speed in traveling
Cooking and diet-related rates
Interest on loans
Commissions received by celebrities' agents
Interest on savings
Prizes for selling (e.g., mail-order greeting cards)
Store discounts
Prize-winning rates on radio giveaway games
Bonus-gift rates for purchases (e.g., number of free music CDs based on the number of purchases)
Physical fitness or growth rates (e.g., percent body fat, rate of improvement in weight training, growth rates)
Sports performance rates (e.g., for foul shots, batting average)

Exhibit 5.21
Diagram Illustrating Ms. Smith's Students' Solution to Ron's Walking-to-School Problem.

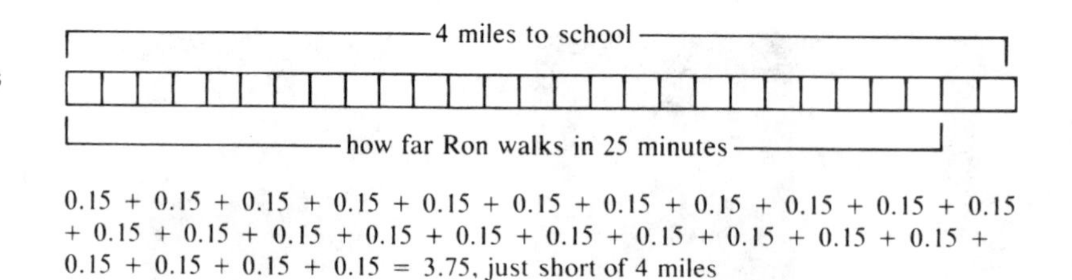

problems. That way, it'll be easier for them to pick up a common pattern when we get into subsequent stages of the lesson. I'd better try that out myself for one of these I've already developed. The distance problems says:

> Ron misses his bus one morning and has 25 minutes to walk the 4 miles to school. Will he make it on time if he covers 0.15 miles each minute?

"How could they possibly diagram that? Let's see. . . ." She illustrates a solution with the diagram in Exhibit 5.21.

She then develops the following problem for the prizes-for-selling-greeting-cards category and draws the diagram appearing in Exhibit 5.22 as she would expect Rachael and others to illustrate their solution:

> For every $25 worth of greeting cards that Mary sells for a mail-order company, she receives 10 points toward prizes from their catalog. For example, it takes 30 points to get a pen and pencil set and 150 points to get an audiocassette player/recorder. How many points would Mary get for selling $200 worth of greeting cards?

Convinced that students will be able to produce comparable diagrams, Ms. Smith develops Exhibit 5.23's list of problems for the initial learning activity.

Pleased with eight problems, Ms Smith thinks: "That takes care of the most difficult part of designing this lesson. Now, to match students to the problems—I'd better see that the less motivated ones are assigned something with which they can identify. Also, there should be at least one who thinks energetically in each group. Okay, Rachael goes to the group with problem 4; I'll call that Team 4. She's bound to be interested in the greeting cards. Bart and Pete should be in separate groups; they get goofy together. Bart to Team 1 and Pete to 2. Let's see, Patrice should . . ."

With the teams of three or four assigned, Ms. Smith prepares to organize and manage the small-group activity efficiently by preparing a direction card that contains the following for each of the 31 students:

- Instructions on how to locate the other members of the task group using a mathematical expression on the card.
- The mathematical expression (e.g., "the sum of the multiplicative inverses of 1, of 2, of 3, and of 4").
- The number of people in the task group that the student is to join (i.e., either 3 or 4).
- Instructions on what the task group is to do (i.e., one of the eight problems is stated, along with what the group is to accomplish relative to the problem).

Exhibit 5.24 displays one of the 31 direction cards.

Just before the students enter the classroom for the period during which she plans to begin the lesson for the discover-a-relationship objective on rates, Ms. Smith tapes each student's card to the underside of her desktop. Initially, the students are not aware the cards are there.

The students are seated at their regular places when Ms. Smith begins the activity with these directions: "For the next 15 minutes or so we will be working in three- or four-person task groups. A card with directions for locating your partners and the task your group is to accomplish is attached to the underside of your desktop. Please find the card now and begin following the directions."

As Ms. Smith walks about the room monitoring behavior, the students quickly locate their partners and begin working out solutions to the problems. At 11:13, Ms. Smith calls for the first of the eight reports to the class.

Explanations of solutions for the first four problems (i.e., those involving Ron's rate of walking, City Bank's interest rate, Sally's interest rate, and the point rate for selling greeting cards) produce diagrams and repeated addition procedures similar to Exhibits 5.21 and 5.22. Team 5 (the one solving Mau-Lin's discount rate problem) simply multiplies 45 by 0.20 for its answer, but Ms. Smith directs them to illustrate the multiplication with a diagram.

Team 6 fails to solve the sports-agent rate problem. They complain that they "couldn't add something 125,000 times." Other students ask, "Why didn't you just multiply?" Team 7 simply multiplies 12 by 7, but also illustrates their logic with a diagram similar to Exhibit 5.22. Similar to the first four groups, Team 8 uses repeated addition and a diagram in its solution.

Exhibit 5.22
Diagram Illustrating Ms. Smith's Students' Solution to Mary's Greeting-Cards Problem.

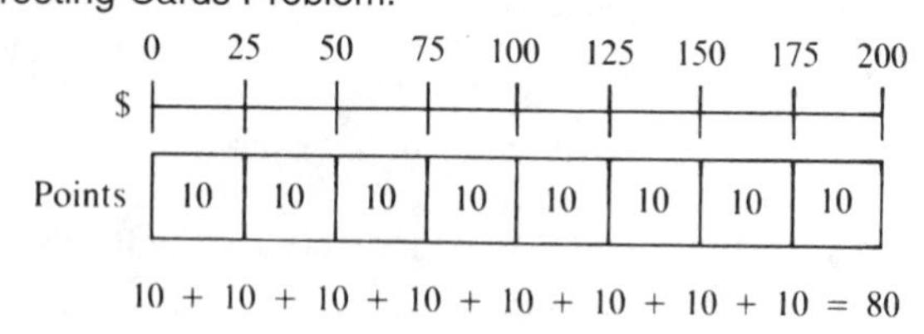

Exhibit 5.23
Ms. Smith's List of Problems for Stage 1 of Her Construct-a-Concept Lesson.

1. Ron misses his bus one morning and has 25 minutes to walk the 4 miles to school. Will he make it on time if he covers 0.15 miles each minute?
2. City Bank is now paying 4¢ for each dollar that remains in a savings account for a whole year. How much interest will Lois make by leaving $60 in a City Bank savings account for a year?
3. Ernie and Sally engage in the following conversation:

   Ernie: "Would you lend me $15?"
   Sally: "Why should I?"
   Erine: "Because I just got my bike fixed and I need to pick it up from the shop."
   Sally: "No, I mean what's in it for me?"
   Ernie: "I'll pay you back in a month when I get paid for my paper route. But I need my bike so I can work."
   Sally: "You still haven't given me a good reason for lending you the money."
   Ernie: "Okay, I'll give you a nickel more on the dollar."
   Sally: "Make that a dime on the dollar for each week until you pay me back, and you've got a deal."

   If Ernie accepts the deal, how much more than $15 will he owe Sally if he waits 4 weeks to pay her back?
4. For every $25 worth of greeting cards that Mary sells for a mail-order company, she receives 10 points toward prizes from their catalog. For example, it takes 30 points to get a pen and pencil set and 150 points to get an audiocassette player/recorder. How many points would Mary get for selling $200 worth of greeting cards?
5. Mau-Lin wants to buy clothes at a store that will cost her $45 today. However, she knows that in a week the store will have a sale marking everything down by 20 percent. How much will she save by waiting a seek to purchase the clothes?
6. Lucy Davis is an agent for professional athletes. For negotiating their contracts, she is paid 5¢ out of every dollar the athletes earn from those contracts. How much will Lucy earn from a $125,000 contract?
7. A mail-order company that sells music CDs advertises that for every 12 CDs purchased, a customer may select a free CD. How many CDs would a customer need to purchase to receive seven free CDs?
8. Every school day, Woody eats a Whiz-O candy bar at lunch. Each Whiz-O contains 312 calories. How many calories does this habit add to Woody's diet during nine weeks of school?

Exhibit 5.24
Direction Card for One of Ms. Smith's Students.

Directions for Patrice Melville

HOW TO FIND THE OTHER MEMBERS OF YOUR WORK GROUP

1. Simplify the following expression:

   $5^2 \div \sqrt{144}$

2. Get up and move about the room until you locate the 3 other people with expressions on their cards that equal yours.

WHAT TO DO ONCE YOU'VE LOCATED YOUR PARTNERS

1. Find a convenient work area without disturbing other groups.
2. As a group, agree to a solution to the following problem:

   Mau-Lin wants to go to a store to buy clothes that will cost her $45 today. However, she knows that in a week the store will have a sale marking everything down by 20%. How much will she save by waiting a week to purchase the clothes?

3. Illustrate your solution with a diagram.
4. At about 11:15, select someone from the group to explain to the whole class your problem and how you solved it.
5. Return to your regular desk when Ms. Smith gives the signal around 11:15.

In the ensuing discussion, similarities among the solutions to the eight problems are pointed out by different students responding to Ms. Smith's probing questions (e.g., "Why did Team 6 want to add five 125,000 times instead of 125,000 five times?"). However, before a general relationship is articulated by the class, Ms. Smith halts the activity and directs everyone to make a copy of each team's illustration and computations. As part of the homework assignment, each student is to develop a formula for solving these types of rate problems.

The following day, the students' formulas are reported and discussed and the class agrees to a multiplication rule for computing the accumulative effects of a rate applied to a frequency. Ms. Smith engages the class in a verification activity in which students apply their formula to other problems and then develop a proof based on multiplication to accomplish the same result as repeated addition.

To deepen your insights into how lessons for discover-a-relationship objectives should be designed, engage in Activity 5.7:

Activity 5.7

Write a discover-a-relationship objective that specifies one of the relationships from the following list as its subject content:

- If A is the lateral surface area of a right cylinder with radius r and height h, then $A = 2\pi rh$.
- The relationship that explains how the graph of $f(x) = ax^2 + c$ is determined by the values of a and c (e.g., if $a < 0$, then the parabola opens downward)
- The Pythagorean relationship
- The complex roots of a quadratic equation $ax^2 + bx + c = 0$ where a, b, and c are real constants can be found via the following relationship:

$$x = \frac{-b \pm \sqrt{b^2 - 4ac}}{2a}$$

- The area of the interior of a rectangle is the product of the rectangle's length and width.
- If C is the circumference of a circle and d is its diameter, then $3.1 < C/d < 3.2$.
- Rate × time = distance.
- $\binom{n}{r} = \frac{n!}{r!(n-r)!}$
- $x^a \cdot x^b = x^{a+b} \forall x \in$ {reals} and $a, b \in$ {whole numbers}.
- $\frac{\sin \alpha}{\cos \alpha} = \tan \alpha$
- $x \in$ {reals} and n is a rational constant $\ni f(x) = x^n \Rightarrow f'(x) = nx^{n-1}$
- $(a_1, a_2, a_3, \ldots)$ is an arithmetic sequence $\Rightarrow \sum_{i=1}^{n} a_i = \frac{n}{2}(a_1 + a_n)$
- $p \in$ {primes} and $p | \prod_{i-1}^{n} a_i \Rightarrow p | a_j$ for some $j \in \{1, 2, \ldots, n\}$
- If $x \in$ {reals} and a and b are real constants, then the graph of $f(x) = ax + b$ is a line.
- If E and F are mutually exclusive events, $P(E)$ is the probability of E and $P(F)$ is the probability of F, then $P(E \text{ or } F) = P(E) + P(F)$.

Develop a plan for a lesson that is designed to lead two or three students to achieve the objective. Discuss your plan with and get feedback from colleagues. Refine your plan as you see fit.

Implement your lesson plan with two or three students. Seek feedback from the students regarding what they learned and how you might make the lesson more engaging for them.

Share your experiences with colleagues who are also engaging in this activity.

Store the lesson plan—annotated with comments about the experience—in the "Cognition, Instructional Strategies, and Planning" section of your working portfolio.

MINIEXPERIMENTS RELEVANT TO STUDENTS' ACHIEVEMENT OF DISCOVER-A-RELATIONSHIP OBJECTIVES

Measuring students' achievement of a discover-a-relationship objective after they have experienced a four-stage lesson for that objective can be a challenging responsibility. Ms. Gaudchaux's students in Case 5.12, for example, will have thoroughly discussed statements of relationships indicated by the objective by the end of the lesson outlined in Exhibit 5.18. Thus, even those who really did not discover the relationships for themselves may have had enough exposure to what their classmates discovered to simply remember statements such as, "If $b < 0$, then the line intersects the y-axis below the origin." Consequently, it is especially important to monitor student achievement during lessons for discover-a-relationship objectives. Observe them experimenting and listen to their conjectures during discussion sessions.

One strategy for gaining some indication of discover-a-relationship level learning after a lesson is to prompt students to describe the experience they had that led them to discover the relationship. Exhibit 5.25 is an example of such a miniexperiment.

To stimulate your thoughts about how to design miniexperiments for discover-a-relationship objectives, engage in Activity 5.8:

Activity 5.8

Design a miniexperiment to measure the two or three students' achievement of the discover-a-relationship

Prompt:

You know that a in the function $f(x) = ax + b$ determines how the line graph of the function is inclined or angled. Write two paragraphs—using diagrams as needed—that explain how experiments you and your classmates performed with calculators helped you to figure that out.

Observer's Rubric:

Maximum score is 8 points depending on how well the response reflects the following 4 criteria:

A. The two paragraphs display comprehension of the task posed by the prompt.
B. The two paragraphs include references to examining specific linear functions, comparing characteristics of the graph to values of a or b.
C. The two paragraphs indicate that inductive reasoning was used to draw a general relationship from specific values of a.
D. Nothing erroneous or extraneous is included in the two paragraphs.

For each of the four criteria, points are awarded as follows:

+ 2 if the criterion in question is clearly met.
+ 1 if it is unclear as to whether or not the criterion in question is met.
+ 0 if the criterion in question is clearly not met.

Exhibit 5.25
One of the Miniexperiments Ms. Gaudchaux Used to Measure Her Students' Achievement of the Objective of Exhibit 5.19's Lesson Plan.

objective for which you designed and implemented a lesson when you engaged in Activity 5.7.

Try out your miniexperiment with the students. Get feedback from them on how it prompted them to think and on the clarity of the directions.

Discuss the experience with a colleague who is also engaging in this activity. Refine the miniexperiment in light of what you learn.

Store the resulting miniexperiment—along with your notes regarding the results of the experiment—as either (a) an attachment to the annotated lesson plan you stored in the "Cognition, Instructional Strategies, and Planning" section of your working portfolio or (b) in the portfolio's "Assessment" section.

In Case 5.14, Jermain and Tina discuss the two miniexperiments one of them designed while engaging in Activity 5.8:

CASE 5.14

Jermain: Whereas I selected the first relationship listed for Activity 5.7, the objective to which my miniexperiments are supposed to be relevant is for the student to explain why the lateral surface area of a right cylinder is $2\pi rh$—discover a relationship.

Tina: So, let's see them.

Jermain: Rather than design miniexperiments for only formative feedback during the lesson, I tried my hand at designing them for summative evaluation near the end of the unit. I decided to give students a problem with the computation already laid out for them, then see if they can explain why the computation will give them the lateral surface area. Here it is.

Jermain shows Tina Exhibit 5.26.

Tina: I like that the prompt directs the student to analyze why the computation works, not to set up or do the computation. It definitely taps a cognitive process that's not the usual focus on the solution. I'm not sure if by itself it tells us if the student actually discovered the relationship.

Jermain: I don't think any one miniexperiment does that by itself. One miniexperiment is just one part of a measurement that, at best, gives us some indicator or evidence of student achievement. We as teachers have to make a judgment based on that evidence. The evidence isn't the answer, it's just something on which to base an answer.

Tina: I agree. We need to keep reminding ourselves that one miniexperiment is only one small piece of a big puzzle. What about your second one?

Jermain: Well I began to worry that this first one may put too much of a premium on students' ability to comprehend the directions as well as their writing skills. So, instead of coming up with an entirely new idea for a second miniexperiment, I concentrated on modifying this first one so that students' abilities to figure out what I wanted and to organize their thoughts in an essay weren't so critical. I wanted to allow them to concentrate more on the mathematics and less on the writing. So, I—

Tina: Show it to me.

Tina examines Exhibit 5.27.

Tina: This version should be easier for most students. You've done some of the cognitive work for them.

Jermain: Which one do you like better?

Exhibit 5.26
The First Miniexperiment Jermain Designed to Be Relevant to How Well Students Discovered the Formula for Lateral Surface Area of a Right Cylinder.

Prompt:

Suppose you wanted to find out how much paint would be needed to cover the outside of the tube pictured (it's open at both ends). Use between one half and a whole page to explain why the size of the surface to be painted can be found by the following computation:

$2 \times 3.14 \times 3.98 \times 15$

3.98 decimeters

15 decimeters

Observer's Rubric:

Maximum score is 8 points depending on how well the response reflects the following 4 criteria:

A. The explanation indicates that the right cylinder can be transformed into a rectangle without changing its lateral surface area.
B. The explanation associates the circumference of the right cylinder with one side of the rectangle.
C. The explanation associates the height of the right cylinder with the other side of the rectangle.
D. The explanation associates the computation with $2\pi rh$.

For each of the four criteria, points are awarded as follows:

+ 2 if the criterion in question is clearly met.
+ 1 if it is unclear as to whether or not the criterion in question is met.
+ 0 if the criterion in question is clearly not met.

Tina: It depends on whether you want them to organize their explanations themselves or you want to do some of the structuring for them. The first reflects a more sophisticated achievement of the objective. But the second one will be easier for you to score.

Jermain: Yeah. If this were only one of many miniexperiments on a test and I had 30 students, I'd be happier with the second when I was up late at night scoring the test.

SYNTHESIS ACTIVITIES FOR CHAPTER 5

1. For each of the following multiple-choice prompts, select the one response that either completes the statement so that it is true or accurately answers the question:
   A. Which one of the following is a concept?
      a) December 10, 1968
      b) 10 hours, 14 minutes, 19 seconds
      c) A point in time
   B. Which one of the following is a specific?
      a) 14 meters
      b) The distance between 2 points
      c) Perimeter
   C. Which one of the following is a relationship from a specific to a concept?
      a) Paul Hoffman wrote *Archimedes' Revenge: The Joys and Perils of Mathematics.*
      b) Paul Hoffman wrote a book.
      c) People write books.
   D. Which one of the following relationships is discoverable?
      a) Among professional mathematicians, a proposition is not accepted to be theorem until a proof of the proposition has been reviewed and judged as valid by a panel of referees.
      b) The relation $\{(\sqrt{4}, 3), (3, -4), (2, \sqrt{9}), (0, -4)\}$ is a function from $\{0, 2, 3\}$ into {reals}.
      c) "NCTM" stands for "the National Council of Teachers of Mathematics."
   E. Which one of the following relationships is conventional rather than discoverable?
      a) A positive integer n is *perfect* iff n is equal to the sum of its proper divisors.
      b) $6, 28, 496, 2^{60}(2^{61} - 1) \in$ {perfect numbers}.
      c) $60 \notin$ {perfect numbers}.
      d) $\{1, 2, 3, 4, 5, 6, 10, 12, 15, 20, 30\} =$ {proper divisors of 60}.
   F. A concept is a _____ .
      a) relationship
      b) constant
      c) variable

Exhibit 5.27

A Miniexperiment Similar to Exhibit 5.26's But With Some of the Cognitive Work Done for the Student.

Prompt:

Suppose you wanted to find out how much paint would be needed to cover the outside of the tube pictured (it's open at both ends). You compute the following:

$2 \times 3.14 \times 3.98 \times 15$

State the formula on which the computation is based:

___

Why does the formula work for the tube in this problem? Explain why in a paragraph. Include a drawing showing how the tube has the same surface size as another—more familiar—figure:

3.98 decimeters

15 decimeters

Paragraph

___

Drawing

Why is 2 in the computation $2 \times 3.14 \times 3.98 \times 15$

___

Why is 3.14 in the computation? ___

Why is 3.98 in the computation? ___

Why is 15 in the computation? ___

Observer's Rubric:

Maximum score is 8 points depending on how well the response reflects the following 4 criteria:

A. The response indicates that the right cylinder can be transformed into a rectangle without changing its lateral surface area.
B. The response associates the circumference of the right cylinder with one side of the rectangle.
C. The response associates the height of the right cylinder with the other side of the rectangle.
D. The response associates the computation with $2\pi rh$.

For each of the four criteria, points are awarded as follows:

+ 2 if the criterion in question is clearly met.
+ 1 if it is unclear as to whether or not the criterion in question is met.
+ 0 if the criterion in question is clearly not met.

G. Students conceptualize when they _____ .
  a) comprehend the definition of a concept
  b) construct a concept in their own minds
  c) explain the conventional meaning of a word naming a concept

H. Direct instructional strategies are appropriate for designing lessons for which one of the following types of objectives:
  a) Construct a concept
  b) Discover a relationship
  c) Algorithmic skill

I. Inquiry instructional strategies are appropriate for designing lessons for which one of the following types of objectives:
  a) Simple knowledge
  b) Discover a relationship
  c) Algorithmic skill

J. Students are stimulated to formulate generalizations from specifics during ________ .
  a) inductive learning activities
  b) deductive learning activities
  c) direct instruction

2. With a colleague, compare and discuss your responses to the multiple-choice prompts from Synthesis Activity 1. Also, check your choices against the following key: A–c, B–a, C–b, D–b, E–a, F–c, G–b, H–c, I–b, J–a.
3. Resurrect the documents resulting from your engagement in Synthesis Activity #6 from the end of chapter 4. In light of your work with chapter 5, revise those objectives if you believe you should. Now select one of your construct-a-concept objectives and one of your discover-a-relationship objectives. For each, devise a lesson plan for a class of students. Discuss your lesson plans with colleagues who are also engaging in this activity. If you are able to, teach your lessons to a class of students while a colleague observes for the purpose of giving you feedback. Insert the lesson plans and other pertinent notes resulting from this activity in the "Cognition, Instructional Strategies, and Planning" section of your working portfolio.

TRANSITIONAL ACTIVITY FROM CHAPTER 5 TO CHAPTER 6

In a discussion with two or more of your colleagues, address the following questions:

1. What mathematical information should students commit to memory?
2. How should lessons for simple-knowledge objectives be designed? What are some strategies for monitoring students' progress during lessons for simple-knowledge objectives?
3. How should lessons for algorithmic-skill objectives be designed? What are some strategies for monitoring students' progress during lessons for algorithmic-skill objectives?

chapter

6

Leading Students to Develop Knowledge and Algorithmic Skills

GOAL AND OBJECTIVES FOR CHAPTER 6

The Goal The goal of chapter 6 is to lead you to develop strategies for designing lessons that lead students to acquire and remember mathematical information and develop algorithmic skills.

The Objectives Chapter 6's goal is defined by the following set of objectives:

A. You will distinguish between examples of (a) mathematical information students need to remember, (b) algorithmic skills students need to develop, and (c) other types of mathematical content specified by learning objectives (construct a concept) 5%.
B. You will describe the reception-overlearning process by which students acquire and retain information (comprehension and communication) 10%.
C. For a given group of middle or secondary school students, you will formulate simple-knowledge objectives that are consistent with *PSSM* (NCTM, 2000b) (application) 5%.
D. You will design lessons for simple-knowledge objectives (application) 20%.
E. You will design miniexperiments that are relevant to your students' achievement of simple-knowledge objectives (application) 10%.
F. You will describe the process by which students develop algorithmic skills (comprehension and communication) 10%.
G. For a given group of middle or secondary school students, you will formulate algorithmic-skill objectives that are consistent with *PSSM* (NCTM, 2000b) (application) 5%.
H. You will design lessons for algorithmic-skill objectives (application) 20%.
I. You will design miniexperiments that are relevant to your students' achievement of algorithmic-skill objectives (application) 10%.
J. You will incorporate the following phrases or words into your working vocabulary: "simple knowledge," "overlearning," "algorithm," "algorithmic skill," and "error pattern" (comprehension and communication) 5%.

MATHEMATICAL INFORMATION TO BE REMEMBERED

In Case 5.7, Ms. Citerelli began a unit targeting the 13 objectives listed in Exhibit 5.8. The "simple knowledge" labels on Objectives C and J indicate students should *remember* certain information (i.e., definitions for "arithmetic sequence" and "geometric sequence" and formulas for finding the nth members of arithmetic and geometric sequences). Whereas constructing concepts and discovering relationships form the basis for learning meaningful mathematics, it is also practical for students to remember conventional names for concepts they have constructed and the statements of relationships they have discovered. Furthermore, they should be informed about and remember certain mathematical conventions and historical names and events (e.g., Nikolai

Ivanovich Lobachevsky and Janos Bolyai invented non-Euclidean geometries in the 19th century (Burton, 1999, pp. 531–545).

Of course, only a minute subset of mathematical information to which you expose your students can be or should be committed to memory. You, like Ms. Citerelli in Case 5.7, need to decide what information students need to know to enable them to do meaningful mathematics. To reflect on the types of information you should consider leading your students to remember, engage in Activity 6.1:

Activity 6.1

Suppose you are planning an introductory unit on trigonometry as part of an algebra II course. The mathematical content of that unit might include some of the following information items:

- Given ΔABC such that $m\angle C = 90°$, and $m\angle A = \theta$, the following six definitions:

$$\sin\theta = \frac{BC}{AB}$$
$$\cos\theta = \frac{AC}{AB}$$
$$\tan\theta = \frac{BC}{AC}$$
$$\cot\theta = \frac{AC}{BC}$$
$$\sec\theta = \frac{AB}{AC}$$
$$\csc\theta = \frac{AB}{BC}$$

- The cotangent, secant, and cosecant are the respective reciprocals of the tangent, cosine, and sine functions.
- Claudius Ptolemy (Egyptian who lived during first century A.D.) contributed to the development of trigonometry in his attempts to solve astronomy and geography problems (e.g., estimating the size of the earth and mapping the heavens).
- The following special angle ratios:

$$\sin 30° = \frac{1}{2} \quad \cos 30° = \frac{\sqrt{3}}{2} \quad \tan 30° = \frac{\sqrt{3}}{3}$$
$$\sin 45° = \frac{\sqrt{2}}{2} \quad \cos 45° = \frac{\sqrt{2}}{2} \quad \tan 45° = 1$$
$$\sin 60° = \frac{\sqrt{3}}{2} \quad \cos 60° = \frac{1}{2} \quad \tan 60° = \sqrt{3}$$
$$\sin\theta = \cos(90° - \theta) \quad \cos\theta = \sin(90° - \theta)$$
$$\tan\theta = \cot(90° - \theta) \quad \cot\theta = \tan(90° - \theta)$$
$$\sec\theta = \csc(90° - \theta) \quad \csc\theta = \sec(90° - \theta)$$

- The following quotient identities:

$$\tan\theta \equiv \frac{\sin\theta}{\cos\theta} \text{ for } \cos\theta \neq 0$$
$$\cot\theta \equiv \frac{\cos\theta}{\sin\theta} \text{ for } \sin\theta \neq 0$$

- The following identities based on the Pythagorean relationship:

$$\sin^2\theta + \cos^2\theta \equiv 1$$
$$1 + \cot^2\theta \equiv \csc^2\theta$$
$$1 + \tan^2\theta \equiv \sec^2\theta$$

For each item of information listed, discuss with a colleague the advantages and disadvantages of having students remember it. Determine which, if any, information items from the list you would likely include among the things for students to remember as part of the unit.

In Case 6.1, Tina and Jermain engage in Activity 6.1:

CASE 6.1

Tina: The first one is the basic definition of the six trigonometric functions—the right triangle version: "opposite over hypotenuse," "adjacent over hypotenuse," and so forth. They are conventional relationships, not discoverable ones; so you would not have a discover-a-relationship objective for them. It's basic vocabulary for doing trigonometry.

Jermain: You have to have a simple-knowledge objective for students to remember those six ratios with their names. You don't want them having to look up the meanings of sine and so forth every time they come across it.

Tina: I agree, but I've got two questions about building this unit. Should these definitions be at the comprehension-and-communication level rather than just at the simple-knowledge level? Because they're so critical to everything we do in trig, students need to be able to interpret the words, not just memorize them.

Jermain: I'm not sure. They seem so simple. Assuming that by the time students are taking trig, they have already constructed the concept of a ratio, what's there to comprehend? Anyway, whether we have a comprehension-and-communication objective or not, students still have to commit the definitions to memory. So for now, let's just say that unless students know these meanings off the tops of their heads, they're not going to be able to communicate about trigonometry. What's your other question?

Tina: When the objectives are listed for the unit, do we want students to memorize all six at once? Or should we bother remembering only "sine," "cosine," and "tangent," because you can do all the trig you want with just those three?

Jermain: Good point! And look at the next item on this Activity 6.1 list—the one about the reciprocals; you really don't need separate definitions for cotangent, secant, and cosecant. I guess you could go either way.

Tina: It'd be nice to cut the number of definitions to be memorized in half before you begin teaching them to do some actual trigonometry.

Jermain: Well, we do agree that at some point in time students need to remember which ratios are associated with which six words. Let's move on to the one about Ptolemy.

Tina: That one is interesting to me because—although I've taken a trig course in high school, had trig units in geometry and algebra courses, and used trig in college courses—I never understood the convenience of using trig for astronomy and geography until last year when I took this math history course. We had to estimate the size of the earth using angle sightings from the moon like Ptolemy did. I gained an appreciation for the value of trig that I never had before!

Jermain: So you're saying that historical events like Ptolemy's contribution to the development of trig should be included. But do historical facts like this need to be committed to memory?

Tina: By remembering just a few such events, students gain a sense of how mathematics developed and it can be useful.

Jermain: But can't you accomplish that through exposure to historical accounts without making it a miserable chore—worrying about what they have to remember for a test? You can hardly enjoy and comprehend historical accounts if you are trying to memorize dates like the time frame between 125 A.D. and 151 A.D. when Ptolemy recorded his astronomical observations and trigonometric inventions in the *Almagest.*

Tina: Wow! I'm impressed the way you just spit out that trivia.

Jermain: You should be. Actually, when I read this Activity 6.1 assignment, the name "Ptolemy" seemed vaguely familiar, but I didn't really remember from where. So I just looked up some stuff in a couple of math history books. Ask me again tomorrow and I won't know "125 to 151 A.D."

Tina: But you will still have a better feel for the history of trigonometry and you will still have an idea of mathematical discovery and invention in the Mediterranean during the first century A.D.

Jermain: I hear us agreeing that students need to comprehend a sample of historical accounts; there's not much advantage to memorizing them.

Tina: Next on the list are these specific values for the so-called special angles—30°, 60°, and 45°. Obviously, those are all discoverable relationships.

Jermain: Clearly, you'd want students to discover each of those for themselves, but then what's the advantage of committing them to memory?

Tina: Because I've used them so much, I remember some of them myself. But not from when my teachers had me memorize them. Back then, I'd memorize them for one test and then rememorize them for the next test. It's convenient to know some of these when you're working on some problems, but now we always have our calculators handy or just look them up when we need the exact ratios. I don't think there's a need to write a simple-knowledge objective for those relationships.

Jermain: I agree, but what about these quotient identities?

Tina: They're so easily discoverable from the definitions of the six basic trig functions. I see no advantage at all in memorizing those. The same is true for the last group of identities on our list. If you understand the six functions and the Pythagorean theorem, you can create these three relationships on the spot as you need them.

Jermain: I guess so, but what about some of the less obvious, but useful, identities like [pause] let's see, let me look one up in the inside cover of my calculus book. Here, one of them is "$(\sin u)(\sin v) = \frac{1}{2}[\cos(u - v) - \cos(u + v)]$." It'd be pretty inconvenient and time consuming to have to rediscover one that complicated every time you needed it.

Tina: I noticed you didn't have it right off the top of your head.

Jermain: But I remembered that it exists.

Tina: Good point. What we should do is have students discover these identities for themselves so they'll make sense out of them and be aware of their existence. Then we just teach them how to look them up when they need them rather than expect them to remember them by heart—like you—they will forget anyway!

Jermain: You are absolutely right; but that's not how I was taught.

THE ACQUISITION AND RETENTION OF INFORMATION

To reflect on the process by which people acquire and remember information, engage in Activity 6.2:

Activity 6.2

Think of an item of information (e.g., your own name) that you remember well but do not specifically recall the event in which you first acquired this knowledge. Speculate on how you first learned this information and explain why you think you will never forget it.

Think of another item of information you also remember well, but this time the information should be something you discovered for yourself. Reflect on the events leading you to discover this information and explain why you think you easily remember the information today.

Think of a third item of information that you remember well, but this time it should be information about which you recall being informed (i.e., you did not discover it for yourself; you read or heard of it). Explain why you think you easily remember the information today.

Now think of a final item of information that you have difficulty remembering (e.g., a formula you once learned for a test in school, but cannot recall right now). Explain why you think you have difficulty remembering this information today.

Share your examples and explanations with a colleague who is also working on this activity. Collaborate with him to formulate a description of a process by which people acquire and retain information. Exchange and discuss your descriptions with other pairs of colleagues who engaged in Activity 6.2.

In Case 6.2, Jermain and Tina engage in Activity 6.2:

CASE 6.2

Jermain: Here, would you read mine (see Exhibit 6.1), while I read yours (see Exhibit 6.2)?

Tina: I have no quarrel with what you've done, especially the part about never forgetting my name.

Jermain: I follow what you've done also, but I have a question about the first one. "Most people enjoy [your] company more when [you] talk about them than when [you] talk about yourself." Is that a form of information or a matter of opinion?

Tina: Thanks for raising that question. It's surely something I believe to be true, but it is a matter of opinion. Does that keep it from being information?

Jermain: I think it's information; it's just difficult to prove. Anyway, it's information that you believe is true. Let's not get hung up on the semantics and get on with the next part of this task.

Tina: So let's come up with a description of a process by which we acquire and retain information.

Jermain: In our examples, the acquisition of the information part varies. For some we found out from an outside source, like hearing or reading about it. Others we figured out for ourselves.

Tina: The ones we figured out for ourselves tend to stay with us longer.

Jermain: But look at the two lists. The information that stuck also tended to be what we continued to use repeatedly. Repetition over time seems to make it indelible.

Tina: Right. Also, your use of a mnemonic device helps hold things in short-term memory.

Jermain: So let's list the steps in the process.

Tina: First, we are either informed of the information or figure it out for ourselves. To get an initial grip on some hard-to-remember things—ones we didn't figure out on our own—we might need to visualize a connection between something already familiar to us and the new, unfamiliar information.

 **Exhibit 6.1**

The Written Part of Jermain's Response to Activity 6.2's Prompt.

Information I'll never forget, but don't remember learning:

My name is "Jermain Jones." Barring a traumatic event or a disease that robs me of my memory, I don't think I'll ever forget my name. I imagine that my mother, aunt, brother, and others who were around me continually in my first few years of life repeatedly referred to me by my name. Gradually, I must have made the association between my name and myself. I easily remember it because nearly every day of my life I have cause to say, write, hear, and read my name repeatedly. The ongoing repetition makes it stick in my memory.

Information I discovered myself and am not likely to forget:

Rainbows typically appear when the sun shines shortly after a rain storm or during a light rain. I discovered that relationship myself from direct and repeated observations. I don't think I'll ever forget it because I discovered it myself and the belief is reinforced every time I see a rainbow.

Information about which I remember being informed and am unlikely to forget:

My friend Tina's name is Tina Huerta. I learned Tina's name when she first introduced herself in class. However, even after I began working with her in cooperative-group activities, I had trouble remembering Huerta. Then I made this silly association: Tina reminds me of "tiny." And something that's tiny is hard to find, so you're always saying "where to" find it; Huerta is pronounced "where-ta." Thus, I used a mnemonic device initially to remember her name.

Information I've forgotten but once knew:

I've heard of the Fundamental Theorem of Algebra. I'm sure I learned what it is for a number of mathematics courses I aced. However, I no longer have any idea as to how to state the Fundamental Theorem of Algebra. It's probably something I use a lot, but I don't know which of the algebraic theorems I've learned over the years is called by that name.

 **Exhibit 6.2**
The Written Part of Tina's Response to Activity 6.2's Prompt.

Something I'll always remember, but don't remember when I learned it:

Most people enjoy my company more when I talk about them rather than when I talk about myself. That is something I discovered to be true sometime between my preadolescent years and as a young adult. But I can't specifically put my finger on when I learned the relationship to be true. I wouldn't forget it because the belief is reinforced again and again in my contact with people.

Note: It just dawned on me that this is all very mathematical in that I've discovered through my experience the following probability relationship: The probability that others will enjoy my company is greater when I focus the conversation on them than it is when I focus it on me.

Something I discovered on my own and will always remember:

What I just wrote fits this category, but because I don't recall the events leading to the realization, I'll select another. It's that the sum of two odd integers is even. I distinctly remember doodling in middle school during a lesson on multiplying signed numbers. "A positive times a positive is positive," Mr. Staple said. That didn't seem surprising, but then I started looking at the sums of odd-integer pairs and I was surprised. I quit paying attention in class and started breaking up odd integers, like $17 = 16 + 1$ and $5 = 4 + 1$. I kept thinking about it and figured out why it's true. It wasn't until I became comfortable with algebraic notation that I expressed odd integers as $2j + 1$ for some integer j, and then proved my conjecture to be a theorem. Although my theorem seems awfully simple and trivial today, it played a significant role in my mathematical development. Whenever I'm struggling with a difficult proof or problem, I think back to how I worked out the explanation for that relationship when I was about 13 years old. It gives me confidence to keep trying no matter how nontrivial the proof or problem seems. My strategies from age 13 still work.

Something about which I was informed and wouldn't forget:

Direct instruction is used to inform students and help them remember; inquiry instruction is used to stimulate them to reason and draw conclusions for themselves. That's something about which I was informed during this mathematics teaching methods course. I wouldn't forget it because we've repeated the idea again and again. Now that I think about it, it's probably an idea I discovered for myself as I recall my own learning experiences. However, it never before stuck with me to use the terms "direct instruction" and "inquiry instruction" in just those ways—although I am pretty sure I was exposed to them before.

Something I no longer remember but once did:

I memorized a massive amount of verbiage dealing with group theory during an abstract algebra course I recently took: "homomorphism," "ideal," "rings," "kernels," "vector spaces," and on and on and on. I even did well on tests regurgitating them. But right now I don't remember what's "ideal" and I still think of a "kernel" as having to do more with corn than with algebra. Although I don't really understand how to apply any of that material from abstract algebra, I do know how to look it up if I ever need it.

Jermain: That's the mnemonic device.

Tina: To retain the information in long-term memory we need to refer to it repeatedly, be exposed to it, or use it over a period of time.

Jermain: That's what we called *"overlearning"* in our educational psychology course.

SIMPLE-KNOWLEDGE OBJECTIVES

Review the definition of "simple-knowledge learning level" and the examples of simple-knowledge objectives in Exhibit 4.24. The three simple-knowledge objectives in Exhibit 4.24 as well as Objectives C and J in Exhibit 5.8 indicate responses for students to remember when presented with certain stimuli. The intent of Objective C is for Ms. Citerelli's students to (a) respond to the stimulus "arithmetic sequence" with the definition they developed when they achieved Exhibit 5.8's Objective A and (b) respond to the stimulus "nth member of an arithmetic sequence" with the equivalent of "$a_n = a_1 + (n - 1)(a_2 - a_1)$."

LESSONS FOR SIMPLE-KNOWLEDGE OBJECTIVES

Facilitating Reception and Retention Through Direct Instruction

Five Stages:

Students achieve construct-a-concept and discover-a-relationship objectives by making decisions based on information they have acquired about examples and nonexamples of a concept or data collected in an experiment. On the other hand, students achieve simple-knowledge objectives by accurately *receiving* and *retaining* information. Reception and retention are accomplished through a five-stage *direct instruction* process: (1st) *exposition,* (2nd) *explication,* (3rd) *mnemonics,* (4th) *monitoring and feedback,* and (5th) *overlearning.*

Stage 1: Exposition

Students are exposed to the information they are to remember. Case 6.3 is an example:

CASE 6.3

Ms. Corbridge's students have already conceptualized Fibonacci sequences. Now she wants them to achieve the following objective:

> The student states the definition of "Fibonacci number" (simple knowledge).

As part of a homework assignment she directs the students to copy the definition of Fibonacci numbers from page 301 of their textbooks and enter it in the glossary file of their computerized notebook. The next day, she displays the definition with the overhead projector, prompts students to read it, and asks them to check the accuracy of the copies in their notebooks.

Stage 2: Explication

The students are given a brief explanation regarding how to respond to a stimulus specified by the simple-knowledge objective (e.g., Ms. Corbridge tells her students, "Anytime you see or hear the words 'Fibonacci number,' you are to think, 'a member of the infinite sequence whose first two members are both 1 and whose subsequent members are the sum of the previous two [i.e., (1, 1, 2, 3, 5, 8, . . .)].' Also, anytime you see that sequence, think 'Fibonacci'.").

Stage 3: Mnemonics

"*Mnemonics*" is a word derived from "Mnemosyne," the name of the ancient Greek goddess of memory. The word means *aiding the memory.* For some—but not all—simple-knowledge objectives, consider providing students with mnemonic devices to enhance retention. Mnemonic devices have proven to be effective in helping students remember new information (Parsons, Hinson, & Sardo-Brown, 2001, pp. 267–269). I, for example, informed you about the derivation of "mnemonics" to help you remember its meaning. Thus, that was a mnemonic aid. However, unless you are already familiar with the goddess Mnemosyne, my mnemonic device is not likely to be effective. The most effective mnemonic devices link new information to be remembered to something already familiar to the student. Cases 6.4 and 6.5 are examples:

CASE 6.4

To help students remember that "whole numbers" refer to {0, 1, 2, . . .}, whereas "natural numbers" refer to {1, 2, 3, . . .}, Mr. Rudd says, "The set of whole numbers is the one with the *hole* in it. The hole is the zero."

CASE 6.5

Had Ms. Corbridge related the history of Fibonacci numbers when her students constructed the concept of Fibonacci sequences, she probably would not need a mnemonic device to help her students associate the sequence with its name. However, because she did not, she decides to use a mnemonic gimmick by telling them, "The name 'Fibonacci' reminds me of 'fibbed on arithmetic' because the Fibonacci sequence is sort of like an arithmetic sequence because you add to find the next member. But not really arithmetic because you do not keep adding the same number."

Usually, it is not necessary to use mnemonics to help students remember definitions of concepts they have already constructed or statements of relationships they have already discovered. Mnemonic devices are more helpful for recalling conventions that are not logically connected to content students have already learned.

Stage 4: Monitoring and Feedback

The accuracy with which students recall what they are supposed to have memorized is monitored. Correct responses are reinforced and errors corrected. Consider Case 6.6:

CASE 6.6

Matt: Ms. Corbridge, is 300 one of those kind of numbers?
Ms. Corbridge: What kind of numbers?
Matt: You know, the kind you were talking about.
Ms. Corbridge: What's the name?
Matt: You know, Fibba-something.
Ms. Corbridge: What's Matt talking about, Riley?
Riley: Fibonacci numbers.
Ms. Corbridge: Oh! Thank you, Riley. Fibonacci numbers! Repeat your question using the words "Fibonacci numbers," Matt.

Stage 5: Overlearning

Students *overlearn* by continuing to practice recalling content even after they have memorized it. Overlearning increases resistance to forgetting and facilitates long-term retention of information (Cangelosi, 2000a, p. 374; Woolfolk, 1993, pp. 317–318). For example, even after the completion of the unit in which she introduced the Fibonacci sequence, Ms. Corbridge continues to confront students with tasks requiring them to use their knowledge of the meaning of "Fibonacci numbers."

A Five-Stage Lesson for a Simple-Knowledge Objective

In Case 6.7, a teacher plans a lesson to include the five stages leading to simple-knowledge-level learning:

CASE 6.7

Ms. Ray designs the lesson for the following objective:

> The student states the definitions of the six trigonometric functions (simple knowledge).

She thinks: "This lesson follows the discover-a-relationship lesson on the association between the size of one of the acute angles of a right triangle and the relative lengths of the triangle's legs. So, the vast majority of the class already understands how the value of θ affects the ratio of one side to another." She draws and muses over the picture in Exhibit 6.3.

She thinks: "All we really need to do here is to get them to associate these six names with the correct ratios. How can I help them keep the names straight? They're familiar with tangents and secants relative to circles, so maybe I could show them these ratios and how y/x is associated with tangent and r/x with secant. [pause] Hmm, then I'd have to get into circular functions and they're not ready for that yet. Besides, it would take more time than I care to spend just so they'll see some logic behind the names. On the other hand, if we took the trouble to do that, I bet they'd never forget the names that go with the ratios. No, I need to keep in mind that this is a simple-knowledge lesson and just give them straightforward, direct instruction. I'll just tell them the names, 'sin θ is y over r' and so forth. Oh, no! that's not the terminology they need to know at this point. We've worked only with right triangles independent of coordinate planes and circles to this point. So I'd better use these." She modifies her notes to read like Exhibit 6.4.

She continues: "That's better, and I think that's the way the book starts off too. [pause] Yep, it does. Okay, the first thing I will do is write each function on the overhead and [pause] No, here's a better idea! I will start by having them list all six possible ratios. That way they will automatically know that the right side of all six equations are permutations of three sides taken two at a time. It'll serve as a mnemonic. Then, after they've listed them, I'll start naming them—getting them to write each one out and saying it aloud. Might as well expose this stuff to as many of their senses as possible. I'd get them to smell these functions if I could. Hmm, actually I could appeal to their sense of touch with some concrete models of each function. Not a bad idea! I'm not going to that trouble this time, but maybe I'll try that with another class.

"Oh! I've also got to make sure they get the abbreviations down. Better include those right from the start. Okay, so I list all the functions and abbreviations for them. Then I better explain just what they're to remember—they're to respond with the name to a given ratio and with the ratio to a given name; I'll test them both ways.

"What are some good mnemonics for these? I ought to check with Frank; he's taught trig for years. I'll bet he's come up with some effective gimmicks. Hmm, how did I ever remember these? My teachers never used mnemonics, but I never forgot. Let's see [pause] I remember! The sine and cosine have the hypotenuse in the denominator, and the sine has opposite for the numerator, so there's nothing left but adjacent for cosine's numerator. Tangent and cotangent are the ones without the hypotenuse and the numerators follow the same order as sine and cosine—opposite for the function and adjacent for the cofunction. I used to remember the cosecant is the reciprocal of sine and the secant is the reciprocal of cosine. I wonder if sharing that with them would work as a mnemonic? Maybe

Exhibit 6.3
Ms. Ray's Drawing as She Begins Designing a Simple-Knowledge Lesson.

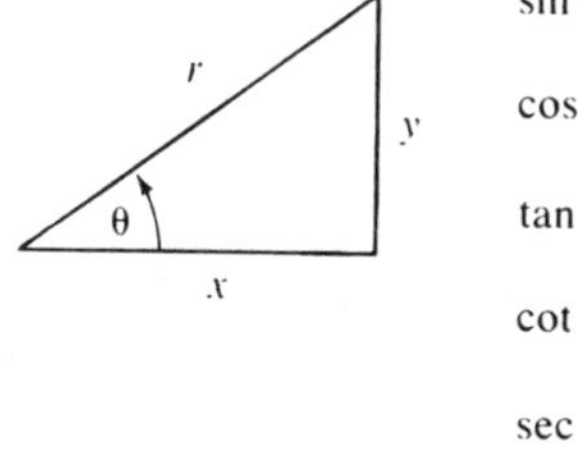

$$\sin\theta = \frac{y}{r}$$

$$\cos\theta = \frac{x}{r}$$

$$\tan\theta = \frac{y}{x}$$

$$\cot\theta = \frac{x}{y}$$

$$\sec\theta = \frac{r}{x}$$

$$\csc\theta = \frac{r}{y}$$

Exhibit 6.4
Ms. Ray Modified Exhibit 6.3.

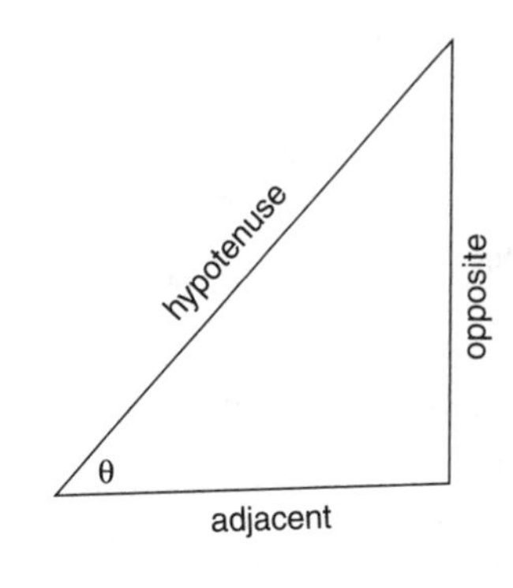

$$\sin\theta = \frac{\text{opposite's length}}{\text{hypotenuse's length}}$$

$$\cos\theta = \frac{\text{adjacent's length}}{\text{hypotenuse's length}}$$

$$\tan\theta = \frac{\text{opposite's length}}{\text{adjacent's length}}$$

$$\cot\theta = \frac{\text{adjacent's length}}{\text{opposite's length}}$$

$$\sec\theta = \frac{\text{hypotenuse's length}}{\text{adjacent's length}}$$

$$\csc\theta = \frac{\text{hypotenuse's length}}{\text{opposite's length}}$$

it would be better to leave them to their own devices. [pause] I don't know. [pause] Think I won't tell this class and see what they come up with. In fourth period I'll tell them and see if they have an easier time remembering—a little action-research project!

"I'll give them some practice exercises for homework and then a quick quiz first thing the next day. We'll go over the quiz right away and correct errors. Overlearning shouldn't be difficult; they'll be using the names with these ratios for the rest of this unit and into the next three units. If they get these correct now, there's no way I'll let them forget them before summer."

To deepen your insights regarding how to design lessons for simple-knowledge objectives, engage in Activity 6.3:

Activity 6.3

Write a simple-knowledge objective that specifies one of the items of information from the following list as its mathematical content:

- (An integer a is a divisor of integer b) $\Leftrightarrow$ ($\exists\, i \in$ {integers} $\ni$ $ai = b$)
- The standard deviation of a sequence of rational numbers, $(x_1, x_2, x_3, \ldots, x_N)$, is σ_x where $\sigma_x = \sqrt{\dfrac{\sum_{i=1}^{N}(x_i - \mu_x)^2}{N}}$ and μ_x is the arithmetic mean of the sequence.
- The distance between two coordinate points, $A(x_1, y_1)$ and $B(x_2, y_2)$, in a Cartesian plane is as follows:
  $AB = \sqrt{(x_1 - x_2)^2 + (y_1 - y_2)^2}$
- {complex numbers} = {reals} $\cup$ {imaginaries}.
- To expand a binomial, use the following formula:
  $(x + y)^n = \sum_{i=0}^{n} \dfrac{n!}{i!(n-i)!} x^{n-i} y^i$
- The volume of a sphere with radius r can be computed as follows: $r = (4/3)\,\pi\, r^3$.
- If $x \in$ {reals} then "$|x|$" means "absolute value of x." And by definition, $|x| = x$ for $x \geq 0$ and $|x| = -x$ for $x \leq 0$.

Develop a plan for a lesson that is designed to lead two or three students to achieve the objective. Discuss your plan with and get feedback from colleagues. Refine your plan as you see fit.

Implement your lesson plan with two or three students. Seek feedback from the students regarding what they learned and how you might make the lesson more engaging for them.

Share your experiences with colleagues who are also engaging in this activity.

Store the lesson plan—annotated with comments about the experience—in the "Cognition, Instructional Strategies, and Planning" section of your working portfolio.

MINIEXPERIMENTS RELEVANT TO STUDENTS' ACHIEVEMENT OF SIMPLE-KNOWLEDGE OBJECTIVES

Stimulus Response

A miniexperiment relevant to students' achievement of a simple-knowledge objective prompts them to respond to the stimulus specified by the objective (e.g., "$\leq$") with the word, image, symbol, name, definition, statement, or other content the objective specifies (e.g., "less than or equal to"). Designing miniexperiments for simple knowledge typically will not tax your creative talents to the same degree as designing ones for construct-a-concept or discover-a-relationship objectives. However, there are pitfalls to avoid. Consider the following objective:

> The student states the Pythagorean Theorem (simple knowledge).

When presented with the stimulus "Pythagorean Theorem," students who achieve this objective remember a response equivalent to: "For any right triangle, the square of the length of the hypotenuse equals the sum of the squares of the lengths of the two legs." Now, examine the four miniexperiments in Exhibit 6.5 and judge how relevant each is to the aforementioned objective.

All four of Exhibit 6.5's miniexperiments appear to have reasonable relevance for the stated objective. However, if your principal purpose for including this objective is to have students remember the Pythagorean Theorem so that they can comprehend messages that use the expression "Pythagorean Theorem," then miniexperiments for the objective should maintain the stimulus-response order. Miniexperiments #1 and #2 do that, but #3 and #4 do not. Unlike #1 and #2, Miniexperiments #3 and #4 require students to remember the name "Pythagorean Theorem" in response to the statement of the relationship. Thus, those miniexperiments require a task that is different from the one students confront when reading or hearing the name "Pythagorean Theorem."

On the other hand, if you also want students to refer to the relationship as "the Pythagorean Theorem" in communications they send, then miniexperiments such as #3 and #4 of Exhibit 6.5 that reverse the stimulus-response order of the objective should also be used. The critical point here is for you to understand the objective well enough so that you design miniexperiments that present students with

Exhibit 6.5
Four Miniexperiments Intended to Be Relevant to How Well Students Can State the Pythagorean Theorem.

Miniexperiment #1

Prompt:

State the Pythagorean Theorem:

Observer's Rubric:

Maximum score is 2 points distributed according to the following criteria:

A. +1 for indicating that the theorem is about right triangles.
B. +1 if the relationship $c^2 = a^2 + b^2$ is clearly implied.

Miniexperiment #2

Prompt:

Multiple choice: Given ΔABC with $m\angle B = 90°$, the Pythagorean Theorem states which one of the following (circle the letter in front of your answer)?

a) $(AB + BC)^2 = AC^2$
b) $AB = BC^2 + AC^2$
c) $AB^2 + BC^2 = AC^2$
d) $(AC + AB)^2 = BC^2$

Observer's Rubric:

+1 for circling "c)" only; otherwise +0.

Miniexperiment #3

Prompt:

What is the name of the following theorem?

For any right triangle, the square of the length of the hypotenuse equals the sum of the squares of the lengths of the two legs.

Answer:

Observer's Rubric:

Maximum score is 2 points distributed according to the following criteria:

A. +1 for "Pythagorean" (spelling need not be exact as long as it is clear what the student meant); otherwise +0.
B. +1 for correct spelling of "Pythagorean;" otherwise, +0.

Miniexperiment #4

Prompt:

Multiple choice: Given ΔABC with $m\angle B = 90°$, then $AB^2 + BC^2 = AC^2$. The name of this statement is which one of the following (circle the letter in front of your answer)?

a) Euclid's Fifth Postulate
b) The Fundamental Theorem of Geometry
c) Archimedes' Principle of Right Triangles
d) The Pythagorean Theorem

Observer's Rubric:

+1 for circling "d)" only; otherwise +0.

tasks remembering appropriate responses to appropriate stimuli.

Avoiding Responses Beyond Simple Knowledge

Students' responses to simple-knowledge prompts should depend only on how well they remember information. Miniexperiments are not relevant to simple-knowledge objectives if students have to use reasoning or higher-order cognitive processes to respond. But, the fact that a miniexperiment is intended to measure simple-knowledge behavior does not guarantee that students will respond at the simple-knowledge level. In Case 6.8 a student responds to what was supposed to be a simple-knowledge prompt with reasoning because of an ambiguity in wording:

CASE 6.8

Early in Mr. Garon's geometry course, Amy engaged in learning activities that led her to discover for herself that the sum of the degree measures of the three angles of any triangle is 180. Six weeks later, as part of a unit on parallel lines in a plane, Amy proves that relationship to be a theorem using Euclid's parallel postulate. Now, Amy is taking the end-of-the-unit test, which includes Exhibit 6.6's miniexperiment that is intended to measure how well students remember—at the simple-knowledge level—statements of the unit's theorems.

When Amy gets to that prompt on the test, she thinks to herself, "We proved that to be true, but we assumed the parallel postulate. So, like Garon said, it's true in Euclidean geometry. But it's not always true because I heard that without the parallel postulate there might not be 180° in a triangle. I think it's less in some geometries. I guess Garon wants us to circle F because the statement isn't always true."

Mr. Garon might argue that Amy should know better because they have been doing Euclidean geometry for nearly the entire course. However, a simple-knowledge prompt does not and should not contain a warning label: "Do not reason on this one." The problem is best avoided by wording prompts so that there is little or no room for misinterpretation.

To stimulate your thoughts about how to design miniexperiments for simple-knowledge objectives, engage in Activity 6.4:

Activity 6.4

Design a miniexperiment to measure the two or three students' achievement of the simple-knowledge objective for which you designed and implemented a lesson when you engaged in Activity 6.3.

Try out your miniexperiment with the students. Get feedback from them on how it prompted them to remember and on the clarity of the directions.

Discuss the experience with a colleague who is also engaging in this activity. Refine the miniexperiment in light of what you learn.

Store the resulting miniexperiment—along with your notes regarding the results of the experiment—as either (a) an attachment to the annotated lesson plan you stored in the "Cognition, Instructional Strategies, and Planning" section of your working portfolio or (b) in the portfolio's "Assessment" section.

ALGORITHMS

An *algorithm* is a multistep procedure for obtaining a result. Algorithms are based on relationships. For example, the algorithm used to find the distance between the two coordinate points $A(8, 7)$ and $B(3, -5)$ in the following computation is based on the Pythagorean Theorem:

$$AB = \sqrt{(8-3)^2 + (7-(-5)^2)} = \sqrt{25 + 144}$$
$$= \sqrt{169} = 13$$

Most people with a relatively unsophisticated understanding of the world of mathematics think of mathematics as nothing more than algorithms to be memorized (Cangelosi, 2001a). Unfortunately, traditional "schoolmath" (Fowler, 1994; Stewart, 1992b) tends to focus almost exclusively on developing algorithmic skills while neglecting construct-a-concept, discover-a-relationship, comprehension-and-communication, application, and creative-thinking learning levels. In an effort to focus attention on aspects of doing mathematics other than executing

Exhibit 6.6
A Miniexperiment Mr. Garon Used in Case 6.8.

Prompt:

Circle "T" if the following statement is true; circle "F" if it is false:

T F The sum of the degree measures of the three angles of any triangle is 180.

Observer's Rubric:

+1 for circling "T" only; otherwise +0.

algorithms, the shapers of curriculum reform movements may have left some teachers with the impression that developing students' algorithmic skills is no longer important. This, of course, was never the intent of those responsible for the NCTM *Standards* (NCTM, 1989a, 1991), for calculus reform (Ferrini-Mundy & Graham, 1991), or *PSSM* (NCTM, 2000b). Even when students use calculators and computers instead of paper and pencil calculations, they are using an algorithm to operate the technology. Furthermore, they need to understand the steps that the machine is helping them execute.

Four types of algorithms are commonly used in mathematics:

- *Arithmetic computations* (e.g., [a] Using paper and pencil, the quotient of 701.333303 and .00567881 is figured via long division. [b] A calculator is used to identify the standard deviation of (37.1, 44.0, 37.0, 44.0, 0.0, 11.8, 37.1, 43.7, 37.1, 28.0, 3.9, 0.0, 41.1, 27.9, 11.6). [c] The Euclidean algorithm is used to find the greatest common divisor of 90075 and 388070).
- *Re-forming symbolic expressions* (e.g., [a] To determine the graphical characteristics of a quadratic function f, a completing-the-square process is used to change the expression of f from "$f(x) = -x^2 + x + 2$" to "$f(x) = -1(x - \frac{1}{2})^2 + \frac{9}{4}$". [b] The polynomial $6x^2 - 34x - 56$ is factored so that it is expressed as "$2(3x + 4)(x - 7)$".)
- *Translating statements of relationships* (e.g., The equation $5x + 15 = 19x - 2x + 43$ is solved so that it reads "$x = -2\frac{2}{3}$".)
- *Measuring* (e.g., [a] A protractor is used to estimate the size of an angle. [b] The frequencies of "Yes" responses and "No" responses to a question on a survey are tallied.)

ALGORITHMIC-SKILL OBJECTIVES

Review the definition of "algorithmic-skill learning level" and the examples of algorithmic-skill objectives in Exhibit 4.24. Algorithmic-skill objectives are concerned with students knowing how to execute the steps in an algorithm. Because you know the answer to the question "What is 9 + 3?" without figuring it out, you have achieved a simple-knowledge objective dealing with arithmetic facts. However, unless you are quite unusual in this regard, you do not know the answer to the question, "What is 168 + 73?" What you do know is how to execute the steps in an algorithm for finding the sum of any two whole numbers such as 168 and 73. This latter skill is indicative of your achievement of an algorithmic-skill objective. Unlike simple-knowledge objectives, the process, not the final outcome, is the target of lessons for algorithmic-skill objectives.

LESSONS FOR ALGORITHMIC-SKILL OBJECTIVES

Facilitating Algorithmic Skills Through Direct Instruction

To gain proficiency with an algorithm, students usually must engage in learning activities that are more tedious and less interesting than learning activities for other types of objectives (e.g., discover a relationship). Sometimes games (e.g., the one integrated into Ms. Saucony's lesson in Case 3.17) can be used to relieve tedium and boredom during drill and practice activities. In any case, algorithmic skill is effected through *direct instruction* in a seven-stage lesson.

Analyzing the Algorithm

Before you are ready to design learning activities that will take students through the seven stages of the lesson, you need to analyze the algorithm, delineating the steps for students to execute. To develop your abilities to delineate the steps students must learn when acquiring an algorithmic skill and to become aware of the number of cognitive steps seemingly simple algorithms involve, engage Activity in 6.5:

Activity 6.5

Select one of the following to analyze:

- An algorithm for factoring a trinomial in the form $ax^2 + bx + c$, where a, b, and c are real-valued constants and x is a real variable
- An algorithm for finding the complex roots of a quadratic equation
- An algorithm for graphing a quadratic function by hand
- An algorithm for adding two fractions
- An algorithm for bisecting an angle with a straightedge and compass
- An algorithm for using trigonometric functions to determine the unknown length of one leg of a right triangle, given the length of the other leg and the measure of one of the acute angles
- An algorithm for measuring the circumference of a circle
- An algorithm for using a graphing calculator to concurrently display the graphs of the following three functions: $f(x) = \sqrt{x}$, $g(x) = x^2$, and $h(x) = x^3$

Exchange your list of steps with that of a colleague. Critique one another's analyses, determining if all the steps students must learn to become proficient with the algorithm are identified.

With your colleague, address and discuss the following questions:

- How is an algorithm like a fractal?
- How is the process by which one executes an algorithm like a fractal?

The first few times you analyze algorithms, you may be quite surprised to discover that algorithms with which you are already proficient involve more steps that you had expected. Schoenfeld (1985, p. 61) pointed out, "It is easy to underestimate the complexity of ostensibly simple procedures, especially after one has long since mastered them." Consider Case 6.9:

CASE 6.9

Mr. Champagne proficiently factors quadratic trinomials with one variable and integer coefficients without much conscious effort. Thus, he does not consider the algorithm to be very complex until he thinks about each step his students will need to remember when they learn it for the first time. He lists the following steps they will need to know in order to factor $x^2 + 10x + 21$:

1. Recognize that this is a quadratic trinomial in standard form.
2. Look for the greatest factor that is common to all three terms.
3. Attempt to factor the trinomial into two first-degree binomials because, in this case, there is nothing to "take out" (i.e., no F so that $Fax^2 + Fbx + Fc$ becomes $F(ax^2 + bx + c)$).
4. Write down: "$(x \quad)(x \quad)$".
5. Note that all coefficients are positive, so insert a "+" sign as follows: "$(x + \quad)(x + \quad)$".
6. Note that the constant term is 21.
7. List each whole number pair whose product is 21.
8. Note that the coefficient for the middle term is 10.
9. Determine which of the pairs listed in Step 7 has a sum of 10.
10. Note that $7 + 3 = 10$.
11. Note that "7" and "3" should be inserted in the blanks in step 5 (in either order).
12. Insert "7" and "3" as follows: "$(x + 7)(x + 3)$".

You may think that enumerating 12 steps for the algorithm Mr. Champagne plans to teach is making the simple complicated. However, please keep in mind that Mr. Champagne's students are just being introduced to this algorithm; consequently, each step represents a potential hurdle that he may have to help them negotiate.

Designing Algorithmic-Skill Lessons

Seven Stages

Once you have the steps in the algorithm clearly delineated, you are ready to design the seven-stage lesson: (1st) explanation of the purpose of the algorithm, (2nd) explanation and practice estimating outcomes, (3rd) general overview of the process, (4th) step-by-step explanation of the algorithm, (5th) trial test execution of the algorithm, (6th) error-pattern analysis and correction, (7th) overlearning.

Stage 1: Explanation of the Purpose of the Algorithm

Algorithms are based on relationships. If your students have experienced a discover-a-relationship lesson relevant to the algorithm, then explaining the purpose of the algorithm is a trivial task. The first stage generally involves nothing more than making an announcement such as, "This algorithm provides us with an efficient way of using our calculators to determine whether a quadratic equation has two, one, or no real roots."

Stage 2: Explanation and Practice Estimating Outcomes

Although algorithmic skills are acquired through direct instruction, students need to get into the habit of estimating or anticipating outcomes before executing the algorithm. This (a) tends to add a little interest to the task as students may be motivated to check their predictions, (b) provides an informal check on the accuracy of the process, and (c) maintains some connection between the algorithm and problem solving.

Traditional algorithmic-skill lessons that are inconsistent with *PSSM* (NCTM, 2000b) do not include this stage. The inclusion of Stage 2 combats the myopic view of mathematical tasks that students typically develop under traditional, follow-the-textbook instruction. Consider Case 6.10.

CASE 6.10

For the purpose of stimulating discussion and obtaining insights about how her students do mathematics on the first day of her algebra I class, Ms. Koa asks Marti to multiply 307 by $\frac{4}{5}$. Exhibit 6.7 displays what Marti wrote on the board.

After Ms. Koa says, "Thank you, Marti," Marti starts to erase her work, but Ms. Koa intervenes, "Please leave it there and work one more next to it. This time, find 80% of 307." Exhibit 6.8 displays Marti's second computation.

Exhibit 6.7
Marti's Initial Computation.

$307 \times \frac{4}{5}$

$\frac{307}{1} \times \frac{4}{5} = \frac{1228}{5}$

$= 245\frac{3}{5}$

```
   245
5)1228
  10
   22
   20
    28
    25
     3
```

Ms. Koa: So you found $\frac{4}{5}$ of 307 to be $245\frac{3}{5}$, but 80% of 307 to be 24.56. Does that seem okay to you?

Marti: I thought I did them right.

Dudley: The second one's not right because you didn't put the decimal in the right place.

Marti: But you're supposed to have as many decimal places in the answer as there are in the problem. See—two here, so I put two here.

Dudley: But that's because you didn't bring the 0 down.

Marti: I thought you were supposed to—

Ms. Koa: Hold on. Doesn't anyone care whether or not the answer makes sense? About what would 50% of 300 be?

Lucy: 150.

Ms. Koa: Then should 80% of 307 be more or less than 150?

The discussion continues with most of the students focusing on the steps of the algorithms while Ms. Koa strives to get them to make sense of the task.

It is not surprising that Ms. Koa's students failed to recognize that 4/5 of 307 is the same as 80% of 307, nor is it surprising that they were conditioned from prior mathematics courses to focus on doing each little step of algorithms without first getting a feel for the results. The primary purpose of including Stage 2 in an algorithmic-skill lesson is to condition students to step back and comprehend the task before diving into the minute steps of the algorithm. As Brenda solved the lightbulb problem in Case 4.7, she estimated outcomes of computations before computing, thus, identifying errors when results did not make sense to her. The student in Case 6.11 had the advantage of working with a mathematics teacher who included Stage 2 in her algorithmic-skill lessons:

CASE 6.11

Jenny thinks as she works the textbook exercises shown by Exhibit 6.9: "Okay, that's just 9×4, which is 36 square meters. No, this is volume, so the answer should come out in cubic, not square, meters. What'd I do wrong? What's the formula again? [pause] Oh yeah, here it is: $\pi r^2 h$. Glad I caught that! So the answer is—punch up π on my calculator times 16 times 9 equals [pause] 452.389 and so. Okay, 452.39 m^3.

"Now for the next one. This is a cone, but the dimensions are the same as the last one. Let's see, this answer should be less than my last one because the top is squeezed together. I bet about 200 m^3. Okay, what's the formula? [pause] $0.333 \times \pi \times 9$ is [pause] 150.645, and so on. So, 150.65 m^3. Why so much smaller? Oh, yeah, this formula is 1/3 of that one, not 1/2."

Stage 3: General Overview of the Process

In this stage, students are provided with an outline of the algorithm they will be executing. This is particularly important for algorithms with so many steps that students may get so involved in details that they lose sight of the overall process. Imagine how confusing it would be to learn each step in the process of shooting a jump shot in basketball if you had never before seen anyone shoot a jump shot. Students need to visualize the overall process before they are ready to follow the detailed steps. The teacher in Case 6.12 conducts Stage 3:

Exhibit 6.9
Two Textbook Exercises Jenny Worked.

Find the volume of the right cylinder:

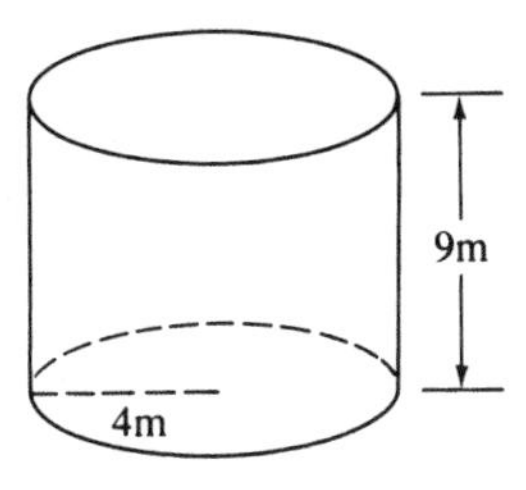

Find the volume of the cone:

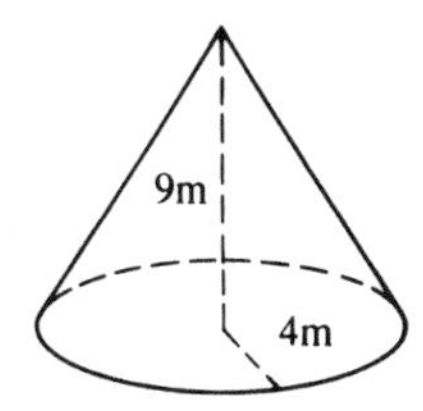

Exhibit 6.8
Marti's Second Computation.

80% of 307

```
  307
x .80
24.56
```

CASE 6.12

Using the overhead projector to illustrate his words, Mr. Anselmo tells his student: "To use this formula to solve quadratic equations, you will first put the equation in standard form so that you can identify *a, b,* and *c* of the formula. Then plug in the values for *a, b,* and *c* and work through the computations with your calculator, making sure to follow the order of operations. The computation will eventually get down to two cases, one for this plus and one for this minus. From there, you'll be working them out as two separate computations. Each result is a potential root."

Stage 4: Step-by-Step Explanation of the Algorithm

This is the paramount stage of the lesson. You begin by explaining the first step of the algorithm and then having the students try it themselves. You then explain how the result of that first step triggers the second. The second step is then explained and tried. Movement to subsequent steps and the steps themselves are each explained and tried in turn.

Consider teaching students three additional steps to be followed after the result or answer from the algorithm is obtained: (a) Compare the result to the previous estimate of the result, as Jenny did in Case 6.11. (b) Check the results. (c) If an error is detected, redo previous steps that might have caused the error.

Stage 5: Trial Test Execution of the Algorithm

Students are assigned exercises selected to demonstrate any error patterns they may have learned. The purpose of this stage is to obtain formative feedback on which aspects of the algorithm students execute correctly and which they do not. This includes how they go about (a) estimating or anticipating outcomes, (b) executing each step in the process itself, and (c) checking for and correcting errors. Exhibit 6.10 illustrates sample exercises—including the teachers' annotations to a student's responses.

Exhibit 6.10
Sample Exercise Completed by a Student With the Teacher's Error-Pattern Analysis.

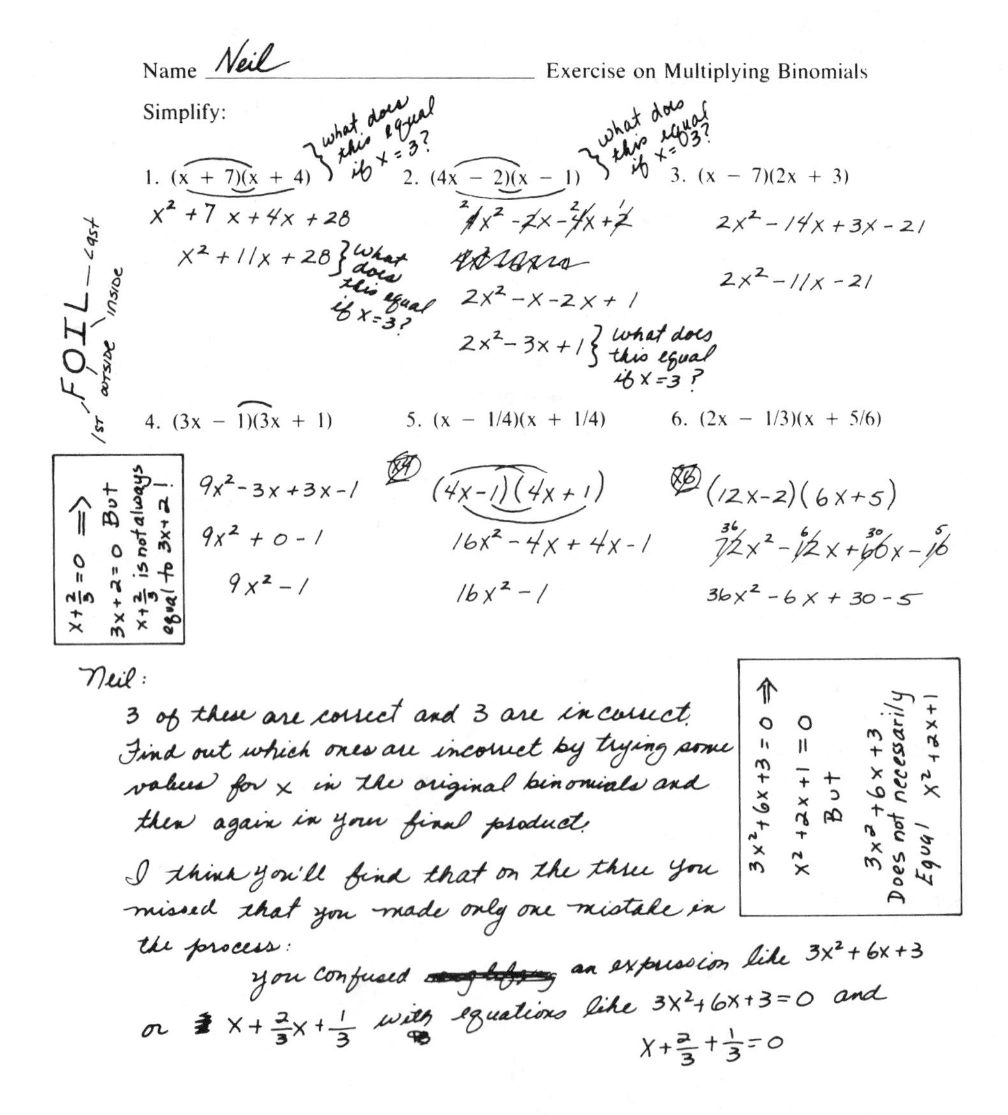

Name Neil Exercise on Multiplying Binomials

Simplify:

1. $(x + 7)(x + 4)$ 2. $(4x - 2)(x - 1)$ 3. $(x - 7)(2x + 3)$

4. $(3x - 1)(3x + 1)$ 5. $(x - 1/4)(x + 1/4)$ 6. $(2x - 1/3)(x + 5/6)$

Stage 6: Error-Pattern Analysis and Correction

Students' responses to the trial exercises are analyzed to diagnose how students are executing the algorithm (e.g., as indicated by the teacher's annotations in Exhibit 6.10). Students are provided with additional explanations as warranted by the error-pattern analysis (Ashlock, 2001).

Stage 7: Overlearning

Students polish their skills and *overlearn* the algorithm through practice exercises. Students' work continues to be monitored as they use the algorithm during subsequent lessons for other objectives.

A Seven-Stage Lesson for an Algorithmic-Skill Objective

A teacher engages students in a lesson for an algorithmic-skill objective in Case 6.13:

CASE 6.13

Ms. Allen is about two thirds through a unit on derivatives of algebraic functions with her calculus class. The previous unit was on limits of functions. In this unit, she already completed lessons designed to lead students to (a) construct the concept of derivatives, (b) discover some relevant relationships involving continuity and tangent lines, (c) know some relevant vocabulary and shorthand notations, and (d) comprehend the proofs of some relevant theorems (e.g., the power rule).

Having already delineated the steps in the algorithm and feeling confident that the students have achieved the prerequisites, Ms. Allen begins a lesson for the following objective:

> Given the continuous function $f \ni f{:}A \to B$ where A, $B \subseteq$ {reals} and $f(x) = a_0x^n + a_1x^{n-1} + a_2x^{n-2} + \ldots + a_{n-2}x^2 + a_{n-1}x + a_n$, the student computes $f'(x)$ using the algorithm based on the following relationship: $f'(x) = na_0x^{n-1} + (n-1)a_1x^{n-2} + (n-2)\,a_2x^{n-3} + \ldots + 2a_{n-2}x + a_{n-1} + 0$ (algorithmic skill).

Tying into the previous lesson, she tells the class: "Recall our definition of derivative of f at x." She displays the following on the overhead:

$$f'(x) = \lim_{h \to 0} \frac{f(x+h) - f(x)}{h}$$

Ms. Allen: From Theorems 4–1 and 4–3, we know two things.
She displays the following:

$$f(x) = a_0x^n + a_1x^{n-1} + a_2x^{n-2} + \ldots + a_{n-2}x^2 + a_{n-1}x + a_n$$

Ms. Allen: If f has a derivative at x, then [pause]
She displays and reads the following:

$$f'(x) = D_x(a_0x^n) + D_x(a_1x^{n-1}) + D_x(a_2x^{n-2}) + \ldots + D_x(a_{n-2}x^2) + D_x(a_{n-1}x) + D_x(a_n)$$

and

$$D_x(bx^m) = mbx^{m-1}$$

Ms. Allen: Those two relationships provide the basis for the algorithm you're about to learn. We'll use this algorithm to find the derivative of f at x. From Theorem 4–1 we get this relationship that allows us to differentiate each term of f one at a time. The relationship from Theorem 4–3 will be used to find each term's derivative. After the algorithm has been executed, what do you think we'll have? Jena.

Jena: f prime of x.

Ms. Allen: I agree. What will f prime of x look like, Michael?

Michael: It will be the derivative of f at x.

Ms. Allen: True, but—I'm sorry I didn't word my question clearly. Let me try again. Whereas $f(x)$ is a polynomial, will $f'(x)$ also be a polynomial, and if so, how will the two differ? Let me give everyone a chance to formulate his own answer. [pause] Okay, Dean.

Dean: f' will be a polynomial of one less degree than f.

Ms. Allen: Okay, let me write that down as you put it in your notes. Anything else we can predict about f', Kayleen?

Kayleen: The constant will go to 0, so you'll have one less term.

Ms. Allen: Okay, let's write that down.

Ms. Allen is now ready to shift into the fourth stage of the lesson, in which she will enumerate, display, explain, and use an example to illustrate each step of the algorithm. See Exhibit 6.11.

She has Randy serve as chalkboard scribe. He is to follow her directions, working through an example at the board as she uses the overhead to list each step. The students know from previous sessions that she expects them to record the steps and work the example in their notebooks.

Ms. Allen: Akeem, when we're done today, would you mind if I duplicate your notes for Randy—at least the ones he'll miss while he's at the board?

Akeem: No problem.

Ms. Allen: Thank you. Okay, get this function down as Randy writes it on the board:

$$g(x) = 2x + 9x^4 - 17.4 - x^2 + \frac{3}{x} + \frac{5}{x^2} - 2x^4 - \frac{1}{x}$$

Our first step is to write the function in standard form, with all variables expressed in numerators with integer coefficients.

Exhibit 6.11
Ms. Allen Lists Steps for the Algorithm on the Overhead Projector as Randy Differentiates g(*x*) on the Board.

She displays the following general form under the associated step on the overhead:

$$f(x) = a_0x^n + a_1x^{n-1} + a_2x^{n-2} + \ldots + a_{n-2}x^2 + a_{n-1}x + a_n$$

Ms. Allen: Let's do that for Randy's example.

Randy writes the following on the board as the students try the task at their places:

$$g(x) = 7x^4 - x^2 + 2x - 17.4 + \frac{2}{x} + \frac{5}{x^2}$$
$$g(x) = 7x^4 - x^2 + 2x - 17.4 + 2x^{-1} + 5x^{-2}$$

Ms. Allen: If your g in standard form doesn't look like Randy's, then change it if you see what you did wrong. If yours doesn't agree and you don't know why, then ask me about it now. [pause] Okay. Next we think about what the derivative of the function should be like. In Randy's example, g is what degree, Ruth?

Ruth: fourth.

Ms. Allen: So g' will have what degree, Ross?

Ross: Third.

Ms. Allen: And how many terms does g' have, Jena?

Jena: I don't know.

Ms. Allen: Count them.

Jena: Oh, okay. six.

Ms. Allen: What's the derivative of a constant, Humula?

Humula: 0.

Ms. Allen: Is one of the six terms of g a constant, Humula?

Humula: Yes, so g' will have five terms.

Ms. Allen: So, now that we have the function in standard form with no variables in denominators, the next step is to apply Theorems 4–1 and 4–3 to find the derivative of each term individually. Let's do it for the first term of g. The rule is to multiply the coefficient by the exponent, reduce the exponent by 1, and then simplify. In general, bx to the mth becomes mbx to the $m - 1$ degree. Work it for the first term for Randy's example. [pause] Does anyone disagree with $28x^3$? Okay, repeat the process for each of the other terms one at a time.

Several minutes later the explanation of the steps is completed and Ms. Allen has the students differentiate a variety of algebraic functions on a tasksheet. She circulates about the room monitoring work and providing help as needed. Rather than reexplaining the algorithm to individual students, she provides very specific directions. For example:

Estes: How do you do this one?

Ms. Allen: What does this step here in your notebook say?

Estes: Find the derivative of each term one at a time.

Ms. Allen: Then do it.

Estes: But this one's funny. What do I do with these negatives?

Ms. Allen: Follow the rule from your notebook.

Estes: You mean you can do that even for the negative of a negative?

Ms. Allen: Yes. Try it and I will check with you after you've had a chance to complete this one and the next three.

From such exchanges and monitoring their work, Ms. Allen is convinced that students generally know the algorithm, but about a third of them tend to get careless in manipulating expressions with negative exponents and negative coefficients, especially when the coefficient is −1. Thus, she decides to include a larger share of functions with such expressions than she originally planned in the practice exercises for homework.

She plans to continue to check on their skill with this algorithm as students use it during the next lesson, which begins the next day and targets the following objective:

> Given a real-life problem, the student determines how, if at all, a solution to that problem is facilitated by finding the derivative of an algebraic function (application).

To deepen your insights regarding how to design lessons for algorithmic-skill objectives, engage in Activity 6.6:

Activity 6.6

Write an algorithmic-skill objective that specifies one of the algorithms listed for Activity 6.5.

Develop a plan for a lesson that is designed to lead two or three students to achieve the objective. Discuss your plan with and get feedback from colleagues. Refine your plan as you see fit.

Implement your lesson plan with two or three students. Seek feedback from the students regarding what they learned and how you might make the lesson more engaging for them.

Share your experiences with colleagues who are also engaging in this activity.

Store the lesson plan—annotated with comments about the experience—in the "Cognition, Instructional Strategies, and Planning" section of your working portfolio.

MINIEXPERIMENTS RELEVANT TO STUDENTS' ACHIEVEMENT OF ALGORITHMIC-SKILL OBJECTIVES

Emphasis on Process, Not Outcome

Students achieve an algorithmic-skill objective by remembering how to carry out a procedure. Thus, an algorithmic-skill prompt should present students with the task of recalling or effecting the algorithm step by step. The nature of algorithms is such that they are remembered via a *sequence* of responses. The first step is triggered by the initial stimulus. The first step serves as the stimulus for the second step, the second for the third, and so on. Thus, the accuracy of subsequent steps in the process depends on the accuracy of previous steps. Because of this phenomenon, algorithmic-skill miniexperiments should be designed to identify which steps in the process are accurately remembered or executed and which are not. Unlike simple-knowledge miniexperiments, more than the final outcome to the initial stimulus needs to be detected. Consider Case 6.14:

CASE 6.14

Ms. Comaneci's students respond to prompts relative to the following objective:

> The student simplifies algebraic expressions with nested parentheses (algorithmic skill).

Ms. Comaneci examines Angel's work on one exercise as it appears in Exhibit 6.12 and thinks to herself: "His answer is off. It should be $-8x^2 + 24x + 104$. What did he do? He's working inside out; he remembered that. Good! Okay, here's his misstep. He missed distributing the negative sign to the 2. But if that's the only mistake, then he'd have 116 there, not +2. [pause] Oh, I see what else he did. He dropped the outside right parenthesis. That was just careless. Okay, I need to remind him of those two things."

Just as Ms. Comaneci analyzes Angel's work to identify which steps in the algorithm were remembered and which were not, algorithmic-skill miniexperiments need to be designed so that what students remember about the algorithm becomes apparent by their responses. Had Ms. Comaneci looked only at Angel's final answer rather than a display of his work, she would not have detected what he did right and what he did wrong—only that he did something wrong. Furthermore, had the exercise not included an example in which the negative sign was to be distributed, Angel's lack of attention to that task would not have surfaced. That particular error with respect to distributing a negative sign is far more common than one in which a negative coefficient other than 1 is to be distributed (Cangelosi, 1984b). Angel may not have made that mistake had the polynomial been $8(3(x + 5) - 4(x^2 + 2))$. The miniexperiments need to sample a variety of situations in which the process is evoked.

Furthermore, the observer's rubrics should reflect the degree to which the steps of the algorithm

Exhibit 6.12

Angel's Responses to an Exercise on Simplifying Algebraic Expressions.

Simplify:

$8(3(x + 5) - (x^2 + 2))$

$8(3x + 15) - x^2 + 2$

$24x + 120 - x^2 + 2$

$-x^2 + 24x + 122$

are remembered—not simply whether or not the final outcome is right or wrong. In Case 6.9, Mr. Champagne analyzed the algorithm, breaking it down into specific steps for students to remember. Such analyses provide the bases for observer's rubrics of miniexperiments relevant to algorithmic-skill objectives. Exhibit 6.13 displays a miniexperiment relevant to Ms. Comaneci's objective from Case 6.14.

Error-Pattern Analysis

Some mistakes students make in executing algorithms are careless oversights that occur inconsistently in their work. Angel's failure to distribute the -1 and to attend to the outside right parenthesis in Case 6.14 may not be recurring types of errors. To determine if each was an isolated incidence of carelessness or evidence of a learned error pattern, Ms. Comaneci needs to examine his work on additional exercises. It is not at all uncommon for students to develop consistent error patterns while attempting to learn an algorithm (Ashlock, 2001; Schoenfeld, 1985, pp. 61–67). Unless these patterns are identified and corrected soon after they are learned, they may become solidified through practice and overlearning. Thus, it is critical for you to devise algorithmic-skill miniexperiments that help identify possible error patterns students might be learning.

In Case 6.15, Ms. Comaneci takes advantage of her experiences observing students' work to devise a miniexperiment:

CASE 6.15

As she did for Angel, Ms. Comaneci looks at Cindy's work shown in Exhibit 6.14, and engages Cindy in the following conversation:

Ms. Comaneci: How did you get from $-8x^2 + 24x + 104$ to $x^2 + 3x - 13$?

Cindy: We are supposed to simplify these all the way—right?

Ms. Comaneci: Yes.

Cindy: So I divided both sides by -8 to get rid of it in front of the x^2.

Ms. Comaneci: But I see only one side.

Cindy: Can't you just factor the -8 out of the equation?

Ms. Comaneci: What equation?

Cindy: This one (as she points to "$-8x^2 + 24x + 104$").

Ms. Comaneci: That's not an equation.

Cindy: It isn't?

Ms. Comaneci: Read this and tell me if . . .

From numerous exchanges similar to this one with Cindy, Ms. Comaneci has developed a wealth of knowledge about incorrect variants of algorithms students accidentally acquire. Her experiences with debugging

Exhibit 6.13
Miniexperiment Designed to Be Relevant to Ms. Comaneci's Objective from Case 6.14.

Prompt:

Simplify; display your work in the box:

$$\frac{1}{3}(c - (c - 3c) - 6(2c + c))$$

Observer's Rubric:

Maximum score is 12 points distributed according to the following criteria (with +2 for each criterion that is met in all phases of the work, only +1 if it is met in some—but not all—of the work, and +0 if it is not met in any of the work):

A. Computations proceed from inside out relative to parentheses.
B. Associative properties are properly applied.
C. Distributive properties are properly applied.
D. Numerical computations are accurate.
E. Final answer is completely simplified.
F. Final answer is equivalent to $-5c$.

Exhibit 6.14
Cindy's Responses to an Exercise on Simplifying Algebraic Expressions.

Simplify:

$8(3(x + 5) - (x^2 + 2))$
$8(3x + 15 - x^2 - 2)$
$8(3x + 13 - x^2)$
$24x + 104 - 8x^2$
$-8x^2 + 24x + 104$
$x^2 - 3x - 13$

students' error patterns have taught her to quickly identify some of the recurring ways students attempt algorithms. She uses this knowledge to devise miniexperiments for algorithmic-skill objectives that cause some of the more common error patterns to surface. Here she designs one such item to test students' proficiency with an algorithm for finding the distance between two points in a Cartesian plane. She thinks to herself, "Okay, this miniexperiment is to help me see how they use the distance formula:

$AB = \sqrt{(x_1 - x_2)^2 + (y_1 - y_2)^2}$

where (x_1, y_1) and (x_2, y_2) are the coordinates for A and B respectively.

"I'll make this one multiple choice with alternatives being results from commonly made errors. First, I'd better come up with a stem for the prompt."

She develops the following stem:

What is the distance between $A(7, -2)$ and $B(0, 3)$?

Next, she computes the correct alternative, writing out the work and examining it:

$AB = \sqrt{(x_1 - x_2)^2 + (y_1 - y_2)^2} =$
$\sqrt{(7 - 0)^2 + (-2 - 3)^2} = \sqrt{7^2 + (-5)^2} =$
$\sqrt{49 + 25} = \sqrt{74} \approx 8.60$

Seeing her answer leads her to rethink the stem, saying to herself, "Should the answer be exact or in decimal form? Because they'll be using calculators, I'll use the decimal form and indicate to the nearest one hundredth in the directions."

Reexamining the computation, she thinks about error patterns students have made in the past, including the following:

- Computing AB as $\sqrt{(x_1 - y_1)^2 + (x_2 - y_2)^2}$, which, if other errors are not made, leads to the answer 9.49.
- Computing AB as $\sqrt{(x_1 + x_2)^2 - (y_1 + y_2)^2}$, which, if other errors are not made, leads to the answer 6.93.
- Computing AB as $\sqrt{(x_1 + x_1)^2 - (y_1 - y_2)^2}$, which, if other errors are not made, leads to an answer of 0 or other possibilities, depending on which values are substituted for (x_1, y_1) and (x_2, y_2), respectively.
- The correct formula is remembered but computed as if it were $\sqrt{(x_1^2 - x_2^2) + (y_1^2 - y_2^2)}$, yielding 6.63 as the answer.
- $\sqrt{(7 - 0)^2 + (-2 - 3)^2}$ is simplified as $\sqrt{7^2 + (-5)^2} = 7 - 5 = 2$.
- $\sqrt{(7 - 0)^2 + (-2 - 3)^2}$ is simplified as $\sqrt{7^2} + \sqrt{(-5)^2} = 7 + 5 = 12$.

She enters the miniexperiment shown by Exhibit 6.15 into her computerized miniexperiment file folders.

To help you begin gaining the kind of experiences Ms. Comaneci used to design miniexperiments in Case 6.15, engage in Activity 6.7:

Activity 6.7

Examine the work with algorithms students displayed in Exhibit 6.16. For each example, describe an error pattern the student might have learned.

Compare your analysis with those of colleagues who are engaging in this activity. Also compare your and your colleague's descriptions to the following ones:

- John's work suggests he is skillful with factoring but does not seem to discriminate between $ab = 0$ and $ab \neq 0$. He seems to blindly use the algorithm without paying attention to the underlying relationship upon which it is based.

Prompt:

To the nearest $\frac{1}{100}$ decimal place, what is the length of AB for $A(7, -2)$ and $B(0, 3)$? Indicate your answer by circling one and only one of the following:

9.49	8.60	6.63	2.00
12.00	6.93	0.00	10.07

Observer's Rubric:

8.60 is the correct response. Note any other response the student makes as evidence of a possible error pattern.

Exhibit 6.15
Multiple-Choice Miniexperiment Ms. Comaneci Developed.

Exhibit 6.16
Identify Students' Error Patterns.

Name *John*

Solve for all real values of x:

1. $x^2 - 6x - 7 = 0$

$(x-7)(x+1) = 0$
$x-7=0$ or $x+1=0$
$x=7$ or $x=-1$

2. $x^2 - 4x - 5 = 16$

$(x-5)(x+1) = 16$
$x-5=16$ or $x+1=16$
$x=21$ or $x=15$

3. $2x^2 - 9x + 7 = 25$

$(2x-7)(x-1) = 25$
$2x-7=25$ or $x-1=25$
$2x=32$ or $x=26$
$x=16$ or $x=26$

4. $3x^2 - 23x - 36 = 0$

$(3x+4)(x-9) = 0$
$3x+4=0$ or $x-9=0$
$3x=-4$ or $x=9$
$x=-4/3$

- -

Name *Jane*

Simplify:

1. $(c^2)^2 =$ c^4 2. $a^2a^4 =$ a^6 3. $(b^3)^2 =$ b^5

4. $(x^3y^5)^0 =$ x^8y^8 5. $b^5b^6 =$ b^{11} 6. $(w^1q^2)r^4 =$ $w^3q^7r^7$

- -

Name *Pat*

Add:

1. $\frac{x+3}{7x} + \frac{2x-8}{7x} = \frac{x+3+2x-8}{7x} = \frac{3x-5}{7x}$

2. $\frac{2x+1}{x-1} + \frac{5}{\sqrt{x-1}} = \frac{2x+1}{x-1} + \frac{25}{x-1} = \frac{2x+26}{x-1}$

Continued

- Jane does not seem to discriminate among a^na^m, a^nb^m, and $(a^n)^m$.
- Pat's work displays two error patterns. First, he seems to think that the value of fractions is unchanged if both the numerator and denominator are raised to the same power. Second, he has some notion of some sort of cross-multiplication process for finding common denominators. He may be confusing the following relationships—the first for adding fractions, the second for re-forming equations:

$$\frac{a}{c} + \frac{b}{d} = \frac{ad + bc}{cd}$$

$$\frac{a}{c} = \frac{b}{d} \Rightarrow ad = bc$$

- Pete appears to add virtually everything in sight to compute perimeters.

To stimulate your thoughts about how to design miniexperiments for algorithmic-skill objectives, engage in Activity 6.8:

Activity 6.8

Design a miniexperiment to measure the two or three students' achievement of the algorithmic-skill objective for which you designed and implemented a lesson when you engaged in Activity 6.6.

Exhibit 6.16
Continued

3. $\frac{\sqrt{x+4}}{x} + \frac{\sqrt{3x^2}}{6} = \frac{x+4}{x^2} + \frac{x\sqrt{3}}{6} = \frac{x+4}{x^2} \; \frac{x^2(3)}{36} = 12(x+4) + x^4$?

4. $\frac{8-x}{9x} - \frac{x}{x^2} = \frac{8-x}{9x} \; \frac{1}{x} = \frac{x(8-x) - 9x}{9x} = \frac{8-x-9}{9} = \frac{-x-1}{9}$

Name *Pete*

For each of the following figures find the perimeter:

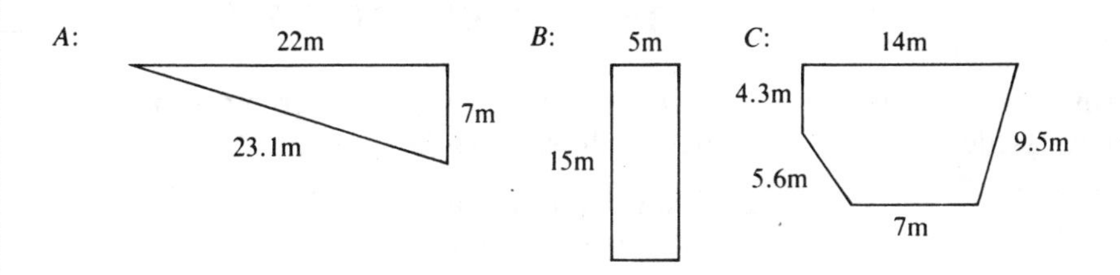

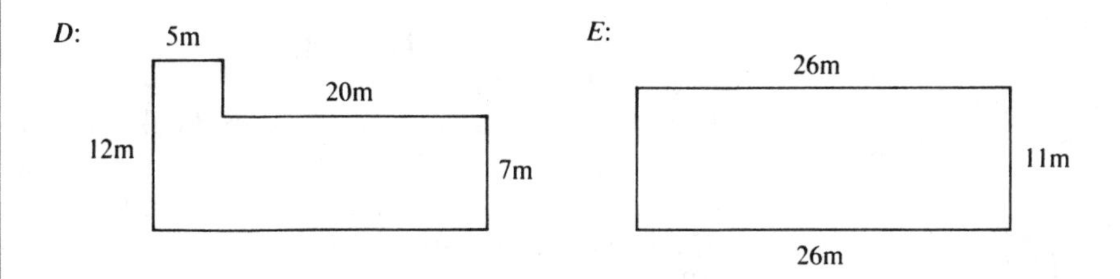

ANSWERS:

A's perimeter = *52.1m* *B*'s perimeter = *20 m*

C's perimeter = *40.4m* *D*'s perimeter = *44m*

E's perimeter = *63 m*

Try out your miniexperiments with the students. Get feedback from them on how it prompted them to attempt to execute an algorithm and on the clarity of the directions.

Discuss the experience with a colleague who is also engaging in this activity. Refine the miniexperiment in light of what you learn.

Store the resulting miniexperiment—along with your notes regarding the results of the experiment—as either (a) an attachment to the annotated lesson plan you stored in the "Cognition, Instructional Strategies, and Planning" section of your working portfolio or (b) in the portfolio's "Assessment" section.

SYNTHESIS ACTIVITIES FOR CHAPTER 6

1. For each of the following multiple-choice prompts, select the one response that either completes the statement so that it is true or accurately answers the question:
   A. Student perplexity is *not* a critical ingredient in lessons for which one of the following types of learning objectives?
      a) Construct a concept
      b) Discover a relationship
      c) Simple knowledge

B. Lessons for simple-knowledge objectives require ______ .
   a) direct instruction
   b) inquiry instruction
   c) error-pattern analysis
C. Lessons for algorithmic-skill objectives require ______ .
   a) direct instruction
   b) inquiry instruction
   c) use of mnemonics
D. Overlearning is associated with what levels of learning?
   a) Construct a concept and discover a relationship
   b) Algorithmic skill and simple knowledge
   c) Construct a concept and simple knowledge
   d) Creative thinking and algorithmic skill
E. Error-pattern analysis is used in learning activities for what type of objective?
   a) Simple knowledge
   b) Discover a relationship
   c) Construct a concept
   d) Algorithmic skill

2. With a colleague, compare and discuss your responses to the multiple-choice prompts from Synthesis Activity 1. Also, check your choices against the following key: A–c, B–a, C–a, D–b, E–d.
3. Resurrect the documents resulting from your engagement in Synthesis Activity #3 from the end of chapter 5. In light of your work with chapter 6, revise those objectives if you believe you should. Now select one of your simple-knowledge objectives and one of your algorithmic-skill objectives. For each, devise a lesson plan for a class of students. Discuss your lesson plans with colleagues who are also engaging in this activity. If you are able to, teach your lessons to a class of students while a colleague observes for the purpose of giving you feedback. Insert the lesson plans and other pertinent notes resulting from this activity in the "Cognition, Instructional Strategies, and Planning" section of your working portfolio.

TRANSITIONAL ACTIVITY FROM CHAPTER 6 TO CHAPTER 7

In a discussion with two or more of your colleagues, address the following questions:

1. What mathematical shorthand notations (e.g., those listed by Exhibit 4.5) should students learn to comprehend and use in their own communications?
2. How can students be taught to communicate with and about mathematics?
3. How is reading a mathematics textbook different from reading a trade book about mathematics (e.g., *Archimedes' Revenge: The Joys and Perils of Mathematics* [Hoffman, 1988])?
4. What strategies should teachers employ to lead their students to achieve comprehension-and-communication objectives?

chapter 7

Leading Students to Communicate With Mathematics

Goal and Objectives for Chapter 7

The Goal The goal of chapter 7 is to lead you to develop strategies for designing lessons that lead students to communicate with the language of mathematics and comprehend mathematical messages.

The Objectives Chapter 7's goal is defined by the following set of objectives:

A. You will explain (a) differences between the language of mathematics and commonly used, nonmathematical language, (b) the value of using the language of mathematics, and (c) difficulties posed by the need for continual transitions between commonly used language and mathematical language (comprehension and communication) 10%.
B. You will explain the role of conversation, speaking, listening, writing, and reading in learning meaningful mathematics (comprehension and communication) 15%.
C. You will develop strategies for teaching students to make use of different types of mathematical literature (e.g., mathematical textbooks, mathematical reference books, and mathematical trade books) for learning and doing mathematics (application) 10%.
D. You will describe the literal and interpretive process by which students comprehend technical mathematical expressions and mathematical messages (comprehension and communication) 10%.
E. For a given group of middle or secondary students, you will formulate comprehension-and-communication objectives that are consistent with *PSSM* (NCTM, 2000b) (application) 10%.
F. You will design lessons for comprehension-and-communication objectives (application) 25%.
G. You will design miniexperiments that are relevant to your students' achievement of comprehension-and-communication objectives (application) 15%.
H. You will incorporate the following phrases or words into your working vocabulary: "comprehension and communication," "mathematical message," "technical mathematical expression," "mathematical trade book," "literal understanding," and "interpretive understanding" (comprehension and communication) 5%.

THE POWER OF MATHEMATICAL LANGUAGE

Consider Case 7.1:

CASE 7.1

As part of a homework assignment in Mr. Rudd's precalculus class, Sunni selects an object from her residence, conceals it in a box, and brings it to class. The object she chose is the salad spinner she is shown holding in Exhibit 7.1. To each student—other than Sunni and LaDaines—Mr. Rudd distributes a copy of the tasksheet shown by Exhibit 7.2; he also displays it on an overhead transparency.

Sunni and LaDaines sit back to back in the center of the room as shown by Exhibit 7.1; the other students observe them from their desks. Sunni removes the salad

Exhibit 7.1
Sunni Describes the Salad Spinner for LaDaines.

Exhibit 7.2
The Tasksheet Distributed to Students and Displayed on a Transparency.

Category V Words	Category S Words

spinner from the box so that it remains out of LaDaines' sight. Mr. Rudd initiates the activity:

Mr. Rudd: In about a minute, LaDaines is going to try to picture the object that Sunni is holding and eventually draw what he pictures but without actually seeing the object with his eyes. He'll be able to use only the descriptions of the objects that Sunni gives him in response to his questions. As they engage in this experiment, I'll occasionally be directing the rest of us to record two kinds of words that they use on our tasksheets: Category V words and Category S words. Yes, Aaron.

Aaron: What are "Category V" and "S" words?

Mr. Rudd: Thanks for asking. I'm not going to reveal the difference right now, but I will tell you what to write on your tasksheets and you'll soon figure out the difference for yourself. Ichiro, would you please display the words on the overhead projector as I call them? [pause] Thank you. Okay, LaDaines, ask Sunni your first question.

LaDaines: What is the object?

Sunni: It's a salad spinner.

Mr. Rudd: Write "salad spinner" in the Category V column.

LaDaines: What's a "salad spinner"? I never heard of that.

Sunni: It's a thing you use in the kitchen to wash and drain lettuce.

Mr. Rudd: Write "wash," "drain," and "lettuce" under Category V.

LaDaines: That's weird. So it's some kind of kitchen utensil.

Mr. Rudd: Write "kitchen utensil" under Category V.

LaDaines: What does it look like?

Sunni: Well, there's a big bowl where you put the lettuce in and there's a top with a crank on it. Actually, there're two bowls—one inside the other.

Mr. Rudd: Write "bowl" and "crank" under Category V and "2" and "big" under Category S.
LaDaines: What shape are the bowls and how big are they?
Sunni: They're round bowls; the bigger one is about like this.
LaDaines: "Like this!" I can't see what you're doing. How big?
Sunni: I'm sorry. The bowl is—
Mr. Rudd: Excuse me, Sunni, before you continue, we're going to list more words. Under Category S, write "round," and "bigger." Please continue, Sunni.
Sunni: It must be about 10 inches across at the top and it tapers down to about 7 inches on the bottom. And then the other bowl fits inside; so it must be just a little smaller.
Mr. Rudd: I see Ichiro has listed "10 inches," "7 inches," and "smaller" under Category S without my asking him to. The rest of us should do the same.
LaDaines: Okay, so what's this crank thing you said?
Sunni: The salad spinner is in three parts: The inside bowl, the outside bowl, and the top which fits over the outside bowl. The crank is part of the top.
Mr. Rudd: Let's list "3," "inside," "outside," and "over" under Category S. Yes, Dominica.
Dominica: I can see why "3" is a Category S, but shouldn't "inside," "outside," and "over" be Category V words? (See Exhibit 7.3.)
Mr. Rudd: Why?
Dominica: I don't know what "V" and "S" stand for, but I figured that the "S" ones refer to numbers or sizes—until you stuck in "inside," "outside," and "over."
Mr. Rudd: What about "round"? Does "round" refer to a quantity like the others?
Dominica: No, but it's a shape like a circle—and that's mathematical.
Mr. Rudd: What area of mathematics does "2," "big," "bigger," "10 inches," "7 inches," "smaller," and "3" fit?
Dominica: What do you mean?
Mr. Rudd: I mean the different kinds of mathematics we study like arithmetic, algebra, geometry, statistics, trigonometry, or calculus.
Dominica: Oh! So, you mean numbers like that could be arithmetic or algebra or I guess anything.
Mr. Rudd: But what about "round," or as you said, "circle," where does that fit?
Dominica: Oh, you mean geometry.
Mr. Rudd: Go ahead, Clarence, what do you want to say?
Clarence: You're going to tell Dominica that "inside," "outside," and "over" are geometry words because geometry isn't only about shapes; it's also about locations and positions.
Mr. Rudd: Keep the discussion going, Clarence.
Clarence: Okay, Susan.
Susan: So, are the S words mathematical and the V words aren't, Mr. Rudd?
Mr. Rudd: Yes. LaDaines.

Exhibit 7.3
The Lists on the Tasksheet When Dominica Raised Her Question.

Category V Words	Category S Words
salad spinner	two
wash	big
drain	round
lettuce	bigger
kitchen utensil	10 inches
bowl	7 inches
crank	smaller
	three
	inside
	outside
	over

LaDaines: Can we get back to the experiment? I'm tired of facing this way. Look what I've drawn so far. Is it right?
Mr. Rudd: Excuse me. LaDaines, without turning around, give your sketch to Lucinda to display on the visual presenter so we can all see it.

Lucinda displays Exhibit 7.4 to the class.

Exhibit 7.4
LaDaines' Initial Drawing.

Mr. Rudd: LaDaines, now that Sunni can see your initial sketch, she should be able to direct your next sketch much more efficiently. Also, I'm going to leave it to the rest of you to continue to categorize words they use without my telling them to you. [pause] Oh, Ichiro, please stay by the overhead so we can all follow along with your lists of words. Thank you.

LaDaines: What's wrong with my drawing?

Sunni: It looks really good.

Mr. Rudd: I'm sorry, I said I wasn't going to list words for you, but I don't want to pass this one up. At the beginning of the experiment, we listed mostly V words, then as Sunni got more specific with her descriptions, we listed more S words. And Sunni just said a V word and I didn't want to miss it. What is it, Aaron?

Aaron: She said "good." I think I know what "V" and "S" stand for.

Mr. Rudd: Go for it.

Aaron: "S" is for "specific"—I'm pretty sure. Is "V" for "value"? Good is a value.

Mr. Rudd: I was thinking of "specific" when I chose "S". I had the opposite of specific in mind when I chose "V". I thought of "vague." But maybe those aren't the best labels I could have chosen. Thanks for getting that out of the way, Aaron. Continue Sunni.

Sunni: LaDaines, you need to move the crank away from the center of the circle and way off near the edge of the top. Also, the handle of the crank is a circular disk that rotates when you crank the handle.

LaDaines: Okay, I'm getting the picture of the top. What else?

Sunni: The bowl itself is not as flat as you drew it; it's longer. And the bottom is flat like you have it, but it's also circular not rectangular like you have it. The bowl is about 7 inches tall and—like I said before—the diameter at the top is about 10 inches and the diameter at the bottom is about 7 inches. The bowl that goes inside is in proportion to the outside bowl—just smaller. I guess in geometry, you'd say the two are similar.

LaDaines: How much smaller? What's the ratio?

Sunni: Let me measure something here. [pause] The inside bowl is 5 inches tall, so the ratio for all the dimensions must be about 5/7. Unlike the outside bowl, the inside bowl is full of holes so water can pass through.

LaDaines: Okay, slow down for a minute while I . . .

The natural conversation (recall chapter 2's section "True Dialogues") continues with LaDaines sketching a very accurate picture and then Sunni using LaDaines' sketch to explain how a salad spinner works. Ichiro's final word lists are shown by Exhibit 7.5.

Mr. Rudd then conducts a question-discussion session that leads students to recognize that we depend on the language of mathematics when we need to communicate an idea or a process as unambiguously and precisely as possible. His objective for the lesson was the following one:

> The student will value the language of mathematics as a tool for precision communications (appreciation).

In Mr. Rudd's next lesson, he plans to lead students to build mathematical models to help explain selected phenomena from students' everyday lives. To successfully do mathematical modeling, they will need to communicate in the precise language of mathematics.

Exhibit 7.5
Ichiro's Final Word Lists.

Category V Words	Category S Words	
salad spinner	two	5/7
wash	big	holes
drain	round	360°
lettuce	bigger	fast
kitchen utensil	10 inches	slow
bowl	7 inches	dry
crank	smaller	
good	three	
water	inside	
clean	outside	
soggy	over	
crisp	center	
hand	circle	
taste	edge	
	flat	
	longer	
	bottom	
	rectangular	
	diameter	
	proportion	
	similar	
	5 inches	
	tall	
	ratio	
	dimensions	

Chapter 3's section "Motivating Students to Do Mathematics" emphasized the advantages of problem-based learning in which students connect mathematics to real-life situations. In Case 7.1, Mr. Rudd attempted to heighten students' awareness of the mathematical words in their everyday language and how their language becomes more mathematical when they formulate precise descriptions and explanations. He also plans to involve them in mathematical modeling in which they will formulate mathematical expressions (e.g., statements of relationships) that reflect real-life phenomena. Mathematical language provides the power to communicate precisely—a necessity for building useful problem-solving models. However, this power has its price: We must learn to shift between the figurative interpretations of ordinary English and literal interpretations of mathematical English.

Although most people find the shorthand symbols (e.g., those in Exhibit 4.5) and technical words peculiar to mathematics (e.g., "homomorphism," "polynomial," or "cosecant") intimidating, the ordinary-sounding English words (e.g., "unique," "most," or "or") may be the main source of the difficulties people have negotiating the language of mathematics (Cangelosi, 2001a). Engage in Activity 7.1:

Activity 7.1

A television personality made the following comments about NBA basketball player John Stockton:

> "Who is more unique than a John Stockton? A 40-year-old point guard who gives a 110% every night! Most 40-year-olds are sitting in their rocking chairs with a cold one right now—not out on the court setting countless back screens on guys twice their sizes and half their ages. In spite of the unlimited punishment, he plays maximum minutes, never complaining about the bumps and bruises. He's one of the greatest players in the history of the game."

With a colleague, examine the quote, discussing how the commentator's use of the underlined words differ from how those words are used in mathematics.

The television personality used the underlined words in her comments about John Stockton the way those words are often used in commonly spoken and written English. But as you and your colleague discussed, they are interpreted quite differently in the world of mathematics:

> In mathematics, something is either unique or not; uniqueness does not exist in degrees. The use of the indefinite article "a" suggests that there is more than one John Stockton. Of course, 110% is impossible and there is not a game every night. In mathematics, "most" refers to at least half. Does the commentator really believe that at least half of the 40-year-olds were all in rocking chairs at the same moment in time? John Stockton frequently sets back screens during games, but the number is far less than countless. It is even less than a countable infinity—let alone an uncountable one. John Stockton is about 6 feet tall; the players on whom he sets back screens are tall, but not 12 feet tall. John Stockton does occasionally come out of games, so he does not play maximum minutes. He once complained and there can be only one greatest.

The point here is not to be critical of the commentator's use of common English. It is colorful and listeners understand what she means. The point is that in order for your students to take advantage of the power of the language of mathematics, they need to learn to discriminate between the most commonly used English and English as it is used in mathematical contexts. Teaching them to make those transitions between the two languages is a complex responsibility.

WHAT IS THE QUESTION?

Return to Case 1.3; reread the dialogue that begins with Harriet's comment, "Well, I'm very happy that your teaching will be . . ." up to Harriet's comment, "I understand why *PSSM*-based . . ."

Note that in the example that Casey worked with Irene in responding to "Solve for x in the equation $8(11 - x)= 56$," Irene failed to focus on the question posed by the prompt. Furthermore, teachers themselves lost sight of the question posed by the multiple-choice prompt Vanessa pulled from a standardized test. My experiences working with students have convinced me that students often fail to respond correctly to prompts on mathematics tests not because they lack the mathematical skills on which they are supposedly being tested, but because they do not comprehend the question they are asked or the task they are directed to perform (Cangelosi, 2001a). Consider Case 7.2:

CASE 7.2

Mr. Rasmussen religiously follows the textbook to instruct his algebra I class. For homework, he directs students to work 15 textbook word problems from the section "Graphing Inequalities." The section includes several examples of worked-out problems that include illustrations with graphs such as those shown by Exhibit 7.6.

Howard begins the assignment with the following word problem as it appears in the textbook:

> CONSUMER AWARENESS: Suppose you are shopping for cassettes and CDs at Music Village. Cassettes cost $8, CDs cost $16, and you have $32 to spend. To show how many different ways you can purchase cassettes and CDs, graph the inequality $8x + 16y \leq 32$.

After looking at "$8x + 16y \leq 32$" and quickly reviewing the worked sample problems from the section, Howard thinks, "It's a less-than-or-equal-to problem, so I shade the part under the line and the line should be solid—not broken. Okay, first find two points on the graph. I'll use (0, 2) and (4, 0). . . . Okay, now for the next one."

Exhibit 7.6

Graphs Included With Worked Sample Problems in the "Graphing Inequalities" Section of Mr. Rasmussen's Algebra I Textbook.

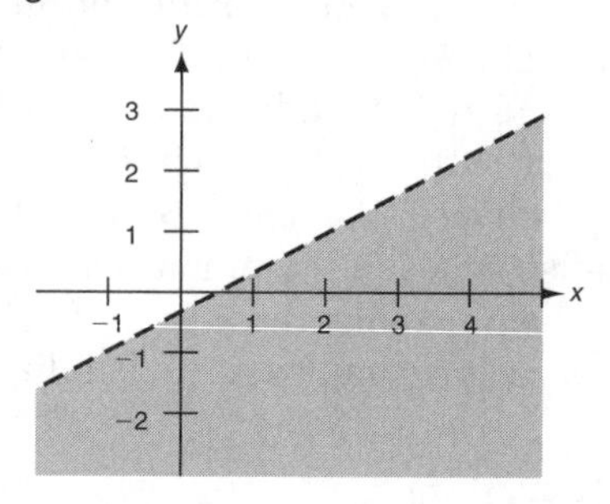

$$3x - 4y > 2$$
$$-4y > -3x + 2$$
$$y < \tfrac{3}{4}x - \tfrac{1}{2}$$

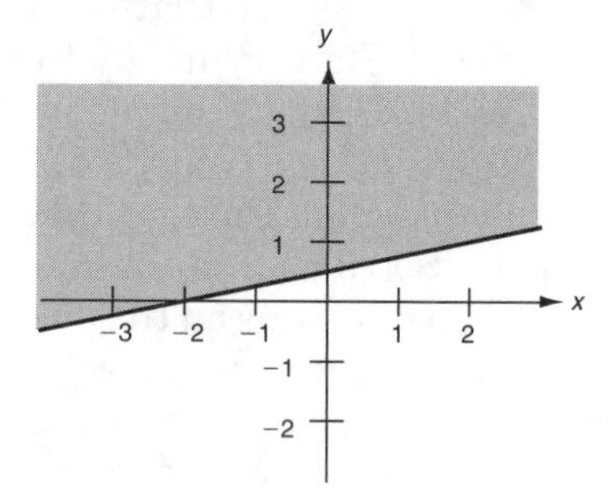

$$x \leq 4y - 2$$
$$x + 2 \leq 4y$$
$$\tfrac{1}{4}x + \tfrac{1}{2} \leq y$$
$$y \geq \tfrac{1}{4}x + \tfrac{1}{2}$$

Exhibit 7.7 displays Howard's "answer" to the first word problem; Exhibit 7.8 displays the "solution" in the teacher's edition of the textbook.

The next day, Ms. Santos, a preservice mathematics teacher who occasionally observes the class and works with students individually, engages Howard in the following conversation regarding his work with that first word problem:

Ms. Santos: So what does this graph tell you about the relationship between the number of cassettes you can buy and the number of CDs you can buy?

Howard: I don't know.

Ms. Santos: I see you included the point (0, 2). What does that tell you about how many CDs you could buy if you bought no cassettes?

Howard: It doesn't tell me anything about that.

Ms. Santos: According to this problem, how much money do you have to spend?

Howard: $32.

Ms. Santos: Okay, so if you didn't buy any cassettes, how many CDs could you buy with your $32 if CDs are $16 a-piece?

Howard: One.

Ms. Santos: But 32 divided by 16 is 2, so wouldn't the answer be 2?

Howard: But what about tax?

Ms. Santos: Oh, okay. You're right. I guess the book wanted you to work this without worrying about sales tax.

Howard: Okay, then it's two. But what has that got to do with this graph?

Exhibit 7.7

Howard's "Answer" to the First Textbook Word Problem.

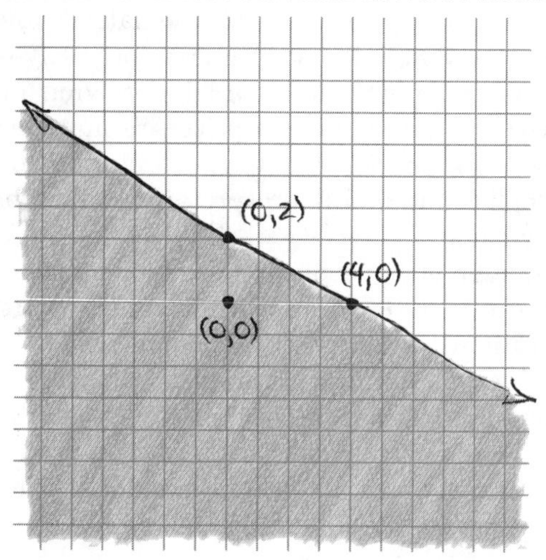

Ms. Santos: Well the graph shows that the more cassettes you buy, the fewer CDs you can buy and vice versa.

Howard: Duhhh! I already knew that.

Ms. Santos: Right! But graphs like this can help you when the relationships aren't so obvious. So let's look at the graph you got. [pause] What does *x* in the inequality represent?

Howard: I don't know. I guess either the cassettes or the CDs.

Ms. Santos: Which one?

Howard: It shouldn't make any difference because Mr. Rasmussen said you could use any letter for a variable. It doesn't make any difference.

Ms. Santos: But in this case it does. You see . . .

The conversation continues with Ms. Santos realizing that Howard completely understands the relationship between number of cassettes and number of CDs, but does

Exhibit 7.8

"Solution" to the First Word Problem as It Appears in the Teacher's Edition of the Textbook.

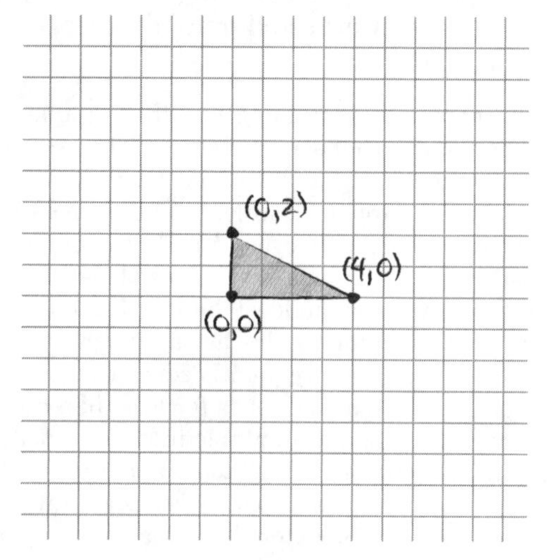

not associate the points on his graph with that relationship. He is not bothered by the fact that his graph indicates that it is possible to purchase an unlimited number of one kind of product as long as you purchase an unlimited negative number of the other type of product. Furthermore, the conversation leads Ms. Santos to realize that the textbook's "solution" shown by Exhibit 7.8 does not make much sense either. Exhibit 7.8's graph indicates that it is possible to purchase fractions of cassettes and CDs—even irrational numbers of them!

One of the factors that sometimes prevents students from comprehending the question posed by prompts like the one in Case 7.2 is that the questions may not make much sense in real-life contexts. There is also a tendency for students who have been taught via textbook-driven curricula to dive into algorithms before stepping back to comprehend the task at hand as suggested by chapter 6's section "Stage 2: Explanation and Practice Estimating Outcomes." Without comprehending the question, many students learn to "play the textbook game" well enough to correctly respond to word problems—but this does not prepare them well to respond to similar prompts on standardized and core-curriculum tests (Cangelosi, 2001a). Consider Case 7.3:

CASE 7.3

Rosario diligently follows his calculus teacher's explanation of how to use the derivative of a quadratic function to find maximum and minimum y values. For homework, Rosario works a number of strictly computational exercises and then gets to the "application" word problems assigned by the teacher. The first one reads:

> If a rock is tossed upward with an initial velocity of 112 decameters per second from an altitude of 700 decameters above the surface of Mars, its altitude above the surface s seconds later is given by $h = -5.6s^2 + 112s + 700$. What is the maximum altitude reached by the rock?

Automatically, Rosario writes "$h(s) = -5.6s^2 + 112s + 700$," but then he pauses and thinks to himself, "I wonder what I'm supposed to do with that first part about velocity of 112 decameters and altitude of 700 decameters. She didn't show us what to do with that. Oh, well, all these are maximum and minimum problems, so I know I've got to take the derivative." He writes as he thinks, "The rule is to multiply the coefficient by the exponent and then reduce the exponent by one and just drop the constants." He writes:

$h(s) = -5.6s^2 + 112s + 700$

$h'(s) = (2)(-5.6)s + 112$

$h'(s) = -11.2s + 112$

He continues, "Now, I've got to equate this to 0 and solve for s, and that should give me the answer." He writes:

$0 = -11.2s + 112$

$11.2s = 112$

$s = \frac{112}{11.2}$

Using his calculator, he divides 112 by 11.2 and writes:

$s = 10$

He thinks, "So my answer is 10; better check it in the back of the book. '1,260 decameters!' Aw, no, I'm not even close! What'd I do wrong? Oh! I wonder if I should have plugged my 10 in for s and solved the function from there. I'll try that. He writes:

$h(s) = -5.6s^2 + 112s + 700$

$h(10) = -5.6(10^2) + 112(10) + 700$

After using his calculator to simplify the right side, he writes "1,260" and exclaims to himself, "Now it's right! So I should work the rest this way. I wonder why I didn't have to use all of the given information. Whatever! This way'll work for all the rest. Okay, next . . ."

The remainder of this chapter focuses on strategies for teaching your students to negotiate the language of mathematics, to use the power of mathematical language, and to clarify the question to be answered or the task to be performed when presented with prompts in mathematics texts (e.g., involving textbook word problems) and on tests. You do this by teaching to appropriate comprehension-and-communication objectives.

NEGOTIATING THE LANGUAGE OF MATHEMATICS

Mathematical Messages

In Cases 5.8–5.10, Ms. Citerelli's students constructed the concept of arithmetic sequence. Afterwards, they formulated a definition for "arithmetic sequence"—thus, generating the following mathematical message:

> A sequence is arithmetic iff there exist real numbers a and d such that the first member is a, the second is $a + d$, the third is $a + 2d$, and so on, with the nth member being $a + (n - 1)d$.

Depending on the success of Ms. Gaudchaux's lesson plan in Exhibit 5.19, her students will hypothesize the following proposition:

> The value of the constant a in the linear function $f(x) = ax + b$ determines the slope of the line graph of f.

The value of the constant b determines where the line graph of f is located but does not influence the slope. $b > 0 \Rightarrow$ the y-intercept is positive. $b < 0 \Rightarrow$ the y-intercept is negative.

In Case 6.13, Ms. Allen's students would hardly be able to follow her explanation of an algorithm for computing derivatives unless they comprehended statements such as the following one:

$$f(x) = a_0x^n + a_1x^{n-1} + a_2x^{n-2} + \ldots + a_{n-2}x^2 + a_{n-1}x + a_n$$

Constructing concepts leads to definitions, discovering relationships leads to statements of propositions, proving propositions leads to statements of theorems and presentations of proofs, inventing algorithms leads to directions for executing step-by-step procedures, and addressing problems leads to the articulation of problems and explanations of work on their solutions. In general, doing mathematics creates messages (e.g., definitions, descriptions, arguments, proofs, directions, explanations) that need to be communicated and comprehended. Of course if mathematics were strictly a solitary activity as it seemed to be for some eccentrics (e.g., Pierre de Fermat in 17th Century France [Singh, 1997, pp. 34–44]), then maybe there would be little need to comprehend and communicate mathematical messages. But for the vast majority of our history, mathematics is a social endeavor with people collaboratively doing mathematics and sharing their discoveries and inventions with others. Thus, your students need to comprehend and communicate mathematical messages via speaking, writing, reading, listening, and observing.

Unfortunately, too many people have developed the mistaken belief that all mathematics resides in textbooks. Mathematics resides in our minds and in our environment. Textbooks are used as one tool for teaching and learning mathematics and teaching students to make use of textbooks is an important aspect of teaching mathematics. But students also need to comprehend mathematical messages expressed in other media besides textbooks (e.g., video programs, oral presentations, discussions, interactive software, and web-based materials). A particularly engaging source of mathematical messages that you and your students will enjoy and learn from is *mathematical trade books.*

Mathematical trade books are books about mathematics or the history of mathematics that are not designed to be used in conjunction with mathematics courses (although I use them in mathematics courses I teach). They include reference books (e.g., *Mathematics: From the Birth of Numbers* [Gullberg, 1997]) and, more importantly, books that inform and entertain (e.g., *Zero: The Biography of a Dangerous Idea* [Seife, 2000]). Some of them are for your younger students to read (e.g., *The Number Devil: A Mathematical Adventure* [Enzensberger, 2000]); others are for you and your older students in more advanced courses (e.g., *Pi in the Sky: Counting, Thinking, and Being* [Barrow, 1992]). Exhibit 7.9 is a relatively minute sample of mathematics trade books that you will find enlightening, entertaining, and helpful to you as you do and teach mathematics.

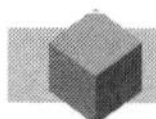

Activity 7.2

Obtain one of the following books listed in Exhibit 7.9 and read the indicated selections: (a) pp. 1–47 of *Archimedes' Revenge: The Joys and Perils of Mathematics,* (b) pp. 1–156 of *Zero: The Biography of a Dangerous Idea,* or (c) all of *Fermat's Enigma* (pp. 1–285).

With a colleague who is also engaging in Activity 7.2, discuss the selections you read. Write a reflection paper (about two pages long) indicating how the reading influenced your thinking about mathematics and how you plan to teach mathematics. Insert the reflection paper in the "Mathematics and the Historical Foundations of Mathematics" section of your working portfolio.

The language of mathematics would have no advantage over commonly used English if it were no different from commonly used English. As suggested by chapter 4's section "Miscomprehension of the Language of Mathematics," those differences create difficulties for students learning mathematics. The language of mathematics, like any other language, depends on people agreeing to certain conventions about what is accepted usage for the meaning of certain technical expressions. The use of technical expressions in mathematics—including common English words with special definitions, technical mathematical words, numerals, shorthand symbols, and the special structures of the language itself—need to be taught.

Technical Expressions

Common English Words With Special Definitions

When you engaged in Activity 7.1, you noted the need for students to discriminate between common usage of words such as "unique," "maximum," "most," and "or" and their use in mathematics. Furthermore the meanings of many of these common English words vary in mathematics depending on the context in which they are used. For example, in common English, the word "power" denotes "capability of doing something." But in an algebraic context (e.g., "x to the nth power"), "power" refers to "the number of times a number is used as a factor." In the context of

Exhibit 7.9

Relatively Minute Sample of Mathematical Trade Books.

Barbeau, E. J. (1997). *Power play.* Washington: Mathematical Association of America.

Barbeau, E. J. (2000). *Mathematical fallacies, flaws, and flimflam.* Washington: Mathematical Association of America.

Barnsely, M. (1988). *Fractals everywhere.* Boston: Academic Press.

Barrow, J. D. (1992). *Pi in the Sky: Counting, Thinking, and Being.* Boston: Back Bay Books.

Bashmakova, I. G., & Silverman, J. J. (1997). *Diophantus and Diophantine equations* (updated ed.). Washington: Mathematical Association of America.

Beautelspacher, A. (1994). *Cryptology.* Washington: Mathematical Association of America.

Beltrami, E. (1993). *Mathematical models in the social and biological sciences.* Boston: Jones and Bartlett.

Benjamin, A., & Shermer, M. B. (1993). *Mathemagics: How to look like a genius without really trying.* Los Angeles: Lowell House.

Bolt, B. (1992). *Mathematical cavalcade.* New York: Cambridge University Press.

Bowers, J. (1988). *Invitation to mathematics.* New York: Basil Blackwell.

Dunham, W. (1994). *The mathematical universe: An alphabetical journey through the great proofs, problems, and personalities.* New York: Wiley.

Dunham, W. (1999). *Euler: The master of us all.* Washington: Mathematical Association of America.

Enzensberger, H. M. (2000) *The number devil: A mathematical adventure.* New York: Henry Holt and Company.

Flato, M. (1990). *The power of mathematics.* New York: McGraw-Hill.

Gardner, M. (1969). *The unexpected hanging: And other mathematical diversions.* New York: Simon & Schuster.

Gibilisco, S. (1990). *Optical illusions: Puzzles, paradoxes, and brain teasers,* #4. Blue Ridge Summit, PA: Tab Books.

Gullberg, J. (1997). *Mathematics: From the birth of numbers.* New York: W. W. Norton.

Hall, N. (Ed.). (1991). *Exploring chaos: A guide to the new science of disorder.* New York: W. W. Norton.

Hardy, G. H. (1992). *A mathematicians apology.* New York: Cambridge University Press.

Hoffman, P. (1988). *The man who loved only numbers: The story of Paul Erdös and the search for practical truth.* New York: Hyperion.

Hoffman, P. (1998). *Archimedes' Revenge: The Joys and Perils of Mathematics.* New York: Fawcett Columbine.

Kasner, E., & Newman, J. R. (1989). *Mathematics and the imagination.* Redman, WA: Tempus Books.

King, J. P. (1992). *The art of mathematics.* New York: Fawcett Columbine.

McLeish, J. (1991). *Number: The history of numbers and how they shape our lives.* New York: Fawcett-Columbine.

Morgan, F. (2000). *The math chat book.* Washington: Mathematical Association of America.

Müller, R. (1989). *The great book of math teasers.* New York: Sterling Publishing.

Paulos, J. A. (1991). *Beyond numeracy: Ruminations of a number man.* New York: Knopf.

Peterson, I. (1988). *The mathematical tourist: Snapshots of modern mathematics.* New York: Freeman.

Pólya, G. (1985). *How to solve it: A new aspect of mathematical methods* (2nd ed.). Princeton, NJ: Princeton University Press.

Poundstone, W. (1992). *Prisoner's dilemma.* New York: Doubleday.

Russell, B. (1993). *Introduction to mathematical philosophy* (revised ed.). London, England, Routledge.

Salem, L. Testard, F., & Salem, C. (1992). *The most beautiful mathematical formulas.* New York: Wiley.

Seife, C. (2000). *Zero: The Biography of a Dangerous Idea.* New York: Viking.

Singh, S. (1997). *Fermat's enigma.* New York: Walker and Company.

Singh, S. (1999). *The code book: The science of secrecy from ancient Egypt to quantum cryptography.* New York: Anchor Books.

Stein, S. (1999). *Archimedes: What did he do besides cry Eureka*? Washington: Mathematical Association of America.

Stewart, I. (1992). *Another fine math you've got me into . . .* New York: Freeman.

Stewart, I. (1992). *The problems of mathematics* (2nd ed.). Oxford, England: Oxford University Press.

Stewart, I., & Golubitsky, M. (1992). *Fearful symmetry: Is God a geometer*? Oxford, England: Birhäuser.

inferential statistics (e.g., "the power of a statistical test"), "power" refers to "the probability of rejecting a null hypothesis."

For students to be able to comprehend the definition of the name of a concept, they need to have previously constructed the concept. Examine, for example, the following definition of "$f{:}A \to B$" (i.e., "f is a function from set A to set B") that is commonly found in advanced high school mathematics books as well as college-level textbooks:

$$f{:}A \to B \text{ iff } (f \subseteq A \times B \ni (\forall x \in A\, \exists y \in B \ni (x, y) \in f) \text{ and } ((x_1, y_1) \in f \text{ and } (x_1, y_2) \in f) \Rightarrow y_1 = y_2)$$

If it has been a long time since you have encountered this definition, you may have forgotten it. However, because you previously constructed the concept of function, you are able to make sense out of this definition—unless, of course, you are unfamiliar with some of the shorthand symbols (e.g., "$f \subseteq A \times B$" means "f is a relation from A to B," "$\ni$" means "such that," "$\forall$" means "for every," and "$\Rightarrow$" means "implies"). However, the definition would make no sense to someone who did not already have the concept of function firmly constructed in her mind—even if she could decode the shorthand symbols.

Technical Mathematical Words

Technical mathematical words and phrases have meanings only within a mathematical structure (e.g., "cosine," "Cauchy sequence," "stochastic," "sexagesimal," and "polyhedron"). As previously suggested, many people are frightened away from the pursuit of mathematics by such foreign-appearing words, but these words are not as likely to be misinterpreted as more familiar-sounding words. Unlike common English words with specialized meanings, if people do not know the meaning of technical mathematical words like "homomorphism" and "harmonic mean," they know they do not know.

Numerals

A *numeral* is a name for a number. Distinguishing between a number itself and its endless list of names is critical to students' comprehending the meaning of equations. For example, there is only one number 2, but {numerals for 2} is infinite. Among the numerals for 2 are "2," "5 − 3," "$\sqrt{4}$," "$16^{1/4}$," and "12 ÷ 6." An equation (e.g., $\sqrt[5]{7-5}^{\,5} = 2$) is a statement indicating that the expression to the left of the equal sign names the same thing as the expression to the right of the equal sign. Thus, 12 ÷ 6 = 2 because "12 ÷ 6" and "2" are numerals for the same number. However, "12 ÷ 6" ≠ "2" because "12 ÷ 6" and "2" are two different numerals. The number of elements in {3, 2 + 1, 14, $\sqrt{9}$} is two just as {George Washington, the first president of the United States, Abraham Lincoln, Martha Washington's husband} is a two-element set. However, the number of elements in {"3," "2 + 1," "14," "$\sqrt{9}$"} is four just as {"George Washington," "the first president of the United States," "Abraham Lincoln," "Martha Washington's husband"} is a four-element set. Quotation marks distinguish references to numerals (e.g., "17") from references to numbers (e.g., 17).

This distinction between numbers (i.e., actual quantities) and numerals (i.e., names for numbers) may seem trivial to you if you have not before dwelled on it. However, making that distinction in your own mind and teaching that distinction to your students—though initially confusing—will serve you and them well with respect to comprehending and communicating with the language of mathematics. Much of what we do in mathematics involves renaming numbers and displaying that two numerals represent the same number.

Shorthand Symbols

Shorthand symbols (e.g., those used in the aforementioned definition of "function") contribute to the mystery of mathematics for those who have not learned to decode them. On the other hand, such symbols serve two valuable purposes: (a) The use of shorthand symbols saves time and space in mathematical communications. (b) By compacting communications, shorthand symbols facilitate the analysis and comprehension of statements of relationships. For example, is comprehension of the definition of "standard deviation" easier when it is stated in ordinary English or when using shorthand symbols?

- *An ordinary English version:* The standard deviation of a sequence of rational numbers is equal to the principal square root of the number derived by adding the squared differences between each member of the sequence (i.e., each data point) and the arithmetic mean of the sequence, and then dividing that sum by the number of numbers in the original sequence.
- *Compact version:* σ is the standard deviation of a sequence of rational numbers X where $X = (x_1, x_2, x_3, \ldots, x_N)$ and μ_X is the arithmetic mean of X iff

$$\sigma = \sqrt{\frac{\sum_{i=1}^{N} (x_i - \mu_x)^2}{N}}$$

The more intricate the relationship, the more useful the shorthand symbols are for comprehending and communicating the idea. As long as a person has been taught the meanings of symbols and has become accustomed to using them, the compact form with the shorthand symbols makes it easier to recognize critical associations (e.g., in the definition of "standard deviation," each difference is squared before the differences are summed).

Communication or Language Structures

Mathematical language includes technical structures such as the Cartesian plane, Euclidean space, and axiomatic systems. The logical underpinnings of how mathematical truths are determined influence how the language is interpreted. For example, within the structure of a geometry, algebra, or number theory, truths are built on a system of undefined terms (e.g., "point"), definitions (e.g., of "line segment"), axioms or postulates (e.g., {integers} is closed under addition), and theorems (e.g., the Pythagorean theorem). As with other conventions, students need to be taught to comprehend and utilize such structures.

As with common-usage English, the meanings of expressions vary depending on the context within which they are used. Think, for example, of the various ways "(a, b)" is used: When you are working within a Cartesian coordinate system, "(a, b)" refers to a point. Most often, "(a, b)" designates an ordered pair or vector, but in number theory where a and b are integers, "$(a, b) = d$" may be read, "The greatest common divisor of a and b is d." Think of how the meaning of the symbol "$|$" depends on context. "$|x|$" could name the absolute value of x. "$|X|$" could name the cardinality of set X. "$a \mid b$" could be read "a is a divisor of b." "$\{(x, y) \mid x \leq y + 6\}$" is read "the set of all ordered pairs x, y such that $x \leq y + 6$."

The dependency of meaning on context—where a single expression is used in more than one way—may initially be a source of confusion with the language of mathematics. But there is a pedagogical advantage: Because you must teach students to interpret mathematical expression in light of contexts, students become conditioned to pausing to make sense out of mathematical messages—much in the same way that you condition them to comprehend the task at hand by engaging them in the second stage of an algorithmic-skill lesson (i.e., explanation and practice estimating outcomes).

COMPREHENSION-AND-COMMUNICATION OBJECTIVES

Review the definition of "comprehension-and-communication learning level" and the examples of comprehension-and-communication objectives in Exhibit 4.24. The mathematical content of a comprehension-and-communication objective can be either a *mathematical message* (e.g., the ϵ, δ definition of the limit of a sequence or a textbook's proof of the Pythagorean theorem), or a *type of technical mathematical expression* (e.g., the summation notation). Comprehension-and-communication objectives specifying a mathematical message are concerned with students being able to interpret the meaning of that message. Comprehension-and-communication objectives specifying a type of technical expression are concerned with students being able to make sense out of messages employing that type of expression and also using such expressions in their own communications about mathematics.

LESSONS FOR COMPREHENSION-AND-COMMUNICATION OBJECTIVES

Language Arts Lessons

Lessons for comprehension-and-communication objectives should be designed following principles suggested by the research-based literature on reading and language arts instruction (Connolly & Vilardi, 1989; Siegel, Borasi, & Fonzi, 1998; Vacca & Vacca, 1999)—an aspect of mathematical instruction that has traditionally been neglected (Lindquist & Elliott, 1996). By engaging students in learning activities that lead them to comprehend and communicate mathematical messages, you obviate one of the more mystifying aspects of formalized mathematics, namely, negotiating its language.

Comprehension and Communication of Mathematical Messages

Literal and Interpretive Understanding

Lessons for comprehension of particular messages focus on two levels of understanding: *literal* and *interpretive*. Students *literally understand* a message if they can accurately translate its explicit meaning, as Cathy demonstrates in Case 7.4:

CASE 7.4

Cathy examines the following definition of the absolute value of a real number:

$|x| = x$ iff $x \geq 0$ and $|x| = -x$ iff $x \leq 0$.

She then displays literal understanding of the definition by formulating the following explanation:

> "The absolute value of a number is the number itself, if and only if, the number is positive or zero. The absolute value of a number is its opposite, if and only if, the number is negative or zero."

Students understand a message at an interpretive level if they can infer implicit meaning and explain how aspects of the communications are used to convey the message, as demonstrated by Cathy in Case 7.5:

CASE 7.5

Cathy examines the following definition of the absolute value of a real number:

$|x| = x$ if $x \geq 0$ and $|x| = -x$ if $x \leq 0$.

She then displays interpretive understanding of the definition by extending her previous explanation from Case 7.4 with the following:

> "This means that the absolute value of any number is nonnegative. The absolute value of 10, for instance, is just 10 because, as the definition says, the absolute value of a positive number is the number itself. But for -10, the absolute value is its negative and the negative of a negative is positive, so the absolute value of -10 is 10. Zero is its own opposite, so when the textbook authors wrote the definition, they included it in both cases."

Designing Learning Activities for Literal Understanding

Interpretive understanding depends on literal understanding. Thus, the initial phase of a comprehension-and-communication lesson should promote literal understanding. To design the learning activities, you will need to analyze the message to be comprehended and identify the following prerequisites:

- *Vocabulary* (What common English as well as technical mathematical words will students need to understand so that they can translate the message? Meanings for words, expressions, and symbols are learned through simple-knowledge lessons. Are there any prerequisite simple-knowledge objectives that should be achieved before students are ready for the comprehension-and-communication level learning activities?)
- *Technical expressions* (What shorthand symbols, communication or language structures, and other technical expressions are used to convey the message that the students need to comprehend? Cathy's explanation in Case 7.4 suggested that she understands how to read mathematical definitions. She seems to understand that "if and only if" serves as a two-way implication. Comprehending a definition requires specialized skills, as does comprehending proofs, word problems, graphs, and other types of communication structures.)
- *Concepts* (What concepts does the author of the message assume the students have constructed prior to receiving the communication? In Case 7.5, Cathy's understanding of the definition depended on her prior construction of the concept of nonnegative numbers.)
- *Relationships* (What relationships does the author of the message assume the students have discovered prior to receiving the message? In Case 7.5, Cathy's understanding of the definition depended on her prior discovery of the relationship: A real number and its opposite are equidistant from zero.)

Once these four prerequisites are achieved, literal understanding of a message is effected through a four-stage *direct instructional* lesson.

Stage 1: The Message Is Sent to the Students

For example, the definition of "absolute value" is stated orally and in writing. Ideally, this is a definition the students formulated themselves near the end of a construct-a-concept lesson, but this is not always practical.

Stage 2: The Message Is Rephrased and Explained

For example, "In other words, the absolute value of a number is the difference between the number and zero, but without concern for whether the number itself is greater than or less than zero."

Stage 3: Students Are Questioned about Specifics in the Message

For example, "Is negative x in the second case a positive or negative number?"

Stage 4: Students Are Provided with Feedback on Their Responses to Questions Raised in Stage 3

For example, "I agree that the absolute value of x can never be negative. So if x is negative, negative x is positive."

Designing Learning Activities for Interpretive Understanding

Interpretive understanding of a message is achieved with learning activities that utilize more inquiry and open-ended questions and discussions than the direct instruction for literal understanding. Students are stimulated to examine the message and extract its main idea, data base or facts, assumptions, and conclusions.

Incorporating the Four Stages Into a Comprehension-and-Communication Lesson

In Case 7.6, note how Mr. Matsumoto shifts from direct instructional activities during the literal understanding phase to activities in which students generate ideas during the phase for interpretive understanding:

CASE 7.6

As part of a unit on aspects of number theory, Mr. Matsumoto would like his precalculus students to discover the following relationship:

$\sqrt{2} \notin$ {rationals}

However, he realizes that such an objective is a bit ambitious for this class, so he settles for helping them comprehend a classic proof for why $\sqrt{2}$ is not rational. Thus, his unit includes this objective:

> The student explains the classic proof by contradiction from Euclid's *Elements* (Burton, 1999, pp. 135–174) of the following theorem: $\sqrt{2} \notin$ {rationals} (comprehension and communication).

He thinks, as he designs the lesson for the objective, "I'd better recreate this proof for myself to make sure I thoroughly understand it." He writes out the proof as shown in Exhibit 7.10.

Looking over his work, he thinks, "That's a nice proof; there's a lot going on for them to understand. It's really too gimmicky for them to discover the theorem from it, but they'll gain a lot just from comprehending the argument.

"I'll present it to them by writing it on the overhead as I explain the logic behind each step. But first, I'd better decide on the form of the presentation. This version (i.e., Exhibit 7.10) is too full of symbols for them. I'll go through my checklist of prerequisites for literal understanding first. Okay, vocabulary: They already understand 'rationals,' 'integers,' 'irrationals,' 'relatively prime'—but not that expression for relatively prime (i.e., '$(p, q) = 1$'). I'll write it out for them. They also know the definitions for 'even' and 'odd;' there's no new vocabulary here.

"Now, for symbols: I write out 'p and q are relatively prime' for '$(p, q) = 1$.' Let's see [pause] all the set symbols are familiar to them and so is the implies symbol; I can leave those. '$\ni$' and '$\exists$' have to go. And I'd better write out the names of the sets or else Jim and Blaine will be distracted.

"Concepts: Let's see [pause] the only concept here that they lack is the big one, irrational numbers. They know the definition, but most haven't conceptualized the set. This lesson won't lead to constructing the concept, but it's a step in that direction. There's no need for a lesson on prerequisite concepts.

"Relationships: Oh, oh! They comprehend the definition of rational numbers, but the fact that a rational number can be expressed as the ratio of relatively prime integers is something that most of them haven't really internalized. So, I need to precede this lesson with a discover-a-relationship one on the following:

$x \in$ {rationals} $\Rightarrow \exists\, p, q \in$ {integers} $\ni (p, q) = 1$ and $x = \frac{p}{q}$

To prove $\sqrt{2} \notin$ {rationals}:

Suppose $\sqrt{2} \in$ {rationals}

$\sqrt{2} \in$ {rationals} $\Rightarrow \exists\, p, q \in$ {integers} $\ni (p, q) = 1$ and $\sqrt{2} = \frac{p}{q}$.

$\sqrt{2} = \frac{p}{q} \Rightarrow 2 = \frac{p^2}{q^2} \Rightarrow 2q^2 = p^2$. Thus, $p^2 \in$ {evens}.

Lemma:

To prove $a^2 \in$ {evens} $\Rightarrow a \in$ {evens}.

Proof: Suppose $a \notin$ {evens}

$a \notin$ {evens} $\Rightarrow a \in$ {odds} $\Rightarrow \exists\, i \in$ {integers} $\ni a = 2i + 1$.

$a = 2i + 1 \Rightarrow a^2 = 4i^2 + 4i + 1 \Rightarrow a^2 = 2(2i^2 + 2i) + 1 \Rightarrow a^2 \in$ {odds}

But $a^2 \in$ {evens} $\otimes$

$\therefore a \notin$ {odds} $\Rightarrow a \in$ {evens}.

Thus, $p^2 \in$ {evens} $\Rightarrow p \in$ {evens}.

$p \in$ {evens} $\Rightarrow \exists\, j \in$ {integers} $\ni p = 2j$.

$\therefore 2q^2 = p^2 \Rightarrow 2q^2 = (2j)^2 \Rightarrow 2q^2 = 4j^2 \Rightarrow q^2 = 2j^2 \Rightarrow q^2 \in$ {evens}.

$q^2 \in$ {evens} $\Rightarrow q \in$ {evens}

$\therefore q, p \in$ {evens}

But $(p, q) = 1$ $\otimes$

$\therefore \sqrt{2} \notin$ {rationals}.

Exhibit 7.10
The Proof as Mr. Matsumoto Initially Wrote It.

"That shouldn't be very difficult because it's just a matter of combining two relationships they've already discovered. Okay, the other relationship they need is the one proved by the lemma (see Exhibit 7.10). I'll also prove the lemma ahead of time, to cut down on the length of the main proof. So, I've got two discover-a-relationship lessons and one comprehension-and-communication lesson on the proof of the lemma to conduct before the main one.

"Communication structures: We're okay here. They're already familiar with paragraph-type formats for proofs, as well as proof by contradiction."

After planning and conducting lessons for the following three prerequisite objectives, Mr. Matsumoto is ready for the comprehension lesson:

- The student explains why any rational number can be expressed as the ratio of two relatively prime integers (discover a relationship).
- The student explains why, if the square of an integer is even, then so is the integer (discover a relationship).
- The student explains a proof of the following theorem: $k^2 \in \{\text{evens}\} \Rightarrow k \in \{\text{evens}\}$ (comprehension and communication).

He begins the literal-understanding phase of the lesson by telling students they will be examining one of Euclid's classic proofs and that their job is to understand it well enough to explain why it is a valid proof. He then directs them to copy the proof as he presents it step by step and to raise any questions to clarify the meaning of any step.

After presenting the proof and responding to students' questions, Mr. Matsumoto summarizes the argument and then queries students with exchanges such as the following:

Mr. Matsumoto: Why could the author of the proof make this statement right here that $\sqrt{2} = \frac{p}{q}$, Wendy?

Wendy: There's p and a q like that for any rational number; we showed that yesterday.

Mr. Matsumoto: Thank you. Why is it okay to rewrite $a = 2i + 1$ as $a^2 = 4i^2 + 4i + 1$, Nadine?

Nadine: He just squared both sides of the equation and that's what you get.

Mr. Matsumoto: What does this circled "*X*" mean right here, Jake?

Jake: It must mean you did something wrong.

Mr. Matsumoto: What do you mean by "wrong," Jake?

Jake: You found a mistake you've go to do over.

Mr. Matsumoto: Actually, this indicates a point where we made a statement that contradicts an earlier statement. That's different from being wrong; it's . . .

After completing the activity in which students are given feedback on their responses to Mr. Matsumoto's questions about meanings regarding different aspects of the proof, Mr. Matsumoto begins raising questions to elicit more in-depth ideas about the proof. He begins this interpretive phase of the lesson with the following:

Mr. Matsumoto: I'd like for everyone to silently take nine minutes to outline on a sheet of paper the main ideas of this proof. Use brief phrases for headings and subheadings that illustrate the logic of the argument. Okay, go.

He walks around the room, monitoring the work and selecting papers for the class to discuss. After nine minutes, he asks several students to put their outlines on the board—including Adam's as shown by Exhibit 7.11. Mr. Matsumoto continues the question-discussion session:

Mr. Matsumoto: I really like the way Adam listed each climactic point in the proof, I through VII. Yes, Chen?

Chen: What does cli-whatever mean?

Mr. Matsumoto: "Climactic." Look at what Adam has pulled from the proof. He didn't list, for example, this

Exhibit 7.11
Adam's Outline of Exhibit 7.10's Proof.

Want: To show that $\sqrt{2}$ is not rational

I. Assume $\sqrt{2}$ is rational

II. So $\sqrt{2} = \frac{p}{q}$ so that p and q are relatively prime

III. p is even

IV. q is even

V. III and IV contradict II

VI. So I is false

VII. Conclusion: $\sqrt{2}$ is not rational

step here where we go from "$\sqrt{2} = \frac{p}{q}$" to "$2 = \frac{p^2}{q^2}$."

But he did list that p is even. Why, Chen, do you think he listed one but not the other?

Chen: Because squaring both sides of the equation is just a step leading to something. But p is even is something we are trying to show.

Mr. Matsumoto: I agree. What do we call a part of a story when ideas or events come together to a conclusion of some sort? It sounds somewhat like "climactic," Grace?

Grace: "Climax."

Mr. Matsumoto: The items Adam listed are climax points of the proof leading to the big conclusion he labels "VII." Okay, Julio.

Julio: Proofs aren't stories.

Mr. Matsumoto: Who would like to debate Julio's point? [pause] Okay, Grace, keep it going with Julio.

Grace: Sure proofs are stories. They tell us the story of why something is true or false—like showing that the square of 2 is irrational.

Julio: You mean the square root of 2 is irrational.

Grace: Right. What did I say?

Mr. Matsumoto: Excuse me. You said "square," even though you meant to say "square root." But I want to hear Julio's argument that proofs aren't stories.

Julio: Well I guess that could be, but stories can be fun; proofs are boring.

Grace: Stories can be boring too, but they're still stories. A story can be boring to one person and fun to another. Mr. Matsumoto acts like he enjoys these proofs—some more than others, just like stories.

Julio: Okay. I can see why you think proofs are like stories, but I don't think of them that way.

Mr. Matsumoto: Excuse me. Your debate helped me gain some new insights on proofs. It's true; I really enjoy some proofs—like this one from Euclid—but find others tedious and boring. You've got me asking myself how the ones I enjoy differ from the ones I don't. Anyway, I'm going to think about that some more. But right now, let's get back to this proof. Grace said that Euclid proved that $\sqrt{2}$ is irrational. But we wrote it as "$\sqrt{2}$ is not rational." Are those two statements equivalent? Okay, let's hear from . . .

The discussion about the proof continues for another 12 minutes and then Mr. Matsumoto moves on to another lesson.

Exhibit 7.12
Graphs of History Test Scores From Two Classes.

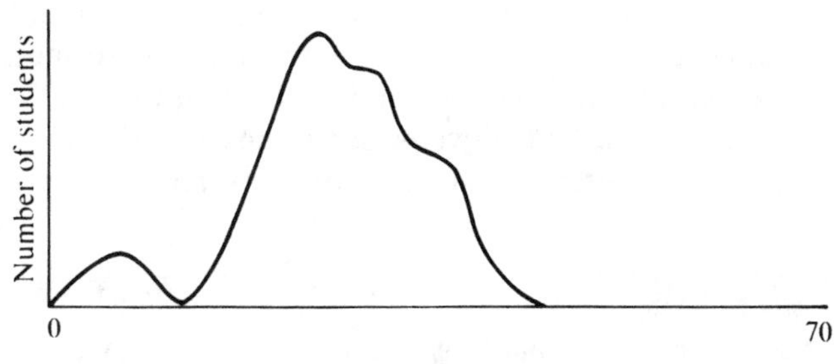

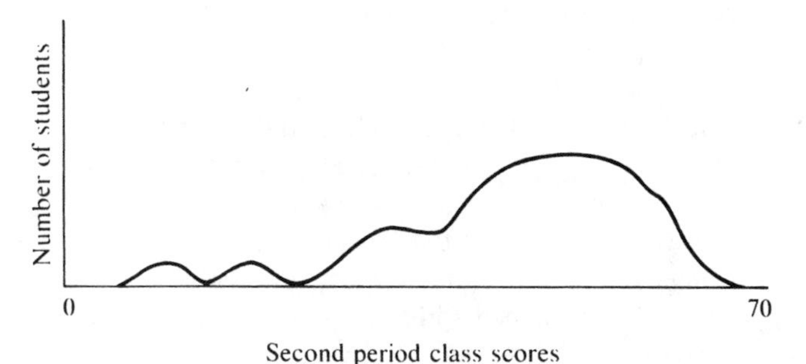

Comprehension of and Communicating With Technical Expressions

Students need to comprehend some of the technical expressions embedded in the language of mathematics so they can receive and send messages about mathematics. Cases 7.7 and 7.8 are examples:

CASE 7.7

Ruth understands the common logic structure of mathematical definitions. Thus, when she confronts the definition of "convex set" for the first time, she knows to look for features common to all convex sets that other sets do not have. Her understanding of mathematical definitions in general serves as an advanced organizer for comprehending unfamiliar ones.

CASE 7.8

Exhibit 7.12 contains the graphs of the scores of two classes of students who took the same history test. By just glancing at it, Edwin reads graphs of real functions in Cartesian planes well enough to note the following: (a) As a group the second-period class scored higher on the test than the first-period class because points on the second period's graph tend to be farther to the right. (b) The two classes contain approximately the same number of students because the areas under the curves are about the same. (c) The first-period scores are more homogeneous than the second-period scores because the scores for the first period are "piled up" and clustered together, whereas the second period scores are spread farther out.

A lesson for comprehension of a technical expression should include learning activities that (a) use direct instruction to inform students about the special conventions of the particular expression specified by the objective, and (b) use inquiry methods to help students develop strategies for using that type of expression. Case 7.9 is an example:

CASE 7.9

Ms. McGiver has planned her prealgebra course so that students confront textbook word problems in every unit. However, after the first two units, she realizes that students are experiencing difficulty solving word problems, not necessarily because they do not understand the particular mathematics applicable to the problem but because they have never developed strategies for solving word problems in general. Thus, she decides to insert a special unit designed to teach general strategies for solving word problems. Her first objective is to get students to comprehend problems as typically presented in textbooks. She states the objective as follows:

> After reading a textbook word problem, the student (a) identifies the question posed by the problem, (b) clarifies the question in her own words, (c) specifies the variable(s) to be solved, and (d) lists facts or data provided in the statement of the problem (comprehension and communication).

Ms. McGiver begins her lesson by displaying the following word problem with the instructions, "Just read this carefully without trying to solve it.":

> Tom is on a television quiz show. He scores five points by correctly answering his first quiz-show question. He then misses the second question and loses 10 points. What is his score after the two questions?

Ms. McGiver knows that everyone in the class can readily solve that problem. So she is not surprised to hear some students blurt, "Aw, that's easy! It's −5." But rather than accept their solutions, she insists that they follow her directions:

Ms. McGiver: Now that you've read the problem, I want you to copy down and answer these questions.
She displays the following on the overhead projector:

1. Solving the problem answers what question?
2. For what variable does answering the question require you to solve?
3. What information are you given with which to work?

After several minutes she begins the discussion:

Ms. McGiver: What is your answer to the first question, Stephanie?
Stephanie: What is the score after the two questions?
Ms. McGiver: Raise your hand if your answer to the first question is significantly different from Stephanie's. Okay, Michelle.
Michelle: I put −5. Isn't that right?

Several students interrupt with, "That's what I got, −5." But Ms. McGiver gives them an icy look, then turns back to Michelle:

Ms. McGiver: What is the question you were to answer?
Michelle: What was the score after the first two questions?
Ms. McGiver: No, read the first question from your paper.
Michelle: "Solving the problem answers what question?"
Ms. McGiver: Now answer that question, Warren.
Warren: What is the score after two questions?
Ms. McGiver: If your answer to the first question doesn't significantly agree with what Stephanie and Warren read, then change it right now so it does. Okay, what is your answer to the second question, Bonita?
Bonita: How many points Bill is behind.
Ms. McGiver: "Behind!" Why behind, Bonita?
Bonita: Because he lost more points than he gained.
Ms. McGiver: Angelo.
Angelo: The variable is how many points he has left. You don't know if he's behind or not until you solve it. She named the variable after she solved for it, so it's no longer a variable.
Ms. McGiver: Angelo, you've made a very insightful point. Thank you. I think Angelo has hit on the reason we're confused. We need to try this on a problem that you won't solve so easily so you'll answer these questions before you know the solution. Here's the problem; answer the same questions for it.

She displays the following:

> An airplane travels 250 miles per hour for 2 hours in a direction of 138° from Albion, NY. At the end of this time, how far west of Albion is the plane?

Some students grumble that the problem is too hard; Ms. McGiver uses a stern look and a gesture to cue them to just answer the three questions. After four minutes, she begins:

Ms. McGiver: Read your answer to the first question, Hartense.
Hartense: How far west of Albion is the plane?
Ms. McGiver: Raise your hand if you disagree. [pause] Okay, what do you have for the second question, Wil?
Wil: How far west the plane is.
Ms. McGiver: Far west from where, Joan?
Joan: From that place in New York.
Ms. McGiver: What kind of variable is it? [pause] Anybody? Okay, Stephanie?
Stephanie: What do you mean?
Ms. McGiver: I mean is it an angle size, distance, weight, or what?
Stephanie: It's a distance.

Ms. McGiver: So what is the variable, Mike?
Mike: The distance from the town.
Ms. McGiver: Okay, what's your answer for the third question, Zeke?
Zeke: I made a list: airplane going 250 mph, 2 hours, at 138°.
Ms. McGiver: Does anyone have more to include? Okay, Joe.
Joe: It's by Albion, New York.
Ms. McGiver: Thank you. Now, let's go back and answer the questions again for the first problem.

After they go through the first problem again, Ms. McGiver explains that they are to answer those three questions for every word problem they work until she notifies them differently. She plans to continue going over this process for comprehending word problems until they appear to do it automatically.

To deepen your insights regarding how to design lessons for comprehension-and-communication objectives, engage in Activity 7.3:

Activity 7.3

Select either a message or a technical expression from the following list:

Messages:

- A passage of your choice from the selection you read when you engaged in Activity 7.2
- A passage of your choice from a middle- or secondary-school mathematics textbook
- The proof of a theorem of your choice from a mathematics textbook
- The definition of a mathematical word or expression of your choice from a mathematics textbook

Technical expressions:

- Histograms
- Box and whisker plots
- Proofs presented in the two-column, statement-reason format
- Summation notation (i.e., "$\sum_{i=a}^{n} f(i)$")

Write a comprehension-and-communication objective that specifies the message or technical expression you selected as its mathematical content. Develop a plan for a lesson designed to lead two or three students to achieve the objective. Discuss your plan with and get feedback from colleagues. Refine your plan as you see fit.

Implement your lesson plan with two or three students. Seek feedback from the students regarding what they learned and how you might make the lesson more engaging for them.

Share your experiences with colleagues who are also engaging in this activity.

Store the lesson plan—annotated with comments about the experience—in the "Cognition, Instructional Strategies, and Planning" section of your working portfolio.

SPEAKING, LISTENING TO, AND WRITING MATHEMATICS

Wanda's behavior in Case 7.10 is similar to behaviors you have observed in classrooms and may remember exhibiting yourself as a student:

CASE 7.10

Ms. Higbee is explaining how two triangles can be similar but not congruent, when Wanda raises her hand:

Ms. Higbee: Yes, Wanda, do you have a question?
Wanda: Yes, why can we [pause] Oh! Never mind, I see now.

Why do you suppose Wanda stopped asking her question? There are a number of possible reasons. One possibility is that the mental activity involved in formulating her question for Ms. Higbee caused her to organize her thoughts in a way that led her to answer her own question. She gained an insight by organizing her thoughts into a *gestalt* (Bourne, Dominowski, Loftus, & Healy, 1986, pp. 24–27; Myers, 1986, pp. 167–169). Articulating words to express thoughts facilitates concept construction, discovery of relationships, comprehension of messages, and problem solving (Coulombe & Berenson, 2001; NCTM, 2000b, pp. 60–63). The role of discourse—verbalizing mathematics—is critical to learning and is an essential aspect of doing mathematics. Teaching students to speak, listen to, and write mathematics involves lessons for comprehension-and-communication objectives that need to be integrated into lessons for construct-a-concept, discover-a-relationship, algorithmic-skill, and application objectives. Consider Case 7.11:

CASE 7.11

Exhibit 7.13 lists the goal and objectives for Mr. Rudd's unit on linear Diophantine equations (i.e., linear equations limited to integer coefficients and solutions) that he is conducting for his precalculus class.

The first time Mr. Rudd designed a unit for the goal listed by Exhibit 7.13, he did not list the three comprehension-and-communication objectives (i.e., B, D and F) that are included in this unit. At the time, he thought it was unnecessary to spell them out because articulating a definition is inherent in the generalizing-and-articulating stage of a construct-a-concept lesson and

Exhibit 7.13

Goal and Objectives for Mr. Rudd's Precalculus Unit on Linear Diophantine Equations.

Unit 11: Linear Diophantine Equations

Goal: Students explore linear Diophantine equations so that they discover fundamental relationships, invent some useful algorithms, and apply them to real-life situations.

Objectives:

A. The student discriminates between linear Diophantine equations and other types of equations (construct a concept).
B. The student formulates a precise definition of "linear Diophantine equation" (comprehension and communication).
C. The student discovers the following relationships regarding two-variable linear Diophantine equations (where a, b and c are constants, x and y are variables, and (a, b) is the greatest common divisor of a and b):
   i) $(a, b) \nmid c \Rightarrow$ the linear Diophantine equation $ax + by = c$ has no solutions.
   ii) $(a, b) \mid c \Rightarrow$ the linear Diophantine equation $ax + by = c$ has a solution.
   iii) $((x_0, y_0)$ is a solution of the linear Diophantine equation $ax + by = c) \Rightarrow ((x_0 + bt, y_0 - at)$ is also a solution for any integer t).
   iv) $((a, b) = 1$ and (x_0, y_0) is a solution to the linear Diophantine equation $ax + by = c) \Rightarrow$ (all solutions of $ax + by = c$ are given by $\{(x, y): x = x_0 + bt, y = y_0 - at \, \forall \, t \in \mathbb{Z}\}$).
   (discover a relationship).
D. The student expresses relationships discovered as propositions in precise mathematical language (comprehension and communication).
E. The student proves propositions she discovers to be theorems (application).
F. The student expresses her proofs of theorems with clarity so that classmates comprehend the arguments (comprehension and communication).
G. The student solves two-variable linear Diophantine equations (algorithmic skill).
H. When presented with a real-life problem, the student determines how—if at all—a solution can be facilitated by linear Diophantine equations (application).

articulating a relationship is inherent in the hypothesizing-and-articulating stage of a discover-a-relationship lesson. However, that initial experience with a previous class taught him that students' struggles with expressing mathematics interfered with their understanding of their own discoveries. Thus, for this current class, he decided to spell out the three comprehension-and-communication objectives to remind himself to explicitly teach for them during the third stages of lessons for Objectives A and C and to emphasize clarity of expression when he teaches for Objective E. He also includes those objectives to remind himself to test for them separately as he monitors students' learning and evaluates their achievement of the unit's goal.

He begins the lesson for Objective A by organizing the class into cooperative-learning teams of about five persons each. Each team collaboratively responds to Exhibit 7.14's tasksheet. How students communicate with one another—articulating and listening to one another's ideas—is essential to the success of the lesson.

One and a half days are spent with the lessons for Objectives A and B with students readily defining "linear Diophantine equation." Mr. Rudd anticipates spending about four days on lessons for Objectives C and D—the centerpiece of the unit.

Mr. Rudd initiates the lesson for Objective C by directing students into the same cooperative-learning teams they were in before to collaboratively respond to the prompts on Exhibit 7.15's tasksheet.

Within each team, students help one another clarify the tasks presented by Exhibit 7.15's prompts. For example:

Sunni: I don't know what we're supposed to do for 1. $14x + 34y = 90$ has two variables. For two variables you need two for a solution.

Dominica: I think he wants us to pick a value for x and then get a y to go with it. Then we get an ordered x, y pair and call that a solution—kind'a like we did yesterday with that other tasksheet with the word problems.

Sunni: Well if it can be just any x, y pair, then all these equations have solutions because you could draw a line graph that would be the solution. All of these equations have lines for solutions.

Roberto: But remember that we're interested only in integer solutions like with Nicole's problem with the shoes for chickens and pigs.

Sunni: And not every point on the line is at a corner on the graph.

Dominica: What do you mean by corner?

Sunni: You know in a graph like this:

Sunni draws Exhibit 7.16 and continues:

Sunni: See, point A would hit at 2, 2 so that'd be a solution. But point B here hits in between—not on a

Exhibit 7.14

The Tasksheet Mr. Rudd's Students Collaboratively Worked With During the Lesson for Objective A.

1. Examine the entries in the following table; determine how those in Column A are alike but different from those in Column B:

A	B
Sunni is in the habit of saving in a jar any 50¢ pieces and silver dollars she gets. Occasionally she rewards herself for work completed (e.g., proving a theorem for precalculus class) by using coins from the jar to buy herself something. *f* is the number of 50¢ pieces and *s* is the number of silver dollars Sunni would have to put in the jar to have a total of $55. What are possible values for *f* and *s*? What would *f* and *s* have to be for the jar to contain exactly 50¢? To contain exactly $3? To contain exactly $7.80?	*l* and *w* are the dimensions of a rectangular field that can be enclosed by 6,000 feet of fencing. What are the possible combinations for *l* and *w*?
How many chickens and pigs can Nicole shoe with a supply of 70 shoes?	At what speed should a car travel for how long to complete a 320.5-mile trip?
How many of each kind of concert tickets must be sold to gross $5,000 if adult tickets sell for $15 apiece and children's tickets sell for $9 apiece?	How many points per minute must Predrag Stojakovic of the Sacramento Kings average and how many minutes must he play per game in order to average at least 15 points per game?

2. Each of the entries in the table includes one or more questions. Each question suggests an algebraic equation. For each entry, formulate an equation or equations suggested by its question. For each equation, determine if it has a solution. If it does, find three.

   Sunni's coin problem:

   The fencing problem:

   Nicole's chickens and pigs problem:

   The traveling car problem:

   The concert ticket problem:

   The Predrag Stojakovic scoring problem:

3. Explain how the equations you formulated for the three problems listed in Column A of the table are like each other but different from the equations you formulated for the three problems listed in Column B of the table.

Exhibit 7.15

The Tasksheet Mr. Rudd's Students Collaboratively Worked With During the Lesson for Objective C.

1. Determine which of the following linear Diophantine equations have solutions:

   $14x + 34y = 90$

   $14x + 35y = 91$

   $14x + 36y = 93$

   $4x + 200y = 10$

   $-4x + 200y = 16$

   $9x + (-12)y = 34$

   $9x + (-12)y = 33$

   $9x + (-12)y = 35$

2. Find two solutions to each of the above linear Diophantine equations that you decided had at least one solution.

3. Formulate (i.e., make up) two linear Diophantine equations that have solutions and two that do not.

4. Make a conjecture (i.e., propose a proposition) about what the relation among a, b, and c needs to be in order for the linear Diophantine equation $ax + by = c$ to have a solution.

5. Consider all the solutions to the following linear Diophantine equation:

   $3x + 5y = 15$

   Fill in the missing cells in the following table of partial solutions for $3x + 5y = 15$ (Note: Ignore *t* for now.):

t	x	y	$3x + 5y$
−4			15
−3			15
−2	−10		15
−1		6	15
0	0	3	
1	5		15
2	10		15
3		−6	15
4		−9	15
5	25		15
6	30	−15	
7			15
8			

6. Using one of the solutions for $3x + 5y = 15$, find a formula for finding all the solutions for $3x + 5y = 15$. (Note: This time, do not ignore *t*.)

7. Consider all the solutions to the following linear Diophantine equation:

   $4x - 3y = 8$

Continued

Exhibit 7.15
Continued

8. Fill in the missing cells in the following table of partial solutions for $4x - 3y = 8$. (Note: Ignore *t* for now.):

t	*x*	*y*	$4x - 3y$
−4			8
−3			8
−2	8		8
−1	5	4	
0	2		8
1	−1	−4	
2		−8	8
3			8
4	−10		8
5	−13		8
6		−24	8
7			8
8			

9. Using one of the solutions for $4x - 3y = 8$, find a formula for finding all the solutions for $4x - 3y = 8$. (Note: This time, do not ignore *t*.)

10. Suppose that you have found one solution (call it "(x_0, y_0)") to a linear Diophantine equation (call it "$ax + by = c$") such that $(a, b) = 1$. Make a up a formula for finding all the solutions to $ax + by = c$.

Exhibit 7.16
Sunni's Drawing to Show That Points in the Cartesian Plane Hit "Corners".

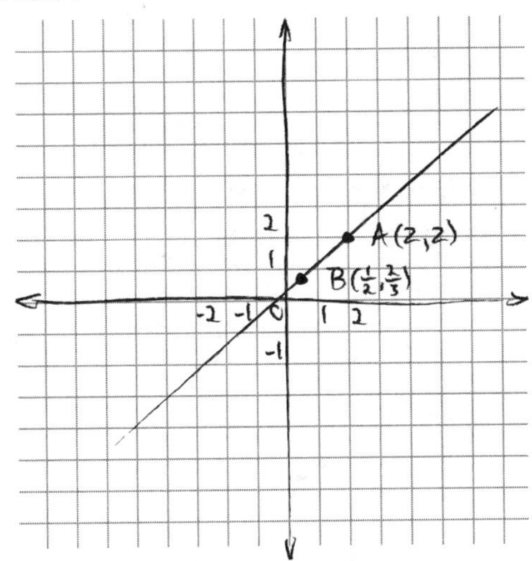

corner—like at about $\left(\frac{1}{2}, \frac{2}{3}\right)$. But I still think every line would hit at least one corner because it's infinitely long.

Roberto: So you're saying all of these equations will have at least one *x*, *y* solution.

Susan: So that takes care of number 1; they all have solutions. Let's go to number 2.

LaDaines: 14 + 34 is 48; that's no good. Let's just try a bunch of numbers on our calculators until we get one.

Dominica: Here, I got 14 × 4 is 56, which is 34 less than 90. So $x = 4$ and $y = 1$ is a solution.

Susan: Good! Let's get the next one.

Roberto: Why don't we just each take a different one. That'll be faster.

Dominica: Okay, Roberto, you take the next one. LaDaines, you get the one after that and . . .

Three minutes later, the group has solutions for only four of the eight equations. They discuss one for which they have not yet solved, $14x + 36y = 93$:

LaDaines: I've tried all kinds of combinations, but none work.
Sunni: There's bound to be one that does. Have you tried combinations with a negative and a positive?
Dominica: Well, maybe it misses all the corners. I think mine does. It's $9x + (-12)y = 34$.
Sunni: It's got to.
Dominica: I don't think so. Look at this: 3 divides 9 and 3 divides -12, but 3 doesn't divide 34.
LaDaines: Sure it does. $34 \div 3 = 11.3333...$
Dominica: Okay, I said it wrong. I mean 3 is not a divisor—you know, like an integer factor—of 34. So it doesn't divide it evenly.
LaDaines: Oh! That's why I couldn't find a solution to mine. 14 and 36 are even, but 93 is odd!
Susan: So maybe not having a solution has something to do with divisors.
LaDaines: Yeah, remember that money problem Rudd gave us last week! We figured out that you couldn't have so many pennies, dimes, and quarters that added up to some amount 'cause it'd make an equation making an even number be an odd number. This is like that.

The discussion continues into the next two days with Mr. Rudd leading students to help one another speak with greater clarity and precision, especially when they began formulating relationships resulting in algorithms for finding all the solutions to linear Diophantine equations. For example, the students are in the process of responding to prompt #6 of Exhibit 7.15's tasksheet and are studying their work reflected by Exhibit 7.17:

Dominica: The *x*s increase by 5s; the *y*s go down by 3s.
Susan: So how do we make a formula out of that?
Sunni: Start with one of the solutions and say, "add by 5s for one and go down by 3s for the other."
Susan: Mr. Rudd is picky; he wants us to say exactly what we mean—like in a formula.
Dominica: It says to use the *t*; he didn't just put that there for nothing. Let's start with 5 and 0. That's the easy one. To get the next solution, you could add 5 to the 5 and -3 to the 0. Then to get the next one you have to add 10 to the 5 and -6 to the 0 and on and on and on.
Roberto: It's always just adding 5 to one *x* to get the next and subtracting 3 from the last *y* to get the next. So the thing that should keep changing in the formula is the *x* and *y* you start with.
Susan: That's going to get too complicated. Let's try to make one formula from just one of the solutions.
Dominica: I still think we've got to use the *t* somehow.

Exhibit 7.17
One Group's Response to Prompt #5 on Exhibit 7.15's Tasksheet.

t	x	y	$3x + 5y$
−4	−20	15	15
−3	−15	12	15
−2	−10	9	15
−1	−5	6	15
0	0	3	15
1	5	0	15
2	10	−3	15
3	15	−6	15
4	20	−9	15
5	25	−12	15
6	30	−15	15
7	35	−18	15
8	40	−21	15

Sunni: I see! If we keep multiplying by *t*s we get 5, 10, 15, and on and on and -3, -6, -12, and on and on. Don't you see?
Dominica: Here it is—just a minute—I've got it. [pause] It's $5 + 3t$ for x and $0 + 3(-t)$ for y.
Roberto: You mean $5 + 5t$ for x and $0 + 3(-t)$ for y.
Dominica: Oh! Right.
Sunni: Why not just say $5 + 5t$ for x and $-3t$ for y? That's the same thing, but shorter.

The discussion continues. The entire unit lasts just over three weeks.

MINIEXPERIMENTS RELEVANT TO STUDENTS' ACHIEVEMENT OF COMPREHENSION-AND-COMMUNICATION OBJECTIVES

Miniexperiments for Comprehension-and-Communication of a Message

A miniexperiment relevant to students' achievement of a comprehension-and-communication objective specifying a particular message as mathematical content prompts students to translate or interpret meanings from the message. What, for example, might be a miniexperiment relevant to the following objective?

> The student explains the following definition of "rational number":

Exhibit 7.18
Example of a Miniexperiment Designed to Be Relevant to How Well Students Comprehend the Definition of Rational Number.

Prompt:

Write "Yes" in the blank for each one of the following statements that can be implied from the definition of "rational number" as given in the textbook's glossary; write "No" in front of each of the other statements:

_______ A. If a number cannot be expressed as a fraction with a numerator an integer and a denominator that is an integer, the number is NOT rational.

_______ B. A whole number is not rational unless it is in the form of a fraction.

_______ C. If $z = \frac{x}{y}$ and x and y are real numbers such that $y \neq 0$, then z is rational.

_______ D. The quotient of two natural numbers is rational.

_______ E. If a number is rational, then it is positive.

Observer's Rubric:

Maximum score is 5 points with +1 for each of the following responses: A–Yes, B–No, C–No, D–Yes, and E–No.

> r is a rational number iff there exist two integers p and q such that $\frac{p}{q} = r$.
> (comprehension and communication).

Exhibit 7.18 displays one possibility.

Miniexperiments for Comprehension and for Communication of a Technical Expression

A miniexperiment relevant to a comprehension-and-communication objective specifying a type of technical expression as mathematical content prompts students to translate or use the technical expression. What, for example, might be a miniexperiment relevant to the following objective?

> The student explains the meaning of expressions using the summation notation (comprehension and communication).

Exhibit 7.19 displays one possibility.

Novelty

Comprehension and communication is a cognitive-learning level involving reasoning and judgment extending beyond what is simply remembered. Similar to miniexperiments for construct-a-concept and discover-a-relationship objectives but dissimilar to simple-knowledge and algorithmic-skill miniexperiments, prompts for comprehension-and-communication miniexperiments present students with tasks that are not identical to ones they have previously encountered. Each prompt needs to have at least some aspect that is novel for the students. The same three questions used in Exhibit 7.19 should not have been previously addressed in class. Of course, during learning activities students should respond to similar but different prompts about the mathematical content of the objective.

To stimulate your thoughts about how to design miniexperiments for comprehension-and-communication objectives, engage in Activity 7.4:

Activity 7.4

Design a miniexperiment to measure the two or three students' achievement of the comprehension-and-communication objective for which you designed and implemented a lesson when you engaged in Activity 7.3.

Try out your miniexperiment with the students. Get feedback from them on how it prompted them to demonstrate how well they comprehend or communicate with the mathematical content of the objective and on the clarity of the directions.

Discuss the experience with a colleague who is also engaging in this activity. Refine the miniexperiment in light of what you learn.

Store the resulting miniexperiment—along with your notes regarding the results of the experiment—as either (a) an attachment to the annotated lesson plan you stored in the "Cognition, Instructional Strategies, and Planning" section of your working portfolio or (b) in the portfolio's "Assessment" section.

Exhibit 7.19
Example of a Miniexperiment Designed to Be Relevant to How Well Students Comprehend the Summation Notation.

Prompt:

Answer each of the following questions, using the area provided:

A. Why is the following statement false?

$$\sum_{i=1}^{3} (i^2 - 1) = (1 - 1) + (4 - 2) + (9 - 3)$$

B. Why can the following expression not be translated?

$$\sum_{j=\frac{1}{2}}^{5} 3j$$

C. Why is the following statement false?

$$\sum_{n=2}^{3} 3^n = 3^2 + 3^3 + 3^4$$

Observer's Rubric:

Maximum score is 12 points distributed according to the following criteria (with +2 for each criterion that is clearly met, only +1 for each criterion that is not clearly met or not met, and +0 for each criterion that is clearly not met):

- For A, the answer indicates that consecutive integers 1, 2, and 3 should be substituted for the index variable only, not for constants.
- For A, the answer includes nothing erroneous (e.g., "The addition should have continued farther.").
- For B, the answer indicates that the function is defined only for integer values for the index variable.
- For B, the answer includes nothing erroneous (e.g., "5 can't be the maximum index value.").
- For C, the answer indicates that the index variable should go from 2 to 3, not 2 to 4.
- For C, the answer includes nothing erroneous (e.g., "The index variable cannot begin with 2.").

SYNTHESIS ACTIVITIES FOR CHAPTER 7

1. For each of the following multiple-choice prompts, select the one response that either completes the statement so that it is true or accurately answers the question:
   A. Which one of the following would NOT be the type of mathematical content specified by a comprehension-and-communication objective?
      a) Statement of a relationship
      b) Definition of a concept
      c) Relationship
      d) Technical expression
   B. Lessons for comprehension-and-communication objectives use _______ .
      a) both direct and inquiry instructional strategies
      b) neither direct nor inquiry instructional strategies
      c) direct instructional strategies only
      d) inquiry instructional strategies only
   C. An acceptable form for presenting the proof of a theorem is an example of a _______ .
      a) constant
      b) discoverable relationship
      c) convention of the Language of Mathematics
      d) shorthand mathematical symbol
   D. Comprehension-and-communication lessons are most appropriate for incorporation into which one of the following stages of a discover-a-relationship lesson?

   a) first
   b) second
   c) third
   d) fourth
2. With a colleague, compare and discuss your responses to the multiple-choice prompts from Synthesis Activity 1. Also, check your choices against the following key: A–c, B–a, C–c, D–c.
3. Examine a section from a mathematics textbook used in a middle, junior high, or high school. Compare that section to a section from one of the mathematical trade books listed by Exhibit 7.9. Discuss with a colleague how the reading-comprehension strategies people need to employ with the textbook differ from those they need with the trade book.
4. Resurrect the documents resulting from your engagement in Synthesis Activity #3 from the end of chapter 6. In light of your work with chapter 7, revise those objectives if you believe you should. Now select one of your comprehension-and-communication objectives. Devise a lesson plan to lead a class of students to achieve that objective. Discuss your lesson plan with colleagues who are also engaging in this activity. If you are able to, teach your lesson to a class of students while a colleague observes for the purpose of giving you feedback. Insert the lesson plan and other pertinent notes resulting from this activity in the "Cognition, Instructional Strategies, and Planning" section of your working portfolio.

TRANSITIONAL ACTIVITY FROM CHAPTER 7 TO CHAPTER 8

In a discussion with two or more of your colleagues, address the following questions:

1. What types of problems do students need to learn to solve in order for them to do meaningful mathematics?
2. How should lessons be designed so that students apply their concepts, discoveries, communication skills, algorithmic skills, and knowledge about mathematics to solve real-life problems? In other words, what strategies should teachers employ to lead their students to achieve application-level objectives?
3. What can teachers do to foster their students' creativity with mathematics?
4. What strategies should teachers employ to lead their students to achieve creative-thinking objectives?
5. What strategies should teachers employ to lead their students to appreciate and enjoy doing mathematics?

chapter

8

Leading Students to Creatively Use Mathematics

Goal and Objectives for Chapter 8

The Goal The goal of chapter 8 is to lead you to develop strategies for designing lessons that lead students to apply mathematics to real-life situations, foster their creativity with mathematics, and develop an appreciation for and willingness to do mathematics.

The Objectives Chapter 8's goal is defined by the following set of objectives:

A. You will distinguish between examples of deductive reasoning and examples of other forms of cognitive behavior (construct a concept) 8%.
B. You will explain how deductive reasoning is used to solve problems (comprehension and communication) 5%.
C. For a given group of middle or secondary students, you will formulate application objectives that are consistent with *PSSM* (NCTM, 2000b) (application) 10%.
D. You will design lessons for application objectives (application) 12%.
E. You will design miniexperiments that are relevant to your students' achievement of application objectives (application) 10%.
F. You will distinguish between examples of divergent reasoning and examples of convergent reasoning (construct a concept) 5%.
G. You will explain strategies for preserving and fostering students' creativity with mathematics (comprehension and communication) 8%.
H. For a given group of middle or secondary students, you will formulate creative-thinking objectives that are consistent with *PSSM* (NCTM, 2000b) (application) 5%.
I. You will design lessons for creative-thinking objectives (application) 5%.
J. You will design miniexperiments that are relevant to your students' achievement of creative-thinking objectives (application) 2%.
K. You will describe how students develop their beliefs about mathematics and acquire the confidence to do mathematics (comprehension and communication) 5%.
L. For a given group of middle or secondary students, you will formulate affective objectives that are consistent with *PSSM* (NCTM, 2000b) (application) 5%.
M. You will design lessons for affective objectives (application) 10%.
N. You will design miniexperiments that are relevant to your students' achievement of affective objectives (application) 5%.
O. You will incorporate the following phrases or words into your working vocabulary: "deductive reasoning," "syllogism," "application," "divergent reasoning," "convergent reasoning," "creative thinking," "synectics," "appreciation," and "willingness to try" (comprehension and communication) 5%.

DEDUCTIVE REASONING FOR PROBLEM SOLVING

Reread Case 4.7.

In Case 4.7, Brenda *applied* certain concepts, relationships, algorithms, and information in an attempt to solve a real-life problem. Problem solving depends on deductive reasoning. When confronted with a question about a specific situation, one *reasons deductively* by deciding how, if at all, a previously learned generality such as a concept or relationship is relevant to that situation. By definition:

> *Deductive reasoning* is deciding that a specific or particular problem is subsumed by a generality. In other words, it is the cognitive process by which people determine whether what they know about a concept or abstract relationship is applicable to some specific situation.

The use of syllogisms is inherent in deductive reasoning. A *syllogism* is a scheme for inferring problem solutions—a scheme in which a *conclusion* is drawn from a *major premise* and a *minor premise.* The major premise is a general rule or abstraction. The minor premise is the relationship of a specific to the general rule or abstraction. The conclusion is a logical consequence of the combined premises. For example:

- *Major premise:* If the discriminant ($b^2 - 4ac$) of a quadratic equation ($ax^2 + bx + c = 0$) is positive and not a perfect square, the equation has two irrational roots.
- *Minor premise:* The discriminate of $x^2 - x - 18 = 0$ (i.e., 73) is positive and not a perfect square.
- *Conclusion:* $x^2 - x - 18 = 0$ has two irrational roots.

Deductive reasoning is the logical foundation of mathematical proofs of theorems. Although they do not typically express their reasoning as formal syllogisms, people use the syllogistic, deductive logic that underlies proofs in real-life problem solving. Consider Case 8.1:

CASE 8.1

Anna is building a playhouse with her children. She has the problem of figuring out how to precut wooden rafter ends so that they will be vertical to the ground when in place. She sketches a diagram with her planned dimensions of the rafters as shown in Exhibit 8.1.

Looking at her diagram, she thinks, "What should α and β be? $\alpha = \beta$ because they're opposite angles of a parallelogram. Okay, so how do I solve for one of them? β has got to be the same as this angle right here; I'll call it Θ." She inserts "Θ" as shown in Exhibit 8.2.

Anna begins to doubt that $\beta = \Theta$, thinking, "Or does it? Let's see, these two lines are parallel, so we have . . .

Exhibit 8.1
Anna's Initial Diagram.

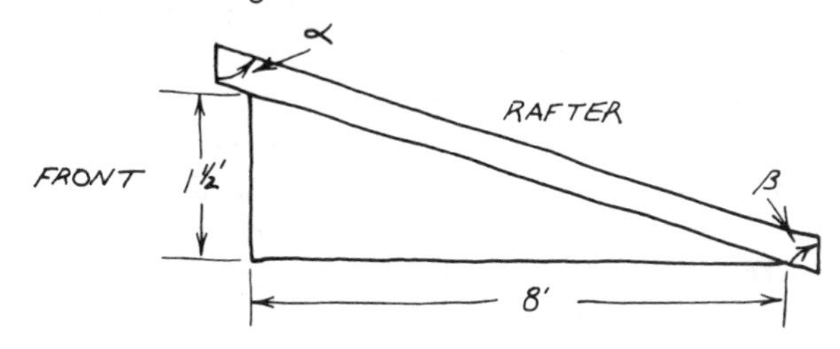

Exhibit 8.2
Anna Inserts Θ.

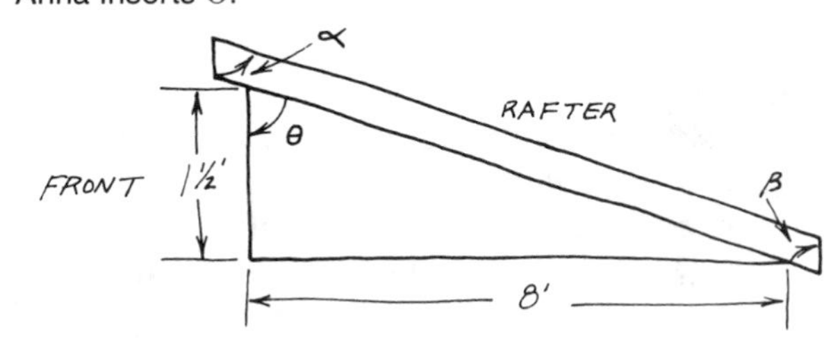

Okay, that's right, $\beta = \Theta$! Alright, I should be able to find Θ, because it's part of this triangle—in fact, it is a right triangle! A right triangle with two known sides—that means I can use a little trig here. Okay, I know Θ's opposite side and its adjacent side. So, tangent is the operable function here. Tangent Θ is 8 over 1.5, so [pause] get this calculator working, arctan, open parenthesis, 8, divide by 1.5, close parenthesis, equals [pause] 1.38544. That's not right! Oh, no wonder, the calculator is set for radians. Okay, switch to degrees and try again. Ahh, [pause] 79.38—that's more like it! Okay, so I cut these two angles at almost 80°."

Anna used several syllogisms to formulate a solution to her problem; here are two of them:

- *Major premise:* Opposite angles of parallelograms are congruent.
- *Minor premise:* The angle with α degrees and the one with β degrees are opposite angles of a parallelogram.
- *Conclusion:* $\alpha = \beta$.

- *Major premise:* The tangent of an acute angle of a right triangle is the ratio of the length of the side opposite to the angle to the length of the side adjacent to the angle.
- *Minor premise:* The angle with Θ degrees belongs to a right triangle whose opposite side measures 8 feet and adjacent side measures 1.5 feet.
- *Conclusion:* $\Theta = \tan^{-1}\frac{8}{1.5}$

APPLICATION OBJECTIVES

Review the definition of "application learning level" and the examples of application objectives in Exhibit 4.24.

When confronted with a problem, a student who has achieved an application-level objective can determine how, if at all, the mathematical content of that objective can be used in a solution to that problem. During application-level lessons, students put into practice previously developed or acquired concepts, relationships, information, and algorithms.

Sometimes teachers confuse algorithmic-skill objectives with application objectives. Compare the following two objectives:

A. Given the velocity of a moving object, the student uses the relationship $rt = d$ (where r = the rate of travel, t = time of travel, and d = the distance traveled) to determine how long it will take to travel a specified distance (algorithmic skill).
B. Given a real-life problem, the student determines how, if at all, using the relationship $rt = d$ (where r = the rate of travel, t = time of travel, and d = the distance traveled) will help solve the problem (application).

Both objectives deal with the same relationship. But Objective B—the application one—requires students to *determine when and how* to use the relationship, whereas Objective A—the algorithmic-skill one—requires students to *remember how to execute the algorithm* based on the relationship. Lessons for application objectives should be designed quite differently from those for algorithmic-skill objectives.

LESSONS FOR APPLICATION OBJECTIVES

Deductive-Learning Activities

A deductive-learning activity is one that stimulates students to reason deductively. An application-level objective is achieved through a four-stage lesson that includes deductive-learning activities: (1st) *initial problem confrontation and analysis,* (2nd) *subsequent problem confrontation and analysis,* (3rd) *rule articulation,* and (4th) *extension into subsequent lessons.*

Four Stages

Stage 1: Initial Problem Confrontation and Analysis

In the initial activity of this stage of the lesson, you confront students with a pair of problems. The pair is chosen so that the problems are very similar except that the mathematical content of the application objective applies to the solution of one of the problems but not the other. Suppose, for example, that the following objective is to be achieved:

> Given a real-life problem, the student decides whether or not a solution is facilitated by computing the area of a polygonal region and, if so, determines how to find that area (application).

The mathematical content of the objective (i.e., area of polygonal regions) is applicable to solving problem A of Exhibit 8.3—but not problem B.

You then engage students in a deductive question-discussion session in which students describe how they would go about solving each problem and then explain why the objective's content was used in one problem but not the other. Mr. Cummings does this in Case 8.2:

CASE 8.2

Mr. Cummings uses problems A and B from Exhibit 8.3 for the aforementioned application objective on polygonal area. He engages in the following discussion with Patsy:

Mr. Cummings: How would you go about solving problem A?
Patsy: I'd divide the wall around the board into rectangles and then find out how many sheets would fit into each rectangle.
Mr. Cummings: But how would you go about finding out the number of sheets to use?
Patsy: I'd have to puzzle in the sheets.
Mr. Cummings: But how would you know how many puzzle pieces to use?
Patsy: By trying it.
Mr. Cummings: But suppose you wanted to find out how many sheets you needed without wasting a bunch of sheets? Is there an easier way?
Patsy: Oh! You mean by computing the area of the wall. Yeah, I could add up the rectangles' areas and divide that by the area of one sheet. Then I'd have it.
Mr. Cummings: Now, what about problem B? How would you solve that one?
Patsy: The same way—compute the area of the wall and divide by the area of one sheet.
Mr. Cummings: Do you really need to compute an area?
Patsy: Oh! No, the length of the crack is what I need to find.
Mr. Cummings: So how would you solve problem B?
Patsy: Just measure the length of the crack and divide that by the height of a sheet and that would give you the number of posters.
Mr. Cummings: Why didn't you compute area for problem B like you did for problem A?
Patsy: B isn't an area problem.
Mr. Cummings: Why not? How is it different from A? They seem pretty similar to me.
Patsy: With A, we had to spread the posters all out. In B, the posters are lined up in just one column.

Exhibit 8.3
Two Contrasting Problems Mr. Cummings Uses in Case 8.2.

A. The front wall surrounding the chalkboard is quite drab. Students suggest that they make decorative posters on standard sheets of cardboard and use them to cover the wall completely, as begun in the picture below on the left. How many posters will be needed to complete the task?
B. There is a nasty-looking crack just above the chalkboard on one side of the wall of the classroom. Students suggest that they make decorative posters and use them to hide the crack, as begun in the picture below on the right. How many sheets will be needed to complete the task?

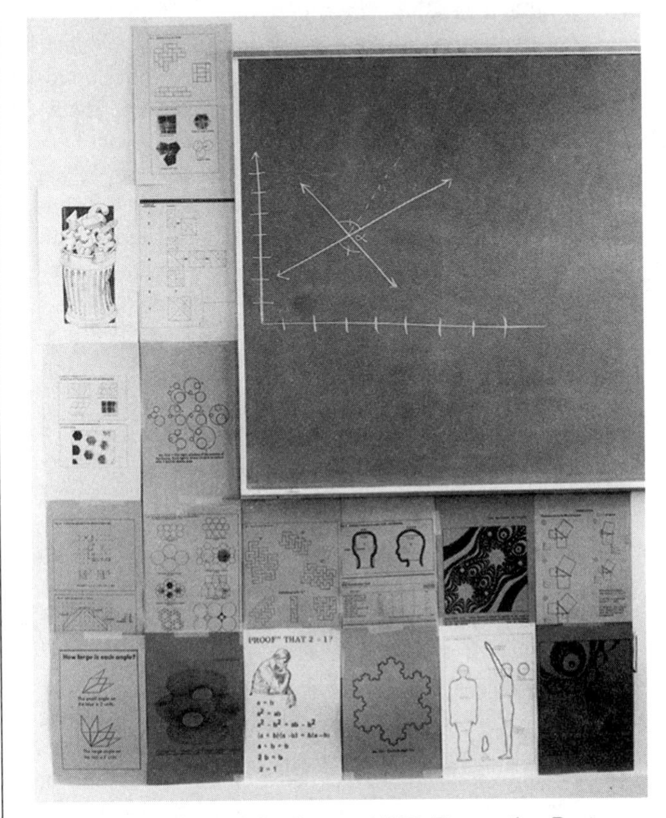

Classroom Wall Partially Covered With Decorative Posters

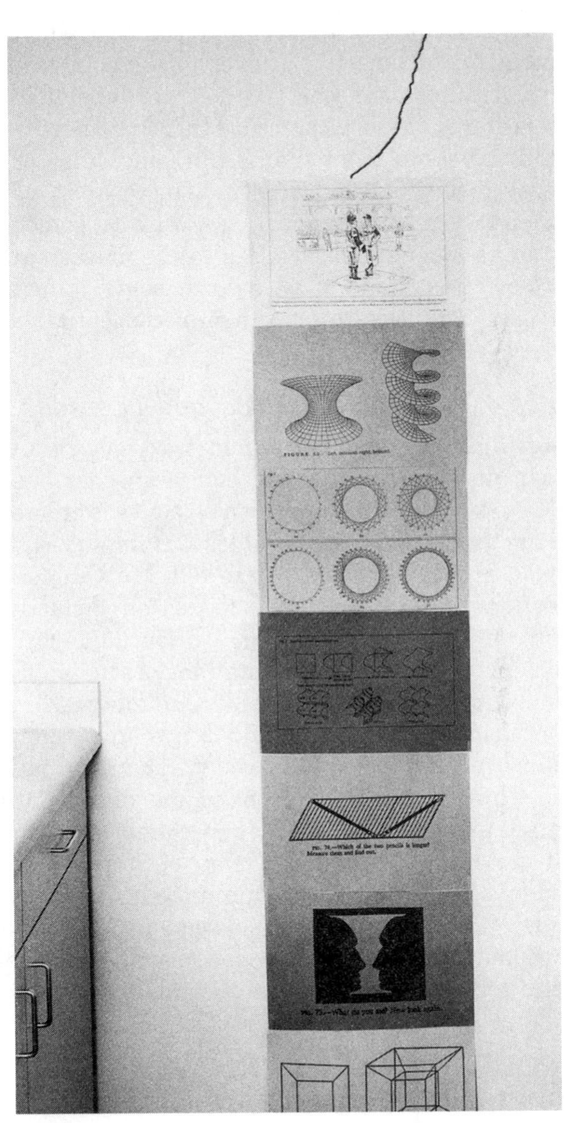

Side Wall of Classroom With Crack Partially Covered by Decorative Posters

Stage 2: Subsequent Problem Confrontation and Analysis

Learning activities for the second stage are similar to those for the first, except students analyze solutions to additional pairs of problems. The pairs are selected so that the students are confronted with a variety of problems each of which they need to determine if the mathematical content is applicable. For the objective on areas of polygonal regions, subsequent problems would include situations other than covering a wall with posters.

How many pairs of problems to include and how difficult each should be depend on how easily students are distinguishing between problems in which the mathematical content applies and those to which it does not. If they readily explain why the content works for one problem but not another, then you should confront them with more difficult problems. On the other hand, select easier problems and delay the start of the third stage if they are experiencing serious difficulties.

Stage 3: Rule Articulation

In this stage, students formulate rules for when the content of the objective applies to the solution of a problem. If relevant construct-a-concept and discover-a-relationship lessons preceded this one, then this often involves no more than rephrasing rules previously constructed or discovered. In cases in which students begin the application level with some conceptual learning gaps relative to the mathematical content, this stage may require some inductive-learning activities to help them reconceptualize or rediscover rules.

Stage 4: Extension Into Subsequent Lessons

Teaching, of course, is an extremely complex art. One lesson does not always end before another begins. Achievement of one application objective is enhanced during the first two stages of lessons for subsequent application objectives. In Case 8.2, Mr. Cummings' questions stimulated Patsy to reason deductively about lengths—a topic from a prior unit—for the problem to which area computation does not apply. In a subsequent unit on volume Mr. Cummings will pair a problem that can be solved via a volume relationship with one that can be solved via an area relationship. Thus, application lessons on one mathematical content extend into those for subsequent mathematical content. A positive side effect of Stage 4 is that students are automatically prompted to do mathematics from prior units in subsequent units. Thus, mathematical content is reviewed and continually connected to new mathematical content.

A FOUR-STAGE LESSON FOR AN APPLICATION OBJECTIVE

Ms. Hundley—whom you first met in Case 3.35—conducts an application lesson in Case 8.3:

CASE 8.3

Exhibit 3.23 depicts what Ms. Hundley calls her "solution blueprint form." She got the idea for the form while attending a NCTM conference (Cangelosi, 1989b); it has proven successful enough for her to incorporate it into most application lessons she now teaches.

Ms. Hundley is beginning a lesson for the following objective:

> Given a real-life problem, the student decides if use of the multiplication principle for independent events, permutation formula, or combination formula will help solve the problem (application).

She directs her students, "Please take out one copy of our solution blueprint form. Now label it 'Jill and Jack's locker security problem' as I display the problem on the overhead." She displays the following:

> Jill and Jack are debating whether it would be better if their school lockers were secured with combination locks or with key locks. Jill says, "Combination locks are better because you don't have to worry about keeping up with your key; you just carry your combination around in your head." Jack says, "Key locks are more secure because a thief could just keep trying your combination until he gets the right one." Jill counters, "The thief would have to try over a hundred combinations before finding yours and, besides, out of hundreds of locks, one key will be able to open more than one."

She begins the question-discussion session:

Ms. Hundley: Let's help settle the debate by finding out whether combination locks or key locks are more vulnerable to being opened by a thief. Everyone, write the question posed by the problem on the form. Read yours, Gene.

Gene: Which type of lock is more secure from a thief, combination or key?

Ms.Hundley: Raise your hand if you disagree with what Gene put. [pause] Okay, then what variables do we have to address, Larry?

Larry: Uhh, wait a minute. [pause] Just how secure the combination lock is and how secure the key lock is.

Ms. Hundley: Everybody, write that down. What other variables are going to affect our results besides those two principal variables, Rembert?

Rembert: Oh, stuff like [pause] ahh [pause] how many thieves there are and a bunch of others.

Ms. Hundley: Thank you. Devon, Marilyn, Lynae, and Wanda each list one. Start, Devon.

Devon: How many keys the thief has.

Marilyn: How many numbers on the combination locks.

Lynae: How many locks out of so many one key will fit.

Wanda: I was going to say the same thing as Marilyn.

Ms. Hundley: Any other variables anyone wants to list? Brent.

Brent: How about the number of numbers in each combination?

Ms. Hundley: Yes, Winne.

Winnie: All our locks use three-number combinations. Let's stick with that.

Ms. Hundley: If it's okay with you, Brent, let's make 3 the constant number for any one lock combination. [pause] Thank you. Considering this list of other variables, I'm going to move us along a little faster by telling you that I did a little research before class. I got Coach Bailey to loan me 100 locks and one randomly selected key. The key opened 5 of the 100 locks. Also, I noted that most locks used in school have 50 numbers on them. So, I suggest that for purposes of the Jill and Jack locker security problem, we control for some of these other variables by delimiting our two principal variables. We solve the problem for one thief operating with one key that'll open 5% of the locks, and that each combination lock has 50 numbers. Assuming you agree to that, what are our two principal variables after they've been delimited, Hank?

Hank: What's "delimited" mean again?

Ms. Hundley: Read the two general principal variables from the second line on your solution blueprint form.

Hank: How secure is the combination lock is one. The other is how secure the key lock is.

Ms. Hundley: Those are two great variables, but there's one difficulty with them if we're going to solve this problem. [pause] What is it, Rembert?

Rembert: They're so general that we can't measure them. We need some mathematical variables.

Ms. Hundley: The delimited variables should be narrower, and—we hope—quantifiable. So, what are the delimited variables, Hank?

Hank: I want to pass. Would you ask somebody else?

Ms. Hundley: Sure. Reed?

Reed: One is the chance that the thief with the one key that opens 5% of the locks will open your key lock. The other one is the chance that the thief who's willing to try a lot of combinations will open your combination lock.

Ms. Hundley: I'll buy that. Everybody get that down. [pause] Could we use a shorthand symbol for those so we can refer to them easily while we work on solutions for each? What do you want to call them, Hank?

Hank: "*K*" and "*C*."

Ms. Hundley: Let's take Hank's suggestion and call the event that the thief opens your key lock "*K*" and opens your combination lock "*C*." And because it's the probability of those events we're interested in, let's use our conventional notation and call our delimited variables "$P(K)$" and "$P(C)$," respectively.

At this point the solution blueprint form has been filled in down to the "solution plan."

Ms. Hundley: After we solve for $P(K)$ and $P(C)$, what should we do next, Winnie?

Winnie: We compare them to see which is greater. If *C* is a more likely event, then Jack is right and key locks should be used. If *K* is more likely, then Jill is right and combinations are better.

Ms. Hundley: Okay, then let's split Jill and Jack's locker security problem into two subproblems: One is to figure $P(K)$ and the other to figure $P(C)$. Let's do $P(C)$ first. How do we go about solving $P(C)$? [pause] Ilone.

Ilone: You've got to figure how many possible combinations there are on the lock.

Ms. Hundley: How many numbers on the lock, Hank?

Hank: 50.

Ms. Hundley: How many numbers in one combination, Tracy?

Tracy: 3.

Ms. Hundley: So, each of you silently figure how many possible lock combinations there are. You have three minutes. [pause] What did you get, Rembert?

Rembert: 19,600.

Ms. Hundley: What did you do to get 19,600, and why did you decide to do it that way?

Rembert: It's a combination of 50 things taken 3 at a time, so I just used the combination function on my calculator.

Ms. Hundley: By the number of hands raised, it looks like a few people want to debate your method. Okay, Kabul, you and Rembert stand up and debate the differences in your approaches.

Kabul: I don't know why they call them "combinations," but I do know the order of the 3 numbers makes a difference. If the lock's combination is 20-30-40, then 40-30-20 won't open it. So it's a permutation, not a combination.

Rembert: So, what's the answer?

Kabul: 117,600.

Ms. Hundley: I see some other folks are anxious to speak. Yes, Marilyn.

Marilyn: But it's not a permutation either. A number can be repeated in a lock combination. What about 20-20-30?

Ms. Hundley: What's the rule for when permutations apply to a situation, Cassandra?

Cassandra: When you want to know how many ways so many things can happen out of so many. But there can't be any repeats in any one way. That's what we said the other day.

Ms. Hundley: And can a lock combination have repeats as 23-14-23, Nancy?

Nancy: Sure; that's why Marilyn is right.

Ms. Hundley: So if permutations don't apply, what does? [pause] Okay, Devon.

Devon: It'd be however many numbers times itself times itself.

Ms. Hundley: Therefore, what? Finish the syllogism for the combination lock problem, Rembert.

Rembert: Therefore, the possible number of lock combinations is $50 \times 50 \times 50$.

Ms. Hundley: So, what's $P(C)$, Kabul?

Kabul: Almost 0; it's $1 \div 50^3$. That's .000008 on my calculator.

Ms. Hundley: Let's move to the other subproblem in Jill and Jack's overall problem. [pause] What's the other variable we need to solve for, Anson?

Anson: $P(K)$.

Ms. Hundley: How should we solve for $P(K)$, Nancy?

Nancy: You said it's 5%. A thief with a key can open 5 out of every 100 key locks. That means key locks are not nearly as secure as combination locks.

Ms. Hundley: Before you compare $P(K)$ to $P(C)$, explain how you got $P(K) = .05$, Nancy.

Nancy: You did it by experimenting.

Ms. Hundley: You mean we couldn't have just used either the permutation formula, the combination formula, or the multiplication principle for $P(K)$ like we did for $P(C)$? [pause] Marie.

Marie: No, it's just one key that fits 5 out of 100 locks. If the thief tries your lock, there's a 5% chance he will get in. There's not so many things to group out of so many—repeating or not repeating.

Ms. Hundley: Oh, grouping! Are you saying if there is no grouping or arrangements, then we don't have a case for permutations, combinations, or the multiplication rule? I understand. Let's go ahead and complete the solution blueprint form for Jill and Jack's overall problem.

Without further discussion, the students complete their forms.

Ms. Hundley continues the lesson by presenting another pair of problems that she considers more difficult than the ones involving $P(C)$ and $P(K)$. However, the students solve them efficiently and appear to distinguish clearly among problems to which combinations apply, permutations apply, the multiplication principle applies, and none of those applies. She thus decides against confronting them with additional problems in class. For homework, she assigns eight more problems. Permutations are applicable to two of them, combinations to two, and the multiplication principle to one other. The other three can be solved by applying mathematics from prior units. She checks the homework to assess if additional learning activities are needed.

The next unit involves binomial distributions. For the application objective on binomial distributions, combinations are needed for binomial distribution problems and permutations are needed for some of the problems that do not involve binomial distributions.

To deepen your insights regarding how to design lessons for application objectives, engage in Activity 8.1:

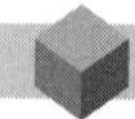

Activity 8.1

Select one of the following objectives:

- Given a real-life problem, the student determines how, if at all, a solution to that problem is facilitated by setting up and solving for a quadratic equation (application).
- Given a real-life problem, the student determines how, if at all, a solution to that problem is facilitated by setting up and solving for a system of linear equations (application).
- Given a real-life problem, the student determines how, if at all, a solution to that problem is facilitated by using probability principles relative to compound events (application).
- Given a real-life problem, the student determines how, if at all, a solution to that problem is facilitated by using relationships derived from triangle congruence postulates and theorems (application).
- Given a real-life problem, the student determines how, if at all, a solution to that problem is facilitated by using the relationship that the ratio of the circumference of any circle to its diameter is π (application).
- Given a real-life problem, the student determines how, if at all, a solution to that problem is facilitated by using the following formula for compound interest where A = the accumulated amount, P = the principal, r = the annual rate, k = the number of times per year the interest is compounded, and n = the number of years:

  $$A = P\left(1 + \frac{r}{k}\right)^{kn}$$

  (application).

Develop a plan for a lesson that is designed to lead two or three students to achieve the objective you chose. Discuss your plan with and get feedback from colleagues. Refine your plan as you see fit.

Implement your lesson plan with two or three students. Seek feedback from the students regarding what they learned and how you might make the lesson more engaging for them.

Share your experiences with colleagues who are also engaging in this activity.

Store the lesson plan—annotated with comments about the experience—in the "Cognition, Instructional Strategies, and Planning" section of your working portfolio.

MINIEXPERIMENTS RELEVANT TO STUDENTS' ACHIEVEMENT OF APPLICATION OBJECTIVES

Deciding How to Solve Problems

A miniexperiment relevant to achievement of an application objective prompts students with a problem and the task of deducing how, if at all, the mathematical content specified by the objective is useful in solving the problem.

Avoiding "Giveaway Words"

Contrast Case 8.4 to Case 8.5:

Prompt:

A group of 3 musicians has 6 instruments that they all can play (a synthesizer, a piano, two different acoustical guitars, a bass, and an electric guitar). How many different instrument combinations can the group play together? Display your work.

Observer's Rubric:

Maximum score is 2 points with +1 for using ${}_nC_r = \frac{n!}{(n-r)!r!}$ and +1 for answering "20."

Exhibit 8.4 One of Ms. Kennedy's Miniexperiments.

CASE 8.4

Exhibit 8.4 displays one of the miniexperiments Ms. Kennedy designed to be relevant to the following objective:

> Given a real-life problem, the student decides if the solution requires determining a combination of n things taken r at a time and, if so, determines how to find that combination (application).

She includes Exhibit 8.4's miniexperiment on a test to measure how well students achieved the goal of a unit involving discrete mathematics. One student, Duane, thinks as he confronts that miniexperiment's prompt, "'How many different instrument *combinations* can [pause] Oh, so this is a combination problem. Let's see, what are the numbers to plug in? Ahh, there's a '3' and a '6.' The rule is the bigger one is n; the smaller one is r—so it's 6! over (6 − 3)!3! which gives me [pause] 20."

Mr. Kennedy records a full two points for Duane's response.

CASE 8.5

Like Ms. Kennedy, Mr. Koebbe includes the same objective stated in Case 8.4 as part of a unit involving discrete mathematics. He includes Exhibit 8.5's miniexperiment on the unit test.

One student, Luanda, thinks as she confronts that miniexperiment's prompt, "'How many different instrument arrangements can the group play together?' [pause] So, what's this instrument arrangement thing? Let's see, it says any set of four instruments they can be playing at once. So, that's like bass, two acoustical guitars, and the piano—that's an instrument arrangement. How many of those are possible? [pause] Aaron is playing piano and Art a guitar, and they switch, does that change the instrument arrangement? No. It's what is being played that counts. So, this isn't a permutation problem. It's a combination. So, okay, how many instruments can they play at once? Seven? No, it's the number of people that count—that's Aaron, Art, Mindy, and Van—four possible out of seven instruments. So, it's a combination of seven instruments taken four at a time, which is [pause] 35. Okay, is that right? Oh, no! Van is the only one that plays the drums. So, drums are in every arrangement of four. I can just ignore drums, so it's really a combination of six instruments taken three at a time, which is [pause] 20."

In Case 8.4 Duane keyed in on the word "combination," remembered the formula, and simply substituted the only data available. He demonstrated algorithmic skill relative to combinations, but he did not have to think deductively to respond correctly to the prompt. Thus, Exhibit 8.4's miniexperiment does not appear relevant to the application-level objective for which it was designed.

On the other hand, Mr. Koebbe in Case 8.5 avoided the word "combination" in the prompt for

Prompt:

Aaron, Art, Mindy, and Van have a musical group with seven instruments (a synthesizer, a piano, two different acoustical guitars, a base, an electric bass guitar, and a drum set). Aaron, Art, and Mindy can play any of the instruments except the drum set. Van can play only drums.

The group refers to any set of four instruments they can be playing together as an "instrument arrangement." How many different instrument arrangements can the group play together? Display your work.

Observer's Rubric:

Maximum score is 2 points with +1 for using ${}_nC_r = \frac{n!}{(n-r)!r!}$ and +1 for answering "20."

Exhibit 8.5 One of Mr. Koebbe's Miniexperiments.

Exhibit 8.5's miniexperiment to increase the chances that students would have to reason deductively—as Luanda did—to make a two-point response. Note, however, that Exhibit 8.5's miniexperiment also taxes students' reading-comprehension skills to a greater degree than Exhibit 8.4's. Because of this unfortunate necessity, it is important to include comprehension-and-communication-level objectives in units and tests for achievement of them separately from application-level objectives.

Extraneous Data

Note that the prompt for Exhibit 8.5's miniexperiment is not only void of giveaway words like "combination," but also includes extraneous information. Having four members and seven instruments in that problem that requires a combination of six things taken three at a time taxes students' abilities to deduce what data are to be used in the formula. If the wording of the problem is such that only exactly what is needed is given, then the miniexperiment is less likely to be at the application level. After all, with a real-world problem, a person is inundated with information, most of which is irrelevant to solving the problem.

Missing Data

Another strategy for designing application-level miniexperiments is to confront students with problems without supplying all of the information needed for the solution. Such miniexperiments help test how well students can deduce what data need to be collected. Keep in mind that in the real world, one must decide what data to collect or what measurements to make to solve problems. In the real world, relevant data are not usually conveniently presented as numerals on a printed page.

To deepen your insights regarding how to design missing-data, application-level miniexperiments, engage in Activity 8.2:

Activity 8.2

Design a missing-data miniexperiment for the following objective:

- Given a real-life problem, the student determines how, if at all, a solution to that problem is facilitated by using the following formula for compound interest where A = the accumulated amount, P = the principal, r = the annual rate, k = the number of times per year the interest is compounded, and n = the number of years:

$$A = P\left(1 + \frac{r}{k}\right)^{kn}$$

(application).

Compare your miniexperiment to those of colleagues who are also engaging in Activity 8.2. Also compare your and your colleagues' work to the miniexperiment shown in Exhibit 8.6.

After refining your miniexperiment if you think it needs to be, store it in the "Assessment" section of your working portfolio.

Mixing Example and Nonexample Problems

As a measure of how well students discriminate between problems to which the objective's mathematical content applies and problems to which it does not, miniexperiments that confront students with example problems as well as miniexperiments that confront students with nonexample problems should be included on tests. The following is an example of a nonexample problem with respect to the application objective on combinations stated in Case 8.4:

> Robbie can play the piano and harmonica. Amanda can play the piano, bass guitar, and lead guitar. Amy can play the piano, harmonica, and lead guitar. The only two instruments that can be played by one person simultaneously are the harmonica and guitar. If the three musicians are to play four different kinds of instruments at the same time, what must Amanda play?

Non-mathematical Topics for Problems

Application miniexperiments confront students with problems to solve. If the nonmathematical aspects of the problem are familiar, students tend to find the miniexperiment easier than one relevant to the same application objective but involving a less familiar nonmathematical topic. For example, students familiar with football will have an easier time visualizing the following problem than students who rarely watch or participate in football:

> At the rate of 9 yards per second, how long will it take a football player to run 10 yards straight down field? At that rate how long would it take the same player to run from the team's 5-yard line to its 15-yard line if the player runs a straight route that makes a 20° angle with the sidelines?

Because you want your application-level miniexperiment to discriminate on the basis of how well students achieved the mathematical learning objective—not how familiar they are with topics such as football—you need to choose nonmathematical topics carefully so that a wide range of interests are included.

To stimulate your thoughts about how to design miniexperiments for application objectives, engage in Activity 8.3:

Exhibit 8.6
Example of a Missing-Data Miniexperiment Designed to Be Relevant to the Objective Listed in Activity 8.2's Prompt.

Prompt:

Several years ago, Riley was shopping for a bicycle. There were two that interested him, one for $650 and another for $375. His dad advised him to buy the one for $375 and put the difference into a savings account. He liked the more expensive one so much more that he bought it anyway. Now, he wonders how much he would have in the bank if he had taken his Dad's advice.

Write a three/four page letter to Riley explaining how to figure out how much he would have in the bank today if he had bought the less expensive bicycle and put the difference in his savings account. Be as detailed as reasonably possible.

Observer's Rubric:

Maximum score is 18 points distributed according to nine 2-point scales with a criterion for each scale so that +2 is awarded if the criterion is clearly met, +1 if it is unclear whether or not it is met, and +0 if it is clearly not met. The nine criteria are as follows:

- The principal is set at $275.
- The question of whether the interest is simple or compound is raised.
- The question of the number of times per year the interest is compounded is raised.
- The process for finding out about how the interest is compounded is explained (e.g., calling the bank or reading a brochure from the bank).
- The question of how long it's been since the bicycle was purchased is raised.
- The process of finding out when the bicycle was purchased is explained (e.g., by looking up the receipt).
- The question of what is the interest rate is raised.
- The process of finding out the interest rate is explained.
- A formula equivalent to the one stated in the objective is suggested.

Activity 8.3

Design a miniexperiment to measure the two or three students' achievement of the application objective for which you designed and implemented a lesson when you engaged in Activity 8.1.

Try out your miniexperiment with the students. Get feedback from them on how it prompted them to reason deductively and on the clarity of the directions.

Discuss the experience with a colleague who is also engaging in this activity. Refine the miniexperiment in light of what you learn.

Store the resulting miniexperiment—along with your notes regarding the results of the experiment—as either (a) an attachment to the annotated lesson plan you stored in the "Cognition, Instructional Strategies, and Planning" section of your working portfolio or (b) in the portfolio's "Assessment" section.

DOING MATHEMATICS CREATIVELY

Some Thoughts on Creativity

Students do mathematics creatively by *reasoning divergently* to originate ideas, conjectures, algorithms, or problem solutions. By definition:

> *Divergent reasoning* is atypical reasoning that deviates from common ways of thinking. It is thought that produces unanticipated and unusual responses.

The antithesis of divergent reasoning is *convergent reasoning*. By definition:

> *Convergent reasoning* is typical reasoning producing predictable responses for most people.

People tend to produce creative ideas in response to dissatisfaction with available resources for dealing with perplexing problems. Contrary to popular belief that aptitude for creative production is found only in rare, exceptional individuals, virtually everyone possesses creative talents (Ormrod, 2000, pp. 133–136; Torrance, 1966). What is rare is for that talent to be recognized and rewarded. Historically, society and its institutions such as schools and churches have frowned upon and generally discouraged creative thinking (Strom, 1969, pp. 222–236; Woolfolk, 1993, pp. 305–310). Divergent reasoning threatens common beliefs. Irrational thought and emotional behaviors are often associated with mental instability. However, Gordon (1961, p. 6) suggested that irrational, emotionally charged thought tends to produce an environment more conducive to creative production than rational, controlled thought. Joyce, Weil, & Calhoun (2000, p. 220) stated, "Nonrational interplay leaves room for open-ended thoughts that can lead to a mental state in which new ideas are possible. The basis for decisions, however, is always rational. The irrational state is the best mental environment for exploring and expanding ideas, but it is not the decision-making stage."

Creativity thrives in an environment in which ideas are valued on their own merit, not on the basis of how they were produced nor who produced them (Strom, 1969, pp. 258–267). In such an environment, irrationally produced ideas are evaluated with the same regard as those resulting from a rational process. The attention afforded an idea should not depend on the eminence of the originator.

Gordon's (1961) studies challenged typical views about creativity with four ideas:

- Creativity is important in everyday circumstances; it should not be associated only with the development of great works.
- Creativity is utilized in all fields, not just the arts.
- Creative thoughts can be generated by groups as well as by individuals alone via similar processes. This is contrary to the common view that creativity must be an intensely personal experience.
- The creative process is not mysterious; it can be described and people can be taught to use it.

Gordon's points are critical to justifying the inclusion of creative-thinking-level lessons in mathematics curricula. However, how best to teach for creativity is still not well understood. One difficulty is resolving the phenomenon that creative thought seems to rise unpredictably (Bourne, Dominowski, Loftus, & Healy, 1986, pp. 9–10).

Preserving Creativity

Studies indicate a steady decline in most students' curiosity and creative activity during their school years; Strom (1969, pp. 259–260) stated:

> Given the great number of children with creative prospect and the fact that it represents a natural evolving process, the first concern among educators ought to be one of preservation. Creativity will develop if allowed to grow, if teachers permit and encourage a course already begun (see Gowan, et al., 1967). A primary clue comes from the process itself—allowing inquiry, manipulation, questioning, guessing, and the combination of remote thought elements. Generally, however, the preferred cognitive style of learning creatively is discouraged (in typical classrooms). Studies indicate that discontinuities in creative development occur at several grade levels and that the losses are accompanied by a decline in pupil curiosity and interest in learning. At the same grade levels at which creative loss occurs, increases are noted in the incidence of emotional disturbances and egregious behavior. Among Anglo-American cultures, the greatest slump in creative development seems to coincide with the fourth grade; smaller drops take place at kindergarten and seventh grade. Children at each of these grades perform less well than they did one year earlier and less well than children in the grade below them on measures of divergent thinking, imagination, and originality. This problem was ignored, since it was judged to be a developmental phenomena instead of man-made or culture-related (Torrance, 1962). Not long ago it was first recognized that in certain cultures the development of creative thinking abilities are continuous. And, even in our own country, under selective teachers who encourage creative boys and girls and reward creative behavior, no slump occurs at grade four.

As a teacher, you can choose not to include creative-thinking objectives in your mathematics curriculum. However, simply managing to preserve students' creativity and allowing it to grow requires some conscious effort on your part.

Fostering Creativity

Consistent with *PSSM* (NCTM, 2000b), you may choose not only to preserve your students' creativity but also to conduct lessons that help them achieve creative-thinking objectives. Creative-thinking lessons can be efficiently interwoven with those for other types of objectives, especially construct a concept and discover a relationship. The strategy is to conduct these other lessons so that students feel free to question, make mistakes, and disagree with ideas—even yours. Particularly important is for students to be positively reinforced for depending on themselves and on their own devices for decision making and problem solving.

Although the creative process is not well understood, some strategies for developing creative thinking appear to succeed (Bourne, Dominowski, Loftus, & Healy, 1986, p. 9: Joyce, Weil, & Calhoun, 2000, pp. 238–240, 314–315). Strom (1969, p. 261) recommended students be exposed to examples of creative production (e.g., through historical accounts of mathematical inventions and discoveries, and through teachers' modeling divergent thinking in think-aloud sessions). Beyer (1987, pp. 35–37) pointed out the importance of heuristic activities such as brainstorming, open-ended question sessions, and discussions in which ideas for consideration are examined regarding purpose, structure, advantages, and disadvantages.

CREATIVE-THINKING OBJECTIVES

What is the fifth term in the infinite sequence (0, 5, 10, 15, . . .)? Most people who comprehend the question reason that the fifth term is 20. They recognize the arithmetic of uniformly increasing multiples of 5 beginning with 0. Such a response requires convergent reasoning because such reasoning produces the expected answer. But suppose a student's reasoning diverges from the usual pattern as in Case 8.6:

CASE 8.6

Ms. Strong: What is the fifth term of the infinite sequence 0, 5, 10, 15, and so forth?
Willie: 26.
Ms. Strong: Why 26?
Willie: Because each number is different from a perfect square by exactly 1.
Willie writes, "$0 = 1^2 - 1$, $5 = 2^2 + 1$, $10 = 3^2 + 1$, $15 = 4^2 - 1$," and continues:
Willie: So the pattern repeats with $n^2 - 1$ once followed by $n^2 + 1$ twice, then another $n^2 - 1$, and so on.

Willie's divergent reasoning justifies 26 for the fifth term just as well as convergent reasoning justifies 20. Do not confuse divergent reasoning with the thinking of the student in Case 8.7:

CASE 8.7

Ms. Strong: What is the fifth term of the infinite sequence 0, 5, 10, 15, and so forth?
Bonnie: 17.
Ms. Strong: Why 17?
Bonnie: I don't know. Did I guess right?

Bonnie's unanticipated answer did not appear to be the result of divergent reasoning.

Review the definition of "creative-thinking learning level" and the examples of creative-thinking objectives in Exhibit 4.24.

The condition of being novel is met as long as the concept, conjecture, algorithm, or solution strategy is completely new to the student himself. A student, for example, displays achievement at the creative-thinking level by originating a method for proving a theorem that is dissimilar to any previous methods to which the students have been exposed—even though the method may have been previously developed unbeknownst to the student.

LESSONS FOR CREATIVE-THINKING OBJECTIVES

Synectics

Metaphors and Analogies

One of the more systematic and researched methods for fostering creativity was referred to by its designer, William J. J. Gordon (1961), as "synectics." *Synectics* is a means by which *metaphors* and *analogies* are used to lead students into an illogic state for situations where rational logic fails. The intent is for students to free themselves of convergent reasoning and to develop empathy with ideas that conflict with their own.

Three types of analogies are used in learning activities based on synectics: (1st) *direct analogies,* (2nd) *personal analogies,* and (3rd) *compressed conflicts.*

Direct Analogies

In the first stage, students make direct analogies by raising and analyzing comparisons between the mathematical content and some familiar idea. For example:

- How is a function like a tossed salad?
- What is the difference between a continuous sequence and frozen yogurt?
- Which is rounder, a hexagon or a television show?

Personal Analogies

In this second stage, students make personal analogies by empathizing with the mathematical content and losing themselves in some imaginary world. For example:

- Imagine yourself as f where $f:(-1, 6) \Rightarrow \{\text{reals}\}$ such that $f(x) = \frac{3}{6 - x}$. Describe how you feel as x moves from 5.75 to 5.83, to 5.97, to 5.99994, nearer and nearer to 6.
- Imagine that you are {rational numbers} and that you must give up one of your infinite subsets to {irrationals}. You get to choose which one to give away. Which one do you give away? Explain why.
- You have just invented a way of constructing an equilateral right triangle. How do you feel about your invention? How will this accomplishment change your life?

Compressed Conflicts

In this third stage, students confront compressed conflicts that involve metaphors with opposing ideas. For example:

- Draw a continuously discrete graph.
- How would mathematics be different if only parallel lines could be perpendicular.
- Show how an infinite set is small.

The metaphors and analogies are used to stimulate students to reconstruct old ideas, thus promoting divergent reasoning.

INCORPORATING SYNECTICS INTO CREATIVE-THINKING LESSONS

Synectics is used in Case 8.8:

CASE 8.8

Ms. Ferney occasionally mixes learning activities for creative thinking with lessons for other types of objectives. Her algebra II class has recently engaged in construct-a-concept lessons on functions and continuity, as well as comprehension-and-communication lessons on the language associated with those concepts. Students have not, however, been introduced to the idea of limit of a function. At this point, she intends to help them achieve the following objective:

> The student generates a variety of novel functions and describes their features including some suggesting the idea of limits (creative thinking).

She directs the students to begin writing down functions of their own design. She insists that at least one of the functions have a domain that is not a set of numbers. After six minutes, she calls a halt to the activity and asks Brook to write one of his functions on the board; he writes the following and returns to his desk:

$q(a) = -\sqrt{|9-a|}$

Ms. Ferney directs the class, "Please take a blank sheet of paper and number it from −3 to 2. Leave about three lines between numerals to write answers in response to prompts I'm about to give you." She lists the following prompts, giving them about two minutes for a response between each:

−3. How is Brook's function like a night light?
−2. If you were Brook's function, why do you suppose you would be accused of being fickle? How would you answer your critics?
−1. Why would anyone call Brook's function a "variable constant"?
0. Write out a question about Brook's function for the class to discuss.
1. Write out another question about Brook's function for the class to discuss.
2. Write out yet another question about Brook's function for the class to discuss.

Ms. Ferney engages the class in a question-discussion session:

Ms. Ferney: What is your answer to question -3, Katrina?
Katrina: A night light is for security and the function is very secure because you can't take the root of a negative number, and by putting in the absolute value, that protects you from having a negative inside the radical.

The session continues with responses for each of the six prompts discussed. Ms. Ferney conducts similar sessions with personal analogies and then compressed conflicts, but she spread them out over several days—interspersing them with other lessons when she judges the class could use a break from more routine activities.

To deepen your insights regarding how to design lessons for creative-thinking objectives, engage in Activity 8.4:

Activity 8.4

Select one of the following objectives:

- The student describes a novel paradigm illustrating the following relationship: $x^a x^b = x^{a+b}$, where $x \in$ {reals} and $a, b \in$ {integers} (creative thinking).
- The student generates novel conjectures about constructions with a straightedge and compass and either proves or disproves them (creative thinking).
- The student invents patterns for novel sequences of numbers (creative thinking).
- The student invents novel structures for illustrating cause-and-effect relationships (e.g., unconventional graphical representations) (creative thinking).
- The student describes a novel arithmetic in which at least some of the conventional algebraic field axioms do not hold (creative thinking).
- The student describes a novel non-Euclidean geometry in which at least some of the fundamental Euclidean postulates are not assumed (creative thinking).

Develop a plan for a lesson that is designed to lead two or three students to achieve the objective you chose. Discuss your plan with and get feedback from colleagues. Refine your plan as you see fit.

Implement your lesson plan with two or three students. Seek feedback from the students regarding what they learned and how you might make the lesson more engaging for them.

Share your experiences with colleagues who are also engaging in this activity.

Store the lesson plan—annotated with comments about the experience—in the "Cognition, Instructional Strategies, and Planning" section of your working portfolio.

MINIEXPERIMENTS RELEVANT TO STUDENTS' ACHIEVEMENT OF CREATIVE-THINKING OBJECTIVES

Unless you devise very unusual curricula for students, relatively few of your objectives specify creative thinking for the learning level. Lessons fostering mathematical creativity tend to be integrated with other lessons and extend beyond the confines of a single teaching unit. You may, for example, include short learning activities based on synectics within most teaching units, but you will detect an increase in students' creative mathematical pursuits only over the course of several units. Consequently, assessing achievement at the creative-thinking level may be more of a long-range endeavor than assessing achievement of other types of cognitive objectives.

Prompts for creative-thinking miniexperiments present students with tasks relative to the objective's mathematical content that require divergent reasoning to accomplish. Rubrics are designed so they reflect divergent rather than convergent reasoning.

Note that the observer's rubrics for simple-knowledge, algorithmic-skill, comprehension-and-communication, construct-a-concept, discover-a-relationship, and application objectives emphasize convergent reasoning (i.e., responses that match previously conceived responses).

Exhibits 8.7 and 8.8 provide examples of creative-thinking objectives, each accompanied by a miniexperiment designed to be relevant to it.

To stimulate your thoughts about how to design miniexperiments for creative-thinking objectives, engage in Activity 8.5:

Activity 8.5

Design a miniexperiment to measure the two or three students' achievement of the creative-thinking objective for which you designed and implemented a lesson when you engaged in Activity 8.4.

Exhibit 8.7

An Example of a Creative-Thinking Objective on Number Patterns and a Miniexperiment Designed to Be Relevant to It.

Objective: The student categorizes numbers in unconventional ways and formulates a rule for each category (creative thinking).

Prompt:

Given $A = \{-\sqrt{3}, \sqrt{-3}, 27, 3, 3.333, \ldots\}$, compose five *distinct* (i.e., no two are equal) subsets of *A* such that each subset contains exactly three elements. For each of the five subsets, write a rule that defines set membership.

Write the rule without actually naming any of the three elements:

1st Subset: ______________________

1st Rule: ______________________

____

____

2nd Subset: ______________________

2nd Rule: ______________________

____

____

3rd Subset: ______________________

3rd Rule: ______________________

____

____

4th Subset: ______________________

4th Rule: ______________________

____

____

5th Subset: ______________________

5th Rule: ______________________

____

____

Observer's Rubric:

Maximum score is 5 points with +1 for each subset-rule pair that fits the criterion established in the directions.

Exhibit 8.8
An Example of a Creative-Thinking Objective on Number Theory and a Miniexperiment Designed to Be Relevant to It.

Objective: The student formulates and proves theorems about subsets of whole numbers (creative thinking).

Prompt:

The number of dots in each of the following figures is called a "triangular number":

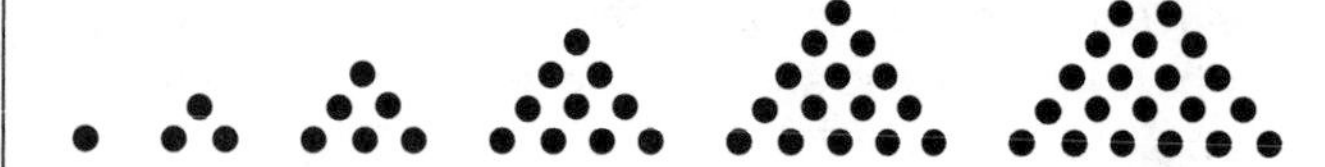

{triangular numbers} is infinite. Take at least 15 minutes to examine triangular numbers. Then make three different statements you think are true about all triangular numbers. These statements should be conjectures that are not immediately apparent from just glancing at the number. For example, "All triangular numbers are positive integers" is too obvious to include. Try to prove your statements. Display your proof or your work toward a proof on a separate sheet and attach it.

1st Statement:

____

____

____

2nd Statement:

____

____

____

3rd Statement:

____

____

____

Observer's Rubric:

The rules are based on comparing responses to those of others. First of all, any blatantly obvious statement (e.g., "No triangular number is imaginary") is eliminated. Then each of the remaining statements is compared to a list of statements compiled from other students who have responded to the prompt. Comparison statements are sequenced from the most frequently occurring to the least frequently occurring. The statement from this student is then ranked in the sequence and given a number of points equal to its rank.

Thus, if there are 50 comparison statements and 20 of them have been made more than once, then if the statement is equivalent to one of the 20, it receives a score from 1 to 20, inclusive. If the statement is equivalent to one of the 30 unique comparison statements, it receives a score of 21. If the statement is not equivalent to any of the 50 statements, it receives a score of 36 (i.e., a three-way tie for 21st place). If the display of the work on the proof demonstrates a discernible line of thought, the statement score is multiplied by 4. If the statement is actually proved, that score is then doubled.

Try out your miniexperiments with the students. Get feedback from them on how it prompted them to reason divergently and on the clarity of the directions.

Discuss the experience with a colleague who is also engaging in this activity. Refine the miniexperiment in light of what you learn.

Store the resulting miniexperiment—along with your notes regarding the results of the experiment—as either (a) an attachment to the annotated lesson plan you stored in the "Cognition, Instructional Strategies, and Planning" section of your working portfolio or (b) in the portfolio's "Assessment" section.

INFLUENCING STUDENTS' ATTITUDES ABOUT MATHEMATICS

Affective Objectives

Unlike cognitive objectives, *affective* objectives are not concerned with students' abilities with mathematical content but rather their attitudes about mathematical content. As indicated in chapter 4, the affective domain includes two learning levels: *appreciation* and *willingness to try.*

Review the definition of "appreciation learning level" and the examples of appreciation objectives in Exhibit 4.24. Achievement of an appreciation-level objective requires students to hold certain beliefs but does not require them to act upon those beliefs.

Review the definition of "willingness-to-try learning level" and the examples of willingness-to-try objectives in Exhibit 4.24. By believing that an understanding of systems of linear equations can help solve problems they care about, students have achieved at the appreciation level. But to learn content at the willingness-to-try level, the student has to act upon that belief by trying to learn about systems of linear equations.

Lessons for Appreciation Objectives

When you teach for an appreciation objective, you are attempting to influence students' preferences, opinions, or desires regarding mathematical content specified by the objective. Students who learn to value mathematical content are intrinsically motivated to increase their skills and abilities with it, and thus achieve cognitive objectives you establish for the unit.

Telling students about the importance and value of certain mathematics is generally ineffectual as a learning activity for an appreciation objective. Consider Case 8.9:

CASE 8.9

Mr. Shaver realizes that if his algebra students appreciate the value of being able to use permutation and combination formulas efficiently, they will be more receptive to achieving the cognitive objectives of his unit on those topics. Thus, his initial objective for the unit is the following:

> The student recognizes the advantages of being able to compute permutations and combinations in real-life situations (appreciation).

In an attempt to lead students to achieve the objective, he tells the class: "Today, we're going to begin studying about permutations and combinations. We need to learn about permutations and combinations so that we can extend our abilities to solve probability problems. Now, I know you'll enjoy working with probabilities because solving probability problems helps us make critical decisions in our lives. Once you understand how to use permutations and combinations, you'll be able to solve some really neat probability problems that'll actually make a difference in your own lives. You're going to enjoy this first activity. First, think about how many ways you can arrange . . ."

In general, students do not learn to appreciate something by being told what they enjoy and will find important (Cangelosi, 2000b, pp. 126–131, 237–245). Rather than wasting time with lip service for his appreciation-level objective, Mr. Shaver should integrate learning activities for the appreciation objective into lessons for his cognitive objectives so that the first few examples used to introduce the content involve situations in which most students have already demonstrated an interest. Questions such as those listed in Exhibit 8.9 can get students' attention and entice them to do mathematics.

At the beginning of a lesson in which a new mathematical content is introduced, tasks to which the mathematical content is applied should be selected so that the value of the new concept, relationship, algorithm, or message is readily demonstrated. For example, if the content is the formula for computing rectangular areas ($A = l \times w$), then which one of the tasks in Exhibit 8.10 would better demonstrate the advantage of having such a formula?

It is just as easy to count the unit squares to find the area of a 2-by-4 rectangle as it is to use the area formula. The value of the formula is apparent for the task of finding the area of the 6-by-16 rectangle, because, with the formula, students need to count only the number of unit squares on two edges rather than all 96 cells.

Similarly, the value of the quadratic formula is demonstrated with the second rather than the first of the two tasks listed by Exhibit 8.11. The first equation is more easily solved via factoring.

Whenever practical, students should construct new concepts and discover new relationships for themselves rather than simply being informed of them. Besides being critical for meaningful learning, construct-a-concept and discover-a-relationship levels of learning have the added benefit of developing in students a feeling of ownership of the mathematical content. Ms. Citerelli's students in Cases 5.7–5.10 are more likely to appreciate arithmetic sequences than students who were simply told about them. Similarly, when Ms. Smith's students from Case 5.13 work with rate relationships, they will be working with mathematics they discovered themselves.

Note how learning activities for an appreciation objective are integrated with a lesson for a discover-a-relationship objective in Case 8.10:

Exhibit 8.9

Questions Identified as Important by Adolescents That Have Been Incorporated Into Problems Addressed in Mathematics Classes (Cangelosi, 2001a).

Art and Aesthetics

- While thinking about how to sketch a picture: At what angles should I make these lines intersect to give the illusion I'm trying to create?
- In deciding how to decorate a room: What color combinations do people tend to associate with being happy?

Cooking

- While planning a meal: How should I expand this recipe so all my guests get enough to eat, but I don't have a lot of food left over?
- What, if any, functions can I formulate (and then write a computer program for) for relating recipe ingredients to output variables such as calories, fat content, nutrients, sweetness, and sourness?

Earning Money

- In considering a fund-raising class project: Would we net more money with a car wash, a bake sale, a "run for donations," used-book sale, or "rent-a-teenager" offer?
- Is this offer to sell greeting cards I just received in the mail a good deal for me?

Electronics

- How can I efficiently link this cable television, videotape recorder, and computer?
- What, if any, functions can I formulate for maximizing amplification of this sound system while minimizing reverberations?

Employment

- Considering time on the job, travel, expenses, opportunity for advancement, security, and benefit from experiences, which of these three jobs should I take?
- Is my paycheck accurate, considering my hours and overtime?

Environmental Concerns

- What's the most efficient way for us to get our message across to the most influential people?
- In preparing for a field trip: How can we minimize our impact on the flora and fauna of the forest?

Family

- In response to the claim that too much time is spent listening to music and watching television and not enough time working on school work and doing chores: How much time do I usually spend a day on each of those four things?
- How can I help my brother manage his time better?

Friends

- Do people really care how their friends dress?
- What factors create friendships?

Gardening and Growing Plants

- What, if any, rules can I formulate (and then write a computer program for) to maximize the growth of beans as a function of soil composition, space, exposure to sun, moisture, etc.?
- What effect does varying the amount and frequency of watering have on plant's health?

Health

- What's the best exercise program for me?
- How should I change my diet?

Managing Money

- How should I go about saving money to buy a car when I'm 16?
- How should I budget my money?

Music

- Who is the hottest music group right now?
- Since I eventually want to work in a rock group, would I be better starting off learning to play the piano, guitar, or drums?

Parties

- How many people should we invite?
- What kind of food should we serve?

Personal Appearance

- What's the best way to treat pimples?
- How do different people respond to "muscular" women?

Personal Planning

- How should I budget my time?
- Would I be better off taking more college prep or business courses in high school?

Pets and Raising Animals

- What kinds and numbers of fish can this aquarium support?
- Is the behavior modification I've started with my cat working?

Politics

- What strategies should we employ to get Allison elected to the student council?
- What can we do to sway people's thinking on this gun-control issue?

Continued

Exhibit 8.9
Continued

School Grades

- What's the relation between the amount of time I study and the grades I get?
- Is it best "cram" the night before a test or spread test preparation out over a longer period of time?

School Subjects Other Than Mathematics

- In response to a problem assigned in science class: How much does it cost to leave a light bulb burning?
- In response to a health and physical education assignment: How many push-ups would I need to do to burn 100 calories?

Social Issues

- Considering the composition of our student body with respect to ethnicity and gender, did ethnic or sex bias influence the outcome of the last school election?
- What can we do to discourage drug abuse in our school?

Sports and Games

- What kind of tennis racquet should I buy?
- What strategy (e.g., regarding lap times) should I use to minimize my time in the 1500-meter run?

Television, Movies, and Videos

- How does gun use in movies compare to gun use in real life?
- In what ways are people influenced by television commercials?

Travel

- What is the most efficient way for me to get from here to Tucson?
- In planning a class trip: Where should we plan to stop along the way?

Vehicles

- Which of these two skateboards is better for speed, control, and durability?
- Regarding a remote-control model car: How are speed, acceleration, maneuverability, and response time affected by battery power and distance between controller and car?

Exhibit 8.10
Which Task Better Demonstrates the Value of the Formula $A = 1 \times w$?

A. Find the area of the following rectangle:

B. Find the area of the following rectangle:

Exhibit 8.11
Which Task Better Demonstrates the Value of the Quadratic Formula?

A. Find the real roots of the following equation:

$x^2 - 4x = 21$

B. Find the real roots of the following equation:

$15x^2 - 7x = 2$

CASE 8.10

The first few lessons of Mr. Polonia's unit on permutations and combinations include learning activities designed to help his algebra students achieve three objectives:

A. The student recognizes the advantages of being able to compute permutations and combinations in real-life situations (appreciation).
B. The student discriminates between examples and nonexamples of each of the following two concepts: permutations and combinations (construct a concept).

C. The student explains why the following relationships hold:

$$_nP_r = \frac{n!}{(n-r)!} \text{ and } _nC_r = \frac{n!}{(n-r)!r!}$$

(discover a relationship).

Mr. Polonia begins the first learning activity by telling the class: "Over the past two weeks, I've kept notes on comments I've overheard students make. Here, I'll show you five of them from students who gave me permission to share their comments with you." He reads each as he displays the following on the overhead projector:

α. "Did you notice that at the [school] dances, they never play two slow songs in a row? I think they're afraid of too much close dancing."

β. "One of us is bound to win the drawing; they pick five winners!"

γ. "Almost every time a teacher picks a group to do something, there are more non-Blacks than Blacks—like Mr. Johnson today, he picked me and two Whites to supervise the drawing."

δ. "Ms. Simmons has never chosen one of my poems for the newspaper."

ε. "You ought to try the lunchroom; there'll always be at least one thing you like."

Mr. Polonia continues: "Tomorrow, we will divide up into collaborative teams with each team assigned to analyze one of these statements for implications and causes. Let's take one now to show you what you'll be doing." He displays Statement γ, then initiates the following discussion:

Mr. Polonia: This statement hints at the possibility of racial bias influencing teachers' selection of student groups and committees. How might we examine the validity of that suggestion?

Theresa: We could keep a record of groups that teachers select over the next month or so and see how often Blacks are in the minority.

Tracy: And if Blacks are in the minority most of the time, then that would show bias.

Eva: I don't think so.

Tracy: Why not?

Eva: We African Americans are a minority in the school, so you'd expect most of the groups would have more non-Blacks.

Milton: I think it's because the teachers always want to have one Black in a group, so they have to spread us out in all the groups.

Mr. Polonia: How often would groups have Black students in the majority if the teachers never considered color when they picked groups?

Don: That's impossible. A few teachers are out-and-out prejudiced, but the others bend over backwards to show they're not.

Mr. Polonia: Maybe so, but if we figured what the numbers would be if the choices were never biased, then we'd have something to compare with the actual choices. Theresa suggests we keep a record.

Tracy: Well, if there were no bias, then the percentage of groups with a majority of African American students should equal the percentage of African American students in school.

Mr. Polonia: Okay, in this class we have 9 Black students and 15 non-Black students. That's—

Eva: 9 out of 24 is 37.5%.

Tracy: So, 37.5% of the groups in this class ought to have an African American majority and the other—

Eva: 62.5%.

Tracy: The other 62.5% should have an African American minority.

Estelle: I don't think it's that simple because . . .

After a few more minutes Mr. Polonia calls a halt to the discussion and directs the students as follows: "I would like for us to continue to work on this problem, but to move us toward developing a model, let's limit the situation for now to selecting groups of three people each from this class. Remember 9 of us are Black and 15 of us are non-Black. The question is, if there is no bias in the selection of a group of three, what are the chances that either two or three of the three will be Black? What's the first thing we need to do to figure that chance?"

Eva: Make a sample space.

Mr. Polonia: The sample space for this problem might be quite long. So, for homework let's divide up the task by having each of you list all possible groups of 3 from the class of which you yourself are a member. Tomorrow, we'll eliminate the duplications, combine the rest, and voilà! We'll have our sample space.

After further clarification of the assignment, the students begin the task, returning the next day to learn that there are many more possible groups of three than they had expected—2,024 in all. In class they complete the arduous task of counting the number of groups for each relevant category and to the surprise of most, discover the following:

> 4% of the groups are all Black, 27% contain 2 Blacks and 1 non-Black, 47% contain 2 non-Blacks and 1 Black, and 22% contain all non-Blacks.

Thus, they conclude that under the no-bias supposition, 31% of the time a group of three would have a black majority. After further discussion regarding the implication of their findings (i.e., how much above or below the 31% figure should be tolerated before the figures are indicative of bias), Mr. Polonia directs their attention to the process by which they obtained the 31% figure. All agree that the process was quite tedious and that they should search for easier ways.

From work with other examples, Mr. Polonia leads the students over the next few days to construct the concepts of permutations and combinations and to discover relationships on which they base algorithms they invent for computing them.

Because Mr. Polonia was concerned with the appreciation objective as well as his cognitive objectives, he carefully chose initial examples that would get students' attention. Once he had them working on a problem, the mathematical content to be taught (i.e., permutation and combination formulas) came as a welcome tool for making their work easier and more efficient.

Lessons for Willingness-to-Try Objectives

Even though students have learned to appreciate certain mathematical content, they still may not attempt to work vigorously with it because they lack confidence that they will use it successfully in situations they find meaningful. Until they have accumulated successful experiences in using mathematics, they tend to be reluctant to pursue problem solutions, as did Mr. Polonia's students in Case 8.10 and Brenda in Case 4.7.

Willingness-to-try objectives, such as the following one, require learning activities similar to appreciation objectives:

> The student attempts to (a) solve problems involving permutations and combinations, and (b) discover models that facilitate efficient solutions to such problems (willingness to try).

But to take students from the appreciation level to the willingness-to-try level, you need to select problem tasks that are interesting enough to maintain their attention and yet easy enough for them to experience success. Keep the following in mind:

- Until students gain confidence in their problem-solving abilities and in the benefits of working on perplexing mathematical tasks, most of the mathematical tasks you assign them should be such that they will experience success before they experience frustration. As their confidence builds, you gradually work in more perplexing and challenging tasks.
- The more a task relates to what already interests students, the more students tend to tolerate perplexity before giving up. It is quite a challenge for you to have to judge that fine line between interest and frustration.
- Achievement of willingness-to-try objectives requires a learning environment in which students feel free to experiment, question, hypothesize, and make errors without fear of ridicule, embarrassment, or loss of status. Recall suggestions from chapter 2's section "Establishing a Favorable Climate for Learning Mathematics."
- By presenting students with problems requiring application of previously acquired mathematical skills and abilities, students not only maintain and improve earlier achievements, they are also afforded opportunities to succeed with mathematics. The four-stage application lessons ensure that students are confronted with problems to which the mathematical content of the objective applies, as well as problems to which mathematical content from previously achieved objectives is applicable. Thus, application lessons provide students experiences with success by including activities in which they apply previously learned mathematics.

To deepen your insights regarding how to design lessons for affective objectives, engage in Activity 8.6:

Activity 8.6

When you engaged in Activities 5.4 and 5.7, you designed a lesson for a construct-a-concept objective and a lesson for a discover-a-relationship objective. Retrieve your plan for one of those lessons. Now, design a lesson for an affective objective that you can integrate into that lesson.

Exchange your integrated lesson plans with that of a colleague who is also engaging in Activity 8.6. Critique one another's work. Store the lesson plan in the "Cognition, Instructional Strategies, and Planning" section of your working portfolio.

MINIEXPERIMENTS RELEVANT TO STUDENTS' ACHIEVEMENT OF AFFECTIVE OBJECTIVES

Choice, Not Ability or Skill

A miniexperiment is relevant to achievement of a cognitive objective when students who have achieved that objective can perform the task presented by its prompt with a higher success rate than those who have not achieved the objective. On the other hand, affective objectives are not concerned with students being able to do anything. Achievement of an appreciation objective is the acquisition of a belief in the value of something. Achievement of a willingness-to-try objective is the acquisition of a tendency to attempt something. The observer's rubric for an affective miniexperiment does not address whether or not students' responses indicate that they *can* perform the task presented, but rather how they *choose* to respond to the task.

Consider the task of designing a miniexperiment for the following objective:

> The student attempts to formulate algebraic open sentences himself when solving word problems before turning to others for help to set up the sentence for him (willingness to try).

To be relevant to this objective, a miniexperiment's prompt must present students with the task of choosing between attempting to formulate an open sentence themselves or having it done for them. One option is to use the *self-report* approach; another is to use the *observational* approach.

The Self-Report Approach

With the self-report approach, you ask students what they would do. Exhibit 8.12 is an example of a miniexperiment designed to be relevant to the aforementioned willingness-to-try objective on formulating algebraic open sentences; it employs the self-report approach.

The value of the self-report approach is limited to situations in which students are confident that they risk nothing by answering honestly. Fortunately, you may want to measure your students' achievement of affective objectives only for formative feedback, not for summative evaluations. Assessments of both their progress relative to appreciating mathematics and willingness to do mathematics provides you with critical formative feedback for regulating lessons. However, consider basing grades only on evaluation of their cognitive mathematical achievements.

The Observational Approach

With the observational approach, you observe students' behavior—either in person or by electronic means—in situations where they are free to make choices that reflect appreciation of or willingness to try mathematics. Exhibit 8.13 is an example of a miniexperiment designed to be relevant to the aforementioned willingness-to-try objective on formulating algebraic open sentences; it employs the observational approach with a computer-administered prompt.

To deepen your insights regarding how to design miniexperiments for affective objectives, engage in Activity 8.7:

Activity 8.7

Design a miniexperiment that you would use as an indicator of students' achievement relative to the affective objective for which you designed a lesson when you engaged in Activity 8.6.

Exchange your miniexperiment with that of a colleague who is also engaging in Activity 8.7. Critique one another's work. Store the resulting miniexperiment as either (a) an attachment to the lesson plan you stored in the "Cognition, Instructional Strategies, and Planning" section of your working portfolio or (b) in the portfolio's "Assessment" section.

Exhibit 8.12
Miniexperiment Employing the Self-Report Approach.

Prompt:

MULTIPLE CHOICE: Suppose that while thumbing through a magazine, you came across one of those brain-teaser type sections in which there was a mathematical word problem to solve. You read the problem and think that with some effort you might be able to solve it by setting up an algebraic equation. You are not sure if you can solve it, but there is a note telling you that a solution is worked out on another page of the magazine. Which one of the following actions are you most likely to take? (Circle the letter in front of your answer.):

A. Work on the problem yourself until you come up with a solution. Only after you come up with the solution do you check with the one given on the other page.
B. You go directly to the solution on the other page rather than try to solve the problem yourself.
C. You neither try to solve the problem yourself nor look at the solution on the other page.
D. You see if it is a kind of problem you already know how to solve; if it is, you solve it yourself before checking with the other page. If it is not one you know how to solve, you do not pay any more attention to the problem.
E. You see if it is a kind of problem you know how to solve; if it is, you solve it yourself before checking with the other page. If it is not one you know how to solve, you do not try to solve it but check with the other page to learn how.
F. You attempt to solve it yourself before checking with the other page. However, if after about five minutes you do not make much progress, you find out how to solve if from the other page.

Observer's Rubric:

Maximum score is 3 points with +3 for selecting A; +2 for selecting F; +1 for selecting D or E; +0 for selecting B or C.

Prompt:

COMPUTER-ADMINISTERED PROMPT: From a bank of word problems, the teacher selects one that he believes the student is capable of solving with some degree of effort (where the equation to formulate is not immediately obvious). The word problem is presented on the computer screen to the student with the following instructions:

> Enter an algebraic equation for solving the given problem. First label the variable. You may request help in setting up the equation anytime in the process by typing HELP to access HELP MODE. After you have received help, the computer will automatically return you to SOLUTION MODE, but you can return to HELP MODE by again typing HELP. Good luck!

Observer's Rubric:

The computer is programed to record the number of seconds spent in SOLUTION MODE and the number of seconds spent in HELP MODE. The score for the item is $S \div H$ where S = the number of seconds in SOLUTION MODE and H = the number of seconds in HELP MODE.

Exhibit 8.13
Miniexperiment Employing the Direct-Observational Approach.

SYNTHESIS ACTIVITIES FOR CHAPTER 8

1. For each of the following multiple-choice prompts, select the one response that either completes the statement so that it is true or accurately answers the question:
   A. With which one of the following tasks do students usually have to deal when solving real-life problems, but not when solving textbook word problems?
      a) Identify and solve for variables
      b) Identify relationships
      c) Remember and execute algorithms
      d) Distinguish between relevant and irrelevant data
   B. Which one of the following do students usually have to determine in order to solve textbook word problems?
      a) The variable to be solved as indicated by the question given in the problem
      b) The implications of the solution outcome
      c) What measurements to make
      d) Whether or not mathematics should be used to solve the problem
   C. Learning activities for which one of the following types of objectives are LEAST likely to be effectively integrated into lessons for other types of objectives?
      a) Willingness to try
      b) Creative thinking
      c) Algorithmic skill
      d) Appreciation
   D. Which one of the following strategies is LEAST likely to enhance students' achievement of an appreciaton objective?
      a) Students use the objective's content to solve problems that concern them.
      b) The teacher tells students how important understanding the content will be for them.
      c) The teacher demonstrates that use of the content can save time.
      d) Students discover and invent mathematics for themselves.
   E. Student perplexity is a critical ingredient in lessons for all BUT which one of the following types of objectives?
      a) Creative thinking
      b) Application
      c) Simple knowledge
      d) Discover a relationship
   F. Lessons for application objectives require _____.
      a) direct instruction
      b) deductive-learning activities
      c) inductive-learning activities
      d) use of mnemonics
   G. Synectics is used in learning activities for what type of objective?
      a) Algorithmic skill
      b) Application
      c) Willingness to try
      d) Creative thinking
2. With a colleague, compare and discuss your responses to the multiple-choice prompts from Synthesis Activity 1. Also, check your choices against the following key: A–d, B–a, C–c, D–b, E–c, F–b, G–d.
3. Resurrect the documents resulting from your engagement in Synthesis Activity 4 from the end of chapter 7. In light of your work with chapter 8, revise those objectives if you believe you should.

Now select one of your application objectives. Devise a lesson plan to lead a class of students to achieve that objective. Discuss your lesson plan with colleagues who are also engaging in this activity. If you are able to, teach your lesson to a class of students while a colleague observes for the purpose of giving you feedback. Insert the lesson plan and other pertinent notes resulting from this activity in the "Cognition, Instructional Strategies, and Planning" section of your working portfolio.

TRANSITIONAL ACTIVITY FROM CHAPTER 8 TO CHAPTER 9

In a discussion with two or more of your colleagues, address the following questions:

1. What strategies should teachers employ to monitor students' learning for the purpose of guiding instruction?
2. What strategies should teachers employ to make accurate summative evaluations of students' achievement of learning goals?
3. How should grades be determined? How should students' progress in mathematics be reported to their parents and to the students themselves?
4. How should individual student portfolios be used to reflect and stimulate mathematical achievement?
5. What role should standardized tests and core-curriculum tests play in mathematics curricula?

chapter

9

Assessing and Reporting Students' Progress With Mathematics

Goal and Objectives for Chapter 9

The Goal The goal of chapter 9 is to introduce you to fundamental principles and strategies for assessing and reporting your students' achievement of learning goals.

The Objectives Chapter 9's goal is defined by the following set of objectives:

A. You will distinguish among examples of measurements, formative judgments, and summative evaluations of students' mathematical progress (construct a concept) 10%.
B. You will explain why a measurement's validity depends on its relevance and reliability and why its usefulness depends on its validity and usability (comprehension and communication) 15%.
C. You will describe common uses and misuses of four types of measurements of students' mathematical achievements: teacher-developed measurements, commercially produced tests, core-curriculum tests, and standardized tests (comprehension and communication) 10%.
D. You will organize a computerized system for miniexperiments to facilitate efficient development of relevant measurements of your students' progress with mathematics (algorithmic skill) 10%.
E. Given the need to make a summative evaluation of students' achievement of one of your unit goals, you will develop a useful measurement of that goal (application) 25%.
F. You will explain the relative advantages of different methods for converting measurement scores to grades (comprehension and communication) 12%.
G. You will develop strategies for using individualized student portfolios as a tool for encouraging and communicating students' mathematical achievements (application) 10%.
H. You will incorporate the following phrases or words into your working vocabulary: "formative judgment," "summative evaluation," "measurement," "planned measurement," "unplanned measurement," "measurement result," "measurement error," "measurement validity," "measurement relevance," "mathematical-content relevance," "learning-level relevance," "measurement reliability," "internal consistency," "scorer consistency," "measurement usability," "measurement usefulness," "traditional percentage grading," "visual-inspection grading," "compromise grading," "individualized student portfolio," "high-stakes test," "standardized test," "standardized test norms," "stanine," "percentile," "grade equivalent," "scaled score," "core-curriculum test," "norm-referenced evaluation," and "criterion-referenced evaluation" (comprehension and communication) 8%.

COMPLEX DECISIONS

A Student's Mind: A Complex of Continua

Think of one of your students; for this example, refer to her as "Amy." Visualize Amy's mind with the following metaphor:

> Exhibit 9.1 is a cutaway diagram of Amy's head displaying the inside of her mind. Within her mind are trillions of networked continua ($T_1, T_2, T_3, \ldots, T_k$). Each continuum (i.e., T_j where $j \in \{1, 2, 3, \ldots, k\}$) is a scale for Amy's level of achievement with respect to a specific learning objective—a learning objective that may or may not have been identified by you, Amy's mathematics teacher. At any one moment in time, Amy's level of achievement with respect to the learning objective associated with j is pinpointed on T_j. See Exhibit 9.2.

Exhibit 9.1
A Cutaway Diagram of the Complex of Continua in Amy's Mind.

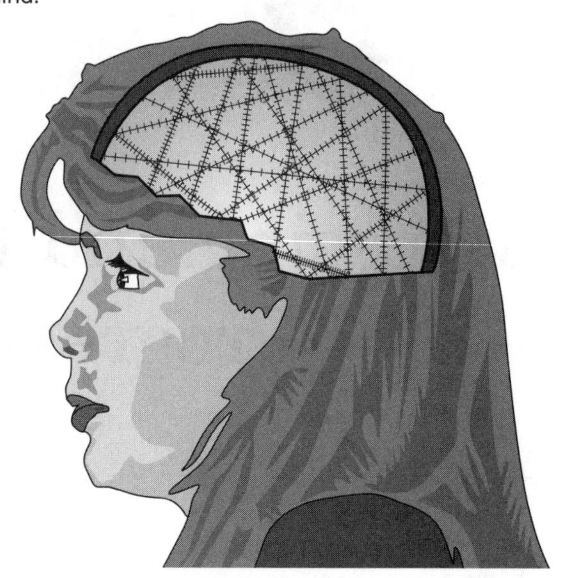

Of course, Amy's achievement level as reflected by T_j varies with time due to influences such as learning and forgetting. To monitor or assess Amy's progress with respect to a specific mathematical learning objective you are targeting with a lesson, you need to examine T_j for the j associated with your objective. That is what you would do to make formative judgments to guide your lesson and to eventually evaluate what Amy learned from the lesson. That is what you would do, if only you could. But even if this were not only a metaphor, performing multiple surgeries to examine and reexamine a T_j for each one of your learning objectives would not be very practical—especially in light of the fact that Amy is not your only student, and achievement of any one learning objective varies from student to student.

However, to monitor the progress of students as you teach and to report what students have learned,

Exhibit 9.2
Each continuum in Exhibit 9.1 Is Associated With a Specific Learning Objective.

T_j indicating Amy's level of achievement with respect to the learning objective associated with j for some $j \in \{1, 2, 3, \ldots, k\}$:

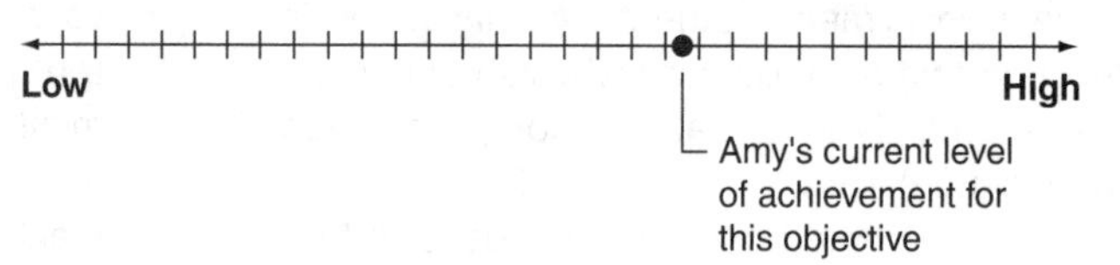

For example:

If the objective associated with $j = 8{,}336{,}504$ is as follows:

The student explains why the following relationship holds:

A is the lateral surface area of a right cylinder with radius *r* and height *h*, then $A = 2\pi rh$.

(discover a relationship).

Then $T_{8,336,504}$ might reflect what Amy learned from your lesson for that objective:

you need to measure achievement levels. Our practical strategies for doing so do not involve surgery, but the metaphor illustrated by Exhibits 9.1 and 9.2 will help us develop those strategies.

Measurements for Formative Judgments

You are the teacher in Case 9.1. But before you read what you do, reacquaint yourself with Ms. Gaudchaux's lesson plan in Exhibit 5.18.

Now you are ready to put yourself in Case 9.1:

CASE 9.1

You have just spent nearly three days conducting the lesson outlined by Exhibit 5.18 for your 26 algebra I students. Regarding the following objective, you have tentatively decided that 18 of the students have achieved it and three have not; you are unsure about the other five:

> The student explains how the values of a and b affect the graph of the linear function $f(x) = ax + b$ (discover a relationship).

You have based this judgment about their achievement on your interactions with and observations of students during the lesson. Here, for example, is an exchange you had in class yesterday:

You: Make up an algebraic formula for a linear function, Naomi.
Naomi: Ahh, how about $3x - 8$?
You: Thank you. May I call it "*n*" for "Naomi?"
Naomi: Sure.
You: So we have $n(x) = 3x - 8$. Now formulate another linear function—lets name this one "*q*" for "Quinton"—with a line graph that's parallel to *n*'s, Quinton.
Quinton: Okay, ahh, $q(x) = -3x + 8$.
You: Yes, Jason.
Jason: That's not going to work because *q*'s *a* is not the same as *n*'s *a*. You've got to have a function like $f(x) = 3x$ plus something.
You: Pick something for *f*'s "something," Jason.
Jason: $f(x) = 3x + 17$.
You: Quinton, please come up to the overhead and use the display calculator to graph *n*, *q*, and *f* simultaneously.

Quinton displays the graphs as shown by Exhibit 9.3, then says:

Quinton: Okay, now I remember, the *a*s have to be the same. But, why?
You: Quinton, please conduct the discussion. Call on someone to answer your question and keep it going.
Quinton: Okay, Dalma.
Dalma: Because the bigger the *a*, the steeper the line.
Quinton: But why?

Exhibit 9.3
Quinton Displays Graphs for $n(x) = 3x-8$, $q(x) = -3x + 8$, and $f(x) = 3x + 17$.

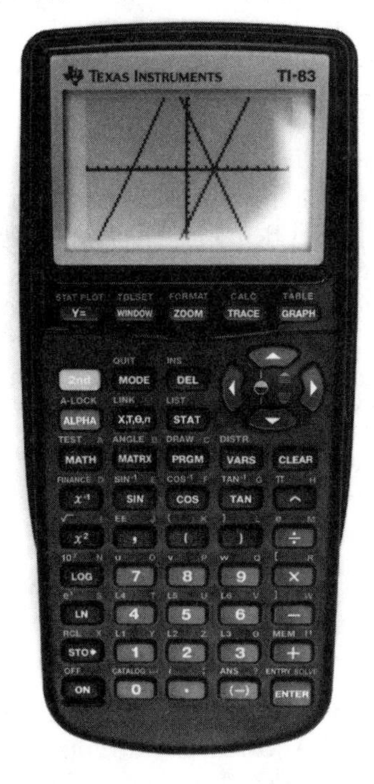

Dalma: Because that's what kept happening when we experimented in our groups the other day. Remember, when *a* was positive, we got a positive slope; when *a* was negative we got a negative slope. And the more negative or positive, the steeper the line. Then we showed that *b* made the line just move up or down.

That exchange left you with the impression that Jason and Dalma had achieved the objective to a greater degree than Quinton, but it did not provide any clues regarding Naomi's progress with the objective.

Now the school day is over and you are thinking about Quinton's question of *why a—x's* coefficient—influences the slope. You think to yourself, "Dalma really didn't explain why *a* influences the line as it does; she simply appealed to the results of our experiment. Although it seems that most students have discovered the influence of *a* and *b* on the line, I doubt that they can explain why. I'm glad Quinton pushed that question. I'll perform an informal miniexperiment at the beginning of class tomorrow to give me a better indication of how well they can explain why *a* and *b* influence the graph as they've discovered they do. If it seems like they need it, I'll jump into a quick lesson to lead them to discover why the relationship is what it is—not just that the relationship exists."

The next day you engage the class in the following question-discussion session:

You: We discovered that for a linear function $f(x) = ax + b$ that the greater $|a|$, the steeper the line that's the graph of *f*. Now, I'd like to return to the question Quinton raised yesterday: Why is that so? Okay, Jason.

Jason: Because every time we jacked up *a*, the line got steeper.

You: I understand. But—assuming now that *a* is positive—why does jacking up *a* cause the slope to increase? Jason.

Jason: Look here on my calculator.

You: Please put it on the overhead so we can all see.

Jason: Okay. [pause] Here's my graph of 2*x*. Now, here's my graph of 10*x*, look how much steeper 10*x* is.

You: Yes, I agree. But what causes that phenomenon to take place? Krystal.

Krystal: Just look at the slope-intercept form in the book. It says "$m = a$," so it's got to be true that the angle of the line depends on *a*.

You then decide to embark on a brief lesson for the following objective:

The student explains why the greater $|a|$, the steeper the line graph of $y = ax + b$ (discover a relationship).

As indicated in chapter 4's section "*PSSM*'s Assessment Principle," a *formative judgment* is a decision made by a teacher that influences how she teaches. As indicated in chapter 5's section "Miniexperiments Relevant to Students' Achievement of Construct-a-Concept Objectives," a *measurement* is a process by which data or information are gathered via empirical observations and those data or information are recorded or remembered. The results of your measurements of student achievement influence your formative judgments. In Case 9.1, you made a number of formative judgments about students' learning that influenced your decision to embark on a lesson for a second discover-a-relationship objective. But what measurement results (i.e., what you heard your students say or saw them do) influenced those formative judgments? To address that question, engage in Activity 9.1:

Activity 9.1

In a discussion with a colleague, address the following questions about your activities in Case 9.1:

1. What are some of the specific measurement results made explicit by Case 9.1 that influenced the judgments you made regarding students' levels of achievement of the objective for which you had just finished conducting a lesson?
2. What are some other possible measurement results that might have influenced your judgment about student achievement of that first objective—measurement results that are not made explicit in the case but which were from measurements you conceivably could have made?
3. What are some of the specific measurement results made explicit by Case 9.1 that influenced the formative judgment you made regarding students' achievement of the second objective—a formative judgment that led you to decide to teach a lesson for that second objective?
4. Although less formal than the miniexperiments you designed while engaging in activities for chapters 5–8, how did the second conversation (i.e., the one beginning with you saying, "We discover that for a linear function . . . Jason.") serve as a miniexperiment that is relevant to students' achievement of the second objective?

Compare your and your colleagues' work with Activity 9.1 to Tina's and Jermain's as reflected by Case 9.2's conversation:

CASE 9.2

Jermain: For Question #1, we need to identify some of the specific measurement results that influenced judgments regarding students' levels of achievement of the first objective. So that refers to the judgments right in the beginning of Case 9.1—the ones about 18 students having achieved it and so forth.

Tina: Okay, so if "measurement results" refer to what I recorded or remembered from empirical observations, then we're talking about things I heard or saw students do during the lesson.

Jermain: So to get specific measurement results, we need to go to this first example where I ask Naomi to give a function.

Tina: But I don't hear anything that influences my judgment about their achievement until Quinton formulates a function that's not going to have a graph parallel to Naomi's.

Jermain: And that influences your—I mean my—judgment that Quinton hadn't achieved the objective as well as some of the others.

Tina: Namely, Jason and Dalma, whose responses to my prompts during the conversation indicated that they understood the relationship.

Jermain: Like Jason picking a function with a graph that's parallel to Naomi's and Dalma saying, "the bigger the *a*, the steeper the line."

Tina: Before now, I really didn't think of just talking to people as making measurements. But I guess what happened here fits the definition of using empirical senses to collect facts—that's like data.

Jermain: It doesn't seem as quantitative as measuring a length with a ruler or giving a test to get scores.

Tina: But it's still data, just not in the form of numerals.

Jermain: As we interact with our students, we pick up memories that influence our judgments and evaluations of what they know and understand.

Tina: Unlike formal tests, they're accidental measurements we didn't plan beforehand to make.

Jermain: Or conduct miniexperiments we designed.

Tina: We're getting ahead of ourselves; it sounds like you're already on Question #4.

Jermain: Okay, Question #2: "What are some other possible measurement results that might have influenced your judgment about student achievement of that first objective—measurement results that are not made explicit in the case but which were from measurements you conceivably could have made?"

Tina: Maybe when I was walking around the room while the class was in collaborative groups working on those tasksheets from that lesson plan [pause] What was it?

Jermain: Exhibit 5.18. I've got it right here.

Tina: Yeah, I observed what they wrote down.

Jermain: But they were just putting down the results of their experiments.

Tina: Well, the more accurate the results of their experiments, the more likely they were to discover the relationships.

Jermain: I guess so, but I'd think what you heard them say during Stage 3 of the lesson would be more indicative of how well they were discovering what you—I mean I—wanted them to discover.

Tina: So, my reading from their tasksheets is still making a measurement; it's just not as relevant to my judgment of their achievement of the objective as my hearing what they said in the discussion.

Jermain: Okay. Some measurements are more valid than others. Let's go to Question #3. For this one we need to identify some of the specific measurement results that influenced the formative judgment I made about students' achievement of the second objective.

Tina: Why are we calling the judgment "formative"?

Jermain: Because it clearly influenced how we taught—deciding to add a new lesson to the unit.

Tina: That helps; thank you. So we need to go to the next day's conversation where I prompt them to say stuff that reveals their understanding or misunderstanding of why the coefficient of *x* influences the graph the way they know it does.

Jermain: So, here you—I mean I—try to prompt students to expose those continua in their heads that are related to the second objective.

Tina: Like in Exhibit 9.1.

Jermain: Right.

Tina: We perform miniexperiments in the classroom—including questions on tests—so that we can infer what those continua in their heads look like without having to cut open their skulls.

Jermain: Getting back to Question #3, I don't think the students had discovered why the value of *a* determined the slope of the line—only that it did. And I believe this because they kept referring back to the results of their experiment when asked why, rather than explaining why *a* determined how much the *y*-coordinate jumps up for each value of *x*.

Tina: Yeah, I was thinking of how a student who really understood *why* would respond to Quinton's initial question. She would probably illustrate the relationship by showing something like this: If a is 3, then the graph goes up 3 for each 1 it moves to the right, whereas if $a = 0.5$, the graph only goes up a half for each 1 it moves to the right.

Jermain: Now, you're the one getting us into Question #4.

Tina: Why?

Jermain: Because you're in the process of creating a rubric to a miniexperiment with a prompt, "Explain why $|a|$ determines the steepness of the line graph of $f(x) = ax + b$."

Tina: Oh, I see what you mean. I was formulating a "correct" response to the prompt.

Jermain: So let's quit resisting Question #4 and answer it.

Tina: I think we already did. In the next day's conversation, I intentionally raised prompts.

Jermain: Like, "Why does jacking up *a* cause the slope to increase?"

Tina: And then I listened to student's responses with a rubric in my mind that I used to categorize what students said as either a positive or negative indicator of achievement of the objective.

Jermain: And that's how we peer into their minds.

As Tina and Jermain pointed out in Case 9.2, the results of *unplanned measurements* as well as *planned measurements* influence your formative judgments of students' learning. By definition (Cangelosi, 2000a, pp. 31–37):

- Unplanned measurements are the continual flow of information from ongoing empirical observations that occur in the course of daily activities. Unplanned measurements influence decision making although they are not deliberately designed for that purpose.
- A planned measurement is a sequence of miniexperiments a teacher deliberately conducts for the purpose of collecting information to be used to make a formative judgment or summative evaluation.

Of course, measurements and miniexperiments do not give us completely accurate indicators of the relevant continua in students' minds as illustrated by Exhibits 9.1 and 9.2. Thus, the results of a measurement of students' achievement of the objective associated with *j* should be interpreted in light of the following relationship:

> $S_j = T_j + E_{S_j}$ where S_j is the result of the measurement, T_j is the student's actual true achievement level of the objective, and E_{S_j} is the error of measurement associated with S_j.

There are practical techniques for estimating the error of measurement (i.e., E_{S_j} of tests you use in evaluating your students' achievement of learning goals). However, those techniques are not within the scope of this book as they are for a text on assessment practices for teachers (e.g., *Assessment Strategies for Monitoring Student Learning* [Cangelosi, 2000a]). For purposes of this chapter, you need to be aware of measurement error in interpreting measurement results and also to conceptualize *measurement validity*—a topic for a subsequent section.

To lead you to develop strategies for conducting lessons so that you are better able to make unplanned measurements as well as conduct miniexperiments for making ongoing formative judgments, engage in Activity 9.2:

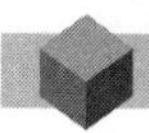

Activity 9.2

With a colleague, reread Case 3.12. Discuss some of the specific strategies Mr. Heaps employed that facilitated his making unplanned measurements and conducting miniexperiments for monitoring students' progress during the learning activity.

Resurrect one of the lesson plans stored in the "Cognition, Instructional Strategies, and Planning" section of your working portfolio. Discuss how you would conduct some of the learning activities outlined by that plan so that you are in a position to make unplanned measurements and conduct miniexperiments that help you make informed formative judgments as you teach. If your reflections suggest that you should modify the lesson plan, then modify it before returning it to your working portfolio.

Measurements for Summative Evaluations

Case 9.3 is a continuation of Case 9.1:

CASE 9.3

In Case 9.1, you were working in the first half of a unit on the behavior of linear functions. You are near the end of the unit, having taught lessons for all of its objectives. Exhibit 9.4 displays those objectives along with the goal they define.

For each student, you are responsible for assigning a letter grade that reflects your summative evaluation of that students' achievement of the unit's goal. Using a procedure explained in a subsequent section of this chapter (i.e., "Designing Measurements of Students' Achievement of Unit Goals"), you develop the test document shown by Exhibit 9.5. Exhibit 9.6 shows the scoring key with the rubrics for the miniexperiment prompts that appear on the test.

You spend three days administering the test. Because of the complexity of the directions and because of the wide variability among your students' reading-comprehension levels, you orally explain the directions for

Exhibit 9.4

Objectives Defining the Goal for Your Unit on the Behavior of Linear Functions.

Unit Goal:

Students discover characteristics of linear functions and apply those discoveries to address real-life problems.

Unit Objectives:

A. The student distinguishes between relationships that are better described by linear functions and relationships that are better described by nonlinear functions (construct a concept) 12%.
B. The student explains why the graph of $f(x) = ax + b$, where a and b are real constants, is a line (discover a relationship) 8%.
C. The student explains how the values of a and b affect the graph of the linear function $f(x) = ax + b$ (discover a relationship) 8%.
D. The student explains why the greater $|a|$, the steeper the line graph of $y = ax + b$ (discover a relationship) 8%.
E. The student explains why b affects the location of the line graph of $y = ax + b$ (discover a relationship) 8%.
F. The student expresses linear functions in the form "$f(x) = ax + b$," and from that expression determines f's slope, y-intercept, and x-intercept (algorithmic skill) 12%.
G. The student incorporates the following terms into her working vocabulary: "linear function," "slope," "y-intercept, and "x-intercept" (comprehension and communication) 9%.
H. The student explains how the behavior of linear functions differs from the behavior of higher degree functions (discover a relationship) 8%.
I. The student interprets among standard English expressions of linear relationships, algebraic expressions of linear relationships, and graphical expressions of linear relationships (comprehension and communication) 12%.
J. Given a problem, the student (i) determines whether or not the solution is facilitated by formulating a linear function and graphing it, and if so, (ii) formulates the function and interprets the graph (application) 15%.

Exhibit 9.5

Test Document With Prompts for Miniexperiments You Designed to Be Relevant to Exhibit 9.4's Objectives.

Algebra I: Opportunity to Demonstrate What You Learned
During Our Unit on Linear Functions
Part I (Monday, October 14)

1. What is your name? ____________________________

2. ORAL AND WRITTEN DIRECTIONS: During our class meeting on Tuesday, Sept. 24, we experimented with different one-variable first-degree equations to show that the graph of some first-degree functions is a line. Formulate four one-variable first-degree equations to demonstrate that the graph of the following function is a line:

   $f(x) = 8x + 3$

   Do not directly graph *f*. Instead use graphs of your four equations for the demonstration.

   State your four equations in the box on the right.

   After graphing your four equations below, write several sentences explaining how the graph of your four equations relates to the graph of *f*.

1)	2)
3)	4)

   Write your explanation in the following box:

   Your explanation:

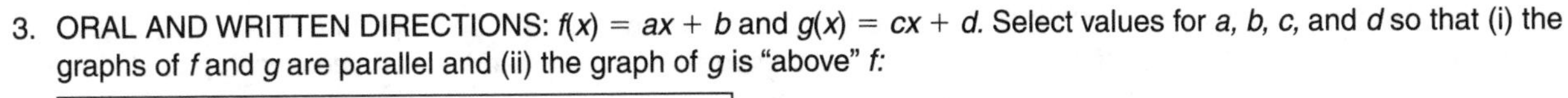

3. ORAL AND WRITTEN DIRECTIONS: $f(x) = ax + b$ and $g(x) = cx + d$. Select values for *a, b, c,* and *d* so that (i) the graphs of *f* and *g* are parallel and (ii) the graph of *g* is "above" *f*:

   $a =$, $b =$, $c =$, and $d =$

   In two sentences, explain your reasons for selecting the values you chose for *a* and *c:*

   Your explanation:

Continued

Exhibit 9.5
Continued

In two sentences, explain your reasons for selecting the values you chose for *b* and *d:*

Write your explanation in the following box:

Your explanation:

4. ORAL AND WRITTEN DIRECTIONS: Write a one-paragraph explanation why the graph of $f(x) = 2x$ is a steeper line than the graph of $g(x) = 0.5x$. To make your explanation clear, draw on and refer to the close-up shot of the coordinate plane that appears just below your explanation:

Your explanation:

	(1,2)		
	(1,0.5)		

(0,0)

5. ORAL AND WRITTEN DIRECTIONS: Write a one-paragraph explanation why the graph of $h(x) = x + 2$ "stays above" the graph of $k(x) = x + 0.5$. To make your explanation clear, draw on and refer to the close-up shot of the coordinate plane that appears just below your explanation:

Your explanation:

Continued

Exhibit 9.5
Continued

(1,3)			
(0,2)			
	(1,1.5)		
(0,0.5)			

6. Smile; you have finished taking advantage of Part I of this opportunity.

Algebra I: Opportunity to Demonstrate What You Learned
During Our Unit on Linear Functions
Part II (Tuesday, October 15)

7. What is your name? ____________________

8. WRITTEN DIRECTIONS: Solve for the slope, y-intercept, and x-intercept of $12x - 3y = 24$:

 The slope =

 The y-intercept =

 The x-intercept =

9. WRITTEN DIRECTIONS: What is the y-intercept and the x-intercept of a linear function whose graph includes the following points: (10, 14102.5302), (3, 14338.1502), (428.97,0), (−3, 14540.1102), (0, 14439.1302)?

 The y-intercept =

 The x-intercept =

10. DIRECTIONS: Using your graphing calculator, graph $f(x) = 2x$ and $g(x) = x^2$ on the same screen.

 A. In a paragraph, explain why the graph of f includes points below the horizontal axis, but the graph of g does not:

 Your explanation:

 B. Compute each of the following:

 $f(1) - f(0.5) =$ _____ $g(1) - g(0.5) =$ _____

 $f(2) - f(1) =$ _____ $g(2) - g(1) =$ _____

Continued

Exhibit 9.5
Continued

C. Notice that for $x > 0$, f increases at the same rate along a straight line. But for $x > 0$, the rate of increase for g changes with the graph curving upward "faster and faster." Write a paragraph explaining why the rate of increase for f remains steady but for g the rate of increase changes as x gets larger. Use the results of your computations from Part B in your explanation:

Your explanation:

D. Once again, compare the two graphs on your calculator. Notice that $g(x) < f(x)$ when $0 < x < 2$. But when $x > 2$, $g(x) > f(x)$. Write a paragraph explaining why this happens.

Your explanation:

11. Smile; you have finished taking advantage of Part II of this opportunity.

Algebra I: Opportunity to Demonstrate What You Learned
During Our Unit on Linear Functions
Part III (Wednesday, October 15)

12. What is your name? ________________________________

13. ORAL AND WRITTEN DIRECTIONS: Read each one of the following four situations and decide which one of the graphs that appears below best fits the situation. No graph may be selected more than once. For each situation, write the Roman numeral corresponding to the graph of your choice in the blank. In a brief paragraph, explain why you chose the graph you chose for the situation.

A. Every Sunday for many weeks Fred deposits the same exact amount of money into his piggy bank. During that time, no money is removed nor are any other deposits made. One of the choices is a graph relating two variables: *Time* on the x-axis and the *amount of dollars in the piggy bank* on the y-axis. Which graph is it? _____

Your explanation:

Continued

Exhibit 9.5
Continued

B. One of the choices is a graph relating two variables: *Time* on the *x*-axis and the *amount a person, Agatha, weighs during the first 30 years of her life* on the *y*-axis. Which graph is it? _____

Your explanation:

C. An automobile is traveling faster and faster at a constant rate of acceleration. One of the choices is a graph relating two variables: *Time during the period of the constant rate of acceleration* on the *x*-axis and the *speed of the automobile* on the *y*-axis. Which graph is it? _____

Your explanation:

D. An automobile is traveling slower and slower at a constant rate of deceleration. One of the choices is a graph relating two variables: *Time during the period of the constant rate of deceleration* on the *x*-axis and the *speed of the automobile* on the *y*-axis. Which graph is it? _____

Your explanation:

Continued

Exhibit 9.5
Continued

THE GRAPHS FOR YOU TO CHOOSE:

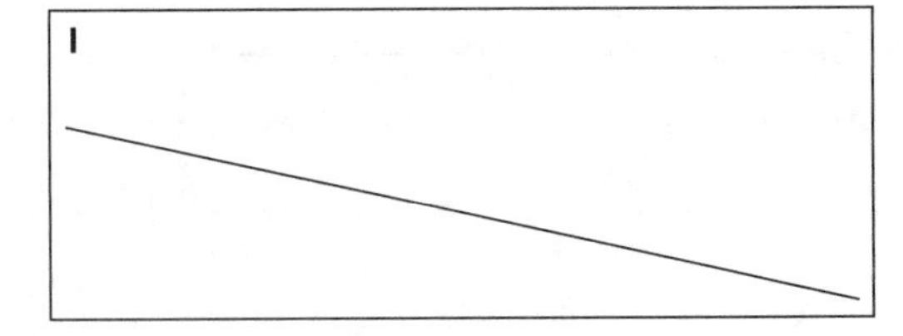

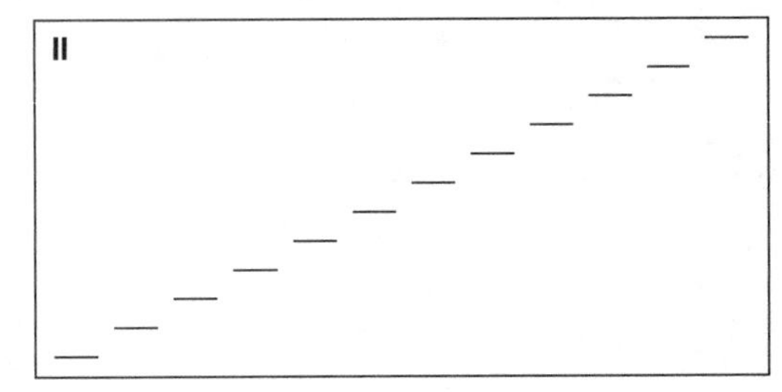

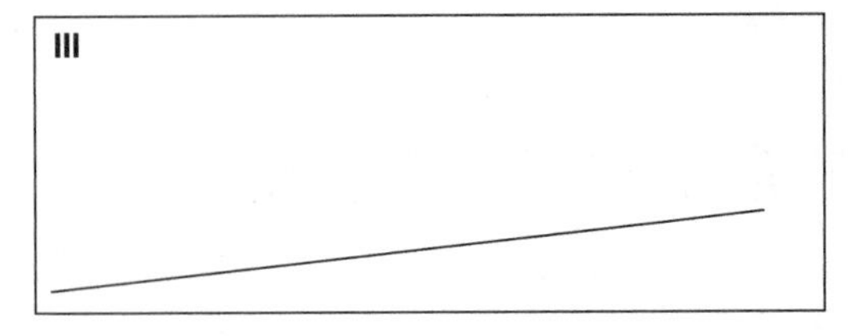

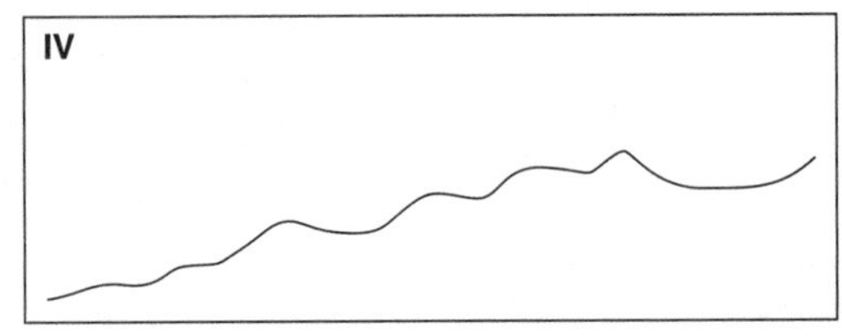

14. ORAL AND WRITTEN DIRECTIONS: It is 2 P.M. when a red hot-air balloon is 150 meters above the ground descending at a steady rate of 20 meters per minute. At the same time, a yellow hot-air balloon is only 10 meters above the ground rising at the steady rate of 15 meters per minute. The temperature of the air just above the ground is holding steady at 24°C. Draw a graph to illlustrate how high the balloons are above the ground between 2 P.M. and when the red balloon lands.

    You may use your graphing calculator, but transfer your graph to the grid below. Label the horizontal axis and the vertical axis and indicate the size of the unit for the horizontal scale and the size of the unit for the vertical scale:

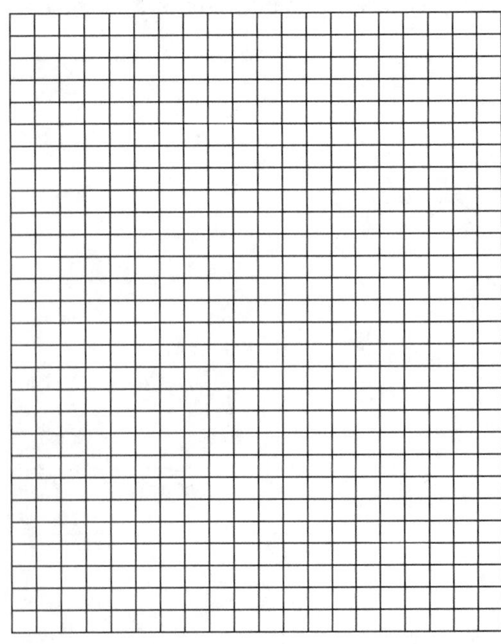

Continued

Exhibit 9.5
Continued

Please answer the following questions:

A. What size unit did you use for the horizontal scale? ____________

B. What size unit did you use for the vertical scale? ____________

C. What variable did you plot on the horizontal axis? ____________

D. What variable did you plot on the vertical axis? ____________

E. At what time will the balloons be the same distance above the ground? ____________

F. How many meters above the ground will the balloons be when thay are at the same altitude? ____________

G. At what time will the yellow balloon land? ____________

H. How high will the yellow balloon be when the red balloon lands? ____________

15. Smile; you have finished taking advantage of this opportunity.

Exhibit 9.6
Scoring Key With Rubrics for Exhibit 9.5's Miniexperiments.

Scoring Form for ____________ (name)

Opportunity Relative to Unit on Linear Functions

Rubric for Response to Prompt #2 (9-point maximum):

+2 or +0 for writing four different equations $y_1 = 8x + 3$, $y_2 = 8x + 3$, $y_3 = 8x + 3$, and $y_4 = 8x + 3$ where y_1, y_2, y_3, and y_4 are any four constants.
+1 or +0 for plotting the solution to the first on the graph of $y = y_1$.
+1 or +0 for plotting the solution to the second on the graph of $y = y_2$.
+1 or +0 for plotting the solution to the third on the graph of $y = y_3$.
+1 or +0 for plotting the solution to the fourth on the graph of $y = y_4$.
+2 or +0 for explaining that the four points associated with the solutions of the four equations lie on the graph of *f*.
+1 or +0 for pointing out that the four plotted points are collinear.

_________ Score for response to Prompt #2

Rubric for Response to Prompt #3 (4-point maximum):

+1 or +0 if $a = c$.
+1 or +0 if $b < d$.
+1 or +0 for explaining why $a = c$ causes the lines to be parallel.
+1 or +0 for explaining why $b < d$ causes the graph of *g* to be "above" the graph of *f*.

_________ Score for response to Prompt #3

Rubric for Response to Prompt #4 (4-point maximum)

+1 or +0 for explaining that each time *x* increases by 1, $f(x)$ becomes twice as large as the new *x*, whereas $g(x)$ only increases to half of the new *x*.
+1 or +0 for associating the 2 to 1/2 relationship to the difference in the steepness of the lines.
+1 or +0 for using the graph to illustrate how the difference in steepness continues.
+1 or +0 if the explanation includes nothing erroneous or extraneous.

_________ Score for response to prompt #4

Rubric for Response to Prompt #5 (3-point maximum):

+1 or +0 for explaining that each time *x* increases by 1 the corresponding values of $h(x)$ and $k(x)$ remain 1½ units apart.
+1 or +0 for using the graph to illustrate how the difference in the locations of the lines remains the same.
+1 or +0 if the explanation includes nothing erroneous or extraneous.

_________ Score for response to Prompt #5

Continued

Exhibit 9.6
Continued

Rubric for Response to Prompt #8 (4-point maximum):

+1 or +0 for indicating that the slope is 4.
+1 or +0 for indicating that the *y*-intercept is −8.
+1 or +0 for indicating that the *x*-intercept is 2.
+1 or +0 for expressing all three answers accurately (e.g., a response of "(0, −8)" would indicate that the student computed the *y*-intercept correctly, but didn't express it accurately possibly because she doesn't comprehend the definition of "*y*-intercept").

_________ Score for response to Prompt #8

Rubric for Response to Prompt #9 (3-point maximum):

+1 or +0 for indicating that the *y*-intercept is 14439.1302.
+1 or +0 for indicating that the *x*-intercept is 428.97.
+1 or +0 for expressing both answers accurately.

_________ Score for response to Prompt #9

Rubric for Response to Prompt #10 (32-point maximum):

For each criterion listed for responses to Prompts A, C, and D, points are distributed as follows: +2 if the criterion is clearly met; +1 if it is unclear as to whether or not the criterion is met; + 0 if the criterion is clearly not met.

For the responses to Prompt B: +1 for each computation result that matches the following:

$f(1) - f(0.5) = 1$ $g(1) - g(0.5) = 0.75$

$f(2) - f(1) = 2$ $g(2) - g(1) = 3$

+4 or +3 or +2 or +1 or +0

Criteria for responses to A, C, and D:

+2 or +1 or +0 A. Writes a paragraph addressing the question.
+2 or +1 or +0 A. Points out that $2x$ can be negative.
+2 or +1 or +0 A. Points out that x^2 cannot be negative.
+2 or +1 or +0 A. Nothing erroneous or irrelevant included.

+2 or +1 or +0 C. Writes a paragraph addressing the question.
+2 or +1 or +0 C. Points out the constant effect of *f* on *x* by the same number—namely 2.
+2 or +1 or +0 C. Points out the effect of *g* on *x* is to multiply *x* by itself and, unlike 2, *x* varies.
Thus, as *x* varies so does the number by which it is being multipied.
+2 or +1 or +0 C. Uses work from B to illustrate points.
+2 or +1 or +0 C. Nothing erroneous or irrelevant included.

+2 or +1 or +0 D. Writes a paragraph addressing the question.
+2 or +1 or +0 D. Uses the fact that $x^2 < x$ for $0 < x < 1$.
+2 or +1 or +0 D. Uses the idea that *g* "catches up" with *f* at $x = 2$.
+2 or +1 or +0 D. Uses the fact that $x^2 > x$ for $x > 1$.
+2 or +1 or +0 D. Nothing erroneous or irrelevant included.

_________ Score for response to Prompt #10

Rubric for Response to Prompt #13 (24-point maximum):

For each criterion listed for responses to Prompts A, B, C, and D, points are distributed as follows: +2 if the criterion is clearly met; +1 if it is unclear as to whether or not the criterion is met; +0 if the criterion is clearly not met.

+2 or +1 or +0 A. Justifies the choice of graph for situation A (which is expected to be "II").
+2 or +1 or +0 A. Recognizes that graph would not be a smooth curve or a straight line.
+2 or +1 or +0 A. Recognizes that graph should show an overall increase.

+2 or +1 or +0 B. Justifies the choice of graph for situation B (which is expected to be "IV").
+2 or +1 or +0 B. Recognizes that graph would be a smooth curve.
+2 or +1 or +0 B. Recognizes that the rate of increase varies.

+2 or +1 or +0 C. Justifies the choice of graph for situation C (which is expected to be "III").
+2 or +1 or +0 C. Recognizes that graph should be a straight line.
+2 or +1 or +0 C. Recognizes that graph should have a positive slope.

Continued

Exhibit 9.6
Continued

+2 or +1 or +0 D. Justifies the choice of graph for situation D (which is expected to be "I").
+2 or +1 or +0 D. Recognizes that graph should be a straight line.
+2 or +1 or +0 D. Recognizes that graph should have a negative slope.

_________ Score for response to Prompt #13

Rubric for Response to Prompt #14 (27-point maximum):

+3 or +0 for displaying any part of the graph of $f(x) = -20x + 150$.
+3 or +0 for displaying any part of the graph of $g(x) = 15x + 10$.
+3 or +0 for displaying the graph of $f(x) = -20x + 150$ for $0 \le x \le 7.5$.
+3 or +0 for displaying the graph of $g(x) = 15x + 10$ for $0 \le x \le 7.5$.
+3 or +0 for selecting convenient units for the two scales (e.g., 1 minute, 10 meters).
+3 or +0 for giving an answer to Question E that's equivalent to "2:04 P.M.".
+3 or +0 for giving an answer to Question F that's equivalent to "70 meters".
+3 or +0 for giving an answer to Question G that's equivalent to "2:07.5 P.M.".
+3 or +0 for giving an answer to Question H that's equivalent to "122.5 meters".

_________ Score for response to Prompt #14

TOTAL SCORE FOR THE OPPORTUNITY______/110

each prompt on Parts I and III. You administer Part II in a more traditional test-taking manner.

Exhibit 9.7 displays one student's responses to the prompts; Exhibit 9.8 shows the results of your scoring her responses. Exhibit 9.9 displays the scores resulting from all your students taking the test and your scoring their tests. Exhibit 9.9 also displays the grades you assigned to students' scores. You converted the scores to grades using the *compromise method*—a technique for asigning grades to scores that is explained in a subsequent section of this chapter (i.e., "Assigning Grades to Measurement Scores").

You spend another day reviewing students' responses using the scoring forms (e.g., Exhibit 9.8) that you return to them.

The test you administered in Case 9.3 is somewhat more complicated than typical tests given to typical algebra I classes. This is necessitated by the fact that the objectives listed in Exhibit 9.4 emphasize higher cognitive-learning levels than typical mathematics units that emphasize simple-knowledge and algorithmic-skill learning levels. You designed the unit this way to lead your students to do meaningful mathematics. Your students are able to follow the complex directions for prompts because of the following: (a) The types of tasks required by the test's miniexperiments are similar to tasks you prompted your students to confront during the unit's lessons. (b) The test is spread over three days so students have time to apply reasoning to formulate their responses. (c) For Parts I and III, you supplement the written directions with your own oral explanations. (d) From the first day of class, you have been conditioning these students to comprehend and communicate with the language of mathematics differently than they comprehend and use the language of common English.

As indicated in chapter 4's section, "A Summative Evaluation of Student Achievement of the Learning Goal," you are expected to make periodic reports of student achievement to students, their parents, and your supervisors. Consequently, most teaching units terminate with measurements of students' achievement of the goal. Your judgments of students' achievement for the purpose of reporting success are referred to as "*summative evaluations.*"

It is important for you to distinguish between your summative evaluations and your formative judgments. Furthermore, your students need to understand that the ongoing unplanned measurements and informal miniexperiments you make during learning activities are strictly for formative purposes and will not be used to influence their grades. Periodically administered tests for purposes of summative evaluations that do influence students' grades should clearly stand apart from continual formative feedback. Otherwise, you will be unable to build that atmosphere conducive to doing meaningful mathematics that is emphasized throughout this book and was the focus of chapter 2's section "Establishing a Favorable Climate for Learning Mathematics" (Cangelosi, 2000a, pp. 115–118). Recall that section's suggestions for using descriptive instead of judgmental language and using true dialogues instead of teacher-student interactions dominated by IRE cycles.

Exhibit 9.7
Test Document With Krystal Bojado's Responses.

Algebra I: Opportunity to Demonstrate What You Learned
During Our Unit on Linear Functions
Part I (Monday, October 14)

1. What is your name? Krystal Bojado
2. ORAL AND WRITTEN DIRECTIONS: During our class meeting on Tuesday, Sept. 24, we experimented with different one-variable first-degree equations to show that the graph of some first-degree functions is a line. Formulate four one-variable first-degree equations to demonstrate that the graph of the following function is a line:

   $f(x) = 8x + 3$

   Do not directly graph f. Instead use graphs of your four equations for the demonstration.

   State your four equations in the box on the right.

   After graphing your four equations below, write several sentences explaining how the graph of your four equations relates to the graph of f.

1) $y_1 = x$	2) $y_2 = x + 3$
3) $y_3 = 8x$	4) $y_4 = -8x - 3$

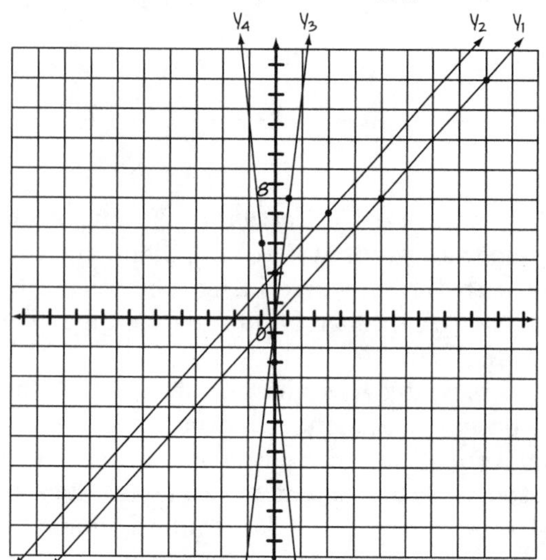

Write your explanation in the following box:

Your explanation:

The graphs of $y_1 - y_4$ being linear suggest that f(x) is linear as well. Consider $y_1 = x$. f(x) differs from y_1 in that f(x) has a steeper slope (a rise of 8 for every run of 1) and each value is 3 greater than that of $y_3 = 8$ (also linear). So, f(x) is linear too since it differs from the above equations by some constant amount for every x-value.

3. ORAL AND WRITTEN DIRECTIONS: $f(x) = ax + b$ and $g(x) = cx + d$. Select values for a, b, c, and d so that (i) the graphs of f and g are parallel and (ii) the graph of g is "above" f:

   $a = 3$, $b = 4$, $c = 3$, and $d = 7$

Continued

Exhibit 9.7
Continued

In two sentences, explain your reasons for selecting the values you chose for *a* and *c:*

Your explanation:
For f and g to be parallel, their slopes must be of equal value. That is why I chose $a = b$.
For g to be above f, its y-intercept must be larger than that of f. That is why I chose $d > b$.

In two sentences, explain your reasons for selecting the values you chose for *b* and *d:*
Write your explanation in the following box:

Your explanation:

4. ORAL AND WRITTEN DIRECTIONS: Write a one-paragraph explanation why the graph of $f(x) = 2x$ is a steeper line than the graph of $g(x) = 0.5x$. To make your explanation clear, draw on and refer to the close-up shot of the coordinate plane that appears just below your explanation:

Your explanation:
The slope of f(x) is 2. This means that the graph is a line which rises 2 units for every run of 1. The slope of g(x) is 0.5, meaning that its graph rises only 1/2 unit for each run of 1. As the diagram below shows, a rise of 2 in a run of 1 (point B) from the origin, is a steeper line than going up 0.5 and over 1 (point C) from the origin. Thus f(x)'s graph is steeper than g(x)'s graph.

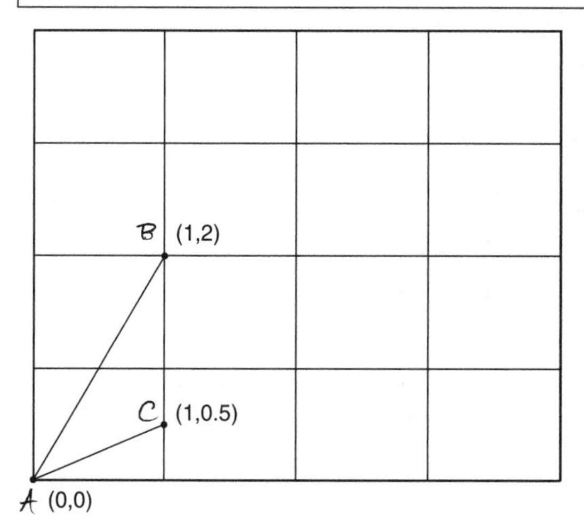

Continued

Exhibit 9.7
Continued

5. ORAL AND WRITTEN DIRECTIONS: Write a one-paragraph explanation why the graph of $h(x) = x + 2$ "stays above" the graph of $k(x) = x + 0.5$. To make your explanation clear, draw on and refer to the close-up shot of the coordinate plane that appears just below your explanation:

Your explanation:
Since both h(x) and k(x) have a slope of 1, they are parallel and thus one will not "cross over" the other. So, we know that the graphs of h(x) and k(x) either lie on the same line or one is above the other. For any given X-value, h(x) is two more than that value, whereas k(x) is only 0.5 more. So this means that h(x) is always 1.5 greater than K(x) for some x. (I drew this below with the brackets.)

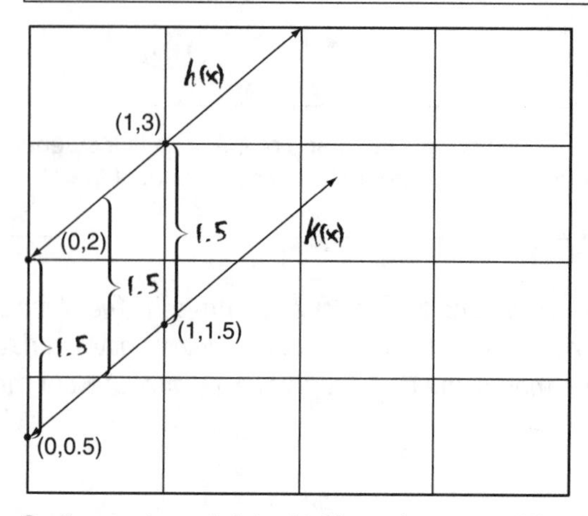

6. Smile; you have finished taking advantage of Part I of this opportunity.

Algebra I: Opportunity to Demonstrate What You Learned
During Our Unit on Linear Functions
Part II (Tuesday, October 15)

7. What is your name? Krystal Bojado

8. WRITTEN DIRECTIONS: Solve for the slope, y-intercept, and x-intercept of $12x - 3y = 24$:

The slope = −4

The y-intercept = (0, 8)

The x-intercept = (2, 0)

$$y = \frac{-12x + 24}{3}$$

$$y = -4x + 8$$

9. WRITTEN DIRECTIONS: What is the y-intercept and the x-intercept of a linear function whose graph includes the following points: (10, 14102.5302), (3, 14338.1502), (428.97,0), (−3, 14540.1102), (0, 14439.1302)?

The y-intercept = (0, 14439.1302)

The x-intercept = (428.97,0)

Continued

Exhibit 9.7
Continued

10. DIRECTIONS: Using your graphing calculator, graph $f(x) = 2x$ and $g(x) = x^2$ on the same screen.

A. In a paragraph, explain why the graph of f includes points below the horizontal axis, but the graph of g does not:

Your explanation:
f is negative whenever x is negative. In contrast, g(x) is positive even when x is negative since any number times itself (i.e., its square) is positive.

$f(x) = 2x,\ g(x) = x^2$

B. Compute each of the following:

$2 - 1$ $f(1) - f(0.5) = $ 1	$1 - .25$ $g(1) - g(0.5) = $ .75
$4 - 2$ $f(2) - f(1) = $ 2	$4 - 1$ $g(2) - g(1) = $ 3

C. Notice that for $x > 0$, f increases at the same rate along a straight line. But for $x > 0$, the rate of increase for g changes with the graph curving upward "faster and faster." Write a paragraph explaining why the rate of increase for f remains steady but for g the rate of increase changes as x gets larger. Use the results of your computations from Part B in your explanation:

Your explanation:
As shown in Part B, when X is small (like 1 or 0.5), the difference between points of g is smaller than that for f. But for larger x-values (like 2) the difference between points of g is greater than that for f. As numbers are squared, they increase in value "faster & faster." But on a line, values increase by a constant amount.

D. Once again, compare the two graphs on your calculator. Notice that $g(x) < f(x)$ when $0 < x < 2$. But when $x > 2$, $g(x) > f(x)$. Write a paragraph explaining why this happens.

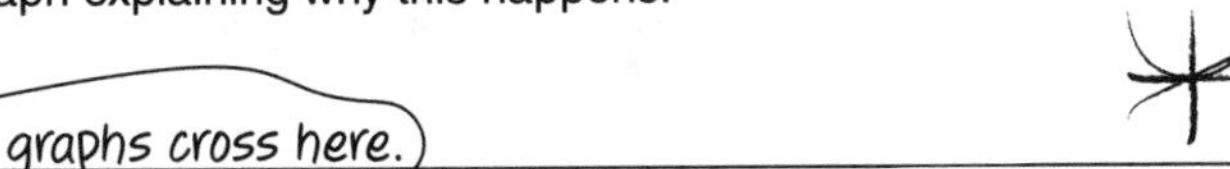

Your explanation:
Like I was saying in part C, g first has small differences between its points (for small x's) but then these differences get bigger than those for f as x gets larger. Here the points of differentiation between "small x's" and "large x's" is $x = 2$. This is because for $x = 2$, $2x = x^2$. Thus the

11. Smile; you have finished taking advantage of Part II of this opportunity.

Continued

Exhibit 9.7
Continued

Algebra I: Opportunity to Demonstrate What You Learned
During Our Unit on Linear Functions
Part III (Wednesday, October 15)

12. What is your name? Krystal Bojado

13. ORAL AND WRITTEN DIRECTIONS: Read each one of the following four situations and decide which one of the graphs that appears below best fits the situation. No graph may be selected more than once. For each situation, write the Roman numeral corresponding to the graph of your choice in the blank. In a brief paragraph, explain why you chose the graph you chose for the situation.

A. Every Sunday for many weeks Fred deposits the same exact amount of money into his piggy bank. During that time, no money is removed nor are any other deposits made. One of the choices is a graph relating two variables: *Time* on the *x*-axis and the *amount of dollars in the piggy bank* on the *y*-axis. Which graph is it? II

Your explanation:
Graph II shows a constant value (horizontal line) for a constant amount of time (length of each line). This could describe the amount of money in the bank and each consecutive week.

B. One of the choices is a graph relating two variables: *Time* on the *x*-axis and the *amount a person, Agatha, weighs during the first 30 years of her life* on the *y*-axis. Which graph is it? IV

Your explanation:
I would expect Agatha's weight to overall increase, but not necessarily at a constant rate. I chose graph IV because there's an unsteady increase displayed.

C. An automobile is traveling faster and faster at a constant rate of acceleration. One of the choices is a graph relating two variables: *Time during the period of the constant rate of acceleration* on the *x*-axis and the *speed of the automobile* on the *y*-axis. Which graph is it? III

Your explanation:
Graph III shows a steady increase if time were on the x-axis & speed on the y-axis. Because it's a line, it displays constant rate of acceleration (i.e., speed increases linearly).

Continued

Exhibit 9.7
Continued

D. An automobile is traveling slower and slower at a constant rate of deceleration. One of the choices is a graph relating two variables: *Time during the period of the constant rate of deceleration* on the *x*-axis and the *speed of the automobile* on the *y*-axis. Which graph is it? I

Your explanation:
Graph I shows a steady decrease. Because deceleration is constant the speed would decrease in linear fashion—to each unit of time, the car would slow by a certain amount. The graph needs to be a line with a negative slope.

THE GRAPHS FOR YOU TO CHOOSE:

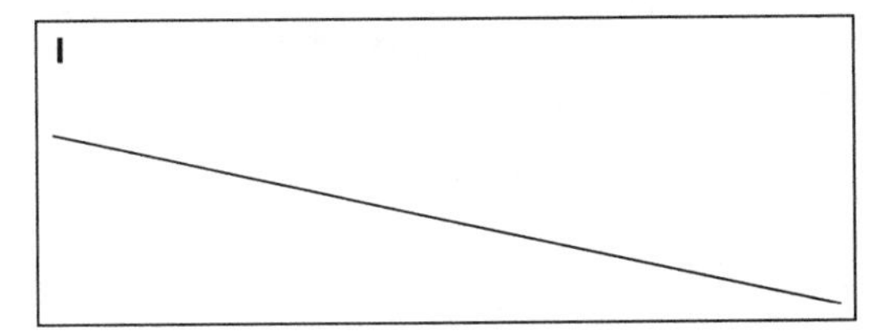

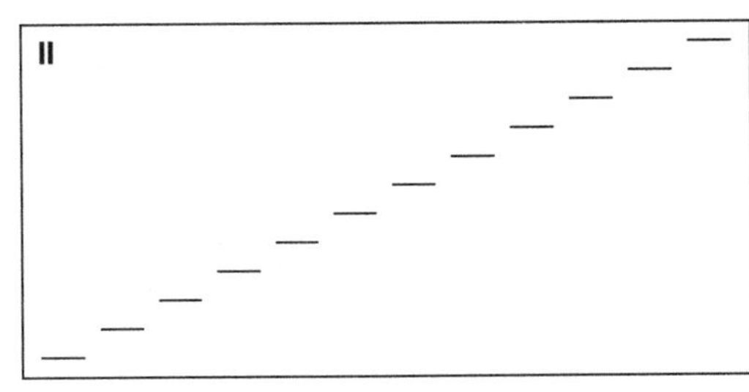

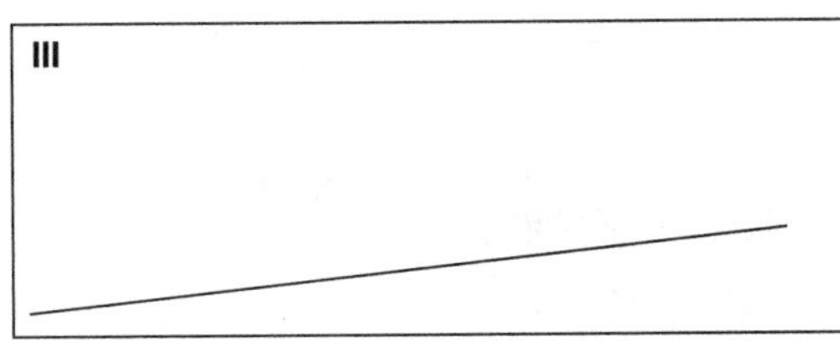

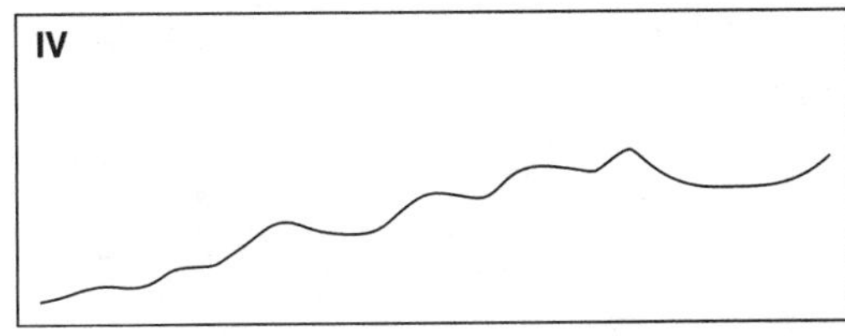

14. ORAL AND WRITTEN DIRECTIONS: It is 2 P.M. when a red hot-air balloon is 150 meters above the ground descending at a steady rate of 20 meters per minute. At the same time, a yellow hot-air balloon is only 10 meters above the ground rising at the steady rate of 15 meters per minute. The temperature of the air just above the ground is holding steady at 24°C. Draw a graph to illlustrate how high the balloons are above the ground between 2 P.M. and when the red balloon lands.

2 p.m. -red 150 m up, @ 20 m/min

Continued

Exhibit 9.7
Continued

You may use your graphing calculator, but transfer your graph to the grid below. Label the horizontal axis and the vertical axis and indicate the size of the unit for the horizontal scale and the size of the unit for the vertical scale:

Yellow 10 m up ↑ 15 m/min, T = 24°C

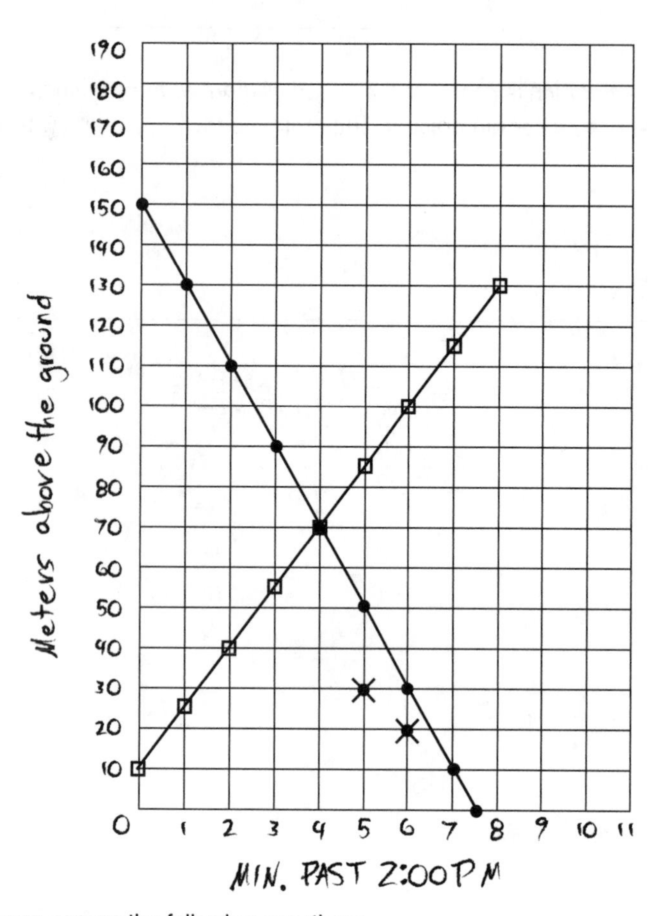

Please answer the following questions:

A. What size unit did you use for the horizontal scale? 1 unit = 1 minute past 2 pm

B. What size unit did you use for the vertical scale? 1 unit = 10 m above ground

C. What variable did you plot on the horizontal axis? minutes past 2 pm

D. What variable did you plot on the vertical axis? meters above ground

E. At what time will the balloons be the same distance above the ground? 2:04 pm

F. How many meters above the ground will the balloons be when they are at the same altitude? 70 m

G. At what time will the yellow balloon land? 2:07 pm (and 30s)

H. How high will the yellow balloon be when the red balloon lands? 125 m

15. Smile; you have finished taking advantage of this opportunity.

Exhibit 9.8
Scoring Form You Completed and Attached to Exhibit 9.7's Document.

Scoring Form for Krystal Bojado (name)
Opportunity Relative to Unit on Linear Functions

Rubric for Response to Prompt #2 (9-point maximum):

+2 or (+0) for writing four different equations $y_1 = 8x + 3$, $y_2 = 8x + 3$, $y_3 = 8x + 3$, and $y_4 = 8x + 3$ where y_1, y_2, y_3, and y_4 are any four constants.
+1 or (+0) for plotting the solution to the first on the graph of $y = y_1$.
+1 or (+0) for plotting the solution to the second on the graph of $y = y_2$.
+1 or (+0) for plotting the solution to the third on the graph of $y = y_3$.
+1 or (+0) for plotting the solution to the fourth on the graph of $y = y_4$.
+2 or (+0) for explaning that the four points associated with the solutions of the four equations lie on the graph of *f*.
+1 or (+0) for pointing out that the four plotted points are collinear.
0 Score for response to Prompt #2

Rubric for Response to Prompt #3 (4-point maximum):

(+1) or +0 if $a = c$.
(+1) or +0 if $b < d$.
(+1) or +0 for explaining why $a = c$ causes the lines to be parallel.
(+1) or +0 for explaining why $b < d$ causes the graph of *g* to be "above" the graph of *f*.
4 Score for response to Prompt #3

Rubric for Response to Prompt #4 (4-point maximum)

(+1) or +0 for explaining that each time *x* increases by 1, $f(x)$ becomes twice as large as the new *x*, whereas $g(x)$ only increases to half of the new *x*.
(+1) or +0 for associating the 2 to 1/2 relationship to the difference in the steepness of the lines.
(+1) or +0 for using the graph to illustrate how the difference in steepness continues.
(+1) or +0 if the explanation includes nothing erroneous or extraneous.
4 Score for response to Prompt #4

Rubric for Response to Prompt #5 (3-point maximum):

(+1) or +0 for explaining that each time *x* increases by 1 the corresponding values of $h(x)$ and $k(x)$ remain 1½ units apart.
(+1) or +0 for using the graph to illustrate how the difference in the locations of the lines remains the same.
(+1) or +0 if the explanation includes nothing erroneous or extraneous.
3 Score for response to Prompt #5

Rubric for Response to Prompt #8 (4-point maximum):

+1 or (+0) for indicating that the slope is 4.
+1 or (+0) for indicating that the *y*-intercept is −8.
(+1) or +0 for indicating that the *x*-intercept is 2.
+1 or (+0) for expressing all three answers accurately (e.g., a response of "(0, −8)" would indicate that the student computed the *y*-intercept correctly, but didn't express it accurately possibly because she doesn't comprehend the definition of "*y*-intercept").
1 Score for response to Prompt #8

Rubric for Response to Prompt #9 (3-point maximum):

(+1) or +0 for indicating that the *y*-intercept is 14439.1302.
(+1) or +0 for indicating that the *x*-intercept is 428.97.
+1 or (+0) for expressing both answers accurately.
2 Score for response to Prompt #9

Rubric for Response to Prompt #10 (32-point maximum):

For each criterion listed for responses to Prompts A, C, and D, points are distributed as follows: +2 if the criterion is clearly met; +1 if it is unclear as to whether or not the criterion is met; + 0 if the criterion is clearly not met.

Continued

Exhibit 9.8
Continued

For the responses to Prompt B: +1 for each computation result that matches the following:

$f(1) - f(0.5) = 1$ $g(1) - g(0.5) = 0.75$

$f(2) - f(1) = 2$ $g(2) - g(1) = 3$

(+4) or +3 or +2 or +1 or +0

Criteria for responses to A, C, and D:

(+2) or +1 or +0 A. Writes a paragraph addressing the question.
(+2) or +1 or +0 A. Points out that $2x$ can be negative.
(+2) or +1 or +0 A. Points out that x^2 cannot be negative.
(+2) or +1 or +0 A. Nothing erroneous or irrelevant included.

(+2) or +1 or +0 C. Writes a paragraph addressing the question.
(+2) or +1 or +0 C. Points out the constant effect of f on x by the same number—namely 2.
(+2) or +1 or +0 C. Points out the effect of g on x is to multiply x by itself and, unlike 2, x varies. Thus, as x varies so does the number by which it is being multipied.
(+2) or +1 or +0 C. Uses work from B to illustrate points.
(+2) or +1 or +0 C. Nothing erroneous or irrelevant included.

(+2) or +1 or +0 D. Writes a paragraph addressing the question.
+2 or +1 or (+0) D. Uses the fact that $x^2 < x$ for $0 < x < 1$.
(+2) or +1 or +0 D. Uses the idea that g "catches up" with f at $x = 2$.
+2 or +1 or (+0) D. Uses the fact that $x^2 > x$ for $x > 1$.
(+2) or +1 or +0 D. Nothing erroneous or irrelevant included.

28 Score for response to Prompt #10

Rubric for Response to Prompt #13 (24-point maximum):

For each criterion listed for responses to Prompts A, B, C, and D, points are distributed as follows: +2 if the criterion is clearly met; +1 if it is unclear as to whether or not the criterion is met; +0 if the criterion is clearly not met.

(+2) or +1 or +0 A. Justifies the choice of graph for situation A (which is expected to be "II").
(+2) or +1 or +0 A. Recognizes that graph would not be a smooth curve or a straight line.
(+2) or +1 or +0 A. Recognizes that graph should show an overall increase.

(+2) or +1 or +0 B. Justifies the choice of graph for situation B (which is expected to be "IV").
+2 or (+1) or +0 B. Recognizes that graph would be a smooth curve.
(+2) or +1 or +0 B. Recognizes that the rate of increase varies.

(+2) or +1 or +0 C. Justifies the choice of graph for situation C (which is expected to be "III").
(+2) or +1 or +0 C. Recognizes that graph should be a straight line.
(+2) or +1 or +0 C. Recognizes that graph should have a positive slope.

(+2) or +1 or +0 D. Justifies the choice of graph for situation D (which is expected to be "I").
(+2) or +1 or +0 D. Recognizes that graph should be a straight line.
(+2) or +1 or +0 D. Recognizes that graph should have a negative slope.

23 Score for response to Prompt #13

Rubric for Response to Prompt #14 (27-point maximum):

(+3) or +0 for displaying any part of the graph of $f(x) = -20x + 150$.
(+3) or +0 for displaying any part of the graph of $g(x) = 15x + 10$.
(+3) or +0 for displaying the graph of $f(x) = -20x + 150$ for $0 \leq x \leq 7.5$.
(+3) or +0 for displaying the graph of $g(x) = 15x + 10$ for $0 \leq x \leq 7.5$.
(+3) or +0 for selecting convenient units for the two scales (e.g., 1 minute, 10 meters).
(+3) or +0 for giving an answer to Question E that's equivalent to "2:04 P.M."
(+3) or +0 for giving an answer to Question F that's equivalent to "70 meters."
(+3) or +0 for giving an answer to Question G that's equivalent to "2:07.5 P.M."
+3 or (+0) for giving an answer to Question H that's equivalent to "122.5 meters."

24 Score for response to Prompt #14

TOTAL SCORE FOR THE OPPORTUNITY 89 /110

Name	Unit Test Score out of 110	Unit Test Grade
Dalma	103	A
Panchito	99	A
Jason	97	A
Melanie H.	90	A−
Rahmin	89	A−
Krystal	89	A−
Qu	89	A−
Mia S.	88	A−
Marian	85	B+ or A−
Marion	84	B+ or A−
Der	82	B+
Charles	81	B+
Quinton	77	B
Naomi	75	B
Tyrone	75	B
Mia L.	75	B
Cyril	75	B
Aaron	74	B− or B
Melanie C.	74	B− or B
Christina	68	C+
Keith	66	C+
Ping	65	C+
Dave	65	C+
Heather	41	D
Chris (Parts I & II only)	55/72	no grade yet
Brady (didn't take test)	no score yet	no grade yet

Exhibit 9.9
Scores and Grades Based on Results of Exhibit 9.5's Tests.

MEASUREMENT USEFULNESS

Both Valid and Usable

As suggested in this chapter's section "Measurements for Formative Judgments," measurements do not yield completely accurate indicators of students' achievement levels; note the relationship $S_j = T_j + E_{S_j}$.

Exhibit 9.9's scores (i.e., $\{S_j\}$)are influenced not only by how well your students achieved the unit goal (i.e., $\{T_j\}$) but also by measurement error (i.e., E_{S_j}). For example, according to Exhibit 9.8, Krystal's response to Prompt #2 on your test merited a score of 0. Miniexperiment #2 was designed to be relevant to how well students achieved Objective B as listed in Exhibit 9.4. Krystal may have responded poorly to Prompt #2 because her actual level of achievement of Objective B was low. In such a case, the result of Miniexperiment #2 was accurate because the result (i.e., S_j) reflects the truth (i.e., T_j). On the other hand, Krystal's poor response to Prompt #2 may have been influenced by factors other than her level of achievement of Objective B. Possibly, she misread or misheard the directions. In such a case, measurement error (i.e., E_{S_j}) may have been significantly negative causing $S_j < T_j$.

Measurement error could also be positive—causing $S_j > T_j$. For example, according to Exhibit 9.8, Krystal's response to Prompt #3 on your test merited a maximum score. Miniexperiment #3 was designed to be relevant to how well students achieved Objective C as listed in Exhibit 9.4. Krystal may have responded well to Prompt #3 because her actual level of achievement of Objective C was high. In such a case, the results of Miniexperiment #3 was accurate because the result (i.e., S_j) reflects the truth (i.e., T_j). On the other hand, Krystal's correct response to Prompt #3 may have been influenced by factors other than her level of achievement of Objective C. Possibly, she simply remembered a "rule" for formulating a function that is parallel and lies "above" the graph of a give equation—a rule she really does not understand.

In such a case, measurement error (i.e., E_{S_j}) may have been significantly positive.

In most cases, the results of measurements are influenced by both the actual levels of student achievement and measurement error. So usually, $E_{S_j} \neq 0$, but—depending on the *validity* of the measurement—it may be close to 0.

A measurement is *valid* to the same degree that its results approximate students' true achievement levels (i.e., its measurement error approaches 0). For a measurement to be *useful,* it must have a satisfactory degree of validity, but how useful the measurement is to you also depends on a second factor: *measurement usability.*

A measurement is *usable* to the degree that it is inexpensive, does not consume time, is easy to administer and score, and does not interfere with other activities (Cangelosi, 2000a, pp. 177–180).

When you examined Exhibits 9.5 and 9.6, you may have worried that such an elaborate test over three days would consume too much time and would be too troublesome for you to score. If you did, you were concerned that the test was not adequately usable. If you were bothered that the test would not accurately reflect students' achievement of Exhibit 9.4's goal, then you were concerned that the test's results would not be satisfactorily valid. To be useful, a measurement must have a satisfactory degree of validity as well as a satisfactory degree of usability.

Measurement Validity

A measurement's validity depends on its *relevance* and its *reliability.* A measurement is *relevant* to the same degree that it is pertinent to the decision influenced by its results. A measurement is *reliable* to the same degree that it can be depended upon to provide noncontradictory information. See Exhibit 9.10.

Measurement Relevance

Both Mathematical-Content and Learning-Level Relevance

For a measurement to be relevant to students' achievement of a unit goal, its miniexperiments must pertain to the *mathematical content* and the *learning levels* specified by the objectives that define that goal. A miniexperiment pertains to the mathematical content and learning level specified by an objective iff students' scores on the miniexperiment depend on how well they operate at the specified learning level (i.e., either construct a concept, discover a relationship, simple knowledge, algorithmic skill, comprehension and communication, application, creative thinking, appreciation, or willingness to try) with the specified mathematical content.

Exhibit 9.10

A Measurement's Usefulness Depends on Its Validity and Its Usability; Its Validity Depends on Its Relevance and Reliability.

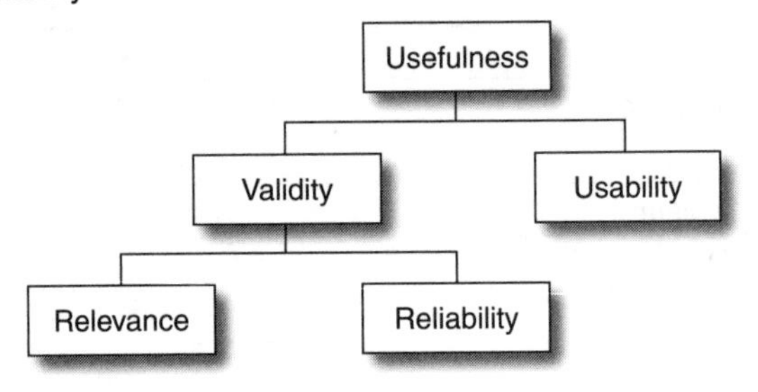

Consider Case 9.4:

CASE 9.4

To help her evaluate her seventh-graders' achievement of the learning goal for a unit on surface areas, Ms. Curry is designing a test. The unit has 10 objectives labeled "A" through "J". Objective J is as follows:

> J. When confronted with a real-life problem, the student determines whether or not computing the area of a surface will help solve the problem (application).

Ms. Curry develops three miniexperiments she intends to be relevant to Objective J. Exhibit 9.11 displays them.

How well do Exhibit 9.11's miniexperiments match Objective J from Case 9.4? Miniexperiment #1 appears to be relevant because it requires students to operate at the application level (which is what Objective J specifies) with surface area (which is the mathematical content specified by Objective J). Of course, students might circle "c" just by guessing, but to increase one's chance from $\frac{1}{3}$ based on random guessing, one must reason deductively about surface area.

Miniexperiment #2 does not seem to match Objective J very well. The correct response (i.e., circling "a") can be determined simply by remembering how to compute a surface area without having to decide whether or not it should be computed. Thus, although Miniexperiment #2 pertains to the mathematical content of Objective J (i.e., surface area), its learning level is algorithmic skill rather than application. Miniexperiment #2 lacks learning-level relevance and, thus, relevance.

Miniexperiment #3 requires students to operate at the application learning level as specified by Objective J. However, the mathematical content of Miniexperiment #3 (i.e., ratios) does not match the mathematical content of Objective J (i.e., surface area). Miniexperiment #3 lacks mathematical-content relevance and, thus, relevance.

Exhibit 9.11
How Relevant Are These Miniexperiments to Objective J From Case 9.4?

Miniexperiment #1

Multiple-Choice Prompt:

Computing a surface area will help you solve one of the following three problems. Which one is it? (Circle the letter in front of your answer.)

a) We have a large bookcase we want to bring into our classroom. Our problem is to determine if the bookcase can fit through the doorway.
b) As part of a project to fix up our classroom, we want to put stripping along the crack where the walls meet the floor. Our problem is to decide how much stripping to buy.
c) As part of a project to fix up our classroom, we want to install new carpet on the floor. Our problem is to decide how much carpet to buy.

Rubric:

+1 for circling "c" only; otherwise +0.

Miniexperiment #2

Multiple-Choice Prompt:

What is the surface area of one side of the sheet of paper from which you are now reading? Use your ruler and calculator to help answer the question. (Circle the letter in front of your answer.)

a) 93.5 square inches
b) 93.5 inches
c) 20.5 square inches
d) 20.5 inches
e) 41.0 square inches
f) 41.0 inches

Rubric:

+1 for circling "a" only, otherwise +0.

Miniexperiment #3

Multiple-Choice Prompt:

As part of our project for fixing up the classroom, we need to buy some paint for the walls. The paint we want comes in two different size cans. A 5-liter can costs $16.85 and a 2-liter can costs $6.55. Which one of the following would help us decide which size can is the better buy? (Circle the letter in front of your answer.)

a) Compare 5 × $16.85 to 2 × $6.55
b) Compare $16.85 ÷ 5 to $6.55 ÷ 2
c) Compare $16.85 − $6.55 to 2/5

Rubic:

+1 for circling "b" only, otherwise +0.

Emphasis According to Relative Importance of Objectives

For a unit test to be relevant to a learning goal, it is necessary—but not sufficient—for each miniexperiment to match one of the objectives. It is also necessary for the objectives to be represented on the test according to their relative importance for attainment of the goal. Suppose, for example, that you want to design a measurement relevant to your students' achievement of the following learning goal:

> Students understand that the ratio of the circumference of any circle to its diameter is π and that they make use of the relationship in real-life situations.

Further suppose that you define this goal with the eight objectives listed in Exhibit 9.12.

Eight objectives define the goal. In your opinion, are some of the objectives more important to goal attainment than others? If not, your unit test should reflect students' achievement of any one of the objectives by 1/8 or 12.5%. Thus, for the case in which you believe each objective is equally important, 12.5% of the maximum test score should relate to Objective A, 12.5% to Objective B, ..., 12.5% to Objective H.

On the other hand, suppose your analysis of the objectives prompted the following thought: "Objectives A and H are more important than the others. Objective A is especially important because if students

Exhibit 9.12
Are Some Objectives More Important to Goal Achievement Than Others?

Unit Goal:

Students understand that the ratio of the circumference of any circle to its diameter is π and that they make use of the relationship in real-life situations.

Unit Objectives:

A. The student makes an inductive argument for concluding that the ratio of the circumference of any circle to its diameter is π (discover a relationship).
B. The student attempts to develop a method for obtaining rational approximation of π (willingness to try).
C. The student explains at least three methods for obtaining a rational approximations of π: (a) an algorithm for averaging measurements that students themselves make, (b) one of the ancient methods (von Baravalle, 1989), and (c) a computer-based method (comprehension and communication).
D. The student states the following: (a) π is the ratio of the circumference of any circle to its diameter. (b) π is an irrational number. (c) $\pi \approx 3.1416$ (simple knowledge).
E. The student explains why $C = 2\pi r$ for a circle with circumference C and radius r (discover a relationship).
F. Given a circle's circumference, the student solves for its radius (algorithmic skill).
G. Given a circle's radius, the student solves for its circumference (algorithmic skill).
H. Confronted with a real-life problem, the student determines how, if at all, a solution to that problem is facilitated by using the relationship $\pi = C/d$ (application).

discover that ratio, they will be able to develop related formulas for themselves. Objective H is the culminating objective of the unit. Objectives B and C—although important—are less important than the rest. All the rest seem about equally important."

If you really thought like that, you might decide to design the test with relative weights for the objectives as indicated by Exhibit 9.13.

Thus, to design a relevant test for student achievement of the goal, you would select miniexperiments and distribute the points according to the values in Exhibit 9.13. If, for example, 50 is the maximum possible score a student could attain on the test, then approximately 10 of those points should pertain to Objective A, 3 to Objective B, 3 to Objective C, 6 to Objective D, 6 to Objective E, 6 to Objective F, 6 to Objective G, and 10 to Objective H.

Now you understand the purpose of the percent values that appear right after objectives listed in this book. Briefly revisit the second paragraph of the section "A Note" near the very beginning of chapter 1.

To lead you to better understand how you might weight objectives in your own teaching, engage in Activity 9.3

Activity 9.3

Resurrect the unit goal and list of objectives you last revised—or at least considered revising—when you engaged in Synthesis Activity 3 from chapter 8.

Carefully examine the unit goal and each objective you formulated to define it. Weight the objectives to reflect the relative importance of each objective with respect to goal attainment.

Exhibit 9.13
Example of Weighted Objectives.

Objective	Relative Weight
A	20%
B	06%
C	06%
D	12%
E	12%
F	12%
G	12%
H	20%

Discuss how you weighted the objectives with a colleague who is also engaging in this activity. After making any modification you think you should in light of your discussion, reinsert the goal and objectives—which are now weighted—in your working portfolio.

Measurement Reliability

Internal Consistency

As indicated in this chapter's section "Measurement Validity," for a measurement to produce valid results, it must not only be relevant to the intended learning goal, it must also be reliable. To be reliable, a measurement must have both *internal consistency* and *scorer consistency.* See Exhibit 9.14.

Suppose a friend tells you, "Math is so boring; I don't know how you can bear studying it!" You respond, "You find mathematics pretty dry, eh?" Your friend: "Absolutely! Of course, I love using numbers

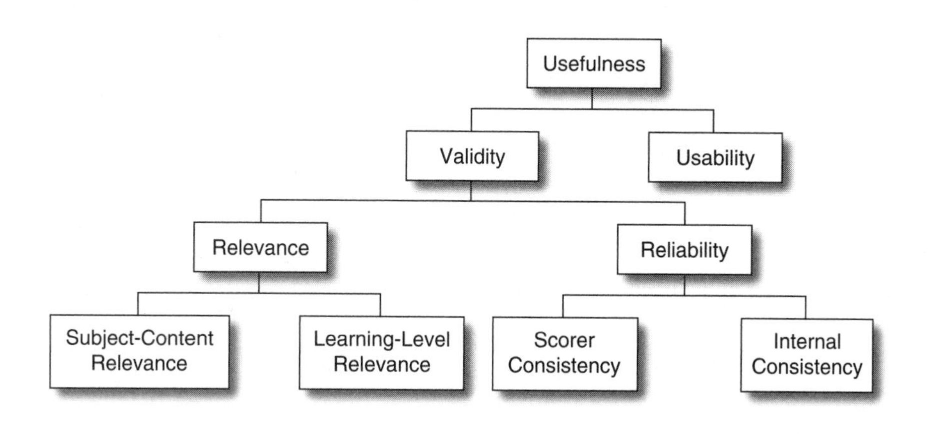

Exhibit 9.14
A Measurement's Relevance Depends on Its Learning-Level Relevance and Its Mathematical-Content Relevance; Its Reliability Depends on Its Internal Consistency and Its Scorer Consistency.

to solve problems—and there's something beautiful about geometry." Does your friend like mathematics or not? Such comments are contradictory. Thus, the results of the unplanned measurement, based on what you heard the friend say, lack internal consistency. By definition:

> A measurement is internally consistent to the degree that results from its miniexperiments are in harmony (i.e., noncontradictory and in agreement).

Consider the internal consistency of the partial test results reported in Case 9.5:

CASE 9.5

Ms. Curry administers a test with 22 miniexperiments to help her evaluate the goal for a unit on surface area. Two of the miniexperiments, #7 and #19, are intended to be relevant to the sixth of the unit's 10 objectives (i.e., Objective F):

> F. Given the dimension of a right triangle, the student computes its area (algorithmic skill).

Exhibit 9.15 displays the two miniexperiments. Curious about how well her students achieved Objective F, Ms. Curry notes five students' scores on Miniexperiments #7 and #19 as displayed in Exhibit 9.16.

Relative to Objective F, the partial results from Exhibit 9.16 suggest that Roxanne's level of achievement is high, Luanne's is low, and Mel's is somewhere in between. However, Ms. Curry is perplexed by Izar's and Jan's results from Miniexperiments #7 and #19. Do Izar and Jan know how to compute areas of right triangles? Miniexperiment #7 suggests that Jan does and Izar does not, whereas Miniexperiment #19 suggests that Izar does and Jan does not.

Relative to Objective F, two of the miniexperiments on Ms. Curry's test yielded contradictory results for Izar and Jan. If the test results are dominated by such inconsistencies, the test lacks internal consistency and is, therefore, unreliable. On the other hand, if there are relatively few contradictions and the results are more in line with what Ms. Curry obtained for Roxanne, Luanne, and Mel, the test is internally consistent.

For a measurement to produce internally consistent results, it must be designed so that the prompts are presented clearly and unambiguously. For students to respond consistently to prompts, they must clearly understand the task required by each. Directions need to be unambiguously communicated. In Case 9.6, students are confronted with a vaguely defined task:

CASE 9.6

One of Mr. Cotrell's teaching units includes the following objective:

> The student interprets the meanings of expressions using the language of sets (comprehension and communication).

Among the miniexperiments for that objective on his unit test are the two in Exhibit 9.17 (i.e., Miniexperiments #9 and #11).

Confronted with Prompt #9, one student, Robin, counts the five elements and writes "5" in the blank. For the next one she thinks, "1, 2, 3, 4—the answer is 4. Oh! Maybe not. *a, b, c,* and *d* could be variables and maybe two or more are equal to one another. If $a = b$, that'd make not more than 3 in the set. The answer could be 1, 2, 3, or 4. So, it's got to be 1, because I know there's at least one element, but I can't tell if there are more."

To fix the prompt for Miniexperiment #11, Mr. Cotrell needs to replace "{*a, b, c, d*}" with an unambiguous expression (e.g., {Sophie Germain, Hypatia, Maria Agnesi, Emmy Noether} or {"a," "b," "c," "d"}). It is understandable that ambiguity crept into Mr. Cotrell's prompt. Hopefully, when he reviews the test results with the class, Robin will explain her reasoning and Mr. Cotrell will turn his "mistake" into a productive learning experience.

Exhibit 9.15

Two Miniexperiments Ms. Curry Designed to Be Relevant to Objective F From Case 9.5.

Miniexperiment #7

Prompt:

What is the area bounded by a right triangle with dimensions 5 cm, 4 cm, and 3 cm? Display your computations in the box and write your answer in the blank:

____

Rubric (4-point maximum):

+1 for using $A = \frac{1}{2}bh$.
+1 if 4 and 3 are used in the computation, but not 5.
+1 if 6 (irrespective of the unit) is given as the area.
+1 if the answer is expressed in square centimeters.

Miniexperiment #19

Prompt:

What is the area of the interior of $\triangle ABC$ if $m\angle B = 90$, $AB = 6$ cm, $BC = 8$ cm, and $AC = 10$ cm? Display your computations in the box and write your answer in the blank:

____

Rubric (4-point maximum):

+1 for using $A = \frac{1}{2}bh$.
+1 if 6 and 8 are used in the computation, but not 10.
+1 if 24 (irrespective of the unit) is given as the area.
+1 if the answer is expressed in square centimeters.

To be internally consistent, a measurement must also include an adequate number of miniexperiments. In a test with a large number of miniexperiments there is a greater likelihood that consistent student response patterns will emerge than in one with only a few. If, for example, Mr. Cotrell's test in Case 9.6 included 10 miniexperiments similar to Miniexperiment #9 and 10 ambiguous ones similar to Miniexperiment #11, then a consistent pattern to Robin's responses would likely surface. There would be a clear distinction between her responses to prompts with well-defined sets and those with ambiguously defined sets.

Furthermore, the more miniexperiments on a test, the less the test results are affected by fortuitous factors. Suppose, for example, that one of a test's objectives is measured by only two true-false miniexperiments. By random guess, a student has a

Exhibit 9.16
Results From Five Students' Responses to Exhibit 9.14's Miniexperiments.

Student	Points from Miniexperiment #7	Points from Miniexperiment #19
Roxanne	4	4
Luanne	1	0
Izar	0	4
Mel	3	2
Jan	4	1

25% chance of scoring 0, a 50% chance of scoring 1, and 25% chance of scoring 2. Results of the two miniexperiments are unlikely to distinguish consistently between students who have and have not achieved the objective. However, if there were 10 such true-false miniexperiments, then random guessing is less of a factor, with a 0.1% chance of scoring 0, a 1.0% chance of scoring 1, a 4.4% chance of scoring 2, a 11.7% chance of scoring 3, a 20.5% chance of scoring 4, a 24.6% chance of scoring 5, a 20.5% chance of scoring 6, a 11.7% chance of scoring 7, a 4.4% chance of scoring 8, a 1.0% chance of scoring 9, and a 0.1% chance of scoring 10.

Tests need to be administered under controlled conditions. A scientific experiment is the application of systematic procedures for the purpose of uncovering evidence that helps answer a specified question. A unit test is a scientific experiment in which the question to be answered is how well students have achieved a particular learning goal. The experimental conditions need to be controlled to minimize distractions to students and assure that directions are followed. Internal consistency is threatened whenever students are distracted during a test.

Exhibit 9.17
Two Miniexperiments Mr. Cotrell Designed to Be Relevant to Objective Stated in Case 9.6.

Miniexperiment #9

Prompt:

How many elements in {8, −9, 0, 11.3, 77}? _____

Rubric:

+1 for "5," otherwise +0

Miniexperiment #11

Prompt:

How many elements in {*a, b, c, d*}? _____

Rubric:

+1 for "4," otherwise +0.

Scorer Consistency

To gain an experience that will heighten your awareness of the need to be concerned with *scorer consistency* when designing miniexperiments, engage in Activity 9.4:

Activity 9.4

Examine Exhibit 9.18. Note that it contains two prompts from a test with Eugene's, Shirley's, and Lydia's responses to each. The maximum possible score for Miniexperiment #1 is one point and four points for Miniexperiment #2. Before consulting with anyone, score the three students' responses to the two prompts.

Rank the three attempts to prove that the sum of any two odds is even from best to worst.

Now, compare your scoring and ranking to those of colleagues who are also engaging in this activity. Discuss the handicap under which you were working because you were not using a detailed rubric for Miniexperiment #2.

Obviously, Eugene's response to Prompt #1 merits the full point. But how many points out of four did you award for his response to Prompt #2? Surely Eugene's response indicates he understands why the proposition is true, but his argument is flawed. For one thing, he demonstrated the proposition to be true only for consecutive odd integers, not any two odd integers.

Shirley's proof seems to be the least complete. But her examples are relevant, so should not some of the four points be awarded? Because Eugene gave a general proof, but only for consecutive odd integers, do you think his work merits more points than Shirley's? Lydia's proof seems complete and clear. I assume you awarded the full complement of points for it. Did you and your colleagues have an easier time agreeing on the score for Eugene's response to Prompt #1 than for his response to Prompt #2? By definition:

> A measurement has *scorer consistency* to the same degree that (a) the teacher (or whoever scores the test) faithfully follows the miniexperiments' rubrics so that the measurement results are not influenced by *when* the measurement is scored; and (b) different teachers (or scorers) who understand the measurement's mathematical content and learning levels agree on the score warranted by each response so that results are not influenced by *who* scores the measurement.

Scorer consistency will be poor if the measurement is dominated by miniexperiments that do not have clearly specified rubrics. There is only one correct response to Prompt #1 of Exhibit 9.18; it is not difficult to score responses to Prompt #1 with consistency. Prompt #2, on the other hand, requires the

Exhibit 9.18
Eugene's, Shirley's, and Lydia's Responses to Two Prompts.

Eugene

1. (1 pt.) Is the sum of any two odd integers odd or even? even
2. (4 pt.) Either prove or disprove that the sum of any two odd integers is even.

a & b is even for every a, b $\in$ Odds because: if a and b are odd, then they can be written as $2i + 1$ for a and $2i - 1$ for b so that i $\in$ integers. Then $a + b = (2i + 1) + (2i - 1) = 2i + 2i + 1 - 1 = 4i + 0 = 4i$ and 4i is even. QED

Shirley

1. (1 pt.) Is the sum of any two odd integers odd or even? even
2. (4 pt.) Either prove or disprove that the sum of any two odd integers is even.

Two odds are even because $7 + 7 = 14$, $3 + 1 = 4$, $9 + 1 = 10$, ...

Lydia

1. (1 pt.) Is the sum of any two odd integers odd or even? even
2. (4 pt.) Either prove or disprove that the sum of any two odd integers is even.

a & b $\in$ {*evens*} *for any a , b* $\in$ {*odds*} *because a, b* $\in$ {*odds*} $\Rightarrow$ *These exist i, j* $\in$ {*integers*} *such that* $a = 2i + 1$ *and* $b = 2j + 1$. *So, a & b* $= (2i + 1) + (2j + 1) = 2i + 2j + 2$. *Thus,* $a + b = 2(i + j + 1)$. *And since* {*integers*} *is closed under addition,* $i + j + 1 \in$ {*integers*} *and so* $a + b$ *is twice some integer which makes a & b even by definition.*

scorer to make judgments about responses. To build scorer consistency into such miniexperiments, observer's rubrics must specifically indicate just how points are to be distributed. Formulating rubrics is a major aspect of miniexperiment design.

Rubrics for most of the miniexperiments appearing throughout this book are specified. Of course, for a miniexperiment to be relevant, both the prompt and the rubric must be developed in light of the mathematical content and the learning level specified by the objective. Hopefully, you are now in the habit of carefully constructing prompts and rubrics from your work with chapters 5–8.

COMMON MALPRACTICE

Measurements provide the basis for teachers' evaluations of student achievement. Unfortunately, studies examining the validity of tests commonly used in schools—both commercially produced and teacher developed—and the assessment methods of many teachers suggest that common malpractice and inaccurate evaluations are widespread (McMillan, 2001). Too often, faith is placed in poorly designed tests that tax students' test-taking skills but do not reflect actual achievement of the learning goals. Of particular concern are incongruences between learning levels specified by the objective and the actual learning levels measured by tests (i.e., learning-level relevance). It is quite common for teachers to include higher cognitive-level objectives (e.g., application) for their units, but to test only for achievement of lower cognitive levels (e.g., algorithmic skill). The consequence of this practice was pointed out by Stiggins (1988, p. 365):

> Teacher-developed paper-and-pencil tests and many tests and quizzes provided by textbook publishers are currently dominated by questions that ask students to recall facts and information. Although instructional objectives and even instructional activities may seek to develop thinking skills, classroom assessments often fail to match these aspirations. Students who use tests to try to understand the teachers' expectations can see the priority placed on memorizing, and they respond accordingly. Thus poor quality assessments that fail to tap and reward higher-order thinking skills will inhibit the development of those skills.

Although testing is widely malpracticed, you can—as many teachers do—manage to collect valid data and make satisfactorily accurate summative evaluations of your students' achievement of learning goals. You begin to do this by generating relevant miniexperiments using measurement-design strategies you practiced while engaging in activities from chapters 5–8. Then you systematically synthesize measurements from those miniexperiments using strategies suggested in the next section. As you gain experiences designing miniexperiments, employing these systematic strategies, and trying out your measurements with students, your talents for assessing students' mathematical achievement will flourish.

DESIGNING MEASUREMENTS OF STUDENTS' ACHIEVEMENT OF UNIT GOALS

A Haphazard Approach

Does the method for developing a measurement employed by Ms. Houlahan in Case 9.7 seem familiar to you?

CASE 9.7

Ms. Houlahan begins the task of developing a test she plans to administer to her geometry class tomorrow. She thinks, "Let's see, this unit was on π and using $\pi = C/d$ in problem solving. I'll start with an easy item—one about the value of π. I could just ask, 'What is π?' But that'd be ambiguous; they wouldn't know if I meant for them to put 3.14 something, or circumference ÷ diameter. Hmmm, okay, first two items." She writes: "1. π is between what two whole numbers? _____ and _____. 2. For any circle with diameter d and circumference C, C/d = _____."

She continues thinking, "Now what should I do for item 3?" she thinks, "I'd better grab some of these problems out of the book, and just change some of the numbers around. Anyone who has been paying attention and keeping up with the homework shouldn't have any trouble with them. Okay, page ahh . . ."

Ms. Houlahan continues in this vein until her unit test is ready.

Ms. Houlahan's haphazard method is unlikely to produce a relevant measurement for her learning goal. She did not seem to pay attention to the objectives that define the goal—if she even defined it at all. This procedure of thinking up prompts in the same order they will appear on the test typically results in tests that overemphasize mathematical skills that are easy to measure and neglect objectives that are more difficult to measure. For example, recall your experience when you engaged in Activity 5.5. Did you not find it easier to design a miniexperiment for Objective C from Exhibit 5.8 (i.e., a simple-knowledge objective) than you did for Objective A from Exhibit 5.8 (i.e., a construct-a-concept objective)?

Consequently, typically developed tests (i.e., using Ms. Houlahan's method) stress memory-level cognitive learning with only little emphasis on reasoning-level learning (Cangelosi, 2000a, pp. 255–257). Even when teachers include reasoning-level objectives in their unit plans, students bother to learn only at the memory level if memory level is emphasized on the tests (Stiggins, 1988; Stiggins, Conklin, & Bridgeford, 1986). Thus, students are unlikely to do meaningful mathematics.

A Systematic Approach

Four Phases

A more systematic approach is needed, one with the following advantages over the haphazard approach:

- You have a system for designing unit tests that reflect objectives according to your judgment of their relative importance to goal attainment. You consciously control the mathematical content and learning levels to which the test is relevant.
- You establish a practical, efficient system for producing valid and usable measurements.

There are four phases to the systematic, research-based method: (a) clarifying the learning goal, (b) designing relevant miniexperiments and storing and accessing them via a computerized folder structure, (c) developing a test blueprint, and (d) synthesizing the test.

Clarifying the Learning Goal

Eavesdrop on the faculty-room conversation in Case 9.8:

CASE 9.8

Mr. Coco: Hey Eva, would you take a look at this test for me? Do you think it's any good?

Ms. Scott: Good for what?

Mr. Coco: For my algebra II class; it's the first draft of my midterm.

Ms. Scott: I can't judge the validity of this test without knowing what you want to evaluate.

Mr. Coco: What my students learned.

Ms. Scott: What did you intend for them to learn?

Mr. Coco: Algebra.

Ms. Scott: You've got to be more specific than that. A test's validity depends on how relevant it is to exactly what its results are used to evaluate. Unless you have the learning goal you want to evaluate clearly spelled out, there's no way to judge the relevance of the test. If this test is reliable—and from the looks of it, I'll bet it is—then it's relevant to something. The question is whether or not that something is what you want.

You can hardly be expected to create an achievement test until you have answered the question: Achievement of what? Thus, the first phase in the test design and development process is to clarify the learning goal to which the test is supposed to be relevant.

Fortunately, you initially defined your learning goal with a set of objectives when you developed the teaching unit. You also weighted each objective according to its relative importance to goal achievement; recall your work with Activity 9.3. So by designing your unit so that the goal is well defined by weighted objectives—each specifying a mathematical content and learning level—you have automatically completed the first phase of developing a unit test.

Computerized Miniexperiment Folders

Return to Exhibit 1.12—Mr. Rudd's computerized folder structure for his courses. Note the subfolder "Miniexperiments" that is subsumed by "Unit 1" that is subsumed by "Algebra I" that is subsumed by "Courses." This subfolder is where Mr. Rudd stores the miniexperiments he develops for Unit 1's objectives—with those for Objective A under "Objective A," those for Objective B under "Objective B," ..., those for Objective K under "Objective K."

When you engaged in Activity 1.3, you set up a similar computerized system. To build a subfolder structure for your miniexperiments within that system, engage in Activity 9.5:

Activity 9.5

Retrieve the latest version of the unit goal and objectives you recently worked with by engaging in Activity 9.3. Now within your computerized management system, create the necessary subfolders so that for each of your unit's objectives, you have a subfolder ready to house files with miniexperiments that you will soon design for that objective.

Designing miniexperiments for each objective is the most difficult phase of developing unit tests. However, once you have begun generating miniexperiments for your courses, you will find that this system not only helps you create more valid and usable tests, but it also saves you valuable time. There are at least five advantages of organizing and maintaining miniexperiment files in this manner:

- Building miniexperiment files requires you to analyze objectives one at a time—stimulating you to deepen your insights into mathematical content and learning levels.
- Each miniexperiment focuses on the mathematical content and learning level of an objective. Thus, a test synthesized from miniexperiments drawn from this computerized system is more likely to be relevant than one with prompts designed while the test is being put together as Ms. Holahan did in Case 9.7.
- Being able to access such files before miniexperiments are selected for a test facilitates constructing the test so that scores reflect objectives as you have weighted them.
- Being able to associate each test point with a particular objective provides a means for you to assess how well specific objectives were achieved. This facilitates using detailed test results as a diagnostic tool.
- It is much easier and efficient to develop and modify tests once such a computerized system for maintaining and expanding miniexperiment files is in place.

Complete the second phase of building a unit test by engaging in Activity 9.6:

Activity 9.6

Retrieve your work from Activity 9.5. For each objective in your unit, design two relevant miniexperiments. Insert the file with prompts and rubrics for those miniexperiments in the subfolder for that objective in your computerized management system.

Print out the miniexperiments for the unit and discuss them with a colleague who is also engaging in this activity. If you think you should in light of your consultations, modify the miniexperiment files.

A Test Blueprint

An Example of a Blueprint

A *test blueprint* is an outline specifying the features you want to build into a unit test you plan to develop from the prompts stored in your miniexperiment files. Typically, the blueprint indicates (a) the title of the unit, (b) anticipated administration dates and times, (c) provisions for accommodating students with special needs, (d) approximate number and types of miniexperiments to be included, (e) an approximation of the maximum possible score for the measurement, (f) how points should be distributed among the objectives that define the goal (based on the weights of the objectives), (g) the overall structure of the measurement, and (h) for summative evaluations, the method for converting scores to grades.

The blueprint serves as a guide for building the measurement the way you want it built. Exhibit 9.19 is an example of a blueprint for a unit test to be relevant to the goal listed in Exhibit 9.12 with the objectives weighted as indicated by Exhibit 9.13.

Administration Time and Dates

A traditional group-administered test consisting solely of written-response prompts with directions read by the students is simpler and less time consuming than one with prompts that are presented

Exhibit 9.19

Sample Blueprint for a Unit Test of Exhibit 9.12's Goal With Objectives Weighted as Indicated by Exhibit 9.13.

Unit: 4: Relationships Involving π

Administration Time and Dates:

Prompts Presented in Three Sessions:

1. Section I of the test is a "take home" to be assigned to students on Mon., 10/7 and due on Wed., 10/9.
2. Sections II and III on Wed., 10/9 from 1:10 to 2:00.
3. Sections IV and V on Thurs., 10/10 from 1:10 to 1:40.

Time for Scoring With the Rubrics:

1. Section II to be scored by aide on Wed. between 2:45 and 3:15.
2. Sections I, III, IV, and V scored by me on Thurs. between 3:15 and 7:00.

Provisions for Accommodating Students With Special Needs:

1. I meet with Ms. Mueller to (a) explain the directions for Sections II–V and (b) schedule Chad and Antonio to be administered Sections II and III under her supervision on Wednesday from 1 P.M. to 3:15 and then Parts IV and V on Thursday from 1 P.M. to 3:00.
2. I meet with Mr. Isley-Vaughn to arrange to have a Spanish version of the measurement documents developed for Lupe and Rosanna as well as a Tongan version for Charlotte.

Maximum Number of Points and Maximum Number Per Objective as Indicated by Objectives' Weights:

Maximum: 60

Objective	Points
A	12
B	3 or 4
C	3 or 4
D	7 or 8
E	7 or 8
F	7 or 8
G	7 or 8
H	12

Measurement Outline:

Session 1: Section I: One "take-home" 5-point essay for Objective E and one take-home 4-point display-product prompt for Objective B.

Session 2: Section II: 19 multiple-choice miniexperiments (1 point each) distributed among the objectives as follows: 7 for D, 5 for F, 5 for G, and 2 for H.

Section III: Three display-computation prompts distributed among the objectives as follows: a 2-point one for C, a 3-point one for F, and a 3-point one for G.

Section IV: Two 2-point brief "explain" prompts: one for Objective C and one for Objective E.

Session 3: Section V: Two 5-point display-solution prompts for Objective H and one 12-point essay prompt for Objective A.

Method for Converting Scores to Grades:

Compromise method with approximate midpoints for grade intervals as follows: 55 → A, 45 → B, 35 → C, 25 → D, 15 → F

individually (e.g., for miniexperiments with interview or performance-observation formats). Although convenient, the more usable, simpler measurements often lack relevance for complex learning goals that lead students to do meaningful mathematics. Thus, the complexity of your blueprint and the time allotted to the administration of the measurement depends on the learning goal and students with whom you work.

Validity considerations must be compromised with usability considerations. Generally speaking, measurements with many complex miniexperiments tend to be more valid than ones with only a few simple miniexperiments. However, the shorter, simpler tests tend to be more usable. You cannot, of course, take more time administering and scoring a measurement than you have available. In deciding how much time to devote to the administration of a measurement, keep two other ideas in mind that are somewhat contrary to the traditional wisdom that simpler is more usable:

- When measurements are used for making formative judgments, they can be integrated with learning activities as Mr. Heaps did in Case 3.12. Miniexperiments that are part of learning activities are more complex than most traditional ones on conventional paper-and-pencil tests, but instead of taking time away from learning, they contribute to learning. Thus, for formative purposes, more complex measurements can be more usable than simpler measurements administered separately from learning activities.
- As previously suggested, the administration of measurements used for making summative evaluations should not be integrated with learning activities because students tend to be guarded and fearful of making mistakes when they think their responses are being "graded." However, reviewing responses with students after their scores are recorded can be an extremely productive learning experience. Having complex miniexperiments included on a measurement may provide richer learning experiences when responses are reviewed than you would expect from a simpler test. Thus, additional complexity and time spent for even measurements used for summative evaluations may pay dividends during subsequent learning activities.

Another factor influencing the time you allot to administering a measurement is the importance of students' making rapid responses to prompts. In general, students' responses to prompts for miniexperiments that are relevant to simple-knowledge or algorithmic-skill objectives should be made more rapidly than those requiring reasoning. Inductive reasoning, deductive reasoning, and creative thinking are deliberate processes for which students should not be hurried to exhibit. Thus, miniexperiments relevant to construct-a-concept, discover-a-relationship, comprehension-and-communication, application, and creative-thinking objectives should allow unhurried time for responses to prompts. For a measurement that includes both miniexperiments relevant to memory-level learning and miniexperiments relevant to reasoning-level learning, consider having a timed session that includes only the memory-level prompts and another session in which students have ample time to respond to the reasoning-level prompts.

Note that Exhibit 9.19's blueprint calls for the test to be administered in three different sessions. Keep in mind that you need not confine the administration of a measurement to a single class period or to one school day. A number of short sessions, sometimes with "take-home" sections, is often preferable to one extended "sit-down" grind.

Just as you must limit the time your students have for responding to prompts, you must also consider the time you can reasonably devote to using rubrics to score their responses. The "Time for Scoring With the Rubrics" slot in your blueprint serves as a reminder that you must factor your time into the design of a measurement.

Accommodations for Students With Special Needs

The reason you employ systematic measurement-development strategies is, of course, to produce valid measurement results relative to specified learning goals. However, the validity of a measurement varies according to individual differences among your students. For example, some students are more easily distracted during a test than others. Reading skills, writing skills, motivation to succeed in school, self-confidence, willingness to follow directions, ability to focus on a task, psychomotor skills, tolerance for perplexity, and visual perception are only a minute subset of the myriad of characteristics with which students differ that influence their scores on tests. Ideally, the results of a measurement used to evaluate students' achievement of a goal should be sensitive only to differences among students relative to how well they achieved that goal. Of course, the complexities of classroom-based measurements make such perfectly valid measurements unrealistic.

Although you can hardly control for all the sources of measurement error due to individual differences among your students, you must concern yourself with accommodating students with special needs or disabilities that influence the way they are able to respond to miniexperiment prompts. Not only are such accommodations necessary to improve the validity of measurements, but they are accommodations that we, as professional teachers, are legally required to make. Referring to the *Individuals with Disabilities Act (IDEA)*, initially enacted by Congress in 1975 as Public Law 94–142 and expanded extensively since then, Turnbull and Turnbull (1998, p. 119) stated:

> . . . tests are to be selected and administered to students with impaired sensory, manual, or speaking skills so that the test results accurately reflect the student's aptitude or achievement level—or whatever other factor the test purports to measure—rather than the student's impaired skill, except when those skills are the factors the test seeks to measure. Section 300.35(b) drives home the point that tests should not be misinterpreted, that undue reliance on general intelligence tests is undesirable, and that tests should be administered in such a way that their results will not be distorted because of the student's disability.

For students with disabilities that will impair their opportunities to respond to prompts on tests in a way that accurately reflects what they have learned, their IEPs should spell out accommodations for modifying test administration procedures to mitigate the effects of the disability contaminating measurement validity. Consider Case 9.9 (adapted from Cangelosi, 2000a, pp. 279–280):

CASE 9.9

Several students with various special-education classifications are included in Mr. Johnson's geometry class. For example, Chad has been diagnosed as having a condition that impairs his ability to process information presented verbally and to concurrently focus on multiple tasks. Chad comprehends messages as well as most students except when word and numeral sequences become distorted by his reception mechanisms. This presents a particular problem when Chad attempts to comprehend directions as presented in the prompts of miniexperiments. Often the problem can be solved simply by having the directions rephrased and presented in a second format (e.g., having written directions rephrased orally) and allowing Chad more time to interpret and respond to prompts than most students need. The inability to focus on multiple tasks prevents Chad from doing such things as listening to a lecture and taking notes at the same time. Chad's concentration while responding to miniexperiments is easily disturbed by the presence of others in the classroom.

When Chad and a special-education teacher first explained this condition to Mr. Johnson, Mr. Johnson recognized the need to accommodate Chad's needs during learning activities as well as for administrations of measurements for purposes of making summative evaluations of Chad's achievement of learning goals. However, Mr. Johnson worried that modifying the administration of tests for Chad would be "unfair" to other students for whom special accommodations would not be provided. Then the special-education teacher had Mr. Johnson view the videotape program "Understanding Learning Disabilities: How Difficult Can This Be? The F.A.T. City Workshop" (Lavoie, 1989). The program led Mr. Johnson to experience what students with perceptual disabilities suffer in classrooms. Mr. Johnson realized that failing to make special accommodations for some students is what is truly unfair. For purposes of evaluating students' learning, he recognized that measurement validity is improved by making such accommodations.

Thus, Mr. Johnson took the lead in adding a provision in Chad's IEP that spelled out how the administration of tests for summative evaluations would be modified. By including this in the IEP, Mr. Johnson increased the chances that he would receive support and resources from the school administration for providing the necessary accommodations. The IEP specified that the school would provide an area for Chad to take tests in an extended time period with a supervisor who would be available to clarify directions for prompts without influencing Chad's responses to those prompts once the directions were understood.

Note the "Provisions for Accommodating Students With Special Needs" section of your blueprint shown by Exhibit 9.19. You use this section of the blueprint to plan how to modify the administration of measurements not only for students with special-education classifications but also for students with limited English proficiency. The bilingual-education program in many schools provides test translators for students whose first language is not English—at least until those students become proficient enough in English so that they can take tests with prompts presented in English without seriously contaminating the validity of the results.

Distribution of Points and Weights of Objectives

The maximum possible score for a measurement is, of course, the sum of the maximum possible scores for the individual miniexperiments from which the measurement is composed. If all the rubrics are dichotomously scored as either +1 or +0, then the maximum possible score for the measurement is equal to the number of miniexperiments. But when miniexperiments are included with rubrics with maximum points greater than one, the number of points possible is greater than the number of miniexperiments. In any case, you need to estimate the maximum possible score to compute how many points you will draw from each miniexperiment file.

Once you have estimated the maximum possible score, you select miniexperiments from your computerized files and make transformations of the points from the rubrics so that scores are influenced more by miniexperiments relevant to heavily weighted objectives and less by miniexperiments relevant to lightly weighted objectives. In Case 9.10, Mr. Johnson demonstrates how he accomplished this, resulting in his entries under "Maximum Number of Points Per Objective as Indicated by Objectives' Weights" in Exhibit 9.19's blueprint.

CASE 9.10

After looking at the weights for the eight objectives of his unit on the relationships involving π and surveying the miniexperiments he could access from his computerized folder for that unit, Mr. Johnson estimated that the maximum possible score for the unit test should be about 60. He then calculated the number of points he needs from each miniexperiment file as shown in Exhibit 9.20.

Exhibit 9.20
Mr. Johnson Computes the Number of Points for Each Objective Based on Exhibit 9.12's Weights and a Maximum Test Score of 60.

Objective	Relative Weight	Computation	Points
A	20%	.20 × 60	12
B	06%	.06 × 60	3 or 4
C	06%	.06 × 60	3 or 4
D	12%	.12 × 60	7 or 8
E	12%	.12 × 60	7 or 8
F	12%	.12 × 60	7 or 8
G	12%	.12 × 60	7 or 8
H	20%	.20 × 60	12

Thus, Mr. Johnson plans to select miniexperiments from File A that contribute approximately 12 points to the measurement's anticipated 60-point maximum. He realizes he can accomplish this task in many ways; for example, by either (a) selecting 12 one-point miniexperiments from File A, (b) selecting one 12-point miniexperiment from File A, (c) selecting four one-point miniexperiments from File A and then multiplying each student's score by 3, or (d) selecting any one of a number of other possible combinations from File A and adjusting the scoring so that the total number of points on the measurement from File A's miniexperiments is 12. A similar process is followed for selecting miniexperiments from the other seven files.

Final Elements of the Blueprint

After designing the overall structure of the measurement and developing an outline, you need to choose the method for converting scores to grades—a topic addressed in a subsequent section of this chapter.

Complete the third phase of building a unit test by engaging in Activity 9.7:

Activity 9.7

Retrieve the latest version of your unit goal defined by weighted objectives (from Activity 9.3). Also look over the miniexperiments you stored in your computer files for those objectives (from Activity 9.6). Now devise a measurement blueprint for a unit test that you will design to be relevant to the goal.

In a discussion with a colleague who is also engaging in this activity, critique the blueprint. Modify it if you think you should. Store a file in the "Tests" subfolder of your computerized management system. Also, insert a hard copy in the "Assessment" section of your working portfolio.

Synthesizing the Measurement

You synthesize the test from miniexperiments selected from your computer folders by following the directions from your measurement blueprint. As you do this, keep in mind that one miniexperiment may interact with another; Case 9.11 is an example:

CASE 9.11

Following Exhibit 9.19's blueprint, Mr. Johnson selects miniexperiments from his computerized folders to synthesize the test and then administer it.

During the second session, one student, Judith, responds to one of the multiple-choice prompts that directs her to select two integers between which π falls. Judith selects "between 4 and 5." Later in the session, Judith is reading the directions for one of the display-computation prompts: "Use 3.14 for π." "Oh!" she thinks, "so π isn't between 4 and 5 like I thought." She returns to the multiple-choice section and changes her response to the aforementioned prompt to "between 3 and 4."

On Friday, when Mr. Johnson reviews responses with the class, Judith reveals how she correctly responded to that multiple-choice prompt. Mr. Johnson thinks to himself, "I'd better be more careful in the future to select prompts that don't interact in ways that contaminate test results."

Besides concern for interactions among miniexperiments, you also need to decide how to sequence the prompts. Look at the measurement outline section of Exhibit 9.19's blueprint. Prompts are grouped by format. The test is organized so that students can respond to prompts that have less time-consuming formats (e.g., multiple choice) before responding to those that are more time consuming (e.g., essay). Grouping prompts together with the same format simplifies the directions and prevents students from having to reorient their thinking frequently due to changes in format.

For timed sections of a measurement (e.g., in which all miniexperiments are relevant to memory-level objectives), prompts using the same format should usually be sequenced from less difficult to more difficult. With this arrangement, students avoid spending so much time responding to difficult prompts that they do not have time to respond to easier ones. The easy-to-hard arrangement is unnecessary for untimed sections emphasizing reasoning-level learning.

You include directions to students as part of the prompts you design for miniexperiments. But when synthesizing the measurement, you collapse directions from miniexperiments with the same format so that students need to hear or read them only once.

Complete the fourth phase of building a unit test by engaging in Activity 9.8:

Activity 9.8

Retrieve the measurement blueprint you devised when you engaged in Activity 9.7. Using a subset of the miniexperiments you stored in your computer files when you engaged in Activity 9.6, synthesize a unit test that is consistent with your blueprint.

In a discussion with a colleague who is also engaging in this activity, critique the unit test. Modify it if you think you should. Store a file in the "Tests" subfolder of your computerized management system. Also, insert a hard copy in the "Assessment" section of your working portfolio.

USING TEST RESULTS FOR FORMATIVE FEEDBACK

Obviously, you and your students depend on achievement test results as indicators of how you should regulate your teaching and they should regulate their studying. Reviewing results from recently taken tests is a learning activity that provides students with formative feedback. Test documents—including completed scoring forms with rubrics (e.g., Exhibit 9.8) should be returned to students as soon as reasonably possible; ordinarily, students are primed for corrective feedback and reinforcement the day after a test. "Why didn't you teach us this before we took the test?" is an oft-heard question in sessions in which recently tested mathematical content is reviewed. The answer, of course, is, "The experience of working with the content under test conditions led you to be more receptive to my teaching."

Here is a routine some teachers successfully follow in the class period after a test has been administered and test documents are returned to students:

1. Test documents are returned with *descriptive* annotations. Descriptive comments provide students with specific information about their performance (e.g., "Your drawing made it easier for me to follow your logic" or "Squaring both the numerator and denominator changed the value of this number"), as opposed to judgmental comments (e.g., "good" or "poor").
2. In a brief large-group session, the teacher makes general comments about the test and, if necessary for the first few tests, explains how to interpret scores and annotations.
3. A small collaborative-group session is held, with each group of about six students going through the test—miniexperiment by miniexperiment—and answering one another's questions.
4. A large-group question-discussion session is held in which the teacher (a) reviews matters that test results indicate needed reviewing, and (b) responds to students' questions that were either not answered in the small-group session or arose as a result of that session.

Such activities are particularly effective when students realize that subsequent tests will include miniexperiments relevant to objectives from prior tests. The use of scoring forms with rubrics (e.g., Exhibit 9.8) helps focus discussions on mathematical content so that the sessions are learning activities rather than "gripe" sessions with questions like, "Why did you take off so many points?"

ASSIGNING GRADES TO MEASUREMENT SCORES

Grades

Test review sessions are used to provide students with formative feedback. Grades are the traditional means for communicating your summative evaluations of achievement of learning goals. Periodically assigning letter grades to students' achievements is a responsibility faced by practically all middle and secondary school mathematics teachers. Data from tests are the primary basis for grades.

Many methods for converting test scores to grades are used; none appears completely tenable (Huetinck & Munshin, 2000, pp. 341 & 344; NCTM, 1995, pp. 113–138). How to establish suitable cutoff scores for grades is a question that has been addressed by evaluation specialists but never satisfactorily answered (Berk, 1986; Cangelosi, 1984a; Plake, Impara, & Irwin, 2000). Research studies of grading methods have been more successful in demonstrating the weaknesses of common practice than in providing practical and effective models.

Traditional Percentage Grading

Among laypersons and teachers who are uninitiated in summative evaluation principles, the most familiar method for converting measurement scores to grades is the traditional percentage method. The "percentage," of course, refers to the score's percentage of the maximum possible points for the test. However, the percentage is not indicative of the percent of content a student has learned or objectives achieved.

Consider Case 9.12:

CASE 9.12

Mr. Nelson, a high school geometry teacher, uses the following scale for percentage grading:

94%—100% for an A

86%—93% for a B

78%—85% for a C

70%—77% for a D

00%—69% for an F

The students and their parents are comfortable with his seemingly "objective" scheme. They think they understand exactly what it takes to make a certain grade in the class.

On the first test of the semester, over half the students' scores are less than 77% of the maximum. Mr. Nelson, holding fast to his "standards," assigns the majority of the scores Ds and Fs. However, either consciously or not, he constructs the next test so that it is much easier than the first. The grade distribution for the second test is more in line with what he had anticipated for the first test. Mr. Nelson commends the class for their efforts, commenting that they must have studied harder for the second test than they did for the first.

Because Mr. Nelson thinks of C as average and because of his grading scale, he tends to design tests so that the average percentage score will be between 78 and 85. This causes his tests to be so easy that they fail to measure higher levels of achievement. If he includes enough miniexperiments with very challenging prompts requiring sophisticated responses, too many students would "fail." Because the students who reach sophisticated achievement levels do not find his tests particularly challenging, they are unlikely to learn at the levels they are capable of achieving (Stiggins, 1988).

Case 9.13—a continuation of Case 9.12—illustrates another weakness of traditional percentage grading:

CASE 9.13

Mr. Nelson administers a test consisting of 60 prompts, each with a simple scoring rubric so that +1 is awarded the "correct" response and +0 otherwise. Three students' numbers of correct responses and percentage scores are as follows:

Milan: 51 correct responses → 85%

Albin: 52 correct responses → 87%

Joyce: 56 correct responses → 93%

As shown by Exhibit 9.21, Mr. Nelson uses his scale to convert Milan's score to a C, Albin's to a B, and Joyce's to a B.

Surely measurement error is a likely cause for the difference of 1 between Albin's 52 and Milan's 51, yet in Case 9.13, they received different grades. On the other hand, Joyce's score received the same grade as Albin's although the difference between 56 and 52 is greater than the difference between 52 and 51.

Visual-Inspection Grading

One of the more commonly recommended processes for assigning letter grades to measurement scores is the visual-inspection method (Smith & Adams, 1972). The process is as follows:

1. Draw a scale that encompasses the range of the scores. For example, if the least score is 5 and the greatest is 44, the segment of the scale in Exhibit 9.22 would suffice.
2. Graph the frequency distribution of the scores onto the scale as shown by Exhibit 9.23.
3. Identify gaps or significant breaks in the distribution.
4. Assign a letter grade to each cluster of scores appearing between gaps. If, for example, you choose to define C as average, the cluster containing the middle score might be assigned the grade C. Or you may choose to sample some of the students' responses from a particular cluster and decide the grade for that cluster based on the quality of the responses in the sample. In this method, every score within the same cluster is given the same grade. Exhibit 9.24 depicts one possible assignment of grades to scores from Exhibit 9.23.

Exhibit 9.21
Comparison of Three Grades Determined by the Traditional Percentage Method.

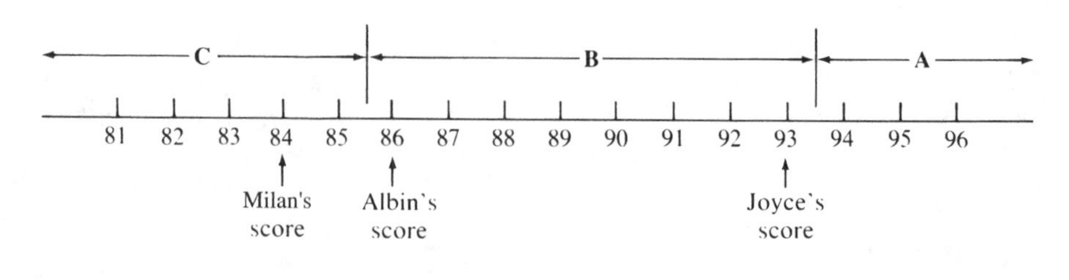

Exhibit 9.22
Segment of Number Line for Use With Visual-Inspection Grading for Scores Between 5 and 44.

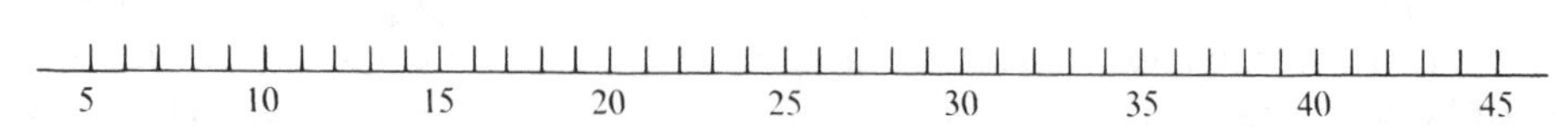

Compromise Grading

Conflict Between the Theoretical and the Practical

Teachers who are introduced to visual-inspection grading readily recognize the following advantages it has over traditional percentage grading:

- Measurements can be designed to include challenging prompts for students with more sophisticated levels of achievement without fear that too many students will receive low grades. The difficulty of the measurement can be factored into the grading scheme.
- Scores that are not markedly different from one another are not assigned different grades; measurement error is acknowledged.

However, teachers tend to reject visual-inspection grading because of the following:

- Establishing criteria for A, B, C, D, and F after a measurement has been administered does not seem to be as objective as having predetermined cutoff points (e.g., 70% for passing) of which students are aware before the test.
- Scores often fail to fall into convenient clusters with significantly large enough gaps between different groupings as required by the visual-inspection method. The distribution in Exhibit 9.25 is a possibility.

A Resolution

The method I suggest has the two advantages of visual-inspection grading while obviating its two weaknesses. This method is a *compromise* between the traditional percentage and the visual-inspection methods. The compromise grading is implemented as follows:

1. As with traditional percentage grading, establish cutoff points for each letter grade before administration of the measurement. However there are two differences:
   A. To allow for the use of a measurement that includes some prompts that will challenge students with sophisticated achievement levels, you are free to set unconventionally low cutoff points for A, B, C, D, and F. Exhibit 9.26, for example, presents possible criteria for a measurement designed to produce an average score of 40 out of a possible 80.
   B. The cutoff point for each letter grade is established with the understanding that there is a buffer or in-between zone between each letter grade category. For example:

   > In Exhibit 9.27, to be assigned a definite B, a score would have to be at least 54 and no greater than 66. For a definite C, the score would be between 34 and 46 inclusive. Scores greater than 46 but less than 54 would fall in the "C or B" category.

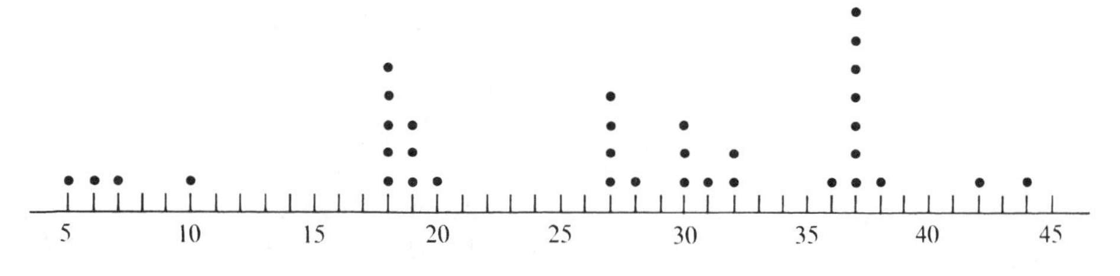

Exhibit 9.23
Sample Scores and Frequency Distribution.

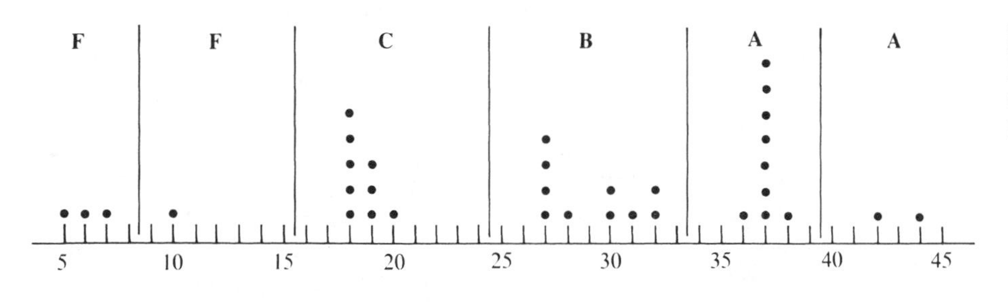

Exhibit 9.24
Example of Scores From Exhibit 9.23 Assigned Grades via Visual Inspection.

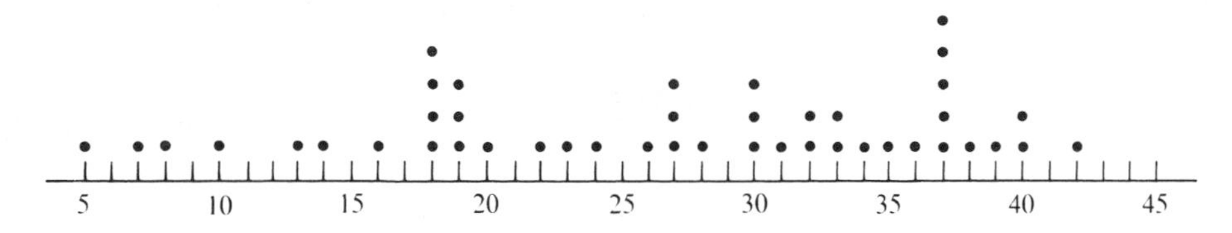

Exhibit 9.25
Inconvenient Score Distribution for Use With Visual-Inspection Grading.

Exhibit 9.26
Example of Letter-Grade Cutoff Points Used for Compromise Grading.

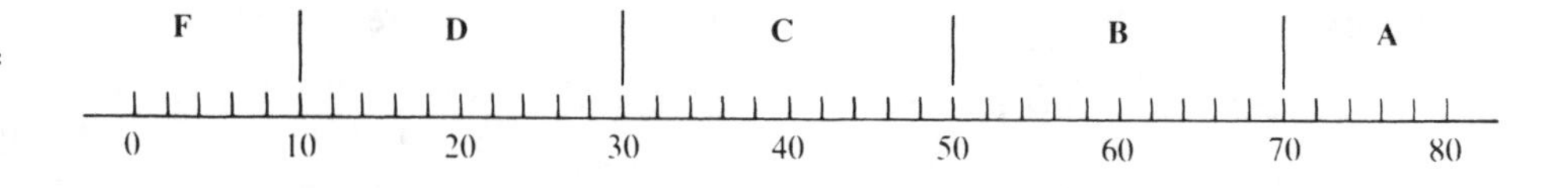

Exhibit 9.27
Example of Letter-Grade Cutoff Points With In-Between Zones for Compromise Grading.

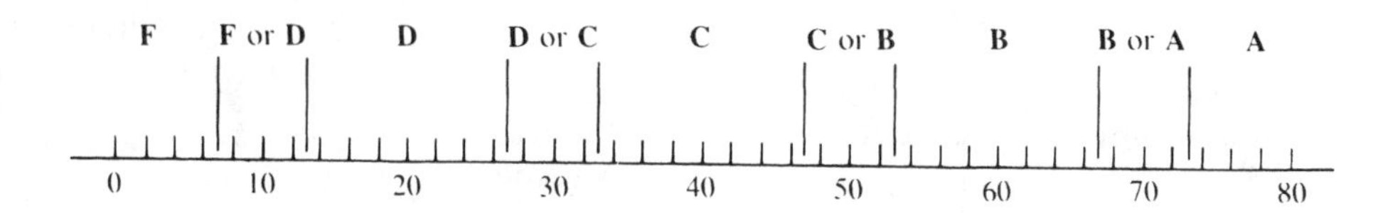

You decide the cutoff points and the size of the in-between zones. The size of these zones can be computed using the standard error of measurement—a statistical routine that is a function of the reliability of the test (see Cangelosi, 2000a, pp. 595–627). However, just creating a little distance to account for at least some measurement error is recommended here.

2. Assign in-between grades to scores that fall between the definite grade intervals. For example:

   Using Exhibit 9.27, a score of 51 falls in the "C or B" category. Final determination of whether the higher or lower grade prevails depends on results from other measurements (e.g., an interview in which the student is prompted to elaborate on some of her responses from the original measurement). You may also have the option of simply letting the grade remain in limbo between the two letter grades and then factoring the in-between grade into your final assignment of a grade on the periodic report.

Point and Counterpoint

The compromise method is criticized for two reasons:

- Some teachers are uncomfortable with scores falling within the buffer zones. They want clear lines of demarcation between grades.
- Unconventionally challenging prompts may lead students to be frustrated because students' scores will be lower than with tests used in conjunction with traditional percentage grading.

It would be convenient to have clear lines of demarcation between letter grades with no scores falling within buffer zones. However, the state of the art of evaluating student achievement is not advanced to the point where precision measurement is possible. Small differences between scores are simply not meaningful; measurement error is too great. The educational community needs to acknowledge this fact.

Regarding the second criticism, there may be some pedagogical disadvantages of challenging prompts and in some circumstances they should not be included. However, once your students become accustomed to responding to challenging prompts, you can use such miniexperiments as tools to lead them to more advanced achievement levels. This is especially true once they have experiences with challenging prompts for purposes of formative feedback. They get used to these challenges in situations where the results are not used for summative evaluations. Then when it is time to test for purposes of making summative evaluations, they are accustomed to struggling with challenging prompts.

PORTFOLIOS FOR COMMUNICATING SUMMATIVE EVALUATIONS OF STUDENT ACHIEVEMENT

The Complexity of Students' Mathematical Achievements

One point on which virtually everyone seems to agree is that meaningful evaluations of student achievement should be based on multiple measurements (American Psychological Association, American Educational Research Association, & National Council on Measurement in Education, 1999; NCTM, 1995; 2000b). In other words, no one type of measurement (e.g., written-response test or performance observation) sufficiently provides the relevant and reliable data that reflect what students have actually learned; a variety of measurements is needed. At least some of the measurements should be individually tailored to students' characteristics, tendencies, and learning styles. Furthermore, a letter grade or other set of symbols can hardly reflect complex learning outcomes (O'Hagan, 1997). Many even suggest that grades interfere with learning (Bauman, 1997; Huetinck & Munshin, 2000, pp. 341 & 344).

Case 9.14 demonstrates the dilemma teachers face in trying to report their summative evaluations of students' achievements and the need for alternatives or supplements to traditional grade reports:

CASE 9.14

After their students depart for the day, Hadezza Robinson and Tyler Longley engage in the following conversation in the faculty workroom of Greystone High School:

Ms. Robinson: You don't look happy, Tyler. Did you have one of "those" days?

Mr. Longley: Actually, I'm really pleased with the way the day went; the kids really got into their work. But thanks for asking.

Ms. Robinson: Well, you don't appear very pleased, sitting there grimacing and moaning to yourself.

Mr. Longley: Because my students had such a productive day, I feel worse about having to boil down all their accomplishments in mathematics to a grade. It's distressing!

Ms. Robinson: So it's about one of those times of year again when we have to get grade reports out. I'm not looking forward to it either, but I'd think your students' successes would make grading a more pleasant chore. It's always easier to report successes than failures.

Mr. Longley: I would've thought so too, but that's not the way things are working. Here's an example of my dilemma. You know Rebecca Powell?

Ms. Robinson: Sure, I have her brother T.J. for precalculus.

Mr. Longley: Rebecca really got turned on to mathematics in our unit on probability. She began to see applications of mathematics she's been exposed to for the past couple of years—something she's never recognized before. She demonstrated such creativity in that unit, coming up with all kinds of ways to apply the relationships and algorithms.

Ms. Robinson: I'm trying to figure out why all this is distressing you. I'd think you'd be jumping at the chance to report her successes!

Mr. Longley: I am, but the report card business has me hamstrung because the grade period covers four units. She got an A for the probability unit, and an A for a unit on sequences. But for the descriptive statistics unit and another on special functions, she didn't learn much of anything. When I average out her grades, she'll end up with a B− at best. B− just doesn't reflect her enthusiasm for what she did with probabilities. If only I had made a more concerted effort to relate real-life situations in the statistics and special functions units like I did for the other two units, she might have gotten turned on to those also!

Ms. Robinson: I understand what you're saying. You want to highlight her accomplishments without having them blended with her lack of success in other areas.

Mr. Longley: And that's not the only example that's bothering me. Sawyer Bond's scores on my written-response tests are really low. But now and then when I catch him in a one-on-one situation and ask him to explain concepts, relationships, and algorithms, he shows sophisticated insights. I think his test performances in mathematics are affected by some major learning gaps relative to reading comprehension. On top of that, he doesn't demonstrate his understanding in large-group class activities because he's shy in those situations. But he's learned a lot about doing mathematics; it just doesn't show up on conventional tests and demonstrations. If I give him a mathematics achievement report based on the usual measures, it's going to send him and his parents the wrong message. On the other hand, I've got to grade equitably.

Ms. Robinson: To compound the problem, I bet you've got students whose test scores exceed their true levels of achievement.

Mr. Longley: You've got that right! I can think of several who are really test wise plus they have exceptional language arts skills.

Ms. Robinson: What we need is to keep an individualized file for each student—one that reflects their special accomplishments and interests as well as reports of more general achievement levels.

Mr. Longley: Which one of us is going to propose that at the next faculty meeting?

Individualized Student Portfolios

How can you efficiently maintain and manage a file on each student? Many teachers do so by having their students develop individualized portfolios reflecting their accomplishments similar to your developing your professional portfolio. Besides being a mechanism for learning and formative feedback, portfolios can be used to communicate some of the details of what students have learned that are missing from grade reports. In the professional literature, "*portfolio assessment*" refers to using individualized student portfolios to communicate summative evaluations of student achievement. Portfolio assessment is commonly used in many language arts and English courses that emphasize students' writing skills (Gallager, 1998, pp. 302–320) and is a mandated component of the grading and reporting systems in some school districts.

In Case 9.15, mathematics teachers Tyler Longley and Hadezza Robinson act on the need to incorporate portfolio assessment in their grade-reporting system that they discussed in Case 9.14:

CASE 9.15

Greystone High School's students have been dismissed for the summer break and Tyler Longley and Hadezza Robinson have just completed their work assigning students' final grades. Their conversation turns toward plans for the next school year:

Mr. Longley: You know how the English faculty went to portfolio assessment this year; they seemed pretty pleased with the way it worked out.

Ms. Robinson: Portfolio assessment—that's where you save examples of students' work in folders or boxes and use them to demonstrate what they've learned and accomplished. Wouldn't that address the problem that upsets you every 10 weeks when grade reports are due? You remonstrate having to boil all students learned down to a single symbol. Why don't you have your students get some folders and stick the best samples of their work in them? Maybe you could even convince Sarah [Greystone High's principal] to let you experiment with supplanting letter grades with portfolios.

Mr. Longley: I don't think of portfolios as simply containers for students' work samples. I'd like to see us do something that significantly impacts the way we teach and students learn mathematics.

Ms. Robinson: What do you mean?

Mr. Longley: Well, we're always talking about how we should do a better job of addressing the *PSSM* (NCTM, 2000b)—you know, getting students to apply mathematics to address real-life problems, communicate mathematics, use technology, see mathematics from a historical perspective, and, most importantly, to make connections among different mathematical subtopics as well as connections between mathematics and other school subject-content areas. But then we complain that we're so busy "covering" material and preparing students to take traditional tests that we don't have time to emphasize these more important aspects of learning mathematics.

Ms. Robinson: So how would portfolios help?

Mr. Longley: If we do it right, individualized portfolios could provide a vehicle for stimulating students to take the mathematics they learn in class and relate it to their special interests, reflect on the connections and history, and do some expository work on topics that pique their interest. But if we move in this direction, it should also be reflected in our summative evaluations of their achievement—in the grade reports.

Ms. Robinson: So why don't you run this idea by Sarah and try it?

Mr. Longley: I'm hoping it'll be something we might develop and try as a department. Sometimes it's tough to be the only one attempting something new. Parents and students start to wonder, "Why is goofy Mr. Longley doing this when nobody else is?" If we're all involved, we'll support one another and share ideas. Sarah can be pretty persuasive with the school board. Someday, we might consider going to the Board with a proposal to supplant grades completely with portfolios.

Ms. Robinson: Not until we've tried it out for a couple of years and gotten the bugs out of the program.

Mr. Longley: Right. But to show that the portfolios are an integral, significant part of our mathematics curriculum, we need to tie them to grades somehow. I'd like to see us factor the portfolios into the grading scheme—maybe make the students' portfolio projects 30% of the 10-week term grades; 70% of each grade can be calculated in the traditional way using test scores, homework, or whatever teachers are doing now.

Ms. Robinson: But shouldn't samples of students' test papers and homework be included in their portfolios?

Mr. Longley: Not the way I'm visualizing the portfolio projects.

Ms. Robinson: What do you mean by a "portfolio project"?

Mr. Longley: I'm not completely sure I know myself. But it should be something that gets students to synthesize, extend, and expound upon the mathematical content we're teaching them in our regular lessons and on which we evaluate their learning. I need you to brainstorm with me. Could we get together and collaborate on a proposal to present to the rest of the department?

Ms. Robinson: Yes, but don't expect everybody in the department to go for the idea. Scott will object to anything that deviates from following his religion.

Mr. Longley: What religion?

Ms. Robinson: His beloved, blessed mathematics textbooks. His religion is to get students through the textbooks page by page and follow the commandment, "Thou shalt not deviate from the sacred textbooks." To him mathematics doesn't exist outside of textbooks.

Mr. Longley: You're right, but at least we might get the majority of the department to contribute their ideas and agree to try something like this.

Ms. Robinson: What time tomorrow should we meet?

Several days later, Mr. Longley and Ms. Robinson draft the proposal displayed by Exhibit 9.28. Exhibit 9.29 shows a scoring rubric for portfolio assessment and Exhibit 9.30 shows a planning form for a portfolio project.

If you decide to incorporate individualized student portfolios into your grading system as planned in Case 9.15, then either the portfolio itself or each portfolio item (e.g., a computer program or the proof of a theorem) will need to be treated as a miniexperiment with a prompt for students and an observer's rubric.

TESTS YOU DID NOT DESIGN

To this point, chapter 9 has focused on measurements for formative judgments and summative evaluations that you design yourself. The ongoing informal miniexperiments you conduct during lessons guide how you teach and how your students learn. The unit tests you develop influence students' grades. Grades and individualized student portfolios communicate mathematical accomplishments. You are in the best position to design valid measurements that reflect your unique students' mathematical accomplishments.

Exhibit 9.28
Portfolio Assessment Proposal Mr. Longley and Ms. Robinson Drafted for Consideration by Greystone High's Mathematics Department.

Proposed Provisions for Greystone High School Mathematics Department's Portfolio Assessment

1. Each student enrolled in a mathematics course at Greystone High School will be required to develop and maintain an individualized mathematics portfolio for the purpose of exhibiting her portfolio projects.
2. For each of the four 10-week-long grading terms in the school year, the student will complete a combination of one to five projects inclusive.
3. The purpose of the combination of projects is for the student to extend and exhibit her ability to do the following:
   A. Connect mathematics learned during the 10-week school term to mathematics learned in previous school terms or in previous mathematics courses. In the first 10-week term, the connection should be between a mathematical topic (e.g., rational functions) from one of the first term's units to a mathematical topic from a previous mathematics course (e.g., rational numbers). In a subsequent term, the connection should be between a mathematical topic from one of the current term's units (e.g., theorems about similar triangles) to a mathematical topic from the prior term (theorems about parallel lines).
   B. Connect mathematics learned during the school term (e.g., nonlinear functions) to subject content learned in nonmathematics Greystone High School courses the student is currently taking (e.g., ecological systems in biology).
   C. Apply mathematics learned during the school term (e.g., nonlinear functions) to address real-life situations the student faces outside of his work at Greystone High School (e.g., personal health and physical fitness questions).
   D. Explain mathematics learned during the school term from a historical perspective (e.g., the pursuit of perfect numbers).
   E. Use technology to discover, invent, or communicate mathematics related to what is learned during the school term (e.g., use *Geometer's sketchPAD* to compare relationships in hyperbolic geometry to those in Euclidean geometry).
4. The combination of projects should be designed so that all five parts (i.e, listed "A" through "E" above) of the purpose as stated in #3 above are realized. This could be accomplished by five separate projects with each project addressing a different part or, preferably, by incorporating the parts in fewer projects (e.g., two). The following is an example of a possible project combination that accomplishes all five parts of the purpose:

   One of the topics ninth-grader Allison studies during the second term of a geometry course is features of triangles in Euclidean geometry (e.g., the sum of the three degree measures of any triangle is 180). One of the topics she studied in the first term was parallel lines. Allison is also taking world history and one of the seocnd-term topics is 19th century philosophy. Allison completes three second-term portfolio projects; one is as follows:
   i) Using *Geometer's sketchPAD,* she develops a computerized simulation demonstrating how the parallel postulate is applied to prove that the sum of the degree measures of the angles of a triangle is 180. This project fulfills the requirements of parts "A," "D," and "E" listed above.
   ii) She writes a library paper explaining how the development of non-Euclidean geometries in the 19th century shook the foundations of classical logic and changed the face of philosophy. This project fulfills the requirements of parts "B" and "D" listed above.
   iii) She develops a detailed floor plan for her "dream house" with a brief narrative explaining how she applied her knowledge of features of triangles in the design. This project fulfills the requirement of part "C" listed above.
5. Each combination of term projects should produce concrete artifacts that demonstrate what the student accomplished. The artifacts that are stored in the student's portfolio can employ a combination of the following modes:
   - Videotaped presentation
   - Expository essay
   - Computer-based presentation
   - Poster display illustration
   - Other presentation mode mutually agreed to by the student and her teacher.
6. The combination of portfolio projects should be scored by the teacher using the attached rubric (see Exhibit 9.29).
7. The score for the combination of portfolio projects should influence the term grade by 30%.
8. By the middle of the term, the student in collaboration with the teacher should complete the attached planning form (see Exhibit 9.30).

Exhibit 9.29
Proposed Scoring Rubric for Portfolio Assessment.

Scoring Form for ______________________________ 's
(student's name)

______________ **Term Mathematics Portfolio**
(1st, 2nd, 3rd, or 4th)

Note: For each criterion listed, points are awarded as follows:

+2 if the criterion is clearly met
+1 if it is unclear as to whether or not the criterion is met
+0 if the criterion is clearly not met

Each encircled numeral (either 0, 1, or 2) indicates the number of points received for the given criterion.

A. The combination of projects clearly demonstrates that the student has made insightful associations between the following two mathematical topics:

____

from the current term

and

____

from the prior term or (if the current term is the first, then a previous mathematics course)

0 1 2

B. The combination of projects clearly demonstrates that the student has made insightful associations between the following mathematical topic:

____

from the current term

and the following topic from the ______________ course that the student is also taking:

____

from the current term

0 1 2

C. The combination of projects clearly demonstrates that the student has applied the following mathematical content:

____

from the current term

to the following situation from her own real life:

____

____

0 1 2

D. The combination of projects clearly demonstrates that the student possesses insights into the historical foundations of the following mathematical topic:

____

from the current term.

0 1 2

E. The combination of projects clearly demonstrates that the student has applied the following class of technology:

____

to accomplish the following mathematical task related to the current term:

____

____

0 1 2

Score for the combination of portfolio projects (10) ________

Exhibit 9.30
Proposed Student-Planning Form for a Term's Portfolio Project.

My name is ______________________________. The date is ______________________.

I plan to accomplish the following tasks via a combination of mathematics portfolio projects during this the _____ term of the school year:

A. Demonstrate that I have made insightful associations between the following two mathematical topics:

____

from the current term

and

____

from the prior term or (if the current term is the first, then a previous mathematics course)

My plan for accomplishing this is to do the following:

____

____

____

____

using the following mode of communication (circle one):

Videotaped presentation **Expository essay** **Computer-based presentation** **Poster display illustration**

Other (explain) ____

B. Demonstrate that I have made insightful associations between the following mathematical topic:

____

from the current term

and the following topic from the _____ course that I am taking:

____

from the current term

My plan for accomplishing this is to do the following:

____

____

____

____

using the following mode of communication (circle one):

Videotaped presentation **Expository essay** **Computer-based presentation** **Poster display illustration**

Other (explain) ____

C. Demonstrate that I have applied the following mathematical content:

____

from the current term

to the following situation from my own real life:

____

____

Continued

Exhibit 9.30
Continued

My plan for accomplishing this is to do the following:

____

____

____

____

using the following mode of communication (circle one):

Videotaped presentation **Expository essay** **Computer-based presentation** **Poster display illustration**

Other (explain) ______________________________

D. Demonstrate that I possess insights into the historical foundations of the following mathematical topic:

____

from the current term

My plan for accomplishing this is to do the following:

____

____

____

____

using the following mode of communication (circle one):

Videotaped presentation **Expository essay** **Computer-based presentation** **Poster display illustration**

Other (explain) ______________________________

E. Demonstrate that I have applied the following types of technology:

____

to accomplish the following mathematical task related to the current term:

____

____

My plan for accomplishing this is to do the following:

____

____

____

____

using the following mode of communication (circle one):

Videotaped presentation **Expository essay** **Computer-based presentation** **Poster display illustration**

Other (explain) ______________________________

Thus, I will be conducting ________ portfolio projects this term.

Signed by me ____________________________ and my teacher:

____________________________/____________ (date)

However, there are commercially available tests that you may sometimes choose to use.

Mathematical achievement tests can be purchased from commercial publishers. Ordinarily, each class-size set of student textbooks is accompanied by a packet of materials for the teacher that includes tests—usually one test per chapter. Most such tests emphasize simple knowledge and algorithmic skills. Because they do not tend to be relevant to higher cognitive-learning levels (e.g., construct a concept, discover a relationship, and application), it is hardly advisable to use them for unit tests. However, tests from textbook publishers can be a source of individual prompts—especially at the algorithmic-skill level—for you to incorporate in tests you design yourself. Computerized files of individual miniexperiments—usually referred to as "test items"—are available commercially. Advertisements for such products are sent to inservice mathematics teachers and members of NCTM; they are also found in professional journals and displayed at NCTM conferences and meetings of some other professional societies.

There are also the so-called "high stakes" tests that (a) your students choose to take (e.g., American College Test [ACT], Scholastic Assessment Test [SAT], Advanced Placement [AP] Calculus, and AP Statistics) that influence their plans for college or (b) state legislatures and school boards mandate that influence public perceptions of your school.

HIGH-STAKES TESTS

Government Mandates

A test is considered *high stakes* if it is administered statewide or district-wide for the purpose of (a) accountability for student achievement or (b) evaluations affecting promotion, retention, or graduation (U.S. General Accounting Office, 1993). Legislative statutes now exist in the vast majority of states that require high-stakes testing in the form of *core-curriculum tests* as well as *standardized achievement tests* (Merrow, 2001). School districts throughout the United States share a common history in which the roles of these tests has been expanded far beyond the use for which they were originally designed (Cangelosi, 2000a, pp. 510–525). The reasons for the increase in mandated high-stakes testing is politically rather than scientifically based (Kohn, 2001; Merrow, 2001).

Exhibit 9.31 is NCTM's position statement on high-stakes testing (NCTM, 2000a).

Core-Curriculum Tests

Almost all state offices of education publish core-curriculum guides in hard copy and on their websites (e.g., www.state.ct.us/sde). The guides list standards or goals to be incorporated in all public school courses in the respective subject-content areas. For example, the Utah State Office of Education core-curriculum guide for sixth-grade mathematics lists 13 standards, explicating each with a statement of purpose, a list of objectives, and a list of skills and strategies. Exhibit 9.32 displays one of those standards as presented in the guide (Utah State Office of Education, 1995).

Most state offices of education provide core-curriculum tests—in some areas referred to as "end-of-level tests"—intended to be relevant to the core standards or goals in their respective core-curriculum guides. In most states, these tests are developed by committees of teachers, state office personnel, and curriculum and testing consultants. In some states, the tests are developed through contracts with commercial testing companies (e.g., Measured Progress). State core tests are usually designed so that scores indicate the percentage of points each student obtained relative to each core-curriculum standard. Thus, unlike standardized tests, these tests are referred to as "criterion referenced" rather than "norm referenced" because they are intended to be used for criterion-referenced summative evaluations of students' achievements.

A summative evaluation of a student's achievement is *criterion referenced* if the evaluation is influenced by how the measurement results relevant to that student's achievement compare to a standard that is not dependent on results obtained from others. On the other hand, a summative evaluation of a student's achievement is *norm referenced* if the evaluation is influenced by how measurement results relevant to that student's achievement compare to the results obtained from others.

Exhibit 9.33 is an example of a student's score report from a state core-curriculum test.

Some local school districts have their own core-curriculum standards and competencies and their own district core-curriculum tests that are administered district-wide. These tests are very similar to state core-curriculum tests. In fact, some of the state-level tests were developed as revisions of existing district-level tests. With the advent of statewide testing, fewer and fewer local school districts are now administering their own core-curriculum tests.

These tests are especially high stakes in states with legislation that ties students' passing subjects (e.g., mathematics) and graduation to their core-curriculum test scores. Because schoolwide and district-wide results are publicized by news media—thus, affecting taxpayers' perceptions of schools—many teachers feel pressured by their administrators to push students to obtain high scores; the net effect on curricula has been destructive (Franklin, 2001; Merrow, 2001).

Exhibit 9.31
NCTM's Position Statement on High-Stakes Testing (NCTM, 2000a).

High-Stakes Testing (Nov., 2000)

Position

The National Council of Teachers of Mathematics believes that far-reaching and critical educational decisions should be made only on the basis of multiple measures. A well-conceived system of assessment and accountability must consist of a number of assessment components at various levels.

Rationale

High-stakes tests are tests that are used to make significant educational decisions about children, teachers, schools, or school districts. To use a single objective test in the determination of such things as graduation, course credit, grade placement, promotion to the next grade, or placement in special groups is a serious misuse of such tests. This misuse of tests is unacceptable. The movement toward high-stakes testing marks a major retreat from fairness, accuracy, and educational equity. When test use is inappropriate, especially in making high-stakes decisions about a child's future, it undermines the quality of education and equality of opportunity.

Just as disturbing as the serious misuse of these tests is the manner in which the content and format of these high-stakes tests tends to narrow the curriculum and limit instructional approaches. Test results may also be invalidated by teaching so narrowly to the objectives of a particular test that scores are raised without actually improving the broader, often more important, set of academic skills that the test is intended to measure.

Assessment should be a means of fostering growth toward high expectations and should support high levels of student learning. When assessments are used in thoughtful and meaningful ways, students' scores provide important information that, when combined with information from other sources, can lead to decisions that promote student learning and equality of opportunity. The misuse of tests for high-stakes purposes has subverted the benefits these tests can bring if they are used appropriately.

Recommendations

- Multiple sources of assessment information should be used when making high-stakes decisions. No single high-stakes test should be used for making decisions about the tracking, promotion, or graduation of individual children.
- Assessment methods must be appropriate for their purposes.
- All aspects of mathematical knowledge and its connections should be assessed.
- Instruction and curriculum should be considered equally in judging the quality of a program.
- Assessment should advance students' learning and inform teachers as they make instructional decisions.
- Assessment should be an open process with everyone knowing what is expected, what will be measured, and what the results imply for what should be done next.
- If tests are used as one of multiple measures in making high-stakes decisions about students, those tests must be valid and reliable for the purposes for which they are used; they must measure what the student was taught; they must provide students with multiple opportunities to demonstrate proficiency; and they must provide appropriate accommodations for students with special needs or limited English proficiency.
- All standardized assessments of mathematical understanding at the national, state, province, district, or classroom level should be aligned with the NCTM Standards.

Standardized Tests

A test is *standardized* if it has been field-tested on a large sample of people—referred to as the "norm group"—to (a) measure reliability of the test and (b) establish norm-referenced standards for use in interpreting scores for subsequent administrations of the test.

The norm-referenced standards (referred to as "test norms") provide averages and standard deviations for different grade levels. Each grade-level average provides the standard to which scores of students in that grade level are compared; the standard deviation serves as the unit of comparison. Thus, a student's score on the test is reported as the number of standard deviations that score fell above or below the average of the norm-group scores. This, of course, is what is commonly known in statistics as a "*z*-score."

Standardized test reports use a variety of methods for reporting a student's score in comparison to the norm-group average and standard deviation. For example, the numbers in Exhibit 9.34, derived from Corina's score on the mathematics test, should be interpreted as follows:

- *Percentile:* A score S_j is reported as a percentile of p if $S_j > p\%$ of the norm group's scores and $S_j < (100 - p)\%$ of the norm group's scores. Thus, Exhibit 9.34 indicates that Corina's mathematics score is greater than 55% of the mathematics scores from the norm group for her grade level—which is 8.2 (i.e., second month of eighth grade).

Exhibit 9.32
Excerpt From *Mathematics Core Curriculum: Level 6* (Utah State Office of Education, 1995).

Standard 5060-02 The students will show understanding and application of mathematical concepts and justification of solutions to problems by communicating in oral, pictorial, and/or written form.

Purpose: This standard highlights the need to involve students individually in cooperative learning groups actively doing mathematics. Exploring, investigating, describing, and explaining mathematical ideas promotes communication. Teachers facilitate this process when they pose probing questions and invite students to explain their thinking.

Objectives

5060-0201. Model situations using oral, written, concrete, pictorial, graphical, and algebraic methods.
5060-0202. Reflect on and clarify thinking about mathematical ideas and situations.
5060-0203. Develop common understandings of mathematical ideas including the role of definitions.
5060-0204. Interpret and evaluate mathematical ideas by using the skills of reading, listening, and viewing.
5060-0205. Discuss mathematical ideas and make conjectures and convincing arguments.
5060-0206. Understand the value of mathematical notation and its role in the development of mathematical ideas.

SKILL AND STRATEGIES

1. Represent mathematical topics studied at this grade level by making sketches, drawing diagrams, or by using objects.
2. Recognize that representing data in a list, table, or graph is a way of translating a mathematical idea into a new form.
3. Discuss, read, and write about mathematical topics presented at this grade level. (Write the steps to follow when multiplying a two-digit number by a two-digit number.)
4. Explain and justify solutions to a given problem.
5. Write about and discuss interpretations and various solutions to open-ended problems.

Because of the way percentiles are scaled, percentiles do not lend themselves to arithmetic operations as do familiar cardinal numbers. Thus, percentiles should not be used to gauge progress over time. The results from subtracting two percentiles, for example, can be very misleading (Cangelosi, 2000a, pp. 553–558).

- *Grade equivalent:* Grade equivalents are reported in the form of "*g.m*" with *m* associated with the *m*th month of grade *g*. They are the most widely misinterpreted ways of reporting standardized test results (Cangelosi, 2000a, pp. 560–563). According to Exhibit 9.34, Corina's grade equivalent result in mathematics indicates that her score is about equal to the average score of those students in the norm group who were in the second month of the eighth grade when they took the test. Arithmetic computations should never be used with grade equivalents, as the scaling characteristics will distort results even more gravely than with percentiles.
- *Scaled score:* Scaled scores are three-digit numbers usually greater than 200 that do not have the scaling problems of other reporting methods. Scaled scores are the only type of standardized test numbers that lend themselves to gauging progress over time. Thus, if on the same standardized test, Corina received scaled-score results of 320, 540, and 580, respectively, from three consecutive years, it would indicate that her performance improved to a greater degree between the first and second years than it did between the second and third years. Such comparisons should not be made with other methods of reporting standardized test results.
- *Stanine:* To encourage people to perceive standardized test scores for what they are—rough comparisons to norm-group scores—scores are collapsed into nine categories or intervals so that each raw score is reported simply as either a "1," "2," . . ., or "9." Scores very near the norm-group average are assigned to the middle interval, which is stanine 5. The length of each of the middle seven stanine intervals is one half a norm-group standard deviation. Thus, the stanine of 5 reported for Corina's mathematics test in Exhibit 9.34 indicates that her score was approximately equal to the average of the scores from the 8.2 grade-level norm group.

Exhibit 9.35 displays a stanine scale stacked with a *z*-score scale and a percentile scale.

Under certain conditions, standardized tests are useful in research studies to compare two or more populations with respect to a particular variable. They are not designed to be used in many of the ways in which school personnel are being mandated to use them (Kohn, 2000). Case 9.16 (adapted from Cangelosi, 2000a, pp. 570–571) is a conversation between Casey Rudd and a parent:

CASE 9.16

Ms. Frese: Just how good are these standardized tests we're giving our children? Are they valid?

Exhibit 9.33
Example of a Student's Score Report From a State Core-Curriculum Test for One Subject-Content Area.

State Core-Curriculum Test
Mathematics Level 6
Individual Student Report
5/17/02

District: South Point

School: Parkway Middle

Student: Sarah Anderson
Teacher: Ms. C. Yanjun

	Core Standard	Subscore	Competency Level
3360–01	Applies math concepts and skills to solve real-life problems	3	Sufficient
3360–02	Communicates, illustrates, and justify solutions to problems	3	Sufficient
3360–03	Uses inductive reasoning to form mathematical relationships	4	Excellent
3360–04	Recognizes connections within mathematics and to other disciplines	3	Sufficient
3360–05	Communicates with and translates among multiple number representations	2	Insufficient
3360–06	Demonstrates an understanding of various subsets of real numbers	2	Insufficient
3360–07	Explains relationship underlying computation and estimation algorithms	3	Sufficient
3360–08	Analyzes, generalizes, and represents functional relationships and patterns	3	Sufficient
3360–09	Informally explores fundamental algebraic concepts and relationships	3	Sufficient
3360–10	Organizes, translate, and represents data in useful ways	4	Excellent
3360–11	Explores real-life situations by experimenting with probability models	3	Sufficient
3360–12	Informally explores and discovers fundamental geometric relationships	4	Excellent
3360–13	Applies appropriate measurement tools for problem solving	4	Excellent

Total Score: 41 Overall Competency Level: Sufficient

Mr. Rudd: Allow me to try to address the question by asking you a question. Do you think that there are times when we should have our temperatures taken—like when we see our physicians for checkups?

Ms. Frese: Sure.

Mr. Rudd: Then do you think measuring body temperature is a valid measure?

Ms. Frese: I've always thought so.

Mr. Rudd: What if you went to your doctor for a checkup and she had her nurse take your temperature, resulting in a score of 98.6°? Based on that test score only, your doctor declared you to be disease free, healthy, and very fit. How valid do you think the temperature test is now?

Ms. Frese: That would be ridiculous. Having a normal temperature doesn't rule out most diseases—like

Exhibit 9.34
Corina Twyman's Scores From a Standardized Test Battery.

Pupil Profile for Corina Twyman Grade level: 8.2

	Language	Math	Social Studies	Science
Percentile	78	55	71	49
Nat. Grade Equivalent	9.9	8.2	9.0	8.2
Scaled Score	566	540	551	494
National Stanine	7	5	6	5

Exhibit 9.35
z-Score, Stanine, and Percentile Scales.

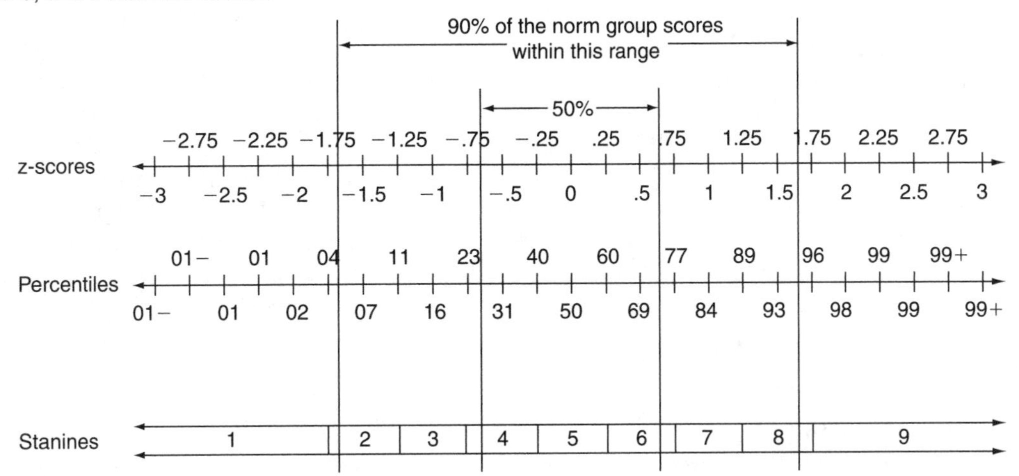

cancer. And even if my temperature were elevated, it would only indicate the possibility that I have an infection; that suggests a need for further tests—such as analyzing a blood sample—to pinpoint the cause of the elevated temperature.

Mr. Rudd: So you're saying that taking your temperature is valid only for a very limited purpose—like to help decide whether to conduct further tests for a possible infection. There are a myriad of other health-related factors to which taking one's temperature is not relevant.

Ms. Frese: Right; it's valid for the one thing, but not others.

Mr. Rudd: The analogy works when we address the question about the validity of various standardized achievement tests. They are valid only for very limited purposes—like to compare two large groups relative to some very general skills. The standardized tests themselves aren't bad, but what is bad is misusing them to make evaluations about questions for which they are not relevant.

AP Calculus and Statistics Tests

If you teach AP calculus or AP statistics, you must concern yourself with preparing students to take advanced-placement tests that can earn them college credit. The AB level of the AP calculus test is designed so that students scoring high enough receive credit for a one-semester beginning calculus course. The BC level of the test can merit credit for a second college calculus course. The maximum for either level of the exam is 5 with 3 considered "passing." However, required cutoff scores for college credit varies depending on the college or university. AP tests are criterion referenced with a specific criterion associated with each possible score. Students can take your AP calculus course but opt not to take the corresponding AP test. The grades they receive from your course are independent from their scores on the AP test.

In 1997, the AP statistics exam was introduced, which affords high school students an opportunity to earn college credit for a noncalculus-based statistics course.

PSSM-Based Practice and High-Stakes Testing

Revisit Case 1.3 in which Casey Rudd interviewed for his position at Rainbow High School. Note how concerned Principal Harriet Adkins was about students' performances on core-curriculum and standardized tests and how Casey—with help from mathematics teacher Vanessa Castillo—demonstrated how *PSSM*-based teaching strategies are more likely to lead to high scores on these tests than the traditional follow-the-textbook approach.

SYNTHESIS ACTIVITIES FOR CHAPTER 9

1. Select the one response to the following multiple-choice prompts that either completes the statement so that it is true or accurately answers the question:
   A. Anytime teachers evaluate student achievement they _____
      a) make value judgments
      b) use valid measurements

c) base evaluations on unit test results
d) determine better ways of teaching

B. Stiggins' research suggested that _____
a) all achievement tests emphasize simple knowledge and algorithmic skills
b) the cognitive levels at which students are tested tend to limit the cognitive levels at which they learn
c) commercially produced tests are superior to teacher-designed measurements
d) valid and usable measures of student achievement are virtually impossible to design

C. Which one of the following is NOT a measurement?
a) Administering a unit test
b) Seeing a student write on the board
c) Hearing a student say, "Math is fun!"
d) Seeing that a student is unable to bisect an angle with a compass and straightedge
e) Seeing that a student is not bisecting an angle with a compass and a straightedge

D. Which one of the following is a *necessary* condition for measurement relevance?
a) Internal consistency
b) Pertinence to the learning goal
c) Validity
d) Usability

E. Which one of the following is a *sufficient* condition for measurement relevance?
a) Internal consistency
b) Pertinence to the learning goal
c) Valibility
d) Usability

F. Which one of the following is a *necessary* condition for reliability?
a) Scorer consistency
b) Usability
c) Relevance
d) Pertinence to the intended mathematical content

G. Which one of the following is a *sufficient* condition for measurement reliability?
a) Usability or relevance
b) Usability and relevance
c) Internal or scorer consistency
d) Internal and scorer consistency

H. Which one of the following is a *sufficient* condition for a measurement to be useful?
a) Usability, internal consistency, and pertinence to the intended mathematical content
b) Relevance, reliability, and validity
c) Relevance, reliability, and usability

I. Which one of the following modifications to a miniexperiment is most likely to improve its scorer consistency?
a) Raise the learning level to which the miniexperiment is relevant from simple knowledge or algorithmic skill to one that requires reasoning (e.g., application).
b) Make the rubric more specific.
c) Change the format of the prompt from multiple-choice to essay.
d) Provide greater latitude for the scorer to use professional judgment.

J. Which one of the following variables depends on the stated purpose of the measurement?
a) Usability
b) Relevance
c) Scorer consistency
d) Internal consistency

K. Which one of the following variables depends on the time it takes to administer a test?
a) Usability
b) Relevance
c) Scorer consistency
d) Internal consistency

L. Out of concern for his students' abilities to read books and articles about mathematics, Mr. Kembloski includes an objective in most of his teaching units for his students to remember definitions of certain mathematical terms. During the course, he plans to build their reading vocabulary of mathematical words. One miniexperiment he intends to be relevant to one of those objectives appears in Exhibit 9.36. Which one of the following is a weakness of the miniexperiment?
a) "Fractal" is not a mathematically well-defined word.
b) The prompt reverses the stimulus-response order of the objective.
c) The miniexperiment requires only simple-knowledge cognition.

M. A miniexperiment with a rubric designed to reflect which steps in a process students do or do not remember is likely to be relevant to which one of the following types of objectives?
a) Discover a relationship
b) Creative thinking
c) Construct a concept
d) Application
e) Algorithmic skill
f) Simple knowledge

N. Why should at least some miniexperiments that are designed to be relevant to an application objective that specifies the Pythagorean relationship as mathematical content confront students with situations in which the Pythagorean relationship does not apply?
a) The Pythagorean relationship is limited; students have broader concerns.

Exhibit 9.36
Mr. Kembloski's Miniexperiment.

Prompt:

Fill in the missing word:

A ________ is a curve or surface (or solid or higher dimensional object) that contains more but similar complexity the closer one looks

Rubric:

+1 for "fractal," otherwise +0.

Exhibit 9.37
Standaedized Mathematics Test Results Reported for Ichiro Sayaki.

Student: Ichiro Syaki
Grade Level: 11.7

	Math Knowledge	Math Computation
Percentile	33	46
Stanine	4	5
Grade Equivalent	10.8	11.2
Scaled Score	325	331

b) Learning the Pythagorean relationship is a prerequisite to learning about other relationships (e.g., the formula for finding the distance between two coordinate points in a Cartesian plane).
c) Students need to also learn about other relationships involving triangles, not only $a^2 + b^2 = c^2$.
d) Achievement of an application objective requires students to discriminate between problems to which the specific mathematical content does and problems to which it does not apply.

O. For miniexperiments to be relevant to creative-thinking objectives, they must _____.
a) provide students with opportunities to reason convergently
b) present students with tasks never previously accomplished by anyone
c) require students to produce novel products
d) have rubrics that discriminate between atypical and typical reasoning

2. With a colleague, compare and discuss your responses to the multiple-choice prompts from Synthesis Activity 1. Also, check your choices against the following key: A–a, B–b, C–d, D–b, E–c, F–a, G–d, H–c, I–b, J–b, K–a, L–b, M–e, N–d, 0–d.
3. Resurrect the unit test you developed when you engaged in Activity 9.7. If reasonably convenient to do so, administer your test to one or more students. In light of the results and discussions with the student(s), revise the test as you see fit. Reinsert the revised test in your portfolio along with notes reflecting on what you learned from the field test.
4. Write an essay, about one page long, explaining (a) the relative advantages and disadvantages of using visual-inspection grading instead of traditional percentage grading and (b) the relative advantages and disadvantages of using compromise grading instead of visual-inspection grading. Exchange your essay with that of a colleague and discuss issues stimulated by reading one another's explanations.
5. Get together with a colleague to practice interpreting the standardized test report shown by exhibit 9.37.
6. Access the website of a state office of education for a state of your choice. Use the site to inform yourself about the state's curriculum standards for mathematics and core-curriculum testing program. Compare your findings to that of a colleague who went to another state's office of education website.

TRANSITIONAL ACTIVITY FROM CHAPTER 9 TO CHAPTER 10

In a discussion with two or more of your colleagues, address the following questions:

1. Cases throughout this book report on teachers implementing lessons that they designed themselves—lessons that included learning activities that extend beyond what is available from typical mathematics textbooks. What do teachers like these use for sources of ideas on teaching mathematics?
2. What technologies are generally available to teachers to enable them to make professional-quality presentations, displays, and tasksheets? How can the internet be used to enhance the quality of mathematics lessons?
3. How should students use graphing calculators and computers to learn and do mathematics?
4. How can teachers take advantage of professional associations such as NCTM to enhance their teaching?

chapter

10

Technology and Resources for Teaching and Learning Mathematics

Goal and Objectives for Chapter 10

The Goal The goal of chapter 10 is to provide you with an overview of resources and technologies available for leading students to do meaningful mathematics.

The Objectives Chapter 10's goal is defined by the following set of objectives:

A. You will describe a variety of resources for stimulating ideas on teaching mathematics, learning mathematics, and doing mathematics (comprehension and communication) 30%.
B. You will describe a variety of technologies and mathematics curriculum materials typically available for use in middle, junior high, and high schools (comprehension and communication) 30%.
C. You will develop your own strategies for incorporating various technologies and curriculum materials into your teaching to enhance learning activities and the efficiency with which you organize and prepare for instruction (application) 25%.
D. You will be an active member of NCTM and take advantage of NCTM's resources and services (willingness to try) 15%.

SELECTION AND USE OF TEXTBOOKS

You, of course, are the person ultimately responsible for designing, developing, and organizing mathematics curricula for your students. Plan to draw your ideas from many sources including other teachers, instructional supervisors, professional journals (e.g., *Mathematics Teacher* and *Mathematics Teaching in the Middle School*), professional conferences, inservice workshops, college or university courses, trade books, websites, electronic media, reference books, textbooks and accompanying supplements for teachers, and personal experiences. The availability, dependability, quality, and sophistication of these sources vary considerably depending on where you teach and how assertive and resourceful you are in seeking them out and taking advantage of them. Not necessarily the most useful, but surely ones that are standard virtually anywhere you teach are adopted mathematics textbooks.

In some school districts, a single publisher's textbook series with a volume for each course may be adopted either school-wide or district-wide. In other school districts, textbook-adoption policies may allow textbooks to be considered on a course-by-course basis so that texts for different mathematics courses might be drawn from a variety of publishers.

For each class-size set of student textbooks for a course, there is typically a special edition of the book for teachers that includes margin notes, answers to exercises, and addenda as an aid in the designing and planning of instruction. A variety of ancillary materials for teachers is also available from most publishers with such items as (a) CD-ROMs with reference materials, miniexperiment files, computer-based learning activities, and electronic manipulatives, (b) bulletin board displays, (c) overhead transparencies or black-line masters for making transparencies, (d) chapter tests, (e) scope and sequence charts for planning a course using the textbook, (f) lists of instructional materials for use with the text including concrete manipulatives and models, and (g) website addresses for instructional support services.

The degree of control you can exercise regarding the selection of textbooks for your courses varies considerably depending on your school situation. Typically, selections are from a state-adopted list of approved textbooks. Some districts depend on panels of mathematics teachers and school administrators to select mathematics textbooks for all the district's schools. In other districts, the selections are made at the school level, usually by mathematics department faculty members. Of course, options for new texts are hardly available until existing texts are worn out or judged to be out of date. In many situations, teachers are required to live with whatever textbooks have been adopted for them.

To increase your familiarity with current secondary and middle school mathematics textbooks and to extend your talent for critiquing texts, engage in Activity 10.1:

Activity 10.1

Obtain two different mathematics textbooks currently being used in secondary or middle schools. If reasonably convenient, select two competing teachers' editions of the texts with accompanying ancillary materials from different publishers that are intended to be used for the same course. Current teachers' editions of textbooks and accompanying ancillary materials can be checked out from many college and university curriculum libraries for teaching majors. Also, you may be able to borrow textbooks from locals schools or school-district resource centers.

Compare the two textbooks with regard to each of the following questions and comments:

1. How well do the book's topics match those specified by relevant curriculum guidelines—including the ones you are supposed to follow at your school—and the ones you have listed for your teaching units? It is not a drawback for the text to include topics you do not plan to include in your teaching units as long as the text's treatment of topics you do include is not dependent on book topics you exclude. Including topics in your units that are not dealt with in the text can be inconvenient.
2. Does the text provide high-quality exercises relevant to the learning objectives you want your students to achieve? More than anything else, textbooks provide teachers with an abundance of exercises for students to practice. Typically, the preponderance of textbook exercises are of the algorithmic-skill variety. Even most textbook word problems require only algorithmic skills once the wording is deciphered and comprehended.
3. How accurate is the mathematics presented in the text? Virtually all textbooks contain a few misprints and report a few incorrect computational results. Sometimes such mistakes provoke healthy discussions that enhance rather than detract from learning. However, conceptual errors or mathematical treatises that conflict with what you want your students to learn can hinder your lessons. For example, many algebra books define a *variable* as "a letter that stands for a number." If you take that definition literally—and I assume you want your students to take mathematical definitions literally—then interest rates, age, speed, time, shape, set, location, length, number of, angle, angle measure, and all the other variables we deal with in problem solving are not variables. These texts say a variable is a *letter*—not what the letter stands for, but the letter itself! Such a restrictive definition precludes connections of mathematics with real-life situations. Furthermore, in the language of a rigorous mathematical system, arithmetic and algebraic operations (e.g., $+$) are defined on numbers, not letters. Set theory operations (e.g., $\cap$) are defined for sets, not letters. Consequently, you need to consider the conceptual treatment of key topics (e.g., variables and functions) before selecting a textbook.
4. How readily can the textbook be used with the technology of your choice, especially graphing calculators and computers? Most textbooks published since the NCTM *Standards* (NCTM, 1989a) include calculator-based and computer-based exercises. Unfortunately, in some texts, the technology-based exercises and technology-related explanations are clearly token inserts presented in addition to the "real mathematical" topics of the chapter. Fortunately, there is a trend to publish textbooks that actually integrate the use of technology in the treatise of mathematical topics, not just as extra add-ons.
5. How consistent are the book's organization and presentations with *PSSM*-based teaching and research-based learning principles suggested herein? The organization, presentation of topics, and pedagogy of typical mathematics textbooks are inconsistent with instructional strategies that lead students to learn meaningful mathematics (American Association for the Advancement of Science, 2000; Cangelosi, 2001a). Thus, you should expect to organize your lessons differently from your textbook's presentations. However, it would be convenient to use a text that is in harmony with teaching strategies you employ. In the supplemental materials accompanying textbooks, most publishers now include a cross-reference list of entries in the texts to either the NCTM *Standards* (NCTM, 1989a) or *PSSM* (NCTM, 2000b). In further response to NCTM's recommendations, suggestions for attending to

cooperative learning, problem solving, connections among mathematical subdisciplines (e.g., algebra and geometry), use of technology, alternative assessments, multicultural education, accommodation of students for whom English is not a first language, and activities for students classified as "gifted and talented" are also commonly inserted into teacher supplements.

6. How readable is the textbook for your students? Will they understand the explanations? Typically, students do not read explanations in mathematics textbooks. Generally, they read only examples and exercises (Kuehl, 2002). However the need for students to read mathematics is well publicized (Friedlander & Tabach, 2001: Miura, 2001).
7. How attractively packaged is the text? The text's aesthetic appeal (e.g., colorful pictures and clever use of white space) may influence the amount of time your students spend with the book. Be careful, however, that you do not let an appealing package obscure substantive factors (e.g., accuracy of the mathematics). Furthermore, the trend toward glossy glitz with pages filled with colorful, eye-catching pictures of celebrities and products may actually interfere with comprehension of the mathematical messages (Cangelosi, 2001a). As you examine texts, note how common it is for the most noticeable feature on a page to be a bright picture of something like a bag of french fries with a cup of soda pop. Ask yourself, what is the pedagogical and mathematical relevance of the pictures?
8. How practical and helpful are the supplements for teachers? Will the software packages, references, tests, and display materials really be useful to talented, qualified teachers like you or are they just gimmicks to make the textbook series more appealing to school administrators and textbook adoption boards? Some sales pitches unscrupulously target lay school board members with the idea that fancy-appearing textbook supplements make it possible for students to learn mathematics despite unqualified or untalented teachers.
9. How much does the textbook package cost? Cost factors may or may not be critical, depending on the particular textbook-acquisition arrangement under which your school administrators operate.

Exchange your comparative critiques with those of colleagues who are also engaging in this activity. Discuss the features of the textbooks and how you might incorporate them in your teaching.

Insert documents resulting from this activity in the "Technology" section of your working portfolio.

When choosing a school to practice your profession, assess the availability of resources and the degree to which you and other teachers exercise control over those resources, as well as the selection of textbooks and other learning materials.

SOURCES OF IDEAS FOR TEACHING MATHEMATICS

Colleagues

Preservice teacher preparation programs provide beginning teachers with necessary, but insufficient, competencies to be successful inservice teachers. Your effectiveness as a mathematics teacher depends on how well you develop those competencies on the job as a professional inservice teacher (Duke, Cangelosi, & Knight, 1988; Smith, 2001). To be consistently effective—especially in the first few years of your career—you need support, guidance, and feedback as you develop curricula, design and conduct lessons, manage students, and assess achievement (Cangelosi, 1991, pp. xi–xii, 121–173; Evans, 1989).

Research findings in the area of instructional supervision suggest that the most useful means of stimulating ideas for teaching involve interactions among teachers. Visiting one another's classrooms, peer coaching, colleague mentoring, sharing responsibilities for students, and think sessions are invaluable ways for you to learn from other teachers as they learn from you (Birman, Desimone, Porter, & Garet, 2000; Danielson, 2001; Feiler, Heritage, & Gallimore, 2000). Recall Casey Rudd's interactions with his colleagues near the end of Case 1.13 and in Case 1.14 from the section "Benefitting From Instructional Supervision."

Some schools have established systems whereby the school week is structured so that teachers have time to work with one another, engage in joint planning, collaborate on developing integrated curricula, and participate in workshops. In other schools such collegial efforts occur only when individual teachers initiate their own informal networks. Unfortunately, teaching becomes a solitary art in some schools with climates that discourage collegiality.

In many geographic areas, collaborative networks of mathematics teachers have been established with support from funding agencies (e.g., through the National Science Foundation and the Eisenhower Program of the U.S. Office of Education); the networks provide a means for mathematics teachers from different schools and school districts to interact, share ideas, and provide assistance to one another (e.g., the *Mathematics Teacher Network* [Wilford, 1993]).

NCTM and Other Professional Societies

Besides supporting teachers' causes, professional organizations (e.g., NCTM, National Education Association [NEA], Mathematical Association of America [MAA], and American Federation of Teachers [AFT]) provide forums for idea sharing and resource materials in the form of conferences, journals, books, newsletters, video programs, websites, and computer software.

Your NCTM membership benefits include a subscription of your choice to one of the following journals:

- *Mathematics Teacher,* published nine times a year, contains articles specifically for the purpose of providing secondary school mathematics teachers with practical ideas they can implement in their classrooms.
- *Mathematics Teaching in the Middle School,* published nine times a year, is similar to *Mathematics Teacher* but focuses on the middle school level.
- *Teaching Children Mathematics,* published nine times a year, is similar to the two aforementioned journals but focuses on the elementary and pre-K school levels.
- *Journal for Research in Mathematics Education,* published five times a year, reports research studies relevant to questions about teaching and learning mathematics.

Besides ideas, suggestions, and learning activities to implement in your classroom, *Mathematics Teacher, Mathematics Teaching in the Middle School,* and *Teaching Children Mathematics* communicate information relevant to professional mathematics teachers (e.g., [a] reviews of publications, software, manipulatives, and technology, [b] notices of professional conferences, workshops, and funding opportunities, and [c] information on how to obtain instructional materials).

To increase your familiarity with *Mathematics Teacher* and *Mathematics Teaching in the Middle School* and to stimulate your ideas for designing lessons for your students, engage in Activity 10.2:

Activity 10.2

Examine at least one recent issue of *Mathematics Teacher* and one recent issue of *Mathematics Teaching in the Middle School.* Familiarize yourself with some of the journals' regular features (e.g., "Activities," "Calendar," "Technology Tips," "Teacher to Teacher," "Menu or Problems," and "Media Clips"). From each journal, select and read an article. Discuss what you learned with a colleague who is also engaging in this activity.

Make a note of any references that provided you with ideas or activities you might implement at some point in your teaching. Insert that note in the appropriate subfolder (e.g., "classroom management," "prealgebra," or "geometry") under "Resources" in the computerized folder structure you set up when you engaged in Activity 1.3. Also insert a copy in the "Professional Development" section of your working portfolio.

NCTM members also receive (a) a subscription to *NCTM News Bulletin*—published six times a year—which reports on current events relevant to the mathematics teaching profession, (b) discounts on subscriptions to the three journals you did not select as part of your regular membership fee, as well as other publications (e.g., NCTM *Yearbooks* and the *PSSM Navigations Series*), (c) opportunities to participate in national regional NCTM conferences consisting of lectures, workshops, seminars, displays, business meetings, and exchanges of ideas among an international group of colleagues, and (d) access to inservice and instructional materials and technology at discounted prices.

If you are not already a member of NCTM, then engage in Activity 10.3:

Activity 10.3

Join NCTM by either accessing www.nctm.org, phoning 1-800-235-7566, or mailing a membership application form from a recent NCTM publication (e.g., *Mathematics Teacher* or *NCTM Catalog*). Configure your membership so that you receive both *Mathematics Teacher* and *Mathematics Teaching in the Middle School.* If you are a full-time student, ask about full-time membership at half the annual fee; it is easier to receive the 50%-off-fees benefit for full-time students by phone than by mail or website applications.

Resource Centers and Inservice Providers

Most, but not all, public school district offices sponsor periodic inservice workshops for teachers at which mathematics-education specialists or teachers present ideas on a variety of topics (e.g., classroom management, alternative assessment, technologies available in the district, or core-curriculum standards and tests). Many districts also maintain resource centers where teachers can borrow instructional materials, professional enrichment literature, software, equipment for classroom use, presentation materials, and mathematical manipulatives. School mathematics departments typically maintain their own collections of such materials for their teachers.

Colleges and universities through regular on-campus programs, evening schools, summer schools, extension and outreach programs, and distance-education networks offer inservice and graduate-level courses for teachers. Mathematics courses and advanced mathematics teaching methods courses are typically included among the offerings.

Government agencies (e.g., the U.S. Office of Education and the National Science Foundation) and research and development centers (e.g., WestEd Laboratories) sponsor projects that provide inservice educational opportunities for teachers as well as professional enrichment materials. They distribute reports and announcements on funded projects for inservice teacher education programs as well as curriculum and professional enhancement materials (e.g., the University of Chicago School Mathematics Project [Zimmer, 2001]). There are also opportunities for you to obtain funding for your own curriculum development, inservice education, or action-research projects. NCTM journals and the *NCTM News Bulletin* keep readers apprized of some of these opportunities; these agencies' websites are another source of information (e.g., www.nsf.gov and www.ed.gov).

Electronic and Literary Sources on Mathematics Teaching

Besides trade books (e.g., those listed in Exhibit 7.9) and professional journals (e.g., *Mathematics Teacher* and *Mathematics Teaching in the Middle School*), massive amounts of literary and electronic sources of information, ideas, and learning activities for mathematics teachers are available. Twenty years ago, when a mathematics teachers was faced with a particular problem or question, it was a challenge to locate any sources at all that addressed the problem or question. Today, the challenge is to select which sources to use from the vast pools of books, reports, websites, CD-ROMs, and video programs. Exhibit 10.1 is a relatively minute sample of these sources. Besides this list, also keep Exhibits 2.8, 2.20, 2.21 4.11, and 7.9 in mind.

To acquaint yourself with some of the many websites for mathematics teachers, engage in Activity 10.4:

Activity 10.4

Access the internet; use a search engine to identify websites for the topic "Teaching Mathematics." Read the descriptors for about 10 of the thousands of sites. Visit several sites that appear interesting; make note of those that you anticipate utilizing as you develop curricula and seek teaching ideas.

HANDS-ON MANIPULATIVES AND CONCRETE MODELS

Learning activities in which students work with hands-on, concrete objects, models, and measuring devices are common in mathematics lessons in the primary grades, but in the past they have been relatively rare at the secondary and middle school levels. Fortunately, the NCTM *Standards* (NCTM, 1989a) and *PSSM* (NCTM, 2000b) have influenced secondary and middle school mathematics teachers to incorporate

Exhibit 10.1

Examples of Sources for Ideas and Activities Relevant to Teaching Mathematics.

Artzt, A. F., & Armour-Thomas, E. (2002). *Becoming a reflective mathematics teacher: A guide for observations and self-assessment.* Mahwah: NJ: Lawrence Erlbaum.

Ashlock, R. B. (2001). *Error patterns in computation: Error patterns to improve instruction* (8th ed.). Upper Saddle River, NJ: Prentice-Hall.

Association for Curriculum and Supervision Development. (2001). *The brain and mathematics* [Videotapes: (1) *Making number sense* & (2) *Classroom applications*]. Baltimore: Author.

Association for Curriculum and Supervision Development. (2001). *The lesson collection: Math strategies* [Videotapes #21–#24: (21) *Pre-algebra—patterns and formulas,* (22) *Algebra I—comparing data,* (23) *Algebra II—series and sequences,* (24) *Geometry—surface area and volume*]. *Baltimore: Author.*

Atweh, B. (Ed.). (2001). *Sociocultural research on mathematics education: An international perspective.* Mahwah: NJ: Lawrence Erlbaum.

Brumbaugh, D. K., & Rock, D. (2001). *Teaching secondary mathematics* (2nded.). Mahwah: NJ: Lawrence Erlbaum.

Burke, M., Erickson, E., Lott, J. W., & Obert, M. (2001). *Navigating through algebra in grades 9–12.* Reston, Va: NCTM.

Cangelosi, J. S. (1988). *Demystifying school mathematics (Episode 1: Mathematics: A discovery or an invention; Episode 2: The mystification of mathematics in schools; Episode 3 Real-life applications of mathematics)* [Videotape series]. Washington: National Science Foundation; Logan: Utah State University Telecommunications Division.

Cangelosi, J. S. (1989). *Using mathematics to solve real-life problems (Episode 1: Variables: Real-life and mathematical; Episode 2: Relations: Real-life and mathematical; Episode 3: Measurements; Episode 4: Numbers; Episode 5: Binary Operations; Episode 6: Interpretations and judgments)* [Videotape series]. Washington: National Science Foundation; Logan: Utah State University Telecommunications Division.

Continued

 **Exhibit 10.1**
Continued

Cangelosi, J. S. (1999). *Leading students to use their mathematical skills to describe biological phenomena (Episode 1: Questioning strategies for student discovery).* [Videotape]. Washington: U.S. Office of Education; Logan: Utah State Multimedia and Distance Learning Services.

Chazan, D. (2000). *Beyond formulas in mathematics and teaching: Dynamics of the high school algebra classroom.* New York: Teachers College Press.

Cooney, T. J., Brown, S. I., Dossy, J. A., Schrage, G., & Wittmann, E. C. (1998). *Mathematics, pedagogy, and secondary teacher education.* Westport, CT: Heinemann.

Countryman, J. (1992). *Writing to learn mathematics.* Westport, CT: Heinemann.

Fennell, F., Bamberger, H., Rowan, T., Sammons, & Suarez, A. (2001). *Connect to NCTM Standards 2000* [volumes for Grades 6, 7, & 8]. Bothell, WA: Creative Publications.

Friel, S., Rachlin, S., & Doyle, D. (2001). *Navigating through algebra in grades 6–8.* Reston, Va: NCTM.

Heddens, J. W., & Speer, W. R. (2001). *Today's mathematics: Part 1: Concepts and classroom methods* (10th ed.). New York: John Wiley & Sons.

Heddens, J. W., & Speer, W. R. (2001). *Today's mathematics: Part 2: Activities and instructional ideas* (10th ed.). New York: John Wiley & Sons.

Hirsch, C. R. (Ed.). *Addenda series: Grades 9–12.* Reston, VA: NCTM.

Huetinck, L. & Munshin, S. N. (2000). *Teaching mathematics for the 21st century: Methods and activities for grades 6–12.* Columbus, OH: Prentice-Hall/Merrill.

Madfes, T. (1999). *Learning from assessment: Tools for examining assessment through standards* [Videotape]. Reston, VA: NCTM.

Mathematics Teacher [a monthly journal] published by NCTM (Reston, VA, www.nctm.org).

Mathematics Teaching in the Middle School [a monthly journal] published by NCTM (Reston, VA, www.nctm.org).

Mathematical Thinking and Learning [a quarterly journal] published by Lawrence Erlbaum (Mahwah, NJ, www.erlbaum.com).

National Council of Teachers of Mathematics. (1989). *Curriculum and evaluation standards for school mathematics.* Reston, Va: Author.

National Council of Teachers of Mathematics. (1991). *Professional standards for teaching mathematics.* Reston, VA: Author.

National Council of Teachers of Mathematics. (1995). *Assessment standards for school mathematics.* Reston, VA: Author.

National Council of Teachers of Mathematics. (2000). *Principles and standards for school mathematics.* Reston, VA: Author.

Posamentier, A. S., & Stepelman, J. (1999). *Teaching secondary mathematics: Techniques and enrichment units.* Upper Saddle River, NJ: Merrill/Prentice-Hall.

Stiff, L. V., & Curcio, F. R. (1999). *Developing mathematical reasoning in Grades K–12: 1999 yearbook.* Reston, VA: NCTM.

Internet Sources:

As you will discover by entering "mathematics teaching" with a search engine, there are thousands of web pages for mathematics teachers. The following are examples:

- Wide variety of resources (e.g., lesson plans) from the U.S. Office of Education: www.thegateway.org
- Eisenhower National Clearinghouse for mathematics and science education: www.enc.org
- Study WEB (math) is a gateway site providing extensive links: www.studyweb.com
- Web Sites and Resources for Teachers provides a well-organized collection of links: www.csun.edu/~vceed009
- Univ. of Tennessee, Knoxville—Mathematics Archives provide access to math resources with an emphasis on teaching materials—particularly educational software: www.archives.math.utk.edu/
- Math Archives—AZ-MATH Software provides resource links to math learning/teaching software on the Internet: www.archives.math.utk.edu/azmath.html
- Florida State University—Mathematics Virtual Library provides a variety of mathematical information categorized by subject: www.euclid.math.fsu.edu/Science/math.html
- Interactive Mathematics provides interactive exercises, online calculators and graphing tools, interactive recreations, archives of teaching documents: www.wims.unice.fr

them in their lessons also. All students—even those in college—need experiences working with manipulatives and concrete models such as those pictured in Exhibits 10.2, 10.3, and 10.5. This is especially critical to promote inductive reasoning during the experimenting stage of lessons for discover-a-relationship objectives. For most mathematical concepts and relationships, students should work from hands-on concrete examples, to numerical examples, to pictorial representations, and finally to symbolic representations (Friedlander & Tabach, 2001; Friel, Rachlin, & Doyle, 2001, pp. 7–17).

Exhibit 10.2
Möbius Strip.

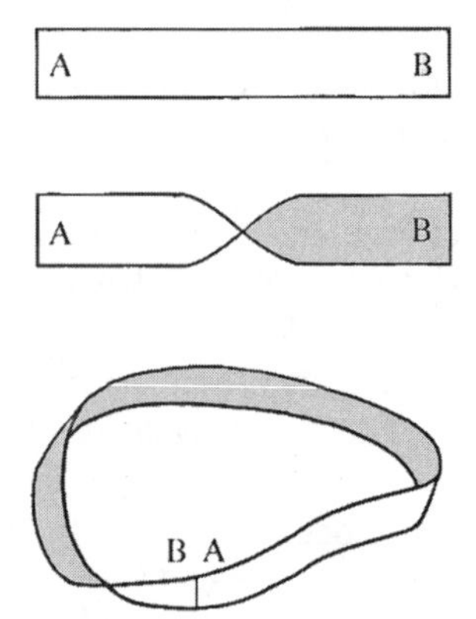

Once mathematics teachers begin involving their students in hands-on activities with concrete objects, they tend to continue doing so. Students prefer these activities to more passive paper-and-pencil exercises and, consequently, are more cooperative, engaged, and easier to manage. However, do not make the mistake of using manipulatives solely for the purpose of keeping students busy and entertained; a teacher in Case 10.1 makes this mistake:

CASE 10.1

In the faculty lounge at the end of the school day, mathematics teachers Lottie Walker and Fred King have the following conversation:

Fred: How'd your day go?

Lottie: Fantastic! We worked with Möbius strips (Hoffman, 1988, pp. 108–121) in all my classes today. They had a ball—got some great discussions going! They were fascinated.

Fred: What's a Möbius strip?

Lottie: Here, I'll show you. See this strip of paper? I'll mark it A and B like this. [pause] Now, I'll twist it and attach *A* to *B* with tape. Now, use your pencil to draw a line from *A* to *B*. (See Exhibit 10.2.)

Fred: There, but I never crossed an edge!

Lottie: Use these scissors to cut down the middle on the line.

Fred: Oh, wow! How can that be? It's only one large continuous band! Why did that happen?

Lottie: That's what the lively discussions in my classes were all about today. I told them if they cooperate during the next unit on polynomial operations, we'll do more of these fun sorts of things afterwards.

Ms. Walker seems to think of her Möbius-strip activity as an aside from her regular lessons—solely for the purpose of entertaining and fascinating students. Such fascinations can lead students to appreciate mathematical phenomena and motivate them to pursue other mathematical relationships. But more is gained from experiences with concrete objects when those experiences are an integral part of lessons that target the learning goal of the unit. When you work with Case 11.9 from chapter 11, note, for example, how Mr. Rudd uses the manipulative activity shown in Exhibit 11.25 to help students discover that $(a + b)^2 = a^2 + 2ab + b^2$. In Case 10.2, Mr. Pitkin incorporates learning activities with manipulatives and concrete models into a unit on Platonic solids. When he first read about the activities in the *Mathematics Teacher* article in Exhibit 10.3, he was not involved in teaching a unit to which the activities would be relevant. But he filed the article for later reference when he might be teaching the unit alluded to in Case 10.2:

CASE 10.2

To provide students with experiences discovering relationships and inventing algorithms as well as achieving the goal of a unit on Platonic solids, Mr. Pitkin integrates activities described in Exhibit 10.6—an article by Hopley (1994).

Concrete materials collected from students' everyday environments usually make manipulatives and mathematical models that are more meaningful to students than prepared instructional materials available from commercial outlets. Such "natural" objects help associate mathematics with real life. However, the commercial products (e.g., attribute blocks, pattern blocks, fraction bars, pentominoes, geoboards, base-ten blocks, geometric solid models, conic section models, algebra tiles, tangrams, and measuring devices and scales) are quite convenient for construct-a-concept and discover-a-relationship lessons. Both natural and contrived objects are useful. The commercial products are advertised in catalogs and websites such as those listed in Exhibit 10.4.

Hands-on manipulatives, concrete models, and other technologies are displayed and demonstrated at NCTM conferences and other professional meetings for mathematics teachers. As an NCTM member, you can expect to receive catalogs listing such products and notices of workshops demonstrating their use.

To stimulate your ideas for incorporating manipulatives and concrete models in lessons you design, engage in Activity 10.5.

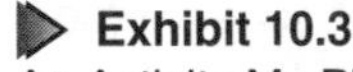
Exhibit 10.3
An Activity Mr. Pitkin Incorporated Into His Unit on Platonic Solids.

Ronald B. Hopley

Nested Platonic Solids: A Class Project in Solid Geometry

Several years ago at a regional NCTM conference in Phoenix, the author was fascinated by a set of cardboard Platonic solids that were nested inside each other. The Platonic solids are polyhedra whose faces are congruent regular polygonal regions, such that the number of edges that meet at each vertex is the same for all vertices; only five are possible. Since the set is no longer commercially available, the author decided to make a nested set for classroom demonstrations and instructions for students to make their own.

Making the nested set is an activity that takes students through many levels of geometric experience. It is a hands-on activity with angle and linear measurement. Students use applications of algebraic equations to geometric relations that result in physical products and gain more experience working with three-dimensional geometry. When they are finished, students have a set of Platonic solids to keep. These solids have interested mathematicians for thousands of years.

Determining the relative sizes of the polyhedra is trivial if the only goal is to make the inner ones smaller than the outer ones. However, the relationships between pairs of these figures is quite interesting if the vertices of the inner figures intersect the outer figures in the right places, as revealed in Pugh's table of relationships (see **table 1**). From 120 possible permutations of the five Platonic

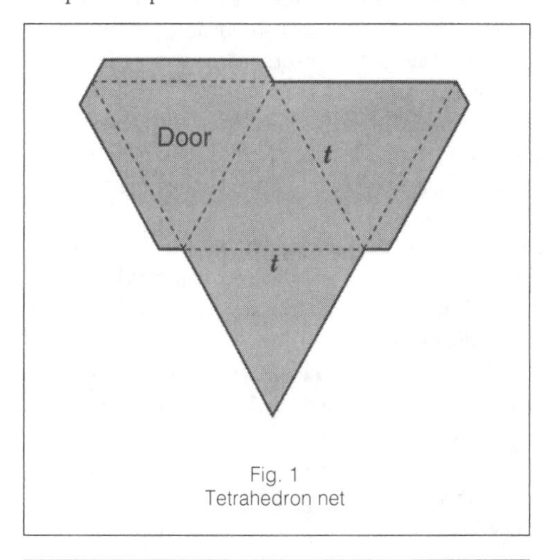

Fig. 1
Tetrahedron net

Ron Hopley teaches at Tucson High Magnet School, Tucson, AZ 85705. He is currently creating files of geometric solids for a three-dimensional computer-graphics program.

Continued

Exhibit 10.3
Continued

TABLE 1
Relationships between Pairs of Platonic Polyhedra

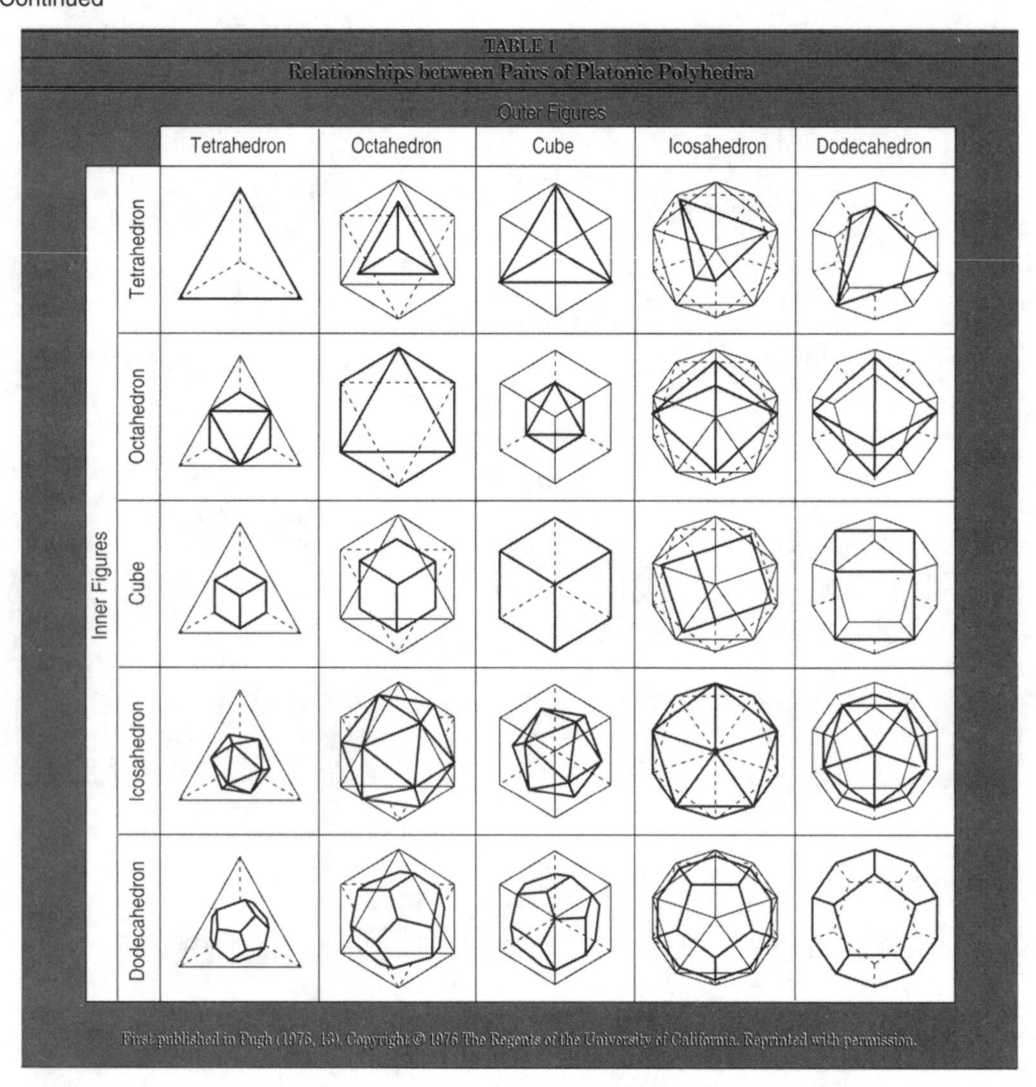

First published in Pugh (1976, 13). Copyright © 1976 The Regents of the University of California. Reprinted with permission.

solids, the set chosen, from innermost out, consists of octahedron, tetrahedron, cube, dodecahedron, and icosahedron. Another octahedron was added on the outside to show that the sequence of solids could continue. The calculations are at the level of high school geometry students. Building the tetrahedron is a good place for students to begin, since that solid has the fewest faces.

Each solid requires approximately one class period to build, except the icosahedron, which takes a little more time for a full discussion of the calculations. Students are supplied with lightweight poster board, scissors, rulers, and glue sticks. Students should have studied right triangles before making the cube and should have been introduced to trigonometry before making the dodecahedron and icosahedron. Thus they do not spend five consecutive class periods building the set. This past year the activity was broken up into two sessions, and next year four or five sessions will be tried.

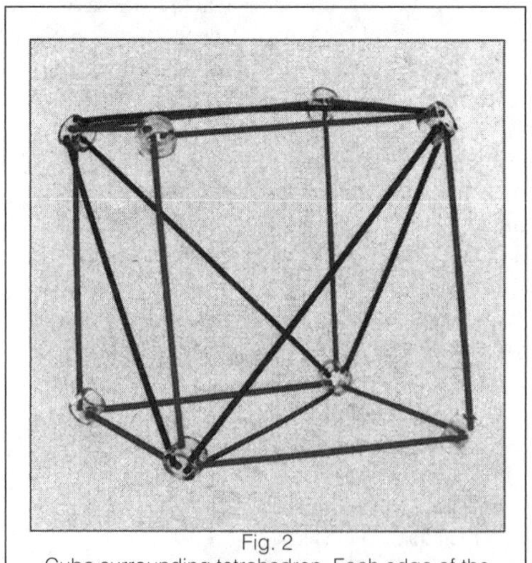

Fig. 2
Cube surrounding tetrahedron. Each edge of the tetrahedron is a diagonal of each face of the cube.

Continued

Exhibit 10.3
Continued

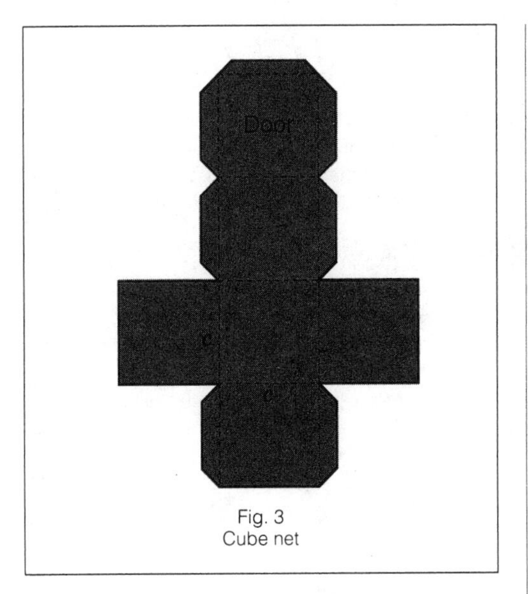

Fig. 3
Cube net

For a follow-up assignment, students make an Archimedean solid on their own.

Several days a year we build the solids until we have five

TETRAHEDRON

The teacher holds up a tetrahedron in front of the class and asks what it would look like if it were cut along the three edges from the "top" vertex to the base and folded open. If none of the responses would work, select one face of the tetrahedron and call it the base. Then point out that each lateral face is attached to the base, which results in the net in **figure 1.** Then draw the net on the chalkboard and discuss with the class the placement of the flaps and the length of the sides of the triangles. An edge length of 8 cm works well for the tetrahedron. Let t be the length of the side used in the calculations that follow.

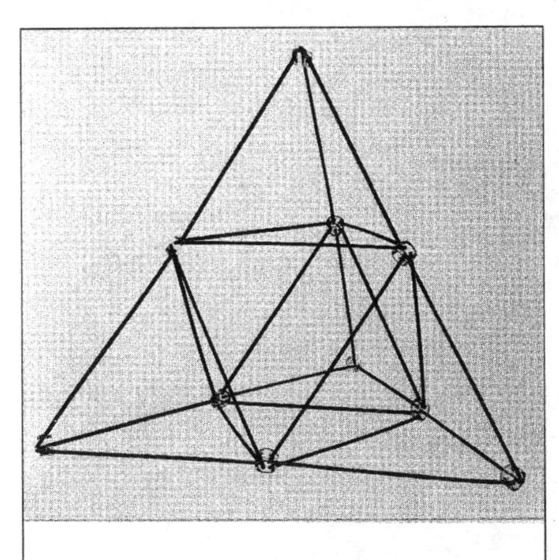
Fig. 4
Tetrahedron surrounding octahedron. Each vertex of the octahedron is the midpoint of each edge of the tetrahedron.

As a class, calculate the dimensions of the smallest rectangle to contain the net without flaps, then add 3 to 5 centimeters for the flaps and minor errors. The calculation requires that the students know how to determine the altitude of an equilateral triangle. Have the students cut rectangles out of the poster board before drawing any of the nets. This step enables every student to get a quick start. With the net on the chalkboard as a guide, students use the rulers and either protractors or compasses to draw their own.

The face with two flaps attached is the door. It should be creased inward to help it stay closed. The single flap attached to the other face is the only flap that should be glued. For crisper edges, use a compass and ruler to score the net before folding.

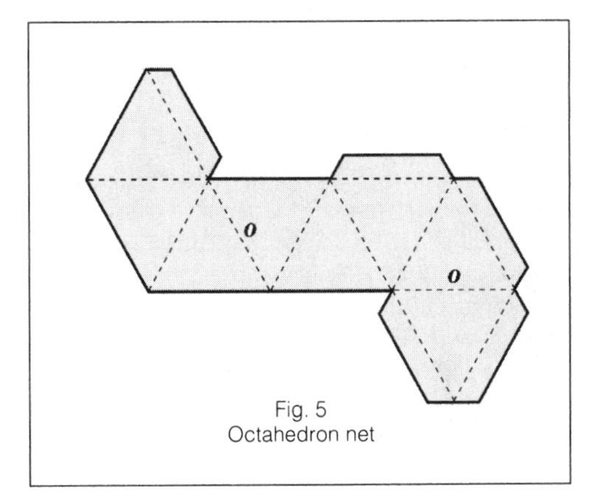

Fig. 5
Octahedron net

CUBE

The cube fits around the tetrahedron with each edge of the tetrahedron a diagonal of each square face of the cube (see **fig. 2**). Since the diagonal of a square is $\sqrt{2}$ times the length of its side, the length of the square's side, c, becomes $(t\sqrt{2})/2$. Thus the edge length of the cube is

$$c = \frac{t\sqrt{2}}{2} \approx 0.707t.$$

Have students add 2 millimeters to the edge length of the tetrahedron before calculating the cube's edge length. This addition takes care of minor errors and the thickness of the poster board.

To determine the net, hold up a cube and have students tell what the net should look like. Most of them are already familiar with it (see **fig. 3**). Leave one face unglued for the door. Make sure that this face has flaps on it to hold it closed.

INNER OCTAHEDRON

The octahedron fits inside the tetrahedron (see **fig. 4**). Its edge length, o, is half the edge length of the

Continued

Exhibit 10.3
Continued

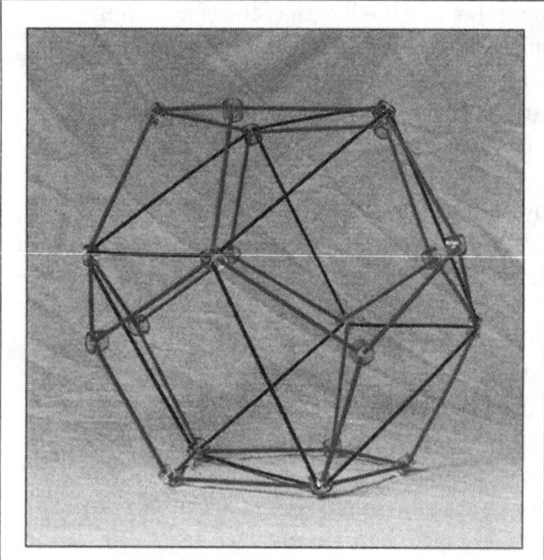
Fig. 6
Dodecahedron surrounding cube. Each edge of the cube is a diagonal of each face of the dodecahedron.

tetrahedron.

$$o = 0.5t$$

By this time, students have had experience visualizing the nets of two of the solids. They can be asked to draw the net of the octahedron as part of a homework assignment the night before they make it. As a start, have the class put four or five of their solutions on the chalkboard, making sure both correct and incorrect nets are selected. Many different correct versions are possible. The most common incorrect net has five triangles sharing a common vertex instead of four. One correct net is presented in **figure 5.**

Because this solid fits inside one that is already made, subtract 2 millimeters from the edge length of the tetrahedron before calculating the octahedron's edge length. It is the innermost figure, so no need arises for a door and thus all flaps can be glued.

DODECAHEDRON

The dodecahedron fits around the cube with each edge of the cube a diagonal of each pentagonal face of the dodecahedron (see **fig. 6**). The interior-angle measure of a pentagon is 108 degrees (see **fig. 7**). Since $\triangle ACB$ is isosceles, if a perpendicular is dropped from C to $\overline{AB}$, both $\overline{AB}$ and $\angle ACB$ are bisected.

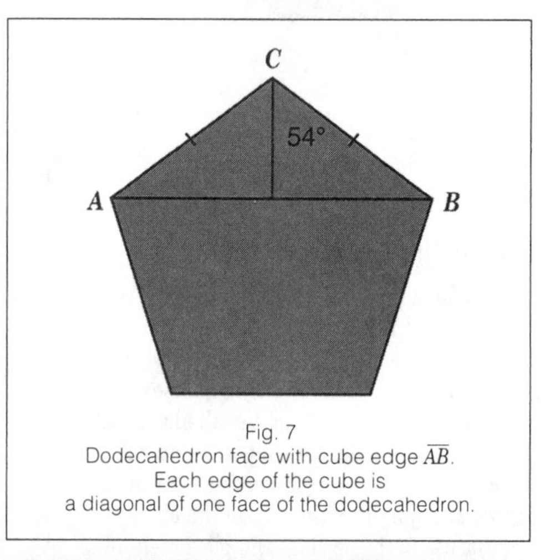

Fig. 7
Dodecahedron face with cube edge $\overline{AB}$.
Each edge of the cube is
a diagonal of one face of the dodecahedron.

Students gain experience in visualizing nets

Continued

Exhibit 10.3
Continued

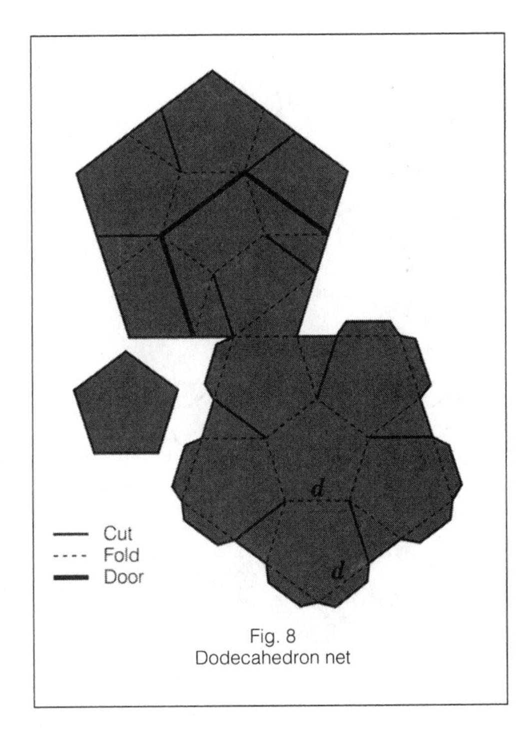

Fig. 8
Dodecahedron net

Elementary trigonometry gives the equation

$$\sin 54° = \frac{0.5c}{d},$$

where c is the length of an edge of the cube and d is the length of an edge of the dodecahedron. So

$$d = \frac{c}{2\sin 54°} \approx 0.618c.$$

The calculation of the edge length of the icosahedron is more involved than earlier calculations

Add 2 millimeters to the edge length of the cube before calculating the edge length of the dodecahedron.

Students are at first reluctant to offer suggestions on the net of a dodecahedron. When they focus on one face as a base surrounded by pentagons and the opposite face as another base also surrounded by pentagons, they quickly come up with the net in **figure 8.**

Students carefully draw one pentagon on scrap poster board to use as a template and then trace out the net. The pentagon template should be saved to glue on the door as a flap to help keep it closed. After cutting out the net, glue the five pentagons around each base first. Then cut the three diagonals for the door. Finally, glue the two halves of the dodecahedron together and add the spare pentagon to the door.

ICOSAHEDRON

The icosahedron fits around the dodecahedron (see **fig. 9**). These solids mirror each other. The dodecahedron's vertices are the centers of the faces of the icosahedron. In **figure 10a**, $\triangle WXY$ and $\triangle XYZ$ are adjacent triangles of the icosahedron with A and B being the circumcenters of the triangles (and two adjacent vertices of the dodecahedron). Point C is the midpoint of $\overline{XY}$; $\overline{AB}$ is an edge of the dodecahedron. The calculation of the edge length of the icosahedron is more involved than the previous calculations. It is worthwhile to work through it with

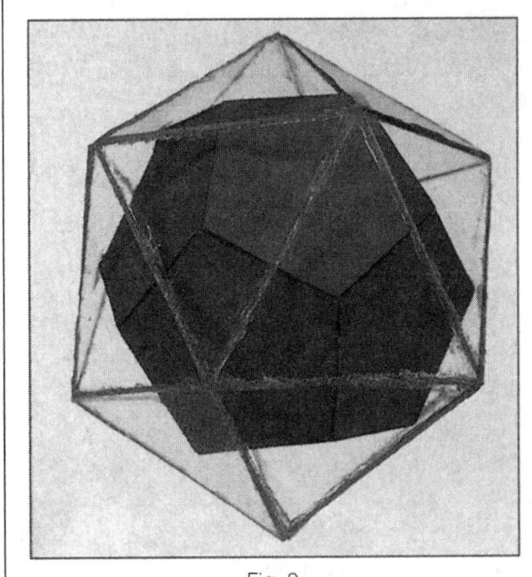

Fig. 9
Icosahedron surrounding dodecahedron.
Each vertex of the dodecahedron is
the circumcenter of each face of the icosahedron.

students because they are accustomed to short solutions to problems.

Calculate the length of $\overline{AC}$, an apothem of $\triangle WXY$. Then calculate the length of $\overline{XY}$, an edge of the icosahedron. AC is calculated by focusing on $\triangle ABC$. Besides being a part of the triangle, $\angle ACB$ is also a

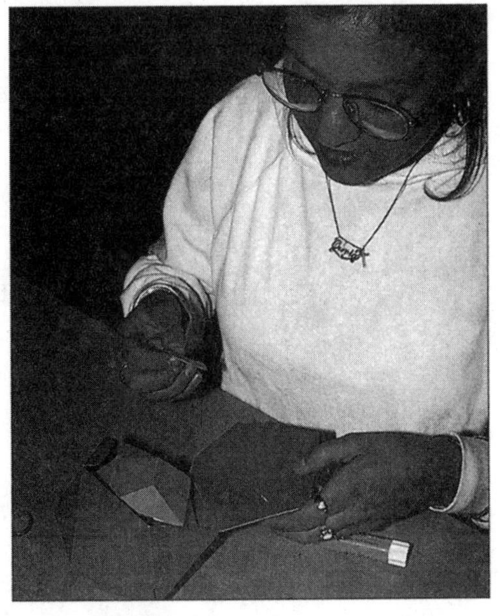

Continued

Exhibit 10.3
Continued

dihedral angle of the icosahedron ($\angle A\text{-}CY\text{-}B$). Its measure is 138 degrees 11 minutes 22 seconds (Williams 1979, 67). As before, $\triangle ABC$ is isosceles, so trigonometry can again be used to determine the length of $\overline{AC}$:

$$AC \approx \frac{d}{2\sin 69^\circ 6'}$$

Now that AC is known, because $\triangle ACY$ is a 30-60-90 triangle, $CY = AC\sqrt{3}$ (**fig. 10b**). And since $XY = 2CY$, then $XY = (2\sqrt{3})AC$. By substituting into the foregoing equation and using i to represent the edge length of the icosahedron,

$$i = \frac{d\sqrt{3}}{\sin 69^\circ 6'} \approx 1.854d.$$

Add only 0.5 millimeters to the edge length of the dodecahedron before calculating the icosahedron.

To help students figure out the net of an icosahedron, point out five triangles meeting at a vertex "on top" of the solid and five meeting "on the bottom." The remaining ten triangles form a belt, alternating up and down, around the middle (see **fig. 11**).

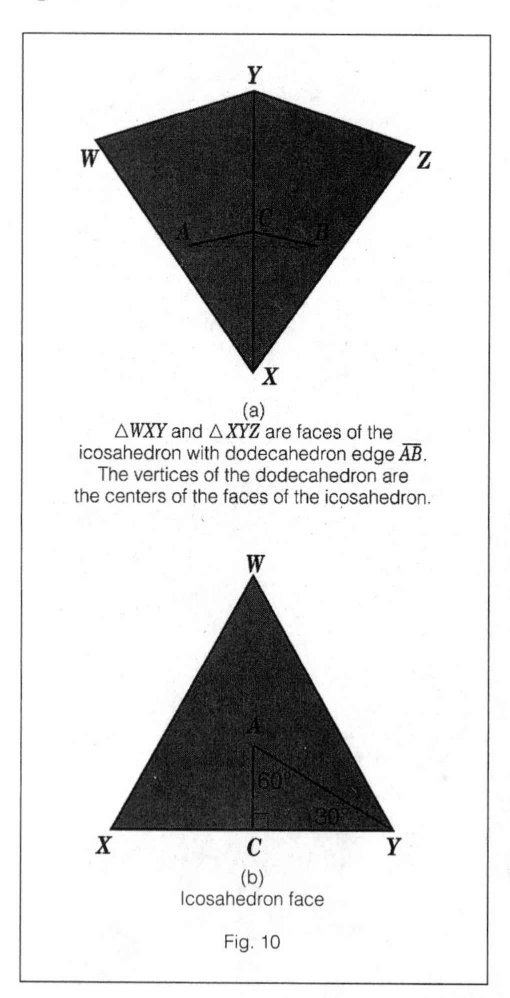

(a)
$\triangle WXY$ and $\triangle XYZ$ are faces of the icosahedron with dodecahedron edge $\overline{AB}$. The vertices of the dodecahedron are the centers of the faces of the icosahedron.

(b)
Icosahedron face

Fig. 10

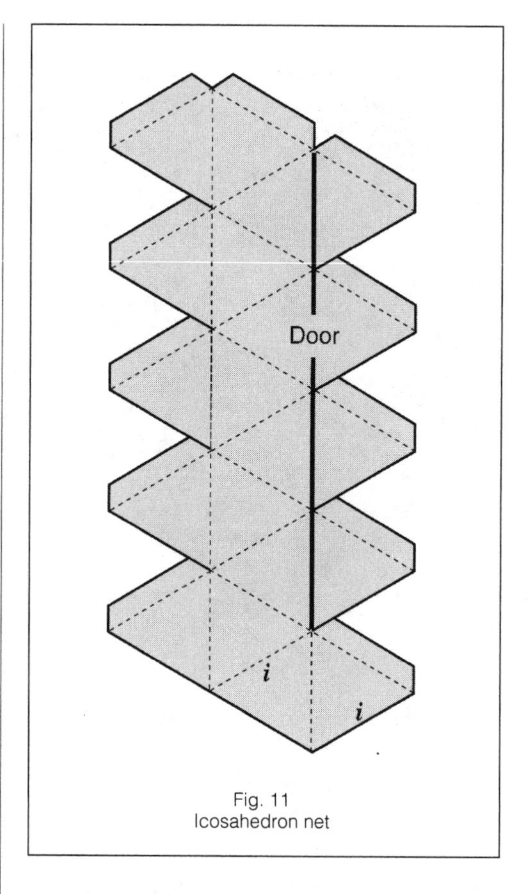

Fig. 11
Icosahedron net

A door for this solid can be cut from its edges. Choose any vertex. The five triangles that share the vertex form the door. Cut along four of the edges and fold along the fifth. Add extra flaps to hold it closed.

OUTER OCTAHEDRON

Students need not necessarily make this second octahedron, but it adds to the set. **Figure 12** shows how the icosahedron nests face to face inside the octahedron. The interesting aspect about this pair is that the vertices of the inner triangle cut the edges of the outer triangle in the golden ratio (Pugh 1976, 13). In the figure,

$$AD{:}DB = BE{:}EC = CF{:}AF = 1{:}\,\frac{1+\sqrt{5}}{2} \approx 1{:}1.618.$$

Let $AD = x$. Then

$$DB = \frac{1+\sqrt{5}}{2}\,x$$

and

$$AB = x + \frac{1+\sqrt{5}}{2}\,x = \frac{3+\sqrt{5}}{2}\,x.$$

Since $DB = AF$, then

$$AF = \frac{1+\sqrt{5}}{2}\,x.$$

Continued

Exhibit 10.3
Continued

TABLE 2
Formulas to Calculate the Edge Length of the Outer Figure as a Function of the Edge Length of the Inner Figure (x)

Inner Figures \ Outer Figures	Tetrahedron	Octahedron	Cube	Icosahedron	Dodecahedron
Tetrahedron	$\frac{x\sqrt{3}}{\sin 35°\,16'}$	$\frac{x\sqrt{6}}{\sin 54°\,44'}$	$\frac{x\sqrt{2}}{2}$		
Octahedron	$2x$		$x\sqrt{2}$		
Cube	$\frac{x\sqrt{6}}{2\sin 35°\,16'}$	$\frac{x\sqrt{3}}{\sin 54°\,44'}$			$\frac{x}{2\sin 54°}$
Icosahedron		$\frac{x(3\sqrt{2}+\sqrt{10})}{4}$			$\frac{x}{(\sin 58°\,17')(\tan 54°)}$
Dodecahedron		$\frac{x\sqrt{3}(\sqrt{5}+1)}{2\sin 54°\,44'}$		$\frac{x\sqrt{3}}{\sin 69°\,6'}$	

Calculations are based on the relationships in **table 1.**

Let $i = FD$. From using the law of cosines,

$$i^2 = x^2 + \left(\frac{1+\sqrt{5}}{2}x\right)^2 - 2x\frac{1+\sqrt{5}}{2}x(0.5)$$
$$= x^2 + \frac{3+\sqrt{5}}{2}x^2 - \frac{1+\sqrt{5}}{2}x^2$$
$$= 2x^2.$$

So

$$x = \frac{i\sqrt{2}}{2}.$$

Then

$$AB = \frac{3+\sqrt{5}}{2}x$$
$$= \left(\frac{3+\sqrt{5}}{2}\right)\left(\frac{i\sqrt{2}}{2}\right)\sqrt{2}$$
$$= \frac{3\sqrt{2}+\sqrt{10}}{4}i$$
$$\approx 1.851i.$$

Thus the edge length of the outer octahedron is approximately 1.851 times the edge length of the icosahedron. Add 1 millimeter here before calculating. Instead of a door, locate four edges that form a square. Cut three of these and use the fourth as a hinge. The octahedron will open up into two square-based pyramids. Add flaps to keep it closed.

A summary of these and additional calculations is included in **table 2.** Nine cells are blank because the calculations haven't been determined yet. The author is interested in seeing these cells filled, with explanations from readers who would like to work on them.

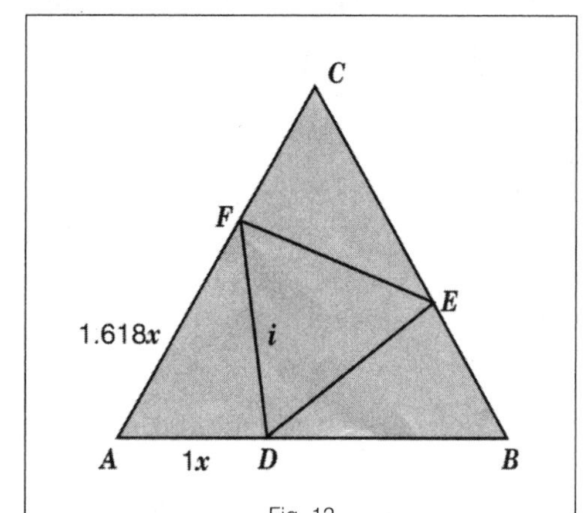

Fig. 12
Icosahedron face $\triangle DEF$ inside octahedron face $\triangle ABC$. The vertices of the icosahedron cut the edges of the octahedron in the golden ratio.

REFERENCES

Pugh, Anthony. *Polyhedra: A Visual Approach.* Berkeley and Los Angeles: University of California Press, 1976.

Williams, Robert. *The Geometrical Foundation of Natural Structure.* New York: Dover Publications, 1979.

Exhibit 10.4
Sources for Commercially Produced Hands-On Manipulatives and Concrete Models.

American Guidance Service (www.agsmath.com)
4201 Woodland Road
P.O. Box 99
Circle Pines, MN 55014-1796

Carolina (www.carolina.com)
2700 York Road
Burlington, NC 27215-3398

Creative Publications (www.creativepublications.com)
19201 120th Avenue NE
Bothell, WA 98011-9512

Didax (www.didaxin.incom)
395 Main Street
Rowley, MA 01969

ETA/Cuisenaire (www.etacuisenaire.com)
500 Greenview Court
Vernon Hills, IL 60061-1862

MindWare (www.mindwareonline.com)
121 5th Ave. NW
New Brighton, MN 55112

NASCO (www.enacco.com)
901 Janesville Ave.
P.O. Box 901
Fort Atkinson, WI 53538-0901

Sunburst Technology (www.sunburst.com)
101 Castelton St., P.O. Box 100
Pleasantville, NY 10570-0100

Wieser Educational (www.wieser.com)
30281 Esperanza
Rancho Santa Margarita, CA 92688-2130

Activity 10.5

Study the descriptions of the learning activities in Exhibit 10.5—an excerpt from *Patterns and Functions* (Phillips, 1991, pp. 60–65).

Now plan a lesson for the following objective so that the lesson incorporates a subset of Exhibit 10.5's activities:

> The student explains various relationships between the volume and height of selected solid figures (discover a relationship).

After obtaining the necessary materials, conduct the lesson with two or more students. Seek feedback from the students regarding what they learned and how you might make the lesson more engaging for them.

Share your experiences with colleagues who are also engaging in this activity.

Store the lesson plan—annotated with comments about the experience—in the "Cognition, Instructional Strategies, and Planning" section of your working portfolio.

COMPUTER-ASSISTED MATHEMATICS TEACHING

Electronic Manipulatives and Computer Simulations

Computer software is now available whereby activities with hands-on, physical manipulatives can be simulated on a computer just as you can play chess or solitaire on a computer without concrete hand-on chess pieces or playing cards. These electronic or virtual manipulatives are an extremely convenient and powerful supplement to hand-on manipulatives—not a complete replacement for the tactile real-world experiences that students need manipulating concrete objects and measurement instruments. But following up physical manipulatives with virtual ones has immense advantages. Besides emulating physical hands-on manipulatives, electronic manipulatives have pedagogical and mathematical features that are impossible with the physical variety. For example, virtual objects can instantaneously change color (e.g., to highlight a specific surface or edge that has been selected). In the midst of an activity (e.g., with virtual geoboards or conic sections), you can interrupt your work, save it, and retrieve it later exactly as you left it. You do not have to worry, for example, about stacks of cubes toppling over.

There are also electronic manipulatives that do not have physical counterparts. For example, in the virtual world, figures can be readily translated, projected, turned inside out, and resized. Of course, the appeal, mathematical capabilities, pedagogical features, user-friendliness, and cost of different software varies tremendously. One of the better designed packages is available to you and your students to use at no cost via the internet. It was developed through the National Library of Virtual Manipulatives for Interactive Mathematics Project sponsored by the National Science Foundation (Heal, Cannon, & Dorward, 2001). To stimulate your ideas for incorporating electronic manipulatives in your lessons, engage in Activity 10.6:

Activity 10.6

Access the following website of the National Library of Virtual Manipulatives for Interactive Mathematics: www.matti.usu.edu.

From the opening home page choose the "Virtual Library" link. Note the grid pairing *PSSM*'s Content Standards to grade bands as shown by Exhibit 10.6. Click on the cell of your choice (e.g., "Number and Operations" for "Grades 9–12"), which brings up a page such as Exhibit 10.7. Now select one of the links by clicking on the icon of your choice (e.g., for "Number Line Bounce," "Polyominoes," "Coordinate Geoboards," "Isometry Transformations," "Polygonal Fractals," or "Scatterplot"), which brings up a page such as the one shown by Exhibit 10.8.

Besides the back-arrow link, note the four icons at the top of the page. Select the "E-Standard" icon, which opens a window relating the learning activities from this page to *PSSM* (NCTM, 2000b). Close the "E-Standard" window and then visit the "Activities," "Lesson Plan," and "Directions" windows. With the "Directions" window still open, select and engage in a subset of the activities. Play around long enough to understand some of the features of the virtual manipulatives and models.

In a similar fashion, sample some of the activities from other links.

Discuss how you might incorporate some of the activities that you tried with colleagues who are also engaging in this activity. Make notes on some of the ideas you gained. Insert those notes in the "Technology" section of your working portfolio.

Numerous software packages exist that include features for incorporating electronic manipulatives into your learning activities (e.g., Geometer's SketchPAD available from Key College Publishing [www.keycollege.com]).

Using Computers to Do Mathematics

Archimedes used a stick to draw and write in the sand (see Exhibit 10.9) (Eves, 1983, pp. 83–95). Euclid executed geometric algorithms with a collapsible compass and straightedge (Retz & Keihn, 1989). Fermat scribbled proofs in the margins of books (Singh, 1997, pp. 34–69). Today, a mathematician's principal tool is a computer equipped with powerful software for (a) executing algorithms, (b) illustrating multivariable, multidimensional relationships, (c) storing and retrieving data and functions, and (d) communicating in the language of mathematics. As emphasized in chapter 4's section "*PSSM*'s Technology Principle," it is essential for you and your students to use computers to do mathematics. What software packages (e.g., *DERIVE, MathCAD,* or *Mathematica*)

Exhibit 10.5
An Activity for You to Incorporate in a Discover-a-Relationship Lesson.

INVESTIGATION 2. THE BOTTLE FUNCTION

Problem: Investigate the relationship between the volume and the height of a bottle.

Teacher Notes

Materials. Supply each group with a straight-sided transparent container (different sizes for different groups); some sand, rice, or water; and a measuring cup, a ruler, and some graph paper. For this first experiment, you may wish to provide some axes already drawn. You will also want to try out the size of the measuring cup to be sure it does not fill up the container too quickly or too slowly.

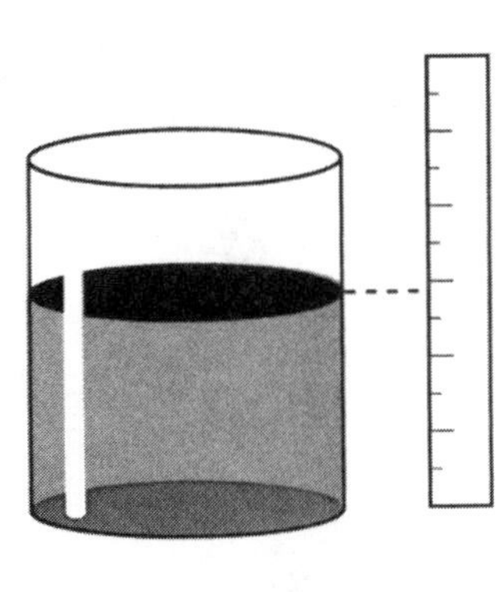

Fig. 5.6

Launch. Demonstrate how to add the same amount of water to a straight-sided transparent container. After each addition of water, measure the height of the water in the container (fig. 5.6). Record the data on a graph. Divide the class into small groups.

Explore. Allow students to work in groups of two or three. One student puts a measure of water into the container; the second student measures the height (in centimeters is simplest); and the third records the data in a table and a graph. If the students are careful about their measurements, they should produce a straight-line graph (fig. 5.7). You will want them to have to use enough measures in filling the bottle to make the "straightness" of the line, the constant slope, clear and convincing.

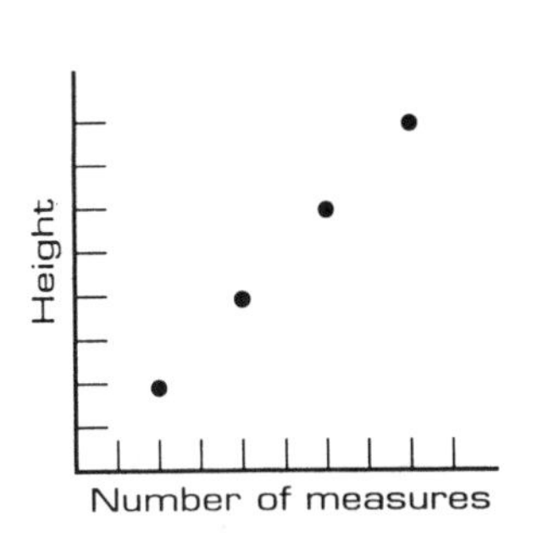

Fig. 5.7

Summarize. What pattern do you notice in the table and in the graph? You can expect students to say something about the increases in height being the same for every measure added—or the slope of the graph being the same—in words appropriate to the age group.

Will this pattern go on forever? [Until the container overflows!]

Continued

Source: Reprinted with permission from E. Phillips, in *Patterns and Functions,* pp. 60–67. Copyright 1991 by National Council of Teachers of Mathematics.

Exhibit 10.5
Continued

♦ ♦ ♦ ♦ ♦ ♦ ♦ ♦

What would the graph look like if we keep on dumping in measures? [See fig. 5.8.] [The height keeps on increasing, up to a limit.]

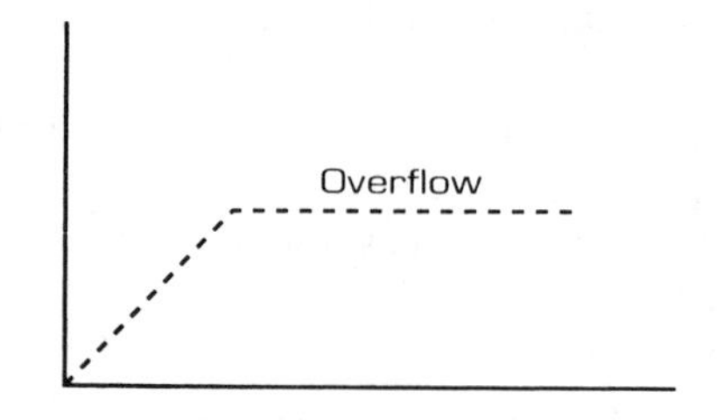

Fig. 5.8

What does it depend on? [The number of measures]

What would happen if we used a different measure? The different groups could compare their experiences. Smaller measures produce smaller increases in height. But the rate of growth is still a constant.

What is the slope of the line? What does this mean? [It is the change in height per unit measure, or the change in height per unit of volume.]

You may ask each group to repeat the experiment using a larger or smaller measure. For additional related activities, see Stewart et al. (1990).

Extension 2.1. Different-shaped containers

Use two different straight-sided containers, one wider than the other. *What will the graph look like?*

Discussion. Give each group two or three different cylindrical bottles. Each group should add measures (*m*), record heights (*h*), and make graphs for each container (fig. 5.9).

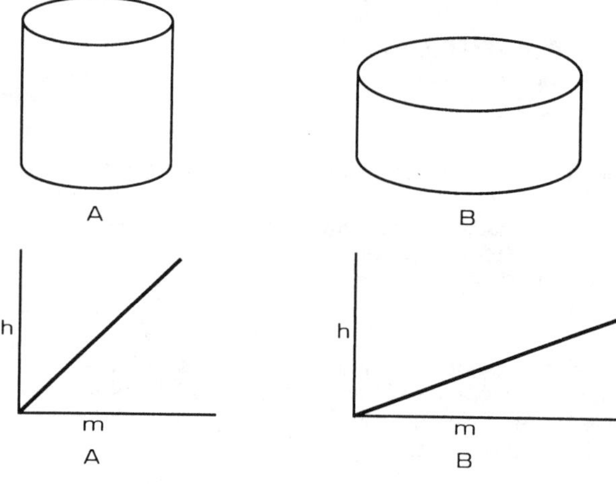

Fig. 5.9

What happens after you add a measure to bottle A? [The height increases.] *Does the same thing happen when you add the same measure to bottle B?* [Yes, but B's height grows more slowly.]

Reasoning

How are the graphs the same? How are they different? In each case the information graphed shows that height depends on the number of measures added. *The graphs are different, so what else does height depend on? What makes it grow quickly? Slowly?* Students will talk about the containers being wide or narrow. Discuss the "cross section" of each bottle. Using the cross sections of containers is a good way to describe the shape or the volume of the containers. The cross sections of a cylinder are congruent circles.

Extension 2.2. Irregular-shaped containers

Repeat the preceding bottle experiments with irregular-shaped transparent containers (fig. 5.10).

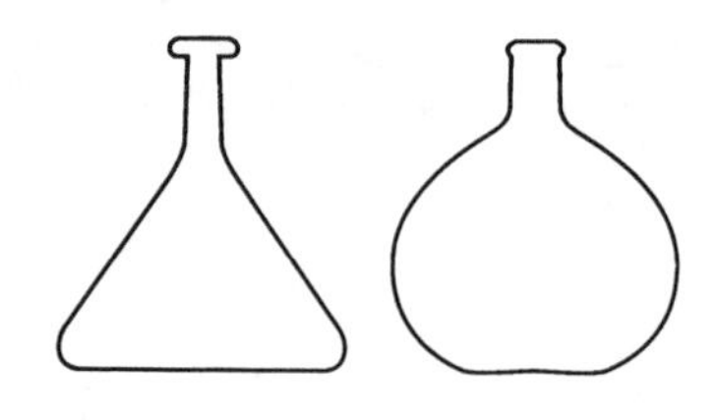

Fig. 5.10

Continued

Exhibit 10.5
Continued

♦ ♦ ♦ ♦ ♦ ♦ ♦ ♦

Discussion. Before beginning, ask students to *think about what will happen as each measure is added. Will this container fill at the same rate as the containers in figure 5.9?*

After the data are graphed, ask questions that focus on the relationship between the growth in the height of the water and the cross section of the containers.

How did your data and graphs differ from those in the experiment you did with the straight-sided containers? [The curved containers sometimes filled quickly, sometimes slowly.]

What is increasing in your experiment? [Height]

When is the height increasing most rapidly? Most slowly? Students should be able to point to the narrow parts of the bottles as the places that filled most rapidly. They may also relate this to the steepest parts of the graph.

Could the graph in figure 5.11 go with the bottle in figure 5.11? [No. The bottle starts out wide and so the height should grow slowly.]

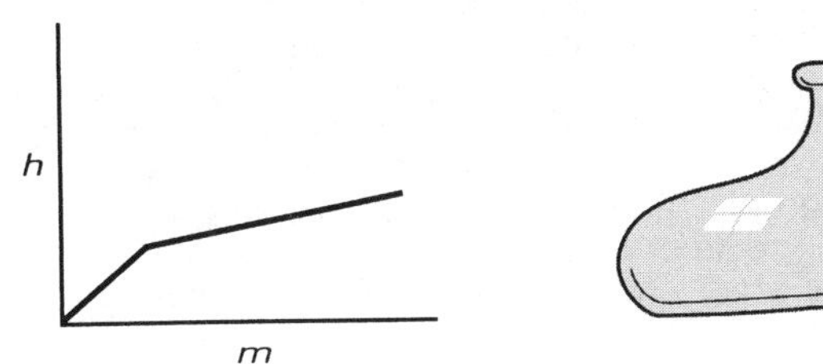

Fig. 5.11

Summarize. The height, the number of measures, and the cross section all relate to each other. Height depends on the number of measures added. The graph is a record of how fast the height of the water grows compared to the cross section of the bottle.

Extension 2.3. Determining the shape of a bottle

Sketch the shape of the bottle if the graph of the water height and the number of measures is given.

Discussion. Give the students graphs (fig. 5.12) and ask them to sketch the shape of the corresponding bottles (fig. 5.13).

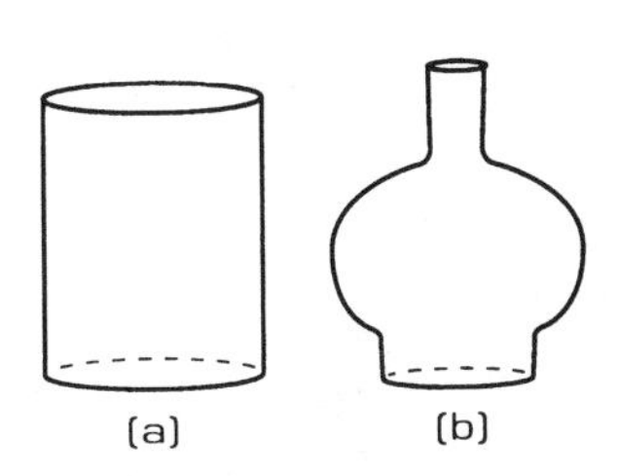

Fig. 5.13

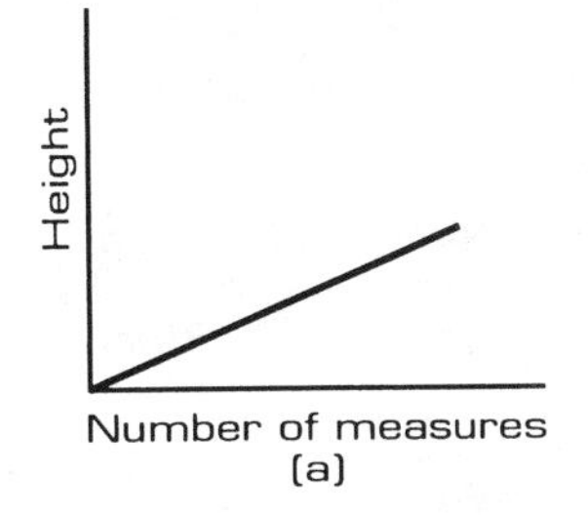

Height
Number of measures
(b)

Fig. 5.12

Continued

Exhibit 10.5
Continued

♦ ♦ ♦ ♦ ♦ ♦ ♦ ♦

Discuss the rate of change. Where is the rate, height/cross section, the greatest? A constant? Let students make up a graph. Have students explain the shapes that go with each graph.

Extension 2.4. Decreasing rates of change

Discussion. From the chemistry lab borrow a bottle that allows water to be drawn off the bottom; attach it to a ring stand (fig. 5.14).

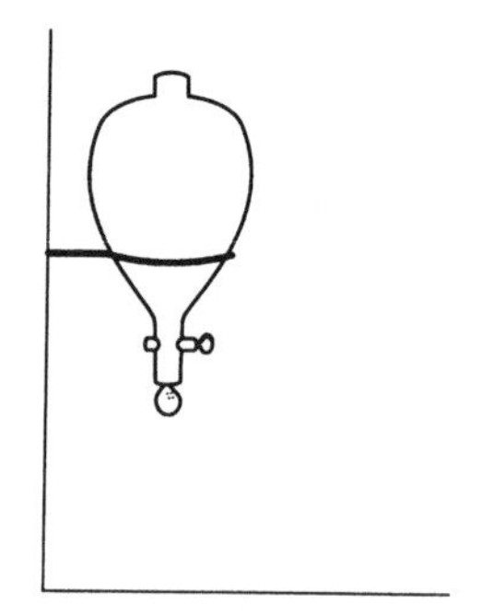

Fig. 5.14

Draw off measures of water and record the height of the water in the bottle. Focus on the height of the bottle. Ignore the distance from the bottle to the table. This will produce a graph similar to figure 5.15. The graph is decreasing from left to right.

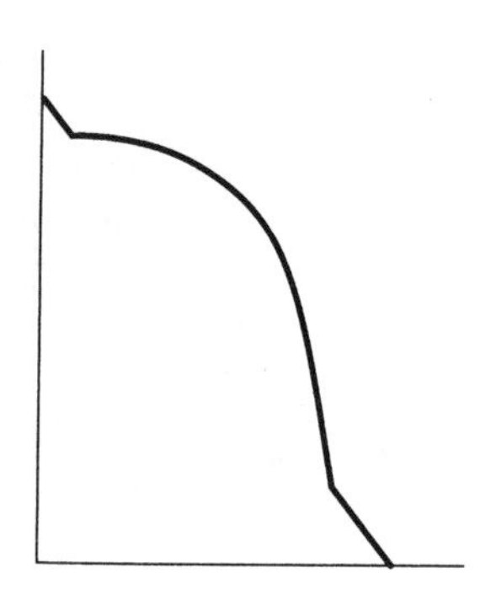

Fig. 5.15

Extension 2.5. More experiments

The following experiments can be conducted in class. For each experiment, allow the groups time to carry out the experiment; collect and organize the data; draw an appropriate graph; answer the questions; and, if possible, determine a rule for the functional relationship between the varying quantities.

Experiment 1. Circumference and pi

Collect several circular lids or cans in different sizes . Older students can measure the circumference (using a string) and the diameter and then plot each point (diameter, circumference) on a graph. Younger students can use a string or colored tape to determine the length of the circumference. They can attach the tape or place the string on the graph at the appropriate diameter (fig. 5.16a). The diameter can be determined by placing the circular lid on the horizontal axis so that the horizontal axis divides the circle into two congruent halves (for younger students). Alternatively, the diameter can be determined by tracing the lid on a piece of paper, cutting out the circle, and folding it in half. The crease line is the diameter.

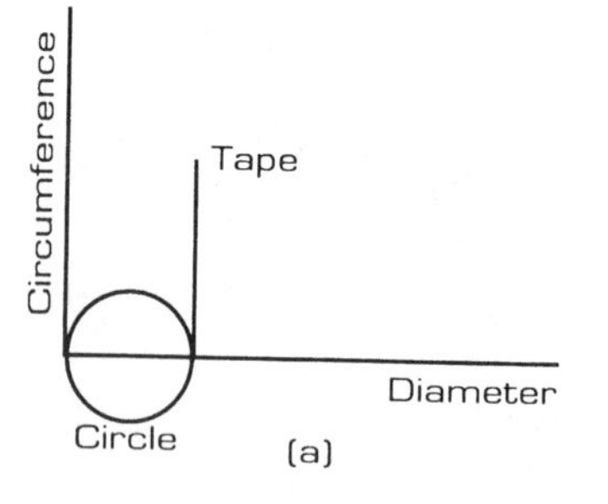

(a)

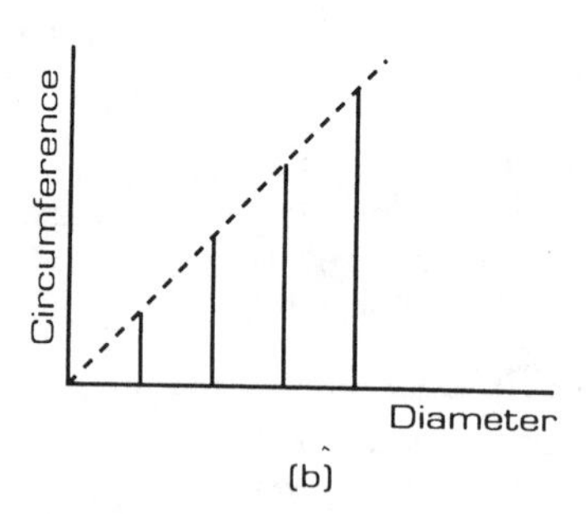

(b)

Fig. 5.16

After each circumference has been determined and plotted on a graph with its corresponding diameter, draw a line through the points (fig. 5.16b). Ask the students to *calculate the vertical and horizontal distances between two points on the line.* This is the process used to find slope (Extension 1.1 in this chapter). *Calculate the ratio of vertical distance to horizontal distance or circumference to diameter.* The ratios come very close to 3.14, which is an approximation for π. This would be an appropriate time to discuss some of the history of π. (See *The Story of Pi* by Tom Apostol [1989].)

Calculator

Connections

What does the graph of the circumference to the diameter look like? [A straight line] *What is the slope of the line?* [Approximately 3.14] We call

Continued

Exhibit 10.5
Continued

♦ ♦ ♦ ♦ ♦ ♦ ♦ ♦

this number π. *Determine a rule that relates the circumference of a circle to its diameter.* [Circumference = diameter × π] The equation $c = \pi d$ is a linear equation, and π is the slope of the line. *Use both your graph and the rule for circumference to find the circumference of a circle with a diameter of 6 centimeters. What is the diameter of a circle whose circumference is 54 centimeters?*

Calculator

Experiment 2. The open box

Give each student a square piece of centimeter paper. Describe how an open box can be made by cutting an identical square from each corner of the large square and folding the edges to form a box (fig. 5.17). This problem is from the *Curriculum and Evaluation Standards for School Mathematics* (NCTM 1989, p. 80).

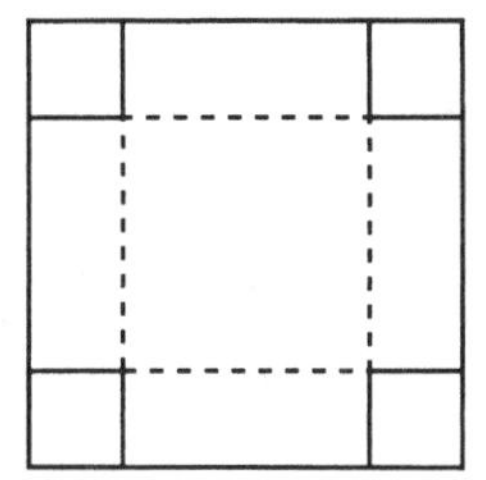

Fig. 5.17

Let students guess which box will have the greatest volume. *If we cut out different-sized squares, which box will have the greatest volume? What is the smallest square we can cut? The largest square?* Review volume for younger students. Centimeter cubes can be stacked in the box to illustrate the concept of volume (Beaumont, Curtis, and Smart 1986). Have each group cut out various-sized squares with a whole-number length and calculate the volume. Organize the data in a table (fig. 5.18). Graph the data (fig. 5.19).

Length of the side of cutout square (in units)	Dimensions of the open box	Volume (in cubic units)
1	1 × 16 × 16	256
2	2 × 14 × 14	392
3	3 × 12 × 12	432 (greatest volume)
4	4 × 10 × 10	400
5	5 × 8 × 8	320
6	6 × 6 × 6	216
7	7 × 4 × 4	112
8	8 × 2 × 2	32
9	9 × 0 × 0	0

Fig. 5.18. Data for original 18-by-18-centimeter square

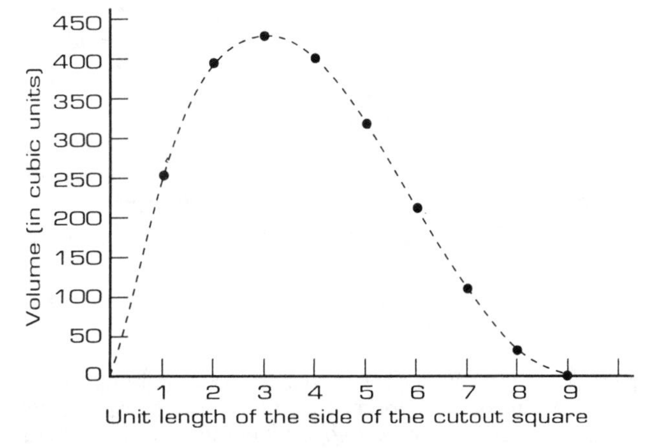

Fig. 5.19

Continued

Exhibit 10.5
Continued

♦ ♦ ♦ ♦ ♦ ♦ ♦ ♦

If we allow the length of the side of the cutout square to be a rational number, can we obtain a larger volume? Try some values very close to 3. [They produce a box with a smaller volume.] *Is the graph a parabola?* [No, it is not symmetric.]

Looking toward Algebra: The pattern for volume can be generalized into an equation that relates the volume, V, and the length, x, of the side of the cutout square (fig. 5.20).

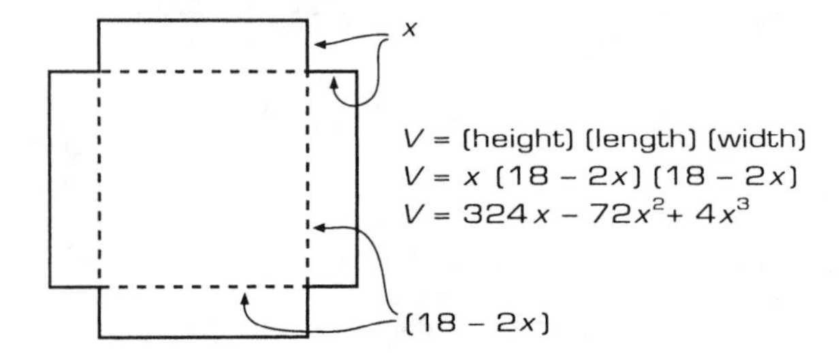

Fig. 5.20

Since the greatest exponent of the variable is 3, this equation is called a *cubic equation*. Older students can extend this problem to an $m \times n$ rectangle.

Experiment 3. Bouncing ball

Drop a ball from a specified height and record the maximum height after each bounce. This will take some practice to read the maximum height between bounces (fig. 5.21a).

The height is approximately half the previous height. The graph is an exponential decay (see decreasing exponential function in chapter 1). See figure 5.21b.

Connections

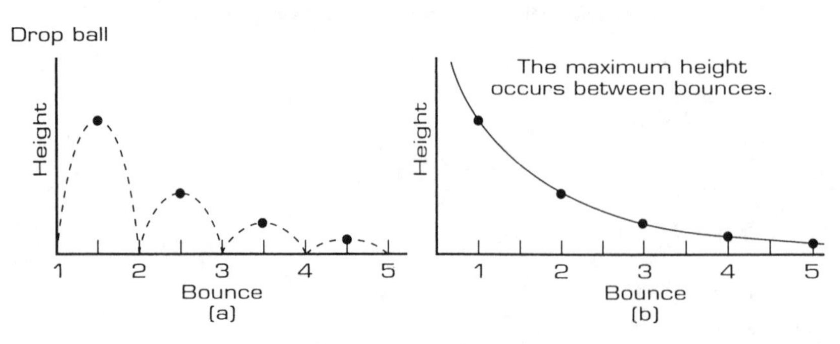

Fig. 5.21

For younger students the relationship between the drop height and the height of the first bounce can be determined for various drop heights. This relationship is linear. *What will happen if we try different balls and different starting heights?*

Continued

Exhibit 10.5
Continued

♦ ♦ ♦ ♦ ♦ ♦ ♦ ♦

Experiment 4. The Ferris wheel

Suppose a person is riding on a Ferris wheel. Record the height of the person from the ground after a specified time interval (fig. 5.22a). Measure the height from the bottom of the chair to the ground. The distance is a vertical distance. As an interval of time, use the time between chairs as they pass through the loading position. A Ferris wheel can be made from a paper circle notched at regular intervals to indicate the seats. Use a tack to position the wheel on a piece of cardboard.

Collect data through two or three complete turns of the Ferris wheel. Graph the data (fig. 5.22b).

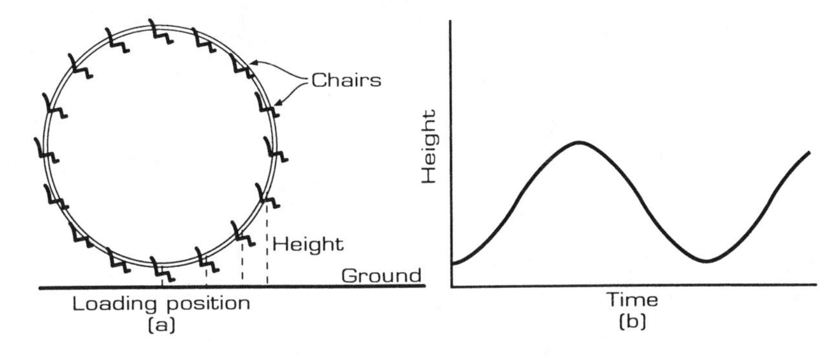

Fig. 5.22

Why does the graph not start at zero? What is the maximum height? When does it occur? Why does the shape of the graph repeat? When does it repeat?

Looking toward Trigonometry: This graph, which looks like a wave, is the graph of the *trigonometric function* $h = \sin x + b$. The variable h is the height of the seat from the ground, and the variable b is the height of the seat in a loading position from the ground.

Experiment 5. The distribution of the sum of two dice

In this experiment we will toss a pair of dice and calculate the sum of the numbers. Ask students, *Guess which sum will occur most often? Least often?* Throw a pair of dice thirty-six times and record the sum of the numbers of the two dice. The sum 7 should occur most often, the sums 2 and 12 least often. The more trials that occur, the more likely it is that these distributions will occur. Collect the data and graph the sum and the number of times each sum occurs (fig. 5.23). For younger students a bar graph can be used.

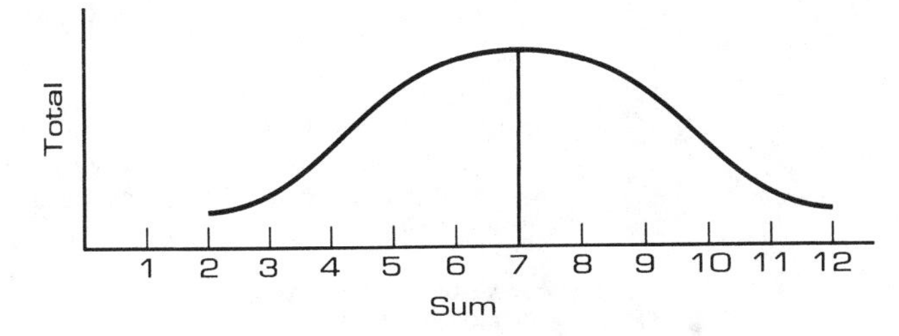

Fig. 5.23. Sum of two dice when rolled a large number of times

Continued

Exhibit 10.5
Continued

♦ ♦ ♦ ♦ ♦ ♦ ♦ ♦

Looking toward Probability and Statistics: The resulting graph is called a *bell-shaped curve*, a normal curve, or the binomial distribution. If the students have studied probability, this problem can be posed as, *Which sum is most likely to occur?* The data can be analyzed theoretically by arranging all the sums in a chart (fig. 5.24). The *probability* of a sum occurring is the number of times the sum occurs divided by the total possible sums.

Sum		Die 1: 1	2	3	4	5	6
	1	2	3	4	5	6	7
	2	3	4	5	6	7	8
Die 2	3	4	5	6	7	8	9
	4	5	6	7	8	9	10
	5	6	7	8	9	10	11
	6	7	8	9	10	11	12

The sum 7 occurs 6 times out of 36 total sums.

Probability (sum of 7 occurs) = $\frac{6}{36}$

Fig. 5.24

Additional sources of experiments that relate mathematics to science can be found in TIMS (Teaching Integrated Mathematics and Science project, University of Illinois, Chicago, IL 60680).

Exhibit 10.6
PSSM's Content Standards Paired With Grade Bands From the Website of the National Library of Virtual Manipulatives for Interactive Mathematics (www.matti.usu.edu).

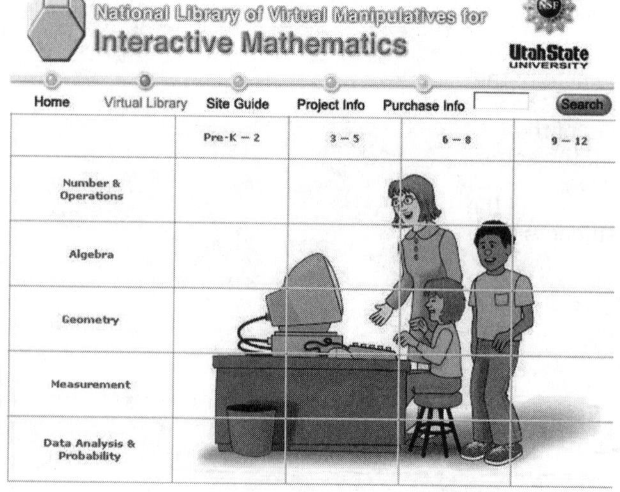

Exhibit 10.7
Number & Operations for Grades 9–12 Web Page From the National Library of Virtual Manipulatives for Interactive Mathematics.

Number & Operations (Grades 9 - 12)

Virtual manipulatives related to the NCTM *Number & Operations* standard for grades *9 - 12*.

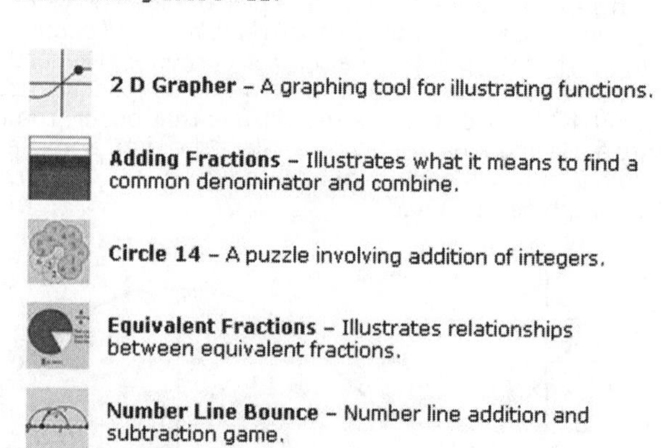

Exhibit 10.8
Conway's Game of Life Page Under "Number & Operations (Grades 9–12)" From the National Library of Virtual Manipulatives for Interactive Mathematics.

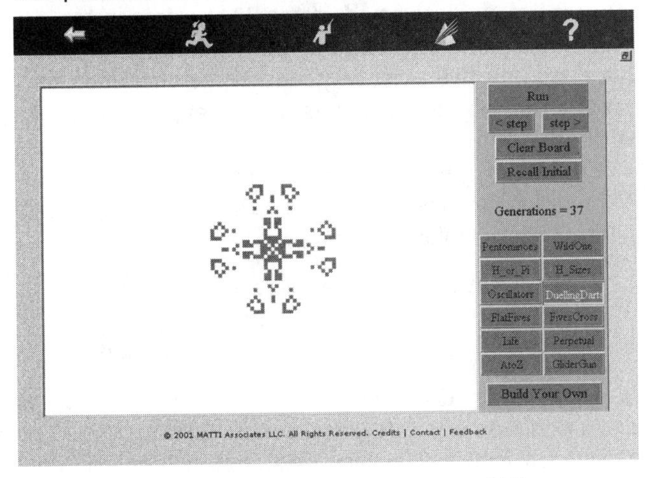

Copied with permission @ 2001 MATTI Associates LLC.

Exhibit 10.9
Archimedes Employing Mathematical Technology.

are right for you? After questioning scores of mathematics teachers and mathematicians, I am convinced that people who use computers to do mathematics prefer whatever they are used to. Once they become acquainted with and use any particular software, they tend to stick with it.

The software makes it possible to easily and quickly (a) generate multitudes of examples with parameters you or your students define, (b) execute algorithms—including algorithms students invented and for which they have written programs, (c) create illustrations, and (d) explore relationships. For example, once students have learned to construct an angle with a compass and straightedge, there is not much gained by having them repeat that process time and time again. However, if you want them to discover relationships involving angles, they may need to construct many angles. The computer software makes it possible for them to experiment with scores of angles of whatever size they choose, move them around, stack them, rotate them, close them, open them, and build figures from them. Thus, they spend time doing mathematics rather than trying to draw and manipulate pictures.

Case 10.3 is an example involving number theory:

CASE 10.3

During an algebra II unit utilizing the Pythagorean theorem, Amanda becomes intrigued by the puzzle of identifying primitive Pythagorean triples (i.e., $a, b, c \in$ {integers} $\ni a^2 + b^2 = c^2$ and 1 is the greatest common divisor of a, b, and c). She engages her teacher, Mr. Johnson, in a conversation:

Amanda: There should be a way of finding three lengths for sides of right triangles that's easier than trial and error and more trial and error!

Mr. Johnson: Why don't you see if you can find a pattern by looking at a number of primitive Pythagorean triples and comparing them to one another and also to integer triples that aren't Pythagorean?

Amanda: That would take me all year to come up with enough to pick out a pattern!

Mr. Johnson: Not if we use a computer to do the work for us. We can write a program to generate both kinds of triples for you to analyze.

Within five minutes after the program is ready, Amanda has two lists of triples to analyze. Not having to generate the examples and nonexamples by hand, she has the time and energy to devote to the sophisticated analytical task of exploring patterns.

Over the next several weeks, with some guidance from Mr. Johnson, Amanda not only develops and tests propositions about primitive Pythagorean triples, she also involves some of her classmates in the effort. Eventually, she becomes convinced—but does not deductively prove—the following:

> $(a, b, c) \in$ {Pythagorean triples} $\Leftrightarrow a = v^2 - u^2$, $b = 2uv$, and $c = u^2 + v^2$ for all positive integers u and v such that $v > u$, u and v are relatively prime, and either u or v is even.

Amanda reports her conjecture to the class; she challenges anyone to find a Pythagorean triple that does not follow her pattern or to find a nonexample that does. Attempts at counterexamples are quickly dispensed with using a computer program that tests whether a given integer triple is a primitive Pythagorean and whether or not it fits Amanda's pattern.

Computer-Based Learning Activities

You can engage students in problem-solving learning activities with electronic manipulatives and models in much the same way that biology students use computers to simulate animal dissection. For example, in the learning activity described in Exhibit 1.3, Mr. Rudd used a trash barrel to lead students to discover a formula for surface area of a right cylinder. As an alternative to getting his students to imagine the barrel being cut and unrolled in the large-group question-discussion session, he might have had students work with a computer-simulation program that allowed them to experiment with a variety of figures emanating from the reshaping of a right cylinder.

Case 10.4 is an example of a teacher conducting computer-based learning activities:

CASE 10.4

Mr. Fernandez's classroom has a computer station for presentations and a cluster of six networked PCs for student use. To accommodate the 29 students in his sixth-grade mathematics class, he frequently organizes the class into five activity groups. While one group takes a turn on the computers, the other four are busy with other work.

As part of a unit titled "Collecting, Analyzing, Interpreting, and Presenting Data," Mr. Fernandez conducts a lesson for the following objective:

> The student uses measures of central tendency and variability to describe characteristics of data sequences (comprehension and communication).

Building on ideas he developed at an NCTM conference and making use of a software package, *Mathematics Exploration Toolkit* (Puhlmann & Petersen, 1992), Mr. Fernandez explains and demonstrates the assignment

Exhibit 10.10
Mr. Fernandez's Tasksheet.

Part I: Measurement

At home, record the height and age of each member of your household.

Part II: Computer

1. From the keyboard, enter the title of your graph in the title box of the chart on the screen.
2. Enter the headings: NAME HEIGHT AGE(yrs)
3. Enter the name, height, and age of each member of your household as prompted on the screen.
4. When the chart is complete, click on the double-bar graph icon in the lower right-hand part of the screen.
5. You should see a bar graph. Now, click on "Average" (lower right of screen). Make sure the average height and age are shown at the bottom.
6. You are ready to print. Make sure the printer is switched to match the letter on your computer (A, B, C, D, E, or F). Push the "Print Screen" key.
7. When the graph is printed, check to see that it looks correct. (It is okay if the last few letters of longer names are chopped off.) If you are satisfied, go back to the screen.
8. Click on "Average" again. Click on the picture of the chart (lower right) to bring back the chart.
9. Click on the picture of the computer. Select "Clear Screen." Do not save your chart. When your screen is cleared, return to your seat.

Part III: Interpreting and Writing

The third part of the assignment is to compute a few more statistics, interpret the data, and write an article about them. Do the following:

1. Compute the mode, median, and range of the heights and of the ages.
2. Think about the following questions before you write your article:
   A. How will your means compare to those of your classmates? Will they be higher or lower? Why?
   B. How do your medians compare to your means? Can you explain why they may be different, close, or far apart?
   C. How many modes are there for heights or ages? Why?
   D. How do you expect your ranges to differ from those of your classmates? Why?
3. Write your article as directed in class.

Exhibit 10.11
Kevin James' Response to Mr. Fernandez's Family-Statistics Assignment.

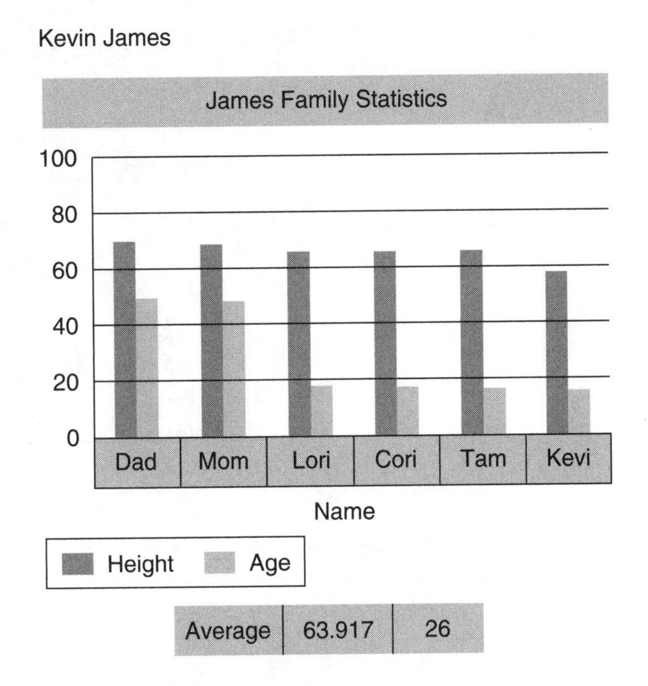

My family consists of mostly teenagers and adults. Therefore our mean (average) height is higher than most of my classmates, because they have younger people in their family. The median age in our family is not very close to the mean because all of the adult kids in our family have moved out. There is a height mode in our family. This is different because most people don't have two people in their family that are the same height. We have a big age range in my family, it is 37 years. We have this range because my parents are a little older than usual parents of seventh graders. The range in my family for age is 37 yrs., and for height 10". There is no age mode, but there is a height mode—64". The median height is 64" and for age is 17 yrs. The mean age for my family is 26 yrs., and for the height 63.917".

reflected by Exhibit 10.10's tasksheet. Exhibit 10.11 is one student's finished product from the assignment.

In a subsequent activity, students work in pairs measuring each others' heights and arm spans to address the question, "Am I square?" Mr. Fernandez collects their data and uses the same software to summarize and display it from his computerized teaching station. The display for 10 of the students is shown in Exhibit 10.12.

To familiarize yourself with computer-based learning activities developed by preservice mathematics teachers, engage in Activity 10.7:

Activity 10.7

Access the following website managed by Joe Koebbe, professor of mathematics at Utah State University: www.j.k.math.usu.edu/koebbe.

Follow the link to Math 4620 Projects. Select and review several project reports of computer-based learning activities developed and field-tested by preservice teachers. Engage in some of the activities.

Discuss how you might incorporate some of the activities you examined with colleagues who are also engaging in this activity. Make notes on some of the ideas you gained. Insert those notes in the "Technology" section of your working portfolio.

CALCULATOR-ASSISTED MATHEMATICS TEACHING

Considering the wealth of research supporting the use of calculators in mathematics curricula at all grade levels and for all courses (Dessart, DeRidder, & Ellington, 1999; Kaput, 1992) and the relatively low cost of powerful handheld calculators, it seems just as critical for all students to have ready access to a calculator as it is for them to have paper and pencils.

Once students understand how and why an algorithm works, calculators can relieve them from the

Exhibit 10.12
Part of the Summarized Data Mr. Fernandez Displayed for the Class.

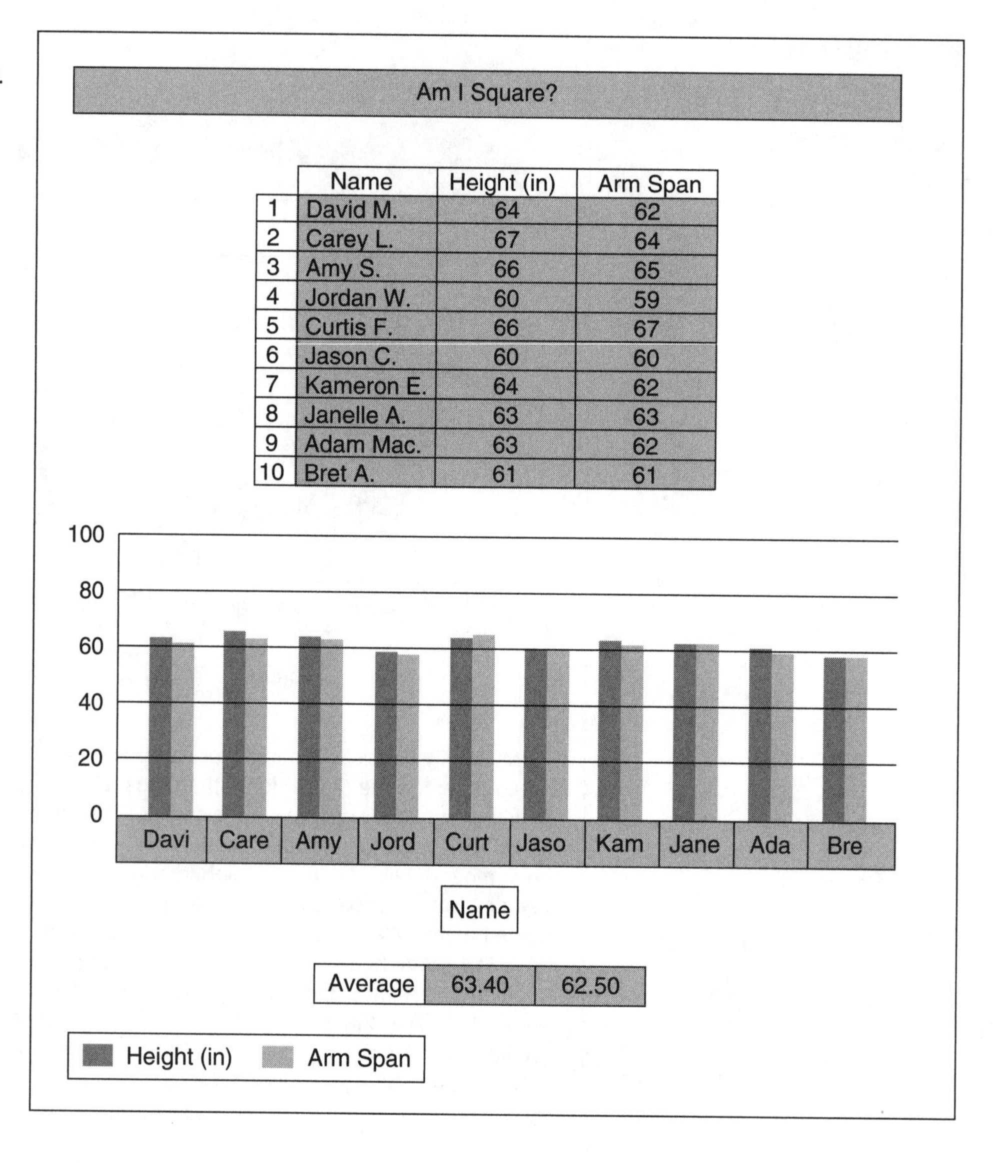

Am I Square?

	Name	Height (in)	Arm Span
1	David M.	64	62
2	Carey L.	67	64
3	Amy S.	66	65
4	Jordan W.	60	59
5	Curtis F.	66	67
6	Jason C.	60	60
7	Kameron E.	64	62
8	Janelle A.	63	63
9	Adam Mac.	63	62
10	Bret A.	61	61

tedium of having to work out each step during learning activities in which their energies should be directed toward more sophisticated cognitive processes (e.g., inductive reasoning, deductive reasoning, comprehension and communication, or divergent reasoning). Calculators free both you and your students to work on problems requiring manipulation of real-world, realistic data; whereas, paper-and-pencil calculations are simply too time consuming unless the problem is contrived so that numbers are easy to work with and result in a pat answer. Because Ms. Citerelli's students in Cases 5.8–5.10 had their calculators in hand as she led them to construct the concept of arithmetic sequence, they were able to find the difference between consecutive numbers in complicated-looking sequences as easily as they could for simple-looking sequences. In Case 8.3, Ms. Hundley and her students were able to try out various algorithms with realistic data in their quest for a solution to the combination-key lock problem.

Besides relieving you and your students from time-consuming, boring algorithms, calculators—like computers—also serve as valuable tools for exploring mathematics. Revisit Case 5.12, for example, and note how Ms. Gaudchaux's students used graphing calculators in collaborative teams to discover the effects of a and b on the graph of $f(x) = ax + b$. Case 10.5 is another example of the calculator as an exploratory tool during stage 1 of a discover-a-relationship lesson:

CASE 10.5

For her algebra II class, Ms. Long is conducting the experimenting stage of a lesson targeting the following objective:

> The student explains how the values of t, f, and g affect y, where t is a real number variable, $f(t) = x$, and $g(x) = y$ (discover a relationship).

Exhibit 10.13

Ms. Long's Students-Use-the Parametric Mode on Their Graphing Calculators.

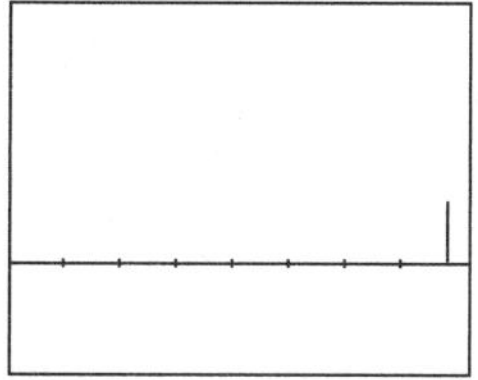

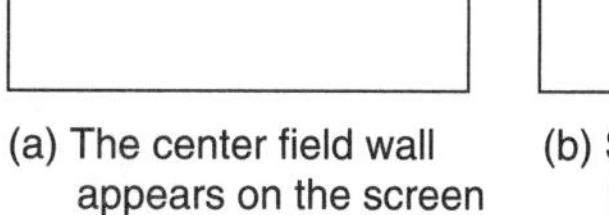

(a) The center field wall appears on the screen

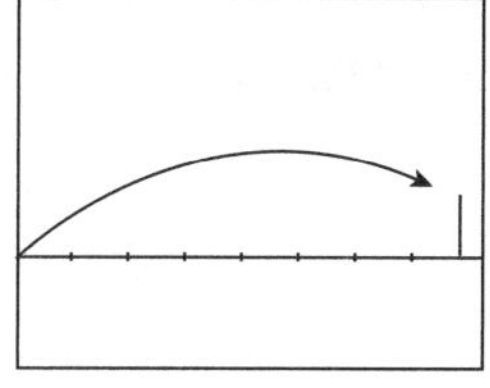

(b) Students watch the ball's flight on their calculator screens

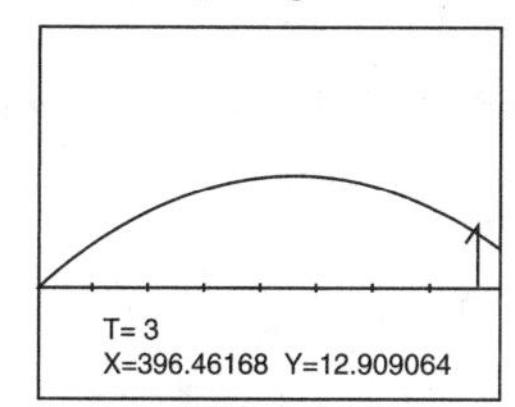

(c) The TRACE function is used to determine how high up the ball strikes the fence

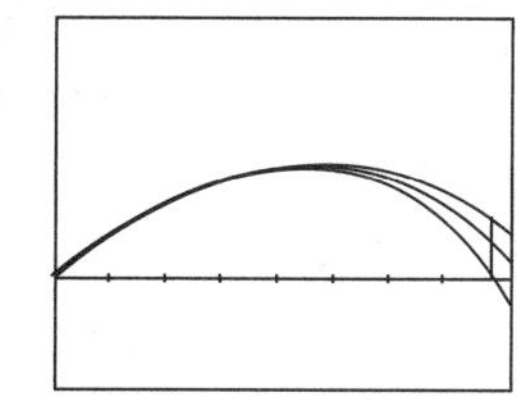

(d) The graphs of three different wind conditions

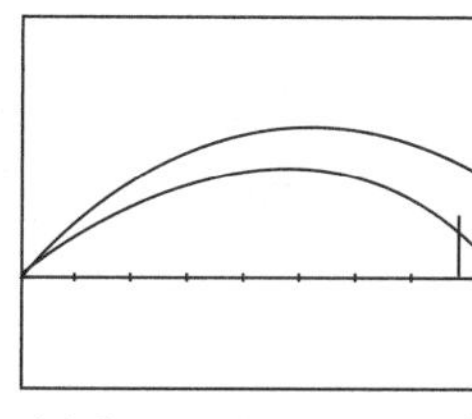

(e) Graphs for hits at 20° and 25°

She uses her display calculator with an overhead projector to demonstrate what she explains as students work along with their calculators.

Ms. Long designed the lesson so that she initially presents students with a physics problem involving motion they could simulate on their graphing calculators:

> Nomar Garciaparra of the Boston Red Sox is at the plate in a game against the Seattle Mariners. The pitcher throws a fast ball at the waist about 3 feet high. Garciaparra hits the ball at a velocity of 150 feet per second at a vertical angle of 20° straightaway toward center field. The center field fence is 20 feet high and 400 feet from home plate. At the moment he hits the ball there is a 6-mph wind blowing straight in from center field. Is this hit a home run? Is the ball catchable?

Using the parametric graphing utility mode, Ms. Long explains the following as she defines the functions on her display calculator:

> The problem will be analyzed in terms of vectors so that the horizontal component of motion is a distance problem. The distance the ball travels equals the rate multiplied by the time of travel. For the purposes of this problem, the ball is assumed to maintain a constant speed during its flight. The horizontal speed is 150 cos 20°, so the horizontal component of motion is (150 cos 20°)t. The effect of the wind is −8.8 feet per second, so the net horizontal component of motion is $x_{1t} = (150 \cos 20°)t - 8.8t$. The vertical component of motion is related to gravity, with an initial velocity of 150 sin 20° and an initial starting height of 3 feet. Then $y_{1t} = -16t^2 + (150 \sin 20°)t + 3$. The graph produced by this set of parametric equations represents the flight path of the ball as it travels toward the center field wall. Use the following RANGE values: $T_{min} = 0$. $T_{max} = 5$. $T_{step} = .05$. $x_{min} = 0$. $x_{max} = 420$. $X_{scl} = 50$. $y_{min} = -25$. $y_{max} = 100$. $y_{scl} = 10$.

She draws the outfield wall on the screen as shown in Exhibit 10.13(a) by returning to the home screen (2nd QUIT) and choosing the line command from the DRAW menu and entering the endpoints of the wall (i.e., "Line (400, 20, 400, 0)") and pressing ENTER.

She executes the function and the students follow the flight of the ball as it moves from Garciaparra's bat toward the wall (see Exhibit 10.13[b]). They use the TRACE function to track the flight of the ball and determine if it is catchable (see Exhibit 10.13[c]).

Having exposed them to a model for the basic problem, Ms. Long uses collaborative-group sessions to lead the students to use the calculators to address the following questions and tasks:

1. What would happen to the ball if the wind suddenly died? What would happen if the wind increased to 12 mph? Graph three possibilities at the same time. (See Exhibit 10.13[d].)
2. What would happen if the ball is hit at an angle of 25°? Compare this hit with the original hit at 20° with a 6-mph wind. (See Exhibit 10.13[e].)
3. A line drive is a ball hit on what appears to be a straight line. If a line drive is hit at a 10° angle, what velocity would it take for it to clear the fence and be a home run?
4. Compare the time of flight of a high fly ball to a line drive using the TRACE function. Graph both simulations at the same time.
5. What is the optimal angle to hit the ball for a home run?
6. What variable is more important to hitting a home run, the angle of the hit or the initial velocity?
7. Change the problem situation to a punter on a football field. What should the punter do to produce the optimal "hang time" for a punt?
8. Change the problem situation to a golfer hitting a golf ball with different types of clubs. For example, a driver has a head angle between 9.5° and 11°. Simulate several different iron shots at the same time.
9. Develop a problem using the same mathematics as in the baseball problem, but make the problem involve something other than sports.

In Case 10.6, a teacher and his students use calculators with a fraction display feature to compare different ways of expressing numbers:

Exhibit 10.14
Mr. Clair-Tresia's Three Tasksheets.

Tasksheet I: How many meters of fencing are needed to enclose the property diagramed below?

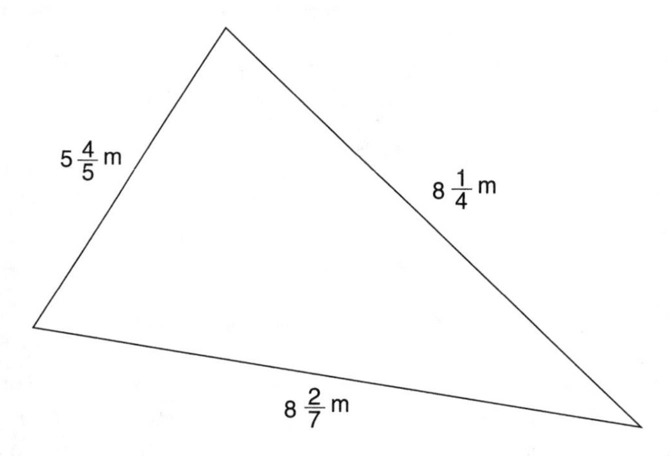

Tasksheet II: How many meters of fencing are needed to enclose the property diagramed below?

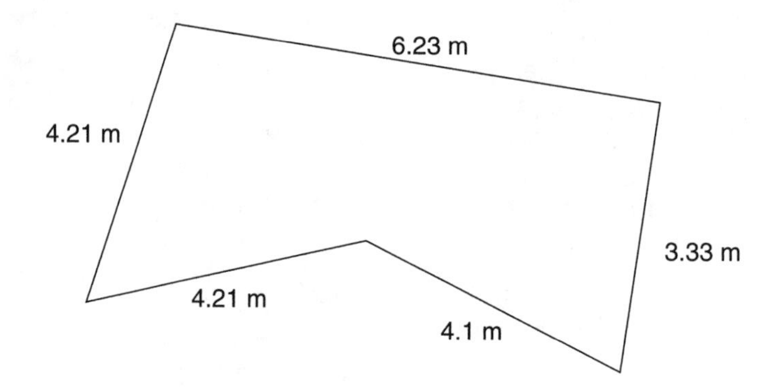

Tasksheet III: How many meters long is the fence that is built along the edge of the stream diagramed below?

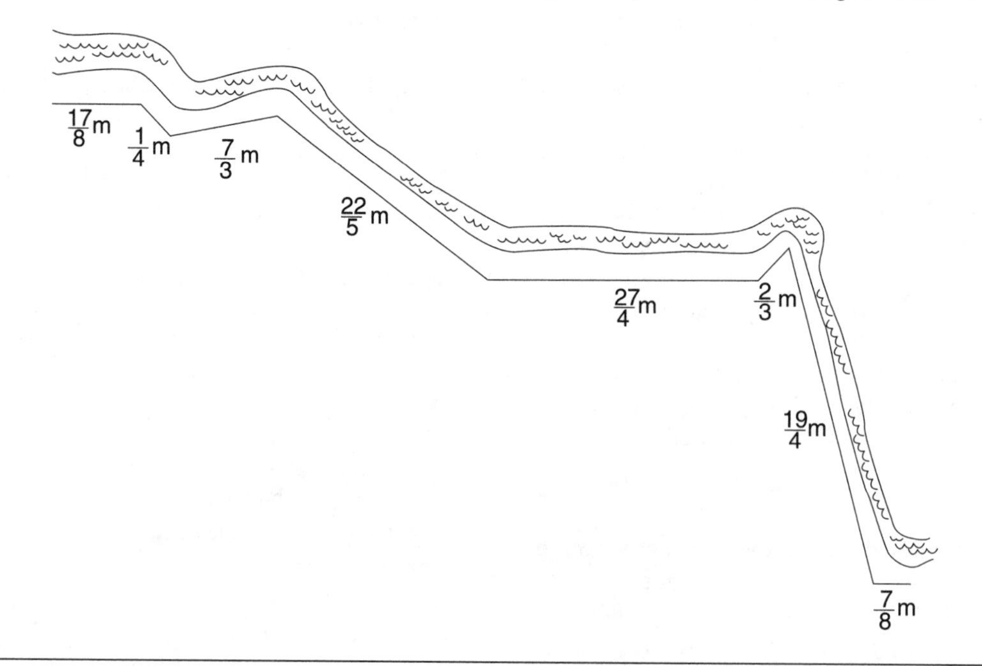

CASE 10.6

After completing a unit on rational numbers, Mr. Clair-Tresia is determined to reinforce the association among the various types of numerals used to express rational numbers (i.e., fractions, mixed numbers, decimals, and percents). So until he is convinced that his sixth graders thoroughly grasp the connections among these forms of expressions and comfortably communicate with one form as well as the others, he incorporates the following strategy in units subsequent to the one on rational numbers: He assigns problems and exercises and presents students with tasks so that some students will be doing the mathematics using one form of expression while others do the same mathematics using another form of expression. Then he has them use their *TI-Explorer* calculators (one of several brands and models with fraction and mixed-number displays and operations in addition to the usual decimal expressions) to compare the different forms.

For example, during a unit on perimeter and area, he directs one group of students to complete Tasksheet I, another group to complete Tasksheet II, and a third group to complete Tasksheet III, as shown in Exhibit 10.14.

For Tasksheet I, the first group reports "$22\frac{47}{140}$m." For Tasksheet II, the second group reports "22.08 m." For Tasksheet III, the third group reports "$\frac{1329}{60}$m." Mr. Clair-Tresia asks the class, "Which piece of land has the least amount of fencing on it? Which one has the most?"

Using the display version of the *TI-Explorer* calculator that he simply sets on the overhead projector, he uses fraction-conversion features to express each of the three answers as a mixed number, as a fraction, and as a decimal. In a discussion session, students answer the two questions.

In Case 10.7, the convenience of students having graphing calculators with them is demonstrated:

CASE 10.7

Nate works on an exercise his algebra teacher, Ms. Van Dusen, assigns as follows:

Factor $2x^2 - 17x + 21$

$(2x - 7)(x - 3)$

He then engages Ms. Van Dusen in the following conversation:

Nate: Is this right?

Ms. Van Dusen: I don't know; check it out on your graphing calculator.

Nate: How?

Ms. Van Dusen: Get into graph mode. Let $Y1 = 2x^2 - 17x + 21$. [pause] Now let $Y2 = (2x - 7)(x - 3)$. [pause] Now graph them on the same screen.

Nate obtains the results in Exhibit 10.15 and the conversation continues:

Exhibit 10.15

Nate Graphs $Y1 = 2x^2 - 17x + 21$ and $Y2 = (2x - 7)(x - 3)$.

Ms. Van Dusen: If $2x^2 - 17x + 21 = (2x - 7)(x - 3)$, what would you expect about the graphs of $Y1$ and $Y2$?

Nate: I don't know.

Ms. Van Dusen: Would $Y1 = Y2$?

Nate: Sure.

Ms. Van Dusen: Look at the two curves on your calculator. Are they the same?

Nate: No. Oh! then $2x^2 - 17x + 21$ isn't the same as $(2x - 7)(x - 3)$ or else they'd be only one curve.

Ms. Van Dusen: You've just invented a test for factoring accuracy.

Seconds later, Nate enters $2x^2 - 17x + 21$ for $Y1$ and $(2x - 3)(x - 7)$ for $Y2$ on his calculator, resulting in the display in Exhibit 10.16.

TECHNOLOGY FOR PRESENTATIONS

Overhead Projector

An overhead projector is a standard classroom feature. Making presentations with an overhead projector is usually preferable to using a chalkboard or dry-erase surfaces because, with the overhead, you can (a) maintain eye contact with students and monitor their behavior, (b) prepare professional-looking illustrations, (c) review and reuse illustrations, (d) focus students' attention on one aspect of your illustrations at a time, and (e) save transition time in class by not having to erase, rewrite, or prepare illustrations while students are waiting. Compare Case 10.8 to Case 10.9:

Exhibit 10.16
Nate Graphs $Y1 = 2x^2 - 17x + 21$ and $Y2 = (2x - 3)(x - 7)$.

CASE 10.8

Mr. Barkin wants to explain the development of a formula for approximating the area of a circular region as depicted in Exhibit 10.17.

His geometry students are poised with pencils and notebooks as he announces, "Let's develop a formula for approximating the area of any circular region." He turns toward the chalkboard and draws a circle as shown in Exhibit 10.18. Looking over his shoulder he says, "Does everyone have a circle in their notes? Okay, let's call the radius of the circle *r*," as he draws on the board. "And circumscribe the circle in a square like this," he continues. Some students—especially those having difficulty seeing his illustrations until he turns around and moves away—entertain themselves with off-task conversations. Mr. Barkin is annoyed by the noise, but because his back is turned, he is not sure who is talking, so—for now—he ignores it.

Facing the class, he asks, "What is the area of the large rectangle?" Some students do not pay attention to the question because they are now busy copying the figure on the board that they could not see while Mr. Barkin was drawing it. Others are cued to stop talking by Mr. Barkin facing the class. Leona answers, "$4r^2$." Mr. Barkin: "Why $4r^2$?" Leona: "Because . . . "

The lecture-discussion continues, with Mr. Barkin eventually constructing the octagon whose area approximates that of the circle and concluding that the circle's area is approximately $(3.111\ldots)r^2$.

Exhibit 10.17
Development of a Formula for Approximating the Area of a Circular Region.

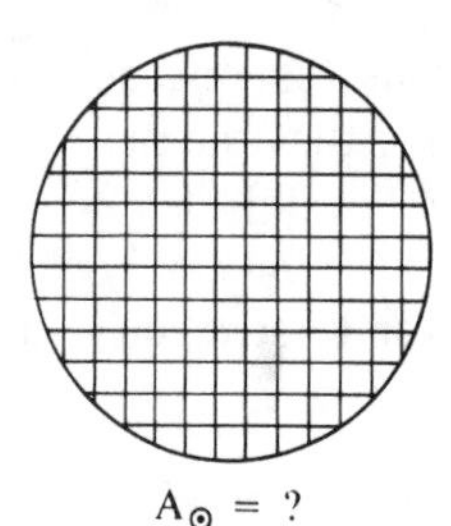

$A_{\odot} = ?$

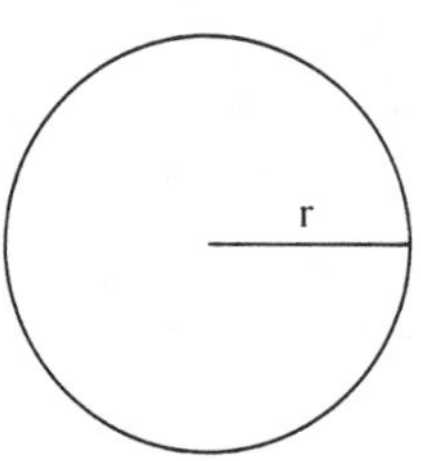

Let the radius = r

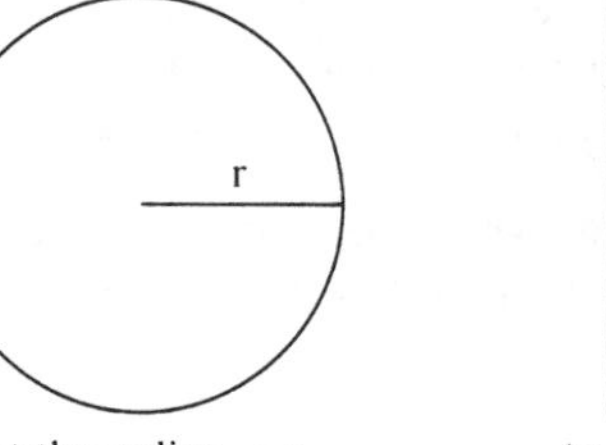

Area of large square = $4r^2$, thus $A_{\odot} < 4r^2$

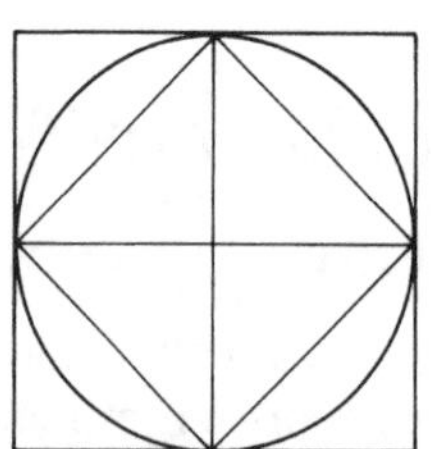

Area of interior square = $2r^2$, so $2r^2 < A_{\odot} < 4r^2$

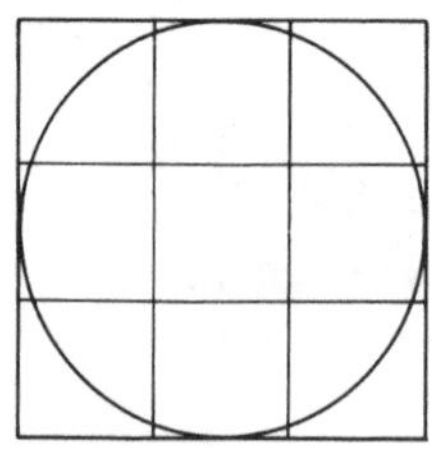

each of 9 squares has an area = $\frac{4}{9}r^2$

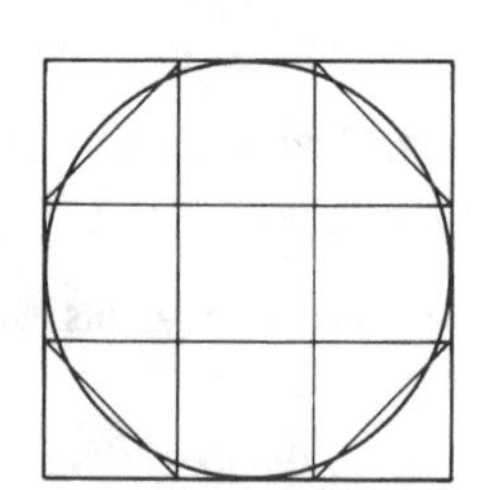

The area of the octagon $\approx A_{\odot}$
And the area of the octagon = $\frac{7}{9}$ of the area of the large square.

Thus, the area of the octagon = $\frac{7}{9}(4r^2) = \frac{28}{9}r^2 = (3.111\ldots)r^2$.
$\therefore A_{\odot} \approx (3.111\ldots)r^2$

Exhibit 10.18
Mr. Barkin Explains the Development of the Formula From Exhibit 10.17 With His Back to the Class.

The session was marred by some students' failing to pay attention whenever he turned his back to write on the board. They were restless as he blocked their view of the board, so they entertained themselves in off-task ways. Some of the other students tried to remain on-task but always seemed to be a step behind, trying to copy what was on the board as Mr. Barkin went on to the next phase of his explanation.

CASE 10.9

Like Mr. Barkin, Ms. Ramos conducts a lecture-discussion session in her geometry class to develop an approximation of a circular region that is depicted in Exhibit 10.17. However, instead of turning her back to the class to illustrate points on the chalkboard, she uses an overhead projector and transparencies she prepared prior to meeting with the class (see Exhibit 10.19). Thus, she is able to monitor the students' behavior throughout the session; students are able to copy the figures and listen at the same time.

There is much less off-task behavior than in Case 10.8. Before moving on, Ms. Ramos is able to see when students are finished writing and whose attention is drifting. Students who appear disengaged are prompted with questions or the sound of their name used during the presentation. For example, "How many small squares do we have now, Nancy?" or "Everyone, quietly count the number of squares there are inside the circle—you too, Scott."

In contrast to Case 10.8, in which students were looking at or copying from the board while Mr. Barkin talked about something else, Ms. Ramos' students see only what she wants them to see because she controls illustrations with a flip of a switch, placement of a transparency, or exposure of only part of a transparency.

The overhead projector is more versatile than simply serving as a superior substitute for a chalkboard or whiteboard:

- An *LCD panel* allows you to project images from a computer screen through the overhead projector.
- *Display calculators* allow you to project images from a calculator screen through the overhead projector.
- You can use *colorful, translucent versions of manipulatives and mathematical models* to illustrate and demonstrate points with the overhead projector.
- An *electronic image writer* allows you to display on the overhead screen what you write and draw, but without being confined to the area near the overhead projector. Thus, you can move about the room while illustrating points for all the class to see. Furthermore, the electronic image writer makes it possible for you to control as well as write on projected images from your computer screen as you move about the room.

Visual Presenter

Like the overhead projector, *visual presenters* display images. However, instead of projecting light through transparencies, a visual presenter uses a small video camera to display the image on its platform onto a television monitor. Unlike an overhead

Exhibit 10.19
Ms. Ramos Faces the Class to Explain the Development of the Formula From Exhibit 10.17 With the Aid of an Overhead Projector.

projector, a visual presenter can project images of objects that are not translucent. Thus, you can use a visual presenter to display—in color—documents directly from the original sources without having to convert them to transparencies—as well as solid, concrete manipulatives and models. Imagine how convenient it is for displaying work students produce independently or in cooperative groups. For example, instead of having students copy their homework onto the board, they simply place it on the platform of the visual presenter for all to see.

Computer-Based, Multimedia Presentations

Word-processing software interfaced with mathematical software (e.g., *Mathematica*) and equation editors (e.g., *MathType*) for writing and illustrating in the language of mathematics allows you to readily produce professional-looking documents for presentation, tasksheets, and tests. Used in conjunction with slide-show software (e.g., *Corel Presentation* or *Microsoft PowerPoint*) that can incorporate video and audio technology (e.g., digital cameras) as well as the internet, you and your students can create engaging, interactive sessions and store them on CD-ROMs or your website.

An impressive variety of videotape and CD-ROM programs can be checked out from libraries or rented or purchased from commercial sources or professional associations.

EXTENDING YOUR MATHEMATICS CLASSROOM BEYOND THE WALLS OF THE SCHOOL

Although it is true that multimedia networking helps bring a global environment into your classroom, you still need to lead your students to do mathematics outside the classroom—making real-world measurements and integrating their mathematics with what they learn in other academic areas. Assignments in which they gather data at banks, real estate offices, engineering firms, sporting events, industrial sites, construction sites, concerts, home, social events, farms, museums, science laboratories, business firms, hospitals, judicial courts, prisons, farms, synagogues and churches, parks, nurseries, and just about anywhere else you can imagine are critical to your efforts to maintain the association between mathematics and the real world.

Traditionally, people are more likely to associate field trips with a science or social studies class than with a mathematics class. But wherever there is science, social studies, physical education, art, language, or music, there is mathematics. Consider taking your mathematics class on a joint field trip with a social studies, science, or English class and use the trip as part of an effort to integrate curricula across academic areas. Involving students in mathematics clubs may help stimulate these types of outside-of-school activities.

SYNTHESIS ACTIVITIES FOR CHAPTER 10

1. You have developed a number of lesson plans and stored them in your working portfolio. Examine a subset of them. For each, determine how, if at all, the lesson might be enhanced by incorporating different or more technologies (e.g., concrete models, electronic manipulatives, computer-based presentations, and internet-based activities) that you had not previously planned. Discuss how you might modify your plans with a colleague who is also engaging in this activity. Modify your lesson plans as you see fit before returning them to your working portfolio.
2. Make plans to participate in a regional or national NCTM conference.

TRANSITIONAL ACTIVITY FROM CHAPTER 10 TO CHAPTER 11

In a discussion with two or more of your colleagues, address the following questions:

1. How should instructional strategies vary according to students' grade levels? How, for example, should teachers interact with sixth-grade mathematics students differently from ninth-grade mathematics students?
2. How should instructional strategies vary according to the mathematical content of courses? How, for example, should an algebra II course be conducted differently from a life-skills mathematics course?
3. What are some strategies you would employ to analyze another teacher's lesson plans and the activities in his classroom for the purpose of learning from one another?

chapter

11

Analyzing Examples of Mathematics Curricula and Instructional Practice

Goal and Objectives for Chapter 11

The Goal The goal of chapter 11 is to lead you to further develop your talent for designing mathematics curricula and conducting lessons for students by prompting you to analyze examples of teachers attempting to implement strategies you learned from your work with chapters 1–10.

The Objectives Chapter 11's goal is defined by the following set of objectives:

A. You will consider variations of students with respect to grade levels when designing mathematics curricula and interacting with students (application) 15%.

B. You will explain fundamental differences among various mathematics courses for middle school, junior high, and high school students (comprehension and communication) 15%.

C. You will analyze examples of mathematics curricula and instructional practice to stimulate your own ideas about how to practice your profession (application) 70%.

BUILDING FROM EXPERIENCES

Often when you implement well-planned lessons that are consistent with research-based principles and *PSSM*'s recommendations using sound classroom management strategies, the lessons do not work out nearly as well as you expected. Why do well-planned lessons go awry sometimes? Ms. Lewis' plan went awry in Case 2.1 because she failed to employ sound classroom management strategies. But you say to yourself, "I didn't botch things like Mr. Lewis. I followed chapters 2's and 3's suggestions, yet the lesson still wasn't as successful as it should have been!" There are always some factors you can not control. More importantly, developing the complex art of teaching meaningful mathematics requires experience practicing the research-based strategies. Chapter 2, for example, suggests that you establish more productive communication patterns with your students by using descriptive instead of judgmental language. But most people are not in the habit of considering their words carefully enough to use descriptive language consistently. Consequently, it may take a concerted effort to practice descriptive language with your students before descriptive phrases flow from your tongue and you begin to reap benefits of this technique.

Chapter 3 illustrated strategies for fully engaging all students in question-discussion sessions. However, most of us spent time in classrooms with mathematics teachers who encouraged us to spit out answers in rapid response to questions. Conditioning yourself and your students to afford opportunities for all to articulate reasoned responses takes practice.

As suggested in chapter 4, sequencing lessons so that students (a) construct a concept before attaching a name to that concept and (b) discover a relationship before learning algorithms based on that relationship is a strategy firmly based in cognitive science. However, planning a unit that is consistent with this strategy is inconsistent with the way typical textbooks are organized and the way most of us were "taught" mathematics. Thus, it may require more experience designing *PSSM*-based units and trying out your designs before you are satisfied with your success rate. Although you may not want to think of your students as subjects of experiments, instructional events do serve as experiments that fine-tune your talent for the complex art.

From chapter 6, you learned the importance of analyzing an algorithm prior to planning a lesson for it and to apply error-pattern analysis during the lesson. Experience observing students struggle with algorithms will lead you to understand the level of detail with which you must break down an algorithm before teaching it. It takes experience analyzing numerous examples of students' work with algorithms to develop your talent for identifying their error patterns.

Chapter 8 emphasized that you lead students to willingly do mathematics by prompting them with problems from their own real worlds. The more you interact with students, the more you are able to integrate problems and situations into your teaching that turns them on to mathematics.

As you work with this chapter, you will be prompted to analyze examples of teachers attempting to implement strategies you learned from your work with chapters 1–10. Although vicarious experiences do not replace actual experiences with students in your own classroom, analyzing others' experiences will accelerate fine-tuning the talent you have developed up to this point. Furthermore, it will help you develop the analytical skills and mentality for learning from your own experiences. Hopefully, you have or will have colleagues (e.g., Casey Rudd and Vanessa Castillo) who will analyze your teaching and whose teaching you will analyze.

In chapters 1–10, hardly any discrimination was made between teaching different grade levels or different courses; suggestions for instructional strategies apply to all middle, junior high, and high school mathematics courses. Choice of mathematical content is not nearly as critical as how the content is taught. This point was illustrated in chapter 4 when you compared Cases 4.1–4.4. Furthermore, mathematical content does not vary from grade to grade and from course to course nearly to the extent that most people expect (Cangelosi, 2001a). Casey Rudd pointed out redundancies throughout mathematics curricula during his interview in Case 1.3. For example, the following topics are introduced as if students had no prior exposure by most mathematics textbooks for grades 6–12: slope, ratio (i.e., rational number), irrational number, rate, absolute value, Pythagorean theorem, greatest common divisor, and prime number. As indicated by Exhibits 4.12, 4.13, 4.14, 4.16, and 4.17, *PSSM*'s (NCTM, 2000b) expectations for grades 6–8 and expectations for grades 9–12 are quite similar for each of the five content standards.

Recommendations for middle-level education (e.g., integrating mathematics with other subject-content areas) and for many special populations (e.g., including creative-thinking lessons for "gifted and talented" students and connecting mathematics to the interests of students "at risk") make perfectly good sense for all students at all grade levels.

Having argued the case for applying suggestions from chapters 1–10 to all grade 6–12 mathematics courses, some attention needs to focus on differences according to students' grade levels and mathematics courses. In the remainder of this chapter, such differences are highlighted.

A SIXTH-GRADE MATHEMATICS COURSE

Most sixth-grade students at most middle schools take a course called "Sixth-Grade Mathematics" or "Math 6." This course typically serves as a transition between the numeric manipulations of arithmetic courses and the symbolic manipulations of algebra courses. Typically, geometry is integrated throughout; units focusing on probability and data analysis are also common. That description also fits courses with the title "Prealgebra" that are more commonly taught to seventh and eighth graders. At many middle schools, Math 6 is considered pre-prealgebra. Students' success with Math 6—as judged by their teachers—often influences whether they take Math 7 (which is very repetitious of Math 6), Prealgebra (which is a hybrid of Math 6 and Algebra I), or Algebra I in seventh grade. Mr. Edginton, whom you met in Cases 4.5 and 4.5.1, plans his sixth-grade mathematics course in Case 11.1:

CASE 11.1

Jalen Edginton is beginning his 22nd year as a mathematics teacher at Capitol Heights Middle School. To the question, "What courses do you prefer teaching?" he responds: "I don't really prefer one over the other. But without a doubt, the courses for seventh graders are the most challenging to manage. Most seventh graders are in the disintegration phase of transescence (Baenen, 2000; Kellough, et al., 1996, pp. 8–12). They are going through so many emotional, physiological, and psychological changes, that doing mathematics and cooperating in a classroom setting are necessarily low on their priority lists. Most sixth graders, on the other hand, bubble over with enthusiasm for learning; their energy seems boundless. The toughest thing about teaching a class of sixth graders is keeping up with them. I gain insights into mathematics from teaching sixth-grade classes; I learn patience and interaction skills from teaching seventh-grade classes. The prealgebra and algebra I classes are usually a mixture of seventh and eighth graders. The eighth graders who have made it past the disintegration phase tend to think the focus of the world is observing and judging them. Are they adults or are they children? Many seem to have learned no math in seventh grade and forgot all they learned in sixth.

"One of the cool things about teaching early adolescents is that within any one classroom you get to work with a huge range of emotional, cognitive, and physical development stages. You have to be really careful in assessing their needs and aptitudes. You may find that this really immature-looking kid that could be mistaken for a third grader is very mature cognitively and emotionally, whereas the mature-looking six footer in the next desk might be the one who thinks like a third grader. The trick is to get to know each student individually. You then have a chance of leading them to do meaningful mathematics."

As Mr. Edginton develops a course outline for the two sections of his upcoming Math 6 course, he refers to the adopted textbook, the state core-curriculum guide, and *PSSM*'s Content Standards and Expectations for Grades 6–8 listed in Exhibits 4.12, 4.13, 4.14, 4.16, and 4.17. He thinks to himself: "This time, I'm not going to be overly concerned with 'covering' all the specific mathematical topics from the textbook. There's nothing here that won't be reintroduced to them next year and the year after when they'll claim it's all new to them. At this stage of their school careers, they mostly need a successful experience doing any kind of challenging, meaningful mathematics. I'll take advantage of all that sixth-grade energy to lead them to discover and invent real mathematics."

"Last year, I promised myself I'd integrate more of what we do in math with other subjects they're taking. Before I finalize this outline, I'll hold a meeting with as many of their other teachers as I can interest in coordinating their units with mine."

Health and physical education teacher Annika Jackson, life science teacher Jason Martin, social studies teacher Corina Knight, English teacher Harmony Rodriguez, and Mr. Edginton meet:

Mr. Edginton: Most students who will be in my two Math 6 sections will be in your sixth-grade classes also. Let's see if we can coordinate some of our units to integrate our curricula wherever convenient.

Ms. Jackson: Because math and history are tied to a linear sequence and P.E. is the least dependent on a fixed sequence, maybe we should start with lists of Jalen's and Corina's units.

Mr. Edginton: I resisted sequencing my units until we had this meeting. Actually, we can be very flexible with Math 6; it's not as lock-step as most people think.

Ms. Knight: It's the same with the history components of the social-studies courses. I've gotten away from sequencing units by chronological events. I plan to have units on topics like conflict in America and struggles in America. So, I have a bit of flexibility too.

Mr. Edginton: I'll build units involving ratios, probability, data analysis, prime numbers, coordinate geometry, measurement, solving open sentences, functions, and a little trig. I can shuffle them around to match with some of yours.

Mr. Martin: What do you have in mind?

Mr. Edginton: Here's an example. Annika teaches students exercises for improving cardiovascular endurance, while you teach them the mechanisms of how that works in a unit on human respiratory and circulatory systems. They'll collect data from endurance tests in P.E. to analyze and interpret in a statistics unit in my class.

Ms. Rodriguez: And in my language arts class, they could be developing a report on the results of the tests as they improve their writing skills.

Mr. Martin: Okay, I see how that'll work out for your math topics like statistics, measurements, and solving equations, but not for a pure math topic like prime numbers.

Mr. Edginton: I don't expect we'll be able to integrate all of our units, but prime numbers is one topic that I'd like to integrate just so students will learn how to put this traditionally pure topic to practical use. Two applications come to mind. One of them fits with something you might be interested in teaching, Jason: different life cycles.

Mr. Martin: Oh! I know what you're getting at. Some animals—especially insects—have peculiar life cycles. Cicadas emerge only every 17 years and that allows most of them to avoid parasites that emerge every x years where x isn't a divisor of 17. (Singh, 1997, pp. 96–97)

Mr. Edginton: And while you've got students questioning why these animals evolved this way, I'll be playing off that question to get them to discover attributes of prime numbers.

Ms. Jackson: You said you had a second example in mind.

Mr. Edginton: Oh, thanks for reminding me, Annika. Because we have no algebraic formula for generating primes, primes play a critical role in cryptography. Codes are built using functions on primes.

Ms. Knight: Now that would really fit nicely with one of my history units on conflict and wars. Oh, better yet! I just thought of a new title for a unit: "The History of Espionage." That would intrigue them.

Mr. Martin: It's too bad the sixth graders don't take Judith's technology course; she could teach them to write programs for coding, encoding, and decoding messages.

Mr. Edginton: I'll speak with her about having our sixth graders team with her eighth graders. That's another way to integrate curricula.

Exhibit 11.1
Teachers Agree to Integrate Curricula by Coordinating Units Around Common Themes.

School Weeks	Theme	Math 6	English 6	H. & P. E. 6	Life Science 6	Social Studies 6
3–4	origins of communications	numeration	origins of language		early discoveries	early civilizations
9–13	making our bodies work better	measurements and statistics	keeping records and writing reports	cardiovascular fitness	respiratory and circulatory systems	making society work
15–16	teaching one another	*Euclid's Elements*	early books	formalizing games		
17–20	engineering	geometric structures	writing structures	strengthening bones and muscles	anatomy	social structures
18–32	cryptography	primes	word searches and presenting reports			espionage
21–23	strange creatures	primes	presenting reports	avoiding dangers to health	life cycles	survival
31–34	decision making	probability	influencing others	game-winning strategies		historical success and blunders

Ms. Rodriguez: My language arts class will fit right in any of these units because whatever you do will require them to read and present reports.

Ms. Knight: Code breaking involves studying word and letter patterns. So wouldn't that be another connection between the cryptography project and your class, Harmony?

Mr. Edginton: Yes, the Bletchley Park team that broke the Nazi Enigma code for the Allies in World War II consisted of linguists and people who were good with crossword puzzles as well as mathematicians. (Singh, 1999, pp. 143–189)

The conversation continues for another hour. They agree to coordinate some of their units around common themes as reflected in Exhibit 11.1.

Energized by the meeting collaborating with colleagues, Mr. Edginton begins developing his course outline, tentatively scheduling the first two weeks for the unit he names "Problem-Solving and Modeling Strategies," the next three weeks for [pause] then as he tries to puzzle in the themes for integrating with the other subjects as reflected in Exhibit 11.1, he suddenly thinks, "Subsequent units always depend on prior units. Why not take that a step further and extend all my units until the end of the school year? We'll still have unit tests right before new units begin, but no unit will actually end until the school year is over. This will facilitate integration because I can plug any of the concurrent units that have already begun into any of the themes. A theme might come too early for a particular unit, but never too late. I like this idea; I should have thought of it years ago!"

Exhibit 11.2 is the Math 6 course outline Mr. Edginton put together. Having overlapping, continuing units reduces the number of units he worked with in the past from about 16 to the 11 listed in Exhibit 11.2. No longer, for example, does he need to list separate units for polygons, compass constructions, and solids because Euclid's Geometry will subsume those topics. Integrating the study of prime numbers with other courses can be ongoing once Euclid's Number Theory begins. This way of organizing the course facilitates long-range projects (e.g., involving cryptography).

In the 17th school week, Mr. Edginton begins Unit 7 by implementing the lesson plan shown by Exhibit 4.4. Once students have constructed the concept of primes, he leads them to invent algorithms (e.g., the sieve of Eratosthenes) for discriminating between primes and composites. The class initiates a rest-of-the-year, ongoing effort to discover "new" primes. A large poster is displayed in the room on which students record their discoveries by circling numerals of primes they discover, slashing numerals for composites they discover, and leaving alone numerals for natural numbers greater than 1 that they have yet to determine to be prime or composite. Exhibit 11.3 shows part of this poster.

Exhibit 11.2
Mr. Edginton's Sequence of Units for Math 6.

School Weeks (Overlapping Units)	Unit #	Unit Title
1–36	1	Problem-Solving and Modeling Strategies
3–36	2	Place-Value Systems and Operations
6–36	3	Applying and Expressing Ratios
9–36	4	Measurements and Data Collection
11–36	5	Analyzing, Displaying, and Interpreting Data
14–36	6	Euclid's Geometry
17–36	7	Euclid's Number Theory
24–36	8	Applications of Integers
26–36	9	Algebraic Equations and Inequalities
31–36	10	Probability
34–36	11	Extending What We Learned

Exhibit 11.3
Wall Poster Showing Primes and Composites Students Have Discovered.

2	25	48	71	94	117	140	163	186	209	232	255	278	301
3	26	49	72	95	118	141	164	187	210	233	256	279	302
4	27	50	73	96	119	142	165	188	211	234	257	280	303
5	28	51	74	97	120	143	166	189	212	235	258	281	304
6	29	52	75	98	121	144	167	190	213	236	259	282	305
7	30	53	76	99	122	145	168	191	214	237	260	283	306
8	31	54	77	100	123	146	169	192	215	238	261	284	307
9	32	55	78	101	124	147	170	193	216	239	262	285	308
10	33	56	79	102	125	148	171	194	217	240	263	286	309
11	34	57	80	103	126	149	172	195	218	241	264	287	310
12	35	58	81	104	127	150	173	196	219	242	265	288	311
13	36	59	82	105	128	151	174	197	220	243	266	289	312
14	37	60	83	106	129	152	175	198	221	244	267	290	313
15	38	61	84	107	130	153	176	199	222	245	268	291	314
16	39	62	85	108	131	154	177	200	223	246	269	292	315
17	40	63	86	109	132	155	178	201	224	247	270	293	316
18	41	64	87	110	133	156	179	202	225	248	271	294	317
19	42	65	88	111	134	157	180	203	226	249	272	295	318
20	43	66	89	112	135	158	181	204	227	250	273	296	319
21	44	67	90	113	136	159	182	205	228	251	274	297	320
22	45	68	91	114	137	160	183	206	229	252	275	298	321
23	46	69	92	115	138	161	184	207	230	253	276	299	322
24	47	70	93	116	139	162	185	208	231	254	277	300	323

As part of his preparations for conducting the cryptography project, Mr. Edginton read *The Code Book: The Science of Secrecy from Ancient Egypt to Quantum Cryptography* (Singh, 1999). Exhibit 11.4 shows the overall plan for the project that he, Ms. Knight, and Ms. Rodriguez developed.

The cooperative learning activities related to the cryptography project were designed to lead students to achieve a subset of Unit 7's objectives. The need to measure students' achievement of those objectives for summative evaluation purposes presents a dilemma for Mr. Edginton. Although students need to be positively reinforced for contributing to their team's efforts, he does not want to award points toward their grades for participation or for successfully creating a code, decoding, or code breaking. Doing so could detract from the spirit of collaboration he is building for cooperative learning. To maintain the link between participation and achievement without students having to worry about "being graded" as they collaborate, he decides to include miniexperiments with prompts such as the following: Explain how prime numbers were used by Team Y's Code Creators to make keys for their code?

How diligently students participated influences how well they achieved objectives to which such miniexperiments are relevant. Thus, participation is positively reinforced without Mr. Edginton putting a damper on cooperative-learning activities by walking around with a clipboard to award "participation points" (Cangelosi, 2000b, pp. 137–145).

To prompt yourself to analyze teaching episodes for the purpose of strengthening your talent for the complex art, engage in Activity 11.1:

Activity 11.1

In a discussion with a colleague, address the following questions about Case 11.1:

1. What are the advantages and disadvantages of integrating mathematics curricula with that of other content areas?

Exhibit 11.4
Overall Plan for the Cryptography Project.

1. On the first day, Ms. Knight shows the videotape program "Spies: Code Breaking" (Columbia House, 1993).
2. On the second day, Ms. Knight establishes rules for the Espionage Game in which the class of 24 is partitioned evenly into two opposing teams. (Refer to them as "Team X" and "Team Y" for now, but students will select more colorful names.) Part of the game will involve Team X exchanging coded messages among its members that Team Y intercepts but does not have the keys to decipher the code. Events follow (e.g., the simulation of a supply line being cut off) the passing of messages from which the opposing team might infer what the message said (after it is too late). Concurrently, Team Y's members will be exchanging coded messages that Team X will attempt to decipher.
3. On the third day, Mr. Edginton teaches relatively simple coding, decoding, and code breaking techniques involving the pattern illustrated below:

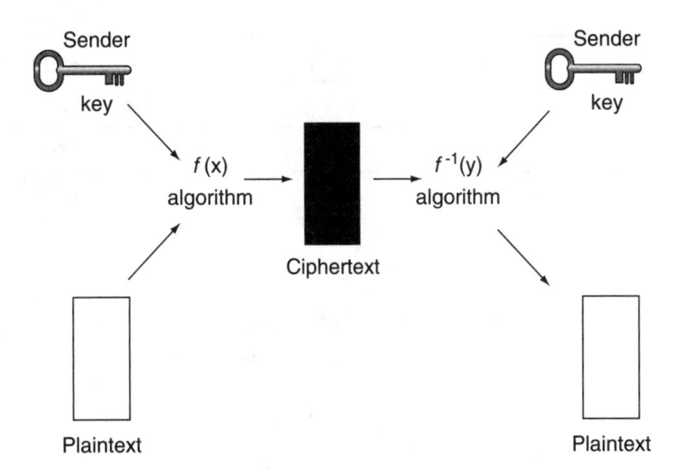

The role of prime numbers is emphasized.

4. On the third and fourth days, Ms. Rodriguez engages students in activities in which they examine word patterns. They also play word-search games.
5. On the fourth day, Ms. Knight organizes the two teams into the following subteams:

Team X's Code Creators: Tyrone, Tanisha C., Allison, & Peggy
Team X's Decoders: Tricia, Bernard, Byron, & Brittany
Team X's Code Breakers: Kellie, Seritta, Saddam, & Alicia
Team Y's Code Creators: Ping, Rosalie, R.J., & Roberta
Team Y's Decoders: Roeberto, Tanisha L., Soo-Shen, & Wansoo
Team Y's Code Breakers: Magnella, Shambay, Yolanda, & Tyler

6. On the fifth day, Ms. Rodriguez teaches students how to organize the reports that they will complete at the conclusion of the game.
7. On the fifth and sixth days in Mr. Edginton's class, code creators secretly build their codes.
8. Between the sixth and 12th days, the students play the game with the coded messages being sent and intercepted and each code-breaking subteam trying to figure out the secret formula for the opposition's code.
9. On the 13th day, a truce is called; peace is declared in all three classrooms. Each of the six subteams prepares a report to reveal all of their work.
10. On the 14th day—using all three classes—reports are presented and all secrets are revealed.

Note: Throughout the 14 school days, students engage in activities for other lessons besides the ones involving the cryptography project.

2. What do you think of the choice of themes and the way they were tied to the various mathematical topics as indicated by Exhibit 11.1?
3. What are the advantages and disadvantages of setting up unit sequences as in Exhibit 11.2 as opposed to the more traditional sequencing patterns in which one unit stops when the next begins (e.g., Exhibit 11.33)?
4. If you were in Mr. Edginton's place, how would you modify Exhibit 11.2's course outline?
5. What are the advantages and disadvantages of the long-range activity using Exhibit 11.3's wall poster?

Exhibit 11.5
Mr. Edginton's Sequence of Units for Prealgebra.

School Weeks (Overlapping Units)	Unit #	Unit Title
1–36	1	Variables, Relationships, Operations, and Algebraic Expressions
4–36	2	Solving Open Sentences and Graphing Solutions
9–36	3	Using Rational Expressions
12–36	4	Discrete Mathematics and Probability
18–36	5	Collecting, Analyzing, Displaying, and Interpreting Data
23–36	6	The Geometry of Algebra
25–36	7	The Algebra of Geometry
31–36	8	Trigonometry
33–36	9	Extending What We Learned

6. With respect to pattern recognition—even unrelated to primes, what are some of the advantages of having Exhibit 11.3's poster constantly hanging on the wall?
7. How prudent is it for teachers to direct students to play competitive games that include simulations of violent acts and opposing teams, considering the need to discourage violence and gang activity at schools? What are some strategies for reaping the benefits of such projects while still fostering a peaceful campus community?
8. What do you think of Mr. Edginton's strategy for measuring objectives achieved through cooperative learning activities?
9. Would courses organized as Mr. Edginton organized the course in Case 11.1 work any differently for seventh graders than it would for sixth graders?
10. What insights did you gain from your work analyzing Case 11.1? How do you plan to apply those insights to your own teaching?

A PREALGEBRA COURSE

Case 11.2 reflects a small slice of the prealgebra course Mr. Edginton also teaches:

CASE 11.2

Mr. Edginton's prealgebra classes are a mixture of seventh and eighth graders. The table of contents for the adopted textbook is shown by Exhibit 1.4—the same one Ms. Castillo displayed during Mr. Rudd's interview in Case 1.3.

Pleased with the idea of overlapping units that he developed for Math 6 in Case 11.1, he outlines the prealgebra course as shown by Exhibit 11.5.

At Capitol Heights Middle School, individual students' class schedules vary more for seventh and eighth graders than for sixth graders. Thus it is not as convenient for Mr. Edginton to integrate his prealgebra curriculum with that of other courses as it is for Math 6. But he has managed some integration, especially with some of Ms. Jackson's H. & P.E. classes.

It is now March and Mr. Edginton's fifth-period prealgebra class is working in units 1–7. He is about to begin a lesson for the following objective:

> When confronted with a real-life problem, the student determines whether or not her knowledge of relationships involving circles is useful in solving the problem (application).

For the lesson, he arranged with Ms. Jackson to substitute for one of her H. & P.E. class sessions that includes most of his fifth-period prealgebra students. He meets the class on the school's track; students are dressed in P.E. uniforms. He announces: "Ms. Jackson asked me to have you race two at a time around the 400-m track. I have her list of who races against whom. But first, Helen, lead us in the warm-up and stretching routine that Ms. Jackson taught you to do before doing any kind of sprinting."

When the warm-up is completed, he directs the two students who are excused from participating in vigorous exercises to operate stopwatches and record times for each pair of competitors. He begins the races by directing Jonathan and Issac to the starting positions as shown by Exhibit 11.6 and emphasizing that the runners are required to stay in their respective lanes for the entire race.

After several races—one in which Rebecca "illegally" switched from the second to the first lane in the first turn—students complain that the races are not fair. They explain to Mr. Edginton that the first-lane runner does not

Exhibit 11.6
Runners Start Side by Side.

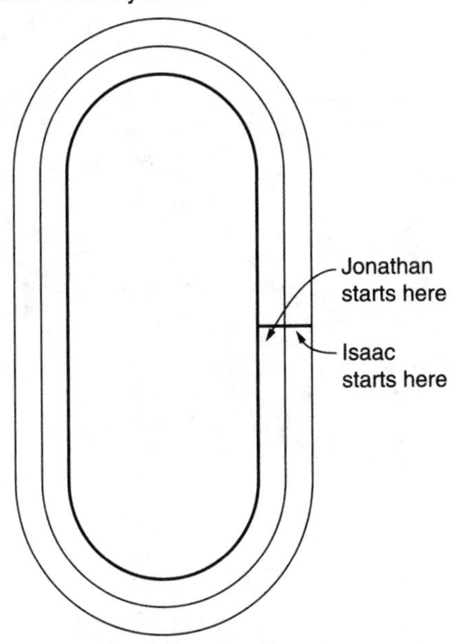

Exhibit 11.7
Runners With an Exaggerated Stagger.

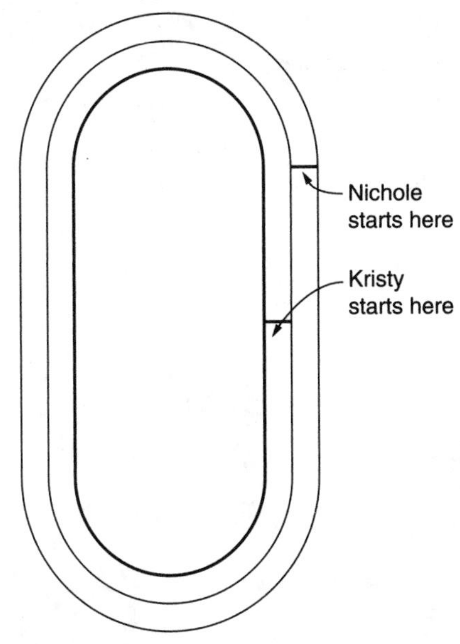

have as far to go, so they should be able to switch lanes as Rebecca did. Mr. Edginton insists that they stay in their lanes as Ms. Jackson told him they should, but agrees to give the second-lane runner a head start. He lines up the next pair, Kristy and Nichole, as shown by Exhibit 11.7.

Kristy complains that the head start or stagger is too big. Others support her contention. Mr. Edginton has them race anyway, resulting in an easy win for Nichole. After more complaints are heard, Mr. Edginton indicates that tomorrow he will present the question of what the length of the stagger should be to his prealgebra class.

The next day, Mr. Edginton has prealgebra students who were also in yesterday's P.E. session with him explain to the others the problem that surfaced on the track. He conducts a question-discussion session leading to several proposals for solutions:

- Return to the track and measure—with a tape or trundle measure or by carefully counting steps—the distance around the first lane and the distance around the second lane (see Exhibit 11.8[a]). The difference should be the length of the stagger.
- Model the problem using string. The difference between the lengths of the two strings as shown in Exhibit 11.8(b) would represent the length of the stagger.
- Ask Ms. Jackson (see Exhibit 11.8[c]).
- Somehow figure it out on paper (i.e., algebraically; see Exhibit 11.8[d]).

Mr. Edginton encourages students to try the measurement, modeling, and algebraic approaches, but asks that they wait until they can solve the problem themselves before asking Ms. Jackson—who he knows will not yet tell because she is in on the plan. He organizes the 31 students into six teams: A, B, C, D, E, and F. Teams A and B address the problem using measuring devices at the track. Teams C and D are provided with string, scissors, and marking pens to develop their modeling approach. Teams E and F attempt to work out a general algebraic formula.

Two days later, the groups report to the class:

- Team A reports that each of the track's two straightaways is about 105 m, each of the first lane's curves is about 95 m, and each of the second lane's curves is about 97.8 m. Team A concluded that the second-lane runner should have a head start of about 5.6 m.
- Team B measured the distance around the first lane to be 400.4 m and around the second lane to be 405.2 m and concluded that the stagger should be 4.8 m.
- Team C modeled the inside edge of the track with a 48″ piece of string in the shape of an oval and then placed a second piece of string outside of it to form two concentric ovals about 1″ apart. Then they measured the length of the second string to be a little more than 54″. Thus, they concluded that the stagger should be a little more than 6″ if the track were 48″ around.
- Some members of Team D went out to the track and measured the width of a lane to be 0.9 m. Others marked 40 "evenly spaced" dots with the pen on a piece of string. Altogether, they put the string in the shape of the oval track to represent the inside edge of the first lane. They then configured a second piece of string outside the first so that it was at a distance of 0.9 of the space between neighboring dots on the first string. They then picked up the strings, stretched them out together and cut off the extra length of the second

Exhibit 11.8

Students Propose Four Different Approaches for Addressing the Track-Stagger Problem.

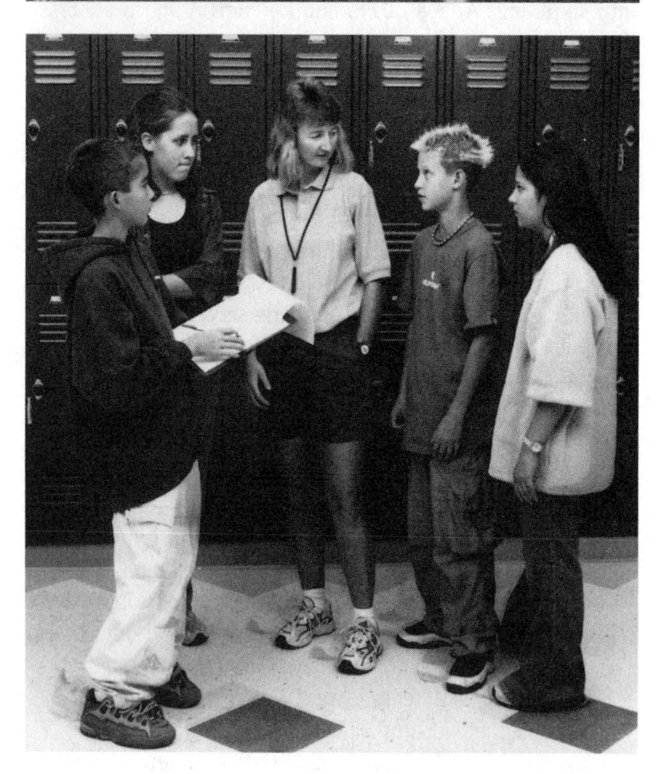

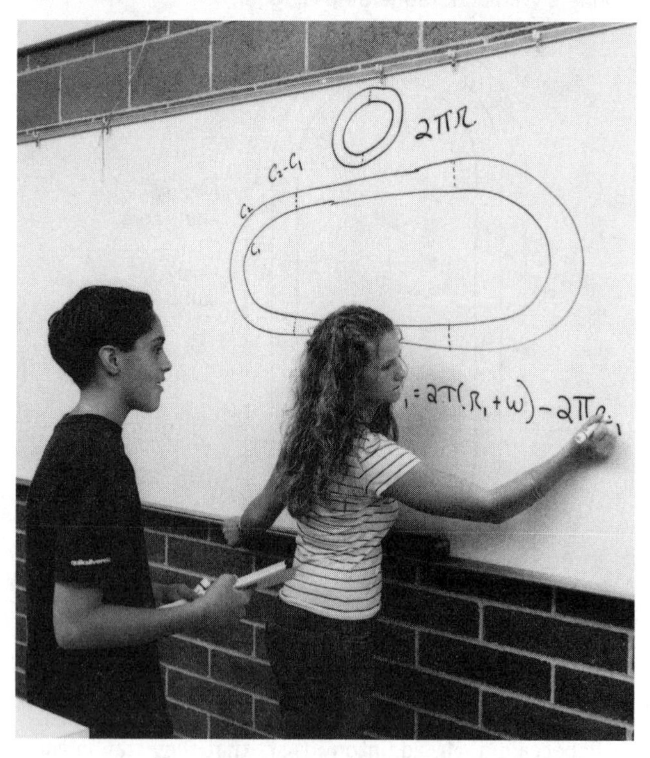

string. The cut-off part—which they marked with dots as they had the first string—represented the difference in distance around the two lanes. Because they marked five dots with about another half of an interval leftover, they concluded the stagger should be about 5.5 decameters. To explain why they used decameters, they said, "because we put only 40 dots on the string instead of 400—but that seems like way too much."

- Team E reports that they "couldn't figure it out." But Mr. Edginton—who consistently focuses on the mathematics students do rather than on bottom-line solutions, prompts them to discuss what they did.

Exhibit 11.9
Group F Discovered That the Straightaways Have No Bearing on the Problem.

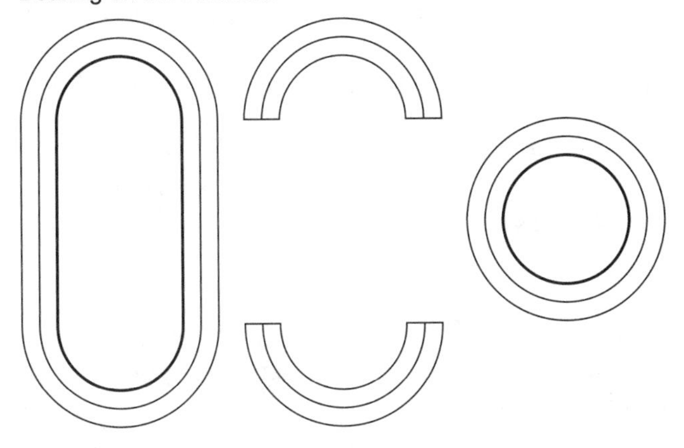

- Team F reports that they did not get a final "answer," because they did not take measurements. They did, however, discover that the difference between the two distances is caused strictly by the two curves. Thus, they eliminated the straightaways and focused on the circle that is left as shown by Exhibit 11.9. They then developed a general formula for finding the length of the stagger. The following reflects what they developed, using crisper mathematical expressions that are more familiar to you than to them—as opposed to the clumsier, inexact expressions they used but which Mr. Edginton understands and helps them translate:

  Let C_1 = the circumference of the inner circle, C_2 = the circumference of the outer circle, r_1 = the radius of the inner circle, r_2 = the radius of the outer circle, and w = the width of the lane. Then the length of the stagger is $C_2 - C_1$, and $r_2 = r_1 + w$. The circumference of a circle is $2\pi \times$ its radius. Therefore, $C_2 - C_1 = 2\pi(r_1 + w) - 2\pi r_1$.

Over the next two days, the six team reports are thoroughly discussed, compared, and related. After simplifying Team F's formula to read "$C_2 - C_1 = 2\pi w$," the class uses Team C's measurement of $w = 0.9$ m to compute that $C_2 - C_1 \approx 5.65$ m, which is analyzed in light of and compared to findings from Teams A, B, C, and D.

To prompt yourself to analyze teaching episodes for the purpose of strengthening your talent for the complex art, engage in Activity 11.2:

Activity 11.2

In a discussion with a colleague, address the following questions about Case 11.2:

1. If you were in Mr. Edginton's place, how would you modify Exhibit 11.2's course outline?
2. What are the advantages and disadvantages of Mr. Edginton "playing dumb" about the need for staggered starts when he substituted for Ms. Jackson?
3. What are the advantages and disadvantages of the way Mr. Edginton conducted the application lesson with respect to achievement of the stated objective?
4. Besides making progress toward the stated objective, what might students learn about doing meaningful mathematics from their engagement in the application lesson?
5. What strategies do you imagine Mr. Edginton employed in the months prior to March for him to have developed a classroom atmosphere in which students discover and invent mathematics as they did while addressing the track-stagger problem?
6. What insights did you gain from your work analyzing Case 11.2? How do you plan to apply those insights to your own teaching?

AN ALGEBRA I COURSE

Some seventh graders take algebra I, but more commonly, algebra I is a course for eighth and ninth graders. High school students who need to fulfill a "requirement for graduation" may take it as 10th, 11th, or even 12th graders. Typically, algebra I courses are very similar to prealgebra courses but with a little more emphasis on symbolic manipulation and less on arithmetic, number theory, geometry, and discrete mathematics.

Unlike three-year high schools, Rainbow High School, where Casey Rudd teaches, includes ninth-grade students. As indicated by his teaching schedule shown by Exhibit 1.5, Mr. Rudd conducts a first-period, two-semester algebra I section. Note the 17 units he anticipated as listed under "What will you learn from this class?" of his Algebra I syllabus shown by Exhibit 1.7.

Cases 11.3–11.13 describe his work in some detail for one of those units:

CASE 11.3

Ten weeks have elapsed since the opening of school. Mr. Rudd's algebra I class is in the early stages of Unit 6, Polynomials. Mr. Rudd anticipates the class beginning Unit 7, Factoring Polynomials, in two weeks, and starts designing it.

It is nearly 10 P.M. as Mr. Rudd sits in front of his computer. Strewn about are the algebra I course syllabus, algebra I textbook, reference books, a calculator, paper, pens, pencils, and his teaching notebook. Since August, as ideas occur to him that might be useful for future lessons, he jots them down in this teaching notebook and transfers some to the "Resources" subfolder of his computerized management system. He reads the goal of Unit 7:

> Students understand why certain factoring algorithms work and can use them in problem solving.

He thinks: "I need to define this goal with objectives. Let's see, why do I want to teach them about factoring polynomials in the first place? Primarily because they need to know how to factor and why factoring works in order to learn how to find roots of higher degree equations, inequalities, and functions. So this unit should really emphasize discovery of relationships underlying certain factoring algorithms, as well as the development of their skills with those algorithms. The real application comes in Units 8, 9, and so forth. But I had better build at least one application objective in here or else I'll lose them before we even get to Unit 8.

"Okay, so I'll end with an application objective. But where do I start? First I'll get them to conceptualize the role of factoring in problem solving—the value of undoing multiplication. This'll almost be like an affective objective—but, I'll stick with discover a relationship."

In a few minutes, Mr. Rudd formulates Objective A as listed in Exhibit 11.10. He then reviews pages 210–243 of the textbook, which will be the primary reference for this unit. He plans to cover the four principal factoring algorithms included in those pages with two objectives each—one discover-a-relationship objective for the students to understand why the algorithm works and one algorithmic-skill objective for the students to execute it efficiently. After coming up with those eight objectives—listed C–J in Exhibit 11.10, he decides to include an objective involving comprehension of the language related to factoring polynomials. He feels it is important to list it to guard against just assuming students follow and use communication conventions about the topic—a mistake he made with previous units. Consequently, he formulates Objective B as listed in Exhibit 11.10. He notes that the vocabulary specified as mathematical content for Objective B are all words and phrases to which students were previously exposed—but only in the context of working with integer constants, not polynomials.

Like Objective B, Objective K is not a central objective of the unit, but computer programming is something he is trying to build into most units. Achievement of the application objective, L, is to be the climax of the unit.

For purposes of the blueprint for the unit test, Mr. Rudd weights the objectives. It is past midnight when he completes his plan for Unit 7 as it appears in Exhibit 11.10.

To prompt yourself to analyze teaching episodes for the purpose of strengthening your talent for the complex art, engage in Activity 11.3:

Activity 11.3

In a discussion with a colleague, address the following questions about Case 11.3:

1. If you were in Mr. Rudd's place, how would you modify the course outline as reflected in Exhibit's 1.7's Algebra I syllabus?
2. Would you organize your own unit plans differently than the way Mr. Rudd organized the one in Exhibit 11.10? If so, how?
3. How consistent is Exhibit 11.10's unit plan with suggestions from chapters 4–8?
4. Do you agree with Mr. Rudd's choice of vocabulary for students to comprehend for Unit 7 as specified by Objective B listed in Exhibit 11.10?
5. If you were in Mr. Rudd's position, would you have included Objective K as listed in Exhibit 11.10? Why or why not?

In Case 11.4, Mr. Rudd reviews the results of the unit test on polynomials and sets the stage for the opening of Unit 7:

CASE 11.4

It is Monday, the final day of Unit 6. After reviewing the unit test, Mr. Rudd calls attention to a prompt he had planted on the test just for this moment. Exhibit 11.11 reflects the way most students responded to it.

Mr. Rudd engages the students in a brief reasoning-level question-discussion session:

Mr. Rudd: I'm just looking at the $-3ab$ and wondering what it means. What is it? Juaquin.

Juaquin: It's a number.

Mr. Rudd: I agree. But what kind of number—large, small, or what? Dustin.

Dustin: It's negative.

Mr. Rudd: What do you want to say, Kevin?

Kevin: It wouldn't be negative if ab is negative or 0.

Mr. Rudd: Raise your hand if you disagree with Kevin. [pause] So, everyone agrees that $-3ab$ could be negative, 0, or positive depending on a and b. Give us a number, Cassandra.

Cassandra: 24.

Mr. Rudd: If $-3ab$ is 24, what would a and b be? Mary.

Mary: You still don't know, because the value of one affects the other.

Mr. Rudd: -3 times what equals 24? Dustin.

Dustin: 8—no! I mean -8.

Mr. Rudd: So, what's ab if $-3ab$ is 24? Mary.

Mary: -8.

Mr. Rudd: If ab is -8, what is a? Omar.

Omar: You don't know unless you know b. If b were 2, a would be -4.

Mr. Rudd: Thank you. [pause] I see quite a few of you still have something to contribute to this matter, but we're nearly out of time, so I'll explain the homework assignment and we'll start Unit 7 tomorrow.

Mr. Rudd distributes the tasksheet shown by Exhibit 11.12 except that numerals, "24," "24," and "10" and "0" are not initially written in the blanks.

Exhibit 11.10
Mr. Rudd's Plan for Unit 7 on Factoring Polynomials.

Course: Algebra I

Unit 7 (of 17): Factoring Polynomials

Goal: Students understand why certain factoring algorithms work and apply them to real-life situations.

Objectives:

A. The student explains how factoring polynomials can facilitate problem solving (discover a relationship) 06%.
B. The student incorporates the following words and phrases into her working vocabulary relative to algebraic polynomial expressions: "factor," "prime factorization," "greatest common factor," "differences of squares," "perfect square trinomial," and "prime polynomial" (comprehension and communication) 06%.
C. The student explains why the distributive property of multiplication over addition can be used to express a polynomial in factored form (discover a relationship) 09%.
D. The student factors polynomial expressible in the form $ax + ay$ (algorithmic skill) 09%.
E. The student explains why polynomials expressible in the form $a^2 - b^2$ can be expressed as $(a + b)(a - b)$ (discover a relationship) 09%.
F. The student factors polynomials expressible in the form $a^2 - b^2$ (algorithmic skill) 09%.
G. The student explains why polynomials expressible in the form $a^2 + 2ab + b^2$ can be expressed as $(a + b)^2$ (discover a relationship) 09%.
H. The student factors polynomials expressible in the form $a^2 + 2ab + b^2$ (algorithmic skill) 09%.
I. The student explains why some polynomials expressible in the form $ax^2 + bx + c$, where x is a real variable and a, b, and c are rational constants, can be expressed in factored form $(dx + e)(fx + g)$, where d, e, f, and g are rational constants, and others cannot (discover a relationship) 10%.
J. Given a polynomial expressible in the form $ax^2 + bx + c$, where x is a real variable and a, b, and c are rational constants, the student determines if it can be expressed in the factored form $(dx + e)(fx + g)$, where d, e, f, and g are rational constants, and, if so, does (algorithmic skill) 10%.
K. The student writes and uses computer programs to execute algorithms for factoring polynomial expressions (comprehension and communication) 04%.
L. Given a real-life problem, the student explains how, if at all, factoring polynomials can be utilized in solving the problem (application) 10%.

Estimated number of class periods: 12

Textbook page references: 210–243 (Unless otherwise indicated, page numbers referred to in the overall plan for lessons are from the course textbook.)

Overall plan for lessons:

I will confront students with a problem either in class or as part of a homework assignment that will be designed to set the stage for an inquiry lesson for Objective A. That lesson should be designed to stimulate students to discover the need to "undo" multiplication and to generalize from constant expressions in factored form to variable expressions in factored form. I'll need to design examples and nonexample problems for the lesson, but some useful follow-up practice exercises, as well as needed definitions, are given in the text on pp. 211–213. (Estimated time needed: 1.5 class periods plus homework)

For Objective B, direct-instructional and comprehension-and-communication activities will be integrated within lessons for the other objectives to help students utilize the vocabulary terms in communications. (Estimated time needed: 2–10 intermittent minutes each class period plus homework)

For Objective C, I will conduct a relatively brief inquiry lesson to help students discover the algorithm using their understanding of the distributive property. Examples on p. 214, the geometric paradigm on p. 216, and the Excursions in Algebra on p. 215 of the text should prove useful. (Estimated time needed: 0.5 of a class period plus homework)

I will conduct a direct-instructional lesson for Objective D as a natural extension to the previous lesson. Adequate practice exercises are on p. 215. (Estimated time needed: 0.4 of a class period plus homework)

For Objective K, direct-instructional and comprehension-and-communication activities will be integrated within lessons for Objectives D, F, H, and J to help students write programs and utilize computers in executing algorithms. (Estimated time needed: Homework plus a total of 40 minutes intermittently distributed among different class periods)

A brief test on objectives A and C and parts of objectives B and K will be given and reviewed with the class for purposes of formative feedback. Using the test review as a lead-in, lessons for Objectives E and F will be conducted following a design similar to that for Objectives C and D. The geometric model and the examples explained on pp. 217–218 will be utilized in the inductive activities, along with the "Excursions in Algebra" section on p. 220. The paper-folding and cutting-up-squares experiments from pp. 172–173 in Sobel & Maletsky (1988) may also be incorporated. For the direct-instructional lesson, exercises on p. 219 as well as the "Using Calculators" section should prove useful. (Estimated time needed: 1.75 class periods plus homework)

Continued

Exhibit 11.10
Continued

For Objectives G and H, I will follow a similar inquiry-direct-instruction sequence as for prior conceptualization/algorithmic-skill pairs. Examples explained on pp. 224–226, as well as the paper-folding experiment on p. 171 from Sobel & Maletsky (1988) will be utilized in the inquiry lesson, and exercises and examples from p. 223 for the direct lesson. (Estimated time needed: 1.75 class periods plus homework)

A brief test on Objectives A, C, D, E, F, G, and H and parts of Objectives B and K will be given and reviewed with the class for purposes of formative feedback. (Estimated time needed: 0.5 of a class period)

Again, the inquiry-direct-instruction sequence will be used, but this time for Objectives I and J. Cardboard models from Sobel & Maletsky (1988, pp. 174–175) for factoring trinomials will be utilized for the inquiry lesson. Examples and exercises, including the "Using Calculators" sections from pp. 224–234, will be used in both lessons. (Estimated time needed: 2.5 class periods plus homework)

I will conduct an inquiry lesson for Objective L, drawing example and nonexample problems from a variety of sources, including prior units, references sources, and my head. (Estimated time needed: 0.75 of a class period plus homework)

Using a practice test, I'll conduct a review session for the unit test for Objectives A–L. (Estimated time needed: 0.5 class period plus homework)

A unit test for Objectives A–L will be administered and the results reviewed with the class. (Estimated time needed: 1.5 class periods)

Extraordinary learning materials and equipment needed:

Cardboard for cardboard cutout models as explained in Sobel & Maletsky (1988, pp. 173–174).

Exhibit 11.11
One Student's Response to a Prompt Mr. Rudd Planted on the Unit 6 Test to Set the Stage for Unit 7.

Simplify the following polynomial:

$21a + 3a(b - 2b - 7)$

$21a + 3a(-b - 7)$

$21a - 3ab - 21a$

$-3ab$

He gives the directions for completing the homework:

Mr. Rudd: Look at Prompt #1 on the tasksheet. Fill in the blank with the number Cassandra chose for us. What was it again, Russell?

Russell: 24.

Mr. Rudd: Thank you. Now, plug the 24 into the two blanks of Prompt #2. For the first prompt you need to find all possible integer pairs for *a* and *b* that make the statement true. That's really only one table; I split it up into two parts just to conserve space. For Prompt #2, find three more pairs that work, but this time you aren't restricted to integers. Yes, Mary?

Mary: We have to use fractions?

Mr. Rudd: Try fractions if you like—as long as they're real numbers. Cassandra chose the number for the blank in the first two prompts, I'm going to choose numbers for the third and fourth. Put "10" in the blank for Prompt #3 and put "0" in for Prompt #4.

Engage in Activity 11.4:

Activity 11.4

In a discussion with a colleague, address the following questions about Case 11.4:

1. What are the advantages and disadvantages of Mr. Rudd's planting Exhibit 11.11's prompt on the Unit 6 test?
2. Why do you suppose Mr. Rudd questioned students so intently on the various possible values of *a* and *b* in the expression $-3ab = 24$? What was he trying to accomplish?
3. What do you suppose Mr. Rudd was attempting to accomplish by assigning Exhibit 11.12's tasksheet for homework?

In Case 11.5, Mr. Rudd plans for Day 1 of Unit 7:

CASE 11.5

After school Mr. Rudd works on lessons for Tuesday's five classes. He starts with plans for the first day of Unit 7 for algebra I. Mr. Rudd thinks: "I hope that little homework assignment will turn their thoughts toward undoing multiplication. But for this first lesson on Objective A, I really need to grab their attention with a real-life problem that'll stimulate them to think why we ever want to look at factors."

He checks his unit plan (i.e., Exhibit 11.10) and rereads the part pertaining to teaching for Objective A. Struggling to come up with the attention-grabbing problem to open the lesson, he considers using a bank interest problem in which the interest earned is given and

Exhibit 11.12
Homework Tasksheet Mr. Rudd Assigned for the First Day of Unit 7.

1. Use the given table to show all possible pairs of integers (a, b) that make the following statement true:

$-3ab = 24$

a	b

a	b

Note: You may not need to use all the spaces in the table.

2. Find three more pairs for (a, b) that work for $-3ab = 24$, but this time a and b do not have to be integers:

$-3ab = 24$

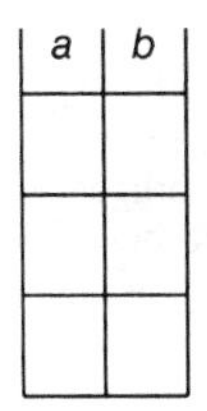

3. Use the given table to show all possible pairs of integers (a, b) that makes the following statement true:

$-3ab = 10$

a	b

a	b

Note: You may not need to use all the spaces in the table.

4. Use the given table to show five possible pairs of integers (a, b) that make the following statement true:

$-3ab = 0$

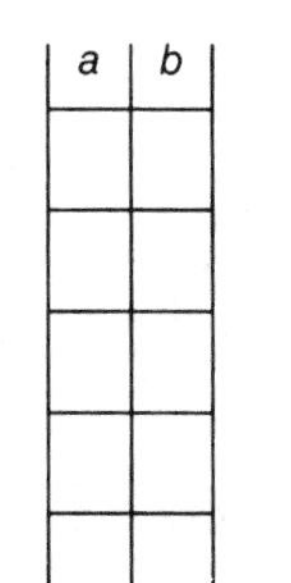

How many possible such pairs for (a, b) are there?

____

questions are raised about possibilities for the variables principal and interest rate. Trying to use that same idea, but shifting it to a more interesting topic, he thinks about a situation in which a politician brags that she has funneled so many thousands of dollars into a project that helps crime victims; he could then have the students figure how that translates into help for individuals in terms of the number of crime victims needing the help. He likes that idea but continues to think, trying to come up with something easier to illustrate.

Problems involving finding possible dimensions of figures given the areas of their interiors are considered, as

Exhibit 11.13
Problem Mr. Rudd Uses in the First Lesson for Unit 7.

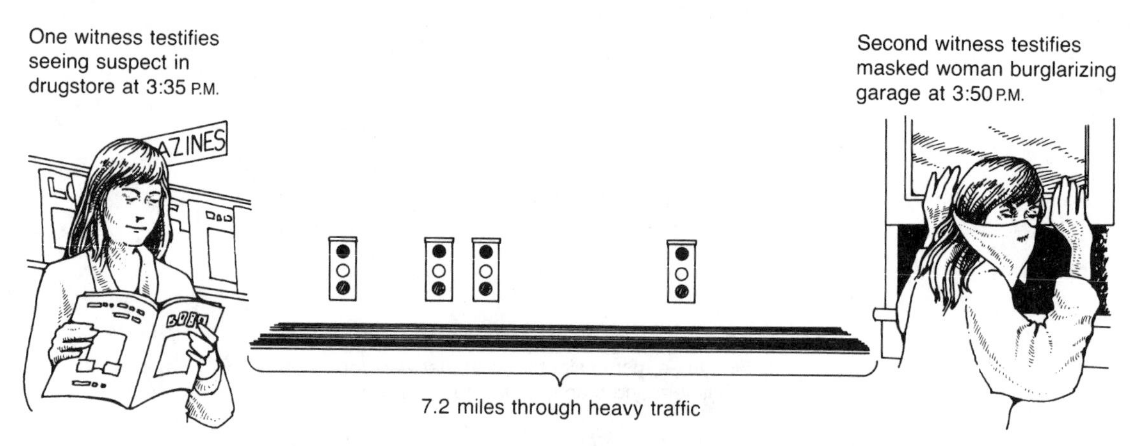

A woman is accused of burglarizing a garage at approximately 3:50 P.M. on a Tuesday. At 3:35 P.M. that day, she was seen in a drugstore 7.2 miles west of the garage. The direct route between the two locations is congested with traffic at that time and has four traffic lights.

A traffic engineer testifies that:

Between 2 P.M. and 4 P.M. on the day of the crime, the fastest average rate a vehicle could travel from a point near the drugstore to the crime scene would be 30 mph. The average rate at which traffic travels there at that time is 20 mph. Unless there is an extraordinary event (e.g., an accident) stopping traffic on that road, traffic will move at 15 mph at the very least. On that day, there was no such traffic stoppage.

The woman's defense attorney claims she could not possibly have been at the crime scene at 3:50 because she was in the drugstore at 3:35.

The question:

Is it possible for her to have been in the drugstore at 3:35 and be burglarizing the garage at 3:50? If so, is it likely she was able to travel the 7.2 miles in the available time?

well as travel problems in which rate and time vary but distance is constant. Thinking about travel problems leads to the problem he finally decides to use for the opening lesson for Objective A. He selects it because it involves the mathematical content of Objective A and is also a topic in which many students have shown interest—namely, solving crimes. The problem is explained in Exhibit 11.13.

With the paramount problem for the day determined, Mr. Rudd designs and schedules the activities. He thinks: "I want them working on this problem right away, but I also should respond to that homework assignment. It was really just to get them thinking about isolating and analyzing factors. I don't want to spend too much time discussing it, just enough to reinforce engaging in homework—and also to plant a few seeds that'll eventually help them discover some relationships down the road. 24, 10, and 0 are going to turn out to be great choices!

"We'll begin with an independent-work session in which they deal with a problem on a tasksheet while I go around checking their homework. Then I can make comments about what I learned from reading the homework before we discuss the problem. But that won't work—some will finish the problem before I could possibly look through all their homework papers, and then they will be ready to discuss the problem and not be interested in the homework. I need another plan.

"I've got it! After quickly going over the homework, we'll deal with the problems in a large-group session. Oh, an even better idea! I'll introduce the problem by having them role-play the characters in the problem. It'll be easy for me to make out a script, pass it out, and assign the parts. As they go through the script, I'll put this diagram on the overhead (see Exhibit 11.13). After the problem is presented, should I have them discuss and solve it as a

Exhibit 11.14
Mr. Rudd's Tasksheet for the Independent-Work Session to Solve the Problem Given in Exhibit 11.13.

Your mission is to solve the problem by answering the following questions:

1. Based on the evidence, was it possible for the suspect to have been in the drugstore at 3:35 P.M. on the day in question and also to be burglarizing the garage at 3:50 P.M.?

   Yes_____ No_____. Explain exactly how the evidence supports your conclusion in the space provided.

2. If you answered Yes to the first question, was it likely or probable that she was able to travel the 7.2 miles in the available time?

   Yes_____ No_____. Explain exactly how the evidence supports your conclusion in the space provided.

3. If you answered No to the second question, how high would the fastest average rate given by the traffic engineer have to be before you would conclude that it is likely or probable to travel the 7.2 miles in the available time?

   _____mi/h. Defend your answer with an explanation.

group or put them into the independent-work session? Maybe I'll have those who don't have parts to read make the judgment; I don't know. Okay, I'm going to go with the independent-work session, but I'll make it a writing exercise—they'll have to explain and argue for what they decide. That way, everybody will have a chance to at least put some thought into the problem before the fast ones with answers get restless. I should probably give them about a 12-minute writing assignment on the problem and then we'll discuss it."

For the in-class writing assignment, Mr. Rudd develops the tasksheet shown in Exhibit 11.14.

He thinks: "There won't be much class time left after that—just time to sum up, review the vocabulary for Objective B, and make an assignment. Better get the script written."

Within 10 minutes, Mr. Rudd has the script written. As it and the tasksheet are being printed and then duplicated, he writes the agenda for Tuesday's class. Exhibit 11.15 contains the script; Exhibit 11.16 shows the agenda.

Engage in Activity 11.5:

Activity 11.5

In a discussion with a colleague, address the following questions about Case 11.5:

1. What was Mr. Rudd attempting to accomplish by confronting students with Exhibit 11.13's crime-solver problem? Would you use such a problem near the beginning of a unit on factoring polynomials? Why or why not?
2. Even though he seemed to prefer to get into the crime-solver problem right away, why do you suppose Mr. Rudd felt obliged to spend time reviewing students' responses to homework prompts?
3. Do you know that Mr. Rudd's use of *inner speech* (i.e., the way he addresses questions in his mind that he poses to himself) is associated with successful problem solving and cognitive development (McCormick & Pressley, 1997, pp. 182–187)?
4. What are the advantages and disadvantages of students playacting Exhibit 11.15's script instead of Mr. Rudd simply explaining the problem?
5. Do you plan to write daily agendas for each class period you conduct (e.g., Exhibit 11.16)? Why or why not?

 **Exhibit 11.15**

Script for Students to Follow in Playacting Situations Given in Exhibit 11.13.

Prosecutor Wilma Jones: (To witness Alvin Smith) Mr. Smith, at approximately 3:50 P.M. on Tuesday, January 26, tell us what you saw just outside the garage located at 821 North Street.

Alvin Smith: I saw a woman in a mask enter the garage. A couple of minutes later she came out of the garage carrying a television set and a small box. She put them into a car parked just outside the door. Then she made two more trips, taking what looked like auto equipment—speakers and tools and stuff.

Wilma Jones: And what kind of car was the suspect—ah, excuse me, I mean the person you observed—using?

Alvin Smith: A light green hatchback of some kind—maybe a Dodge or something.

Wilma Jones: (To suspect Alice Brown) Ms. Brown, what kind of car do you drive?

Alice Brown: A green 1987 Dodge hatchback, but I didn't—

Wilma Jones: That's all Ms. Brown; thank you.

Defense Attorney Willie Adams: (To witness Irene Johnson) Ms. Johnson, where do you work?

Irene Johnson: At Sitman's Drugstore on 5980 North Street.

Willie Adams: Ms. Brown, please stand up so this witness can take a good look at you. Thank you. Now Ms. Johnson, have you ever seen this person at any time before today?

Irene Johnson: Only once, and that was at Sitman's Drugstore where I work. She bought some gloves and I rang up the sale for her.

Willie Adams: And at what time and on what day did this occur?

Irene Johnson: At exactly 3:35 P.M. on January 26 of this year.

Willie Adams: You seem so sure of the date and time. How can you be so sure?

Irene Johnson: The date and time are right here on this cash register slip.

Willie Adams: (To traffic engineer Bob Moore) Mr. Moore, what position do you hold with our great city?

Bob Moore: I'm a traffic engineer. I study traffic patterns and basically work to keep traffic on our city streets flowing as smoothly as possible.

Willie Adams: Mr. Moore, how far is it between Sitman's Drugstore at 5980 North and the garage, allegedly burglarized at 821 North?

Bob Moore: 7.2 miles.

Willie Adams: Please tell us, based on your scientific studies, how long it would take someone to travel those 7.2 miles between 3:35 and 4 o'clock on Tuesday, January 26, a workday.

Bob Moore: The average rate at which traffic travels between those two points on a Tuesday at that time is 20 miles per hour. It's quite congested and there are four major traffic lights.

Willie Adams: That's the average. What's the fastest a car could average over that 7.2 miles?

Bob Moore: 30 miles per hour max.

Willie Adams: You say that's the most. Is it realistic to expect someone to average as high as 30 miles per hour over that 7.2-mile stretch at that time of day on a Tuesday?

Bob Moore: It would be a rare occurrence, to say the least.

Exhibit 11.16

Mr. Rudd's Agenda for the First Day of Unit 7.

1. Transition period: Start class and direct students to take out the tasksheets they completed for homework (i.e., Exhibit 11.12).
2. Question-discussion session in which students share homework responses leading to the idea of undoing multiplication.
3. Transition period: Crime-solver problem scripts are distributed and roles are assigned.
4. Interactive lecture-discussion session: Designated students play their characters from the crime-solver script and the problem is clarified.
5. Transition period: Distribute tasksheets (i.e., Exhibit 11.14) and give directions for independent-work session.
6. Independent-work session: Students complete the tasksheet as I circulate, selecting sample responses to be read during the follow-up question-discussion session.
7. Transition period into the question-discussion session.
8. Question-discussion session:
    - Selected contrasting responses are read aloud and analyzed.
    - Strategies for solving the problem are articulated.
    - Types of problems with solutions requiring undoing multiplication are characterized.
    - Additional example and nonexample problems are compared.
9. Transition period: Students are directed to open the textbook to p. 210.
10. Interactive lecture session:
    - Review notebook glossary definitions of "factoring," "prime number," "composite number," and "greatest common factor."
    - Explain the examples and directions for the textbook exercises on pp. 212–213.
11. Transition period: Assign homework:
    - Think up and write out a real-life problem. Make it one with a solution that requires the undoing of multiplication.
    - Work the following exercises from p. 213: 11, 15, 19, 20, 21, 25, 49, & 55.
12. Independent-work session: Begin homework.
13. Transition period into second period.

Exhibit 11.17
Students' Initial Whiteboard Entries for the Activity Reviewing the Homework Assignment From Exhibit 11.12's Tasksheet.

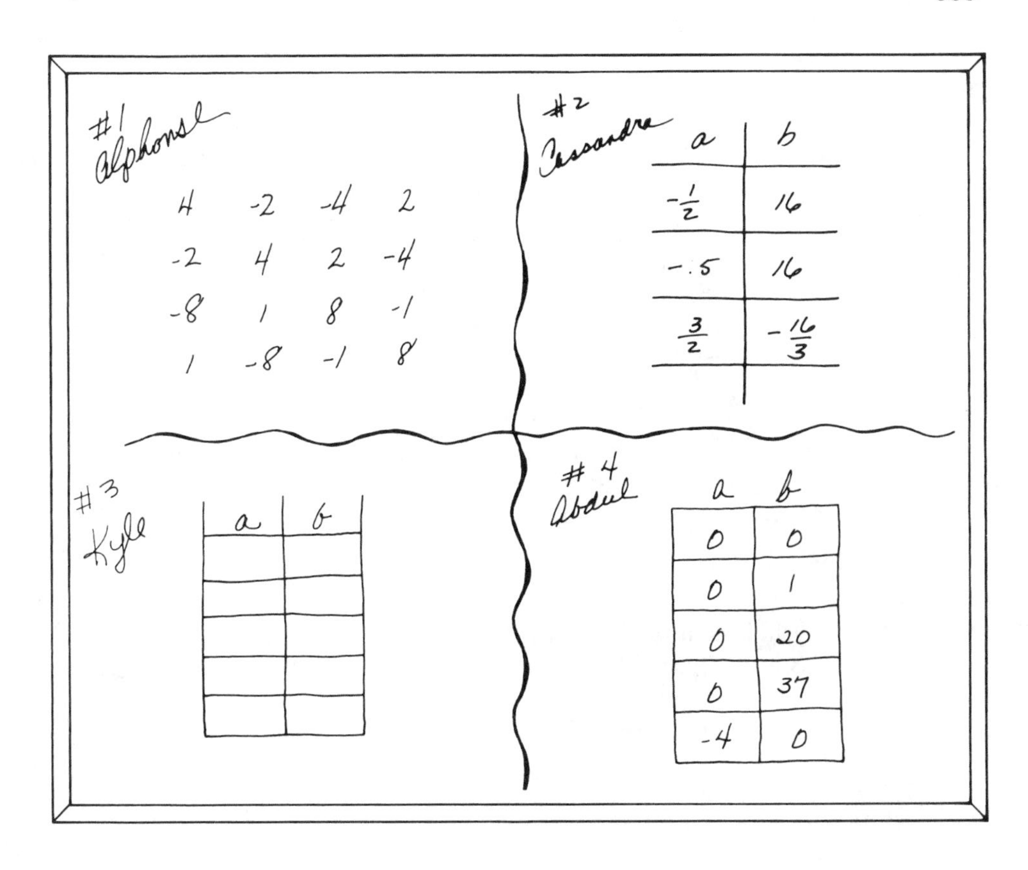

In Case 11.6, Mr. Rudd conducts the class on Day 1 of Unit 7:

CASE 11.6

On Tuesday, 24 of Mr. Rudd's 26 algebra I students are in class. Four minutes into the class period, the students have their homework sheets (i.e., Exhibit 11.12) in front of them as he conducts the planned reasoning-level question-discussion session:

Mr. Rudd: How many pairs did you come up with for the first prompt, Alphonse?

Alphonse: 16.

Mr. Rudd: Quickly, list them on the board for us—and while Alphonse is doing that, let's have Cassandra list hers for Prompt #2, Kyle for Prompt #3, and Abdul for #4.

Exhibit 11.17 displays what the students write.

Mr. Rudd: Let's turn our attention to Alphonse's entries for the first. Where are the rest of them, Alphonse?

Alphonse: What do you mean?

Mr. Rudd: You said you had 16; I only see eight pairs.

Alphonse: That's what I meant; I was counting numbers, not pairs.

Mr.Rudd: Any comments or questions for Alphonse? Do his entries match your own? Go ahead, Dick.

Dick: Did we have to repeat them like he did? 4 and -2 is the same as -2 and 4.

Mr. Rudd: Direct your questions to Alphonse. He may either answer you himself or call on a volunteer.

Alphonse: I don't know. Okay, Dustin.

Dustin: You've got to do them both, because if like, ah, $a = 4$, then b has to be -2. But if a is -2, b has to be 4. What I did was to solve for all the as; then b had to be whatever it had to be.

Mr. Rudd: Do you have a response to Dustin's explanation, Dick? No? Okay, anybody else? Put your hand up if you don't understand why Alphonse's eight pairs are all possible ordered pairs of integers a, b in the case where $-3ab = 24$. Questions or comments for Cassandra? Ron.

Ron: You've only got two pairs up there, not three.

Cassandra: What do you mean—see, 1, 2, 3?

Ron: The first and second are the same one; $-1/2$ is $-.5$.

Cassandra: Oh! Sorry, just switch the two—make the second row (16, $-.5$) instead. Edwin.

Edwin: I thought they all had to be fractions.

Cassandra: Here, does this make you happy?

Cassandra modifies her entries so that they now appear as in Exhibit 11.18.

Mr. Rudd: We'll allow any pair of real numbers that works as long they're different from what she had for #1. All the numbers that Cassandra has can be expressed as fractions, but she doesn't have to express them as fractions. Marlene.

Marlene: Are her answers right? None of mine agree.

Exhibit 11.18
Cassandra's Modified Entry.

#2 Cassandra

a	b
$-\frac{1}{2}$	$\frac{16}{1}$
$\frac{16}{1}$	$-\frac{.5}{1}$
$\frac{3}{2}$	$\frac{-16}{3}$

Mr. Rudd: Raise your hand if you have exactly the same entries as Cassandra. [pause] Nobody agrees with Cassandra! But her answers are exactly correct; so nobody else got correct answers. I'm shocked! Read yours, Sid.

Sid: (8/7, 7), (8/9, 9), and (8/10, 10).

Mr. Rudd: That would work if Sid sticks in some negative signs here and there, but then it still wouldn't be the same as Cassandra's. Read yours, Omar.

Omar: (32, −1/4), (−1/4, 32), and (16, −1/2).

Mr. Rudd: Well that surely works, but it's different from Cassandra's. Marlene.

Marlene: So you're really saying that there's more than one correct answer to this one.

Mr. Rudd: How many possible *a, b* pairs work if we don't say *a* and *b* do not have to be integers, Xavier?

Xavier: I don't know.

Mr.Rudd: What three pairs did you write down for #2, Xavier?

Xavier: I didn't do it.

As Mr. Rudd redirects his initial question for Xavier to another student, he quickly writes on a slip of paper, "Xavier, please check with me later on in class today right after I assign the homework for tomorrow. I want to speak with you briefly." He hands the note to Xavier as he continues.

Mr. Rudd: How many possible *a, b* pairs will work if we don't say *a* and *b* have to be integers, Juaquin?

Juaquin: There's no stopping; you can keep going and going and going.

Mr. Rudd: Let's go to #3 and look at what Kyle put. [pause] Blanks! All blanks! You were supposed to fill in the blanks, not leave them blank, Kyle! What do have to say about this?

Kyle: There are no answers.

Mr. Rudd: Do you think I'd give you an exercise with no answers?

Kyle: You do it all the time.

Mr. Rudd: Explain to the class why this one has no answers.

Kyle: 3 doesn't go into 10, so there's no integer times 3 that'll give you 10. Mary?

Mary: But you could put in fractions, right?

Kyle: But you weren't supposed to for this one.

Mr. Rudd: If nonintegers were allowed, how many answers would there be, Juaquin?

Juaquin: You could keep going and going and going.

Mr. Rudd: Okay, that brings us to the last one. Do you agree with Abdul's entries? Alena.

Alena: I agree, but they're not the same as mine.

Mr. Rudd: Super quick, write yours on the board, and while Alena is doing that, write yours also, Terri. Okay, does anybody want to quarrel with anything that's on the board? How many possible correct pairs could we have? Salvador.

Salvador: Infinite.

Mr. Rudd: Describe what must be true for a pair to work. You tell us, Salvador.

Salvador: You've got to have a 0.

Mr. Rudd: We're going to be okay as long as *a* or *b* is 0. If *a* is 0, then what is *b*, Mary?

Mary: It can be anything. And if *b* is 0, *a* can be anything.

Mr. Rudd distributes copies of the crime-solver script and appoints individuals to play the roles. With Exhibit 11.13's illustration displayed on the overhead, the students read their parts. As planned, they complete the tasksheet shown in Exhibit 11.14 while Mr. Rudd circulates about the room selecting sample responses to use in the follow-up question-discussion session. Students enthusiastically engage in that session—listed as item 8 in Exhibit 11.16. With only seven minutes remaining in the period Mr. Rudd terminates the discussion. Thinking quickly, he tries to summarize and explain the type of problem on which they will be focusing in this unit (i.e., ones that involve situations in which they already have a product and want to look at different factor combinations that could have resulted in that product).

Lacking time to get to agenda item 10, Mr. Rudd shifts to item 11, explaining and assigning the part of the homework in which they think up and write out a sample real-life problem. He postpones the part of the assignment from the textbook.

Engage in Activity 11.6:

Activity 11.6

In a discussion with a colleague, address the following questions about Case 11.6:

1. During question-discussion sessions, Mr. Rudd hardly ever responds to students' responses with comments such as "right," "wrong," or "good." What does he do instead? What are the advantages and disadvantages of responding as he does rather than as most teachers

Exhibit 11.19
Mr. Rudd's Agenda for the Second Day of Unit 7.

1. Transition period: I start class and distribute an overhead transparency and transparency (erasable) pen to each student and direct them to take out the problems they formulated for homework.
2. Independent-work session: each student quickly copies her homework problem onto the transparency (make sure they display their names).
3. Transition period: I collect completed transparencies and the pens.
4. Interactive lecture session:
   - I quickly display each transparency on the overhead as the student who wrote it reads the problem aloud.
   - After each problem is shown, the students independently and quietly classify the problem on their worksheet as either an example or a nonexample.
   - After all the problems are shown, I select two example problems and two nonexample problems (assuming they exist in the sample—if not, use my own as a last resort) and use the contrast between the two types to sum up the need to undo multiplication to solve some types of problems.
5. Transition period: students are directed to open the textbook to p. 210.
6. Interactive lecture session:
   - Review the textbook and notebook glossary definitions of factoring, prime number, composite number, and greatest common factor.
   - Explain the examples and directions for the textbook exercises on pp. 212–213.
   - Call their attention to the purpose of the textbook section Factoring Using the Distributive Property on p. 214.
7. Transition period: assign the following homework:
   - Work the following exercises from p. 213: 11, 15, 19, 20, 21, 25, 28, 33, 49, 53, 54, and 55.
   - Study page 214, working through the four examples with the authors.
   - Work the following exploratory exercises from p. 215: 1, 2, 3, and 9.
   - Work the following written exercises from p. 215: 1, 2, 3, 5, 9, and 15.
8. Independent-work session: begin homework.
9. Transition period into the second period.

do? Do the conversations in his classroom have fewer IRE cycles that the conversations in most classrooms?

2. What do you suppose Mr. Rudd did in the 12 weeks of class preceding Unit 7 to condition students to wait for the floor to speak aloud during question-discussion session and to listen to whoever has the floor?
3. What are the advantages and disadvantages of the way Mr. Rudd responded to Xavier not having done the homework assignment? What would you need to know about Xavier as an individual before recommending a strategy for dealing with his off-task behavior?
4. Why do you suppose Mr. Rudd included Prompt #3—the one without an "answer"—in Exhibit 11.12's tasksheet?

In Case 11.7, Mr. Rudd plans for Day 2 of Unit 7:

CASE 11.7

It is 4:30 when Mr. Rudd gets around to reflecting on the day's classes. Regarding algebra I, he thinks: "I like the way they got into the two inquiry activities today. The trouble is, we didn't leave time to really get things summed up at the end and to go through the vocabulary. I'm sorry I didn't get to work in at least two more example problems and at least one nonexample problem. My attempt to tie things up at the end didn't work because we had only one problem to think about. They loved that crime-solver business—think I'll try that in the geometry classes. That was a good idea. Was it a mistake to postpone the textbook assignment? Earlier in the year, I would've assigned it even though I hadn't gone through the vocabulary, but I've learned that usually doesn't work for most of this group—they see a word they don't know and just quit.

"Okay, so where do I go with this tomorrow? That'll be Wednesday, so three more days this week. Oh! we've got plenty of time; the extra time spent on the discover-a-relationship objective will make the rest of the unit go efficiently. Okay, back to tomorrow—we need to get to the vocabulary. But first I need to hit them with example and nonexample problems and then try to do a better job of summing things up. Better jot down these problems I thought of at the end of class. Let's see, one was on area, the other on money spent by the politician. Oh! It would be better to use the problems they bring to class from the homework assignment. I almost forgot about that. Some are bound to come up with super examples; others won't actually involve factoring—I can use some of those for nonexamples. I'm glad I made that assignment; it will work out well. Then, we tie up Objective A and go straight to the textbook and the vocabulary stuff planned for today."

With the unit plan (i.e., Exhibit 11.10), Tuesday's agenda (Exhibit 11.16), and page 212 of the textbook in front of him, Mr. Rudd calls up the computer file with Tuesday's agenda and begins modifying it into Wednesday's agenda, resulting in Exhibit 11.19. As Mr. Rudd looks at it, he thinks, "We probably won't get through this one either,

Exhibit 11.20
Assignment From Mr. Rudd's Textbook.

Examples

1 **Use the distributive property to write $10y^2 + 15y$ in factored form.**

First, find the greatest common factor of $10y^2$ and $15y$.

$10y^2 = 2 \cdot (5) \cdot (y) \cdot y$
$15y = 3 \cdot (5) \cdot (y)$ *The GCF is 5y.*

Then, express each term as a product of the GCF and its remaining factors.

$10y^2 + 15y = 5y(2y) + 5y(3)$
$= 5y(2y + 3)$ *Use the distributive property.*

2 **Factor: $21ab^2 - 33a^2bc$**

$21ab^2 = (3) \cdot 7 \cdot (a) \cdot (b) \cdot b$
$33a^2bc = (3) \cdot 11 \cdot (a) \cdot a \cdot (b) \cdot c$ *The GCF is 3ab.*

Express the terms as products.

$21ab^2 - 33a^2bc = 3ab(7b) - 3ab(11ac)$
$= 3ab(7b - 11ac)$ *Use the distributive property.*

3 **Factor: $12a^5b + 8a^3 - 24a^3c$**

$12a^5b = 2 \cdot 2 \cdot 3 \cdot a \cdot a \cdot a \cdot a \cdot a \cdot b$
$8a^3 = 2 \cdot 2 \cdot 2 \cdot a \cdot a \cdot a$
$24a^3c = 2 \cdot 2 \cdot 2 \cdot 3 \cdot a \cdot a \cdot a \cdot c$ *The GCF is $4a^3$.*

$12a^5b + 8a^3 - 24a^3c = 4a^3(3a^2b) + 4a^3(2) - 4a^3(6c)$
$= 4a^3(3a^2b + 2 - 6c)$

4 **Factor: $6x^3y^2 + 14x^2y + 2x^2$**

$6x^3y^2 = 2 \cdot 3 \cdot x \cdot x \cdot x \cdot y \cdot y$
$14x^2y = 2 \cdot 7 \cdot x \cdot x \cdot y$
$2x^2 = 2 \cdot x \cdot x$ *The GCF is $2x^2$.*

$6x^3y^2 + 14x^2y + 2x^2 = 2x^2(3xy^2 + 7y + 1)$

but I'd much rather have too much planned than not enough and be left with dead time."

When Mr. Rudd first begins writing the agenda, he does not plan for the activities related to Objective A to be as elaborate as agenda items 1–4 suggest. Initially, he planned to quickly tie together the end of Tuesday's activities for that objective. However, as he thinks of how to make efficient use of the homework and the need to involve problems besides the crime-solver one, he ends up with a more elaborate plan. This should provide a stimulating start to the period, but will not leave enough time to do all the following: (a) Go through the vocabulary and set the stage for the textbook activities, (b) provide students with needed exposure to and practice factoring constant and variable expressions from page 213 (e.g., "Find the prime factorization of 112" and "Find the greatest common factor of $18a^2b^2$, $6b$, $42a^2b^3$"), and (c) conduct the brief inquiry lesson relative to Objective C as suggested in his unit plan and the follow-up lesson for Objective D.

He estimates that stopping with only the first two tasks completed will leave unused class time but not enough to complete the third task. Exhibit 11.19 represents a compromise; he will plan to have the students get a jump on Objectives C and D by working ahead on textbook page 214 shown by Exhibit 11.20. He does not really like the idea of students being introduced to steps in an algorithm before experiencing a discover-a-relationship lesson on why it works, but he goes ahead with the agenda (i.e., Exhibit 11.19) because it seems to be an efficient way to use the available time. Besides, he does not consider this particular algorithm—which is based on the familiar distributive property—to be conceptually complex for the students.

In Case 11.8, Mr. Rudd conducts the class on Day 2 and plans for Day 3 of Unit 7:

CASE 11.8

Wednesday's algebra I class smoothly followed the agenda of Exhibit 11.19. Mr. Rudd reflects on the day's events and makes plans for Thursday's classes. He thinks to himself: "That first part really went well. Flashing every one of their problems on the overhead screen reinforced their engagement in homework and exposed them to a variety of problems—examples and nonexamples—with them both seeing and hearing the problems. I'm going to use that tactic more often, but next time I need to think of it before I assign the homework. That way they can save class time by taking transparencies home to prepare them before class. Too bad I don't have a visual presenter.

"Even agenda item 6 went well—switching to direct instruction using the textbook was a change of pace after all the lively inquiry activities. I wonder just how much they got from studying page 214. I'd better start off tomorrow with a brief test to get some formative feedback on where they are and to reinforce their work on the assignments. The test shouldn't take more than 20 minutes. I should have one or two miniexperiments on Objective A, a few for the vocabulary we went over today, and some similar to the textbook exercises relative to Objective D. A pretty easy miniexperiment on Objective C should tell me if that little explanation at the top of page 214 helped them understand that the distributive property in reverse is the basis for the algorithm. I'm afraid this brief test may not be as brief as I'd like!"

Mr. Rudd spends nearly an hour entering miniexperiments for Objectives A, B, C, and D into his computerized miniexperiment files. He longs for the day when he will not have to create new miniexperiments whenever he puts together a test—the second year of teaching will be easier than this one. After synthesizing the test, Mr. Rudd works out Exhibit 11.21's agenda.

Exhibit 11.21
Mr. Rudd's Agenda for the Third Day of Unit 7.

1. Transition period: Start class and initiate a brief test.
2. Brief test relevant to Objectives A, B, C, and D.
3. Transition period from the test to the review of the results.
4. Interactive lecture and question-discussion session:
   - I call out the scoring key for each miniexperiment while students check and score their own responses.
   - Miniexperiments about which students raise questions are discussed.
   - I explain any topics the test results suggest need to be explained.
5. Transition period: Students are directed to take out their homework papers, open textbooks to p. 214, and open their notebooks to Part 3, where they keep computer programs and flowcharts developed during the course.
6. Interactive lecture session: I walk the students through writing a basic program for factoring using the distributive property.
7. Transition period: Distribute homework tasksheets (Exhibit 11.22); explain tasksheets and assign homework as follows:
   - Respond to the prompts on the tasksheet.
   - ____________________ (to be determined in class depending on feedback from brief test).
8. Independent-work session: Begin homework.
9. Transition period into second period.

In Case 11.9, Mr. Rudd conducts the class on Day 3, plans for Day 4, conducts the class for Day 4, and plans for Day 5 of Unit 7:

CASE 11.9

Feedback from the brief test suggests that most students failed to achieve Objectives C and D well enough to be ready for Objectives E and F. Thus, Mr. Rudd spends the majority of the class period on item 4 of Exhibit 11.21's agenda. Because he needs to have them complete the tasksheet of Exhibit 11.22 before he spends class time on Objectives E and F, he postpones agenda items 5 and 6 and begins explaining and assigning the tasksheet immediately after item 4. He does not give any other homework assignment because the period ends before he gets to item 8.

For Friday, he plans quick coverage of items 5 and 6 from Thursday's agenda and then use of their responses to Exhibit 11.22's homework tasksheet to lead students to discover that $a^2 - b^2 = (a + b)(a - b)$. He expects them to bring in a variety of square and rectangular regions, and he will use inductive-questioning strategies to get them to generalize from their own examples to the relationship for factoring the difference of squares. He plans to assign written and calculator exercises from pages 215 and 218–219 in which they practice factoring algorithms from both Objectives D and F.

On Friday Mr. Rudd spends more time with the computer-programming activities than he had planned. The reasoning-level question-discussion session is quite lively, with students abstracting the general relationship for factoring the difference of two squares (see Exhibit 11.23) from the examples produced in the homework assignment. However, Mr. Rudd is able to get the students through only the first two and part of the third stage of the four-stage lesson. He is displeased with having to stop the lesson with just enough time remaining to assign the homework. Most students appear disappointed that the class period is over.

The high point on which the first period ends leaves Mr. Rudd so energized that he begins planning for Monday's algebra class during second period, instead of collecting his thoughts for third and fourth periods, as he does on most days.

He locks his classroom door, gathers the unit plan and the day's agenda, sits at his computer, and thinks to himself: "I should've given them a computer assignment to follow up that programming session. They could've handled it with the weekend break. Wow! They really got into discovering why $a^2 - b^2 = (a + b)(a - b)$. I sure hate to wait until Monday to tie it all down. Let's see,

Exhibit 11.22
Tasksheet Mr. Rudd Uses for the Third Day of Unit 7.

DIRECTIONS:

1. Take a sheet of cardboard (preferably) or a sheet of paper and cut it into a square region no smaller than 8″ by 8″.

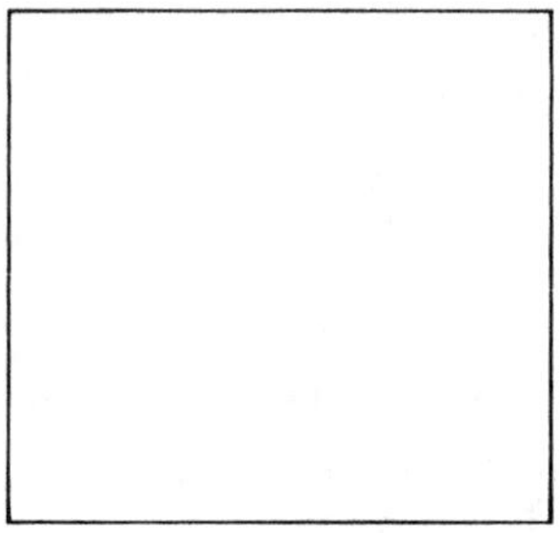

2. Measure your square carefully and record the dimensions here:

   Length = ________ Width = ________ Area = ________

3. Now, measure a small *square* region out of the top right corner of your original square region as shown below:

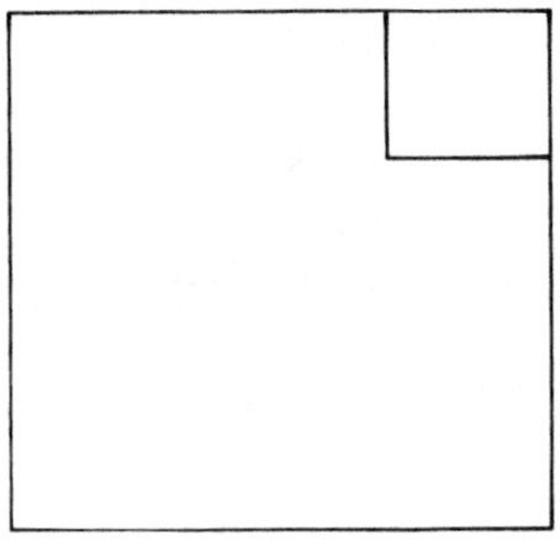

4. Record the dimensions of the small square region here:

   Length = ________ Width = ________ Area = ________

5. Now, cut the small *square* region out and remove it from your original square region as shown below:

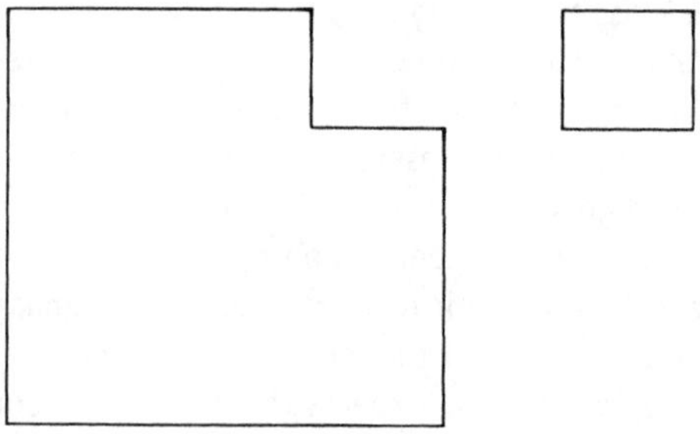

Continued

Exhibit 11.22
Continued

6. Now, divide the region pictured on the left below into two rectangular regions as shown:

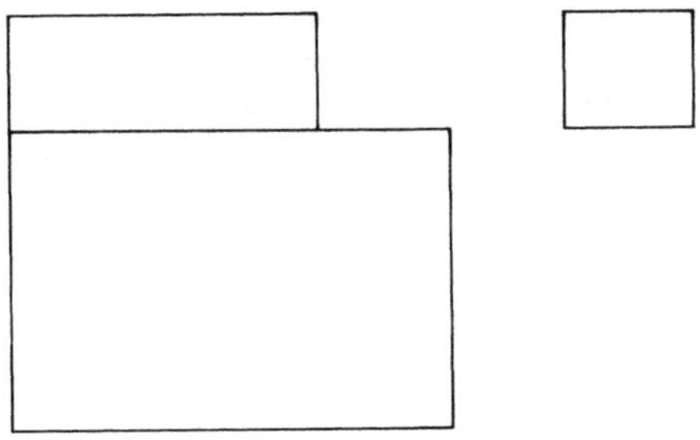

7. Carefully measure the two rectangular regions, both the top one and the bottom one. Record the dimensions here:

   Top rectangle:

   Length = ________ Width = ________ Area = ________

   Bottom rectangle:

   Length = ________ Width = ________ Area = ________

8. Carefully cut off the top rectangular region as shown below:

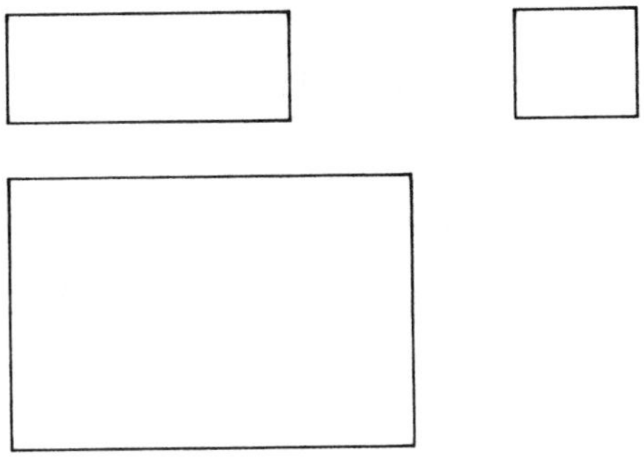

9. Now rotate the top rectangular region and attach it to the left side of the bottom rectangular region as shown below (it should fit exactly):

10. Carefully measure the rectangular region you just formed. Record the dimensions here:

    Length = ________ Width = ________ Area = ________

Exhibit 11.23

General Model for Factoring the Difference of Squares Casey's Students Abstracted From Their Examples Based on Exhibit 11.22's Tasksheet.

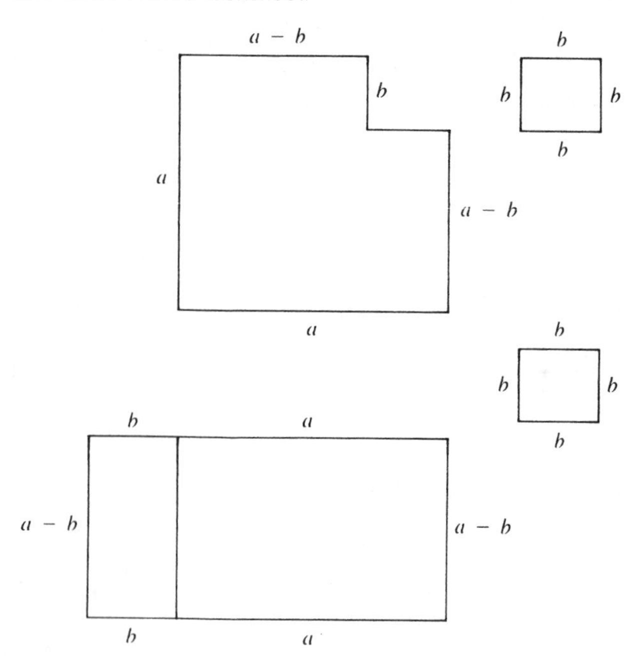

page 217 gives them a pretty good explanation of a geometric model that's isomorphic to the one we discovered in class. Not many of them would understand it if I hadn't put them through the tasksheet (i.e., Exhibit 11.22) experience first. Page 218 lays out the algorithm quite clearly with five examples and a variety of coefficients. Then the written and calculator exercises I assigned will give them skill work for this and the previous algorithm. I shouldn't feel so bad about having had to cut the session short. It'll come together for those who put an effort into this homework. I'd better start Monday's class with a brief test to reinforce those who make the effort."

Mr. Rudd develops Monday's agenda as shown in Exhibit 11.24, but he does not get the brief test completed until Monday morning.

Exhibit 11.24

Mr. Rudd's Agenda for the Fifth Day of Unit 7.

1. Transition period: Start class and initiate a brief test.
2. Brief test relevant to Objectives C, D, E, and F.
3. Transition period from the test to the review of the results.
4. Interactive lecture and question-discussion session:
   - I call out the scoring key for each miniexperiment while students check and score their own responses
   - Miniexperiments and homework exercises about which students raise questions are discussed.
   - I explain any topics the test results and questions suggest need to be explained. (This may require a direct-instructional learning activity on the algorithm for factoring the difference of perfect squares.)
5. Transition period: I distribute a 10″ by 10″ square-shaped sheet of colored paper to each student.
6. Interactive lecture session:
   - I demonstrate and work through the paper-folding experiment (see Exhibit 11.25 illustrating why $a^2 + 2ab + b^2$ can be expressed as $(a + b)^2$).
   - Use direct instruction to explain and provide practice with the algorithm for factoring $a^2 + 2ab + b^2$.
7. Transition period: Assign the following homework:
   - Read the textbook section "Perfect Squares and Factoring" on pp. 221–223.
   - Complete the following Exploratory Exercises from p. 222: 1–6 and 9–12.
   - Complete the following Written Exercises from p. 223: 5, 7, 9, 10, 12, 25, 34, 42, 47, and 48.
8. Independent-work session: Begin homework.
9. Transition period into second period.

Engage in Activity 11.7:

Activity 11.7

In a discussion with a colleague, address the following questions about Cases 11.7–11.9:

1. Do you know that the geometric representations of factoring algorithms such as those shown by Exhibit 11.23 were used in the early Greek mathematics (500–200 B.C.) of the Pythagoreans, Euclid, Archimedes, and Apollonius to avoid theological difficulties with irrational numbers and practical difficulties of working with Greek numerals (Burton, 1999, pp. 135–161; Pratt, 1989)?
2. What do you think about the relevance of Mr. Rudd's homework assignments? Do you think the workload was too heavy or too light?
3. Besides providing formative feedback to him and his students, what other purposes are served by the brief test Mr. Rudd plans to administer on Day 5 of Unit 7?
4. The prompts for the brief test are shown by Exhibit 11.26. How relevant does the test appear to be with respect to students' progress toward Objectives C, D, E, and F as listed in Exhibit 11.10?

In Case 11.10, Mr. Rudd conducts Day 5 and plans for Day 6 of Unit 7:

CASE 11.10

Monday's first period follows the schedule of Exhibit 11.24. Twelve minutes into the period, six of the 25 students have responded to the prompts on Exhibit 11.26's test and are waiting at their places for the others to finish. Mr. Rudd

Exhibit 11.25
Paper-Folding Experiment in Which Mr. Rudd Engaged Students During the Fifth Day of Unit 7.
Source: Max A. Sobel/Evan M. Maletsky, *TEACHING MATHEMATICS: A Sourcebook of Aids, Activities, and Strategies, 2e,* © 1988, pp. 171–172. Reprinted by permission of Prentice Hall, Inc. Englewood Cliffs, New Jersey.

Paper Folding for $(a + b)^2 = a^2 + 2ab + b^2$

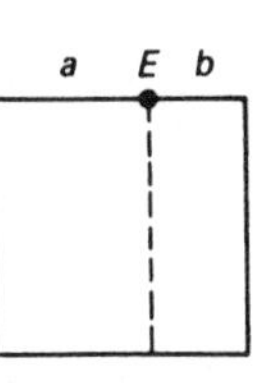

Material

One square piece of paper per student.

Directions

1. Fold one edge over at a point E to form a vertical crease parallel to the edge. Label the longer and shorter dimensions a and b.
2. Fold the upper right-hand corner over onto the crease to locate point F. Folding this way, point F will be the same distance from the corner as point E.
3. Now fold a horizontal crease through F and label all outside dimensions.
4. Find the areas of the two squares formed. Find the areas of the two rectangles formed. Show that these four areas together must equal $(a + b)^2$.

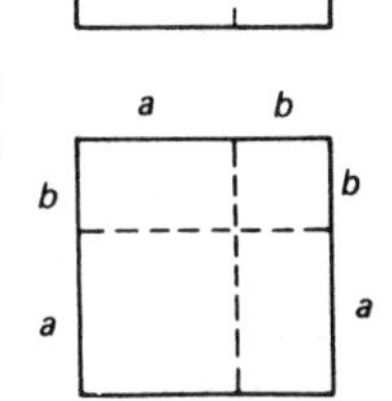

Exhibit 11.26
Brief Test for Objectives C, D, E, and F Mr. Rudd's Uses on the Fifth Day of Unit 7.

Algebra I — Brief Test 7–2

1. Write $18x^2 - 12x$ in factored form:

   $18x^2 - 12x =$ ________________

   Use the distributive property to prove that your answer is correct by indicating what you let a, b, and c equal in the following axiom: $ab + ac = a(b + c)$:

   $a =$ ________ , $b =$ ________ , $c =$ ________

2. What is the GCF of $14ac$ and $21a^3$?

   ____

3. Factor $3x^3y^2 + 9xy^2 + 36xy$

   ____

4. Factor $25w^2 - 81q^4$

   ____

5. Factor $\frac{2}{3}x^2 - \frac{8}{3}$

   ____

6. Illustrate why $100 - x^2 = (10 + x)(10 - x)$ for $x < 10$ by writing in dimensions for the region shown below and then by drawing a rectangle with the same area:

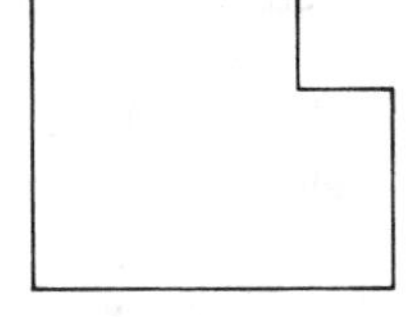

Exhibit 11.27

Mr. Rudd's Display in Response to Prompt #6 of Exhibit 11.26.

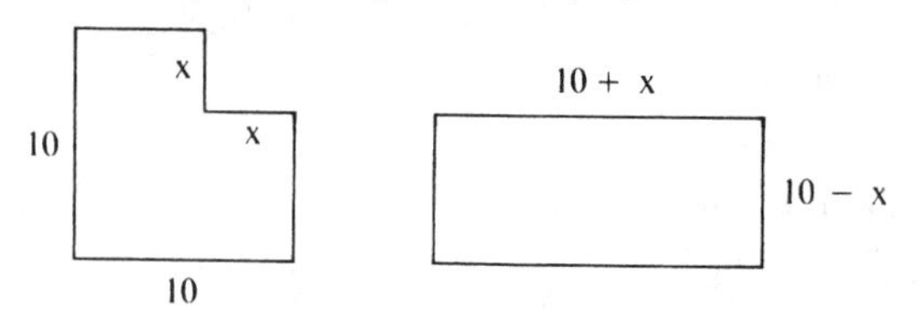

regrets not having made accommodations by having a standby assignment for students finishing at different times. Seven minutes later, most students are finished; he halts the test and begins calling out responses to the prompts:

Mr.Rudd: For #1: $6x(3x - 2)$. To prove that answer, let $a = 6x$, $b = 3x$, and $c = -2$. For #2: the greatest common factor is $7a$. For #3: $3xy(x^2y + 3y + 12)$. For #4: $(5w + 9q^2)(5w - 9q^2)$. For #5: $2/3\ (x + 2)(x - 2)$. And for #6, your dimensions and rectangle might look something like this.

He displays Exhibit 11.27 on the overhead.

Sid: I don't get #5.
Mr. Rudd: When you see "factor $2/3\ x^2 - 8/3$," what's the first thing you think of doing, Sid?
Sid: Look for common terms, but first I got rid of the number at the bottom.
Mr. Rudd: How?
Sid: By multiplying through by 3. That gives me $2x^2 - 8$ and now 2 is common.
Mr. Rudd: But doesn't multiplying by 3 change the value of the expression?
Sid: It's okay to do it to both sides.
Mr. Rudd: Both sides of what?
Sid: The equation.
Mr. Rudd: I don't see an equation, just an expression standing in for a number.
Sid: Oh, yeah.
Mr. Rudd: What did Sid tell us he thinks first when factoring expressions like this, Charlene?
Charlene: Look for a common term.
Mr. Rudd: And is there a common term in this case, Sid?
Sid: They both have 3's on the bottom.
Mr. Rudd: Divided by 3 is the same as multiplied by what, Sid?
Sid: 1/3.
Mr. Rudd: Using the distributive property, take 1/3 out and you have 1/3 of what?
Sid: $2x^2 - 8$. [pause] And now, 2 is common also, so you get $2/3\ (x^2 - 4)$. But that's not the answer you gave us.
Mr. Rudd: Okay, Juaquin, you've been patiently waiting to tell us something. You're up.
Juaquin: $(x^2 - 4)$ is the difference of two perfect squares, so you can factor that too to end up with your answer.
Sid: I see. Thank you.

Twelve minutes later, Mr. Rudd has just talked the class through the paper-folding experiment shown by Exhibit 11.25 and completed item 6 of Exhibit 11.24's agenda by listing the steps in the algorithm for factoring expressions of the form $a^2 + 2ab + b^2$. He talks the students through two examples and then assigns the homework. He leaves the algorithm's steps and illustrative examples displayed on the board as students begin the homework. Although only seven minutes are left for the independent-work session, Mr. Rudd efficiently provides individual help to 12 students by applying strategies suggested in chapter 3's section "Engaging Students in Independent-Work Sessions." Mary, for example, requests help with the following prompt from the textbook:

> Determine if each trinomial is a perfect square trinomial. If it is, state its factors: $n^2 - 13n + 36$.

Mary: Mr. Rudd, I can't do this one.
Mr. Rudd: Is the expression written in standard form?
Mary: Yes.
Mr. Rudd: Are the first and third terms perfect squares?
Mary: Yes.
Mr. Rudd: That means you are on which step listed on the board?
Mary: Ahh, five.
Mr. Rudd: Okay, go with step 5. I'll be back in 33 seconds.

Mr. Rudd returns just in time.

Mary: So this middle term has to be $2ab$, but I still don't get it.
Mr. Rudd: What's a in this case?
Mary: n.
Mr. Rudd: n or $-n$, and what's b in this case?
Mary: 6 or -6.
Mr. Rudd: Is there any way $-13n$ could be $2ab$ for any of those values of a and b?

Mr. Rudd moves to another student as Mary completes the exercise and moves on to #6.

That evening, Mr. Rudd plans for Tuesday's classes. He thinks about the algebra I class: "So up to this point, everyone except for Dick, Blair, and Delia seem to be progressing nicely regarding discovering the relationships underlying the three algorithms we've covered. Skill-wise most of them are coming along. Let's see, next we go to the biggie—factoring $ax^2 + bx + c$ for Objectives I and J. I'd better wait on those until they've had more practice with the distributive one, the difference of squares, and perfect-square trinomials. Also, we haven't hit on any kind of real-life problem solving since the second day. I shouldn't go any longer without some application work, or else they'll lose that real-world connection we started with the crime-solver problem. Okay, better build some applications for tomorrow for the three algorithms, which means we'll begin working on parts of Objective L before getting to Objectives I and J. Also, I should work computer utilization into the practice and review lessons."

Exhibit 11.28
Homework Tasksheet Mr. Rudd Plans to Assign on the Sixth Day of Unit 7.

1. Answer the following questions, assuming that a and b are two numbers such that $ab = 0$.
   If a is 19, what is b? ______________
   If b is -3.74, what is a? ______________
   If a is positive, what must be true about b? ________
   If b is negative, what must be true about a? ________
   If $a = 0$, what must be true about a? ____________
2. Answer the following questions, assuming that c and d are two numbers such that $cd \neq 0$.
   What can you conclude about c? ______________
   What can you conclude about d? ______________
3. There are two numbers for which the following statement is true:
   Seven times the square of the number equals 14 times the number.
   For what two numbers is that statement true? Display your work and write the numbers in the blanks.
   ______________ and ______________
4. Sam's age times itself is the same as 30 times his age less 225. How old is Sam? Display your work and write Sam's age in the blank.
   ____
5. According to a principle of physics:
   If an object is launched from the ground into the air and allowed to fall back to the ground, its height above the ground during the trip is governed by the following formula:
   $$h = vt - 16t^2$$
   where h is the number of feet above the ground the object is t seconds after launch and v is the velocity at which it left the ground in feet per second.
   Use this formula to assess how many seconds a golf ball will be in flight if it is hit so that it leaves the ground at a velocity of 64 feet per second.
   Hint: First solve for h by answering the following question: How high off the ground is the ball when it lands after being hit?
   Draw a picture illustrating the path of the ball, display your work and write your answer in the blank.
   ____
6. Enter, debug, and save on a disk the computer programs for factoring polynomials that we wrote in class.
7. For each of the following use the programs to determine if the polynomial can be expressed in factored form with integer coefficients; if so, factor it and print out the results.
   $$3a + 54a^2 + 81a$$
   $$70x^2 - 7x^2 - 2y(500y + 4y)$$
   $$9b^2 + 16c^2 + 24bc$$
   $$5x^2 - 24x + 10$$

As Mr. Rudd develops the detailed plan, he is frustrated attempting to formulate or locate real-life problems in which the three factoring algorithms apply. Any problems he thinks will interest students and in which factoring is truly useful require the upcoming algorithm for $ax^2 + bx + c$. He avoids problems solvable by finding roots to equations such as $x^2 - 25 = 0$ because he would prefer students use common sense to reason that "x^2 is 25, so x can be only 5 or -5," rather than blindly adhere to a factoring algorithm for solving such equations. Consequently, he reluctantly uses contrived problems for his lesson (e.g., the two word problems in Exhibit 11.28's tasksheet).

The agenda Mr. Rudd develops is displayed as Exhibit 11.29.

Engage in Activity 11.8:

Activity 11.8

In a discussion with a colleague, address the following questions about Case 11.10:

1. What are the advantages and disadvantages of the way Mr. Rudd had students check their responses to the prompts on the brief test? Would it have been more efficient to have students go over them in small collaborative groups?
2. How common is it for students to fail to discriminate between expressions like $2/3\ x^2 - 8/3$ and equations like $2/3x^2 - 8/3 = 0$ as Sid did? What are the advantages and disadvantages of the way Mr. Rudd responded to Sid's "mistake" with a question rather than immediately correcting the mistake?
3. During the independent-work session, Mr. Rudd tells Mary, "Okay, go with step 5. I'll be back in 33 seconds." Why do you suppose he said "in 33 seconds" rather than "in a minute," "soon," "in a sec," or even "in a half a minute"?
4. Rather than using a factoring algorithm, Mr. Rudd prefers to have students solve equations like $x^2 - 25 = 0$ by reasoning "25 from what number is 0? Either 5 or -5." Why do you suppose he feels that way?

Exhibit 11.29
Mr. Rudd's Agenda for the Sixth Day of Unit 7.

1. Transition period: I start class and direct students to do the following:
   - Take out homework and display it on desktops.
   - Begin working the minireview exercises on p. 223 (a short exercise on topics from prior units).
2. Independent-work session on the minireview: I quickly check homework and direct some students to display selected homework exercises on the board.
3. Transition period: Work on the minireview is halted and attention is directed to the completed homework exercises displayed on the board.
4. Interactive lecture and question-discussion session:
   - Students individually explain to the class how they completed homework exercises displayed on the board.
   - Homework and minireview exercises about which students raise questions are discussed.
5. Transition period: I direct students to put away homework and to do the following:
   - Turn their notebooks to the pages relating to the first two days of the unit, during which we worked on the crime-solver problem and in which they developed problems and displayed them on transparencies.
   - Retrieve their completed homework tasksheets for the first day (see Exhibit 11.12).
6. Interactive lecture session:
   - Play upon their work during the first two days of the unit to get them to deduce how factoring can be used to solve open sentences.
   - Walk them through writing computer programs for factoring differences of perfect squares and perfect-square trinomials.
7. Transition period: Distribute and assign the homework tasksheet (i.e., Exhibit 11.28).
8. Independent-work session: Begin homework.
9. Transition period into second period.

5. What do you suppose Mr. Rudd was attempting to accomplish with Exhibit 11.28's tasksheet? What do you like and what do you dislike about it as a homework assignment?

In Case 11.11, Mr. Rudd conducts Day 6 of Unit 7:

CASE 11.11

Twenty-five minutes into the period, the class has completed the first five agenda items and Mr. Rudd initiates the interactive-lecture session. He displays the transparency shown in Exhibit 11.13 that he used a week ago when the class discussed the crime-solver problem.

Mr. Rudd: Someone, please volunteer to briefly summarize the problem this reminds you of. Marlene.

Marlene: This woman was accused of stealing stuff from a garage. But she was seen at the drugstore just before that. Did she have enough time to get from the drugstore to the garage to commit the crime? That's the question.

Mr. Rudd: How far would she have had to travel?

Marlene: 7.2 miles.

Mr. Rudd: Thank you. The day after we solved that problem, we classified problems we wrote for homework as either solvable by factoring or not. Here's a sample of those problems.

Mr. Rudd begins showing a few problems on transparencies one at a time:

Mr. Rudd: This one was a factoring problem. [pause] This one was not. [pause] This one, yes. [pause] And this one, yes. [pause] This one, no. [pause] This one, yes. [pause] Now, what is special about the problems where the solutions required undoing multiplication—that is, factoring? Omar.

Omar: In all those we were given some product, like 7.2, and we had to find out different combinations that got you to it.

Mr. Rudd: So, if you know the dimensions of a polygon and you wanted to find the area, would you have a factoring problem? Megan.

Megan: Yes, sir.

Mr. Rudd: Why would it be a factoring problem?

Megan: I don't know.

Mr. Rudd: What did Omar tell us we're given in a factoring problem?

Megan: I don't remember.

Mr. Rudd: Thanks for waiting; go ahead, Kyle.

Kyle: It wouldn't be a factoring problem because you have to multiply—not unmultiply.

Mr. Rudd: Turn it around so it becomes a factoring problem, Kyle.

Kyle: You would want to find the dimensions and you know the area.

Mr. Rudd: How did I originally describe the problem, Megan?

Megan: Something about a polygon.

Mr. Rudd: Cassandra?

Cassandra: You wanted the polygon's area and you knew its sides. Kyle reversed it, where you had the area but not the sides.

Mr. Rudd: Omar, write your statement on the upper right corner of the board—the one describing the kinds of problems that are solvable using factoring. [pause] Thank you. We're now going to discuss the use of factoring to solve equations. We aren't switching topics; we're still going to be dealing with the kinds of problems Omar's statement describes. What we'll be doing is developing a more systematic approach to solving those kinds of problems. To start, turn your attention to the homework tasksheet for the first day of this unit; it looks like this.

Mr. Rudd displays Exhibit 11.12 with the overhead projector.

Mr. Rudd: You don't have yours, Xavier. Magnolia, would you please allow Xavier to look on with you? Thank you. Everybody, look at what you have for Prompt #4. Magnolia, read yours, giving us one a, b pair at a time.

Magnolia: (0, 7), (7, 0), (0, 0), (11, 0), (1, 0).

Mr. Rudd: How is yours different from Magnolia's, and how is it the same? Alena.

Alena: Well, I had (1, 0) and then—

Mr. Rudd: I'm sorry; I didn't say what I meant. Don't give us your list. Describe how yours differs and how it's the same as Alena's without actually listing pairs.

Alena: Okay. [pause] Just a sec, let me think. [pause] This is hard. Okay! Mine's different because my numbers aren't all the same as hers. But mine's the same because—like Magnolia's—every pair has a 0—it's got to!

Mr. Rudd: Why does it have to, Megan?

Megan: You're picking on me today.

Mr. Rudd: Possibly. Why does a or b have to be 0 if $-3ab = 0$?

Megan: Because the only thing you can multiply 3 by to get 0 is 0. So, ab is 0, and 0 times anything is 0.

Mr. Rudd: Megan makes an important point for us. Zero times any real number is 0. For our purposes, let's turn that around. Okay, Salvador.

Salvador: The only way the product of two numbers can be 0 is if one of the numbers is 0.

Mr. Rudd: Write that down for us under Omar's statement on the board. [pause] Thank you. Consider how we might solve this equation.

Mr. Rudd displays "$3x(x - 8) = 0$" with the overhead projector.

Mr. Rudd: Jot down the solution. [pause] What did you put, Alphonse?

Alphonse: 0.

Mr. Rudd: If $x = 0$ as Alphonse indicates, what is $x - 8$? Delia.

Delia: I don't know.

Mr. Rudd: Look at this equation.

Mr. Rudd writes "$ab = 0$" right under "$3x(x - 8) = 0$."

Mr. Rudd: What is b if a is 0, Delia?

Delia: It can be any number.

Mr. Rudd: So, let $b = x - 8$. Go back to the original equation where $3x = 0$. So what is $x - 8$?

Delia: Any number.

Mr. Rudd: But if $3x = 0$, then what's x, Megan?

Megan: 0.

Mr. Rudd: If $x = 0$, as Megan and Alphonse said, what is $x - 8$? Delia.

Delia: Anyth–Oh! Then $x - 8$ is 0, which is just -8! I get it.

Mr. Rudd: So for $3x(x - 8) = 0$, we have numbers multiplied together giving us a product of what? Xavier.

Xavier: 0.

Mr. Rudd: So for the equation to be true [pause] Juaquin?

Juaquin: One of the factors has to be 0. So either $3x = 0$, which makes $x = 0$, or the other one can be 0.

Mr. Rudd: What's true if the other one is 0, that is, if $x - 8 = 0$? Juaquin.

Juaquin: Then $x = 8$.

Mr. Rudd: Dustin.

Dustin: I don't get it. How can $x = 0$ and $x = 8$?

Mr. Rudd: It can't. But x can *either* equal 0 *or* 8. This equation happens to have two solutions. Kyle.

Kyle: I get it! It's like $ab = 0$. a is $3x$ and b is $x - 8$. Right?

Mr. Rudd: You've got the right idea. But technically, what you just said is not exactly right. And I'm not sure how to help clear it up in the amount of time we have left today. Let me try something. I don't know if it's going to help, but it's worth a try. We didn't run into this difficulty with $ab = 0$ because we were dealing with two different variables. So when we took the case $a = 0$, b could be any number. Now, as Kyle pointed out to us, $3x(x - 8) = 0$ works the same way but with one difference. $3x$ and $x - 8$ are related, so the value of one affects the value of the other. So in the case of $3x = 0$, x must equal 0 and there's only one number x—it's the same x.

Looking at the sea of blank faces, Mr. Rudd feels ambivalent. On one hand, some of the students who tend to struggle with mathematics are on the threshold of conceptual-level understanding; on the other hand, he realizes that the subtleties of what he is trying to explain are not coming across to them. Noticing that only 14 minutes remain in the period, he continues:

Mr. Rudd: We have a point of confusion that we have yet to clear up. But instead of my rattling on about it, I think we should look at two more equations and then assign the homework. The experience should help the explanation make more sense tomorrow. Look at this equation.

Mr. Rudd displays "$x^2 - 7x = 0$."

Mr. Rudd: Let's find what, if any, values for x makes this statement true. With the previous equation $3x(x - 8) = 0$, we had a polynomial in factored form equal to 0. So, we knew that the equation held for an x value where one of the factors equaled 0. But with this one, we have a binomial equal to 0. We're looking at the difference of two terms, not the product. Any suggestions? Do you see a way we can rewrite this binomial so that it's in factored form? Then we could find solutions by finding values of x that make a factor equal 0. [pause] Okay, Mary.

Mary: Just factor the binomial.

Mr. Rudd: Everybody, take Mary's suggestion—quietly, on the paper in front of you. Shhh, just do it please. [pause] Tell us what the equation looks like on your paper, Edwin.

Edwin: $x(x - 7) = 0$.

Mr. Rudd: If one of two things is true, Edwin's statement is true. What are those two things, Magnolia?

Magnolia: If $x = 0$ or if $x - 7 = 0$.

Mr. Rudd: And if $x - 7 = 0$, what is x in that case, Sid?

Sid: 7.

Mr. Rudd: We've got time for one or two more quick ones.

Mr. Rudd writes "$12x^2 = 4x$."

Mr. Rudd: How about this one? Juaquin.

Juaquin: Just divide both sides by $4x$ and you get $3x = 0$.

So x has to be 0.

Mr. Rudd writes and displays the following:

$$12x^2 = 4x$$

$$\frac{12x^2}{4x} = \frac{4x}{4x}$$

$$3x = ?$$

Mr. Rudd: What's $4x$ over $4x$, Juaquin?

Juaquin: Oh, yeah! It's 1, not 0. So then, $3x = 1$ and $x = 1/3$.

Mr. Rudd: Everybody, quickly try 1/3 for x in the original equation to see if it works. Put your hand up if and when you find it does. [pause] Okay, so $x = 1/3$ works. Does any other value for x work? Sid.

Sid: $x = 0$.

Mr. Rudd: Everybody, try it. Raise your hand if and when you find it works. [pause] Okay. So we've got two cases for x that work. I need to raise a caution flag about solving equations like this. Be careful when you divide through by the unknown variable for two reasons: First, you might lose a solution. Second, you need to watch that you don't try to divide by 0. By the way, we did nicely on that one, but I was surprised because I expected you to do this.

Mr. Rudd writes and displays the following:

$$12x^2 = 4x$$

$$12x^2 - 4x = 0$$

$$4x(3x - 1) = 0$$

Either $4x = 0$ or $3x - 1 = 0$

If $4x = 0$, then $x = 0$

If $3x - 1 = 0$, then $x = 1/3$

Mr. Rudd: Let's squeeze in one more. I'm going to write it out quickly. You copy it and think about it as part of your homework.

He writes and displays the following:

$$x^2 + 25 = 10x$$

$$x^2 + 25 - 10x = 0$$

$$x^2 - 10x + 25 = 0$$

$$(x - 5)^2 = 0$$

$$x = 5$$

He then distributes the homework tasksheet (i.e., Exhibit 11.28) but assigns only Prompts #1–5 because he did not get to the computer programming part of the day's agenda (i.e., the second part of item 6 in Exhibit 11.29). The bell sounds; Casey is left pondering what happened in the class. However, he quickly turns his thoughts to the next business at hand—third period geometry.

Engage in Activity 11.9:

Activity 11.9

In a discussion with a colleague, address the following questions about Case 11.11:

1. What are the advantages and disadvantages of Mr. Rudd's directing Xavier to look on Magnolia's tasksheet when Xavier did not have his own?
2. At one point, Megan seemed reluctant to engage in the question-discussion session. What are the advantages and the disadvantages of the way Mr. Rudd attempted to lead her to engage in the activity?
3. After one of Megan's answers in the question-discussion session, Mr. Rudd comments to the class, "Megan makes an important point for us. Zero times any . . ." instead of telling Megan "Good!" or "That's right." What are the advantages and disadvantages of this strategy?
4. In the question-discussion session, Delia answered, "I don't know" to one of Mr. Rudd's questions. How did Mr. Rudd respond? What are the advantages and disadvantages of responding as he did rather the way most teachers do (e.g., by asking another student to "help Delia out")?
5. Mr. Rudd wrote "$ab = 0$" under "$3x(x - 8) = 0$" on a transparency. How did this move backfire? How did it help? In the long run, do you think this strategy will turn out to be helpful or hurtful to students' understanding of mathematics?

6. Do you think Mr. Rudd managed to engage enough of the students in the reasoning-level question-discussion session? How do his tactics compare to those used by Ms. Citerelli in Cases 5.8 and 5.9, Mr. Heaps in Case 3.12, Mr. Grimes in Case 3.18, Mr. Smart in Case 3.19, and Ms. Cramer in Case 3.20?
7. What are the advantages and disadvantages of the way Mr. Rudd responded to Juaquin rewriting the equation $12x^2 = 4x$ as "$3x = 0$"?
8. Mr. Rudd usually presents prompts to students (e.g., questions for all to consider) before calling a name for an aloud response. However, when he directed Omar to write a statement on the board, he used Omar's name before presenting the prompt. What was different about this directive to Omar as compared to most of the other oral prompts he presented during that question-discussion session?
9. Do you think Mr. Rudd attempted to communicate too much information near the end of Day 6's session? Would it have been more efficient to postpone his attempt to clear up the confusion between "$x = 0$ and $x = 8$" and "either $x = 0$ or $x = 8$" for another day?

In Case 11.12, Mr. Rudd reflects on Day 6 and plans for and conducts Days 7-15 of Unit 7:

CASE 11.12

After school, Mr. Rudd thinks about the day's algebra class and plans for Wednesday's. He thinks: "That session was exciting, but it sure went a bit haywire when I tried to parallel $3x(x - 8) = 0$ with $ab = 0$. Next time, I'll anticipate some of them thinking that x in $3x$ is not equal to x in $x - 8$. I learn something every day! I don't think there's any way I can explain that nuance to them until they have more experience in factoring to solve real-life problems. Then they'll be able to make the connections. That's why I wanted to get into applications before we got to the really useful stuff—$ax^2 + bx + c = 0$. I hate waiting too long without touching real-life situations. As things went, my planned application lesson turned out to be more of a discover-a-relationship lesson!

"I hadn't realized that the little business we did with that homework tasksheet (i.e., item 4 of Exhibit 11.12) on the first day didn't do the discovery trick for them. It seems like such a straightforward idea—if $ab = 0$, then a or b must be 0! But I've got to remember these kids are just being introduced to this stuff; I've been using that relationship for years.

"The unit is going to work out fine. I didn't spend enough time with discovery in the beginning, but I'm making up for it now. There's still time to work on application—especially after we get through Objective J. Maybe today's activities, coupled with some review and practice tomorrow, will help them see that factoring has real-world applicability. Of course, part of the difficulty is that the algorithms we've covered up to this point in the unit are ones included because they lead to something that's useful, not necessarily because they are useful themselves. At least, the kids seems to be hanging in with me.

"Where do we go with the rest of this unit? First, I need to use their work with the tasksheet (i.e., Exhibit 11.28) for further discovery and applications of factoring. Most everybody will have done okay with Prompts #1 and #2. Work with Prompt #3 will help clear up that business about x being either one number or another rather than one number and another. I doubt if too many of them will understand how to respond to Prompt #5, but it gets us into application. The progression in the tasksheet from the discovery work of Prompts #1 and #2 to the thought problems of #3 and #4 followed by the application word problem of #5 should do the trick.

"After we get through the homework, I should walk them through writing the computer programs as planned for today. Then I should give them practice with the algorithms to this point and wait until Friday to go for the discovery lesson on factoring trinomials. This unit may take a day or two longer than planned, but if I rush things, it'll just create difficulties in Unit 8 and on down the road."

On Wednesday, Mr. Rudd reviews the homework and then completes the programing writing activity previously planned for Tuesday (i.e., the second part of agenda item 6 in Exhibit 11.29). For homework he assigns Prompts #6 and #7 from Exhibit 11.28, as well as textbook exercises on factoring algorithms.

After reviewing the homework on Thursday (i.e., Day 8), Mr. Rudd divides the class into three independent-work groups. Some students go to the computer lab to complete the factoring exercises using programs they had entered and debugged for homework. Others use the classroom computers to work on the same exercise. The remaining students work on word problem in collaborative-group sessions.

On Friday, Mr. Rudd conducts the lesson for Objective I. In a reasoning-level question-discussion session, he takes advantage of students' prior knowledge of the FOIL method for multiplying binomials (i.e., first terms are multiplied, then the two outer terms, followed by the two inner terms, and then the last two terms) to lead them to formulate an algorithm for factoring trinomials.

In a collaborative group session, four subgroups of about six students each work with cardboard rectangular regions to develop geometric models for factoring different trinomials. The activity for one example (i.e., $2x^2 + 3x + 1 = (x + 1)(2x + 1)$) is illustrated by Exhibit 11.30.

Over the weekend, Mr. Rudd develops the unit by applying procedures you learned by working with chapter 9's section "Designing Measurements of Students' Achievement of Unit Goals." Exhibit 11.31 is a copy of the prompts for that test. He also synthesizes a practice test to use in a review session he plans to conduct the day before the unit test is administered.

Exhibit 11.30
An Activity Mr. Rudd Used on Day 9 of Unit 7.

Demonstration: Factoring a Trinomial

Materials for demonstration:

A set of large squares measuring x by x, labeled by their areas as $x \times x$ or simply x^2

A set of small squares measuring 1 by 1, labeled by their areas as 1×1 or simply 1

A set of rectangles measuring x by 1, labeled by their areas as $x \times 1$ or simply x

Notes:

Monomials, binomials, and trinomials in x can now be represented by the appropriate geometric figures.

In factoring a trinomial, the various monomial parts are arranged in a rectangular shape. The dimensions of the rectangle give the factors.

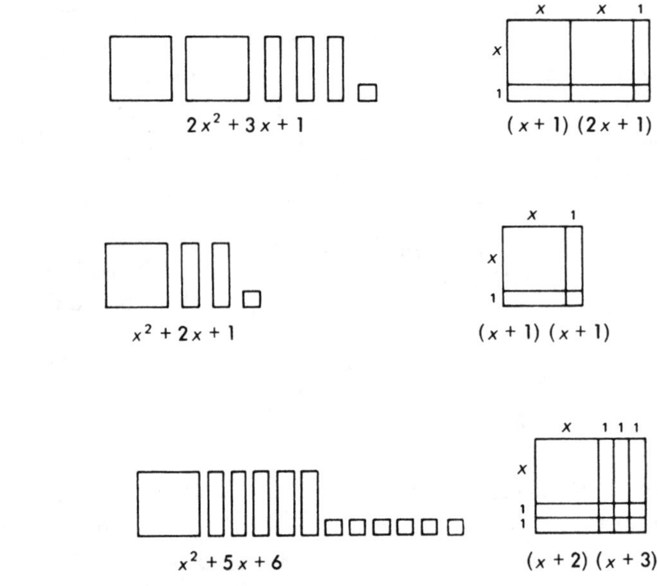

Tasks:

1. Show how to factor $3x^2 + 4x + 1$ using a model.
2. Show how to factor $4x^2 + 8x + 3$ using a model.
3. Show some other trinomials that can be factored using a model. Show some that cannot be factored.
4. Imagine a rectangular array of pieces measuring $5x + 1$ by $2x + 3$. How many pieces are in it? How many of each size are there? What trinomial factorization does it represent?

Source: Max A. Sobel/Evan M. Maletsky, TEACHING MATHEMATICS: A Sourcebook of Aids, Activities, & Strategies, 2e, © 1988, pp. 174–175. Reprinted by permission of Prentice Hall, Inc., Englewood Cliffs, New Jersey.

Monday's class meeting is devoted to direct-instructional activities on the algorithm for factoring trinomials (i.e., Objective J from Exhibit 11.10). Case 6.9 includes a list of the steps in the algorithm Mr. Rudd explains during the fourth stage of the lesson. Students engage in an independent-work session and a homework assignment for the fifth stage.

On Tuesday (i.e., Day 11), Mr. Rudd leads the students to write computer programs for factoring trinomials. He then begins the lesson for Objective L using a tasksheet with word problems. As suggested by chapter 8's section "Lessons for Application Objectives," the tasksheet includes example problems solvable by factoring as well as nonexample problems solvable by

Exhibit 11.31
Mr. Rudd's Unit 7 Test.

Algebra I
1st Period
Unit 7 Test

1. What is your name? ______________________
2. Following is a list of questions. For each, determine whether or not analyzing the factors of a number would help to answer the question. If so, write Yes in the blank. If not, write No. Write one sentence explaining the reason for your answer.

   ________ $5,378 was donated for cause. How many people donated to the cause and what was the average donation?
   Reason: ______________________

   ________ How large an area can I enclose with 60 meters of fencing?
   Reason: ______________________

   ________ If I save $175 in a bank at an interest rate of 6%, compounded monthly for a year, how much interest will I have earned after 2 years?
   Reason: ______________________

   ________ How long will it take me to travel 300 miles by car?
   Reason: ______________________

3. List all the factors of -18 that are integers.
   ____
4. Express 72 in a factored form using only prime numbers.
   ____
5. Which of the following two expressions represents a prime polynomial? Put an X in the blank in front of the prime polynomial and then use the space provided to explain why it is prime and the other one is not.
   ________ $80a^3b - 4ab^2$ ________ $80a + 17b$
   ____
6. Write $21x^3 + 6x$ in factored form. ________
   Prove your answer with the distributive property $ab + ac = a(b + c)$ by indicating what you let a, b, and c equal:
   $a =$ ________, $b =$ ________, $c =$ ________
7. Illustrate that $x^2 + 4x + 3$ can be expressed in factored form as $(x + 3)(x + 1)$ by drawing a puzzle constructed from the following eight puzzle pieces:

   Label your finished puzzle to indicate the dimensions of the two-dimensional region it forms.

Continued

methods from prior units. During the remainder of the period, the class engages in an independent-work session with some students working on computers to enter and debug their programs while others work on the tasksheet. For homework, all students are directed to complete both tasks.

On Day 12, Mr. Rudd conducts a reasoning-level question-discussion session as part of the lesson for Objective L. His deductive questions focus on problems students addressed for homework, leading them through the first three stages of the lesson. For homework, he assigns the practice test he developed over the weekend.

On Thursday, students' responses to prompts from the practice test assigned for homework are reviewed—stimulating questions and issues that help them prepare for Friday's test.

Exhibit 11.31
Continued

8. For each of the following that is factorable, factor completely; write not factorable under the others:

a. $11w^3 - 44w$

b. $2x^2 + 11x - 30 + x^2 + 10$

c. $t^2 + t + 2$

d. $5x^2 - 80$

9. A rectangle is 4 meters wide and 7 meters long. If the length and the width are increased by the same amount, the area is increased by 26 square meters. What are the dimensions of the new rectangle? Display your work and write your final answer in the blank.

____

10. The difference between the square of a particular whole number and itself is 12. What is that whole number? Display your work and write your final answer in the blank.

____

11. Smile.

Over the weekend, Mr. Rudd scores students' responses to the test prompts and uses his computer to generate the detailed results shown by Exhibit 11.32.

Using the compromise method explained in chapter 9, he converts scores to grades for Unit 7 according to the following scale where s is a test score: $74 \leq s \rightarrow$ A, $65 < s < 70 \rightarrow$ B, $40 < s < 60 \rightarrow$ C, $s \in \{22, 24\} \rightarrow$ D, and $s = 15 \rightarrow$ F.

Conclusions he makes from his analysis of the results of the miniexperiments include the following: (a) In general, students tended to have more trouble expressing themselves about mathematics than executing mathematics. (b) Overall, students achieved at the application level (i.e., Objective L) better than expected. (c) Students in general—especially those who scored below 60—tended to have difficulty with Miniexperiment #2, which was designed to be relevant to Objective A. (d) He should consider strategies for (i) providing Magnolia and Juaquin with more advanced work, (ii) keeping Delia and Xavier engaged, as their lack of attention and effort seemed to be a major factor in their lack of progress toward Unit 7's goal, and

Exhibit 11.32
Detailed Results From Mr. Rudd's Unit 7 Test.

Miniexperiment (Max. Points)	2 (8)	3 (3)	4 (3)	5 (3)	6 (8)	7 (12)	8 (28)	9 (8)	10 (8)	Test Score (81)
Magnolia	8	3	3	3	8	12	28	8	8	81
Juaquin	8	2	3	3	8	12	27	8	8	80
Marlene	6	3	3	3	8	10	28	8	8	77
Alena	7	2	1	3	8	12	28	8	8	77
Cassandra	6	3	3	2	8	12	25	8	8	75
Kevin	5	3	2	3	7	10	28	8	8	74
Omar	3	3	3	3	3	10	28	8	8	69
Tawny	8	3	3	3	8	2	28	6	8	69
Edwin	7	1	3	3	8	12	20	6	8	68
Megan	4	3	3	3	8	12	28	3	4	68
Sid	5	3	2	3	4	9	26	8	8	68
Chevron	8	2	3	3	8	12	19	5	8	68
Salvador	8	3	3	3	4	10	24	6	6	67
Dustin	8	3	3	3	8	4	25	6	6	66
Mary	6	1	2	2	4	7	28	8	8	66
Kyle	4	3	3	1	6	10	22	4	6	59
Mylinn	1	1	1	0	8	10	22	8	8	59
Amad	2	2	2	2	7	10	19	6	6	56
Ron	3	3	3	3	8	10	15	4	6	55
Alphonse	2	3	3	3	4	6	26	4	4	55
Russell	6	3	3	3	8	2	12	4	5	46
Blair	5	3	3	3	4	4	13	4	5	44
Delia	2	2	3	0	1	6	5	2	3	24
Xavier	4	2	3	0	3	1	7	1	1	22
Dick	2	1	0	1	2	3	4	1	1	15
Ann	Has not taken test									

$n = 25$ $\mu = 60.32$ $\sigma = 17.51$ $r = 0.78$ SEM = 8.21

(iii) remediating what appears to be significant gaps in Dick's understanding of mathematics.

He notes individual responses to certain prompts that he wants shared with the class on Monday when the test is reviewed. Because he plans to use small collaborative groups as part of the review session, he organizes subgroups so that each one has at least two students with high scores.

On Monday (i.e., Day 15), Mr. Rudd reviews the test results with the class by following the four-stage routine explained in chapter 9's section "Using Test Results for Formative Feedback." During the latter part of the session, he plays off discussions of Miniexperiments #9 and #10 to lead into Unit 8 on solving quadratic equations.

Engage in Activity 11.10:

Activity 11.10

In a discussion with a colleague, address the following questions about Case 11.12:

1. In light of the weighted objectives in Exhibit 11.10, how relevant to Unit 7's goal does the unit test shown by Exhibit 11.31 appear to be?
2. What are the advantages and disadvantages of Mr. Rudd's using the compromise method to convert Exhibit 11.32's test scores to grades? Does his scale seem appropriate? Why or why not?
3. What insights did you gain from your work analyzing Cases 11.3–11.12? How do you plan to apply those insights to your own teaching?

A GEOMETRY COURSE

A course in geometry is commonly taken by students the year after they complete algebra I. In the past, only high schools offered year-long geometry courses. Today, with the increase in sixth and seventh graders taking prealgebra and algebra I, geometry courses are offered by many middle and junior high schools. Typically, a student is expected to take only one geometry course during her precollege school

Exhibit 11.33
Mr. Rudd's Sequence of Units for Geometry.

School Weeks	Unit #	Unit Title
1–2	1	Inductive Reasoning, Deductive Reasoning, and Truth
3–4	2	The History of Geometry
5	3	Points, Distance, and Line Segments
6–7	4	Lines, Rays, and Angles
8–9	5	Planes and Relationships Among Lines
10–11	6	Polygons
12	7	Similarity and Congruence
13–16	8	Triangles and Congruence
17–18	9	Triangles and Similarity
19–21	10	Special Polygons
22–25	11	Circles
26	12	Area
27–28	13	Solid Figures
29–30	14	Surface Area and Volume
31–32	15	Transformation Geometry
33–34	16	Non-Euclidean Geometries
35–36	17	Extending What We Learned

career. However, the study and computation of one-dimensional geometric variables (e.g., lengths of line segments, perimeters of polygons, and circumferences of circles), two-dimensional geometric variables (e.g., areas of polygonal regions), and three-dimensional geometric variables (e.g., volumes of solids) are included in virtually every school mathematics course from fourth grade through algebra I. Typically, geometry courses provide a much more formal, axiomatic treatise of Euclidean geometry than the exposure-to-and-application-of-geometric-formula approach of other courses (e.g., sixth-grade mathematics and prealgebra). In the tradition begun with *Euclid's Elements* around 300 B.C. (Calinger, 1999, pp. 130–149), many high school graduates think of geometry as "that course where we had to do proofs." It is the course in which the logical underpinnings of mathematics are most often emphasized—albeit this may be an unfortunate tradition (Cangelosi, 2001a).

Case 11.13 is a brief glimpse into Mr. Rudd's geometry course:

CASE 11.13

Although just preparing for, conducting, and evaluating outcomes for his algebra I class could occupy all of Mr. Rudd's working days, he must partition his time among five classes. Exhibit 11.33 is an outline for his geometry course. Exhibit 11.34 is one of his unit plans. Exhibit 11.35 is one of his lesson plans.

Engage in Activity 11.11:

Activity 11.11

In a discussion with a colleague, address the following questions about Case 11.13:

1. If you were in Mr. Rudd's place, how would you modify Exhibit 11.33's course outline?
2. If you were in Mr. Rudd's place, how would you modify Exhibit 11.34's unit plan?
3. Exhibit 11.35's tasksheets include prompts for students to attempt the impossible (e.g., draw a triangle with two 95° angles and use the same three sticks to make triangles with different sizes or shapes). What are the advantages and disadvantages of presenting students with such tasks? What types of discussions do you imagine will be stimulated by students' responding to those prompts?
4. If you were in Mr. Rudd's place, how would you modify Exhibit 11.35's lesson plan?

Exhibit 11.34
Mr. Rudd's Plan for Unit 8 on Triangles and Congruence.

Course: Geometry

Unit 8 (of 17): Triangles and Congruence

Goal: Students discover and apply congruence relationships involving triangles and prove them to be theorems.

Objectives:

A. Given that *A, B,* and *C* are three noncollinear points and *D, E,* and *F* are three noncollinear points, the student explains why the following relationships are true:
   i) $(AB = DE, AC = DF, \text{and } BC = EF) \Rightarrow (\triangle ABC \cong \triangle DEF)$
   ii) $(AB = DE, m\angle A = m\angle D, \text{and } AC = DF) \Rightarrow (\triangle ABC \cong \triangle DEF)$
   iii) $(m\angle A = m\angle D, AB = DE, \text{and } m\angle B = m\angle E) \Rightarrow (\triangle ABC \cong \triangle DEF)$
   iv) $(m\angle A = m\angle D, AB = DE, \text{and } m\angle C = m\angle F) \Rightarrow (\triangle ABC \cong \triangle DEF)$
   v) $AB = AC \Rightarrow m\angle B = m\angle C$
   vi) $m\angle B = m\angle C \Rightarrow AB = AC$

(discover a relationship) 35% C,

B. Accepting the second relationship listed for Objective A (i.e., ii) as a postulate, the student proves the other five (i.e., i, iii, iv, v, and vi) to be theorems (application) 15%.
C. The student explains why the proofs the class developed as a result of the lesson for Objective B are valid (comprehension and communication) 15%.
D. The student uses the six relationships listed for Objective A to develop compass-and-straightedge construction algorithms (e.g., to "copy" a triangle or bisect an angle (application) 10%.
E. Given a real-life problem, the student explains how, if at all, postulates, theorems, and algorithms formulated in this unit can be used to address the problem (application) 20%.
F. The student experiments with postulates that are inconsistent with axioms of Euclidean geometry to generate novel and unusual conjectures involving congruence relationships with triangles (creative thinking) 05%.

Estimated number of class periods: 20

Textbook page references: 186–250 & 115–127 (but only for exercises—not the explanations or logic structure)

Overall plan for lessons:

For Objective A, I'll partition the class into 8 collaborative teams of 3 or 4 students each for the 1st stage. Team I will use manipulatives in experiments leading to the discovery of relationship i as stated in Objective A. Teams II, III, V, VII, & VIII will engage in similar activities for relationships ii, iii, iv, v, and vi respectively. Team VI will engage in similar activities but its experiment will lead to the discovery that a correspondence between two triangles is not necessarily congruent even though the two triangles have two pairs of corresponding sides and one pair of corresponding angles congruent. Team IV will engage in similar activities but its experiment will lead to the discovery that a correspondence between two triangles is not necessarily congruent even though the two triangles have three pairs of corresponding angles congruent. Students will engage in question-discussion sessions for the 2nd and 3rd stages. Homework exercises as well as in-class experiments with *Geometer's SketchPAD* (Bennett, 1993) and virtual manipulatives from the National Library of Virtual Manipulatives for Interactive Mathematics (www.matti.usu.edu) will be used to analyze additional examples and search for counter examples during the 4th stage. The lesson should culminate with acceptable statements of the six conjectures; however, I'll stay relaxed even if all six don't work out as planned. As long as the class formulates an adequate subset, we can move on and then come back and finish the lesson later in the unit. (Estimated time needed: the equivalent of 3.5 class periods spread over about a week's time—plus homework)

I'll provide much more direct guidance during the lesson for Objective B that I did for Objective A. Reasoning-level question-discussion sessions will be interspersed over a two-week period in which I'll pull from their ideas to build an overall plan and outline for writing proofs for each of the five theorems. The deductions and justifications for the proofs must come from them; however, I'll clean up the wording and presentation for them. Presumably, not all students will contribute significant ideas to the development of the proofs. However, I will strive to prompt each student to come up with at least one idea that contributes to at least one of the five proofs. In any case, I know I can at least get all students to comprehend the logic of most of the proofs during the lesson for Objective C—and that will be an important accomplishment. (Estimated time needed: the equivalent of 4 class periods spread out over 2-weeks' time—plus homework)

The lesson for Objective C will be integrated with that for Objective B in that as each theorem is proven, we will work on comprehending and communicating the logic of the proof and why it is valid before the next theorem is proven. (Estimated time needed: the equivalent of 2 class periods spread out over 2-weeks' time—plus homework)

Continued

Exhibit 11.34
Continued

A brief test on objectives A, B, and C will be given and reviewed with the class for purposes of formative feedback. The test review will serve as a lead-in to lessons for Objectives D and E. (Estimated time: 1 class period)

To keep the historical and compass-straightedge construction connections from Units 2 and 4, I'll use a four-stage application lesson to lead them to develop algorithms for duplicating triangles, bisecting angles, and other tasks they see as application of this unit's postulate and five theorems. This lesson for Objective D—like that for Objective E, but unlike those for the other four objectives—will be intact (i.e., not integrated with other lessons). (Estimated time needed: 2 class period plus homework)

The lesson for Objective E will follow my usual 4-stage application model. (Estimated time needed: 3 periods plus homework)

I plan to conduct the lesson for Objective F—using synectics—after the lesson for Objective F. However, if students appear to need a jolt to their conventional thinking (e.g., to stimulate ideas for developing proofs or discover a relationship they're struggling to discover), I'll call a halt to whatever we're doing at any point in the unit to engage them in activities involving direct analogies, personal analogies, or compressed conflicts. (Estimated time needed: 1.5 class periods)

Using a practice test, I'll conduct a review session for the unit test for Objectives A–F. (Estimated time needed: 1 class period plus homework)

A unit test for Objectives A–F will be administered and the results reviewed with the class. (Estimated time needed: 2 class periods plus homework)

Extraordinary learning materials and resources needed:

various-size sticks, tape, various-size sticks connected with wing-nut bolts, geoboards with rubber bands, *Geometer's SketchPAD* from our computer network, calculators with software for *Geometer's SketchPAD*, rulers, internet access to virtual manipulatives from the National Library of Virtual Manipulatives for Interactive Mathematics (www.matti.usu.edu), compasses, and protractors.

Exhibit 11.35
Mr. Rudd's Lesson Plan for Objective A Listed in Exhibit 11.34.

Objective: Given that A, B, and C are 3 noncollinear points and D, E, and F are 3 noncollinear points, the student explains why the following relationships are true:

i) $(AB = DE, AC = DF, \text{and } BC = EF) \Rightarrow (\triangle ABC \cong \triangle DEF)$
ii) $(AB = DE, m\angle A = m\angle D, \text{and } AC = DF) \Rightarrow (\triangle ABC \cong \triangle DEF)$
iii) $(m\angle A = m\angle D, AB = DE, \text{and } m\angle B = m\angle E) \Rightarrow (\triangle ABC \cong \triangle DEF)$
iv) $(m\angle A = m\angle D, AB = DE, \text{and } m\angle C = m\angle F) \Rightarrow (\triangle ABC \cong \triangle DEF)$
v) $AB = AC \Rightarrow m\angle B = m\angle C$
vi) $m\angle B = m\angle C \Rightarrow AB = AC$

(discover a relationship)

Note 1: During Stage 1, I will organize the class into collaborative teams as follows:

Team I	Team II	Team III	Team IV
Nashauna*	Tyson*	Catherine*	Ha*
Carmella	Donyell	Antonio R.	Don
Leroy	Angel	Michael	Michelle

Team V	Team VI	Team VII	Team VIII
Nancy R.*	Gayton*	Kathyrn*	Kramer*
Candice	Rahmin	Qu-Ping	John T.
John B	Miguel	Davilon	Janiece
Jima	Manunsell	Antonio M.	

*-Indicates the team's chairperson.

Note 2: Each of the eight teams will complete a tasksheet during Stage 1; following are the eight different tasksheets:

Continued

Exhibit 11.35
Continued

Team I's Tasksheet

1. Pick up the box of supplies labeled "Team I" and bring it to your work location.
2. Note that your box contains colored sticks of varying lengths so that same-colored sticks have the same length and different-colored sticks have different lengths.
3. Select three sticks of the same length.
4. Position the three sticks so that they form a triangle but so they touch only at their ends.
5. With those same three sticks try to form another triangle—again with only ends touching—so that the second triangle has a different shape or size than the first.
6. Answer: What are the results of the experiment through Step 5? ____________
____
7. Repeat Steps 3–5, but this time select a different length for your three same-size sticks.
8. Answer: What are the results of the experiment through Step 7? ____________
____
9. Repeat Steps 3–5, but this time use three sticks that do not have the same lengths.
10. Answer: What are the results of the experiment through Step 9? ____________
____
11. Experiment with three sticks of various lengths to make triangles of various sizes and shapes.
12. Write a conjecture about how the lengths of the three sides of a triangle affect its size and size of its angles: ____________
____
____

Team II's Tasksheet

1. Pick up the box of supplies labeled "Team II" and bring it to your work location.
2. Note that your box contains several pairs of sticks of varying lengths with each pair joined at ends by a wingnut bolt that can be tightened to stabilize the angle.
3. Select one pair of sticks and tighten the wing nut so that the size of the angle is fixed at some point between 0° and 180°. Now imagine a line segment opposite the angle with one of its ends on the other end of one of the two sticks and its other end on the opposite end of the other stick.
4. Answer: As long as that angle stays fixed, can a line segment longer or shorter than the one you imagined serve as the third side of a triangle? ____________
Explain your answer: ____________
5. Now loosen the wing-nut just enough so that you can change the size of the angle between the two sticks. Answer: Every time you change the size of the angle, what happens to the length of the imagined line segment opposite the angle? ________
6. Repeat Steps 3–5 except with another pair of sticks with lengths different from the first pair.
7. Answer: What are the results of the experiment through Step 6? ____________
____
8. Write a conjecture about how the lengths of two sides and the angle that includes the two sides affect the shape and size of a triangle: ____________
____
____

Continued

Exhibit 11.35
Continued

Team III's Tasksheet

1. Pick up the box of supplies labeled "Team III" and bring it to your work location.
2. Note that your box contains protractors, rulers, paper, pencils, and a three-stick contraption joined with two wingnut bolts that can be tightened to stabilize two of the three angles.
3. Use protractors and rulers to draw as many different sizes and shapes of triangles as you can with a 40° angle and a 60° angle that share an 11-cm side.
4. Answer: How many triangles with different shapes or sizes were you able to draw in response to Prompt #3? ____________
5. Use protractors and rulers to draw as many different sizes and shapes of triangles as you can with a 25° angle and a 45° angle that share a 15-cm side.
6. Answer: How many triangles with different shapes or sizes were you able to draw in response to Prompt #5? ____________
7. Experiment with the three-stick contraption to help you develop an explanation as to why the following two statements are true: (a) Every triangle you drew with a 40° angle and a 60° angle that share an 11-cm side had the same size and shape. (b) Every triangle you drew with a 25° angle and a 45° angle that share a 15-cm side had the same size and shape.
8. Write your explanation: ____________
9. Write a conjecture about how two angles and an included side affect a triangle's size and shape: ____________

Team IV's Tasksheet

1. Pick up the box of supplies labeled "Team IV" and bring it to your work location.
2. Note that your box contains protractors, rulers, paper, pencils, and geoboards with rubber bands.
3. Use protractors and rulers to draw four different size triangles so that all four triangles have angles with the following measures: 40°, 60°, and 80°.
4. Use protractors and rulers to draw four different size triangles so that all four triangles have angles with the following measures: 25°, 45°, and 110°.
5. Experiment with the geoboards and rubber bands to help you develop a conjecture about the effect of the sizes of a triangle's three angles on the triangle's shape and size.
6. Write a conjecture about the effect of the sizes of a triangle's three angles on the triangle's shape and size: ____________

Continued

5. What are the advantages and disadvantages of students' experimenting with physical manipulatives and paper-and-pencil drawings before using computer simulations for similar experiments?
6. What insights did you gain from your work analyzing Case 11.13? How do you plan to apply those insights to your own teaching?

ALGEBRA II, TRIGONOMETRY, AND PRECALCULUS

Algebra II courses—which usually include trigonometry—are typically filled with high school students who have completed algebra I and geometry and intend to pursue college degrees. Case 11.14 is an interview in which a teacher explains her ideas of how

Exhibit 11.35
Continued

Team V's Tasksheet

1. Pick up the box of supplies labeled "Team V" and bring it to your work location.
2. Note that your box contains protractors, rulers, paper, pencils, and geoboards with rubber bands.
3. Use protractors and rulers to draw as many different sizes and shapes of triangles as you can with a 50° angle and a 13-cm side attached to it and also a 40° angle that is opposite the 13-cm side.
4. Answer: How many triangles with different shapes or sizes were you able to draw in response to Prompt #3? ______
5. Use protractors and rulers to draw as many different sizes and shapes of triangles as you can with a 70° angle and an 8-cm side attached to it and also a 22° angle that is opposite the 8-cm side.
6. Answer: How many triangles with different shapes or sizes were you able to draw in response to Prompt #5? ______
7. Experiment with the geoboards and rubber bands to help you develop an explanation as to why the following two statements are true: (a) Every triangle you drew with a 50° angle and a 13-cm side attached to it and also a 40° angle that is opposite the 13-cm side had the same size and shape. (b) Every triangle you drew with a 70° angle and an 8-cm side attached to it and also a 22° angle that is opposite the 8-cm side had the same size and shape.
8. Write your explanation: ______
9. Write a conjecture about how two angles and a side affect a triangle's size and shape: ______

Team VI's Tasksheet

1. Pick up the box of supplies labeled "Team VI" and bring it to your work location.
2. Note that your box contains protractors, rulers, paper, pencils, and geoboards with rubber bands.
3. Use protractors and rulers to draw six different-shaped triangles so that all six triangles have an angle that measures 65°, a 10-cm side, and 16-cm side.
4. Use protractors and rulers to draw six different-shaped triangles so that all six triangles have an angle that measures 100°, a 20-cm side, and 10-cm side.
5. Experiment with the geoboards and rubber bands to help you develop a conjecture about the effect of two sides and one angle on the triangle's shape and size.
6. Write a conjecture about the effect of two sides and one angle on the triangle's shape and size: ______

Continued

algebra II should be organized and taught. However, what should happen in algebra II is influenced by the school's offerings in trigonometry, precalculus, and other so-called "college-preparatory" mathematics courses. Thus, the interview includes discussions about trigonometry, precalculus, and other related courses:

CASE 11.14

Mr. Rudd's colleague Vanessa Castillo teaches algebra II at Rainbow High School. I interviewed her about the course:

Me: What is the main thing students need to get out of algebra II?

Exhibit 11.35
Continued

Team VII's Tasksheet

1. Pick up the box of supplies labeled "Team VII" and bring it to your work location.
2. Note that your box contains protractors, rulers, paper, pencils, and geoboards with rubber bands.
3. Use protractors and rulers to draw six different isosceles triangles that differ from one another in shape and size.
4. For each of the six isosceles triangles, use protractors to measure its three angles. Record the results just below:
   The measures of the three angles of the first triangle: _____, _____, & _____.
   The measures of the three angles of the second triangle: _____, _____, & _____.
   The measures of the three angles of the third triangle: _____, _____, & _____.
   The measures of the three angles of the fourth triangle: _____, _____, & _____.
   The measures of the three angles of the fifth triangle: _____, _____, & _____.
   The measures of the three angles of the sixth triangle: _____, _____, & _____.
5. Experiment with the geoboards and rubber bands to help you develop a conjecture about the effect of having two congruent sides of a triangle on the sizes of the triangle's angles.
6. Write a conjecture about the affect of having two congruent sides of a triangle on the sizes of the triangle's angles: ________________________________
   ____
   ____
   ____
   ____

Team VIII's Tasksheet

1. Pick up the box of supplies labeled "Team VIII" and bring it to your work location.
2. Note that your box contains protractors, rulers, paper, pencils, and geoboards with rubber bands.
3. Use protractors and rulers to draw six different triangles so that the first triangle has two 40° angles, the second has two 45° angles, the third has two 25° angles, the fourth has two 60° angles, the fifth has two 75° angles, and the sixth has two 95° angles.
4. For each of the six triangles, use rulers to measure its three sides. Record the results just below:
   The lengths of the three sides of the first triangle: _____, _____, & _____.
   The lengths of the three sides of the second triangle: _____, _____, & _____.
   The lengths of the three sides of the third triangle: _____, _____, & _____.
   The lengths of the three sides of the fourth triangle: _____, _____, & _____.
   The lengths of the three sides of the fifth triangle: _____, _____, & _____.
   The lengths of the three sides of the sixth triangle: _____, _____, & _____.
5. Experiment with the geoboards and rubber bands to help you develop a conjecture about the effect of having two congruent angles of a triangle on the lengths of the triangle's sides.
6. Write a conjecture about the effect of having two congruent angles of a triangle on the lengths of the triangle's sides: ________________________________
   ____
   ____
   ____
   ____

Continued

Exhibit 11.35
Continued

Note 3: I will set up the eight work stations and eight labeled boxes of material prior to the start of class.

The Four-Stage Lesson Plan:

1. Experimenting: I'll partition the class into the 8 teams and conduct the cooperative-learning session as indicated in Notes 1–3. As they work, I'll move among the groups to clarify tasksheet directions, deal with any disengagement, monitor progress, and gauge when to make the transition to a large-group question-discussion session of Stage 2. During the transition, I'll direct the chairpersons to return the materials to the boxes while the rest of the class rearranges the desks for large-group activities.
2. Reflecting and explaining: I'll direct one student to record team findings and conjectures on the overhead as I move about the room conducting the question-discussion sessions. Each team will spend about 5 minutes reporting the results of its experiment and sharing its conjectures. We won't complete this activity before the end of the 1st day, but we will pick it up wherever we left off on the 2nd day.
   There will be no homework assignment directly related to this lesson; however, students who are revising their responses to Unit 7's test prompts will do that for homework.
   We'll complete the question-discussion session on the second day and move to Stage 3 as soon as the final team report is made.
3. Hypothesizing and articulating: On the 2nd day of the lesson, the question-discussion session will continue into Stage 3 with my overhead-projector scribe recording conjectures as they develop during the session. It's okay if we formulate only three or four instead of all six relationships. We'll move on to the 4th stage with whatever propositions do get formulated by the end of the second day and then return to this stage for missing propositions when it seems appropriate later in the unit.
4. Verifying and refining: We'll use *Geometer's SketchPAD* as well as virtual manipulatives from the National Library of Virtual Manipulatives for Interactive Mathematics (www.matti.usu.edu) to experiment with many more examples to inductively test our conjectures and search for counterexamples. We'll begin this work in class and continue it with a homework assignment. For the homework assignment, I'll check out the available calculators with the *SketchPAD* software to students who don't have other access to the technology for the assignment (i.e., those who don't have their own programmed calculators, can't get to the computer lab, and don't have at-home computers with internet access or *SketchPAD).*
   The following day we'll use the results of the homework assignment to finalize our propositions in a question-discussion session. Then we'll make the transition into the lesson for Objective B with whatever propositions we have in hand at that point.

Ms. Castillo: Many people think that high school mathematics curricula have two purposes: to prepare college-bound kids for calculus and provide some fundamental everyday useful skills to kids who don't plan to continue with mathematics in college. I don't particularly agree with that, but for kids who do need to be ready for calculus—either AP calculus in high school or beginning calculus in college, algebra II is where it's at.

Me: What do you mean?

Ms. Castillo: From my experience teaching calculus at the university and AP calculus here, students struggle with calculus—not because they don't understand the new content like limits and derivatives—but because their algebra skills are atrocious. If someone really understood the topics from any good algebra II textbook that also includes trigonometry as they all do now, they would do fine in any reasonably well-taught beginning calculus course.

Me: What about precalculus?

Ms. Castillo: Because of its name, I used to think of precalculus as the bridge between algebra and calculus, just as prealgebra is supposed to be the bridge between arithmetic and algebra. In spite of the name, precalculus courses—if they're offered at all—are more of a hodgepodge of topics like probability and combinatorics, graph theory, iteration and fractals, matrices, mathematical induction, truth tables, sequences and series, and statistics—all useful, important mathematics, but not as directly related to calculus as the more focused algebra II courses.

Me: Then precalculus includes more topics from discrete mathematics than algebra II. And because calculus is the mathematics of continuous functions, algebra II is more directly related to calculus.

Ms. Castillo: I agree, but I'm not saying students don't need discrete mathematics. It's far more relevant to everyday life than conic sections and the behavior of continuous functions—just not as relevant to the study of calculus. Precalculus is the most enjoyable course for me to teach, because I have more freedom to select topics. But with algebra II, I have the pressure of knowing that what I teach them will directly affect their success with calculus.

Me: You prefaced one of your statements about precalculus courses with "if they're offered at all;" please elaborate.

Ms. Castillo: Many high schools offer a course called "algebra III," "college algebra," or "advanced high school mathematics" instead of one called "pre-calculus." "Advanced high school mathematics" is just another name for the traditional precalculus courses like we teach here at Rainbow with that hodgepodge of topics. The algebra III or college mathematics courses are usually more of an extension of algebra II—or in some cases—the content of algebra II is just extended to two courses over 1 1/2 or 2 school years.

Me: What about trigonometry?

Ms. Castillo: Students are now picking up bits and pieces of trig in middle school; but I don't think they are getting much out of it. Within my algebra II course, I include all the trig anyone ever needs to know for calculus, physics, surveying—you name it.

Me: I'm getting the impression that you're not a big fan of trigonometry.

Ms. Castillo: I'm a huge fan of trig. I just think there's much less trig to be learned than typical textbooks lead people to think. It's ridiculous to have a full semester devoted to trig as do some schools.

Me: Please elaborate.

Ms. Castillo: Trig courses or sequences of trig units in other courses usually begin with the six fundamental trigonometric functions—either with stand-alone right triangles or in a rectangular coordinate plane. Then we look at some manipulations of those functions and then apply them to physics problems—so far, so good. Then we get into laws of sines and cosines, more applications, and then graphing trig functions—still okay. But then we lose our minds with trigonometric identities supposedly getting kids to "prove" all kinds of relationships involving the six trig functions and the Pythagorean theorem. Again, the content is okay, but no purpose is served by memorizing identities. I don't remember them myself, but I can generate the identities because I have algebra skills and I remember the six trig functions and the Pythagorean theorem. Instead of memorizing them, my students make a list of the identities as we develop them in class. Then they refer to the list whenever they need them. It's the most ridicu–

Me: You're getting carried away with how trig is often mis-taught. Let's talk about how it should be taught.

Ms. Castillo: Thank you. Okay, so we've taught them the manipulations with the basic functions, graphing—which they really need for calculus courses—and some work with identities all mixed in with lots of good application. But then, if we follow the textbooks, we teach it all over again as trigonometric inverses and then again when we move from rectangular to polar coordinates.

Me: Do you suggest that we stop teaching inverses of trigonometric functions?

Ms. Castillo: Of course not. But just as multiples should be taught right along with divisors because it's simply two ways to express the same idea, saying "$\cos \theta = a$" is the same as saying "$\arccos a = \theta$." So I teach trig inverses right along with straight trig functions. The kids never miss a beat—just view it as two ways to say the same thing.

Me: So do you also teach polar graphing and radians right along with rectangular graphing and degrees?

Ms. Castillo: Even though "$\theta = 90°$" says the same thing as "$\theta = \pi/2$," I begin with the more familiar degree measures. As soon as we move from stand-alone right triangles to rotating angles in a unit circle, they learn radians as another way of measuring angles right along with degrees. And although polar coordinates are just another way of expressing relationships just like rectangular coordinates, I don't bring up polar graphing until they're thoroughly familiar with rectangular graphing. It's just a complication I delay. However, when we do get into polar graphing, I make sure they maintain the connection to the rectangular system by having them frequently translate between the two. Graphing calculators really facilitate that. They get into the habit of asking themselves, "What would that look like in the rectangular system?"

Me: Let me make sure I understand one of your major points about trig courses. The content is critically important, but by reorganizing typical textbook presentations—like mixing trigonometric functions with their inverses and not making such a big deal of identities—we can lead students to understand what they need to understand for physics and calculus in a way that's more efficient than the way it's done in most trig courses. Am I understanding your suggestion?

Ms. Castillo: Yes, but with one more point: Real-life applications must be included throughout. One of the reasons I enjoy teaching trig units is that they readily

Exhibit 11.36
Ms. Castillo's Sequence of Units for Algebra II.

School Weeks	Unit #	Unit Title
1–3	1	A Review of Algebra I and Geometry
4–5	2	Functions and Other Relations
6–8	3	Systems of Linear Relationships and Matrices
9–11	4	Quadratic and Higher Degree Equations
12–15	5	Part I of Quadratic Functions and Other Relations
16–21	6	Part II of Quadratic Functions and Other Relations
22–26	7	Part I of Trigonometric Functions
27–31	8	Part II of Trigonometric Functions
32–34	9	Exponential and Logarithmic Functions
35–36	10	Extending What We Learned

lend themselves to problems in which kids can use on-site measurements to address questions like, "If a storm knocks over that tree outside our window, will it come crashing through our classroom?" It helps me with classroom management to get students out of their chairs to gather data and confer with one another; it's easy to do that with trig.

Me: May I see the sequence of units for your algebra II course?

She shows me Exhibit 11.36.

Me: You have 10 weeks allotted for trigonometry. Is that adequate?

Ms. Castillo: It's enough time for them to do the trig they need to be successful with their physics course and to be ready for calculus.

Me: I see you've deviated considerably from the textbook's table of contents. It looks like you're leaving out quite a few topics.

Ms. Castillo: At Rainbow, the mathematics department decided to put all our get-ready-for-calculus eggs into our algebra II basket. There isn't an algebra or trigonometry topic in the textbook that's not included. It just looks like less because I've combined so many topics to maintain connections.

Me: Like trigonometric functions with their inverses.

Ms. Castillo: The textbook's topics—such as combinatorics—that are actually excluded are included in precalculus at Rainbow.

Me: Doesn't deviating from the textbook create a lot more work for you?

Ms. Castillo: Not as much work as cleaning up the mess that following the textbook creates in kids' minds. I learned a long time ago to use textbooks for the exercises and examples they provide, but not to depend on the explanations or follow their sequences.

Me: Earlier, you emphasized trigonometric applications. I assume you do the same with algebra.

Ms. Castillo: Absolutely! I find the most efficient way to do real-life applications is to integrate non-mathematics.

Me: But with huge variations in high school students' schedules, that's not as easy to do as it is in middle schools.

Ms. Castillo: So true—so I've begun doing more integration with courses they're not taking. I first got the idea from a video you gave me (Cangelosi, 2001b; Lindahl & Cangelosi, 2000). Mindy Bleir teaches beginning biology for mostly ninth graders. For her unit on ecosystems, my algebra II students served the role of applied mathematicians collaborating with biologists. The biologists—that's Mindy's students—would collect data from experiments with creatures like flour beetles and then bring the data to my algebra II class to create functions that modeled population growth.

Me: That must really fit well with your Unit 9 on exponential and logarithmic functions.

Ms. Castillo: It fits well with all but the trig units. The only difficulty was that my students also wanted to be involved in the data collection. They weren't satisfied just using Mindy's students' data. So, for the second semester we got Harriet [the principal] to lean on the scheduling people so that the two classes are held the same period. My students and hers formed teams that occasionally met together in her lab for data-collection activities (see Exhibit 11.37). It works so cool!

Me: For your students to be able to apply mathematics to such real-life situations—formulating functions to model phenomena and make predictions, you must really hit the construct-a-concept and discover-a-relationship objectives hard.

Exhibit 11.37
Ms. Castillo's Algebra II Students Collaborate With Ms. Bleir's Biology Students.

Ms. Castillo: Absolutely. For example, when we do conic sections in Unit 5, I have students work with flashlights in the experimenting stage so that they discover that a parabola is an ellipse with its two foci infinitely far from each other.

See Exhibit 11.38.

Engage in Activity 11.12:

Activity 11.12

In a discussion with a colleague, address the following questions about Case 11.14:

1. If you were in Ms. Castillo's place, how would you modify Exhibit 11.36's course outline?
2. How consistent with yours is Ms. Castillo's perception of the relationships among algebra I, geometry, algebra II, precalculus, algebra III, and calculus?
3. Do you agree with Ms. Castillo's comments about the role of trigonometry in mathematics curricula? Why or why not?
4. Have you tried or do you plan to try the type of integrating between courses that Ms. Castillo and Ms. Bleir use?
5. What insights did you gain from your work analyzing Cases 11.14? How do you plan to apply those insights to your own teaching?

A LIFE-SKILLS MATHEMATICS COURSE

Unfortunately, there has been a long-standing tradition in high schools that courses with names such as "consumer mathematics," "mathematics for life," "basic mathematics," "applied mathematics," "applied algebra," "applied algebra and geometry," and "life skills" are the dumping grounds for students who neither like nor do well in mathematics but who need mathematics credit to graduate. Students, as well as some thoughtless faculty members, referred to these courses as "dummy math." A more appropriate name may have been "terminal mathematics" because they were not intended to prepare students for subsequent mathematics courses. They were thought of as "the last math course you will ever have to take."

Fortunately, this tradition is dying throughout the country and is already dead in many high schools with enlightened mathematics departments. The demise of this ugly tradition has been hastened by NCTM's push toward meaningful mathematics for all students—not only the so-called "college bound" and "mathematically inclined" (NCTM, 1989a; 2000b). Typically, these courses emphasize real-life applications, use of technology, and preparation for careers. With the phrases "real-life applications" and "use of technology" appearing repeatedly in recommendations for leading students to do meaningful mathematics (e.g., *PSSM* [NCTM, 2000b]), the status of courses such as life-skills mathematics has been raised, as they now seem to have been ahead of their time. The *School-to-Careers* movement (Goldberger & Kazis, 1996; Gray, 1996; NCTM, 1997) has also provided financial support to curriculum-reform projects that focus high school curricula more directly on preparing students for occupations than has been traditionally the case.

Some high schools have supplanted courses such as life-skills mathematics with algebra and geometry courses that emphasize applications more than their standard algebra and geometry courses. Other high schools have introduced mathematics courses that specifically focus on technical careers. This second approach is more in line with the *School-to-Careers* movement and is reflected by some of the newer textbook titles (e.g., *Basic Mathematics for Occupational and Vocational Students* [Spangler, 2001] and *Mathematics Applied to Electronics* [Harder, 2001]).

In Case 11.15, Mr. Rudd reflects on the one-semester life-skills mathematics course he teaches at Rainbow High School:

CASE 11.15

Unlike textbooks for his other courses, the life-skills mathematics text is organized more by daily-living topics (e.g., "Earning Money" and "Housing, Taxes, and Insurance") than by mathematical topics (e.g., "Solving Open Sentences" and "Ratios and Proportions"). He considers this arrangement inconsistent with *PSSM* and his conception of how the course should be organized. However, as he prepares the course, he is not as comfortable with the content as he is with his other courses, so he feels more dependent on the textbook. He

Exhibit 11.38

Ms. Castillo's Students Use Flashlights During the Experimenting Stage of a Discover-a-Relationship Lesson on Conic Sections.

A circle is an ellipse with foci 0 apart.

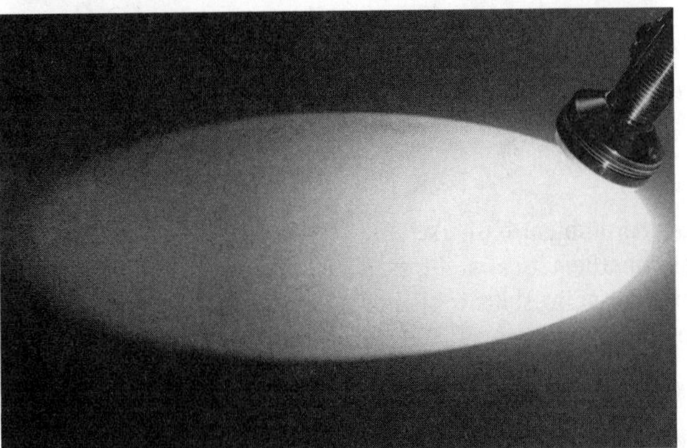

The foci of an ellipse that is neither a circle nor a parabola are a positive distance apart.

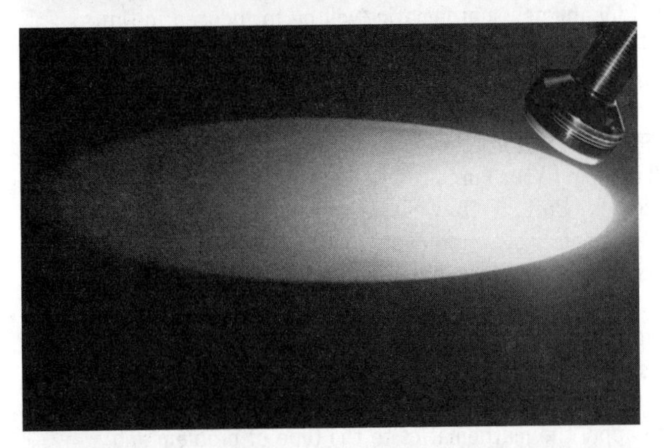

A parabola is an ellipse with foci infinitely far apart.

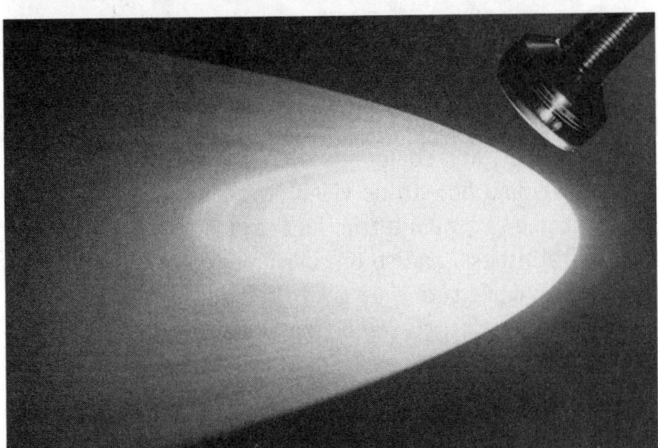

Exhibit 11.39
Mr. Rudd's Sequence of Units for Life-Skills Mathematics.

School Weeks	Unit #	Unit Title
1	1	Using Mathematics to Make Decisions
2	2	Measurements and Numbers
3	3	Calculating, Organizing Data, and Using Formulas
4–6	4	Earning, Borrowing, Saving, and Investing Money
7–8	5	Spending Wisely
9–10	6	Acquiring and Maintaining a Place to Live
11–12	7	Taxes
13–14	8	Insurance
15	9	Transportation and Travel
16–17	10	Sports and Entertainment
18	11	Extending What We Learned

plans a problem-based strategy to lead students to build their mathematical abilities, skills, and attitudes using the textbook's life-skills topic headings to motivate them to do mathematics.

Exhibit 11.39 is the course outline he develops.

Two weeks into the course, Mr. Rudd is quite pleased with the class' progress. However, he is concerned that hardly any new mathematics will be introduced in Units 5–10; they focus on applying previously learned mathematics to new situations. He re-previews some of the textbook's section headings (e.g., "Doing Income Tax," "Buying a Motorcycle," "Planning a Vacation," and "Financing Your Purchase") and thinks: "If I follow the book for Units 5–10, we'll be repeating the same mathematics over and over. That could get monotonous. Maybe I should farm out some of these topics to collaborative teams. With this class, teams of three work pretty well—no more than four for sure. With 31 students, I should pick on 9 or 10 types of problems. Each team would be responsible for searching out information on how to apply the mathematics to the type of problem and teach the rest of the class. I'd like to spice it up with some kind of project. But what?

"I've got it! Some of them love to perform in front of the video camera and almost all of them are interested in televison, videos, and computers. I'll assign each team a topic to develop a how-to-do video program—like how to do your income tax, plan a trip, buy car insurance, assess your physical fitness, and so on. I'll have to set tight parameters to make sure they get to the application of relevant mathematics, clearly define the audience—that stuff. All the presentations must emphasize how the mathematics is done."

In his initial excitement with the idea, Mr. Rudd thinks of using the approach for Units 5–10. However, his excitement is soon tempered by the realization that such a plan requires more long-range preparation than he has time left in the course. Thus, he decides to try out the idea with only one unit to be run concurrently with other units that he will teach as previously planned. He selects Unit 9, "Transportation and Travel," for the experiment. Exhibit 11.40 is the overall unit plan that will run concurrently with Units 6–8 and 10.

At the conclusion of the unit, Mr. Rudd decides that the video program projects worked so well that he will expand the approach next time he teaches the course and he will also try it with other courses.

Engage in Activity 11.13:

Activity 11.13

In a discussion with a colleague, address the following questions about Case 11.15:

1. If you were in Mr. Rudd's place, how would you modify Exhibit 11.39's course outline?
2. How would you design a sequence of mathematics courses for high school students who either do not plan to attend college or whose college plans do not require calculus?
3. What are the advantages and disadvantages of the way Mr. Rudd decided to implement a problem-based approach with the "how-to-do" video projects?
4. How consistent is Exhibit 11.40's unit plan with suggestions from chapters 4–8?
5. What insights did you gain from your work analyzing Case 11.15? How do you plan to apply those insights to your own teaching?

Exhibit 11.40
Mr. Rudd's Plan for Unit 9 to Run Concurrently with Other Life-Skills Mathematics Units.

Course: Life-Skills Mathematics

Unit 9 (of 11): Transportation and Travel

Goal: Students apply their talents for doing mathematics to situations involving transportation and travel.

Objectives:

A. The student incorporates into her working vocabulary words and phrases relevant to communicating information, analyses, and relationships relative to transportation-and-travel projects on which students work as part of this unit* (comprehension and communication) 10%.

   * - Tentatively each project will develop a "how-to" video program to accomplish one of the following:

   i) Efficiently plan your trip.
   ii) Determine the best place to travel for your purposes.
   iii) Maximize the cost effectiveness of your travel by automobile.
   iv) Optimize the safety-to-efficiency factors when traveling by automobile.
   v) Obtain the best lodging deal for your purposes when traveling.
   vi) Obtain the best deal for your purposes when traveling by train.
   vii) Obtain the best deal for your purposes when traveling by bus.
   viii) Obtain the best deal for your purposes when renting an automobile.
   ix) Secure the best deal for your purposes when traveling by airplane.
   x) Budget for your trip.

B. The student explains the rationale underlying fundamental mathematical formulas used in solving transportation-travel related problems (discover a relationship) 30%.

C. The student executes algorithms based on fundamental relationships used in solving transportation-travel related problems (algorithmic skill) 20%.

D. Given a transportation-travel related problem, the student explains how, if at all, mathematical relationships and algorithms studied in this course can be used to solve the problem (application) 40%.

Estimated number of class periods: Parts of about 12 periods spread over the final 10 weeks of the semester

Textbook page references: 263–301 and 336–339

Overall plan for lessons:

The class will be subdivided into 10 teams of three or four each, with each group being assigned one of the 10 transportation-travel topics. Each team is to undertake a project in which they produce a videotape program (approximately 25 minutes long) designed to help the rest of the class achieve the four objectives of the unit relative to the assigned topic. From design through production and presentation, I will supervise each group's project.

In a large-group presentation, I'll (a) provide the class with an overview of the plans for the unit, (b) organize the 10 teams (based on selections I make prior to class—ideally, each team will have a student who has had a relatively easy time with mathematics and one who does not but who is knowledgeable about the travel-transportation area), (c) distribute resources, and (d) explain and have them begin the initial assignment for each team (which will involve reading the relevant section of the textbook and gathering background information).

Each team will report results of the initial assignment to the class and then I will specify the parameters for the projects, explaining what is to be included and outlining each of the following phrases: information gathering, analyzing, designing, script writing, evaluating/refining, and video program production.

In small-task-group sessions, each team will develop, evaluate, and refine a data gathering, development, and production plan and have it critiqued and approved by me.

Students will engage in data-gathering activities, including visiting off-campus sites as necessary and practical.

In small-task-group sessions, each team will complete its production according to its approved plan. Productions are scheduled and presented to the class as a whole as soon as reasonably possible upon completion. Students observing a production are responsible for learning from the presentations in order to achieve the unit's objectives. I will conduct a question-discussion session on each of the 10 topics after it is presented. Brief tests will be intermittently scheduled for formative feedback.

A unit test will be administered and the results reviewed with the class.

Extraordinary learning materials and resources needed:

(a) Informational resources (e.g., travel and buyers' guides), (b) arrangements for students to visit automobile dealers, motorcycle dealers, recreational vehicle dealers, travel bureaus, auto rental outlets, and motor vehicle safety office, and (c) equipment and supplies for videotaping in the classroom over an eight-week period.

AP AND CONCURRENT-ENROLLMENT MATHEMATICS COURSES

AP calculus and AP statistics courses are included in high schools to lead students to do meaningful calculus and statistics as well as to prepare them for the exams alluded to chapter 9's section "AP Calculus and Statistics Tests." The content of the tests largely determines the mathematical topics addressed by the courses. Typical AP calculus textbooks mimic the first two thirds of standard calculus textbooks used for the three-semester sequence of beginning calculus courses taught at colleges and universities. AP statistics textbooks are quite similar to those used in introductory, noncalculus-based college statistics courses.

By passing an AP exam, students are able to fulfill certain requirements for a college degree. Through cooperative arrangements among some colleges and high schools, students can also earn credit directly at those colleges by taking concurrent-enrollment courses. College-level algebra, trigonometry, pre-calculus, and statistics are the more common mathematics courses that high schools offer under concurrent-enrollment arrangements. Typically, college mathematics departments control the mathematical content of these courses; under some arrangements, the high school students may be required to take the same exams as college students who take the college's version of the course. Although you are required to include the same mathematical topics as the college-based version of the course, you are not required—nor should you—teach your students in the same manner that college courses are typically taught. The research-based strategies consistent with *PSSM* (NCTM, 2000b) and explained in this text apply to any mathematics course.

SYNTHESIS ACTIVITIES FOR CHAPTER 11

1. You developed and revised mathematics curriculum materials (e.g., lesson plans and miniexperiments) and stored them in your working portfolio as you have worked with chapters 1–10. In light of your work with Chapter 11, revise them as you think you should.
2. Consider sharing and discussing some of the mathematics curriculum materials you have developed or some of your innovative ideas on teaching mathematics by preparing a proposal for a presentation at a regional NCTM conference or a local meeting of mathematics teachers.
3. Enjoy a productive career as a professional mathematics teacher.

Glossary

Affective domain: If the intent of the objective is for students to develop a particular attitude or feeling (e.g., a desire to prove a theorem or willingness to work toward the solution of a problem), the learning level of the objective falls within the *affective domain.*

Algorithm: An *algorithm* is a multistep procedure for obtaining a result.

Algorithmic-skill learning level: Students achieve an objective at the *algorithmic-skill* level by remembering and executing a sequence of steps in a specific procedure.

Allocated time: The time periods during which a teacher intends for students to be involved in learning activities are referred to as *"allocated time."*

Application learning level: Students achieve an objective at the *application level* by using deductive reasoning to decide how to utilize, if at all, a particular mathematical content to solve problems.

Appreciation learning level: Students achieve an objective at the *appreciation level* by believing the mathematical content specified in the objective has value.

Assertive communication: A person's communications are *assertive* when he sends exactly the message that he wants to send, being neither *passive* nor *hostile.*

Blueprint, measurement: A *measurement blueprint* is an outline specifying the features a teacher wants to build into a unit test. Typically, the blueprint indicates (a) the title of the unit, (b) anticipated administration dates and times, (c) provisions for accommodating students with special needs, (d) approximate number of and types of miniexperiments to be included, (e) an approximation of the maximum possible score for the measurement, (f) how points should be distributed among the objectives that define the goal (based on the weights of the objectives), (g) the overall structure of the measurement, and (h) for summative evaluations, the method for converting scores to grades.

Businesslike classroom: A learning environment in which the students and teacher conduct themselves as if achieving learning goals takes priority over other matters is *businesslike.*

Classroom management: *Classroom management* is the complex art teachers practice for the purpose of controlling their learning environments and teaching students to be on-task.

Cognitive domain: If the intent of the objective is for students to be able to do something mentally (e.g., remember a formula or deduce a method for solving a problem), the learning level of the objective falls within the *cognitive domain.*

Comprehension-and-communication learning level: Students achieve an objective at the *comprehension-and-communication level* by (a) extracting and interpreting meaning from an expression, (b) using the language of mathematics, and (c) communicating with and about mathematics.

Concept: A *concept* is a category people mentally construct by creating a class of specifics possessing a common set of characteristics (i.e., a concept is an abstraction).

Concept attribute: A *concept attribute* is a characteristic common to all examples of a particular concept; a concept attribute is a necessary requirement for a specific to be subsumed within a concept.

Construct-a-concept learning level: Students achieve an objective at the *construct-a-concept level* by using inductive reasoning to distinguish examples of a particular concept from nonexamples of that concept.

Constructivism: *Constructivism* subsumes a variety of perspectives emphasizing the importance of learners constructing knowledge (e.g., constructing concepts and discovering relationships) for themselves from their own experiences.

Convergent reasoning: *Convergent reasoning* is typical reasoning producing predictable responses for most people.

Cooperative learning: *Cooperative learning* is students engaging in activities in which they collaborate for the purpose of learning from one another.

Core-curriculum test: A *core-curriculum test* is a measurement administered statewide or districtwide that is used to make criterion-referenced evaluations about how well statewide or districtwide curriculum goals or standards have been achieved.

Creative-thinking learning level: Students achieve an objective at the *creative-thinking level* by using divergent reasoning to view mathematical content from unusual and novel ways.

Criterion-referenced evaluation: A summative evaluation of a student's achievement is *criterion referenced* if the evaluation is influenced by how the measurement results relevant to that student's achievement compare to a standard that is not dependent on results obtained from others.

Curriculum:

- A *school curriculum* is a system of planned experiences (e.g., coursework, school-sponsored social functions, and contacts with school-sponsored services [e.g., the library]) designed to educate students.
- A *course curriculum* is a sequence of *teaching units* designed to provide students with experiences that help them achieve specified learning goals.
- A *mathematics curriculum* is a sequence of mathematics courses as well as other school-sponsored functions (e.g., a mathematics club) for the purpose of encouraging students to do mathematics.
- A *school district curriculum* is the set of all school curricula within the school district.
- A *state-level curriculum* is the set of all school district curricula within a state.

Deductive-learning activity: A *deductive-learning activity* is a part of an inquiry lesson that leads students to reason deductively.

Deductive reasoning: *Deductive reasoning* is deciding that a specific or particular problem is subsumed by a generality. In other words, it is the cognitive process by which people determine whether what they know about a concept or abstract relationship is applicable to some specific situation.

Descriptive language: *Descriptive language* verbally portrays a situation, behavior, achievement, or feeling.

Destructive positive reinforcer: A positive reinforcer for one behavior is *destructive* if it has undesirable side effects on another behavior.

Destructive punishment: A punishment for one behavior is *destructive* if it has undesirable side effects on another behavior.

Direct-instructional strategy: A teacher employs *direct-instructional strategies* by exposing students to the information or algorithm to be remembered and then engaging repetitive activities to commit the information or algorithm to memory.

Discover-a-relationship learning level: Students achieve an objective at the *discover-a-relationship level* by using inductive reasoning to discover that a particular relationship exists or why the relationship exists.

Disruptive behavior: A student's behavior is *disruptive* if it is off-task in such a way that it interferes with other students being on-task.

Divergent reasoning: *Divergent reasoning* is atypical reasoning that deviates from common ways of thinking. It is thought that produces unanticipated and unusual responses.

Engagement, student: A student exhibits *engaged* behavior by being on-task during allocated time. In other words, whenever a student is attempting to participate in a learning activity as planned by the teacher, the student is *engaged in the learning activity*.

Error-pattern analysis: *Error-pattern analysis* is a strategy mathematics teachers employ to detect exactly how students are executing an algorithm.

Example noise: *Example noise* is any characteristic of an example of a concept that is not an attribute of that concept.

Extrinsic motivation: A student's motivation for engaging in a learning activity is *extrinsic* if the engagement is prompted by a desire to receive the rewards that have been artificially associated with engagement or to avoid consequences artificially imposed on those who are off-task rather than because he recognizes value in experiencing the activity.

Formative judgment: A *formative judgment* is a decision made by a teacher that influences how he teaches.

Goal, learning or Unit: The *learning goal* is the overall purpose of a teaching unit. It indicates what students are expected to gain if the teaching unit is successful.

Grade-equivalent score: A *grade-equivalent score* of *g.m* is associated with a raw score x_i if x_i is approximately equal to the mean of the raw scores of those students in the norm group who were in the *g.m grade level* when they took the test.

Grading, compromise: With the *compromise grading method* for converting measurement scores to grades, cutoff points for each letter grade are established prior to the administration of the measurement. All scores that are only insignificantly below or insignificantly above a grade cutoff point according to some criterion (e.g., within one *standard error of measurement*) are assigned the grade associated with that cutoff point. A significant difference between two scores is required in order for the two scores to be assigned different grades.

Grading, traditional percentage: With the *traditional percentage grading method* for converting measurement scores to grades, each score from a measurement is converted to a percentage of the maximum possible score and translated to a letter grade by some predetermined scale (e.g., 94% to 100% for an A, 86% to 93% for a B, 78% to 85% for a C, 70% to 77% for a D, and 00% to 69% for an F).

Grading, visual inspection: With the *visual-inspection grading method* for converting measurement scores to grades, the frequency distribution of the scores from a measurement is graphed so that grades are assigned according to gaps in the distribution. Every score within a single cluster of scores is assigned the same grade with clusters of higher scores assigned higher grades and clusters of lower scores assigned lower grades.

High-stakes testing: A test is considered *high stakes* if it is administered statewide or districtwide for the purpose of (a) accountability for student achievement or (b) evaluations affecting promotion, retention, or graduation (U.S. General Accounting Office, 1993).

Hostile communication: A person's communication is *hostile* rather than assertive if it is intended it to be intimidating or insulting.

IEP (Individualized Education Program): An *IEP* is an evolving written agreement between the parents of a student with a special-education classification and a school that specifies an assessment of the student's present level of functioning, long- and short-term goals, services to be provided, and plans for delivering and evaluating those services.

Inductive-learning activity: An *inductive-learning activity* is a part of an inquiry lesson that leads students to reason inductively.

Inductive reasoning: *Inductive reasoning* is generalizing from encounters with specifics. It is the cognitive process by which people discover commonalities among specific examples, thus, leading them to formulate abstract categories (i.e., concepts) or discover abstract relationships.

Inquiry instructional strategy: A teacher employs *inquiry instructional strategies* by engaging students in activities in which they interact with information, make observations, and formulate and articulate ideas that lead them toward discovery, concept construction, or invention.

Inservice mathematics teacher: A person who holds a position as a mathematics teacher in a middle, junior high, or high school is an *inservice mathematics teacher.*

Internal consistency, measurement: A measurement is *internally consistent* to the degree that results from its miniexperiments are in harmony (i.e., noncontradictory and in agreement).

Interpretive understanding: Students understand a message at an *interpretive level* if they can infer implicit meaning and explain how aspects of the communications are used to convey the message.

Intrinsic motivation: Students are *intrinsically motivated* to engage in learning activities if they recognize that by experiencing the activity they will satisfy a need. The learning activity itself is perceived to be valuable.

IRE (initiate-respond-evaluate) cycle: An *IRE cyle* is part of a conversation in which one person (e.g., a teacher) initiates by prompting others (usually students) to respond. The response is then evaluated by the initiator (e.g., the teacher).

Judgmental language: *Judgmental language* verbally summarizes an evaluation of a person, achievement, or behavior with a characterization or label.

Learning level, objective's: An objective's *learning level* is the manner in which students will mentally interact with the objective's mathematical content once the objective is achieved.

Learning-level relevance, measurement: A measurement has *learning-level relevance* to the degree that its miniexperiments require students to operate at the learning levels specified by the learning objectives.

Lesson: A *lesson* is a sequence of learning activities designed to lead students to achieve an objective.

Literal understanding: Students *literally understand* a message if they can accurately translate its explicit meaning.

Mathematical content, objective's: The mathematical concept, relationship, information, or algorithm on which a learning objective focuses is the *objective's mathematical content.*

Mathematical-content relevance, measurement: A measurement has *mathematical-content relevance* to the degree that its miniexperiments involve students in the mathematics specified by the learning objectives.

Meaningful learning of mathematics: Learning mathematics is *meaningful* to students if students apply that mathematics to situations they consider important.

Measurement: A *measurement* is a process by which data or information are gathered via empirical observations and those data or information are recorded or remembered.

Measurement, planned: A *planned measurement* is a sequence of miniexperiments a teacher deliberately conducts for the purpose of collecting information to be used to make a formative judgment or summative evaluation.

Measurement error: *Measurement error* is the difference between the results of a measurement and what the measurement results would be if the measurement had been perfectly valid. In a measure of student achievement, measurement error relates to measurement results and actual student achievement as follows:

$S_j = T_j + E_{s_j}$ where S_j is the result of the measurement, T_j is the student's actual true achievement level of the objective, and E_{s_j} is the error of measurement associated with S_j.

Measurement result: The memory or record of the information or data yielded by a measurement is a *measurement result.*

Measurement, unplanned: *Unplanned measurements* are the continual flow of information from ongoing empirical observations that occur in the course of daily activities. Unplanned measurements influence decision making although they are not deliberately designed for that purpose.

Memory-level question: Questions that can be answered by remembering previously learned responses are *memory-level questions.*

Message, mathematical: A *mathematical message* is a specific communication transmitted among people that is intended to convey information, ideas, directions, arguments, or meanings involving mathematics.

Miniexperiment: A *miniexperiment* is a component of a measurement used to monitor student learning that consists of a prompt for students' responses and an observer's rubric for recording or quantifying those responses.

Modeling, behavioristic principle of: The *principle of modeling* states that individuals tend to imitate behaviors—especially attitudes—that they frequently observe.

Norm group, standardized test: A *norm group* for a standardized test is the sample of students with whom the test is field-tested and whose scores provide the standards for interpreting scores from subsequent administrations of the test.

Norm-referenced evaluation: A summative evaluation of a student's achievement is *norm referenced* if the evaluation is influenced by how measurement results relevant to that student's achievement compare to the results obtained from others.

Norms, standardized test: *Standardized test norms* are statistics computed from the norm group's scores that provide the (a) standard to which subsequent scores are compared (i.e., an average of scores from the norm group) and (b) unit of measure for making those comparisons (i.e., a measure of the variability of the scores from the norm group).

Objective, learning: A *learning objective* is a statement specifying what students will do or be able to do if a lesson is successful. The learning objective is the purpose of a lesson.

Off-task behavior: A student's behavior is *off-task* whenever the student fails to be on-task during either transition or allocated time.

On-task behavior: A student's behavior is *on-task* whenever the student is attempting to follow the teacher's directions during either transition or allocated time.

Overlearning: Students *overlearn* by continuing to practice recalling mathematical information or executing an algorithm even after they have memorized it. Overlearning increases resistance to forgetting and facilitates long-term retention.

Passive communication: A person's communication is *passive* rather than assertive if it fails to send the message the person wants to convey because he is intimidated or fearful of the recipient's reaction.

Percentile score, standardized test: A raw score x_i converts to a *percentile* of p if x_i is greater than $p\%$ of the norm group's raw scores and x_i is less than $(100 - p)\%$ of the norm group's raw scores.

Portfolio assessment: *Portfolio assessment* is the use of individualized student portfolios as devices for communicating summative evaluations of student achievement.

Portfolio, individualized student: An *individualized student portfolio* is a collection of products, artifacts, and demonstrations of the student's schoolwork assembled for the purpose of providing a representative sample of the student's achievement. Three types of portfolios are commonly identified in the teacher-education literature: The (a) *working portfolio,*(b) *presentation portfolio,* and (c) *record-keeping portfolio.* The *working portfolio* is a mechanism for documenting the student's ongoing progress. The *presentation portfolio* is a mechanism for reflecting the student's significant and prominent accomplishments. The *record-keeping portfolio* is a mechanism for maintaining a comprehensive record of the student's schoolwork.

Portfolio, professional: A *professional working portfolio* is a collection of items that reflect an individual teacher's professional talents, aptitude for teaching, style of teaching, and teaching performances. A teacher selects a proper subset of the items from his working portfolio to create a *professional presentation portfolio* for use when applying for faculty positions or exhibiting his teaching accomplishments and capabilities.

Positive reinforcer: A *positive reinforcer* is a stimulus presented after a response that increases the probability of that response being repeated in the future.

Preservice mathematics teacher: A *preservice mathematics teacher* is a person who is currently enrolled in a professional teacher-preparation program for the purpose of becoming qualified and certified to teach mathematics in middle, junior high, and high schools.

Principles and Standards for School Mathematics (PSSM): *PSSM* (NCTM, 2000b) is a resource and guide developed by the National Council of Teachers of Mathematics intended to be used by all who make decisions that affect the mathematics education of students in prekindergarten through grade 12. *PSSM* is available as a 401-page book as well as on a CD-ROM. The full text can also be accessed electronically through NCTM's website (www.nctm.org). It includes six principles for school mathematics (i.e., principles for equity, curriculum, teaching, learning, assessment, and technology), five content standards and expectations (i.e., for number and operations, algebra, geometry, measurement, and data analysis and probability), and five process standards (i.e., for problem solving, reasoning and proof, communication, connections, and representation).

Problem-based teaching unit: A teaching unit is *problem based* if it is designed so that students are confronted with problems they perceive a need to solve and are then motivated to engage in the unit's lesson by their desire to address those problems.

Prompt, miniexperiment: A *prompt of a miniexperiment* is the component of the miniexperiment that stimulates students to behave or respond in a manner that is indicative of what the miniexperiment is designed to measure.

Raw score: The *raw score* a student obtains from a test is the sum of the points from the individual miniexperiments composing the test.

Reasoning-level question: A question that requires respondents to reason and make judgments to answer is *reasoning level.*

Relationship: A *relationship* is an association between either (a) concepts (e.g., {irrationals} $\subseteq$ {reals}), (b) a concept and a specific (e.g., $x^2 > -4$ $\forall$ $x \in$ {reals}), (c) a specific and a concept (e.g., 5,981 is prime), or (d) specifics (e.g., $\sqrt{13} \geq 1.1$).

Relationship of convention: A *relationship of convention* is a relationship that exists because it has been established through tradition or agreement.

Relationship, discoverable: A relationship is *discoverable* if one can use reasoning or experimentation to find out that the relationship exists.

Relevance, measurement: A measurement is *relevant* to the same degree that it is pertinent to the decision influenced by its results. For a measurement to be relevant to students' achievement of a unit goal, its miniexperiments must pertain to the *mathematical content* and the *learning levels* specified by the objectives that define that goal. A miniexperiment pertains to the mathematical content and learning level specified by an objective if students' scores on the miniexperiment depend on how well they operate at the specified learning level (i.e., either construct a concept, discover a relationship, simple knowledge, algorithmic skill, comprehension and communication, application, creative thinking, appreciation, or willingness to try) with the specified mathematical content.

Reliability, measurement: A measurement is *reliable* to the same degree that it can be depended upon to provide noncontradictory information. To be reliable, a measurement must have both *internal consistency* and *scorer consistency.*

Rubric for a miniexperiment, observer's: A miniexperiment's *observer's rubric* is the set of rules, key, or procedures a teacher or other observer follows to record an analysis of a student's response to the miniexperiment's prompt.

Scaled score: *Scaled scores* are derived scores from a standardized test that allow results from different levels of the test to be compared. Scaled score values are positive numbers less than 1,000 with raw scores from lower grade levels (e.g., 1.2) associated with lower scaled scores (e.g., 109) and raw scores from higher grade levels (e.g., 11.7) associated with higher ones (e.g., 722). The particular correspondences between scaled scores and other types of scores as well as grade levels differ depending on the test and the edition of the test.

Scorer consistency, measurement: A measurement has *scorer consistency* to the same degree that (a) the teacher (or whoever scores the test) faithfully follows the miniexperiments' rubrics so that the measurement results are not influenced by *when* the measurement is scored and (b) different teachers (or scorers) who understand the measurement's mathematical content and learning levels agree on the score warranted by each response so that results are not influenced by *who* scores the measurement.

Simple knowledge: Students achieve an objective at the *simple-knowledge level* by remembering a specified response (but not multiple-step process) to a specified stimulus.

Specific: A *specific* is a unique entity, something that is not abstract.

Standardized test: A test is *standardized* if it has been field-tested on a large sample of people—referred to as the "norm group"—to (a) measure reliability of the test and (b) establish norm-referenced standards for use in interpreting scores for subsequent administrations of the test.

Stanine: To convert raw scores to *stanine scores,* the raw scores are collapsed into nine categories or intervals so that each raw score is reported simply as either a "1," "2," . . ., or "9." Raw scores very near the norm group mean (i.e., μ) are assigned to the middle interval, which is stanine 5. The length of each of the nine stanine intervals (except for the two most extreme ones) is one half a norm group standard deviation. With z_x = z-score associated with a raw score of x, x is assigned a stanine as follows: 1 if $z_x < -1.75$, 2 if $-1.75 \leq z_x < -1.25$, 3 if $-1.25 \leq z_x < -0.75$, 4 if $-0.75 \leq z_x < -0.25$, 5 if $-0.25 \leq z_x < 0.25$, 6 if $0.25 \leq z_x < 0.75$, 7 if $0.75 \leq z_x < 1.25$, 8 if $1.25 \leq z_x < 1.75$, and 9 if $1.75 < z_x$.

Subconcept: A concept that is subsumed by a broader concept (e.g., irrational number is a subconcept of real number) is a *subconcept.*

Summative evaluation: Your judgments of students' achievement for the purpose of reporting success are referred to as "*summative evaluations.*"

Supervision of teaching, administrative: *Administrative supervision* is the art of controlling the quality of instruction in a school.

Supervision of teaching, instructional: *Instructional supervision* is the art of helping teachers improve their teaching performances.

Supportive reply: A reply to an expression of feelings is *supportive* if the response clearly indicates that the feelings have been recognized and not judged to be right or wrong.

Syllogism: A *syllogism* is a scheme for inferring problem solutions—a scheme in which a *conclusion* is drawn from a *major premise* and a *minor premise.* The major premise is a general rule or abstraction. The minor premise is the relationship of a specific to the general rule or abstraction. The conclusion is a logical consequence of the combined premises. For example:

- *Major premise:* If the discriminant ($b^2 - 4ac$) of a quadratic equation ($ax^2 + bx + c = 0$) is positive and not a perfect square, the equation has two irrational roots.
- *Minor premise:* The discriminate of $x^2 - x - 18 = 0$ (i.e., 73) is positive and not a perfect square.
- *Conclusion:* $x^2 - x - 18 = 0$ has two irrational roots.

Synectics: *Synectics* is a means by which *metaphors* and *analogies* are used to lead students into an illogic state for situations where rational logic fails. The intent is for students to free themselves of convergent reasoning and to develop empathy with ideas that conflict with their own. Three types of analogies are used in learning activities based on synectics: (1st) *direct analogies,* (2nd) *personal analogies,* and (3rd) *compressed conflicts.*

Technical mathematical expression: Communication devices (e.g., shorthand symbols and graphs) that are peculiar to the language of mathematics are *technical mathematical expressions.*

Test, unit: A *unit test* is a scientific experiment (consisting of a sequence of miniexperiments) in which the question to be answered is how well students have achieved a particular learning goal.

Trade book, mathematical: *Mathematical trade books* are books about mathematics or the history of mathematics that are not textbooks designed to be used in conjunction with mathematics courses.

Transition time: The time students spend between learning activities is *transition time.*

True dialogue: A *true dialogue* is a conversation that is not dominated by IRE cycles.

Unit, teaching: A *teaching unit* consists of (a) a learning goal defined by a set of specific objectives, (b) a planned sequence of lessons—each consisting of learning activities designed to lead students to achieve the lesson's objective, (c) mechanisms for monitoring student progress and using formative feedback to guide lessons, and (d) a summative evaluation of student achievement of the learning goal.

Usability, measurement: A measurement is *usable* to the degree that it is inexpensive, does not consume time, is easy to administer and score, and does not interfere with other activities.

Usefulness, measurement: To be *useful,* a measurement must have a satisfactory degree of validity as well as a satisfactory degree of usability.

Validity, measurement: A measurement is *valid* to the same degree that its results approximate students' true achievement levels (i.e., its measurement error approaches 0). A measurement's validity depends on its *relevance* and its *reliability.*

Willingness to try: Students achieve an objective at the *willingness-to-try level* by choosing to attempt a mathematical task specified by the objective.

Withitness: Jacob Kounin (1977) coined the term "*withitness*" to refer to a teacher's awareness of what students are doing.

Z-score: The *z-score* associated with a particular raw score is the number of standard deviations that the raw score falls above the mean. If x is a raw score from a test, the z-score associated with x (i.e., z_x) is given by the following:

$z_x = \frac{x - \mu}{\sigma}$ where μ is the arithmetic mean and σ is the standard deviation of the norm-group scores.

References

Alberti, R., & Emmons, M. (1995). *Your perfect right* (7th ed.). San Luis Obispo, CA: Impact.

American Association for the Advancement of Science. (2000). *Project 2061* [Website: www.project2061.org]. Washington: Author.

American Psychological Association, American Educational Research Association, and National Council on Measurement in Education. (1999). *Standards for educational and psychological testing* (6th ed.). Hanover, PA: Author.

Amundson, H. E. (1989). Percent In *Historical topics for the mathematics classroom* (pp. 146–147). Reston, VA: National Council of Teachers of Mathematics.

Arnold, D., Atwood, R., & Rogers, V. (1974). Questions and response levels and lapse time intervals. *Journal of Experimental Education, 43,* 11–15.

Ashlock, R. B. (2001). *Error patterns in computation: Error patterns to improve instruction* (8th ed.). Upper Saddle River, NJ: Prentice Hall.

Baenen, J. (2000). *Transescent seminar* [Videotape]. Platteville: WI: Center of Education for the Young Adolescent.

Ball, D. L. (1988). Unlearning to teach mathematics. (Issue paper 8–1). East Lansing: Michigan State University National Center for Research on Teacher Education.

Barrow, J. D. (1992). *Pi in the sky: Counting, thinking, and being.* Boston: Back Bay Books.

Battista, M. T. (1999). The mathematical miseducation of America's youth: Ignoring research and scientific study in education. *Phi Delta Kappan, 80,* 424–433.

Bauman, M. (1997). What grades do for us, and how to do without them. In S. Tschudi (Ed.), *Alternatives to grading student writing* (pp. 162–178). Urbana, IL: National Council of Teachers of English.

Baumgart, J. K. (1989). The history of algebra. In *Historical topics for the mathematics classroom* (pp. 233–260). Reston, VA: National Council of Teachers of Mathematics.

Begle, E. G. (1958). The School Mathematics Study Group. *Mathematics Teacher, 51,* 616–618.

Benjamin, A., & Shermer, M. B. (1993). *Mathemagics: How to look like a genius without really trying.* Los Angeles: Lowell House.

Bennett, D. (1993). *Exploring geometry with the Geometer's SketchPAD.* Berkeley, CA: Key Curriculum Press.

Berk, R. A. (1986). Performance standards on criterion-referenced tests. *Review of Educational Research, 56,* 137–172.

Beyer, B. K. (1987). *Practical strategies for teaching of thinking.* Boston: Allyn and Bacon.

Beyer, W. H. (1987). *CRC standard mathematical tables* (28th ed.). Boca Raton, FL: CRC Press.

Birman, B. F., Desimone, L., Porter, A. C., & Garet, M. S. (2000). Designing professional development that works. *Educational Leadership, 57,* 28–33.

Black, P., & Wiliam, D. (1998). Inside the black box: Raising standards through classroom assessment. *Phi Delta Kappan, 80,* 139–148.

Bloom, B. S. (Ed.). (1984). *Taxonomy of educational objectives: The classification of educational goals. Book I: Cognitive domain.* New York: Longman.

Boers-van Oosterum, M. A. M. (1990). Understanding of variables and their uses acquired by students in traditional and computer-intensive algebra. Dissertation. University of Maryland, College Park.

Bolt, B. (1992). *Mathematical cavalcade.* Cambridge, UK: Cambridge University Press.

Bourne, L. E., Dominowski, R. L., Loftus, E. F., & Healy, A. F. (1986). *Cognitive processes* (2nd ed.). Englewood Cliffs, NJ: Prentice Hall.

Bowers, C. A., & Flinders, D. J. (1990). *Responsive teaching: An ecological approach to classroom patterns of language, culture, and thought.* New York: Teachers College Press.

Bransford, J. D., Brown, A. L., & Cocking, R. R. (Eds.). (1999). *How people learn: Brain, mind, experiences, and school.* Washington: National Academy Press.

Brenner, M. E., Herman, S., Ho, H., & Zimmer, J. (1999). Cross-national comparison of representational competence. *Journal for Research in Mathematics Education, 30,* 541–557.

Bridges, E. M. (1986). *The incompetent teacher.* Philadelphia: The Falmer Press.

Brophy, J. (1986). Teaching and learning in mathematics: Where research should be going. *Journal of Research in Mathematics Education, 17,* 323–346.

Brownell, W. A. (1947). The place of meaning in the teaching of arithmetic. *Elementary School Journal, 47,* 256–265.

Burton, D. M. (1997). *Elementary number theory* (3rd ed.). New York: McGraw-Hill.

Burton, D. M. (1999). *The history of mathematics: An introduction* (4th ed.). Boston: WCB/McGraw-Hill.

California Institute of Technology. (2000). *Early history of mathematics* [Videotape]. Pasadena, CA: Author.

Calinger, R. (1999). *A contextual history of mathematics.* Upper Saddle River, NJ: Prentice Hall.

Campbell, D. M., Cignetti, P. B., Melenyzer, B. J., Nettles, D. H., & Wyman, R. M. (1997). *How to develop a professional portfolio: A manual for teachers.* Boston: Allyn and Bacon.

Cangelosi, J. S. (1980). Four steps in teaching for mathematical application. *Mathematics and Computer Education, 14,* 54–59.

Cangelosi, J. S. (1982). *Measurement and evaluation: An inductive approach for teachers.* Dubuque, IA: W. C. Brown.

Cangelosi, J. S. (1984a). Another answer to the cutoff score question. *Educational Measurement: Issues and Practice, 3,* 23–25.

Cangelosi, J. S. (1984b, May). *Teaching students to apply mathematics.* Paper presented at the meeting of the Research Council of Diagnostic and Prescriptive Mathematics, San Francisco, CA.

Cangelosi, J. S. (1989b, April). *A video inservice program for underprepared mathematics teachers.* Presentation at the annual meeting of the National Council for Teachers of Mathematics, Orlando, FL.

Cangelosi, J. S. (1990). *Using mathematics to solve real-life problems* [Videotape series]. Logan, UT: National Science Foundation and the Utah State University Telecommunications Division.

Cangelosi, J. S. (1991). *Evaluating classroom instruction.* New York: Longman.

Cangelosi, J. S. (2000a). *Assessment strategies for monitoring student achievement.* New York: Addison-Wesley/Longman.

Cangelosi, J. S. (2000b). *Classroom management strategies: Maintaining and gaining students' cooperation* (4th ed.). New York: John Wiley & Sons.

Cangelosi, J. S. (2001a). *Mathematics teaching and learning practices in middle and secondary school classrooms: An update of the Jesunathadas study.* Unpublished study. Logan: Utah State University.

Cangelosi, J. S. (2001b, April). *A video program for helping college and university instructors develop effective questioning and discussion strategies.* A presentation at the 2001 annual meeting of the American Educational Research Association, Seattle.

Canter & Associates. (1994a). *Intervening safely during fights.* [Videotape]. Los Angeles: Author.

Canter & Associates. (1994b). *Preventing conflict and violence in the classroom.* [Videotape]. Los Angeles: Author.

Canter, L. (2001). *Assertive discipline: Positive behavior management for today's classrooms.* (rev. ed.). Los Angeles: Canter & Associates.

Cazden, C. B. (1988). *Classroom discourse: The language of teaching and learning.* Portsmouth, NH: Heinemann.

Cobb, P., Wood, T., & Yackel, E. (1994). Discourse, mathematical thinking, and classroom practice. In *Contexts for learning: sociocultural dynamics in children's development.* New York: Oxford University Press.

Cole, R. S. (1993). Why should we care about teaching calculus? *Washington Center News, 7,* 4–6.

Conference Board of the Mathematical Sciences. (1983a). *The mathematical sciences curriculum K–12: What is still fundamental and what is not.* Report to the National Science Board Commission on Precollege Education in Mathematics, Sciences, and Technology. Washington: Author.

Conference Board of the Mathematical Sciences. (1983b). *New goals for mathematical education.* Report. Washington: Author.

Conolly, P., & Vilardi, T. (Eds.), (1989). *Writing to learn mathematics and science.* New York: Teachers College Press.

Coulombe, W. N., & Berenson, S. B. (2001). Representation of patterns and functions. In A. A. Cuoco & F. R. Curcio (Eds.), *The roles of representation in school mathematics: 2001 Yearbook* (pp. 166–172). Reston, VA: NCTM.

Coxford, A. F. (1995). The case for connections. In P. A. House & A. F. Coxford (Eds.), *Connecting mathematics across the curriculum: 1995 Yearbook* (pp. 3–12). Reston, VA: NCTM.

Culotta, E. (1992). The calculus of education reform. *Science, 255,* 1060–1062.

Cunningham, P. A., Moore, S. A., Cunningham, J. W., & Moore, D. W. (1995). *Reading and writing in elementary classrooms: Strategies and observations.* New York: Longman.

Danielson, C. (2001). New trends in teacher evaluation. *Educational Leadership, 58,* 12–15.

Davis, R. B., Maher, C. A., & Noddings, N. (Eds.), (1990). *Constructivist views on the teaching and learning of mathematics.* Reston, VA: NCTM.

Dence, J. B., & Dence, T. P. (1999). *The elements of the theory of numbers.* San Diego, CA: Harcourt Academic Press.

Dessart, D. J., DeRidder, C. M., & Ellington, A. J. (1999). The research backs calculators. *Mathematics Education*

Dialogues [supplement to *NCTM News Bulletin, 35*], 6.

Devlin, K. (1993). Computers and mathematics. *Notices of the AMS, 40,* 1352–1353.

Duke, C. R., Cangelosi, J. S., & Knight, R. S. (1988, February). *The Mellon project: A collaborative effort.* Colloquium presentation at the annual meeting of the American Colleges for Teacher Education, New Orleans.

Dunham, P. H., & Dick, T. P. (1994). Research on graphing calculators. *Mathematics Teacher, 87,* 440–445.

Dunican, P. (2000). Drowning in a deluge of paper. *English Journal, 89,* 27–28.

Eisner, M. P. (1986). An application of quadratic equations to baseball. *Mathematics Teacher, 79,* 327–330.

Emmer, E. T., & Stough, L. M. (2001). Classroom management: A critical part of educational psychology, with implications for teacher education. *Educational Psychologist, 36,* 103–112.

Enzensberger, H. M. (2000). *The number devil: A mathematical adventure.* New York: Henry Holt and Company.

Evans, R. (1989). The faculty in midcareer: Implications for school improvement. *Educational Leadership, 46,* 10–15.

Evertson, C. M. (1989). Classroom organization and management. In M. C. Reyonlds (Ed.), *Knowledge base for the beginning teacher* (pp. 59–70). Oxford, UK: Pergamon.

Evertson, C. M., Emmer, E. T., Clements, B. S., & Worsham, M. E. (1997). *Classroom management for elementary school teachers.* (5th ed.). Boston: Allyn & Bacon.

Eves, H. (1983). *Great moments in mathematics before 1650.* Washington: Mathematical Association of America.

Feiler, R., Heritage, M., & Gallimore, R. (2000). Teachers leading teachers. *Educational Leadership, 57,* 66–69.

Ferrini-Mundy, J., & Graham, K. G. (1991). An overview of the calculus reform effort: Issues for learning, teaching, and curriculum development. *American Mathematical Monthly, 98,* 627–635.

Fisher, C. W., Berliner, D. C., Filby, N. N., Marliave, R., Cahen, L. S., & Dishaw, M. M. (1980). Teaching behaviors, academic learning time, and student achievement: An overview. In C. Denham & A. Lieberman (Eds.), *Time to learn* (pp. 7–32). Washington: National Institute of Education.

Fleischman, H. L., & Hopstock, P. J. (1993). *Descriptive study of services to limited English proficient students.* Arlington, VA: Development Associates.

Foster, A. G., Winters, L. J., Gordon, B. W., Rath, J. N., & Gell, J. M. (1992). *Merrill algebra 2 with trigonometry: Applications and connections* (Teacher's wraparound edition). Westerville, OH: Glencoe Divison, Macmillan/McGraw-Hill.

Fowler, D. (1994). What society means by mathematics. *Focus, 14,* 12–13.

Fraleigh, J. B. (1999). *A first course in abstract algebra* (6th ed.). Reading, MA: Addison-Wesley/Longman.

Franklin, J. (2001). Trying too hard? How accountability and testing are affecting constructivist teaching. *Education Update, 43,* 1, 4–8, & 8.

Friedlander, A., & Tabach, M. (2001). Promoting multiple representations in algebra. In A. A. Cuoco & F. R. Curcio (Eds.), *The roles of representation in school mathematics: 2001 Yearbook* (pp. 173–185). Reston, VA: NCTM.

Friel, S., Rachlin, S., & Doyle, D. (2001). *Navigating through algebra in grades 6–8.* Reston, VA: NCTM.

Gallager, J. D. (1998). *Classroom assessment for teachers.* Upper Saddle River, NJ: Prentice Hall.

Gibilisco, S. (1990). *Optical illusions: Puzzles, paradoxes and brain teasers, Number 4.* Blue Ridge Summit, PA: Tab.

Glassman, M. (2001). Dewey and Vygostsky: Society, experience, and inquiry in educational practice. *Educational Researcher, 30,* 3–14.

Gödel, K. (1943). *On undecidable propositions of formal mathematics.* Princeton, NJ: Princeton University Press.

Goldberger, S, & Kazis, R. (1996). Revitalizing high schools: What the school-to-career movement can contribute. *Phi Delta Kappan, 77,* 547–554.

Goldin, G., & Shteingold, N. (2001). Systems of representations and the development of mathematical concepts. In A. A. Cuoco & F. R. Curcio (Eds.), *The roles of representation in school mathematics: 2001 Yearbook* (pp. 1–23). Reston, VA: NCTM.

Goodlad, J. I., & Su, Z. (1992). Organization of the curriculum. In P. W. Jackson (Ed.), *Handbook of research on curriculum* (pp. 327–344). New York: Macmillan.

Gordon, T. (1974). *Teacher effectiveness training.* New York: David McKay.

Gordon, W. J. J. (1961). *Synectics.* New York: Harper & Row.

Gowan, J. C., Demons, G. D., & Torrance, E. P. (1967). *Creativity: Its educational implications.* New York: Wiley.

Gray, K. (1996). The baccalaureate game: Is it right for all teens? *Phi Delta Kappan, 77,* 528–546.

Groves, S. (1994, April). *Calculators: A learning environment to promote number sense.* A paper presented at the American Educational Research Association, New Orleans.

Guilford, J. P. (1959). *Personality.* New York: McGraw-Hill.

Gullberg, J. (1997). *Mathematics: From the birth of numbers.* New York: W. W. Norton.

Guzzetti, B. J., Snyder, T. E., Glass, G. V., & Gamas, W. S. (1993). Prompting conceptual change in science: A comparative meta-analysis of instructional intervention from reading education and science education. *Reading Research Quarterly, 28,* 117–159.

Harder, P. (2001). *Mathematics applied to electronics.* Upper Saddle River, NJ: Prentice Hall.

Hargreaves, A. (2001). Beyond anxiety and nostalgia: Building a social movement for educational change. *Phi Delta Kappan, 82,* 373–377.

Hatano, G., & Inagaki, K. (1991). Sharing cognition through collective comprehension activity. In L. B. Resnick, J. M. Levine, & S. D. Teasley (Eds.), *Perspectives on socially shared cognition* (pp. 331–348). Washington: American Psychological Association.

Heal, E. R., Cannon, L. O., & Dorward, J. (2001, April). *Hands-on explorations of new web-based electronic manipulatives and tools.* Presentation at the annual meeting of the National Council of Teachers of Mathematics, Orlando.

Heddens, J. W., & Speer, W. R. (2001). *Today's mathematics: Part 1: Concepts and classroom methods* (10th ed.). New York: John Wiley& Sons.

Hidi, S., & Harackiewicz, J. M. (2000). Motivating the academically unmotivated: A critical issue for the 21st century. *Review of Educational Research, 70,* 151–159.

Hiebert, J., & Carpenter, T. P. (1992). Learning and teaching with understanding. In D. A. Grouws (Ed.), *Handbook of research on mathematics teaching and learning* (pp. 65–97). New York: Macmillan.

Hoffman, P. (1988). *Archimedes' revenge: The joys and perils of mathematics.* New York: Fawcett Columbine.

Hopley, R. B. (1994). Nested Platonic solids: A class project in solid geometry. *Mathematics Teacher, 87,* 312–318.

Huetinck, L. & Munshin, S. N. (2000). *Teaching mathematics for the 21st century: Methods and activities for grades 6–12.* Columbus, OH: Prentice Hall/Merrill.

Ifrah, G. (2000). *The universal history of numbers: From prehistory to the invention of computers.* New York: John Wiley & Sons.

James, W. (1890). *The principles of psychology.* Vols. I and II. New York: Holt, Rinehart, & Winston.

Jesunathadas, J. (1990). *Mathematics teachers' instructional activities as a function of academic preparation.* Dissertation, Utah State University, Logan.

Johnson, D. W., & Johnson, R. T. (1999). *Methods of cooperative learning: What can we prove works?* Edina, MN: Cooperative Learning Institute.

Johnson, J. R. (1997, October). *Geometry portfolio projects.* Presentation at the Western Regional Conference of NCTM, Salt Lake City, UT.

Jones, F. H., Jones, P., & Jones, J. T. (2000). *Tools for teaching: Discipline, instruction, motivation.* Santa Cruz, CA: F. H. Jones & Associates.

Joyce, B., Weil, M., & Calhoun, E. (2000). *Models of teaching* (6th ed.). Boston: Allyn and Bacon.

Kamii, C., & Warrington, M. A. (1999). Teaching fractions: Fostering children's own reasoning. In L. V. Stiff & F. R. Curcio (Eds.), *Developing mathematical reasoning in grades K–12: 1999 Yearbook* (pp. 82–92). Reston, VA: NCTM.

Kaput, J. J. (1992). Technology and mathematics education. In D. A. Grouws (Ed.), *Handbook of research on mathematics teaching and learning* (pp. 515–556). New York: Macmillan.

Kellough, R. D., Cangelosi, J. S., Collette, A. T., Chiappaetta, E. L., Souviney, R. J., Trowbridge, L. W., & Bybee, R. (1996). *Integrating mathematics and science for intermediate and middle school students.* Englewood Cliffs, NJ: Prentice Hall/Merrill.

Khisty, L. L. (1997). Making mathematics accessible to Latino students: Rethinking instructional practice. In J. Tretacosta & M. J. Kenney (Eds.), *Multicultural and gender equity in the mathematics classroom: The gift of diversity: 1997 Yearbook* (pp. 92–101). Reston, VA: NCTM.

Kinney, L. B., & Purdy, C. R. (1952). *Teaching mathematics in secondary schools.* New York: Holt, Rinehart, & Winston.

Kohn, A. (2000). *The case against standardized testing: Raising the scores, ruining the schools.* Portsmouth, NH: Heinemann.

Kohn, A. (2001). Fighting the tests: A practical guide to rescuing our schools. *Phi Delta Kappan, 82,* 348–357.

Kounin, J. (1977). *Discipline and group management in classrooms.* New York: Holt, Rinehart & Winston.

Kramer, M. C. (2001). Triumph out of the wilderness: A reflection on the importance of mentoring. *Phi Delta Kappan, 82,* 411–412.

Krathwhohl, D., Bloom, B. S., & Masia, B. (1964). *Taxonomy of educational objectives: The classification of educational goals. Handbook 2: Affective domain.* New York: Longman.

Kreindler, L., & Zahm, B. (1992). *MathFinder sourcebook: A collection of resources for mathematics reform.* Armonk, NY: The Learning Team.

Kuehl, B. B. (2002). *Improving reading comprehension of mathematical texts.* Thesis, Utah State University, Logan.

Lavoie, R. D. (1989). *Understanding learning disabilities: How difficult can this be? The F.A.T. city workshop* [Videotape]. Greenwich, CT: Eagle Hill Outreach (A Peter Rose Production distributed by PBS Video).

Lewinter, M., & Widulski, W. (2002). *The sage of mathematics: A brief history.* Upper Saddle River, NJ: Prentice Hall.

Lindahl, A. M., & Cangelosi, J. S. (2000, October). *Biomath labs: Getting beyond cookbook.* Paper presentation at the National Convention of the National Association of Biology Teachers, Orlando.

Lindquist, M. M., & Elliott, P. C. (1996). Communication—an imperative for change: A conversation with Mary Lindquist. In P. C. Elliot & M. J. Kenney (Eds.),

Communication in mathematics, K–12 and beyond: 1996 yearbook (pp. 1–10). Reston, VA: NCTM.

Manning, M. L., & Bucher, K. T. (2001). Revisiting Ginott's congruent communication after thirty years. *The Clearing House, 74,* 215–218.

Marks, H. M. (2000). Student engagement in instructional activity: Patterns in elementary, middle, and high school years. *American Educational Research Journal, 37,* 153–184.

Martin, G., & Pear, J. (1996). *Behavior modification: What it is and how to do it* (5th ed.). Upper Saddle River, NJ: Prentice Hall.

McCormick, C. B., & Pressley, M. (1997). *Educational psychology: Learning, instruction, and assessment.* New York: Longman.

McLeish, J. (1991). *Number: The history of numbers and how they shape our lives.* New York: Fawcett-Columbine.

McLeod, B. (1996). *Exemplary schools for language minority students.* Report of the Student Diversity Study by the Office of Bilingual Education, University of California, Santa Cruz, CA.

McLeod, D. B. (1992). Research on affect in mathematics education: A reconceptualization. In D. A. Grouws (Ed.), *Handbook of research on mathematics teaching and learning* (pp. 575–596). New York: Macmillan.

McMillan, J. H. (2001). Secondary teachers' classroom assessment and grading practices. *Educational Measurement: Issues and Practices, 20,* 20–32.

Meel, D. E. (2000). Sumgo here and sumgo there. *Mathematics Teaching in Middle School, 6,* 236–239.

Merrow, J. (2001). Undermining standards. *Phi Delta Kappan, 82,* 653–659.

Miller, L. (1989). Radical symbol. In *Historical topics for the mathematics classroom* (pp. 147–148). Reston, VA: National Council of Teachers of Mathematics.

Miura, I. T. (2001). The influence of language on mathematical representations. In A. A. Cuoco & F. R. Curcio (Eds.), *The roles of representation in school mathematics: 2001 Yearbook* (pp. 53–62). Reston, VA: NCTM.

Moore, R. C. (1994). Making the transition to formal proof. *Educational Studies in Mathematics, 27,* 249–266.

Mueller, W. (2001). Reform now, before it's too late! *American Mathematical Monthly, 108,* pp. 126–143.

Myers, D. G. (1986). *Psychology.* New York: Worth Publishers.

National Commission on Mathematics and Science Teaching for the 21st Century. (2000). *Before it's too late: A report of the National Commission on Mathematics and Science Teaching for the 21st Century.* Washington: U.S. Department of Education.

National Commission on Teaching and America's Future. (1996). *What matters most: Teaching for America's future.* New York: Author.

National Council of Teachers of Mathematics. (1940). *The place of mathematics in general education: 15th yearbook.* New York: Teachers College Press.

National Council of Teachers of Mathematics. (1980). *An agenda for action: Recommendations for school mathematics for the 1980s.* Reston, VA: Author.

National Council of Teachers of Mathematics. (1989a). *Curriculum and evaluation standards for school mathematics.* Reston, Va: Author.

National Council of Teachers of Mathematics. (1989b). *Historical topics for the classroom.* Reston, VA: Author.

National Council of Teachers of Mathematics. (1991). *Professional standards for teaching mathematics.* Reston, VA: Author.

National Council of Teachers of Mathematics. (1995). *Assessment standards for school mathematics.* Reston, VA: Author.

National Council of Teachers of Mathematics. (2000a). *High stakes testing: NCTM position statement* [Website: www.nctm.org] Reston, VA: Author.

National Council of Teachers of Mathematics. (2000b). *Principles and standards for school mathematics.* Reston, VA: Author.

National Council of Teachers of Mathematics. (2001a). *Spring 2001 NCTM catalog: Resources for mathematics educators.* Reston, VA: Author.

National Council of Teachers of Mathematics. (2001b). What can we learn from the TIMSS-repeat? *NCTM News Bulletin, 37,* Issue 6. Author.

National Science Board Commission on Precollege Education in Mathematics, Science, and Technology. (1983). *Educating Americans for the 21st century: A plan for improving the mathematics, science, and technology education for all American elementary and secondary students so that their achievement is the best in the world by 1995.* Washington: National Science Foundation.

Noble, T., Nemirovsky, R., Wright, T., & Tierney, C. (2001). Experiencing change: The mathematics of change in multiple environments. *Journal of Research in Mathematics Education, 32,* 85–108.

O'Brien, T. C. (1999). Parrot math. *Phi Delta Kappan, 80,* 434–438.

O'Hagan, L. K. (1997). It's broken—fix it! In S. Tschudi (Ed.), *Alternatives to grading student writing* (pp. 3–13). Urbana, IL: National Council of Teachers of English.

Oliva, P. F. (2001). *Developing the curriculum* (5th ed.). New York: Addison-Wesley/Longman.

Oliva, P. F., & Pawlas, G. E. (2001). *Supervision for today's schools* (6th ed.). New York: Wiley.

Ormrod, J. E. (2000). *Educational psychology: Developing learners* (3rd ed.). Columbus: Merrill/Prentice Hall.

Painter, B. (2001). Using teaching portfolios. *Educational Leadership, 58,* 31–34.

Parsons, R. D., Hinson, S. L., & Sardo-Brown, D. (2001). *Educational psychology: A practitioner-researcher model of teaching.* Belmont, CA: Wadsworth.

Phillips, E. (1991). *Patterns and functions.* Reston, VA: NCTM.

Plake, B. S., Impara, J. C., & Irwin, P. M. (2000). Consistency of Angoff-based predictions of item performance: Evidence of technical quality of results from the Angoff standard setting method. *Journal of Educational Measurement, 37,* 347–355.

Posamentier, A. S., & Stepelman, J. (1999). *Teaching secondary mathematics: Techniques and enrichment units.* Upper Saddle River, NJ: Merrill/Prentice Hall.

Post, T. R. & Cramer, K. A. (1989). Knowledge representation, and quantitative thinking. In M. C. Reynolds (Ed.), *Knowledge base for the beginning teacher* (pp. 221–231). Oxford, UK: Pergamon.

Powell, R. R., McLaughlin, H. J., Savage, T. V., & Zehm, S. (2001). *Classroom management: Perspectives on the social curriculum.* Upper Saddle River, NJ: Merrill/Prentice Hall.

Pratt, G. V. (1989). Early Greek algebra. In *Historical topics for the mathematics classroom* (pp. 289–301). Reston, VA: NCTM.

Price, G. (1997). Quantitative literacy across the curriculum. In L.A. Steen (Ed.), *Why numbers count: Quantitative literacy for tomorrow's America* (pp. 155–160). New York: College Entrance Examination Board.

Puhlmann, N. A., & Petersen, M. L. (1992, April). *The electronic teaching station of the future—today.* Presentation at the annual meeting of the National Council of Teachers of Mathematics, Nashville.

Quina, J. (1989). *Effective secondary teaching: Going beyond the bell curve.* New York: Harper & Row.

Quinn, R. J. (2001). Using attribute blocks to develop a conceptual understanding of probability. *Mathematics Teaching in the Middle School, 6,* 291–294.

Retz, M., & Keihn, M. D. (1989). Compass and straightedge constructions. In *Historical topics for the mathematics classroom* (pp. 192–195). Reston, VA: NCTM.

Rojano, T. (1996). Developing algebraic aspects of problem solving within a spreadsheet environment. In N. Bednarz, C. Kieran, & L. Lee (Eds.), *Approaches to algebra: Perspectives of research and teaching.* Boston: Kluwer Academic Publishers.

Romberg, T. A. (1992). Problematic features of the school mathematics curriculum. In P. W. Jackson (Ed.), *Handbook of research on curriculum* (pp. 749–788). New York: Macmillan.

Rosenshine, B. (1987). Direct instruction. In M. J. Dunkin (Ed.), *The international encyclopedia of teaching and teacher education* (pp. 257–262). Oxford, UK: Pergamon.

Rowley, E. R. (1996). Alternative assessments of meaningful learning of calculus content: A development and validation of item pools. Dissertation, Utah State University, Logan.

Santrock, J. W. (2001). *Educational Psychology.* Boston: McGraw-Hill.

Schank, R. (1987). Let's eliminate math from schools. *Whole Earth Review, 55,* 58–62.

Schmuck, R. A., & Schmuck, P. A. (2001). *Group processes in the classroom* (8th ed.). Boston: McGraw-Hill.

Schoenfeld, A. H. (1985). *Mathematical problem solving.* San Diego, CA: Academic Press.

Schoenfeld, A. H. (1988). When good teaching leads to bad results: The disaster of well-taught mathematics classes. *Educational Psychologist, 23,* 145–166.

Seife, C. (2000). *Zero: The Biography of a Dangerous Idea.* New York: Viking.

Sheets, C. (1993). Effects of computer learning and problem-solving tools on the development of secondary school students' understanding of mathematical functions. Dissertation. University of Maryland, College Park.

Shuell, T. J. (1990). Phases of meaningful learning. *Review of Educational Research, 60,* 531–547.

Siegel, M., Borasi, R., & Fonzi, J. (1998). Supporting students' mathematical inquiries through reading. *Journal for Research in Mathematics Education, 29,* 378–413.

Silver, E. A., Kilpatrick, J., & Schlesinger, B. G. (1990). *Thinking through mathematics: Fostering inquiry and communication in mathematics classrooms.* New York: College Entrance Examination Board.

Silver, E. A., Smith, M. S., & Nelson, B. S. (1995). The QUASAR Project: Equity concerns meet mathematics education reform in the middle school. In W. G. Secada, E. Fennema, & L. Byrd (Eds.), *New directions for equity in mathematics education* (pp. 9–56). New York: Cambridge University Press.

Singh, S. (1997). *Fermat's enigma.* New York: Walker and Company.

Singh, S. (1999). *The code book: The science of secrecy from ancient Egypt to quantum cryptography.* New York: Anchor Books.

Skemp, R. R. (1976). Relational understanding and instrumental understanding. *Mathematics Teaching, 77,* 20–26.

Smith, F., & Adams, S. (1972). *Educational measurement for the classroom teacher* (2nd ed.). New York: Harper & Row.

Smith, M. S. (2001). *Practiced-based professional development for teachers of mathematics.* Reston, VA: NCTM.

Sobel, M. A., & Maletsky, E. M. (1988). *Teaching mathematics: A sourcebook of aids, activities, and strategies* (2nd ed.). Englewood Cliffs, NJ: Prentice-Hall.

Spangler, H. (2001). *Basic mathematics for occupational and vocational students.* Upper Saddle River, NJ: Prentice Hall.

Steen, L. A. (1999). Twenty questions about mathematical reasoning.

In L. V. Stiff & F. R. Curcio (Eds.), *Developing mathematical reasoning in grades K–12: 1999 Yearbook* (pp. 270–285). Reston, VA: NCTM.

Stewart, I. (1992a). *Another fine math you've got me into . . . .* New York: Freeman.

Stewart, I. (1992b). *The problems of mathematics* (2nd ed.). Oxford, UK: Oxford University Press.

Stiggins, R. J. (1988). Revitalizing classroom assessment: The highest priority. *Phi Delta Kappan, 69,* 363–368.

Stiggins, R. J., & Conklin, N. F., & Bridgeford, N. J. (1986). Classroom assessment: A key to effective education. *Educational Measurement: Issues and Practice, 5,* 5–17.

Stigler, J. W., & Hiebert, J. (1999). *The teaching gap: Best ideas from the world for improving education in the classroom.* New York: The Free Press.

Strom, R. D. (1969). *Psychology for the classroom.* Englewood Cliffs, NJ: Prentice Hall.

Tang, E. P., & Ginsburg, H. P. (1999). Young children's mathematical reasoning: A psychological view. In L. V. Stiff & F. R. Curcio (Eds.), *Developing mathematical reasoning in grades K–12: 1999 Yearbook* (pp. 45–61). Reston, VA: NCTM.

Tate, W. F., & Johnson, H. C. (1999). Mathematical reasoning and educational policy: Moving beyond the politics of dead language. In L. V. Stiff & F. R. Curcio (Eds.), *Developing mathematical reasoning in grades K–12: 1999 Yearbook* (pp. 221–233). Reston, VA: NCTM.

Tattersall, J. J. (1999). *Elementary number theory in nine chapters.* Cambridge, UK: Cambridge University Press.

Thompson, D. R., & Senk, S. L. (2001). The effects of curriculum on achievement in second-year algebra: The example of the University of Chicago School Mathematics Project. *Journal for Research in Mathematics Education, 32,* 58–84.

Thompson, S. (2001). The authentic standards movement and its evil twin. *Phi Delta Kappan, 82,* 358–362.

Thorndike, E. L., & Woodworth, R. S. (1901). The influence of improvements in one mental function upon the efficiency of other functions. *Psychology Review, 8,* 247–256.

Tobin, K., Tippins, D. J., & Gallard, A. J. (1994). Research on instructional strategies for teaching science. In D. L. Gabel (Ed.), *Handbook of research on science teaching and learning* (pp. 45–93). New York: Macmillan.

Torrance, E. P. (1962). *Guiding creative talent.* Englewood Cliffs, NJ: Prentice Hall.

Torrance, E. P. (1966). Fostering creative behavior. In R. D. Strom (Ed.), *The inner city classroom: Teacher behavior* (pp. 57–74). Columbus, OH: Merrill.

Trafton, P. R., Reys, B. J., & Wasman, D. G. (2001). Standards-based mathematics curriculum materials: A phrase in search of a definition. *Phi Delta Kappan, 83,* 259–264.

Turnbull, H. R., & Turnbull, A. P. (1998). *Free appropriate public education* (5th ed.). Denver: Love Publishing.

U.S. General Accounting Office. (1993). *Student testing: Current extent and expenditures, with cost estimates for a national examination* (GAO/PEDM-93-8). Washington: Author.

Utah State Office of Education. (1995). *Mathematics core curriculum: Grades 7–12.* Salt Lake City, UT: Author.

Utah State Office of Education. (2001). *Utah core assessment series: Mathematics Test Development Project.* Salt Lake City, UT: Author.

Vacca, R. T., & Vacca, J. L. (1999). *Content area reading: Literacy and learning across the curriculum* (6th ed.). New York: Addison-Wesley/Longman.

Veljan, D. (2000). The 2500-year-old Pythagorean theorem. *Mathematics Magazine, 73,* 259–272.

Von Baravelle, H. (1989). The number π. In *Historical topics for the mathematics classroom* (pp. 148–154). Reston, VA: NCTM.

Ward, C. D. (2001). Under construction: On becoming a constructivist in view of the *Standards. Mathematics Teacher, 94,* 94–96.

Western, D. W. (1989). Fundamental theorem of algebra. In *Historical topics for the mathematics classroom* (pp. 316–318). Reston, VA: NCTM.

Whitney, H. (1987). Coming alive in school math and beyond. *Educational Studies in Mathematics, 18,* 229–242.

Wilford, P. (1993). *Peer collaboration in the Mathematics Teacher Network Project: A qualitative study.* Dissertation, Utah State University, Logan.

Woolfolk, A. E. (1993). *Educational psychology* (5th ed.). Boston: Allyn and Bacon.

Zimmer, T. S. (Ed.), (2001). *UCSMP Newsletter, 29* (Spring). Chicago: The University of Chicago School Mathematics Project.

Index

About the Author

James S. Cangelosi (Ph.D., Louisiana State University, 1972) has extensive experience teaching mathematics at the middle school, high school, and university levels. He specializes in mathematics education, data collection, assessment of student achievement, and behavior management at Utah State University where he serves as a professor in the Department of Mathematics and Statistics. Among his publications are articles in journals (e.g., *Journal for Research in Mathematics Education, Mathematics Teacher, Arithmetic Teacher, SIAM Review, Phi Delta Kappan, Mathematics and Computer Education, Educational Measurement Issues and Practices, Contemporary Education, The Clearing House, NASSP Bulletin,* and *Delta Pi Epsilon*), books (e.g., *Measurement and Evaluation: An Inductive Approach for Teachers* (1982), *Cooperation in the Classroom: Students and Teachers Together* (1984, 1986, 1990), *Classroom Management Strategies: Gaining and Maintaining Students' Cooperation* (1988, 1993, 1997, 2000), *Designing Tests for Evaluating Student Achievement* (1990), *Evaluating Classroom Instruction* (1991), *Systematic Teaching Strategies* (1992), and *Assessment Strategies for Monitoring Student Learning* (2000)), and videotape programs for mathematics teachers (e.g., "Demystifying School Mathematics," "Using Mathematics to Solve Real-Life Problems," and "Leading Students to Use Mathematics Skills to Describe Biological Phenomena"). Funded research and development efforts (e.g., the *Mathematics Teacher Inservice Project, Calculators in the Classroom Study, Mathematics Teacher Network, Calculator-Based Calculus Project, Mathematics Core Curriculum Project for Elementary and Middle School Teachers, Regional Institute for Mathematical Sciences,* and *BioMathLab Project*) reflect Dr. Cangelosi's concern for narrowing the gap between typical teaching practices and research-based teaching practices.